Chemistry
for the clinical laboratory

Chemistry
for the clinical laboratory

WILMA L. WHITE

Graduate Fellow, Department of Pathology,
St. Louis University School of Medicine,
St. Louis, Missouri

MARILYN M. ERICKSON

Chemist, Office of the Medical Examiner, St. Louis County;
Department of Pathology,
St. Louis University School of Medicine,
St. Louis, Missouri

SUE C. STEVENS

Director, Division of Endocrine Chemistry, Jewish Hospital;
Assistant Professor of Pathology,
Washington University School of Medicine,
St. Louis, Missouri

FOURTH EDITION

with 205 illustrations

The C. V. Mosby Company

Saint Louis   1976

FOURTH EDITION

Copyright © 1976 by The C. V. Mosby Company

All rights reserved. No part of this book may be reproduced in any manner without written permission of the publisher.

Previous editions copyrighted 1958, 1965, 1970

Printed in the United States of America

Distributed in Great Britain by Henry Kimpton, London

Library of Congress Cataloging in Publication Data

White, Wilma L
    Chemistry for the clinical laboratory.

  First-3d eds. published under title: Chemistry for medical technologists.
    First-2d eds. by C. E. Seiverd.
    Bibliography: p.
    Includes index.
    1. Chemistry, Medical and pharmaceutical.
I. Erickson, Marilyn M., joint author.
II. Stevens, Sue Cassell, joint author.
III. Seiverd, Charles Edward, 1913-   Chemistry for medical technologists. IV. Title.
RS403.W58   1976   616.07′56   75-31617
ISBN 0-8016-5432-7

CB/CB/B 9 8 7 6 5 4 3 2 1

Contributors

C. F. GUTCH, M.D.

Associate Professor of Medicine, University of Arizona College
of Medicine; Chief, Dialysis Programs, Arizona Medical Center
and Veterans Administration Hospital, Tucson, Arizona

CARL R. JOLLIFF

Clinical Biochemist and Microbiologist, Laboratory Section,
Lincoln Clinic and Lincoln Medical Research Foundation;
Visiting Lecturer, Department of Microbiology, University of Nebraska;
Lecturer, Nebraska Wesleyan University, Lincoln, Nebraska

MARY ANN MACKELL

Toxicologist, Office of the Medical Examiner, St. Louis County;
Department of Pathology, St. Louis University School of Medicine;
Clinical Instructor, St. Louis University School of Nursing and
Allied Health Professions, St. Louis, Missouri

ROBERT C. ROSAN, M.D.

Associate Pathologist, Cardinal Glennon Memorial Hospital for Children;
Associate Professor, Departments of Pathology and Pediatrics,
St. Louis University School of Medicine, St. Louis, Missouri

JOSEPH R. SIMMLER

Simmler & Son, Inc., St. Louis, Missouri

NANCY VanDILLEN

Supervisor, Division of Endocrine Chemistry, Jewish Hospital,
St. Louis, Missouri

Preface

The expanding use of the text has given evidence to the need for changing the title of the fourth edition to *Chemistry for the Clinical Laboratory*. The aim of this edition is to update methods, present new information, and fulfill recommendations submitted by our constructive critics. There is no apology for methods retained or deleted or for procedural format. The revision is directed toward satisfying needs of activity, training, and reference in clinical chemistry laboratories of all sizes.

Chapters have been revised and expanded. A chapter on toxicology from a major toxicology laboratory has been added. The rapid growth of radioassay techniques for laboratory analysis and the requirement that users pass a qualifying examination prompted the inclusion of a basic chapter on radioassay from a large diagnostic laboratory where there is extensive application of the methodology. A unique progressive chapter on the clinical pathology of hydrogen ion homeostasis is presented by a clinical pathologist.

In attempting to maintain a usable size text, subjects of interest that have been covered by the companion text, *Practical Automation for the Clinical Laboratory*, have not been included.

It is a pleasure to acknowledge the assistance of Dorothy Harrison, MT(ASCP), Chief Technologist of St. Francis Hospital, Washington, Missouri, in reviewing methods for small laboratories.

Wilma L. White
Marilyn M. Erickson
Sue C. Stevens

Contents

1. Blood collection and sample preparation, 1
2. Quality control, 9
3. AutoAnalyzer methods, 36
   AutoAnalyzer I, 36
   AutoAnalyzer II, 73
4. Glucose and nonprotein nitrogen compounds, 92
   Glucose, 92
   Nonprotein nitrogen compounds, 101
5. Electrolytes, 115
6. The clinical pathology of hydrogen ion homeostasis, 141
   ROBERT C. ROSAN
7. Hepatobiliary system, 177
8. Atomic absorption spectroscopy, 202
9. Enzymes, 214
   Methods, 220
10. Renal function, 242
    Renal function tests, 242
    Renal insufficiency and dialysis, 262
    C. F. GUTCH
    Routine urine examination, 274
11. Gastrointestinal system, 308
12. Cerebrospinal fluid, 322
13. Radioassay, 328
    NANCY VanDILLEN
14. Hormones, 358
    Pituitary gland (hypophysis), 359
    Thyroid, 366
    Parathyroid glands, 371
    Islets of Langerhans of the pancreas, 372
    Adrenal gland, 374
    Other hormones and glands, 393
    Hormone assays, 401
15. Toxicology, 449
    MARY ANN MACKELL
16. Chromatography, 485
    JOSEPH R. SIMMLER
17. Electrophoresis, 528
    CARL R. JOLLIFF
18. Immunoelectrophoresis and immunoquantitation procedures, 592
    CARL R. JOLLIFF

Appendixes

A. Tables, 657
B. Fire precautions, chemical hazards, and antidotes of poisons, 676
C. Basic information for the student, 678
D. Reagents, 690
E. Company directory, 712

Chemistry
for the clinical laboratory

1

Blood collection and sample preparation

In health the composition of body fluids is maintained within narrow physiologic limits. The human body consists of 50% to 70% fluid by weight. The fluids, which are classified as intracellular or extracellular, maintain a balance in the various compartments and in their constituents.

Blood, one of the extracellular fluids, makes up approximately 7.5% and 7% by weight of normal men and women, respectively. The concentrations of the wide variety of substances in the blood remain relatively constant for the normal individual, with the evaluation of the acceptable normal range dependent on methodology and patient population. In physiologic disorders or pathologic states, there may be a decrease or an increase in the amount of the regular constituents, or some abnormal factor may appear in the blood, where either a diagnosis or the extent of an abnormal condition may be determined.

Whole blood, plasma, and serum are utilized in correlating the patient's clinical condition to that of needed therapy. While the hematology laboratory studies the cellular structure and concentration in relation to disease, immunology and chemistry delve into the biochemical structure.

With children and adults, the blood is usually secured from finger or ear and veins of the forearm, wrist, or ankle. With infants, the blood is most often taken from heel or toe and veins on the forearm, neck, or head. The standard practice is to obtain blood from the finger or from veins of the forearm. Before collection of the blood is started, however, any apprehension on the part of the patient should be dispelled.

BLOOD COLLECTION
Capillary puncture

An extremely old technique for obtaining blood is the puncture wound of the capillaries. Although the finger puncture is most used, you should be familiar also with the ear puncture and its advantages and disadvantages. While the differences in chemistry values between the capillary blood of the ear and that of the finger are for most purposes insignificant, the following notations must be taken into consideration if blood from the ear is used for hematologic work.

Unless the patient has been in the cold or under undue stress, the red cell counts and hemoglobin values are essentially the same as values obtained from a finger stick or from venous blood. The white cell count (WBC) from the ear is slightly higher than the count from the finger. Also, larger white cells may be present in greater numbers.

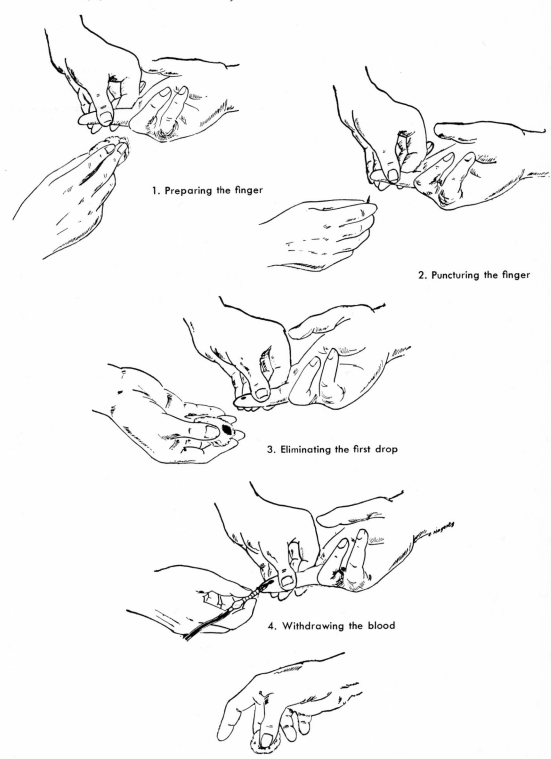

Fig. 1-1. Finger puncture.

There is a striking difference in results obtained from earlobe blood versus blood from finger or vein in certain pathologic conditions. The difference is seen in the WBC and the differential counts of infectious mononucleosis, leukemias, and bacterial endocarditis. The increase of white cells from the earlobe involves all types of white cells, especially the large cells such as monocytes and histiocytes that are seen in large numbers in bacterial endocarditis. This has been explained largely by the greater filtering capacity of the vascular bed of the earlobe in comparison to that of the fingertip.

In preparation of blood films for the study of leukopenias and for certain infections, the earlobe is the most desirable source of blood. The first two drops of blood from an earlobe puncture may contain more large cells. After several drops, the cells from this source are similar to those from the finger stick or the venipuncture.

The finger or earlobe is cleansed with a 70% alcohol sponge, the area permitted to dry, and an incision made with a disposable sterile lancet. The blood will form full rounded drops. If the patient has been outdoors in cold weather, the area is gently warmed to enhance circulation and then cleansed with an alcohol sponge. Cold or cyanotic skin will result in falsely elevated white and red blood cells.

While the finger is more convenient, the earlobe is less sensitive. The choice is usually determined by the patient's condition (bedridden, obese, edematous).

The finger is punctured with a sterile disposable lancet using a loose wrist motion. The puncture is made by firmly grasping the finger and making a quick deliberate stab to a depth of about 3 mm. It is better to induce a good puncture rather than make the patient apprehensive by having to repeat it. A deep puncture is no more painful than a superficial one. The first drop of blood, containing tissue juices and foreign matter, is always wiped off. Free-flowing blood should be used for examination, although gentle pressure may be applied if needed to enhance the flow of blood. Never exert heavy pressure, because that may cause the flow of tissue juices into the blood.

After the required amount of blood has been taken, an antiseptic (70% alcohol pad) is placed on the puncture. The patient may be instructed to apply pressure to the wound until the bleeding has ceased (Fig. 1-1).

Venipuncture

Larger volumes of blood needed for chemical analyses are obtained from the veins of the forearm, wrist, or ankle. For convenience and because of the larger size of the veins in the forearm, this area is usually chosen for a venipuncture. With good technique and sharp needles, the same vein may be used many times with little discomfort to the patient or damage to the vein. However, bandages, soreness, obesity, or scar tissue formed from poor techniques may sometimes make it necessary to use the veins of the wrist or ankle.

Obtaining a venipuncture is an art that demands compassion, knowledge, and skill to minimize discomfort to the patient. If the patient is not confined to a bed, he should be seated comfortably in a chair with his arm placed upon a table, counter, or some other solid form. The chair and table should be on a level such that the arm will be extended downward at an angle and not raised above the shoulder level. Low chairs and high tables should be avoided.

If the patient has a tendency to bend his arm, place a support under the elbow. Use books, a package of paper towels, or anything solid enough to hold the arm straight.

A patient may feel faint from apprehension, so keep a fresh bottle of smelling salts available, or moisten a piece of cotton with dilute ammonium hydroxide and wave this under his nose. If there is no immediate response, call a physician.

Place the blood-collecting tray at the bedside where it will not be upset by either yourself or the patient. Try not to disturb the patient any more than is absolutely necessary.

Take the arm available and gently place it toward you. If the patient shows apprehension, calmly explain what is going to take place and what is expected of him. If there is still a slight pulling away of the arm, two things may be done. Fold a towel about four times and place it under his elbow to keep it from bending; then very gently place your elbow on the patient's hand and wrist. The second recommendation is to hold the patient's elbow gently with your left hand and introduce the needle with the right—or vice versa if you are left-handed. A combination of the two steps is also successful. The object of this is to obtain blood without trauma.

The actual technique of obtaining venous blood is presented in Figs. 1-2 and 1-3.

Equipment

The diameter or bore of the needle is indicated by its gauge number. The smaller the number, the greater the diameter. Gauges 20 and 21 are usually used for the forearm and 25 is used for the veins of the wrist and ankle. The size of the needle chosen is regulated by the size of the vein to be entered.

The choice in size of a syringe is regulated by the amount of blood needed for analysis. A 10-ml syringe is most frequently used. The syringe and needle must be sterile. Do not use a syringe in which there is any dampness, since moisture will hemolyze the blood. Plastic syringes and tips are now widely used.

When the needle and syringe have been removed from their sterile packages, carefully place the needle firmly onto the syringe. Do not handle the cannula of the needle or the plunger of the syringe. Use the hub of the needle (Fig. 1-2) and the head of the plunger.

Vacutainer tubes are widely accepted. Rubber stoppers are color-coded to identify the anticoagulant present, or the tube may be left plain and used for blood that can be allowed to clot. The vacuum within the tube is sufficient to draw a predetermined volume of blood. The blood-drawing set contains a disposable double-pointed needle that screws into the holder. When the needle is in place, the Vacutainer tube is inserted into the holder, with the tube's rubber stopper touching the needle. It is then pushed forward until the top of the stopper reaches the guideline. The needle is partially embedded in the stopper without breaking the vacuum in order to prevent leakage of blood at the time of venipuncture. Once the blood set has been assembled, the venipuncture needle is inserted into the arm in the routine manner and the needle in the Vacutainer stopper is pushed through the diaphragm, breaking the vacuum and thus allowing blood to flow into the tube. When more than one tube of blood is needed, the first tube may be removed from the needle and another inserted while the needle is still in the vein.

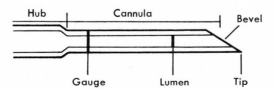

Fig. 1-2. Diagram of the needle.

Technique

Many antiseptic solutions are available, but the one most commonly used is 70% ethyl alcohol. Lundy[2] has expressed the necessity of warming the sponge and antiseptic solution, at least to body temperature, before applying them to the area of injection. Because the arm is warm, it is more sensitive to the cold, and contact with a cold antiseptic will cause the vein to contract. While the alcohol is drying, the tourniquet may be applied.

It is most desirable to enlarge the veins of the forearm to make them more prominent. With the area for puncture located, utmost care must be used for the proper distention of the vein. The arm should hang down for a while. The veins will become more apparent for proper evaluation.

A piece of soft rubber tubing is most commonly used for a tourniquet, although tourniquets may range from pieces of cloth to elaborate mechanisms. A blood pressure cuff is an excellent tourniquet because of the ability to adjust the venous flow without interfering with the arterial circulation.

The tourniquet is placed above the elbow to increase venous pressure. The vein becomes more prominent, therefore making an easier entry possible. NOTE: The tourniquet must be released as soon as possible to prevent hemoconcentration. The patient then continually opens and closes his hand to aid in filling or distending the vein. During the time that the blood is being drawn the hand is kept closed. If the veins are poorly dilated, even after pumping the hand, release the tourniquet, place a hot wet towel over the arm, and cover with a dry towel. Keep this heat pack on the arm for 10 or 15 minutes. Electric blankets, heating pads, and hair dryers have been used successfully to dilate the veins. CAUTION: Do not use the hair dryer around any explosives. When the circulation is improved, reapply the tourniquet.

Normally, when the size of the vein to be entered is sufficiently large in relation to the needle size, place the bevel of the needle upward. If a large needle is all that is available for a small vein, however, place the needle into the vein with the bevel facing down. This will prevent a hematoma that might occur from puncturing the wall of the vein. Occasionally a small vein will collapse, covering the opening of the needle. In attempting to adjust the needle to clear the passage, the blood drawer may puncture the wall, causing a hematoma. Many veins have a tendency to "roll" when stuck with a needle. However, the vein may be "fixed" by holding the vein or grasping the arm below the elbow and pulling the skin taut.

The needle approach is 25 degrees into the long axis of the vein. Never approach the vein from the side. Hold the syringe by both the plunger and the barrel, pierce the skin, and after the blood has entered the syringe, pull the plunger of the syringe slowly back. If the blood flow stops, the needle has either slipped out of the vein or penetrated through the wall. If it has slipped out, carefully reinsert it. If it has gone through the vein, slowly withdraw the needle until it is once again in the vein. This ability comes with experience. In remedying these problems, be sure to avoid pushing the plunger forward in the syringe.

In the case of a collapsed vein, if little or no blood has been obtained but there is a large air space within the syringe, obtain a new syringe and connect it to the needle. This, of course, is unpleasant for the patient, but it must be done. Before the needle is withdrawn from the vein, the tourniquet must be released. Have the patient relax his fist before releasing the tourniquet. Otherwise, a blood flow will continue through the puncture. Withdraw the needle after the tourniquet has been removed or released. Place a dry, sterile pad or adhesive bandage over the puncture wound immediately.

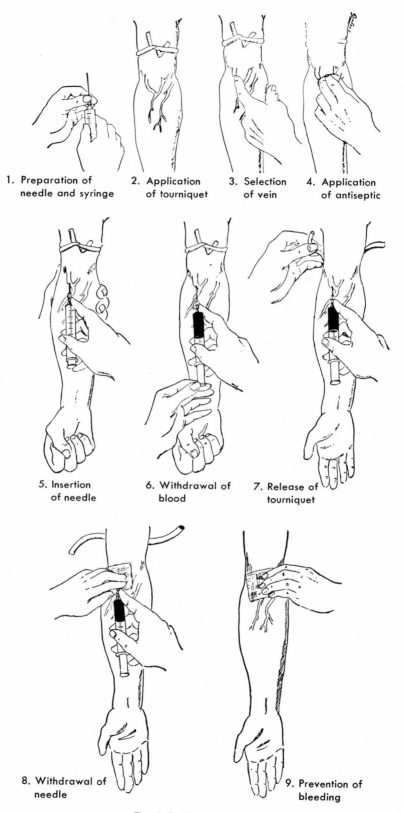

Fig. 1-3. Venipuncture.

The patient may play a role in this procedure by gently applying pressure to the sterile pad covering the wound. Folding the arm up toward the shoulder, combined with the finger pressure, will aid in stopping the flow of blood. The patient should be watched carefully for any reaction before being dismissed. If fainting, shock, or anxiety is evident, the patient should lie down and a physician should be summoned. A hematoma is sometimes formed by the blood seeping into subcutaneous tissue, leaving a large black and blue mark painful to the patient.

Technical problems. Hematomas may be caused by several factors: using too large a needle for the vein, therefore puncturing the wall; failing to have the needle completely in the vein, thus allowing blood to enter the tissues; forgetting to release the tourniquet before withdrawing the needle, with the result that the pressure expels blood from the vein puncture into the tissues; failing to apply finger pressure to the wound for a sufficient time; and not taking the time and care and then having to do repeated punctures of the vein.

The steps in the venipuncture are illustrated in Fig. 1-3.

Failure in obtaining free-flowing blood into the syringe may result from not completing entry into the vein (withdraw needle slightly and reenter the vein); pulling plunger faster than blood flow, resulting in a collapsed vein (carefully move needle back and forth with a gentle pulling of the plunger); and circulatory failure (no corrections can be made on this).

Disposable equipment such as needles, syringes, and lancets have eliminated many problems of contamination, infections, and hepatitis.

SAMPLE PREPARATION

The blood usually requires some preparation before analysis. Serum is preferred to plasma or whole blood for most chemical analysis. However, in general, heparinized plasma may be used in place of serum, since the basic difference between them is the fibrinogen in plasma. Serum or plasma must be separated from the blood cells by centrifugation within an hour after collection. A few procedures require a protein-free filtrate. This can be prepared by using either whole blood (anticoagulated), serum, or plasma.

Serum

The reaction in blood clotting is a conversion of the soluble protein fibrinogen to the insoluble fibrin. This may be simply described as progressing in steps or phases. In the first phase, ionized calcium and thromboplastic factors (cephalin, tryptases, thromboplastins) activate prothrombin, forming thrombin. In the second phase, the thrombin converts fibrinogen into fibrin. This is the actual formation of the clot.

Plasma

In order to obtain plasma, the blood must be kept from clotting. Substances that prevent clotting are called anticoagulants. Some chemical anticoagulants remove the ionized calcium from the blood. Heparin prevents clotting by inactivating prothrombin. The more popular anticoagulants and the amounts required per milliliter of blood are given below:

Anticoagulant	*mg/ml of blood*
Ammonium oxalate	2
Potassium oxalate	2
Sodium fluoride	2
Heparin	0.2

Anticoagulant	mg/ml of blood
Lithium oxalate	1
EDTA	2

Potassium oxalate, ammonium oxalate, and trisodium citrate remove calcium from blood plasma by binding in un-ionized form, preventing coagulation.

A mixture of potassium oxalate and sodium fluoride is used as an anticoagulant and preservative for some glucose methods. However, since sodium fluoride inhibits enzyme activity, it cannot be used as an anticoagulant for any enzyme determination. Heparin, a physiologic anticoagulant, and EDTA (ethylenediamine tetra-acetate), a chelating agent binding calcium, are very popular anticoagulants used in chemistry and hematology laboratories, since there is no destruction of the cells.

Heparin, 0.1 to 0.2 mg/ml of blood, does not affect the corpuscular size and hematocrit. It is the best anticoagulant for prevention of hemolysis and for osmotic fragility tests. However, it is not satisfactory for leukocyte counts or when blood films are to be prepared; it produces a blue background in Wright's-stained slides.

Vacutainers with color-coded tops identifying the anticoagulant are very popular today; it takes only the effort of knowing the analysis compatibility with the anticoagulant. The manufacturer has thoughtfully provided such a coded chart.

Gross examination

The appearance of the blood specimen may give an insight to the patient's condition. Examination of the specimen after centrifugation may reveal a low or high cell volume or normality of the buffy coat, which is a thin white layer separating the red cells and the plasma. A pink or red color to the plasma may be indicative of hemoglobulinemia, or the color may be a hemolysis due to poor technique in blood drawing. A milky or cloudy plasma may indicate cryoglobulinemia, other hyperglobulinemias, or nephrosis, unless a fatty meal had been ingested 1 or 2 hours before the blood drawing. Green or orange coloring of the plasma suggests increased bilirubin. Strict attention to such details with proper notation is a service to the physician and the patient.

References

1. Abbott Laboratories, Medical Department: Parenteral administration, Chicago, 1959.
2. Lundy, J. S.: Suggestions to facilitate venipuncture in blood transfusion, intravenous therapy and intravenous anesthesia, Mayo Clin. Proc. **12:** 122, 1937.
3. Page, L. B., and Culver, P. J.: A syllabus of laboratory examinations in clinical diagnosis: cricital evaluation of laboratory procedures in the study of the patient, revised edition, Cambridge, Mass., 1966, Harvard University Press.

2

Quality control

HISTORY AND FUNCTION OF QUALITY CONTROL

The history of quality control dates back to the Middle Ages. When man learned to build and to see his talents as a craftsman, the quality of his efforts determined his success. Long periods of apprenticeship were required for learning a skill, and this training had to meet certain standards before an apprentice could become a master craftsman. In an effort to maintain quality of output, various methods for quality control have been introduced into the manufacturing market. Today, production quality in industry can be checked by on-line inspection.

STATISTICS

Statistics, a means of grouping and interpreting facts by use of numbers, was introduced about 3 centuries ago.[12] Statistical application was first employed in astronomy and physics and later utilized in biologic and social sciences. In the eighteenth and early nineteenth century three mathematicians—de Moivre, Laplace, and Gauss—independently devised a mathematic displacement distribution resembling a bell-shaped curve (Fig. 2-1), now commonly referred to as the gaussian curve. Such a distribution of values from the average or true value has been called the *law of errors*.

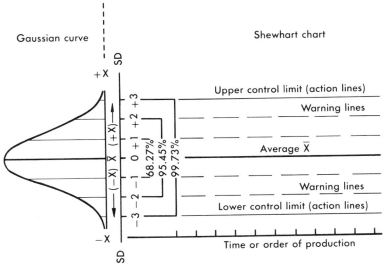

Fig. 2-1. Normal distribution: curve and chart.

Statistical quality control

Statistical quality control (SQC) is a relatively new application. The first sketch of a quality control chart was made by Walter A. Shewhart, Bell Telephone Laboratories, in 1924.[8] Development of this new technique of utilizing charts to evaluate quality continued, and in 1931 Shewhart[14] published a book on statistical quality control. This eventually set the pattern for statistical methods as applied to process control. Dodge and Romig, also of the Bell Telephone Laboratories, adopted the statistical method for sampling inspection. The conclusion of their work resulted in the Dodge-Romig sampling inspection tables. The investigations carried out by Shewhart, Dodge, and Romig constitute the basis for the major part of statistical control.[3-5, 14]

Industrial development

British industry readily accepted statistical quality control and by 1937 its use was widespread. In the United States, however, enthusiasm was lacking. American industry considered that the quality control measures already being applied were adequate and that introducing unnecessary mathematics would be an excessive burden.

With the outbreak of World War II the armed forces became a large consumer of industrial products and influenced the adoption of statistical quality control. The Bell Telephone Laboratories were invited to join the government in developing a sampling inspection program for Army Ordnance. The combined efforts resulted in sampling inspection tables for the Army Ordnance and the Army Service Forces.

The demand for this new method of evaluating both production and products and the shortage of personnel trained to use it led to the establishment of educational programs for teaching statistical quality control. In July, 1942, an intensive 10-day course was given at Stanford University for representatives of war industries and procurement agencies of the armed forces. The training sessions were sponsored by the Engineering Science and Management War Training Program of the U. S. Office of Education. A shorter (8-day) training period was given in Los Angeles. Courses were then provided in various centralized areas throughout the country by the Office of Production Research and Development of the War Production Board.

Clinical development

Compared with industry's application of statistical control, the clinical diagnostic laboratories' adoption of this tool for routine analysis control has been slow. The need for quality control in the medical diagnostic laboratory has long been recognized. Belk and Sunderman[1] in 1947 and Wootton and King[18] in 1953 reported results obtained from surveys in which blood specimens containing constituents that had been determined by reference laboratories were sent to a number of routine hospital laboratories. Wide variation in the analytic data obtained by the different laboratories confirmed the need for control procedures. In 1953 the first commercial control serum of known value was made available. It was in liquid form and was marketed under the name of Lab-Trol. Currently there are lyophilized sera, control urines, and spinal fluids commercially prepared to enable the analyst to maintain a monitoring program of laboratory assay results.

Automation will not eliminate analytic problems in chemical determinations. If a procedure for measuring a specified constituent is poor, improvement will come only with a better chemical method. Statistical quality control will include variables inherent within a chemical determination. The following conditions are monitored by SQC: (1) reagent purity, (2) standard reliability, (3) accuracy of pipets, autodilutors, and analysts, (4) cleanliness of all glassware and equipment, (5) calibration of the readout apparatus, (6) mechanical condition of the analytic hardware, and (7) proper calculation of the data obtained.

In recent years there has been a worldwide realization within medical laboratories that the needed degree of accuracy is not being achieved. Surveys carried out in routine hospital laboratories have revealed wide discrepancies between the various laboratory areas. With the introduction of statistical quality control into the laboratories the whole application has assumed the appearance of a "whodunit?" between the supervisory or managerial staff and the technical staff. For development of a productive, efficient laboratory, members of the technical staff must be oriented in the importance of instrument care, methodology limitations, analysis or quantitation demands, and purpose of a quality control program. They must be trained to utilize quality control rather than be policed by it. There certainly is a need for unknown constituents to be placed in the runs and for active participation in surveys or multiple-hospital control programs.

Purpose of quality control in laboratories

Many factors contribute to the quality of work produced in the laboratory. Management; professional and technical personnel; collection, delivery, and storage of specimens; analytic procedures; and transcription of records are a few of the variables related to the final result of an analysis. Team effort with understanding communications between all areas is needed for a proper program. Quality control is not a gathering of data for display or policing of the various work areas. It is a method of obtaining technical control by stimulating a team objective, with mutual assistance toward success in achieving a goal.

Starting a program

In setting up a program or revamping an existing one, it is important to establish simple rules that are easily understood. Consideration must also be given to the following:

1. Determine whether the program will be sectional (chemistry, hematology, microbiology, serology, etc.), with individual quality control supervisors, or will function as one overall program, including each laboratory and using one quality control team.

2. Choose the individual(s) who will assume the responsibility of inspector and define to all employees the range of authority invested in the inspector. If the laboratory is small, each person may assume the quality control responsibility with his duties on a rotation basis for a month.

3. Train all personnel in (a) proper methods of specimen collection, delivery, preparation, and storage and (b) correct performance of analytic methods and transcription of results. Doing this will include use of teaching sessions, clearly written directions, and the sending of individuals to certified workshops.

4. Determine the acceptable limits of each method.

5. For each procedure add the volumes of control needed in the normal and the abnormal ranges and choose the material to be used as the control.

6. Determine the type of tabulation to be carried out and the charts to be used in each case.

LABORATORY APPLICATION OF QUALITY CONTROL
Specimen collection

Collecting a specimen correctly is a vital step for an analytic method; yet it is practically impossible to eliminate all the errors of collection. A list of the tests and the right methods of specimen collection, including also the proper preservatives or anticoagulants, should be posted in every nursing and blood-drawing station. The instructions should be carried out as written. Physiologic sources, such as medication and foods, and improper collection of specimens may cause misleading laboratory results to be given the physician. If the error is subtle—in that it cannot be detected by the laboratory—such as incorrect labeling of a tube or combination of the urine of two patients in the same bottle, the effect could be harmful to the patient.

A few basic rules for specimen collection follow:
1. Read the directions and follow them.
2. Properly label all timed collections.
3. Do not draw blood from the arm receiving an intravenous fluid or transfusion of blood.
4. Collect a 24-hour urine during a period of exactly 24 hours and add to the designated bottle as soon as a specimen is voided. Each urine bottle may have a color-coded tag to identify the preservative and the group of tests for which its contents may be used.

A moment of care and observation may save the patient—and all concerned—grief, money, or time.

Specimen delivery and storage

Another area that can limit laboratory efficiency and patient care is the transport of specimens to the laboratory. Assuming that the specimens are drawn correctly and are contained in the proper tubes or bottles, there is still the problem of delivery. Smaller hospitals or medical centers have the advantage of utilizing laboratory employees for blood drawing and ward personnel for delivery of other specimens. The use of employees such as pages or dispatch messengers, who have no other contact with the specimen than its delivery, introduces a source of error difficult to control. Specimens that require processing within a limited time are adversely affected by delayed delivery. Thus errors may be compounded by prolonged delays in delivery, improper storage, and deterioration due to incorrect preservative before the analytic procedure is initiated.

Laboratory errors

The specimen's arrival in the laboratory does not guarantee infallibility in handling. Specimen identification must be carefully controlled throughout the analysis and during transcription of the results to the proper requisitions. Manual manipulations of the specimens and work sheets allow many possibilities for error.

Automation does not prevent erroneous handling of specimens. For example, sample plates may be loaded in a sequence different from that recorded on the work sheet, stats may be inserted during a run in the place of removed specimens and the latter not reentered in the series later, or charts may be incorrectly numbered.

Controls

Quality control includes more than tabulating the results from a series of standards and controls. Factors to be taken into consideration include specificity of method, time required for analysis, the available and the needed equipment, cost of analysis, and proficiency of personnel. There must be a commonsense approach in choosing a reliable method and in establishing a continuing program for monitoring both the precision and the accuracy of test results obtained. The latter should be kept as consistent as possible for an accurate monitoring of the area.

Various controls may be used:

1. Solutions of pure chemicals. Depending on the method, these controls may or may not be processed through an entire analysis. In prolonged procedures a solution containing a known concentration of the constituent is frequently analyzed along with the unknown specimen.

2. Patient controls. A specimen from one patient may be carried over in each batch run of a given test. This is an excellent method for checking multiple series of analyses throughout the day. The standard deviation may be calculated by dividing the average difference of the carryover specimen by 1.128, or on a long-term evaluation by using this formula[16]:

$$SD = (\Sigma d^2 / 2\ N)^{\frac{1}{2}}$$
where d = difference between duplicates
N = number of duplicates performed

3. Controls of pooled sera, cerebrospinal fluid (CSF), or urine. Sera and CSF with values in the normal range may be pooled separately. From this range, dilutions can be made for lower concentrations. For abnormal levels of constituents, in which the collection of material is not so prevalent, a group of laboratories could share in the pooling and in a monitoring program. A 24-hour collection of urine will supply the needs for most programs.

4. Commerically prepared sera and CSF. This is a more expensive program. Controls can be purchased as unassayed material or as analyzed products with target values. Quality control programs are now available commercially. With the wide selection of material, it is recommended that dependence not be placed on any one control and that after selection of the primary control other products be maintained for cross-checking.

5. Statistical evaluation of daily results. Such an application of statistics can be extremely useful in monitoring analytic precision. This method will detect a subtle shift in the total operation. It is a rapid technique for observing a curve.

The shelf-life of a control depends largely on refrigeration. There is difficulty in maintaining pooled controls free from bacterial contamination. Frozen material should be held at a temperature of $-20°$ C. Lyophilized material must be kept refrigerated. Constituents in biologic fluids vary in their stability, and pools should be cross-checked to validate the assay results.

Controls for monitoring hormone analyses are usually prepared from pooled collections of plasma and urine. Each new lot of pooled specimen is analyzed along with the previous batch and with any available commercially assayed controls. After such standardizing, the pooled material is divided into small aliquots and stored at $-20°$ C.

Standards

A standard is a criterion for comparison that has been established by custom or authority. It may be a rule governing quality, quantity, or conduct. Standards for weights and measures such as the gram, pound, liter, gallon, meter, and yard are fixed by the National Bureau of Standards (NBS). Since its establishment in 1901, NBS has made available standard reference materials having certified properties and has provided specifications for weights and other units of measurement. The clinical chemist may need either to obtain from the Bureau a certified weight for calibrating weights available in the laboratory or send the weights to the Bureau for calibration, to request volumetric glassware that has been calibrated by the Bureau, and to purchase various chemicals for standardization. NBS has available salts for buffers, such as potassium hydrogen phthalate and phosphates, and chemicals for standards, such as glucose, cholesterol, urea, uric acid, and creatinine.

Manufacturers who provide drugs or substances used in clinical practice must conform to the standards for these products as set forth in the *United States Pharmacopeia*. Included are assay procedures for determining compliance with pharmacopeial standards of purity and strength. Some professional organizations also publish standard or official procedures of analysis. *Official Methods of Analysis* is published by the Association of Official Agricultural Chemists; *Standard Methods for the Examination of Dairy Products,* by the American Public Health Association; and *Standard Methods for the Examination of Water and Sewage,* by the American Public Health Association and the American Water Works Association. Many other organizations are active in programs for control and standardization of laboratory methods.

In an analytic chemistry laboratory a primary standard is defined as a chemically pure compound of known composition that can be accurately weighed and used in the preparation of a solution of known concentration. A secondary standard is one that can be prepared in approximate concentration, standardized against a primary standard, and then used to determine the concentration of another compound for which a primary standard is not available. For example, potassium hydrogen phthalate, a primary standard, may be used to determine the exact concentration of a solution of sodium hydroxide. The latter may then be used to determine the exact concentration of a solution of hydrochloric acid. In this case, the sodium hydroxide acts as a secondary standard.

The clinical chemistry laboratory is concerned with analyzing biologic fluids for various constituents, some of which are not available in pure form. When these compounds are to be used as standards, a known weight of the highest purity obtainable dissolved in a suitable medium is accepted. The solution with compounds of known concentration or primary standard is used as a basis for comparison with the constituents being measured in the unknown sample. Depending on the procedure, the standard may or may not be processed through all the steps of the method in the same manner as the unknown and controls. Regardless of the stage at which the standard is introduced into a procedure, the purpose is to make possible a comparison of the reaction of the unknown and that of the standard.

As a result of claims that standards in an aqueous system react differently from those in serum, commercial preparations have been made available that contain weighed amounts of various constituents added in a manner to duplicate human serum. The known levels of the added chemicals are indicated with each lot purchased. Since known absolute concentrations of desired constituents have been weighed into these products, they are considered, by the manufacturers, primary standards. However, the presence of unknown factors as a part of the medium negates this claim for such a product.

A standard solution must not be confused with a control sample. The latter is a specimen (serum, plasma, spinal fluid, urine) having approximately the same composition as the unknown that is to be analyzed. It is carried through the entire procedure in the same manner as the unknown and serves to monitor the precision of the determination. Concentrations of the constituents should approximate the levels of those in the unknown samples being assayed. Control samples are usually taken from pooled fluids that have been purchased commercially or prepared in the laboratory. They are preserved by lyophilizing, freezing, or sterilizing. The commercially available product is accompanied by a certificate listing the assayed values determined by reference laboratories and also the methods employed in the testing. A control sample can never be used as a standard, but a standard may be employed as a control.

Another type of standard used in clinical laboratories is called an internal standard. This term refers to the addition of a known amount of a pure compound to a routine sample being analyzed.

Data

Collecting

Data accumulated for control purposes should be kept in a special notebook. A serial listing of the daily monitoring makes possible a rapid comparison with previous values. At each work area a daily register of blanks, standards, working control, and baseline readings of the inverse colorimetric methods (AutoAnalyzer) should be maintained. This presents a continual check on chemistry, standards, and instrument conditions. The settings for the AutoAnalyzer colorimeter will indicate to the analyst if the photocells need to be replaced. With the lists available, control calculations can be performed at any time.

The unknown controls are tabulated and charted for visual display. At least 30 analyses should be tabulated for a consistent monitoring of a method. A large quality control program will exceed 30 analyses each month. Therefore the standard deviation is calculated after the desired population (total test) is obtained.

A small laboratory cannot maintain a large quality control program. The processing of control samples along with tests that are prolonged and time-consuming is limited. When there are fewer controls (less than 10 samples), it is more practical to compute the mean deviation than the standard deviation. The relative efficiency of mean deviation is high when measuring dispersion of the small group of samples, 91% for samples up to 10. However, with normal variations, the standard deviation when corrected for bias is more efficient.

Tabulating and studying

Preparing or reviewing data requires understanding of some basic terminology. When symbols or unfamiliar terms are used, the meaning must be explained.

Rates, ratios, and percentages

Rates, ratios, and percentages are used daily in the routine work. The AutoAnalyzer may be employed in an example for defining rate and ratio. *Rate* refers to a numerical proportion between two variables, such as the different sampling speeds per hour (20, 40, or 60/hr). *Ratio* denotes a quantitative relation between similar measurements such as the length of time a sample probe remains in the sample cup and its stay in the wash cup. A comparison of the time interval for each sampling step is the ratio—2:1, 1:1, or 1:2. Symbols, shown here in parentheses and commonly used to compare numbers, are (:) meaning *is to* and (::) meaning *as*. Therefore the statement may be made that 4 is to 2 as 2 is to 1 or 4:2::2:1. *Percentage* usually designates the frequency or quantity in proportion to a sum or total amount of 100 units. Therefore, a percentage is no more than a ratio that employs a base of 100 units.

A rule for computing a ratio is to *always* divide by the base. The base is the reference figure to which the other number is to be compared. For computing percentages, remember that the base should be 100 or more. The percentage of increase may be infinitely large but the percentage of decrease, unless negative values are possible, does not exceed 100. The negative values usually occur in business where profit and loss data are involved.

In any computation, the decimal point is a source of potential error. A 48.9 numerical value, for example, might be recorded as 489 (ten times the acutal figure) or as 4.89 (one tenth the designated value).

To assist in calculating the standard deviation and making a control chart for the daily plotting of tests, the following example is given. The tests being calculated are potassium results that were run routinely in the laboratory. The specimens used were obtained from pooled sera prepared and previously assayed. The potassium value is above the normal range for potassium. Specimens of the order listed last are called *outliers*. Quality control is needed for elevated and abnormal ranges, as well as for checks within the normal range. Table 2-1 illustrates also the results of careless and inferior technical application in analysis.

Table 2-1
Sample work sheet for tabulating test results

A Number of tests	B Test results	C Difference from the mean	D Difference from the mean squared
1	5.8	0.0	0.00
2	5.6	0.2	0.04
3	5.9	0.1	0.01
4	5.8	0.0	0.00
5	5.7	0.1	0.01
6	5.8	0.0	0.00
7	5.8	0.0	0.00
8	5.8	0.0	0.00
9	5.4	0.4	0.16
10	5.7	0.1	0.01
11	5.7	0.1	0.01
12	5.6	0.2	0.04
13	5.8	0.0	0.00
14	5.7	0.1	0.01
15	5.8	0.0	0.00
16	5.8	0.0	0.00
17	5.8	0.0	0.00
18	5.9	0.1	0.01
19	5.7	0.1	0.01
20	5.8	0.0	0.00
21	5.9	0.0	0.00
22	5.7	0.1	0.01
23	5.7	0.1	0.01
24	5.8	0.1	0.01
25	(9.3)	(Specimen contaminated by cigarette ashes)	
Total	138.0		0.34

Standard deviation

Referring to Table 2-1, calculate the standard deviation.

1. Total column B and divide by the number of tests done. Test 25 is eliminated from the averages because of a technical error.

$$\text{Average or mean value} = \frac{138.0}{24} = 5.8$$

2. Determine the absolute difference of each test from the average value (5.8). The difference is recorded in column C, which is *not* added. Do not bother placing plus or minus signs with the differences.
3. In column D, the difference from the mean is squared (multiplied by itself: e.g., $0.4 \times 0.4 = 0.16$). This column is totaled, giving a result of 0.34.
4. With these data, calculate the standard deviation. With n = number of tests done, the formula for calculating SD in this example is as follows:

$$\text{Standard deviation} = \sqrt{\frac{\text{Sum of squared differences (column D)}}{n - 1}}$$

$$SD = \sqrt{\frac{0.34}{24 - 1}}$$

$$SD = \sqrt{\frac{0.34}{23}}$$

$$SD = \sqrt{0.0148}$$

$$SD = 0.12$$

Table 2-2
Squares and square roots

n	n²	√n	n	n²	√n	n	n²	√n	n	n²	√n
1	1	1.000	13	169	3.605	25	625	5.000	37	1369	6.082
2	4	1.414	14	196	3.741	26	676	5.099	38	1444	6.164
3	9	1.732	15	225	3.872	27	729	5.196	39	1521	6.244
4	16	2.000	16	256	4.000	28	784	5.291	40	1600	6.325
5	25	2.236	17	289	4.123	29	841	5.385	41	1681	6.403
6	36	2.449	18	324	4.242	30	900	5.477	42	1764	6.480
7	49	2.645	19	361	4.358	31	961	5.567	43	1849	6.557
8	64	2.828	20	400	4.472	32	1024	5.656	44	1936	6.634
9	81	3.000	21	441	4.582	33	1089	5.744	45	2025	6.708
10	100	3.162	22	484	4.690	34	1156	5.830	46	2116	6.782
11	121	3.316	23	529	4.795	35	1225	5.916	47	2209	6.856
12	144	3.464	24	576	4.898	36	1296	6.000	48	2304	6.928

n = the number.
n² = the number squared.
√n = the square root of the number.

The divisor (n − 1) is called the degrees of freedom in the sample. This term as applied to statistics refers to the number of independent differences possible with a series of observations. For example, given one sample with one result there is no variation possible. If there are ten results from a sample, then there are nine possible differences.

The symbol "$\sqrt{}$" means taking the square root of a number, 0.0148 in the example shown. Values may be obtained from Table 2-2 or from Table A-6, in Appendix A, for the square roots of small numbers. To check the accuracy, multiply the square root by itself. A slide rule may be used to obtain either a square of a number or a square root.

To find the square of a number

Set the hairline to the number on the D scale and read its square under the line on the A scale. The square may also be found by setting the hairline on the C scale and reading the square on the B scale.

To find the square root of a number

Set the hairline on the number of the A scale and read the square root under the line on the D scale. Reading from the B scale to the C scale may also be used.

NOTE: Use the left half of the slide rule A scale for odd numbers left of the decimal point (1.30). Use the right half of the A scale for the even numbers before the decimal point (24.00).

Whether it is found by means of chart, slide rule, or calculation, the standard deviation of 0.12 in this example represents ± 1 SD. In other words, 68% of the tests will vary 0.12 from the average or mean value of 5.8. In view of the fact that controlling a test this closely is highly improbable, 3 SD has been set as the "confidence limits." With some methods, it is desirable to hold to 2 SD. A simple multiplying by 2 or by 3 will give the control limits desired:

	Lower	Average	Upper
± 1 SD = 0.12 mEq/L	5.7	5.8	5.9
± 2 SD = 0.24 mEq/L	5.6	5.8	6.0 (Warning lines)
± 3 SD = 0.36* mEq/L	5.4	5.8	6.2 (Action lines)

Percent of the mean

It is now understood that standard deviation is the unit variation from an average or mean value. In the case of the example given, the unit variation is ± 0.12 mEq/L. The percent of the mean or average value called "coefficient of variation" is the percentage variation. To obtain this coefficient of variation (CV), the standard deviation value is divided by the mean or average value and then multiplied by 100. The following results are obtained by this formula:

$$\frac{SD}{Mean\ value} \times 100 = Coefficient\ of\ variation$$

$$\pm 1\ SD\ of\ the\ potassium = 0.12$$
$$Average\ (mean)\ value\ of\ the\ potassium = 5.8$$

$$\frac{0.12}{5.8} \times 100 = 2.06\% \quad 1\ SD$$

---

*0.36 is rounded off to 0.4.

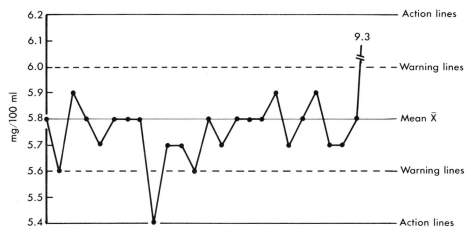

Fig. 2-2. Shewhart chart, using Table 2-1 data on potassium.

$$\frac{0.24}{5.8} \times 100 = 4.13\% \quad 2\ \text{SD}$$

$$\frac{0.36}{5.8} \times 100 = 6.20\% \quad 3\ \text{SD}$$

Control charts

There are at least four ways to present a chart for daily plotting of results:
1. The Shewhart control chart, with the mean and the deviations' boundaries drawn
2. The bell curve, inverted 90 degrees
3. The target control chart, with the rings representing the various deviations (not discussed here)
4. The cusum chart, discussed on p. 21

A convenient method of keeping the chart reusable from month to month is placing it with its numerical values under a plastic cover. The average or mean value and the confidence limits are drawn on the plastic cover with a wax pencil. Marks can be wiped off with an alcohol sponge. An example of the Shewhart chart, with Table 2-1 data, is given in Fig. 2-2.

STATISTICAL METHODS
Control chart techniques
Shewhart chart

Shewhart[15] has suggested that control charts may have these three functions:
1. Specification—define the goal that management wishes to attain.
2. Production—act as an instrument to attain that goal.
3. Inspection—serve as a judge to verify whether the goal has been obtained.

His test on quality control set the pattern for later procedures and for application of statistical methods to process control. In this scheme of process control, samples of a fixed size are used and the statistics of the sample are plotted on a chart. When the point falls outside the chart, corrective action is initiated. The standard deviation (SD) lines may be called action lines.

Dudding and Jennett[6] in 1942 proposed that warning lines be drawn in less extreme positions than the action lines. Therefore the occurrence of data appearing between the warning and the action lines should be regarded as sufficient evidence of need for corrective measures. If the chart range is ± 3 SD, the warning lines would be at ± 2 SD; if the test limits are held at ±2 SD, the warning lines should be at ± 1.5 SD. With a normal distribution of errors (gaussian), the warning limit will be exceeded for only 5% of the control results and the action limit for 0.3%. The control limits may be looked upon as follows:

SD	Percent of observations	Beyond limits (approximate)
± 1	68.27	1 out of 3
± 2	95.45	1 out of 20
± 3	99.73	1 out of 300

Levy and Jennings[11] in 1950 recommended the use of control charts in the clinical laboratory. Therefore some investigators who are medically oriented refer to the control chart as the Levy-Jennings chart.

When enough control samples have been analyzed, it is possible to determine the mean and the far points of distribution for the group. Sample fluctuations in a large group are distributed in a definite statistical pattern known as a normal distribution, illustrated by the bell or gaussian distribution curve. The bell curve is turned on its side, and the confidence limits are shown with SD lines of the control chart in Fig. 2-1.

The proper preparation of a control chart includes more than the grouping of a series of numbers. Methodology must be understood to prevent stringent demands being made on one method and overtolerance permitted for another. Establishment of warning lines requires that consideration be given to work performance under the influence of regular stress and the limitations of a routine day. Other important factors are listed below.

1. Accuracy and repeatability of the test must be reviewed with respect to low, normal, and elevated ranges that the method will have to cover, to identify the physiologic and pathologic limits.

2. Warning and action lines should be adjusted only after the method is under control and any large discrepancies have been corrected.

3. For a new method, duplicate analyses should be carried out and the charting should include both the difference between and the average of the pairs (this is an excellent method for checking variations in multiple batch work).

4. Limits of individual test noise need to be determined and the difference between this noise and trouble signals recognized.

Allowable limits of error in methods form an area of debate. Tonks' formula[17] has set a maximum limit of ± 10% for determinations:

$$\text{Allowable limits of error } (\%) = \frac{\frac{1}{4} \text{ of normal range}}{\text{Mean of normal range}} \times 100$$

Some investigators insist on closer limits whereas others allow more than ± 10%. Without standardized methodology, agreements on limits of error for test procedures is not likely. It is important to recognize that close control must be maintained. A different confidence limit is important at the normal and at elevated levels.

Hoffman and Waid[7] have presented a method for using daily values of the normal range and calculating the mean and coefficient of variation. Each daily tally represents a controlled group or a shift in the method. This approach may be used as a supplement to the regular program.

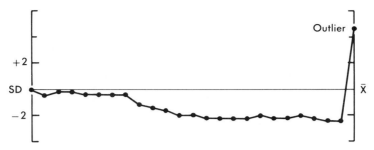

Fig. 2-3. Cusum chart of Table 2-1 data on potassium. Test out of control, as result of inferior technical application.

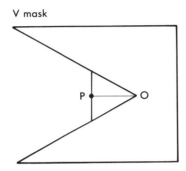

Fig. 2-4. V mask.

Cumulative sum chart

The cumulative sum or cusum chart (CSC) is normally used in industry to maintain current control of a process.[2] It was proposed in 1954 by a British statistician, E. S. Page.[13] The medical laboratories have started utilizing the chart to supplement the Shewhart chart.

The cumulative sum chart has an immediate advantage over other charts in that the work quality can be observed earlier by visual inspection.[10, 13] The other charts display results plotted independently of each other. As observed in Fig. 2-3, the CSC has no warning or action lines. An accumulative variation on the CSC will travel in a horizontal pattern until it is out of control; then there will be a trend away from the previous pattern. The CSC can also be used to check blanks.

A V-shaped mask (Fig. 2-4) can be used to determine any process changes.[9] It also can be utilized to estimate the change in the process average. As each point is plotted on the CSC, it is evaluated to observe any shift from the $\overline{X}$ value. The point P of the mask is placed over the last point plotted on the chart, and the mask is leveled so that the PO of the mask is horizontal. If any of the previously plotted points fall under the mask, there has been a process change. If the points stay within the angle of the V, the process is considered in control. Dade has a cusum control chart with the mask lines on the chart. A ruler also may be used as a diagonal guide.

Tabulation of data on the cusum chart. A point of reference must be established for plotting the CSC. In clinical application the best reference to use is the mean ($\overline{X}''$) of the previous pool if the control material is still the same. With $\overline{X}''$ as the target, the data on the CSC will be the cumulative sums of $\overline{X} - \overline{X}''$. For example, if $\overline{X}''$ is 9.2 and $\overline{X}_1 = 9.2$, $\overline{X}_2 = 9.4$, and $\overline{X}_3 = 9.3$, the chart would be as follows:

Table 2-3
Tabulation of data for cusum chart

	Results	Difference	Plot		Results	Difference	Plot
1.	5.8	0	0	14.	5.7	−0.1	−1.1
2.	5.6	−0.2	−0.2	15.	5.8	0	−1.1
3.	5.9	+0.1	−0.1	16.	5.8	0	−1.1
4.	5.8	0	−0.1	17.	5.8	0	−1.1
5.	5.7	−0.1	−0.2	18.	5.9	+0.1	−1.0
6.	5.8	0	−0.2	19.	5.7	−0.1	−1.1
7.	5.8	0	−0.2	20.	5.8	0	−1.1
8.	5.8	0	−0.2	21.	5.9	+0.1	−1.0
9.	5.4	−0.4	−0.6	22.	5.7	−0.1	−1.1
10.	5.7	−0.1	−0.7	23.	5.7	−0.1	−1.2
11.	5.7	−0.1	−0.8	24.	5.8	0	−1.2
12.	5.6	−0.2	−1.0	25.	9.3	+3.5	+2.3
13.	5.8	0	−1.0				

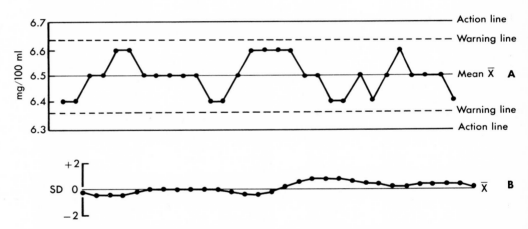

Fig. 2-5. A, Normal distribution. Shewhart chart. Total protein. **B,** Normal pattern. Cusum chart of **A.** Total protein.

$\overline{X}(9.2) - \overline{X}''(9.2) = 0$, $9.4 - 9.2 = +0.2$, $9.3 - 9.2 = +0.1$. The cusum points would be 0, +0.2, and +0.3. Refer to Table 2-3 for the tabulation scheme.

Interpretation of the charts

Normal. With a normal distribution there is an almost even distribution on both sides of the mean line (Fig. 2-5, *A*). The cusum chart in Fig. 2-5, *B*, shows a normal horizontal path within the 95% confidence limits or ± 2 SD.

Systematic trend. Problems within the analytic procedure may show a systematic trend either up or down (Fig. 2-6). A trend upward may be caused by (1) photocells or lamps that need replacing, (2) a deteriorating standard, (3) deterioration of one or more reagents, or (4) contamination in reagents or within the system. A trend

Quality control 23

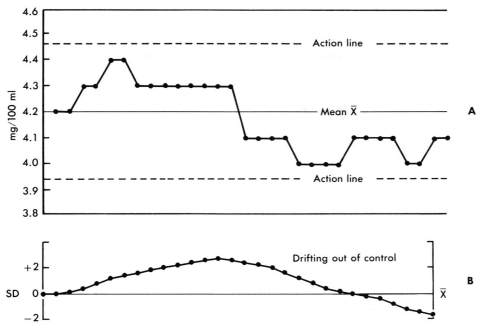

Fig. 2-6. A, Systematic trend resembling a shift. Shewhart chart. Albumin. **B,** Systematic trend more obvious. Cusum chart of **A**. Albumin.

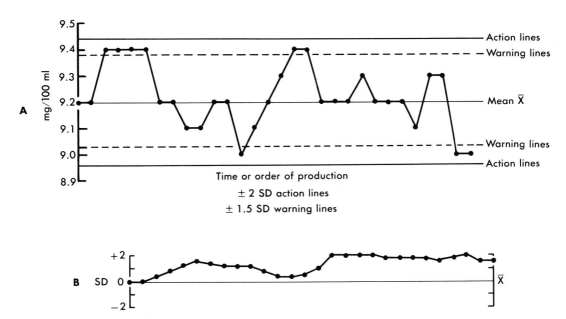

Fig. 2-7. A, Systematic shift. Shewhart chart. Calcium. **B,** Systematic shift. Cusum chart of **A**. Calcium.

downward may be caused by (1) need for rebuilding an AutoAnalyzer manifold, (2) standards that are too concentrated, and (3) contamination within reagents or systems.

Systematic shift. With the systematic shift, the points do not rise gradually, as with the trend. There is a distinct distribution accumulation on either the plus side or the negative side of the mean (Fig. 2-7). Upward shifts may be caused by any of the following: (1) partial electrical failure in the instrument, (2) overdiluted new standard, (3) deteriorated reagent that has stabilized, (4) improperly prepared reagent, too dilute, (5) contaminated reagent, (6) inaccurate timer or one in need of repair, (7) deteriorated indicator, (8) dirty glassware, and (9) overheating in a temperature-sensitive analysis. A downward trend may result from (1) increased concentration of standards and reagents by evaporation of water (keep all bottles covered or stoppered), (2) contaminated reagents, (3) inaccurate timer, too rapid, (4) underheating in analysis, (5) new indicator or new standard that is too concentrated, or (6) contaminated glassware.

Outlier. A single point may go suddenly out of range (Figs. 2-2 and 2-3). The outlier may be due to an individual condition: (1) contamination of the single specimen, (2) faulty pipet, (3) incorrect dilution of the test, or (4) contamination of a tube, pipet, or AutoAnalyzer cup.

Frequency distribution

No two units are ever alike even though a product may be said to be uniform. The variability of the units may be tabulated to observe the characteristics of the total data (universe) from which the sample data came.

Two types of data may be separated into the categories of discrete data and continuous data.

Discrete data

Discrete data differs from cumulative data in that they are an integral unit—they either exist or they do not. There is no variable, or *half-wrong*. Of course, if averages of the data are taken into consideration, it is feasible to derive results in terms of fractions. The data are expressed in the form of a whole number, and variation in this number from result to result is in multiples of 1. This is demonstrated in Tables 2-4 and 2-5. The example given is one method of tabulating results that are out of the acceptable range. It may represent a total day's run of all analyses, one group of analyses, or a report taken from the standard deviation charts. The frequencies may be represented by a histogram. However, with discrete data, remember that although the bars touch, the frequencies apply only to points on the X axis, not to a range of values. If a histogram is used, separate the bars from each other to emphasize the discrete nature of the variable. If histograms are used for the continuous and discrete data, there will be less confusion if the discrete data are plotted using vertical lines at the abscissa values. Refer to Fig. 2-8 for the comparison.

Graphic forms

The frequency distribution may be displayed graphically by a histogram or curve. The average (mid-value) of each interval is plotted on the graph.

When the frequencies of the distribution are cumulated from the low end to the other end of the scale (low → high), it is called a cumulative distribution. When the distribution is continuous, it can be best represented with a smoothline curve called an orgive, rather than by a broken line (Fig. 2-9).

Table 2-4
Number of outliers in 29 groups of calciums (30 tests/group, 870 total)

Group number	Number of outliers	Group number	Number of outliers
1	0	16	1
2	0	17	1
3	5	18	1
4	4	19	1
5	3	20	1
6	0	21	0
7	0	22	0
8	2	23	1
9	1	24	1
10	2	25	2
11	2	26	1
12	1	27	1
13	0	28	0
14	2	29	1
15	2	30 in process	

Table 2-5
Distribution of the number of outliers in 870 calciums

Number of outliers	Groups with designated number of outliers	
	Total	Fraction
0	8	0.2759
1	12	0.4137
2	6	0.2069
3	1	0.0345
4	1	0.0345
5	1	0.0345
Grand total	29	1.0000

$$\frac{\text{Total}}{\text{Grand total}} = \text{Fraction}$$

$$\frac{8}{29} = 0.2759$$

Purpose of the graph

The frequency distribution gives an overall picture of variation in a set of data. The characteristic pattern will give several bits of information.

1. Location of the data concentration will show if there is a dispersion (central tendency). An area with a single concentration is called a unimodal distribution. However, not all distributions are unimodal. There may be more than one concentration in a distribution, which is then called bimodal (Fig. 2-10).

26 *Chemistry for the clinical laboratory*

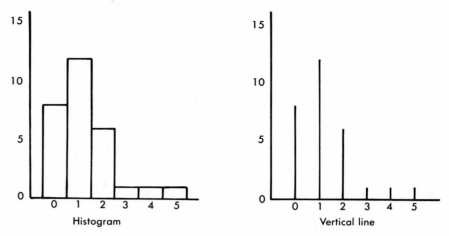

Fig. 2-8. Comparative graphs of discrete frequency distribution.

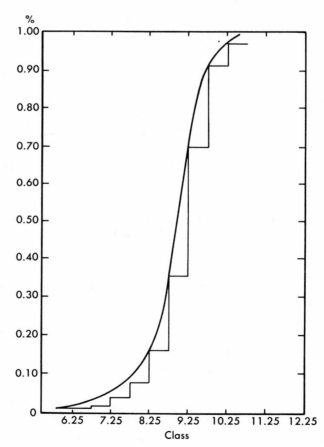

Fig. 2-9. Orgive curve and broken line showing cumulative frequency of calcium data in Fig. 2-16.

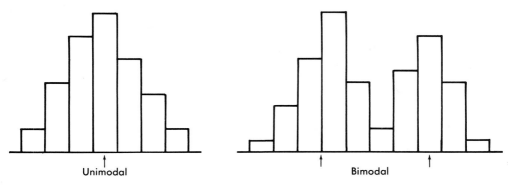

Fig. 2-10. Central tendency.

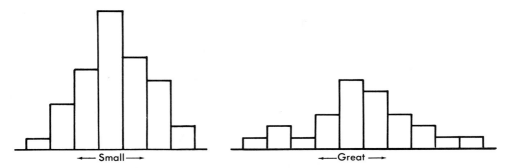

Fig. 2-11. Variability.

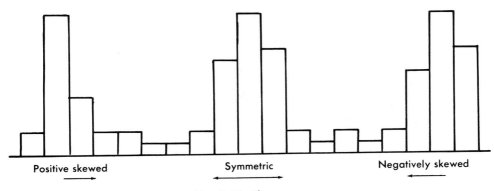

Fig. 2-12. Skewness.

2. Another answer that is needed concerns how much variability there is. Refer to Fig. 2-11. If the two figures are drawn on the same scale, then one obviously demonstrates a greater variability. Variations should be judged on the scale of the graph rather than the visual picture of the graph.

3. The graph shows the symmetry of the distribution. When it is lopsided and departs from symmetry, it is said to be skewed. With concentration of the data on the right side, the distribution is negatively skewed, and a positively skewed graph shows concentration on the left side. When there is no skewness, the graph is symmetrical (Fig. 2-12).

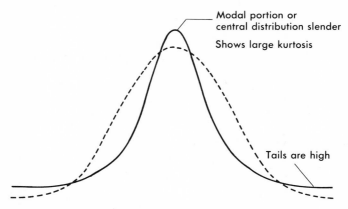

Fig. 2-13. Leptokurtic curve against a mesokurtic (normal) curve.

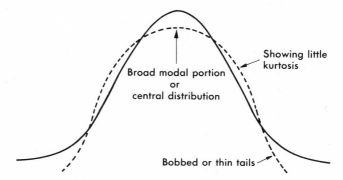

Fig. 2-14. Platykurtic curve against a mesokurtic (normal) curve.

4. The next characteristic to observe is the kurtosis or peakedness of the distribution. The graph may be symmetrical yet differ from the normal graph in the following way. The smooth-line drawing demonstrates variations from the normal curve better than the histogram. Fig. 2-13 illustrates the normal curve (dotted line) and the leptokurtic curve (solid line). The leptokurtic curve has a slender central distribution (modal) and the tails are higher. This is said to have a large kurtosis. The platykurtic curve (dotted line) is shown in Fig. 2-14. The central distribution area is broader and the tails appear to be bobbed or thin. This is said to have little kurtosis.

In some problems or evaluations, all that is needed is the graphic plot of the frequency distribution. For those who wish to measure the skewness and kurtosis, the following formulas are offered:

1. A measure of skewness as coefficient of skewness, based on all data. This can be handled mathematically. However, it is affected by extreme variations.

$$\gamma_1 = \frac{\mu^3}{\sigma^3}$$

2. A measure of kurtosis as coefficient of kurtosis. It is approximately zero for a bell-shaped distribution.

$$\gamma_2 = \frac{\mu^4}{\mu_2^2} - 3 \text{ or } \beta_2 = \frac{\mu^4}{\mu_2^2}$$

Measures of central tendency

Three different measures of central tendency are the mean, the median, and the mode.

The arithmetic mean ($\overline{X}$) is used so often that it is sometimes just called "average" or "mean." It is considered the center of gravity, easy to compute, with a relatively small sampling error. It is based on all cases and is affected by extreme cases. When this is used, it should be designated as the arithmetic mean, since there are other means: the harmonic mean, which is defined as "the reciprocal of the arithmetic mean of the reciprocal of a series of values"; the geometric mean, which is "the Nth root of the product of the values"; and the quadratic mean, which refers to "the square root of the arithmetic mean of the square of the values." So! Still further details may be found in Croxton and Cowden.* Now, see the importance of specifying *arithmetic* mean?

The median (Mi) divides the total area in half. It is difficult to handle mathematically although it is not affected by extreme cases. If the universe is normal, there will be a greater sampling error than that of the arithmetic mean.

The mode (Mo) is the value in a series that appears most frequently. The mode is difficult to find in the universe.

Continuous data

Continuous data (collectively) refer to information that can be of value within an acceptable range. Gathering continuous data requires grouping into different intervals. When raw data are accumulated, there is no order in their arrangement. Therefore, to do anything statistical with the data, you must place them in order (array). The data are listed then according to numerical magnitude. In grouping the data into intervals or classes, it is simpler to keep the intervals a constant size. Two factors to take into consideration are (1) what size interval is to be used and (2) how the intervals are to be located.

If there are too few intervals, the data will be bunched together and information lost. However, if the intervals are too wide, the distribution will be irregular and also of no value in studying the characteristics of the distribution.

In determining the format of the distribution, keep a few ideas in mind. Do not

---

*Croxton, F. E., and Cowden, O. J.: Applied general statistics, New York, 1939, Prentice-Hall, Inc., pp. 176-180.

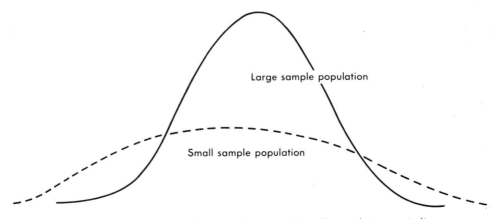

Fig. 2-15. Frequency distribution of large and small population sampling.

30 Chemistry for the clinical laboratory

Class interval	Data range	Class midpoint	5	15	25	35	45	55	65	75	85	95	105	115	125	135	145	155	Frequency total	Cumulative total	Fraction
6.0- 6.5	6.06- 6.55	6.25	‖																2	2	0.0040
6.5- 7.0	6.56- 6.95	6.75	⎮																1	3	0.0060
7.0- 7.5	6.96- 7.55	7.25	⧚																5	8	0.0170
7.5- 8.0	7.56- 8.05	7.75	⧚ ⅠⅠⅠⅠ																9	17	0.0370
8.0- 8.5	8.06- 8.55	8.25	⧚ ⧚ ⧚ ⧚ ‖																22	39	0.0850
8.5- 9.0	8.56- 9.05	8.75	⧚ ⧚ ⧚ ⧚ ⧚ ⧚ ⧚ ⅠⅠⅠ																38	77	0.1688
9.0- 9.5	9.06- 9.55	9.25	⧚ ⧚ ⧚ ⧚ ⧚ ⧚ ⧚ ⧚ ⧚ ⧚ ⧚ ⧚ ⧚ ⧚ ⧚ ⧚ ⧚ ⧚ ⅠⅠ																91	168	0.3688
9.5-10.0	9.56-10.05	9.75	⧚ ⅠⅠⅠⅠ																154	322	0.7072
10.0-10.5	10.06-10.55	10.25	⧚ ⅠⅠⅠ																103	425	0.9335
10.5-11.0	10.56-11.05	10.75	⧚ ⧚ ⧚ ⧚ ⧚ ⅠⅠⅠ																28	453	0.9950
11.0-11.5	11.06-11.55	11.25	⎮																1	454	0.9971
11.5-12.0	11.56-12.05	11.75	⎮																1	455	0.9992
12.0-12.5	12.06-12.55	12.25																			
12.5-13.0	12.56-13.05	12.75																			
13.0-13.5	13.06-13.55	13.25																			
13.5-14.0	13.56-14.05	13.75																			
14.0-14.5	14.06-14.55	14.25	⎮																		

Fig. 2-16. Frequency distribution tally (calcium).

make a frequency distribution tally on less than 30 bits of data. See example of distribution shown in Fig. 2-15. This group may be studied in array. Five or six classes can be utilized for data of 100 and increased according to the volume of material to be tabulated. It is unusual to have as many as 20 intervals, unless the data are extensive. The analytic situation will determine the number of intervals used. The best compromise is to take 10 to 15 intervals with sampling over 100 bits of data. Unless a specific range is required for the subgroup, use the following as a guide:

$$\frac{\text{High value} - \text{Low value}}{\text{Number of intervals}} = \text{Subgroup size}$$

Tabulation of data

When the format has been decided, a tally of the frequency distribution can be made from the raw data. Data may be placed on index cards and sorted manually or may be keypunched and placed on the computer. Use of the computer is an efficient method if the data processing staff has been forewarned and a program has been written for frequency distribution.

An entry form designed for column listing of numerical values according to class is an effective means of checking incorrect entries. Now, a tally form (Fig. 2-16) is drawn, and the values falling within the class interval are recorded with tally marks. The frequency column is a final count; the cumulative frequency gives the total, and the percent column is obvious. If the tally marks are kept in neat order, the general picture of the distribution can be observed.

The percentages and the class intervals are plotted on probability paper (Fig. 2-17). Intersecting lines are drawn at the 2.5% and at the 97.5% acceptance range. This gives the lower and upper limits of distribution.

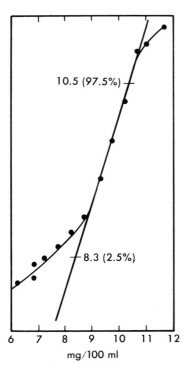

Fig. 2-17. Distribution curve from Fig. 2-16.

The t test

The comparing of lots of material—such as test kits, reagents, or controls—or two different methods has to be accomplished with individual measurements, and this next question raised. Is the difference between the measured items significant? The average or the standard deviation may be within acceptable limits of a single evaluation, when one lot or test is not being compared to another.

For example, however, the PBI method of test kit A can be compared to an automated method, B. Refer to Table 2-6. The results of the first column are squared in the second. Both columns are totaled. To obtain the sum of the x^2 (Σx^2), sums of A and B are totaled (130.15 + 137.30 = 267.45). The sum of the test columns (x_1 and x_2) are squared and divided by the number of tests performed (5).

$$A \text{ column } (25.5)^2 = 650.25 \div 5 = 130.05$$
$$B \text{ column } (26.2)^2 = 686.44 \div 5 = 137.28$$

Now, $\Sigma \dfrac{(\Sigma x)^2}{N}$ is equal to 130.05 + 137.28, or 267.33. The degrees of freedom for A and B are 5 − 1 = 4 (n − 1) individually and 4 + 4 = 8 total. The standard deviation is as follows:

$$\sqrt{\frac{267.45 - 267.33}{8}} = \sqrt{0.015} = 0.123 \text{ SD}$$

Is the difference between the methods due to chance or are they the same?

The average of the test for A = 5.10 and for B = 5.24, and the difference between these two averages is: 5.24 − 5.10 = 0.14.

The difference of 0.14 is not out of precision range for the test methods. Now, determine whether 0.14 is significant or not. The ratio of the 0.14 (difference between the two averages) to the standard deviation of 0.123 will be the determining factor.

This ratio to be calculated is called the t value.

Table 2-6
Comparison of PBI test results from two different methods

	Method A		Method B	
	$x_1(\mu g/dl)$	x_1^2	$x_2(\mu g/dl)$	x_2^2
	5.0	25.00	5.3	28.09
	5.3	28.09	5.2	27.04
	4.9	24.01	5.3	28.09
	5.2	27.04	5.2	27.04
	5.1	26.01	5.2	27.04
Total	25.5	130.15	26.2	137.30
Average	5.1		5.24	

$$t = \frac{\text{Difference between averages}}{\text{SD of difference between averages}}$$

$$= \frac{\overline{x} - \overline{y}}{\sigma} \sqrt{\frac{N_1 \cdot N_2}{N_1 + N_2}}$$

$$= \frac{0.14}{0.123} \sqrt{\frac{5 \cdot 5}{5 + 5}} = 1.138 \sqrt{\frac{25}{10}} = 1.14 \sqrt{2.5}$$

$$= 1.14 \cdot 1.58 = 1.8 \text{ t value}$$

If the t value is large the difference between averages is much greater than chance variation. Table A-8 in Appendix lists the t values, which must be expected in order to forewarn that the method differences are significant.

The degrees of freedom of A + B was 8. Refer to Table A-8 in Appendix. For this degree of freedom, the t value of 2.31 must be exceeded at the 5% probability level. The t value holds good for the 10% level. Therefore the tests are assumed to be identical. If the 10% level had been exceeded, then at least 10 more tests should have been performed and evaluated.

CONCLUSION

Analytic conditions and quality controls must be under continuous review. The need for more specific and improved methodology is evident.

A few rules that will serve advantageously in a good quality control program are (1) keep instruments in good working order with a scheduled preventive maintenance program, (2) properly ground electric appliances, (3) do not use equipment beyond its capabilities, (4) keep all equipment routinely calibrated, (5) maintain clean glassware and use disposable materials when economically feasible, (6) train personnel to utilize good technique, (7) rotate workers weekly or every 2 weeks to prevent boredom and lethargic reactions, and (8) maintain an active program for quality control as a successful team activity.

GLOSSARY

Symbols   **Definition**

accuracy Deviation of an estimate from the true value. The word used in this sense does not mean precision. It may mean bias.

AD $\frac{\Sigma FX}{\eta}$

average deviation Deviation divided by the number of determinations run. The method is rapid for obtaining the magnitude of the standard deviation and is a rough check on calculations. Is valid only on long runs of data. The relationship between the average and the standard deviation should show the SD greater than the AD by a factor of $\sqrt{\frac{\pi}{2}} = 1.25$.

bias Tendency of sample statistic to deviate in one direction from a true value, as by reason of nonrandom sampling.

BIPP Binominal probability paper.

block Group of test or experiments.

column mean square Column sum of squares divided by $\eta - 1$ (degrees of freedom).

confidence coefficient Chance that a confidence interval has of including the universal value.

confidence limits High and low limits of a confidence interval.

continuous sampling plan Continuous production without a change of lots or groups.

±1 SD 68.27%
±2 SD 95.45%
±3 SD 99.73%

control limits Standard deviations based on the normal distribution (68.27%, 95.45%, and 99.73%) of the total observation or universe.

correction factor A constant or adjustment factor applied to observed values for small sample size to obtain a better fit of data to the theoretical χ^2 distribution.

$n-1$

degrees of freedom Number of independent differences or variations possible among individual observations. If only one observation is made, there is zero degree of freedom, no other observation being available for comparison.

duplicates Unit or divided specimens produced or analyzed under the same conditions. This term is used whether there are two or more duplicates.

expected value Arithmetic mean value of a variable.

experimental error Deviation from the expected value, due to the limitations of methodology or production.

fixed factors, variables, or effects Factors, variables, or effects that are constant or the same for all samples.

f

frequency distribution Grouping of the values of a variable in order of size and plotting of the frequencies according to size.

$\bar{x}$

histogram Bar diagram representing a frequency distribution.

homoscedastic Having a uniform scatter.

interaction Tendency of two variables to combine, giving a result that differs from the sum of their independent activities.

γ_2

kurtosis Degree of peakedness in the curve. It is approximately zero for a bell-shaped distribution. $\dfrac{\mu_4}{\mu_2^{\,2}} - 3$

$\bar{x}$

mean (arithmetic) Sum of a series of tests divided by the total number.

mean error Difference between the true result and the mean obtained.

Me
Mi

median Line of equality where there are as many cases above the line as there are below.

50% of cases > Mi
50% of cases < Mi

Mo

mode Class of data with the greatest frequency, indicating the point of central tendency.

normal distribution Distribution of data in a frequency tabulation resembling a bell-shaped curve.

$\bar{x}$

outlier Result in an extreme position, beyond acceptable limits. An obviously wild error not relating to analytic variations; therefore it is not considered in tabulating data.

parameter Constant of a universe, describing an individual characteristic.

precision Exactness in the repeated determination of a test. The smaller the standard deviation, the higher the precision. Precision does not refer to accuracy; it denotes reproducibility.

P

probability Relative frequency of chance. Varies from 0 to 1.

random Condition of disorder or disarray without prediction of individual results.

R

range Simplest measure of general variability describing the difference between the highest value and the lowest value.

$$x_H - x_L$$

relative error Mean error of a series of tests, representing a percentage of the true result.

relative standard deviation Reference to coefficient of variation.

reproducibility Ability to repeat or reproduce the same data on different occasions.

n **sample** Group of items or data from a larger group (universe). Unless specified, samples are considered to be in a random state.

sampling plan Plan that designates the sample size and acceptance criteria.

sampling scheme Coordinated group of sampling plans utilized toward a certain goal.

$\bar{x}$

sensitivity Ability of a test to detect deviations from a target value or goal.

shift Congregating of well-distributed values on either side of the mean.

significant Test result deviating from the confidence limits.

γ_1

skewed Asymmetric curve based on total data. Affected by extreme variation. Can be handled mathematically. $\gamma_1 = \dfrac{\mu^2}{\sigma^3}$

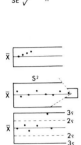

standard deviation Square root of the average of the squared deviations from the arithmetic mean.

standard error Error around the mean for a series of tests.

statistics Collection and use of numerical facts to estimate a universe parameter.

subgrouping Dividing a group of items into subgroups such that, if assignable causes are present, these will become common to all elements of the subgroups.

trend Values increase or decrease in succession from the mean.

universe Group (either infinitely or extremely large) from which samples are taken for tabulation.

variance Square of the standard deviation.

V mask V-shaped mask or overlay used with the cusum chart.

warning limits Limits on the control chart (usually at ± 2 SD) Data exceeding these limits alert for immediate attention.

References

1. Belk, W. P., and Sunderman, F. W.: A survey of the accuracy of chemical analyses in clinical laboratories, Am. J. Clin. Pathol. **17**:853, 1947.
2. van Dobben de Bruyn, C. S.: Cumulative sum tests: theory and practice, New York, 1968, Hafner Publishing Co., Inc.
3. Dodge, H. F.: Using inspection data to control quality, Manufact. Industries **16**:517, 613, 1928.
4. Dodge, H. F.: Statistical aspects of sampling inspection plans, Mech. Eng. **57**:645, 1935.
5. Dodge, H. F., and Romig, H. G.: Sampling inspection tables, single and doubling sampling, New York, 1959, John Wiley & Sons.
6. Dudding, B. P., and Jennett, W. J.: Quality control charts (British Standard 600R: 1942), London, 1942, British Standards Institution.
7. Hoffman, R. G., and Waid, M. E.: The average of normals: method of quality control, Am. J. Clin. Pathol. **43**:134, 1965.
8. Industrial Quality Control **4**:23, July, 1947.
9. Kemp, K. W.: The average run length of the cumulative sum chart when a V-mask is used, J. R. Stat. Soc. **23**:149, 1961.
10. Kemp, K. W.: The use of cumulative sums for sampling inspection schemes, Appl. Stat. **11**:16, 1962.
11. Levy, S., and Jennings, E. R.: The use of control charts in the clinical laboratory, Am. J. Clin. Pathol. **20**:1059, 1950.
12. Moroney, M. J.: Facts from figures, Harmondsworth, England, 1951, Penguin Books, Ltd.
13. Page, E. S.: Continuous inspection schemes, Biometrika **41**:100, 1954.
14. Shewhart, W. A.: Economic control of quality of manufactured product, New York, 1931, Van Nostrand Co., Inc.
15. Shewhart, W. A.: Statistical method from the viewpoint of quality control, Washington, D. C., 1939, Graduate School, Department of Agriculture, chap. 1.
16. Snedecor, G. W.: Queries, Biometrics **8**:85, 1952.
17. Tonks, D. B.: A study of the accuracy and precision of clinical chemistry determinations in 170 Canadian laboratories, Clin. Chem. **9**:217, 1963.
18. Wootton, I. D. P., and King, E. J.: Normal values for blood constituents: interhospitals differences, Lancet **1**:470, 1953.

3

AutoAnalyzer methods

AutoAnalyzer I
ADAPTING MANUAL METHODS TO THE AUTOANALYZER

Most manual methods of analysis can be adapted to the AutoAnalyzer. The following steps are presented as a guide to automating a manual procedure.

1. Consider the kind of specimen required. Is it a simple dilution or a protein-free filtrate? Is the constituent to be measured dialyzable?

2. Consider the chemistry involved. What product of the reaction is measured? Can a different product be measured? Is a specimen blank correction necessary and, if so, can blanks be avoided by dialysis? Example:

$$p\text{-Nitrophenyl phosphate} \xrightarrow{\text{Alkaline phosphatase}} p\text{-Nitrophenol} + \text{Phosphate}$$

Either p-nitrophenol or phosphate could be measured. Both are dialyzable; therefore blank correction is not needed. In this case it is preferable to measure the p-nitrophenol because it has color and can be measured directly. The phosphate would have to be reacted with a reducing agent to produce a color that could be measured.

3. Plan a preliminary flow diagram. Decide what modules are needed. Can pump I handle the number of pump tubes necessary or will pump II be needed? Is a heating bath required? What temperature? Is a dialyzer needed? What will be used to measure the end product?

    a. Colorimeter: Photocell or phototube? What filters?
    b. Fluorometer: What filters?
    c. Flame photometer: What filters?

Will any special glass fittings or coils be required? Solvaflex or Acidflex tubing?

4. Consider the proportions used in the manual method and match the proportions with the pump tubes (Table 3-1). Example:

	Sample	*Substrate*
Manual	0.1 ml	0.9 ml
Automatic	0.015 inch I.D.	0.045 inch I.D.

In systems using dialysis some adjustments of proportions are necessary. Figure on a 10% dialysis ratio (some constituents have a higher dialysis ratio) and adjust tubing sizes accordingly. Match the flow rate through the dialyzer so that it comes as close as possible to being the same on both sides of the membrane. This can be accomplished by adjusting the size of the air tubes.

5. Build and test the system.

6. Evaluate results. Has the desired sensitivity been achieved? If not, readjust the sample and reagent proportions, the incubation time, or the color development time.

Table 3-1
Pump tube sizes and delivery volumes*

Tube I.D. (inch)	Shoulder colors	Clear standard delivery (ml/min)			Clear standard delivery (ml/hr)	Solvaflex delivery (ml/min)			Solvaflex delivery (ml/hr)	Acidflex delivery (ml/min approx)	Acidflex delivery (ml/hr)
		Min	Nom	Max	Max	Min	Nom	Max	Max		
0.005	Orange-black	0.005	0.015	0.029	1.74	0.005	0.015	0.029	1.74	n.a.†	
0.0075	Orange-red	0.016	0.03	0.048	2.88	0.016	0.03	0.048	2.88	n.a.	
0.010	Orange-blue	0.032	0.05	0.072	4.32	0.032	0.05	0.072	4.32	n.a.	
0.015	Orange-green	0.075	0.10	0.128	7.68	0.075	0.10	0.128	7.68	n.a.	
0.020	Orange-yellow	0.13	0.16	0.19	11.4	0.13	0.16	0.19	11.4	n.a.	
0.025	Orange-white	0.19	0.23	0.27	16.2	0.19	0.23	0.27	16.2	n.a.	
0.030	Black	0.28	0.32	0.36	21.6	0.28	0.32	0.36	21.6	0.34	20.4
0.035	Orange	0.37	0.42	0.47	28.2	0.37	0.42	0.47	28.2	0.43	25.8
0.040	White	0.54	0.60	0.66	39.6	0.51	0.56	0.62	37.2	0.53	31.8
0.045	Red	0.73	0.80	0.87	52.2	0.64	0.70	0.76	45.6	0.64	38.4
0.051	Gray	0.92	1.00	1.08	64.8	0.81	0.88	0.93	55.8	0.78	46.8
0.056	Yellow	1.12	1.20	1.28	76.8	0.99	1.06	1.13	67.8	0.92	55.2
0.060	Yellow-blue	1.31	1.40	1.49	89.4	1.14	1.21	1.29	77.4	1.06	63.6
0.065	Blue	1.50	1.60	1.70	102.0	1.29	1.37	1.45	87.0	1.19	71.4
0.073	Green	1.90	2.00	2.10	126.0	1.60	1.69	1.78	106.8	1.44	86.4
0.081	Purple	2.37	2.50	2.63	157.8	1.92	2.02	2.12	127.2	1.71	102.6
0.090	Purple-black	2.77	2.90	3.03	181.8	2.31	2.42	2.53	151.8	2.03	121.8
0.100	Purple-orange	3.26	3.40	3.54	212.4	2.77	2.89	3.01	180.6	2.39	143.4
0.110	Purple-white	3.75	3.90	4.05	243.0	3.26	3.39	3.52	211.2	2.76	165.6

*Courtesy Technicon Instruments Corp., Tarrytown, N. Y.; from White, W. L., Erickson, M. M., and Stevens, S. C.: Practical automation for the clinical laboratory, ed. 2, St. Louis, 1972, The C. V. Mosby Co.
†Not available.

7. If sensitivity is satisfactory, determine the optimum sampling rate. A proper rate is one in which a sample has an absorbance (A) or % T reading equal to 95% of the steady state reading of the same sample. Also, a proper rate will keep the sample interaction at a minimum.

8. Check the automated method against the manual method. If results do not agree, investigate the cause.

9. Readjustment of reagent concentrations may be necessary, or advantageous, in some procedures.

NOTES ON GENERAL OPERATING PROCEDURE

1. Before turning a colorimeter on, be sure that proper filters are in place. Never allow the full intensity of the lamp to strike the photocells. The effect

would be similar to that of a person's looking directly into the sun. Eyes require some time to refocus. If staring at the sun is done repeatedly, the eyes will suffer permanent damage and even blindness. The same type of damage is done to the photocells unless a set of filters is in place. If it is necessary to change filters for another procedure, turn the lamp off just long enough to change the filters.

In reading the following flow charts, assume that the standard 15-mm tubular flowcell is used, unless specified otherwise.

2. Before starting the pump, clean the platen and the manifold pump tubes with a cloth dampened with methanol. Wipe the rollers free of any grease at the same time. This will prolong the life of the pump tubes and contribute to a smooth flow through the tubes. When engaging the roller head assembly on pump I, be sure that the pump tubes are not overlapping.

3. Never release the roller head (pump I) or lift the platen (pumps II or III) while reagents are being pumped. Mixed reagents will back up through the lines into the reagent bottles and cause contamination of the reagents. Clamp the lines with hemostats before releasing the roller head.

4. Allow sufficient time for a stable flow of reagents through the flowcell before setting the reagent baseline. This will lessen the chance of baseline drift during the test run.

5. Compensate for baseline drift by adjusting the standard curve to a standard place in every tenth position in the sample plate.

6. When transferring reagent lines from one bottle to another (reagent or water), wipe the outside of the lines with a tissue to avoid contaminating the other bottle.

7. Each time a dialyzer membrane is changed, the bath water should be changed also. It is not enough to put an antibacterial agent in the water and forget about it. Invariably, something spills or leaks into the bath water. As acids or salts accumulate, corrosion of the metal parts takes place. Always use distilled or deionized water in the dialyzer bath. Be sure the temperature has stabilized at 37° C before running any procedure. If it becomes necessary to change a membrane in the middle of a run, the bath water may be changed when the run is completed. Always run fresh standards after the membrane has been changed before finishing the run.

8. Change the oil in the heating bath at least every 3 months. Remove all old oil from the coil, heating element, and thermoregulator by soaking the top part in the pot filled with a hot detergent solution. Before reassembling the heating bath, be sure there is no charred material coating the heating element. If necessary, scrape it off with a knife.

9. Be sure the circulating motors on the dialyzer and heating baths are working. If the liquid is not circulating, the temperature will vary considerably in different parts of the bath as the heating element cycles on and off. This can cause wide variation in test results.

10. If waste from the colorimeter is being collected in a container instead of draining into a sink, be sure that the debubbler line does not hang down below the surface of the liquid in the container. An all-liquid line would have no trouble in draining, but air bubbles would have to go down below the surface of the liquid to escape. Pressure is required to push the air down and out the end of the drain tube. The farther down below the surface the end of the tube is, the more pressure is required to push the air out. This buildup of pressure will affect the flow through the flowcell.

11. Check the recorder gain each day or before starting a different procedure.

12. Always wait until the last specimen has recorded before removing lines from reagent bottles. If the lines are pulled out of reagents too soon, the last one or two specimens may be lost because there were no reagents with which to react. The sample has a longer distance to travel than do most reagents.

13. An "in-line" filter is now available that is very helpful. Each filter unit is self-contained and is complete with nipples for connection into the reagent lines. It can be attached at any convenient point between the manifold tubing and the reagent bottle. The filter matrix is chemically inert and is unaffected by strong acids or bases. The manufacturer* states that the filter will completely remove any debris from the reagents (down to a 6-μ particle size).

14. One of the ways to adjust the sensitivity of a procedure that requires dialysis is to change the size of the sample, sample diluent, or recipient stream pump tube. When this is done, however, take care to see that the flow rates on each side of the dialyzer membrane are as close as possible to being equal. This can usually be accomplished by changing the air line to balance the change in a reagent or sample line. When calculating flow rates, the air lines must be included.

15. To determine the reagent consumption for a particular procedure, simply note the size of the tube pumping the reagent and refer to Table 3-1, where the approximate ml/min and ml/hr values are given.

16. Detailed information on the individual AutoAnalyzer modules has been published.[83]

WETTING AGENTS

A wide variety of wetting agents or surfactants is now being used in analytic chemical procedures. The main purpose is to lower the surface tension of a liquid and thus permit a smoother flow through small-diameter tubing.

Sterox SE

One of the earliest wetting agents used in the clinical laboratory was Sterox SE,† the brand name for p-oxyethylene thioether. Sterox SE is a 100% nonionic type of surface-active agent and is also a wetting agent, emulsifier, and detergent. It has a relative density of 1.0495 at 25° C. This surfactant has been widely used in manual flame photometry. A concentration of 0.02% Sterox SE in standards and unknown solutions provides a smooth flow through the atomizer into the flame and helps to keep the burner clean.

Brij 35

The surfactant most commonly used in AutoAnalyzer methods is Brij 35,‡ a polyoxyethylene (23) lauryl ether. It is a nonionic, white waxy solid with an approximate relative density of 1.05 and 25° C. Its pour point is approximately 33° C. The solution of Brij 35 supplied by the Technicon Corporation for use with their systems has a concentration of about 30%.

---

*UniChem Corp.
†Harleco.
‡Atlas Chemical Industries, Inc.

Triton X-100

Triton X-100* is a water-soluble isooctyl phenoxy-polyethoxy ethanol containing 10 moles of ethylene oxide. It is a biodegradable, anhydrous liquid, 100% active, nonionic surface-active agent. It is useful as a wetting agent, an emulsifier, and a detergent. The relative density of Triton X-100 is 1.065 at 25° C and its pour point is 7° C.

Triton X-405

Triton X-405* is a preparation of p-tertiary octylphenoxy-polyethoxy ethanols with about 40 moles of ethylene oxide. It is a liquid, 70% active, nonionic surface-active agent. The relative density of Triton X-405 is 1.11 at 25° C and its pour point is −4° C.

Both Triton X-100 and X-405 are ideal surfactants for enzyme determinations in continuous flow systems, since they do not interfere with enzyme activity.

Aerosol 22

Aerosol 22† is a 35% aqueous solution of tetrasodium N-(1,2-dicarboxyethyl)-N-octadecyl-succinamate. It is an anionic surface-active agent with a relative density of 1.12 and a boiling point of 95° C (at 760 mm). Aerosol 22 is soluble in hot or cold water and saturated salt solutions and is insoluble on most organic solvents. Its surfactant properties are not as good as those of the other surface-active agents listed but it has proved of value in systems where other surfactants have been found to interfere in some way. One of these is the Technicon N-13b uric acid procedure.

FC-134

FC-134‡ is one of 3M's Fluorad group of surfactants. It is a cationic brown waxy solid. FC-134 is soluble in acetone and in ethyl, isopropyl, and methyl alcohols but is only slightly soluble in water.

ARW-7

ARW-7§ is an anionic surfactant composed of a 40% by weight aqueous solution of sodium 2-ethylhexyl sulfate. It has a specific gravity of 1.144 at 20° C and is completely soluble in water. Commercially, it is used in caustic solutions for washing fruits and vegetables. In the laboratory, it is used in the biuret reagent for the SMA 12/60 and the AutoAnalyzer II. Care should be used in handling ARW-7 because it will defat the skin and can damage the eyes in its undiluted form. It is corrosive to such materials as steel, aluminum, galvanized iron, and tin. ARW-7 should be stored in polyethylene containers in a place where the temperature does not go below 10° C or above 38° C (50° to 100° F); otherwise, phase separation will occur.

Ultrawet 60L

Ultrawet 60L surfactant,‖ also known as wetting agent A,§ is a 65.5 solution of triethanolamine linear alkylate sulfonate. It is an anionic biodegradable detergent

---

*Rohm & Haas Co.
†American Cyanamid Co.
‡3M Co.
§Technicon Instruments Corp.
‖Arco Chemical Co.

with a relative density of 1.14 at 25° C. Commercially, it is used in formulas for shampoo, bubblebath, and specialty detergents. In the laboratory, it is used in an automated phosphorus procedure.

METHODOLOGY
Electrolytes (Na, K, CO_2, Cl)
Principle

The AutoAnalyzer electrolyte system determines sodium potassium, carbon dioxide, and chloride simultaneously from a single specimen.[20]

Serum or heparinized plasma is mixed with a CO_2-free air–segmented stream of acid lithium nitrate solution. The diluted sample enters the dialyzer where the sodium, potassium, chloride, and lithium ions dialyze into a CO_2-free air–segmented stream of water. The lithium serves as an internal standard in the determination of sodium and potassium. The acid in the lithium nitrate solution liberates free CO_2 gas from the carbonate and dissolved CO_2 in the sample. As the sample stream leaves the dialyzer, it is joined and mixed with an acid antifoam reagent. The stream then enters a gas-liquid phase separator (CO_2 trap), where the liquid goes to waste and the gas phase is resampled. The resampled CO_2 gas is mixed with a weakly buffered phenolphthalein solution. The CO_2 decolorizes the phenolphthalein solution and the loss of color is measured at 550 nm in a 15-mm tubular flowcell.[71] This is called an inverse color reaction.

As the recipient stream leaves the dialyzer, a portion of the stream is resampled and mixed with chloride color reagent. The chloride ions (Cl^-) combine with the mercuric ions (Hg^{++}) in mercuric thiocyanate, $Hg(SCN)_2$, to form $HgCl_2$. The released $(SCN)^-$ reacts with Fe^{+++} to form the red complex–ferric thiocyanate, $Fe(SCN)_3$.[86] The intensity of this color is measured at 480 nm in a 15-mm tubular flowcell.

The remainder of the recipient stream is pumped to the flame photometer, where the major portion is aspirated into the burner for the simultaneous determination of sodium and potassium. The excess liquid and air segments go to waste.

Modules needed

  One sampler II
  One proportioning pump II or two pump I's
  One dialyzer with one set of dialyzer plates
  Two colorimeters with 480- and 550-nm filters
  One flame photometer III or III A
  Two double-pen recorders

Flow diagrams

  See Figs. 3-1 and 3-2.

Reagents

  Water recipient (0.5 ml Brij 35/L water)
  Lithium nitrate stock, 1000 mEq Li/L in 2 N H_2SO_4 (D-1)
  Lithium nitrate working solution, 125 mEq Li/L in 0.25 N H_2SO_4 (D-2)
  Mercuric thiocyanate, saturated (D-3)
  Ferric nitrate, 0.5 M (D-4)
  Mercuric nitrate, 0.2 M in 0.2 N HNO_3 (D-5)
  Chloride color reagent (D-6)
  Acid antifoam diluent (D-7)
  Sodium hydroxide wash solution (D-8)

42 Chemistry for the clinical laboratory

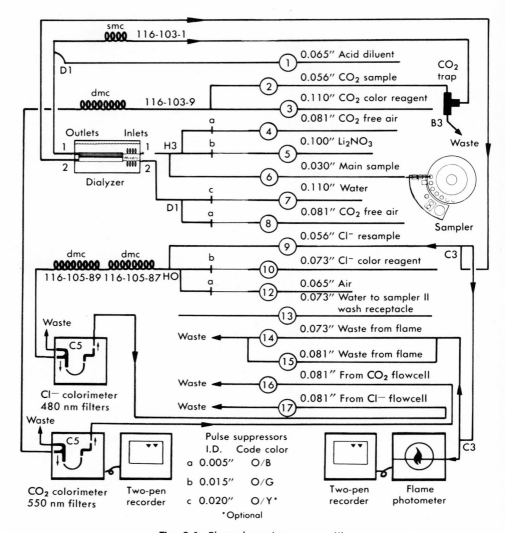

Fig. 3-1. Electrolytes (one pump II).

Phenolphthalein indicator, 1%	(D-9)
Sodium carbonate, 1 M	(D-10)
Sodium bicarbonate, 1 M	(D-11)
Carbonate-bicarbonate buffer	(D-12)
CO_2 color reagent	(D-13)
Sodium chloride stock standard, 1 M	(D-14)
Potassium chloride stock standard, 0.1 M	(D-15)
Ammonium chloride stock standard, 1 M	(D-16)

Dilute stock standards

Prepare dilute stock standards as follows:

Standard A. Measure 100 ml of 1 M sodium carbonate stock, 200 ml of 0.1 M potassium chloride stock, and 100 ml of 1 M ammonium chloride stock standards and transfer to a 1-L volumetric flask. Dilute to volume with water and mix thoroughly. Store in a Pyrex or polyethylene bottle. This standard

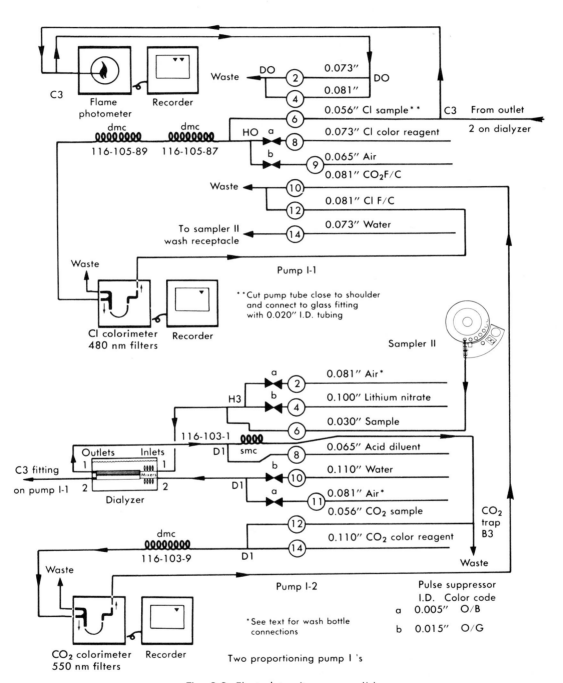

Fig. 3-2. Electrolytes (two pump I's).

contains 200 mEq of Na, 20 mEq of K, 120 mEq of Cl, and 100 mEq of CO_2/L.

Standard B. Dilute 800 ml of 1 M sodium chloride to 1000 ml. This standard contains 800 mEq of Na and 800 mEq of Cl/L.

Dilute working standards

Measure the following amounts of dilute standards A and B into 100-ml volumetric flasks. Dilute to volume and mix.

Working standard	Standard A (ml)	Standard B (ml)	Values in mEq/L of working standards			
			Na	K	Cl	CO_2
1	5	10	90	1	86	5
2	10	10	100	2	92	10
3	15	10	110	3	98	15
4	20	10	120	4	104	20
5	25	10	130	5	110	25
6	30	10	140	6	116	30
7	35	10	150	7	122	35
8	40	10	160	8	128	40

Procedure

1. If flame III, III A, or IV is used, heparinized plasma or serum samples may be run at a rate of 60/hr (2:1 sample-wash ratio). This rate is possible only with flame III, III A, or IV, which operates on natural gas (or propane) and compressed air. The recommended rate for flame I (propane and oxygen) and for flame II (natural gas and oxygen) is 40 specimens/hr.
2. Because of the number of manifold pump tubes required it is recommended that either one pump II or *two* pump I's be used.
3. It is necessary for the air that segments the diluent and recipient streams to be bubbled through a sodium hydroxide wash solution to remove atmospheric CO_2. To accomplish this, the sodium hydroxide wash bottle must have an airtight cap with an air inlet tube that goes down below the level of the sodium hydroxide. The bottle should be no more than two thirds full. The outlet tube should be well above the level of the wash solution to avoid the possibility of sodium hydroxide being carried over into the CO_2 color reagent or to the sample stream. There is now available a carbon dioxide–absorbing cartridge (T11-0415-56)* that may be used in place of the sodium hydroxide wash solution. The cartridge contains a color indicator that is pink while the absorbent is still active and turns tan/yellow when it is spent. The cap or stopper in the CO_2 color reagent should also be airtight. The tube carrying CO_2-free air into the bottle should just go through the stopper but not reach the level of the liquid. The outlet tube, connected to the color reagent pump tube, should be long enough to reach the bottom of the bottle. To avoid interference from atmospheric CO_2, the caps or stoppers of both the sodium hydroxide wash bottle and the CO_2 color reagent bottle *must* be airtight.
4. To maintain good bubble patterns and low noise levels, add 0.5 ml of Brij 35/L to the lithium nitrate solution, the water recipient, and the CO_2 color reagent.

---

*Technicon Instruments Corp.

Operating sequence

1. Turn on gas and air to the flame photometer and ignite the flame according to the instructions in the flame photometer manual. Allow 30 minutes for the instrument to warm up and stabilize. Do *not* allow any liquid to be aspirated into the burner until the flame is ignited. Turn on colorimeters.
2. While the flame photometer is warming up, start pumping water through all the lines. Connect the appropriate line to the flame photometer. Check to see that all parts of the manifold are connected properly and that both streams coming from the dialyzer are flowing properly.
3. Pump reagents through all lines except the CO_2 color reagent line. Leave it in water along with the sample line.
4. Adjust the CO_2 water baseline to 97% or 98% T. Check 0% T and, after the pen returns to the water baseline, check the recorder gain.
5. Connect the CO_2 color reagent line to the output tube on the color reagent bottle.
6. After the reagents have been pumping for about 10 minutes, adjust the chloride baseline to 97% or 98% T. Set the 0% T. Readjust the reagent baseline if necessary. Check the recorder gain.
7. Observe the reagent baseline (RBL) of CO_2. It should be about 20% T but not lower than 10% T nor higher than 30% T. If the RBL is lower than 10% T, dilute the color reagent with water. If the RBL is higher than 30% T, add a few drops more of phenolphthalein indicator and mix.
8. Place the sample line into a tube that contains working standard 8 (160 mEq of Na, 8 mEq of K, 40 mEq of CO_2, and 128 mEq of Cl/L).
9. Adjust the sodium pen to 95% T and the potassium pen to 90% T. Check the recorder gain on both pens.
10. Observe the steady-state position of the CO_2 pen on the chart while aspirating standard 8. It should be between 80% and 90% T—ideally, close to 90% T—for maximum sensitivity. If it reads less than 80% T, dilute the color reagent with water. If it reads greater than 90% T, add 2 to 3 drops of carbonate-bicarbonate buffer to the color reagent.
11. Place the sample line into a tube containing standard 2 (92 mEq of Cl/L). Note the position of the chloride pen on the chart when the steady state of this standard records. It should be between 60% and 65% T. Mercuric nitrate is added to the chloride color reagent to control the sensitivity. The 3.5 ml/L volume called for in the directions binds about 60 mEq/L of chloride ions in the sample. Consequently, only samples with chloride concentrations greater than 60 mEq/L will move the pen from the reagent baseline. This brings the physiologic range of blood into a more readable and accurate portion of the chart. If the steady-state reading of the 92 mEq/L standard is lower (toward 0% T) than 60% T, add more mercuric nitrate in 0.1-ml steps. Excessive use of mercuric nitrate will result in higher noise levels. If the standard reads higher than 65% T, dilute the color reagent with a batch of chloride color reagent containing no mercuric nitrate.
12. Place the sample line in water and load standards on the sample plate in the following order: Three cups of standard 4 followed by standards 1 through 8. Place a cup containing standard 5 after standard 8 and in every tenth position on the plate. Fill the remaining positions with unknown samples to be analyzed. *Do not* use any water cups or leave any empty spaces. Fill any blank spaces with a standard cup.

NOTE: Do not pour standards into cups until the instrument is ready to start

sampling. Standards exposed to air will absorb CO_2 from the atmosphere.
13. When the last sample has recorded, place all the reagent lines in water. After pumping water for 2 minutes, place the chloride color reagent line in a 25% bleach solution. Leave the other lines in water. Continue pumping for 10 minutes; then place the chloride color reagent line in water and flush for 10 minutes. Disconnect the line going to the flame photometer *before* turning the flame off.

Calculations

Values for the unknown samples may be determined by using Technicon comparators or by reading from a curve prepared by plotting standard readings versus concentration on graph paper. The percent transmission of CO_2 and chloride standards should be plotted versus concentration on semilog paper. The percent transmission of sodium and potassium standards may be plotted versus concentration on linear graph paper.

System maintenance and troubleshooting

1. The dialyzer membrane should be changed at regular intervals, depending on use. Before changing the membrane, pump a 25% Clorox solution through the sample and reagent lines for 10 minutes. Follow with a 10-minute water wash and then change the membrane. This flushes out any deposits of precipitated protein in the tubing. Daily rinsing of the chloride color reagent line with bleach prevents a film buildup in the flow cell and keeps the mixing coils clean.

2. The acid antifoam diluent must be shaken each day before use because the Antifoam B has a tendency to settle out. If there is not enough antifoam in the solution, bubbling will occur in the CO_2 trap. When this happens, a portion of the acid solution is resampled through the CO_2 line, mixes with CO_2 color reagent, and decolorizes the purple phenolphthalein. This is recorded on the CO_2 chart as a very erratic pattern, usually spiking toward 95% T.

3. The small capillary leading into the CO_2 trap or gas-liquid phase separator is the one spot on the manifold most likely to accumulate small particles of precipitated protein. When this capillary becomes completely obstructed, pressure builds up back along the diluted sample line. Eventually one of the connections will "pop" but that connection may be far removed from the CO_2 trap. It just happened to be the "weakest link." In other words, if a connection "pops" anywhere along the diluted sample line, look first for an obstruction in this small capillary.

4. Filter the chloride color reagent to avoid noisy recordings.

5. Clean the flame photometer glass chimney and mirror each day before lighting the flame.

6. Try to maintain a constant temperature in the room where the instrument is located. A changing room temperature will cause considerable drift in standardization, particularly for CO_2 and sodium.

Simultaneous glucose and urea nitrogen
Principle

The simultaneous AutoAnalyzer procedure for the determination of glucose and urea nitrogen[69] is based on the individual procedures for glucose and urea nitrogen. Two dialyzer plates are used, connected in series for the sample stream and in parallel for the recipient streams.

The automated glucose procedure was adapted from the manual procedure of

Hoffman.[37] It is a method utilizing the potassium ferricyanide–potassium ferrocyanide oxidation reduction reaction. The yellow potassium ferricyanide solution is reduced by the glucose to the colorless ferrocyanide. The loss of color is measured at 420 nm using a flowcell with a 15-mm light path. The diffusion rate of glucose in biologic materials is similar to that of aqueous standards.

The urea nitrogen method is a slightly modified version of the one described by Marsh and co-workers.[53] It is a modification of the carbamido-diacetyl reaction. In acid solution, diacetyl monoxime (2,3-butanedione-2-oxime) is hydrolyzed to diacetyl, which reacts directly with urea in the presence of acidic ferric ions. The pink color of the reaction product is intensified by the presence of thiosemicarbazide. This enables the determination to be run without the need of concentrated acid reagents. The colored reaction product is measured in a 15-mm flowcell at 520 nm.

In the simultaneous procedure, the sample is diluted with saline, segmented with air, and pumped into a 37° C dialyzer bath where it goes through a mixing coil before entering the top side of a set of dialyzer plates. Glucose in the diluted sample dialyzes through a membrane into an air-segmented stream of alkaline potassium ferricyanide. Upon leaving the dialyzer, the recipient stream enters the 95° C heating bath (40-foot coil, 1.6 mm I.D.) for the color reaction to take place. The glucose reduces the ferricyanide (yellow) to ferrocyanide (colorless) in proportion to the amount of glucose present in the dialysate. The reaction stream then enters the colorimeter, where the loss of color is measured at 420 nm. This is called an inverse color reaction.

The diluted sample stream emerging from the first dialyzer plate (which in a single glucose method would be pumped to waste) is pumped into the top of a second set of dialyzer plates. Urea is dialyzed into an air-segmented stream of working BUN color reagent. As the recipient stream leaves the dialyzer, it is joined by a stream of working BUN acid reagent, passed through a mixing coil, and then pumped into a second coil in the 95° C heating bath. As the acidified solution passes through the heating bath coil, the urea reacts with diacetyl and the color formation is intensified by thiosemicarbazide. The reaction stream then enters the colorimeter where the color intensity is measured at 520 nm. This is a direct color reaction.

Modules needed
  One sampler II
  One proportioning pump (I or II)
  One dialyzer bath with two sets of dialyzer plates
  One double-coiled heating bath at 95° C
  Two colorimeters with 420- and 520-nm filters
  One double-pen recorder or two single-pen recorders

Flow diagrams
  See Fig. 3-3.

Reagents
  Saline, 0.9%   (D-17)
  Alkaline potassium ferricyanide   (D-18)
  Diacetyl monoxime stock, 2.5%   (D-19)
  Thiosemicarbazide stock, 0.5%   (D-20)
  Working BUN color reagent   (D-21)
  Ferric chloride–phosphoric acid stock   (D-22)
  Sulfuric acid stock, 20%   (D-23)
  Working BUN acid   (D-24)
  Benzoic acid, saturated   (D-25)

48 *Chemistry for the clinical laboratory*

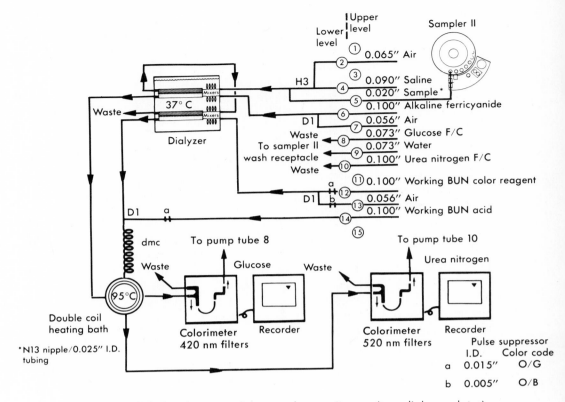

Fig. 3-3. Simultaneous glucose and urea nitrogen (two dialyzer plates).

Glucose stock standard, 10 mg/ml	(D-26)
Standard diluent, 0.01 N H_2SO_4 containing 40 mg/L PMA	(D-27)
Urea nitrogen stock standard, 10 mg urea N/ml	(D-28)

Working standards

Prepare the working standards in 100-ml volumetric flasks. Using volumetric pipets, measure the amounts of the two stock standards indicated below into each of the flasks and dilute to volume with standard diluent (D-27).

Working standard	Glucose stock (ml)	Urea N stock (ml)	Glucose (mg/dl)	Urea N (mg/dl)
1		0.5	0	5
2	5	1	50	10
3		2	0	20
4	10	3	100	30
5		4	0	40
6	15	5	150	50
7		6	0	60
8	20	7	200	70
9		8	0	80
10	25	9	250	90
11		10	0	100
12	30	15	300	150

NOTE: If preferred, the glucose standards may be grouped together in either the first or last six working standards.

Procedure

1. The simultaneous glucose/BUN procedure may be run at a rate of 60 specimens/hr with a 2:1 sample-wash ratio. However, the amount of sample interaction or carryover can be reduced considerably by using a a rate of 50/hr with a 1:1 sample-wash ratio.
2. Glucose and urea nitrogen may be determined in serum, plasma, spinal fluid, or anticoagulated whole blood. If whole blood is used, a mixer should be used on the sampler to maintain a homogeneous suspension of cells in the sample. Unless a fluoride preservative is used, all cells should be removed as soon as possible after the specimen is obtained. Glucose cannot be determined on urine by this method because of the presence of high concentrations of uric acid, creatinine, and other reducing substances. Urea nitrogen can be determined on urine without prior treatment because the reaction is not affected by the presence of ammonia. The urine specimen need only be diluted with water to bring the urea nitrogen concentration within the range of the standards.
3. The bubble patterns of the system are improved and recorder noise is reduced by adding 0.5 ml/L of Brij 35 to the saline diluent, potassium ferricyanide, and working BUN color reagents. If surging occurs in the glucose heating bath coil, add more Brij 35 to the potassium ferricyanide (up to 2 ml/L).
4. Since glucose is determined by an inverse colorimetric procedure, it is necessary to establish a "reference" (colorless) baseline before getting a reagent baseline. This is accomplished by setting the pen to 97% or 98% T while pumping water through all the lines. The reagent baseline is then determined by putting the saline and potassium ferricyanide reagent lines in their respective bottles. This reagent baseline (RBL) should be at 13% T ± 3%. The RBL can be raised by adding a solution of 2% sodium carbonate in 0.9% sodium chloride or lowered by adding a solution of 5% potassium ferricyanide in 0.9% sodium chloride to the potassium ferricyanide solution. It would be advisable to calculate, at the time this adjustment is being made, the amount of potassium ferricyanide required in future batches of reagent to obtain the desired RBL.
A 300 mg/dl standard should read between 80% and 90% T. If it goes higher than 90% T, the RBL may be adjusted down toward 10% T but not below 10% T. If the 300 mg/dl standard still reads above 90% T, the ferricyanide reagent manifold pump tube should be increased to 0.110 inch I.D.
5. BUN color reagent should not be kept in a cold place or precipitation may occur. However, a gentle warming should redissolve the precipitate. If too much sensitivity is obtained, the concentration of sulfuric acid in the BUN acid may be reduced to 15%.

WARNING: Do not use the high speed of a two-speed pump while reagents are going through the lines. There is the danger of "popping" a line and spurting acid over everything in the near vicinity.

Operating sequence

1. Insert appropriate filters in each colorimeter *before* turning on the colorimeter lamps. Allow 20 minutes for optic systems to stabilize.
2. While colorimeters are stabilizing, start pumping water or saline with Brij 35 through sample and reagent lines. Check manifold connections and bubble patterns to see whether everything is functioning properly.
3. When the colorimeters have stabilized, adjust the glucose water (or saline) baseline to 97% or 98% T. Check the gain of the glucose pen.

4. Place all the reagent lines in the appropriate reagent bottles and after about 8 minutes adjust the BUN baseline to 97% or 98% T. Check the the gain of the BUN pen. Note the position of the glucose RBL.
5. Start sampler. Place a cup of water between the last standard and the first specimen.
6. When the last specimen has recorded, place all lines in water and rinse for 10 minutes.

Calculations

The standard curve may be drawn on a Technicon plastic comparator or on graph paper by plotting % T versus concentrations on semilog paper.

System maintenance and troubleshooting

1. When using pump I, four helper springs at each end of the platen will facilitate adequate and even pumping.
2. When lowering the roller head assembly of pump I, be sure there is no overlapping of pump tubes of the manifold. If there is, raise the rollers, press the tubes down into proper position with one finger, and lower the rollers again.
3. Sample interaction can be reduced by using glass transmission tubing wherever possible, especially between the heating bath and the colorimeter.
4. If a noisy recorder pattern is obtained even though the reagents contain the proper amount of Brij 35, filter those reagents into clean bottles. One of the older urea nitrogen procedures called for the addition of urea to the diacetyl monoxime reagent. This was found to cause excessive noise on many systems and is no longer used.
5. The dialyzer membranes should be changed regularly according to use. Many problems can be avoided if a 25% Clorox solution is pumped through the entire system for about 15 minutes just before the membranes are changed. Follow the bleach with a 10-minute water wash before dismantling the dialyzer. The bleach will remove any protein buildup in the lines and any film that might have accumulated in the flowcell.
6. The hot alkaline potassium ferricyanide will gradually corrode the inside of the heating bath coil until it eventually perforates the wall of the coil. As the inside diameter increases, the transit time through the coil increases. The longer dwell time in the heating bath results in higher glucose sensitivity and a shift in the calibration curves.[18, 83]

Creatinine

Principle

The automated determination of creatinine is based on the classic manual method of Folin and Wu.[22, 24] It was adapted for automation by Stevens and Skeggs[76] and by Chasson and colleagues.[13]

The sample is diluted with saline and pumped into the dialyzer where creatinine dialyzes into a water recipient stream. Picric acid and sodium hydroxide are mixed to form alkaline picrate and this stream joins with the recipient stream as it emerges from the dialyzer. They mix in a single mixing coil and then pass into a 40-foot time delay coil immersed in water at room temperature for color development. The reaction mixture then goes to the colorimeter, where the color intensity is measured at 505 nm in a tubular flowcell.

Modules needed
   One sampler II
   One proportioning pump I

AutoAnalyzer methods 51

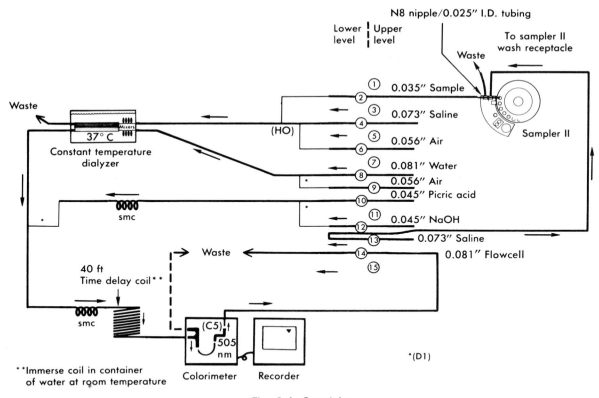

Fig. 3-4. Creatinine.

One dialyzer with one set of dialyzer plates
One colorimeter with 505-nm filter
One single-pen recorder

Flow diagram
See Fig. 3-4.

Reagents

Saline, 0.9%	(D-17)
Sodium hydroxide, 0.5 N	(D-29)
Picric acid, saturated	(D-30)
Creatinine stock standard, 1 mg/ml	(D-31)
Hydrochloric acid, 0.1 N	(D-32)
Hydrochloric acid, 0.02 N	(D-33)

Working standards

Measure the following amounts of creatinine stock standard, 1 mg/ml, into 100-ml volumetric flasks and dilute to volume with 0.02 N HCl.

Creatinine stock standard (ml)	Creatinine working standard (mg/dl)
1	1
2	2
3	3
5	5
7	7
10	10

Procedure

1. Samples may be run at a rate of 60/hr with a 2:1 sample-wash ratio. It may be advisable, however, to place a cup of saline between samples on the plate to provide better wash between samples.
2. Samples may include both serum and urine. Urines should be diluted 1:10 or 1:20 with water. Serum samples with creatinine values greater than 10 mg/dl should be diluted 1:2 or 1:3 with saline.
3. The bubble pattern is improved by using 0.5 ml of Brij 35 per liter of saline and water recipient.
4. To reduce noise level and baseline drift, filter all reagents and use saline instead of water in the sampler wash receptacle.
5. Wash the picric acid and sodium hydroxide lines, coils, and flowcell with 10% acetic acid for 5 minutes at the end of each day's run.

Operating sequence

1. Turn on colorimeter and allow for a 20-minute warmup.
2. Begin pumping water through all lines and check the system for leaks.
3. Place lines in reagent bottles. Leave the sample line in the sampler wash receptacle.
4. Pump reagents for 10 minutes; then adjust reagent baseline. Set the pen to 97% or 98% T and check 0% T. When the pen returns to the reagent baseline check the recorder gain.
5. Begin sampling.
6. When the last sample has recorded, place the picric acid and sodium hydroxide lines in 10% acetic acid for 5 minutes. Then rinse with water for 10 minutes.

Calculations

The standard curve may be drawn on a Technicon plastic comparator or on graph paper by plotting % T versus concentration of semilog paper.

System maintenance and troubleshooting

1. The dialyzer membrane should be changed regularly according to use. Pump 25% Clorox through all lines for 5 minutes and rinse with water for 10 minutes before changing the membrane.
2. Since the picric acid is a saturated solution, some precipitation may occur. If this precipitate is pumped through the reagent line, it will cause excessive noise. If the precipitate will not go back into solution with agitation or gentle warming, the solution should be filtered.

Uric acid

Principle

The automated uric acid procedure[80] is adapted from the manual method of Brown[9] as described by Hawk and co-workers.[32]

The color reaction involves the reduction of the phosphotungstate complex to a phosphotungstite complex. Cyanide improves sensitivity and prevents turbidity.

The sample is diluted with saline and pumped into the dialyzer, where uric acid dialyzes into a saline recipient stream. As the dialysate leaves the dialyzer, it is joined and mixed with a cyanide-urea stream. This mixture is then joined by a phosphotungstic acid stream. The reaction mixture passes through two double mixing coils for color development and on to the colorimeter where the intensity of the color is measured at 660 nm in a 15-mm flowcell.

AutoAnalyzer methods 53

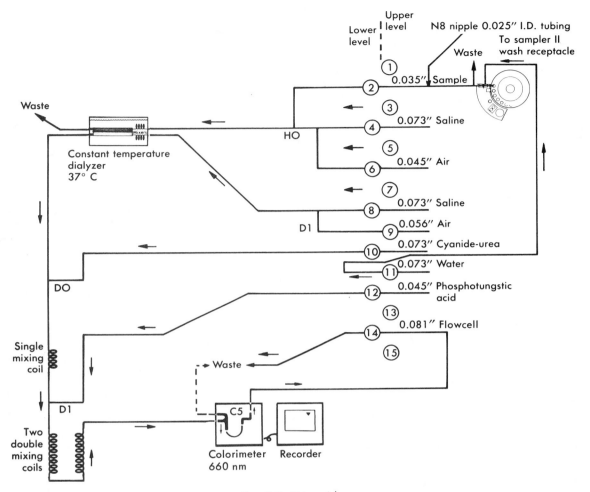

Fig. 3-5. Uric acid.

Modules needed
    One sampler II
    One proportioning pump (I or II)
    One dialyzer with one set of dialyzer plates
    One colorimeter with 660-nm filters
    One single-pen recorder

Flow diagram
    See Fig. 3-5.

Reagents
Saline (with 0.5 ml of Aerosol 22/L)	(D-17)
Phosphotungstic acid	(D-34)
Sodium cyanide stock solution, 10%	(D-35)
Urea stock solution, 20%	(D-36)
Cyanide-urea working solution	(D-37)
Uric acid stock standard, 1 mg uric acid/ml	(D-38)

Working standards

Pipet the following volumes of uric acid stock standard into 100-ml volumetric flasks, dilute to volume with water, and mix thoroughly.

Uric acid stock standard (ml)	Uric acid working standard (mg/dl)
2	2
4	4
6	6
8	8
10	10
12	12

Store the working standards in the refrigerator.

Procedure
1. Samples may be run at a rate of 60/hr with a 2:1 sample-wash ratio. For better wash between samples, use a 50/hr cam with a 1:1 sample-wash ratio or place a saline cup between samples on the sample plate.
2. Samples may include nonhemolyzed serum or plasma and urine. Urines usually require a 1:5 or 1:10 dilution with water.
3. Serum or plasma samples with a uric acid concentration greater than 12 mg/dl should be diluted with saline.
4. Use 0.5 ml of Aerosol 22 per liter of saline to improve the bubble pattern and avoid noise.
5. Waste from the system should be flushed down a sink with running water. The mixing of an acid with the sodium cyanide will generate hydrogen cyanide gas, which is extremely poisonous. If the waste liquid must be collected in a bottle, a neutralizing compound such as trisodium phosphate powder should be used in the bottle.

Calculations

The standard curve may be drawn on a Technicon plastic comparator or on graph paper by plotting % T versus concentration on semilog paper.

Operating sequence
1. Turn on colorimeter and allow a 20-minute warmup.
2. Begin pumping water through the sample and reagent lines. Check for leaks in the system and observe the nature of the bubble pattern coming from the dialyzer.
3. Place the reagent lines in their respective reagent bottles. Leave the sample line in water.
4. After about 8 minutes set the reagent baseline to 97% or 98% T. Check the 0% T and when the pen returns to the baseline, check the recorder gain.
5. Begin sampling standards and specimens.
6. When the last specimen has recorded, place the reagent lines in water and rinse for 10 minutes.

System maintenance and troubleshooting

1. The dialyzer membrane should be changed regularly according to use. Just prior to changing the membrane pump 25% Clorox solution through the sample and reagent lines for 5 minutes. Follow with a 5-minute water wash.
2. If turbidity is observed in the mixing coils it is probably caused by old sodium cyanide solution. Prepare a new sodium cyanide stock solution and a new cyanide-urea working solution.

Calcium

Principle

In this method[11] a metal-complexing dye, cresolphthalein complexone (phenolphthalein, 3'3" bis-{[bis-(carboxymethyl)amino]methyl}-5',5"-dimethyl; metalphalein; phthalein komplexon), is used to determine calcium. The use of this dye was introduced by Anderegg and co-workers.[5] The automated method was adapted by Kessler and Wolfman[46] from a spectrophotometric method of Pollard and Martin[62] and Stern and Lewis.[74] Gitelman[27] further modified the procedure by adding 8-hydroxyquinoline to minimize interference from magnesium.

The sample is mixed with 0.25 N hydrochloric acid and pumped into a dialyzer. The acid releases the protein-bound calcium and the calcium ions dialyze into a 0.25 N hydrochloric acid recipient stream. As the dialysate leaves the dialyzer it is joined by the dye and the base reagent streams. The mixture passes through two 3-mm double mixing coils for mixing and for color development. It then goes to the colorimeter, where the color intensity is measured at 580 nm in a 15-mm flowcell.

Modules needed
>One sampler II
>One proportioning pump (I or II)
>One dialyzer with one set of dialyzer plates
>One colorimeter with 580-nm filters
>One single-pen recorder

Flow diagram
>See Fig. 3-6.

Reagents
>Hydrochloric acid, 0.25 N (D-39)
>Cresolphthalein complexone, 0.01%, with 8-hydroxyquinoline (D-40)
>Calcium base, 3.75% diethylamine (D-41)
>Calcium stock standard, 50 mg Ca/dl (D-42)
>Magnesium stock standard, 100 mg Mg/dl (D-43)

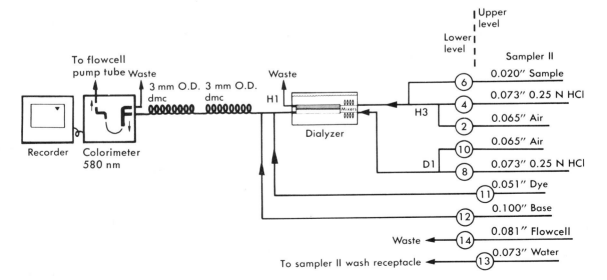

Fig. 3-6. Calcium.

Working standards

Prepare dilute working standards in 100-ml volumetric flasks. Pipet the following volumes of calcium and magnesium stock standards into the flasks, add 1 drop of concentrated HCl to each flask, dilute to volume with water, and mix thoroughly.

Calcium stock standard (ml)	Magnesium stock standard (ml)	Calcium working standard (mg/dl)	Magnesium working standard (mg/dl)
5	2	2.5	2
10	2	5.0	2
15	2	7.5	2
20	2	10.0	2
25	2	12.5	2
30	2	15.0	2

Procedure
1. Calciums may be run at a rate of 60 samples/hr with a 2:1 sample-wash ratio. If a better wash between samples is desired, use a 50/hr cam with a 1:1 sample-wash ratio.
2. Determine calcium directly in serum, heparinized plasma (but no other anticoagulated plasma), urine, and spinal, pleural, or ascitic fluids. Collect urine specimens in bottles containing glacial acetic acid or 6 N hydrochloric acid to keep the calcium in solution. The amount of acid depends on the expected volume of urine. Use 10 ml of acid for a 24-hour collection. Adjust the urine to pH 2, using pH indicator paper.
3. Magnesium (2 mg/dl) is added to the working standards equivalent to the level normally present in serum.
4. Be sure that water rather than air is being aspirated through the sample line when the reagent baseline is being set. Sampling air will give a slightly higher % T reading.
5. Use 0.5 ml of Brij 35 per liter of 0.25 N hydrochloric acid to assure a good bubble pattern and low noise level.

Operating sequence
1. Turn on colorimeter and allow a 20-minute warmup.
2. Begin pumping water through reagent and sample lines. Check for leaks and irregular flow from dialyzer.
3. Place reagent lines in their respective reagent bottles. Leave the sample line in water.
4. After about 8 minutes, adjust the reagent baseline to 97% or 98% T. Check 0% T, and when the pen returns to the reagent baseline check the recorder gain.
5. Begin sampling standards and specimens.
6. When the last specimen has recorded, place the reagent lines in water and rinse for 10 minutes.

Calculations

The standard curve may be plotted either on a Technicon plastic comparator or on semilog paper.

System maintenance and troubleshooting

1. The dialyzer membrane should be changed regularly according to use. Pump 25% Clorox solution through sample and reagent lines for 10 minutes, followed by a 5-minute water rinse before changing the membrane.
2. Run the waste lines into a sink with running water. The mixture of acid with potassium cyanide in the base reagent will generate a small amount of highly poisonous hydrogen cyanide gas.

3. If it is not possible to set the reagent baseline up (toward 100% T) to 97% or 98% T, even with a No. 1 aperture, be sure first that the flowcell is clean. Disconnect it from the system and rinse with full-strength Clorox or chromic acid cleaning solution followed by water. Reconnect the flowcell and, if the baseline still cannot be set to 98% T, add a neutral-density filter to the reference side or partially cover the aperture with black tape.

Inorganic phosphate
Principle

This procedure[41] is a modification of an earlier method[40] that used aminonaphthol sulfonic acid as a reducing agent. It is based on the formation of phosphomolybdic acid which is then reduced to the blue phosphomolybdous acid. This procedure uses the more stable reducing agent, stannous chloride–hydrazine sulfate. The use of this reducing agent in phosphate determinations was first reported by Hurst.[39] The automated procedure was adapted by Kraml.[48] No heating bath is required for this method as it was for the earlier method.

The sample is mixed with saline and pumped into the dialyzer where phosphorus ions dialyze into an air-segmented recipient stream. As the dialysate leaves the dialyzer, it is joined by an acid molybdate stream and mixes. This mixture is then joined by a stannous chloride–hydrazine sulfate stream and mixes again. The color produced is then measured at 660 nm in a 15-mm tubular flowcell.

Modules needed
    One sampler II
    One proportioning pump (I or II)
    One dialyzer with one set of dialyzer plates
    One colorimeter with 660-nm filters
    One single-pen recorder

Flow diagram
    See Fig. 3-7.

Reagents

Saline (with 0.5 ml Brij 35/L)	(D-17)
Saline (with 2 ml wetting agent A/L)	(D-17)
Hydrazine sulfate, 0.2%, in 1 N H_2SO_4	(D-44)
Stannous chloride–hydrazine sulfate working solution	(D-45)
Acid molybdate	(D-46)
Phosphorus stock standard, 1 mg P/ml	(D-47)

Working standards
Measure the following volumes of phosphorus stock standard into 100-ml volumetric flasks, dilute to volume with water, and mix thoroughly.

Phosphorus stock standard (ml)	*Phosphorus working standard (mg/dl)*
1	1
3	3
5	5
7	7
10	10

Procedure
1. Phosphorus determinations may be run at a rate of 60 samples/hr with a 2:1 sample-wash ratio. For a better wash between samples, use a 50/hr cam with a 1:1 sample-wash ratio.

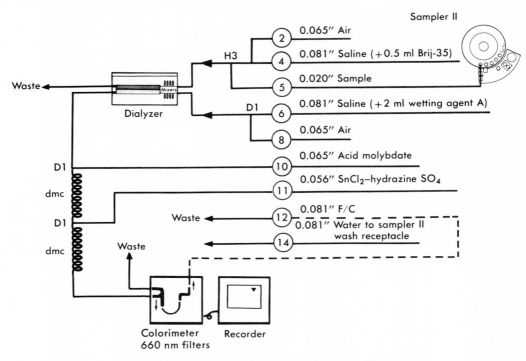

Fig. 3-7. Phosphate (via SnCl$_2$–hydrazine sulfate).

2. Inorganic phosphorus may be run on serum or heparinized plasma. Urine specimens usually require a 1:5 or 1:10 dilution with water.
3. The saline used for sample dilution should contain 0.5 ml of Brij 35 per liter. The saline used as the recipient stream should contain 2 ml of wetting agent A per liter. If other wetting agents are used on the recipient side, a blue precipitate will form and cause a very noisy pattern on the recorder.

Operating sequence

1. Turn on the colorimeter and allow for a 20-minute warmup.
2. Begin pumping water through the sample and reagent lines. Check for leaks in the system and observe the nature of the bubble pattern coming from the dialyzer.
3. Place the reagent lines in their respective reagent bottles. Leave the sample line in water.
4. After about 8 minutes set the reagent baseline to about 97% or 98% T. Check 0% T and when the pen returns to the baseline check recorder gain.
5. Begin sampling standards and specimens.
6. When the last specimen has recorded, place the reagent lines in water.
7. After a 2-minute water wash place the acid molybdate and the stannous chloride–hydrazine sulfate lines in 1 N NaOH. Wash for 5 minutes with 1 N NaOH to remove traces of blue precipitate in the mixing coil and flowcell. Follow with a 10-minute water wash.

Calculations

The standard curve may be drawn on a Technicon plastic comparator or on graph paper by plotting % T versus concentration on semilog paper.

System maintenance and troubleshooting

1. The dialyzer membrane should be changed regularly according to use. Pump 25% Clorox solution through sample and reagent lines for 5 minutes followed by a 5-minute water wash before changing the membrane.

2. If the acid molybdate and stannous chloride–hydrazine sulfate lines are not washed with 1 N NaOH, a blue film will build up on the inside surface of the flow-cell, resulting in a loss of sensitivity.

Serum glutamic-oxaloacetic transaminase (SGOT) (aspartate aminotransferase, EC 2.6.1.1)

Principle

The basic reaction for this procedure is the same as it is in the manual method. However, the oxaloacetic acid produced is dialyzed into a buffer and then coupled with the diazonium salt of N-butyl-4-methoxymetanilamide (azoene fast red [PDC] or fast ponceau L). The intensity of the color produced is measured at 460 nm in a 15-mm tubular flowcell. This method[67] is based on the procedure of Morgenstern and associates.[56] By measuring dialyzed oxaloacetic acid, interference from other diazo-reacting serum constituents is avoided and it is not necessary to run individual serum controls.

A serum sample is aspirated and mixed with an air-segmented substrate stream. It then enters one of two 40-foot coils in a 37° C heating bath for incubation. Upon leaving the heating bath, the substrate mixture enters a dialyzer, where the enzymatically generated oxaloacetic acid is dialyzed into a citrate buffer. As the substrate stream leaves the dialyzer, it continues on to waste. As the citrate buffer recipient stream leaves the dialyzer, it is joined and mixed with the diazonium salt solution. This mixture then returns to the 37° C heating bath to the second coil for color development. Upon leaving the heating bath, the reaction mixture goes to the colorimeter, where the intensity of color produced is measured at 460 nm in a 15-mm flowcell.

Modules needed

    One sampler II
    One proportioning pump I
    One heating bath, 37° C, with two coils
    One dialyzer, one set of dialyzer plates
    One colorimeter with 460-nm filters
    One single-pen recorder

Flow diagram

    See Fig. 3-8.

Reagents

SGOT substrate	(D-48)
Phenyl mercuric acetate, 0.04%	(D-49)
Citrate buffer	(D-50)
Azoene fast red dye diluent	(D-51)
Azoene fast red dye solution	(D-52)
Alcoholic KOH wash solution, 2%	(D-53)

Working standards

    Enzyme "standards" are prepared from a frozen serum pool with high enzyme activity or from a lyophilized reference serum. Dilute as follows:

60 *Chemistry for the clinical laboratory*

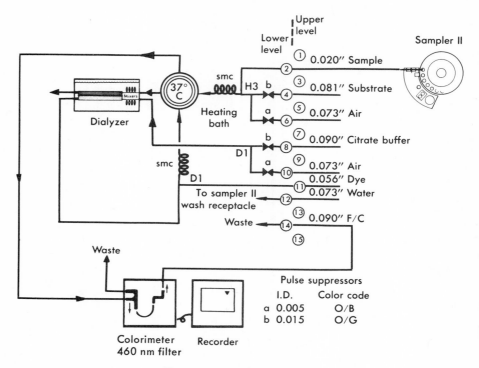

Fig. 3-8. SGOT by dialysis.

Reference serum (RS) (ml)	+	*Saline (ml)*	=	*Karmen units*	*Example Karmen units*
0.1		0.9		RS value ÷ 10	35
0.2		0.8		RS value ÷ 5	70
0.3		0.7		RS value × 3/10	105
0.4		0.6		RS value × 2/5	140
0.5		0.5		RS value ÷ 2	175
0.7		0.3		RS value × 7/10	245
1.0		0.0		RS value	350

These dilutions should be made fresh daily in small test tubes and mixed thoroughly before pouring into sample cups. CAUTION: Before accepting the given value of any lot of lyophilized control serum, the enzyme activity should be determined by a manual spectrophotometric method.

Procedure

1. Samples may be run at a rate of 60/hr with a 2:1 sample-wash ratio. If a number of elevated values are expected it is advisable to use a 1:1 sample-wash ratio to reduce the amount of sample interaction. Use a microsample probe (0.016 inch I.D.) on the sampler.
2. Prepare the dye fresh each day, protect it from light by placing it in an amber glass bottle, and keep the bottle in ice during use. The use of premixed dye diluent is recommended. The inadvertent mixing of the dye with the undiluted alcohol or HCl will cause decomposition of the dye.
3. Wash the dye line, color development coil, and flowcell daily, at the end of the run, with 2% alcoholic KOH solution. Place the dye line in the KOH solution and allow the other reagent lines and the sample line to

pump air. Wash for 2 minutes and then rinse all lines with water for 10 minutes.

Operating sequence

1. Turn on colorimeter and allow for a 20-minute warmup.
2. Start pumping water through all the lines and check for any leaks in the system.
3. Place the substrate and citrate buffer lines in their respective bottles. Leave the dye line in water.
4. After about 5 minutes place the dye line into the dye.
5. Wait about 8 minutes and then adjust the reagent baseline. Set the pen to 97% or 98% T and check the 0% T. When the pen returns to the reagent baseline, check the recorder gain.
6. Begin sampling.
7. At the end of the run place all lines in water for 5 minutes and then proceed to wash the dye line as directed in step 3 under Procedure.

Calculations

The standard curve may be plotted either on a Technicon plastic comparator or on semilog paper.

System maintenance and troubleshooting

1. The dialyzer membrane should be changed regularly according to use. A 25% Clorox wash for 5 minutes before changing the membrane will clear all the lines of any film buildup or protein particles in the sample line.
2. Unless the dye line is washed regularly with 2% alcoholic KOH the flowcell will become coated with the dye product and sensitivity will be decreased. Also, the coating built up in the lines and color development coil may flake off small particles that will cause a noisy pattern on the chart.
3. Filter all reagents to avoid noisy recordings.
4. Be sure the substrate and citrate buffer contain 1 ml of Triton X-405 per liter. Some commercially prepared reagents do not contain wetting agents.
5. Changing ambient temperature will cause considerable baseline drift.

Alkaline phosphatase, EC 3.1.3.1
Principle

This automated method[3] is based on the manual procedure of Bessey and associates[6] using a *p*-nitrophenyl phosphate substrate, pH 10.25. It was modified for automation by Morgenstern and co-workers.[55] The phosphatase enzyme catalyzes the hydrolysis of the substrate while incubating at 37° C. The free *p*-nitrophenol is then dialyzed into a 2-amino-2-methyl-1-propanol (AMP) recipient buffer. Since *p*-nitrophenol is highly colored (yellow) in an alkaline solution, the color intensity can be measured directly at 410 nm. Since dialyzed *p*-nitrophenol is measured, there is no need for specimen blank correction.

A sample is mixed with substrate and goes into a 37° C heating bath for incubation and hydrolysis. The reaction mixture goes into a dialyzer where free *p*-nitrophenol dialyzes into the AMP recipient buffer. The yellow-colored dialysate then goes to the colorimeter, where the color intensity is measured at 410 nm in a 15-mm tubular flowcell.

Modules needed

One sampler II
One proportioning pump (I or II)

62 Chemistry for the clinical laboratory

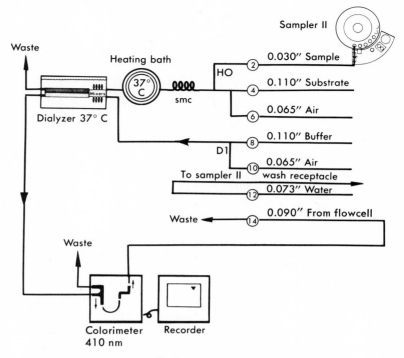

Fig. 3-9. Alkaline phosphatase (N-6b).

    One heating bath at 37° C
    One dialyzer bath with one set of dialyzer plates
    One colorimeter with 410-nm filters
    One single-pen recorder

Flow diagram
    See Fig. 3-9.

Reagents

AMP stock buffer, 50% (w/v)	(D-54)
AMP working buffer, 0.625 M, pH 10.25	(D-55)
Magnesium chloride, 1 M	(D-56)
p-Nitrophenyl phosphate substrate	(D-57)

Working standards

    The International Unit, as proposed by the Enzyme Commission of the International Union of Biochemistry, is used as a measure of phosphatase activity by this method. One unit of enzyme activity is defined as the amount of enzyme that will catalyze the transformation of one micromole of substrate per minute under specified reaction conditions. It has been recommended that concentration be expressed as units per milliliter. Since values in the normal range are less than one International Unit, milliunits (mU) are used. The suggested normal range for this method is 30 to 85 mU/ml.

    Saline dilutions of a reference serum with a known alkaline phosphatase activity may be used as working standards. These dilutions should be made just prior to use. Example:

Reference serum (ml)	Saline (ml)	Working standard (mU/ml)
0.2	0.8	29
0.4	0.6	58
0.6	0.4	87
0.8	0.2	116
1.0	0.0	145

Procedure
1. Samples may be run at a rate of 40/hr, using a 2:1 sample-wash ratio.
2. Serum specimens should not be hemolyzed.
3. Serum specimens with a phosphatase activity greater than the highest standard should be diluted with saline.

Operating sequence
1. Turn on colorimeter and allow a 20-minute warmup.
2. Begin pumping water through the sample and reagent lines. Check the system for leaks and observe the bubble pattern coming from the dialyzer.
3. Place the reagent lines in their respective reagent bottles. Leave the sample line in water.
4. After about 10 minutes set the reagent baseline to 97% or 98% T. Check 0% T and when the pen returns to the baseline check the recorder gain.
5. Begin sampling standards and serum specimens.
6. When the last specimen has recorded, place all lines in water and rinse for 10 minutes.

Calculations
The standard curve may be prepared by using a Technicon plastic comparator or by plotting % T versus mU/ml activity on semilog paper.

System maintenance and troubleshooting
1. The dialyzer membrane should be changed regularly according to use. Pump a 25% Clorox solution through all lines for 10 minutes followed by a 10-minute water rinse before changing the membrane.

Lactic dehydrogenase (LDH), EC 1.1.1.27
Principle
This automated procedure for the determination of lactic dehydrogenase activity[50] is based on a method described by Hochella and Weinhouse.[36] The oxidation of *l*-lactate by NAD is catalyzed by LDH to give pyruvate and reduced NAD (NADH). The enzyme diaphorase catalyzes the coupling of NADH to an electron-acceptor dye (INT). The electron from NADH is transferred to the tetrazolium dye, resulting in the production of a colored formazan.

$$\text{Lactate} + \text{NAD} \xrightleftharpoons{\text{LDH}} \text{Pyruvate} + \text{NADH}$$

$$\text{NADH} + \text{INT} \xrightleftharpoons{\text{Diaphorase}} \text{NAD} + \text{Reduced dye (colored product)}$$

An air-segmented stream of substrate is joined by an INT dye stream and an NAD-diaphorase stream. This mixture is then joined by the sample stream and passes through a double mixing coil to a 37° C heating bath for color development. The reaction mixture then goes to the colorimeter, where the color intensity is measured at 505 nm in a 15-mm tubular flowcell.

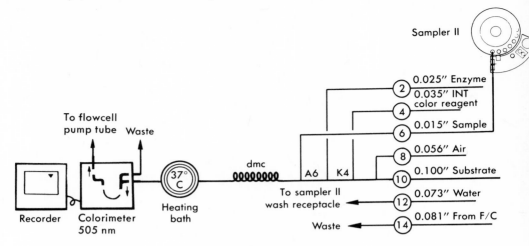

Fig. 3-10. LDH (N-60).

Modules needed

    One sampler II
    One proportioning pump (I or II)
    One heating bath at 37° C
    One colorimeter with 505-nm filters
    One single-pen recorder

Flow diagram

    See Fig. 3-10.

Reagents

LDH substrate	(D-58)
Phosphate buffer, 0.067 M, pH 7.4	(D-59)
NAD-diaphorase	(D-60)
INT dye	(D-61)

Working standards

    The Enzyme Commission of the International Union of Biochemistry defines one International Unit of enzyme activity as the amount of enzyme which will catalyze the transformation of one micromole of substrate per minute under specified reaction conditions. It is recommended that the concentration of enzyme solutions be expressed as Units per milliliter. Since the normal range of LDH activity in human serum is below one International Unit, activity is expressed in milli–International Units per milliliter. The suggested normal range for this procedure is 90 to 200 mU/ml.

    Dilutions of a reference serum of known LDH activity may be used as working standards. Example:

Reference serum (ml)	Saline (ml)	Working standard (mU/ml)
0.5	1.5	90
1.0	1.0	180
1.5	0.5	270
2.0	0.0	360

Procedure

    1. Samples may be run at a rate of 60/hr, using a 2:1 sample-wash ratio.

2. Add Triton X-100 to the substrate as a wetting agent. Do not use any other wetting agents.
3. The tubing that carries the INT and NAD-diaphorase reagents from the bottles to the manifold should be of narrow bore. Use either 0.025- or 0.030-inch I.D. tubes.
4. A microsample probe (0.016 inch I.D.) should be used and should be connected to the sample pump tube with 0.015-inch I.D. tubing.
5. Serum samples must be free of any fibrin particles that could plug the microsample line.
6. Serum samples with LDH activity greater than the highest standard should be diluted with saline.

Operating sequence
1. Turn on the colorimeter and allow a 20-minute warmup.
2. Begin pumping water through all lines and check the system for leaks.
3. Place lines in reagents and leave sample line in water.
4. After about 7 minutes, set the reagent baseline to 97% or 98% T. Check 0% T and when the pen returns to baseline check the recorder gain.
5. Begin sampling standards and serum specimens.
6. When the last specimen has recorded, place all lines in water and rinse for 10 minutes.

Calculations
Prepare a standard curve on a Technicon plastic comparator or plot % T versus concentration on semilog paper.

System maintenance and troubleshooting

1. Eventually a red coating will deposit on the glass coils and flowcell. Clean these periodically by flushing with dichromate cleaning solution followed by a water rinse. The discolored Tygon tubing should then be replaced.
2. Pump a 50% Clorox solution through the INT, NAD-diaphorase, and sample lines for 10 minutes at least once a week. Follow with a 10-minute water rinse.

Cholesterol (direct)
Principle

This procedure[14] employs a direct Liebermann-Burchard reaction for the determination of serum cholesterol. The manual procedure of Huang and associates[38] was adapted for automation by Levine and co-workers.[49]

Serum is mixed with the Liebermann-Burchard color reagent and pumped into a 40-foot glass coil in a 37° C heating bath for color development. When the reaction mixture leaves the heating bath it goes to the colorimeter where the color intensity is measured at 630 nm in a 15-mm flowcell.

Modules needed
One sampler II
One proportioning pump (I or II)
One heating bath at 37° C
One colorimeter with 630-nm filters
One single-pen recorder

Flow diagram
See Fig. 3-11.

Reagent
Liebermann-Burchard color reagent (D-62)

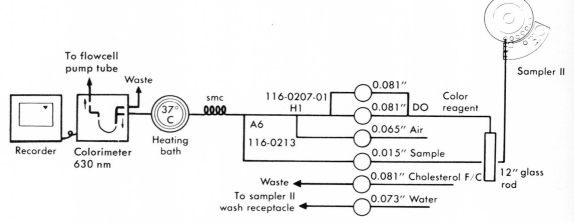

Fig. 3-11. Cholesterol via Liebermann-Burchard.

Working standards

Commercially prepared control sera in the elevated and normal ranges are used as standards. A third standard is prepared by mixing equal volumes of the normal and the elevated serum and calculating the value from the given values of the two. Example:

Normal serum control = 180 mg cholesterol/dl

Elevated serum control = 420 mg cholesterol/dl

$$1 \text{ part normal} + 1 \text{ part elevated} = \frac{180 + 420}{2} = 300 \text{ mg/dl}$$

Procedure

1. The Liebermann-Burchard color reagent should not be exposed to atmospheric moisture any more than is absolutely necessary. The reagent is stable at least 3 months when stored in a refrigerator. Bring to room temperature before use. If it becomes discolored it should be discarded.
2. Precipitated protein may accumulate in the steel nipple in the A-6 fitting where the serum meets the color reagent. This should be cleaned daily with a fine steel probe.
3. The relative density of the color reagent should be 1.185 at room temperature (20° C). If the relative density is above 1.190, the reagent should be diluted with a solution made up of equal parts of acetic acid and acetic anhydride.
4. The reagent and flowcell pump tubes should be changed daily.

Operating sequence

1. Turn on colorimeter and allow a 20-minute warmup.
2. Place the reagent line in the color reagent bottle and the sample line in the sampler wash receptacle. Begin pumping.
3. After about 7 minutes set the reagent baseline to 97% or 98% T. Check 0% T. When the pen returns to the baseline, check the recorder gain.
4. Begin sampling standards and specimens.
5. When the last sample has recorded, remove the reagent bottle. Wipe the glass rod with several layers of tissue. Allow the line to aspirate air for about 2 minutes and then place the line in water and wash for 10 min-

utes. Then remove all lines from water and continue pumping air until all water is pumped out.

Calculations

The standard curve may be drawn on a Technicon plastic comparator or on graph paper by plotting % T versus concentration on semilog paper.

System maintenance and troubleshooting

1. As stated previously, the steel nipple in the A-6 fitting should be cleaned daily.
2. Watch the entire system closely for leaks. The acids in the color reagent could cause serious damage if a leak should go undetected.

Simultaneous direct and total bilirubin
Principle

This procedure for the simultaneous determination of direct and total bilirubin[68] in serum is an adaptation by Gambino and Schreiber[26] of the manual method of Jendrassik and Grof.[44] The method is based on the formation of colored azobilirubin when bilirubin reacts with diazotized sulfanilic acid.

Total bilirubin (free and conjugated) is determined by reacting the serum sample with diazo reagent in the presence of a caffeine–sodium benzoate solution. Since free bilirubin is insoluble in water below pH 8, direct (conjugated) bilirubin is determined by reacting the serum sample with diazo reagent in an acidic aqueous medium.

The sample is aspirated and split into two streams—one for total bilirubin and the other for direct bilirubin. The total bilirubin sample stream is mixed with a caffeine solution, while the direct bilirubin sample stream is mixed with 0.05 N hydrochloric acid. Each stream is then mixed successively with diazo reagent, ascorbic acid, and tartrate buffer. The reaction mixtures then go to the colorimeters, where the color intensity of each is measured at 600 nm in 15-mm flowcells. When running serum blanks, diazo I is used in place of the combined diazo reagent.

This method has high sensitivity and excellent linearity; it is insensitive to hemoglobin and to wide variations in protein concentration and in sample pH.

The caffeine–sodium benzoate accelerates the coupling of bilirubin to diazotized sulfanilic acid. The sodium acetate in the caffeine diluent buffers the pH of the diazo reaction. The ascorbic acid eliminates hemoglobin interference in the total reaction and stops additional color formation in the direct reaction. The strongly alkaline tartrate buffer brings the final pH to 13.4, which solubilizes the protein. It eliminates the effects of variation in sample pH, increases the molar absorptivity of azobilirubin, and increases specificity by shifting the absorbancy maximum to 600 nm.

Modules needed

One sampler II
One proportioning pump II
Two colorimeters with 600-nm filters
One double-pen recorder with absorbance (A) chart paper

Flow diagram

See Fig. 3-12.

Reagents

Caffeine diluent (D-63)
Sulfanilic acid (diazo I) (D-64)

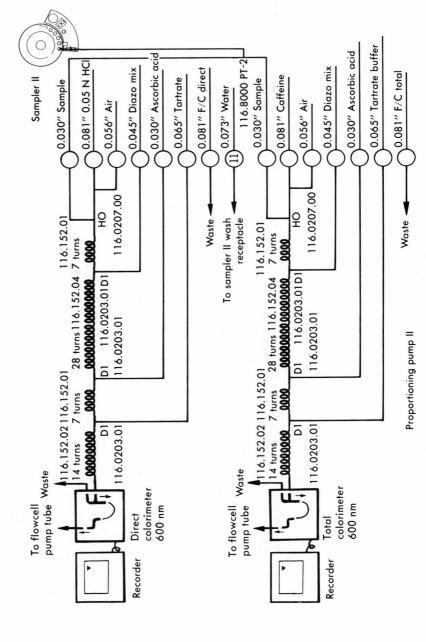

Fig. 3-12. Simultaneous direct and total bilirubin, pump II.

Sodium nitrite, 0.5% (diazo II)   (D-65)
Diazo reagent, combined   (D-66)
Tartrate buffer   (D-67)
Ascorbic acid, 4%   (D-68)
Hydrochloric acid, 0.05 N   (D-69)

Working standards

Working standards can be prepared by diluting a bilirubin-in-serum standard with water. Numerous commercially prepared serum controls with elevated bilirubin values are available. A serum standard may be diluted up to fiftyfold without deleterious effects. However, some protein must be present. Aqueous bilirubin standards without protein are not reliable.[26]

Working standards should be prepared just prior to use and protected from light. The following is an example of how a bilirubin-in-serum standard with a value of 6.0 mg of bilirubin/dl would be diluted:

Serum standard (ml)	H_2O (ml)	Bilirubin working standard (mg/dl)
0.1	0.9	0.6
0.2	0.8	1.2
0.4	0.6	2.4
0.6	0.4	3.6
0.8	0.2	4.8
1.0	0.0	6.0

When running standards, use caffeine diluent in both channels. When the last standard has been aspirated, remove the 0.05 N HCl line from the caffeine diluent and place in the HCl reagent bottle. Readjust reagent baseline and run unknown serum samples. All commercially prepared bilirubin standards are made with free (unconjugated) bilirubin. Conjugated (direct) bilirubin is not available as a stable standard.

Procedure

1. Samples may be run at a rate of 60/hr with a 2:1 sample-wash ratio.
2. Protect serum samples from light as much as possible. Bilirubin is altered by exposure to direct sunlight and fluorescent lights, giving lower results. An amber-colored sample plate cover is available to protect samples in the sampler tray.
3. Serum samples with bilirubin values above the highest standard should be diluted with water.

Operating sequence

1. Turn on colorimeters and allow a 20-minute warmup.
2. Begin pumping water through all the lines and check for leaks.
3. Place the reagent lines (except the 0.05 N HCl line) in their respective reagent bottles. Place the HCl line in the caffeine diluent bottle. Leave the sample line in water.
4. After about 5 minutes set the reagent baseline to 0.1 A on both channels. Check 2.1 A (0% T). When the pens return to baseline, check the recorder gain on both pens.
5. Begin sampling standards.
6. One minute after the last standard has been aspirated, remove the 0.05 N HCl line from the caffeine diluent, wipe the outside of the line with tissue, and place the line in the 0.05 N HCl bottle.
7. Readjust the reagent baseline when the HCl has had time to reach the colorimeter.

8. Begin sampling unknown sera.
9. When the last sample has recorded, place the diazo reagent lines in a bottle containing diazo I (sulfanilic acid) and rerun the standards and unknown serum specimens.
10. When the last serum blank has recorded, place all reagent lines in water and rinse for 10 minutes.

Calculations

From each standard A reading subtract the corresponding blank A reading. Plot these A readings versus bilirubin concentration on linear graph paper. Subtract blank A from test A for each sample and determine bilirubin value from standard curve. Each channel requires its own calibration curve.

System maintenance and troubleshooting

1. This system is relatively trouble-free. Problems can usually be traced to an improperly prepared reagent.
2. As the manifold ages, sensitivity decreases. After a month of daily use sensitivity may drop 12% to 15%.

Simultaneous total protein and albumin
Principle

Total protein is determined by the biuret reaction. This reaction produces a purple-colored complex of copper, in an alkaline solution, with two or more carbamyl groups (—CO—NR—) that are joined together directly or through a single atom of nitrogen or carbon.

The manual procedure of Weichselbaum[82] was adapted for the AutoAnalyzer by Stevens.[75]

The albumin procedure[1] is based on the quantitative binding of an anionic dye, 2-(4′-hydroxyazobenzene) benzoic acid (HABA), specifically to serum albumin. The manual procedure of Ness and co-workers[58] was adapted to the AutoAnalyzer by Nishi and Rhodes.[60]

The sample is aspirated from the sample cup and goes to a PT-2 fitting, where the sample is split into two streams. One sample stream joins an air-segmented biuret reagent stream and more biuret reagent is added. The reaction mixture then goes through three double mixing coils where color development takes place. It then goes to the colorimeter, where the color intensity is measured at 550 nm in a 15-mm tubular flowcell.

The other sample stream joins an air-segmented HABA dye stream and more dye is added. The reaction mixture goes through three double mixing coils where color development takes place. It then goes to the colorimeter, where the color intensity is measured at 505 nm in a 15-mm tubular flowcell.

Serum blanks are obtained by rerunning the samples, using a tartrate-iodide solution in place of the biuret reagent and a phosphate buffer in place of the HABA dye.

Since the albumin portion of the manifold is identical to the total protein portion, one may run simultaneously either total protein and serum protein blanks or albumin and serum albumin blanks if two sets of filters of the same wavelength are available.

Modules needed

One sampler II
One proportioning pump (I or II)

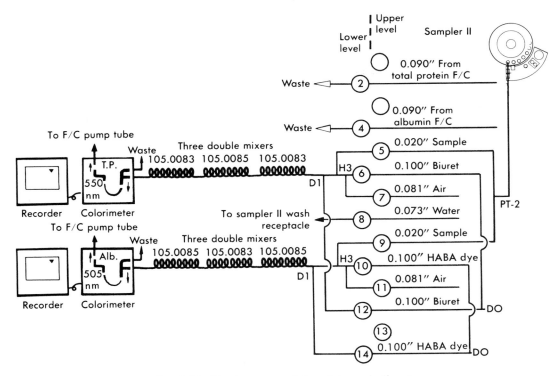

Fig. 3-13. Simultaneous total protein and albumin.

Two colorimeters with 550- and 505-nm filters
One double-pen recorder

Flow diagram
    See Fig. 3-13.

Reagents

Alkaline iodide	(D-70)
Biuret stock solution	(D-71)
Biuret working solution	(D-72)
Protein blank solution	(D-73)
HABA dye stock solution, 6×10^{-3} M	(D-74)
Phosphate buffer, pH 6.05	(D-75)
HABA dye working solution, 6×10^{-4} M	(D-76)
Albumin stock standard, 10 g/dl	(D-77)

Working standards

    Unless a laboratory has the facilities and the inclination to standardize its own total protein and human albumin standard, it is strongly recommended that a stock standard be obtained from Technicon Instruments Corporation (T23-0204). Their standard is made from crystalline bovine albumin, fraction V. The total protein content and a factor with which to calculate the human albumin equivalent are given on the label. It is a good practice to verify the calibration curves with serum controls.

    Pipet the following volumes of standardized albumin stock standard into 25-ml volumetric flasks and dilute to volume with saline containing 50 mg of sodium azide/dl. Mix thoroughly.

Albumin stock (ml)	Total protein (g/dl)	X factor	Albumin (g/dl)
5	2	(e.g. 0.9)	1.8
10	4		3.6
15	6		5.4
20	8		7.2
Undiluted stock	10		9.0

These standards are stable for a year if kept refrigerated.

Procedure

1. This procedure can be run at a rate of 60 samples/hr, using a 2:1 sample-wash ratio.
2. Serum should be separated from the clot within 4 hours after blood is drawn. Care should be taken to avoid hemolysis. Refrigerated serum is stable for 2 days.
3. Serum blanks should be run on all specimens. This is especially important for turbid or icteric specimens.
4. The bovine albumin standard curves should be checked periodically with a human albumin standard or a commercially prepared control serum.
5. Since the dye-binding capacity of bovine albumin is about 90% that of human albumin, the stock bovine albumin solution (if a source other than Technicon is being used) should be standardized by comparing it to a human albumin fraction V standard or to a primary mercaptalbumin standard.[1, 61]
6. The dye-binding capacity of serum albumin obtained from other species will be different from that of human albumin. Inaccurate results will be obtained if, for example, rabbit albumins are calculated by using human albumin standards.

Operating sequence

1. Turn on colorimeters and allow a 20-minute warmup.
2. Pump water through all lines and check for leaks.
3. Place reagent lines in their respective bottles.
4. After 5 minutes set reagent baselines for both total protein and albumin to 97% or 98% T. Check 0% T.
5. When pens return to baseline, check recorder gain on both pens.
6. Begin sampling standards and specimens.
7. When the last specimen has recorded, replace the biuret reagent with the protein blank solution and the HABA dye with the phosphate buffer.
8. After 5 minutes, reset the baselines and begin sampling the specimens again.
9. When the last blank has recorded, place all lines in water and wash for 10 minutes.

Calculations

Plot one curve for total protein and another for albumin—either on a Technicon plastic comparator or on graph paper (% T versus concentration on semilog paper). Determine from the curve the g/dl value for each test and subtract the g/dl value for each corresponding serum blank.

System maintenance and troubleshooting

1. The two sample pump tubes usually wear out before the other tubes. When this happens either the sample will not pass through the tube at all or the sampling is not reproducible. Always replace *both* sample pump tubes at the same time so that the sample flow rate will be about equal through both tubes.

2. After every 50 to 60 hours of operating time pump a 25% Clorox solution through the sample and reagent lines for 10 minutes to clean out the coils and the flowcells. Follow with a 10-minute water rinse.

AutoAnalyzer II

The second generation of basic AutoAnalyzers is known as the AutoAnalyzer II systems. These systems incorporate the miniature dialyzers and heating baths and the improved flowcells used in the SMA 12/60 systems. As a result of the miniaturization and improved flowcell, the reagent consumption and sample size have been reduced considerably. The methods are linear over a wider range and the sample interaction or carryover has been reduced.

The user should be aware, however, that linearity, maximum precision and accuracy, and a minimum of sample interaction are obtained only when a system is properly maintained. Worn pump tubes and dirty dialyzer membranes and tubing will have adverse effects.

METHODOLOGY
Albumin (BCG)
Principle

The automated bromcresol green (BCG) method for the determination of albumin[2] is a modification by Doumas, Watson, and Biggs[19] of the original manual method of Rodkey.[63] The method is based on the binding of bromcresol green dye to albumin. The absorbance of the color produced is directly proportional to the concentration of albumin.

Sample dilution of about 1:441 eliminates the need for a serum blank.

The serum sample is mixed with an air-segmented stream of water containing 1 ml of Brij 35 per liter. An aliquot of the diluted sample is then resampled into an air-segmented stream of 0.105% BCG in pH 4.20 ± 0.1 succinate buffer. Brij 35 is added to the BCG reagent to minimize the absorbance of the reagent blank, to prevent turbidity, and to provide linearity. The absorbance of the analytic stream is measured at 630 nm.

Range of expected values

Age	Outpatients (g/dl)	Inpatients (g/dl)
20	2.7-5.4	2.8-5.2
30	3.6-5.3	2.6-5.1
40	3.5-5.2	2.6-5.0
50	3.4-5.1	2.5-4.9
60	3.3-5.0	2.5-4.8
70	3.2-4.9	2.4-4.7
80	3.1-4.8	2.7-4.6

System performance specifications

The following specifications apply to the albumin (BCG) method as performed on the Technicon AutoAnalyzer II system:

Rate of analysis	60 samples/hr
Sample-wash ratio	9:1
Chart speed	1 inch/min
Linearity range	0 to 6 g/dl
Sensitivity coefficient	0.10 absorbance unit per unit of concentration
Minimum sample required	90 µl (approximate)

Sample dilution

Serum samples with albumin concentrations greater than 6 g/dl should be diluted with distilled water or 0.9% saline and rerun.

Interfering substances

Bilirubin, up to 80 mg/dl, and sodium salicylate, up to 60 mg/dl, do not interfere in the measurement of albumin by this method. No drugs in concentrations used in medical treatment have been found to interfere. Hemolysis has been reported to interfere; however, the high sample dilution (about 1:441) used in this method minimizes the effect of hemolysis on the final test result.

Sample interaction (carryover)

According to Technicon's sample interaction study, there is no clinically significant sample interaction between samples at varying levels of concentration throughout the range of the albumin method. At no time does sample interaction exceed 5%, even when a sample with a very high value is followed by a sample with a very low value.

This, of course, depends on the cleanliness of the system and is best determined by the user.

Alkaline phosphatase, EC 3.1.3.1
Principle

The alkaline phosphatase method of Bessey, Lowry, and Brock[6] was modified and automated by Morgenstern and associates.[55] Both methods use p-nitrophenyl phosphate (PNPP) as a substrate. Alkaline phosphatase hydrolyzes PNPP to form p-nitrophenol. In an alkaline buffer, p-nitrophenol is bright yellow and this color can be measured at 410 nm. The p-nitrophenol concentration is proportional to the alkaline phosphatase activity.

A glycine buffer was used in the Bessey-Lowry-Brock method to maintain the reaction at pH 9.8 to 10.1. Glycine buffer is now considered undesirable because it inhibits alkaline phosphatase activity. Lowry and co-workers[51] introduced the use of 2-amino-2-methyl-1-propanol (AMP) as an alkaline phosphatase buffer. AMP buffer increases enzyme activity through transphosphorylation. The hydroxyl group on the buffer attracts the phosphate group that is split off the PNPP, thereby increasing the rate of reaction.

In this automated method,[4] the serum sample is added to an air-segmented stream of alkaline phosphatase substrate at pH 10.25. The reaction stream enters a heating bath at 37.5° C, where the color-producing reaction proceeds. Following incubation, the free p-nitrophenol is dialyzed into an AMP buffer stream. The dialysis is performed to separate the p-nitrophenol from serum bilirubin and protein. Bilirubin absorbs light at approximately the same wavelength as p-nitrophenol and would cause a positive error in the determination of alkaline phosphatase activity. The absorbance of the dialyzed p-nitrophenol is measured at 410 nm.

Range of expected values[4]

The range of expected values for alkaline phosphatase activity determined by this method is 30 to 100 mU/ml. This range is based on a statistical analysis of an ambulatory, adult population of both sexes, ages 20 to 60 years.

System performance specifications

The following specifications apply to the alkaline phosphatase method as performed on the AutoAnalyzer II system:

Rate of analysis   60 samples/hr

Sample-wash ratio	9:1
Chart speed	1 inch/min
Linearity range	0 to 350 mU/ml
Sensitivity coefficient	0.0011 absorbance unit per unit of enzyme activity (approximate)
Minimum sample required	288 µl (approximate)

Sample stability

Massion and Frankenfeld[54] have studied the lability of alkaline phosphatase activity in human fresh and frozen sera and in lyophilized controls. They found that the activity increases with time and temperature. The greatest increase, 1.5%/hr, was found in frozen and thawed pooled human serum. Individual fresh frozen and thawed sera increased in activity at an average rate of 0.25%/hr.

It is recommended that serum specimens be frozen if they cannot be assayed on the day of drawing.

Sample dilution

Serum specimens with an alkaline phosphatase activity greater than 350 mU/ml should be diluted either with AMP buffer or with inactivated pooled sera.

Interfering substances

The following substances, if present in the sample, reagents, or system, may interfere in the manner described:

Arsenicals[85]	Arsenates are inhibitors of enzyme activity and may cause a decrease in alkaline phosphatase activity.
Beryllium salts[85]	May inhibit enzyme activity and cause a decrease in alkaline phosphatase activity.
Fluorides[85]	May inhibit enzyme activity and cause a decrease in alkaline phosphatase activity.
Bromsulfalein[85]	Has an appreciable absorbance at 410 nm and would be measured along with p-nitrophenol, thus causing an apparent increase in alkaline phosphatase activity.
Manganese salts[85]	May increase enzyme activity and cause an increase in alkaline phosphatase activity.
Magnesium salts[85]	May increase enzyme activity and cause an increase in alkaline phosphatase activity.
Protein	Will decrease test results by approximately 2% per g/dl of protein.
Anticoagulants	May complex the magnesium necessary to catalyze action, thereby greatly decreasing test results.

Sample interaction (carryover)

Within the range of expected values for the alkaline phosphatase method, the amount of sample interaction is less than 5%. In cases where a sample with a very high concentration is followed by a sample with a low concentration, the amount of sample interaction may result in a clinically significant (greater than 5%) elevation of enzyme activity in the low-level sample.[4]

Calcium

Principle

The automated calcium method of Kessler and Wolfman[46] employed the metal complexing dye, cresolphthalein complexone. This was a rapid and reproducible procedure; however, the accuracy of calcium determinations on biologic samples varied with the magnesium concentration. Metals such as magnesium, iron, copper, and zinc also complexed with the dye. The magnesium concentration in serum caused significant interference.

Gitelman[27] eliminated the magnesium interference by adding 8-hydroxyquinoline to the reaction. Potassium cyanide was added to the reaction mixture to eliminate heavy metal interferences.

In the Technicon AutoAnalyzer II calcium method,[12] the sample is added to an air-segmented stream of 0.3 N hydrochloric acid containing 0.25% 8-hydroxyquinoline. The acid releases the protein-bound calcium and the 8-hydroxyquinoline binds the free magnesium ions. The calcium ions are then dialyzed into an air-segmented stream of 0.007% cresolphthalein and 0.25% 8-hydroxyquinoline. A solution containing 3.75% diethylamine and 0.05% potassium cyanide is added to the analytic stream and a colored complex is formed between the calcium and the dye. The absorbance of the colored complex is measured at 570 nm.

Range of expected values

The expected range of values for this calcium method is 8.5 to 10.5 mg/dl.

System performance specifications

The following specifications apply to the calcium method as performed on Technicon AutoAnalyzer II systems:

Rate of analysis	60 samples/hr
Sample-wash ratio	9:1
Chart speed	1 inch/min
Linearity range	0 to 15 mg/dl
Sensitivity coefficient	0.0234 absorbance unit per unit of concentration
Minimum sample required	90 μl (approximate)

Sample stability

The serum should be separated from the red cells within about 2 hours of the time the blood is drawn. If the serum cannot be assayed within 30 minutes of centrifugation, it should be refrigerated.

Sample dilution

Serum samples with calcium concentrations greater than 15 mg/dl should be diluted with distilled water or saline and reassayed.

Interfering substances

The following substances, if present in the serum, reagents, or system, may interfere with serum calcium determinations in the manner described:

Calcium salts[85]	Possible contamination of distilled water will increase calcium values.
Fluorides[85]	May precipitate calcium to form an insoluble salt, thereby decreasing calcium values.
Oxalates[85]	Precipitates calcium to form an insoluble salt, thereby decreasing calcium values.

Sample interaction (carryover)

In Technicon's study of sample interaction there was no evidence that interaction exceeded 5% even when a sample with a very high calcium value was followed by a sample with a very low value.[12]

Cholesterol (direct)
Principle

The manual cholesterol method of Huang and others[38] was modified and adapted to the AutoAnalyzer system by Levine, Morgenstern, and Vlastelica.[50]

The method is based on the classic Liebermann-Burchard reaction in which concentrated sulfuric acid is added to a solution of cholesterol and acetic anhydride. During the ensuing reaction, color changes of red to violet to green may be observed.

The main problems in earlier cholesterol methods based on the Liebermann-Burchard reaction were the instability of the acid reagent and of the final product. In 1961, Huang and associates formulated a stable Liebermann-Burchard reagent and recommended that this reagent be mixed with the unknown cholesterol sample and be allowed to stand at room temperature for 20 minutes before measuring the absorbance at either 550 or 610 nm.

In 1964, Ness, Pastewka, and Peacock[59] suggested that the cholesterol sample be added to chilled (0° C) color reagent and then be allowed to slowly reach room temperature. This manner of controlling the reaction prevented broad and nonspecific absorption curves.

Levine and co-workers introduced other modifications when they automated the method. The formulation of the cholesterol color reagent was modified to prolong pump tube life, and the absorbance of the analytic stream was measured at 630 nm to minimize the effects of serum pigments.

The method is limited, however, by the large number of substances known to interfere. (See interfering substances.)

In the automated method,[15] the sample is diluted with distilled water to about 30% of its original concentration. The diluted sample then is resampled and mixed with an air-segmented stream of chilled cholesterol color reagent. The reaction mixture then flows through a thirty-one turn coil for mixing and color development. The following reactions[35] take place:

(1) Cholesterol + H_2SO_4 → Bis-cholestadienyl monosulfonic acid
(2) Bis-cholestadienyl monosulfonic acid + H_2SO_4 → Bis-cholestadienyl disulfonic acid

The absorbance of the analytic stream is measured at 630 nm.

Range of expected values

The following "normal limits" have been suggested by Fredrickson, Levy, and Lees[25]:

Age	Cholesterol (mg/dl)
0-19	120-230
20-29	120-240
30-39	140-270
40-49	150-310
50-59	160-330

System performance specifications

The following specifications apply to the cholesterol (direct) method as performed on the Technicon AutoAnalyzer II system:

Rate of analysis	60 samples/hr
Sample-wash ratio	9:1
Chart speed	1 inch/min
Linearity range	0 to 500 mg/dl

Sensitivity coefficient	0.007 absorbance unit per unit of concentration
Minimum sample required	380 μl (approximate)

Sample dilution

Samples with cholesterol concentrations greater than 500 mg/dl should be diluted with distilled water or saline and reassayed.

Interfering substances

Several substances have been reported to interfere with the direct cholesterol method and, if present in the serum, reagents, or system, may interfere in the manner described:

Amphotericin B[70]	High concentrations (1 mM/L) of amphotericin B have been reported to cause elevated serum cholesterol values.
Bilirubin	For each mg/dl of bilirubin present in a serum sample, an absorbance equivalent to approximately 4 mg/dl of cholesterol occurs.
Hemolysis[85]	Has been reported to cause elevated serum cholesterol values.
Lipemia[85]	Has been reported to cause turbidity in nonextraction cholesterol methods, thus causing an apparent elevation of the cholesterol values.
Lipochrome[85]	Has an appreciable absorbance at 630 nm and can be detected as cholesterol, thus causing an apparent increase in cholesterol concentration.
Tryptophan[85]	Has been reported to elevate cholesterol values.

Sample interaction (carryover)

According to Technicon's studies, at no time does sample interaction exceed 5% even when a sample with a very high cholesterol value is followed by a sample with a very low value.[15]

Creatinine

Principle

Over the years, many methods have been described to measure creatinine in whole blood, serum, or plasma, but those most frequently used today are based on the method described by Jaffe[43] in 1886. In this method a red color is produced when creatinine is added to an alkaline picrate solution.

Chasson, Grady, and Stanley[13] automated the Jaffe procedure with modifications. The modifications included a change in concentration of the alkaline picrate and in reaction time, the use of dialysis to separate creatinine from protein and other unwanted substances, and the analysis of serum or plasma rather than whole blood. The automated procedure has been shown to be more accurate, more precise, and more specific than the corresponding manual method.

In this method[17] the serum sample is diluted with a 1.8% sodium chloride solution and the creatinine in the diluted sample is dialyzed into water. After dialysis, the recipient stream is mixed with a stream of 0.5 N sodium hydroxide solution. A saturated solution of picric acid is then added to the analytic stream and mixed. The reaction proceeds at room temperature and the absorbance is measured at 505 nm.

Range of expected values

The range of expected values for the creatinine method as performed on the Technicon AutoAnalyzer II system is 0.7 to 1.4 mg/dl.

System performance specifications

The following specifications apply to the creatinine method as performed on the Technicon AutoAnalyzer II system:

Rate of analysis	60 samples/hr
Sample-wash ratio	9:1
Chart speed	1 inch/min
Linearity range	0 to 20 mg/dl
Sensitivity coefficient	0.023 absorbance unit per unit of concentration
Minimum sample required	380 μl (approximate)

Sample dilution

Samples with creatinine concentrations greater than 20 mg/dl should be diluted with distilled water or saline.

Interfering substances

The following substances, if present in the serum, reagents, or system, may interfere with the method by falsely elevating the true creatinine level of the sample[85]:

Acetoacetic acid	Glucose
Acetone	Glycocyamidine
Aminohippurate	Levodopa
Ascorbic acid	Methyldopa
Bromsulphalein	Pyruvate*
Fructose	Resorcinol

Sample interaction (carryover)

Within the normal range of serum creatinine, the amount of sample carryover does not exceed 5%. However, when a sample with an abnormally high concentration is followed by a sample with a normal concentration, the amount of carryover may be significantly greater than 5%. In such a case, the specimen following the one with the high value should be repeated.

Glucose (glucose oxidase)
Principle

The idea of using a glucose oxidase-peroxidase procedure for the determination of glucose was introduced by Keston[47] in 1956. During the ensuing years, a number of compounds, such as o-dianisidine, o-anisidine, o-toluidine, and indophenol, have been used as O_2 acceptors for the enzymatic reaction, but these chromogens were colloidal and unstable.[33]

Problems also were encountered in the direct application of these reagents to serum or plasma because of the presence of inhibitors, such as ascorbic acid, uric acid, bilirubin, catechols, glutathione, and cysteine, or because of hemolysis. These compounds inhibited the determination by competing with the chromogen as proton donors.[33]

The Technicon glucose oxidase method[28] is based on the procedure of Gochman and Schmitz.[31] In this method the specificity of glucose oxidase is combined with a new peroxide indicator couple, 3-methyl-2-benzothiazolinone hydrazone (MBTH) and dimethylaniline (DMA). Glucose catalyzes the reaction in which hydrogen per-

---

*It has been reported[85] that 200 mg/dl of pyruvate will elevate the true creatinine level of the sample by 7.4 mg/dl.

oxide is formed. Peroxidase catalyzes the oxidative coupling of MBTH and DMA to form indamine dye.

The reactions are as follows:

(1) Glucose + O$_2$ + H$_2$O $\xrightarrow{\text{Glucose oxidase}}$ Gluconic acid + H$_2$O$_2$

(2) MBTH + DMA + H$_2$O$_2$ $\xrightarrow{\text{Peroxidase}}$ Indamine dye + H$_2$O + OH$^-$

In the Technicon AutoAnalyzer II method, the sample is mixed with saline and dialyzed into a buffered glucose oxidase reagent. This reaction mixture is incubated briefly in a 37.5° C heating bath and then is mixed with a solution of 3-methyl-2-benzothiazolinone hydrazone and dimethylaniline (MBTH/DMA). Peroxidase reagent is added and mixed with the indicator-H$_2$O$_2$ solution. The peroxidase catalyzes coupling of MBTH and DMA and forms the soluble blue indamine dye. The amount of dye formed is proportional to the amount of glucose present in the original sample. The absorbance of the stable blue color is measured at 600 nm.

Range of expected values

The range of expected values for the glucose (glucose oxidase) method, as performed on the Technicon AutoAnalyzer II system, is 70 to 100 mg/dl.

System performance specifications

The following specifications apply to the glucose (glucose oxidase) method as performed on Technicon AutoAnalyzer II systems:

Rate of analysis	60 samples/hr
Sample-wash ratio	9:1
Chart speed	1 inch/min
Linearity range	0 to 500 mg/dl
Sensitivity coefficient	0.009 absorbance units per unit of concentration
Minimum sample required	90 μl (approximate)

Sample stability

Serum samples are stable for 2 to 3 days when stored at 5° C.[28]

Sample dilution

Serum samples with glucose concentrations greater than 500 mg/dl should be diluted with distilled water or saline and reassayed.

Interfering substances

The following substances, if present in the sample, reagents, or system, may interfere in glucose determinations in the manner described[28]:

Diethreitol **Galactose** **Hydrazine sulfate** **Maltose**	May cause an elevation (5%) in the determination of glucose levels.
	Can cause a significant elevation (30%) in the determination of glucose levels.
Phenylhydrazine HCl	May inhibit the glucose oxidase reaction, resulting in a significant depression (65%) of glucose levels.
Ascorbic acid	May inhibit the glucose oxidase reaction, resulting in a depression (10%) of glucose levels.

Gochman and Schmitz[31] reported that the following compounds (10 mg/dl), when added to a serum pool, had no measurable effect: gentisic acid, acetylsalicylic acid, L-3,4-dihydroxyphenylalanine, D,L-cysteine, 4-hydroxy-3-methoxy mandelic acid, vanillin, epinephrine, norepinephrine, L-ergothioneine, glutathione, phenol, 3,4-dihydroxyphenyl acetic acid, 3,4-dihydroxybenzoic acid, and uric acid or creatinine (100 mg/dl).

Sample interaction (carryover)

Within the range of expected values for the glucose (glucose oxidase) method, the amount of sample carryover is not clinically significant (less than 5%). In cases where a sample with an extremely high concentration is followed by a sample with a low concentration, the amount of sample carryover may result in a clinically significant (greater than 5%) elevation of the concentration in the sample.[28]

Glucose (neocuproine)
Principle

In 1961, Brown[10] described a modified Folin-Wu glucose method that was claimed to be 30 times more sensitive than the original method. The amount of crystalline copper sulfate used to prepare the alkaline copper solution was reduced and the phosphomolybdic acid was replaced with neocuproine (2,9-dimethyl-1,10-phenanthroline hydrochloride) solution.

In 1963, Bittner and McCleary[8] modified Brown's method by decreasing the amount of sodium carbonate used in the alkaline copper solution and by eliminating the chelating agent (the tartaric acid in the alkaline copper solution).

In 1966, Bittner and Manning[7] described an automated version of Bittner's neocuproine glucose method. Manual preparation of a protein-free filtrate was replaced by the use of a dialyzer, thereby eliminating the need for blank correction. The dialysate was pretreated with 0.25 M sodium carbonate to minimize interference from ascorbic acid and sulfhydryl groups.

In the automated method,[29] the serum sample is mixed with saline and is pumped into the dialyzer where glucose diffuses across the semipermeable membrane into a strongly alkaline solution of 0.25 M sodium carbonate. The amount of glucose that is dialyzed into the recipient stream is about 10% of the total glucose concentration. Along with the glucose that diffuses across the membrane, there is a small amount of dialyzable reducing substances including glutathione, creatinine, uric acid, and ascorbic acid. The presence of these substances in the reaction mixture affects the result to some degree. Interference of sulfhydryl groups and ascorbic acid is reduced by treating the dialysate with sodium carbonate prior to the addition of the color reagent.

After dialysis, the recipient stream is mixed with the neocuproine copper reagent and the mixture then is heated to 80° C. The cupric-neocuproine chelate is reduced by glucose in the alkaline medium, resulting in an intense yellow-orange cuprous-neocuproine complex. The absorbance of the colored complex is measured at 460 nm.

Range of expected values

Based on a study of 170 ambulatory adults, the expected ranges of glucose in mg/dl (by age and sex) for the glucose neocuproine method are as follows[29]:

	Males	*Females*	*Males and females*
Under 40	61.1-117.8	70.9-111.1	
Over 40	66.8-126.7	57.7-138.4	
Mean	65.0-125.2	60.3-131.1	62.9-127.8

System performance specifications

The following specifications apply to the glucose (neocuproine) method as performed on the Technicon AutoAnalyzer II system:

Rate of analysis	60 samples/hr
Sample-wash ratio	9:1
Chart speed	1 inch/min
Linearity range	0 to 500 mg/dl
Sensitivity coefficient	0.0012 absorbance units per unit of concentration
Minimum sample required	90 µl (approximate)

Sample stability

According to a study by Ruiter, Weinberg, and Morrison,[65] the glucose level in serum remains constant for at least 48 hours at room temperature, provided the serum is removed from the clot immediately after collection and centrifugation. However, if the serum is not assayed on the day it is collected, it should be stored in a refrigerator to assure stability.

Sample dilution

Serum or plasma samples with glucose concentrations greater than 500 mg/dl should be diluted with distilled water or saline and rerun.

Interfering substances

Several conditions and substances have been reported to interfere with the neocuproine glucose method and, if present in the sample, reagents, or system, may interfere in the manner described[29]:

Contact with red cells	Allowing the serum sample to remain in contact with the cells will cause a decrease in the glucose value.
Glutathione	Has been reported to cause elevated glucose values.
Ribose Mannose Xylose	Since the method is not specific for glucose, when these sugars are present in the sample, elevated glucose values will be obtained.

One millimole per liter of the following drugs have been reported to produce elevated glucose values when specimens are assayed by the neocuproine method:

Aminophenol	Levodopa
Aminosalicylic acid	Mercaptopurine
Ascorbic acid	Methimazole
Epinephrine	Methyldopa
Hydralazine	Propylthiouracil
Isoproterenol	

Sample interaction (carryover)

Technicon's experimental evidence has indicated that sample interaction does not exceed 5% even when a sample with a very high glucose concentration is followed by a sample with a low value.

The amount of sample carryover should be determined by the user on the individual AutoAnalyzer II system.

Glutamic-oxaloacetic transaminase (GOT)
(aspartate aminotransferase, EC 2.6.1.1)
Principle

The Technicon method for the determination of glutamic-oxaloacetic transaminase (GOT)[30] is based on the work of Kessler and associates,[45] who modified the procedure of Schwartz, Kessler, and Bodansky.[66] Both of these automated procedures were adapted from the manual method of Henry and co-workers.[34]

Sample, L-aspartic acid, and α-ketoglutaric acid are incubated and then the rate of formation of oxaloacetic acid is determined by the NAD-NADH system. Malic dehydrogenase (MDH) is added in excess, thus making GOT the rate-limiting enzyme in the following reactions:

$$\text{Aspartic acid} + \alpha\text{-Ketoglutaric acid} \xrightarrow{\text{GOT}} \text{Oxaloacetic acid} + \text{Glutamic acid}$$

$$\text{Oxaloacetate} + \text{NADH} \xrightarrow{\text{MDH}} \text{Malate} + \text{NAD}$$

The rate of decrease in absorbance at 340 nm is proportional to the GOT activity in the sample.

In the Technicon automated method, the serum sample is mixed with an air-segmented stream of GOT substrate and incubated at 37.5° C. The reaction mixture is then pumped into a dialyzer where the oxaloacetate formed is dialyzed into an air-segmented stream of MDH-NADH in a phosphate buffer.

The absorbance of the analytic stream is measured at 340 nm. Since the result of the two reactions is a decrease in absorbance, a signal inversion switch is used to obtain a direct graphic presentation of the test results.

Range of expected values

The range of expected values for the GOT method, as performed on the Technicon AutoAnalyzer II system, is 7 to 40 mU/ml.

System performance specifications

The following specifications apply to the GOT method as performed on the Technicon AutoAnalyzer II system:

Rate of analysis	60 samples/ hr
Sample-wash ratio	9:1
Chart speed	1 inch/min
Linearity range	0 to 300 mU/ml
Sensitivity coefficient	0.0013 absorbance unit per unit of enzyme activity
Minimum sample required	360 μl (approximate)

Sample stability

Serum samples are stable for 1 week when stored at 4° C and for 1 month when stored at −20° C.[85]

Sample dilution

Serum specimens with GOT activities greater than 300 mU/ml should be diluted with saline and reassayed.

Interfering substances

The following may interfere in GOT determinations in the manner described:

    Renal dialysis   Sera from patients undergoing chronic renal dialysis were found to have little or no GOT activity. In this study by Wolf and associates,[84] the analyses were done by both a colorimetric and the UV method of Henry and others.

A number of drugs are listed in Technicon's method (No. SE4-0010FH4) as causing an increase in GOT activity. However, the investigators referenced used a colorimetric method in their studies and their findings would not necessarily apply to this UV method.

Sample interaction (carryover)

Technicon states that, within the range of expected values for the GOT method, the amount of carryover is less than 5%. However, in cases where a sample with a very high value is followed by a sample with a low value, the amount of carryover may exceed 5%. In such cases, the low-valued sample should be repeated.

Inorganic phosphorus
Principle

Most methods for the determination of phosphate are based on the reduction of phosphomolybdic acid by a reducing reagent to molybdenum blue. Through the years several different reducing reagents have been used. Some presented problems such as poor sensitivity and poor stability, and some produced interfering colors.

In 1964, Hurst[39] described a phosphorus method using stannous chloride as the reducing reagent. The stannous chloride in dilute solution was stabilized by combining it with hydrazine sulfate. This permitted the rapid development of the molybdenum blue color at room temperature.

In 1966, Kraml[48] modified the method of Hurst for use on the AutoAnalyzer.

Technicon further modified the method by replacing the sulfuric acid in the ammonium molybdate reagent with hydrochloric acid. This modification has resulted in reduced carryover. The stannous chloride concentration was doubled to ensure linearity.

In the automated method,[42] the sample is mixed with an air-segmented stream of dilute sulfuric acid (0.36 N) and pumped into a dialyzer where the inorganic phosphorus is dialyzed into dilute sulfuric acid (0.36 N). The phosphate-containing dialysate then is mixed with an acid solution of ammonium molybdate to form phosphomolybdic acid. The stannous chloride–hydrazine reagent reduces the phosphomolybdic acid to molybdenum blue. The absorbance of the analytic stream is measured at 660 nm.

Range of expected values

The range of expected values for the inorganic phosphorus method, as performed on the Technicon AutoAnalyzer II system, is 2.5 to 4.5 mg/dl. This range is based on a statistical analysis of an ambulatory, adult population of both sexes, ages 20 to 60 years.[42]

System performance specifications

The following specifications apply to the inorganic phosphorus method as performed on Technicon AutoAnalyzer II systems:

Rate of analysis	60 samples/hr
Sample-wash ratio	9:1
Chart speed	1 inch/min
Linearity range	0 to 10 mg/dl
Sensitivity range	0.062 absorbance unit per unit of concentration
Minimum sample required	207 µl (approximate)

Sample stability

Serum samples are stable for about 8 hours at room temperature, 24 hours at 4° C, and 1 year when frozen.[85] Serum should be separated from the cells as soon as possible after blood is drawn, taking care to avoid hemolysis.

Sample dilution

Serum samples with inorganic phosphorus concentrations greater than 10 mg/dl should be diluted with distilled water or saline and reassayed.

Interfering substances

The following substances, if present in the sample, reagents, or system, may interfere with inorganic phosphorus determinations in the manner described[85]:

Hemolysis	Red blood cells contain a high concentration of organic phosphate esters. The inorganic phosphate increases rapidly as these esters are hydrolyzed by serum phosphatases.
Detergents	Detergents containing phosphates will cause falsely elevated inorganic phosphorus values.
Mannitol	May cause decreased phosphorus values.

Sample interaction (carryover)

Technicon states that at no time does sample interaction exceed 5%, not even when a sample with a very high phosphate concentration is followed by a sample with a very low concentration.

Total protein

Principle

The Technicon total protein method is based on the biuret reaction in which any organic compound containing two or more carbamyl groups ($-CO-NH_2$) that are joined together directly or by a single nitrogen or carbon atom gives a purple-colored complex when added to an alkaline copper salt solution.

In the manual method of Weichselbaum,[82] the biuret reagent contained sodium potassium tartrate as a complexing agent and potassium iodide to prevent autoreduction.

Skeggs and Hochstrasser[72] automated the manual method of Weichselbaum. They included a blank channel consisting of sample mixed with a solution containing potassium iodide and sodium hydroxide. This was run concurrently with the biuret-sample channel to compensate for any endogenous materials that would tend to elevate the total protein value. Blank subtraction was performed automatically by differential colorimetry.

The Technicon total protein method[78] requires the use of two independent but interrelated test channels. Also required is the digital printer, if automatic blank subtraction is desired, or a double-pen recorder if manual blank subtraction is preferred.

An alternative method, requiring only a single test channel, is to run the sample blanks first and then the sample tests (with the biuret reagent). The blank values are then subtracted from the test values.

When a two-channel system is used, the serum sample in one channel is added to an air-segmented stream of biuret reagent. During the ensuing reaction, the protein in the sample combines with the copper in the biuret reagent to form a purple complex.

In the blank channel, the sample is added to an air-segmented stream of total protein blank solution. The blank solution contains all the constituents of the biuret reagent except the copper sulfate and the sodium potassium tartrate. Their absence from the blank channel will prevent any color formation in that channel. Therefore, the absorbance determined in the blank channel is caused by endogenous serum pigments.

The absorbance of both channels is determined at 550 nm. Blank subtraction is performed by differential colorimetry when a digital printer is used.

Range of expected values

The range of expected values for the total protein method, as performed on the Technicon AutoAnalyzer II system, is 6.0 to 8.0 g/dl. This range is based on a statistical analysis of an ambulatory, adult population of both sexes, ages 20 to 60 years.

System performance specifications

The following specifications apply to the total protein method as performed on the Technicon AutoAnalyzer II systems:

Rate of analysis	60 samples/hr
Sample-wash ratio	9:1
Chart speed	1 inch/min
Linearity range	0 to 10 g/dl
Sensitivity coefficient	0.056 absorbance units per unit of concentration
Minimum sample required	180 µl (approximate)

Sample stability

The total protein in serum samples is stable for 1 week at room temperature and for 1 month at 4° C.[85]

Sample dilution

Serum samples with total protein concentrations greater than 10 g/dl should be diluted with saline and reassayed.

Interfering substances

The following substances, if present in the sample, reagents, or system, may interfere in total protein determinations in the manner described:

Dextran[85]	When dextran is present in serum samples, it reacts with biuret reagent to form a turbid solution. Since the biuret reagent is used only in the sample channel, the turbidity will appear only in this channel and, therefore, cannot be corrected for by blank subtraction. The result is elevated total protein values.
Ammonium ions[85]	Ammonium ions react with the copper in the biuret reagent forming a cupric-ammonium complex. The resulting decrease in copper concentration may cause lower total protein values.
Hemolysis[85]	Hemolysis will cause increased total protein values as a result of the added hemoglobin.

Sample interaction (carryover)

Technicon states that there is no clinically significant sample interaction (less than 5%) between samples at varying levels of concentration throughout the entire range of the total protein method.

Urea nitrogen (UN)

Principle

In 1939, Fearon[21] found that the reaction of diacetyl-monoxime followed by oxidation produced colors with urea, creatinine, methylurea and urea derivatives, allantoin, and proteins. These compounds give different chromogens, but only urea produces a yellow color.

In 1957, Marsh, Fingerhut, and Kirsch[52] described an automated urea nitrogen method using diacetyl-monoxime and ferric alum reagents.

In 1963, Coulombe and Favreau[16] demonstrated the effect of thiosemicarbazide on the reaction between diacetyl-monoxime and urea. This compound was shown to intensify the color.

In 1965, Marsh, Fingerhut, and Miller[53] presented both a manual and an automated method for the determination of urea. Both methods are highly sensitive due to the combined use of thiosemicarbazide and ferric ions. This modification of the BUN color reagent made possible the reduction of the acid requirement to about one tenth of that originally needed. The urea nitrogen method[79] as performed on the Technicon AutoAnalyzer II system is a modification of the procedure described by Marsh and co-workers.

In this method, the serum sample is diluted in an air-segmented stream of distilled water. The urea then is dialyzed across the semipermeable membrane into a recipient stream of BUN color reagent. After dialysis, a stream of BUN acid reagent is added to the recipient stream. The reaction mixture passes through a 90° C heating bath where the chromogen forms. The stream then flows to the colorimeter, where the absorbance is measured at 520 nm.

Range of expected values

The range of expected values for the urea nitrogen method, as performed on the Technicon AutoAnalyzer II system, is 10 to 26 mg/dl. This range is based on a statistical analysis of an ambulatory, adult population of both sexes, ages 20 to 60 years.[79]

System performance specifications

The following specifications apply to the urea nitrogen method as performed on the Technicon AutoAnalyzer II system:

Rate of analysis	60 samples/hr
Sample-wash ratio	9:1
Chart speed	1 inch/min
Linearity range	0 to 150 mg/dl
Sensitivity coefficient	0.004 absorbance unit per unit of concentration
Minimum sample required	90 µl (approximate)

Sample stability

Serum samples for urea nitrogen determinations are stable for at least a week when stored at 4° C.[85]

Sample dilution

Serum samples with urea nitrogen concentrations greater than 150 mg/dl should be diluted with distilled water or saline and reassayed.

Interfering substances

The only drug presently known to interfere with the diacetyl-monoxime urea nitrogen method is acetohexamide, which causes a positive interference.[79]

Sample interaction (carryover)

Technicon states that at no time does sample interaction exceed 5%, even when a sample with a very high urea nitrogen value is followed by a sample with a very low value.

Uric acid
Principle

The first quantitative method for determining uric acid in serum was introduced in 1912 by Folin and Denis.[23] This method and the modifications that followed presented problems with turbidity and/or poor sensitivity. In 1945, Brown[9] introduced a method using a cyanide-urea reagent and an improved phosphotungstic acid reagent. Of the methods using cyanide, his has given the best results and is still in use.

In 1965, Sobrinho-Simões[73] introduced a uric acid method that uses sodium tungstate to provide the required alkalinity and hydroxylamine to intensify the color of the final solution. The method has the precision and accuracy of Brown's method and does not require the use of the toxic cyanide-urea reagent.

Musser and Ortigoza[57] modified the reagent concentrations and automated the Sobrinho-Simões method. As a result of these reagent modifications, the time required for color development was greatly decreased.

In the Technicon uric acid method,[81] the serum sample is diluted with saline and is pumped into a dialyzer where uric acid dialyzes into a sodium tungstate-hydroxylamine solution. Phosphotungstic acid is added to the recipient stream as it emerges from the dialyzer. The phosphotungstic complex is formed in proportion to the amount of uric acid present. The absorbance of the reaction stream is measured at 660 nm.

Range of expected values

The range of expected values for the uric acid method as performed on Technicon AutoAnalyzer II systems is 3.5 to 9.0 mg/dl. This range is based on a statistical analysis of an ambulatory, adult population of both sexes, ages 20 to 60 years.[81]

System performance specifications

The following specifications apply to the uric acid method as performed on the Technicon AutoAnalyzer II system:

Rate of analysis	60 samples/hr
Sample-wash ratio	9:1
Chart speed	1 inch/min
Linearity range	0 to 12 mg/dl
Sensitivity coefficient	0.024 absorbance unit per unit of concentration
Minimum sample required	90 µl (approximate)

Sample stability

Serum samples for uric acid determinations are stable for 3 days at room temperature and for 6 months when frozen.[85]

Sample dilution

Serum samples with uric acid concentrations greater than 12 mg/dl should be diluted with saline and reassayed.

Interfering substances

The following substances, when present at the maximum concentration likely to occur in serum or plasma, have been shown to cause falsely elevated uric acid values[70]:

	Concentration ($\mu g/ml$)	Apparent uric acid concentration (mg/dl)
N-acetyl-p-aminophenol	52.3	+0.5
Ascorbic acid	250.0	+2.0
L-Dopa	83.3	+3.5
α-Methyldopa	62.5	+2.5
6-Mercaptopurine	14.0	+0.6

Sample interaction (carryover)

Technicon states that at no time does sample interaction exceed 5%, even when a sample with a very high value is followed by a sample with a very low value.

References

1. Albumin, Technicon Laboratory method file N-15c.
2. Albumin (BCG), Technicon method No. SE4-0030FD4, April, 1974.
3. Alkaline phosphatase, Technicon Laboratory method file N-6b.
4. Alkaline phosphatase, Technicon method No. SE4-0006FC4, March, 1974.
5. Anderegg, G., Flaschka, H., Sallmann, R., and Schwarzenbach, G.: Ein auf erdalkaliionen ansprechendes Phthalein und seine analytische Verwendung, Helv. Chim. Acta 37:113, 1954.
6. Bessey, O. A., Lowry, O. H., and Brock, M. J.: Method for rapid determination of alkaline phosphatase with 5 cubic millimeters of serum, J. Biol. Chem. 164:321, 1946.
7. Bittner, D. L., and Manning, J.: Automated neocuproine glucose method: critical factors and normal values. In Automation in analytical chemistry, Technicon symposia, 1966, White Plains, New York, 1967, Mediad Inc., pp. 33-36, Vol. I.
8. Bittner, D. L., and McCleary, M. L.: The cupric-phenanthroline chelate in the determination of monosaccharides in whole blood, Am. J. Clin. Pathol. 40:423, 1963.
9. Brown, H.: The determination of uric acid in human blood, J. Biol. Chem. 158:601, 1945.
10. Brown, M. E.: Ultra-micro sugar determinations using 2, 9-dimethyl-1, 10-phenanthroline hydrochloride (neocuproine), Diabetes 10:60, 1961.
11. Calcium, Technicon Laboratory method file N-3b.
12. Calcium, Technicon method No. SE4-0003FJ4, September, 1974.
13. Chasson, A. L., Grady, H. J., and Stanley, M.A.: Determination of creatinine by means of automatic chemical analysis, Am. J. Clin. Pathol. 35:83, 1961.
14. Cholesterol (direct), Technicon Laboratory method file N-77.
15. Cholesterol (direct), Technicon method No. SE4-0026FC4, March, 1974.
16. Coulombe, J. J., and Favreau, L.: A new simple semimicro method for colorimetric determination of urea, Clin. Chem. 9:102, 1963.
17. Creatinine, Technicon method No. SE4-0011FH4, August, 1974.
18. Davis, H. A., Sterling, R. E., Wilcox, A. A., and Waters, W. E.: Investigation of a potential source of difficulty in the use of the AutoAnalyzer, Clin. Chem. 12:428, 1966.
19. Doumas, B. T., Watson, W. A., and Biggs, H. G.: Albumin standards and the measurement of serum albumin with bromcresol green, Clin. Chem. Acta 31:87, 1971.
20. Electrolytes, Technicon Laboratory method file N-21b.
21. Fearon, W. R.: The carbamido diacetyl reaction; a test for citrulline, Biochem. J. 33:902, 1939.
22. Folin, O.: On the determination of creatinine and creatine in urine, J. Biol. Chem. 17:469, 1914.
23. Folin, O., and Denis, W.: A new colorimetric method for the determination of uric acid in blood, J. Biol. Chem. 13:469, 1912-1913.
24. Folin, O., and Wu, H.: A system of blood analysis, J. Biol. Chem. 38:81, 1919.
25. Fredrickson, D. S., Levy, R. I., and Lees, R. S.: Fat transport in lipoproteins—an integrated approach to mechanisms and disorders, N. Engl. J. Med. 276:148, 1967.
26. Gambino, S. R., and Schreiber, H.: The measurement and fractionation of bilirubin on the AutoAnalyzer by the method of Jendrassik and Grof, presented at Technicon International Symposium, Automation in Analytical Chemistry, no. 54, New York, 1964.
27. Gitelman, H. J.: An improved automated procedure for the determination of calcium in biological specimens, Anal. Biochem. 18:521, 1967.
28. Glucose (Glucose-oxidase), Technicon method No. SE4-0036FJ4, September, 1974.
29. Glucose (Neocuprione), Technicon method No. SE4-0002FF4, June, 1974.
30. Glutamic-oxaloacetic transaminase (GOT),

Technicon method No. SE4-0010FH4, August, 1974.
31. Gochman, N., and Schmitz, J. M.: Application of a new peroxide indicator reaction to the specific, automated determination of glucose with glucose oxidase, Clin. Chem. 18:943, 1972.
32. Hawk, P. B., Oser, B. L., and Summerson, W. H.: Practical physiological chemistry, ed. 13, New York, 1954, McGraw-Hill Book Co., pp. 564-565.
33. Henry, R. J., Cannon, D. C., and Winkelman, J. W.: Clinical chemistry—principles and technics, ed. 2, New York, 1974, Harper and Row, Publishers, pp. 1281-1282.
34. Henry, R. J., Chiamori, N., Golub, O., and Berkman, S.: Revised spectrophotometric methods for the determination of glutamic-oxalacetic transaminase, glutamic-pyruvic transaminase, and lactic acid dehydrogenase, Am. J. Clin. Pathol. 34:381, 1960.
35. Hewitt, T. E., and Pardue, H. L.: Kinetics of the cholesterol-sulfuric acid reaction: a fast kinetic method for serum cholesterol, Clin. Chem. 19:1128, 1973.
36. Hochella, N. J., and Weinhouse, S.: Automated assay of lactate dehydrogenase in urine, Anal. Biochem. 13:322, 1965.
37. Hoffman, W. S.: A rapid photoelectric method for the determination of glucose in blood and urine, J. Biol. Chem. 120:51, 1937.
38. Huang, T. C., Chen, C. P., Weffler, V., and Raftery, A.: A stable reagent for the Liebermann-Burchard reaction, Anal. Chem. 33:1405, 1961.
39. Hurst, R. O.: The determination of nucleotide phosphorus with a stannous chloride-hydrazine sulfate reagent, Can. J. Biochem. 42:287, 1964.
40. Inorganic phosphate, Technicon Laboratory method file N-4b.
41. Inorganic phosphate, Technicon Laboratory method file N-4c.
42. Inorganic phosphorus, Technicon method No. SE4-0004FH4, August, 1974.
43. Jaffe, M.: Über den Niederschlag, welchen Pikrinsäure in normalen Harn erzeugt, und über eine neue Reaction des Kreatinins, Z. Physiol. Chem. 10:391, 1886.
44. Jendrassik, L., and Grof, P.: Vereinfachte photometrische Methoden zur Bestimmung des Blutbilirubins, Z. Biochem. 297:81, 1938.
45. Kessler, G., Rush, R., Leon, L., Delea, A., and Cupiola, R.: Automated 340 nm measurement of SGOT, SGPT, and LDH. In Advances in Automated Analysis, Technicon International Congress 1970, Vol. 1, Miami, Florida, 1971, Thurman Associates, pp. 67-74.
46. Kessler, G., and Wolfman, M.: An automated procedure for the simultaneous determination of calcium and phosphorus, Clin. Chem. 10:686, 1964.
47. Keston, A. S.: Specific colorimetric enzymatic analytical reagents for glucose, abstract, American Chemical Society (Division of Biology and Chemistry), Dallas meeting, p. 31C, 1956.
48. Kraml, M.: A semi-automated determination of phospholipids, Clin. Chim. Acta 13:442, 1966.
49. Levine, J., Morgenstern, S., and Vlastelica, D.: A direct Liebermann-Burchard method for serum cholesterol, Technicon Symposia 1967, New York, 1968, Mediad, Inc., pp. 25-28.
50. LHD, Technicon Laboratory method file N-60.
51. Lowry, O. H., Roberts, N. R., Wu, M. L., Hixon, W. S., and Crawford, E. J.: The quantitative histochemistry of brain. II. Enzyme measurements, J. Biol. Chem. 207:19, 1954.
52. Marsh, W. H., Fingerhut, B., and Kirsh, E.: Determination of urea nitrogen with the diacetyl method and an automatic dialyzing apparatus, Am. J. Clin. Pathol. 28:681, 1957.
53. Marsh, W. H., Fingerhut, B., and Miller, H.: Automated and manual direct methods for the determination of blood urea, Clin. Chem. 11:624, 1965.
54. Massion, C. G., and Frankenfeld, J. F.: Alkaline phosphatase: lability in fresh and frozen human serum and in lyophilized control material, Clin. Chem. 18:366, 1972.
55. Morgenstern, S., Kessler, G., Auerbach, J., Flor, R. V., and Klein, B.: An automated p-nitrophenyl phosphate serum alkaline phosphatase procedure for the AutoAnalyzer, Clin. Chem. 11:876, 1965.
56. Morgenstern, S., Oklander, M., Auerbach, J., Kaufman, J., and Klein, B.: Automated determination of serum glutamic oxaloacetic transaminase, Clin. Chem. 12:95, 1966.
57. Musser, A. W., and Ortigoza, C.: Automated determination of uric acid by the hydroxylamine method, Am. J. Clin. Pathol. 45:339, 1966.
58. Ness, A. T., Dickerson, H. C., and Pastewka, J. V.: The determination of human serum albumin by its specific binding of the anionic dye, 2-(4'-hydroxybenzeneazo)-benzoic acid, Clin. Chim. Acta 12:532, 1965.
59. Ness, A. T., Pastewka, J. V., and Peacock, A. C.: Evaluation of a recently reported stable Liebermann-Burchard reagent and its use for the direct determination of serum total cholesterol, Clin. Chim. Acta 10:229, 1964.
60. Nishi, H. H., and Rhodes, A.: An automated procedure for the determination of albumin

in human serum, Technicon Symposium, 1965, Automation in Analytical Chemistry, New York, 1966, Mediad Inc., pp. 321-323.
61. Pastewka, J. V., and Ness, A. T.: The suitability of various serum albumin products as standards for the quantitative analysis of total protein and albumin in human body fluids, Clin. Chim. Acta 12:523, 1965.
62. Pollard, F. H., and Martin, J.: The spectrophotometric determination of the alkaline-earth metals with murexide, Eriochrome Black T and with o-cresolphthalein Complex-one, Analyst 81:348, 1956.
63. Rodkey, F. L.: Direct spectrophotometric determination of albumin in human serum, Clin. Chem. 11:478, 1965.
64. Romano, A. T.: Automated glucose methods: evaluation of glucose oxidase-peroxidase procedure, Clin. Chem. 19:1152, 1973.
65. Ruiter, J., Weinberg, F., and Morrison, A.: The stability of glucose in serum, Clin. Chem. 9:356, 1963.
66. Schwartz, M. K., Kessler, G., and Bodansky, O.: Automated assay of activities of enzymes involving the diphosphopyridine nucleotide ⇌ reduced diphosphopyridine nucleotide reaction, J. Biol. Chem. 236:1207, 1961.
67. Serum glutamic oxalacetic transaminase (SGOT), Technicon Laboratory method file N-25b.
68. Simultaneous direct and total bilirubin, Technicon Laboratory method file N-51a.
69. Simultaneous glucose and urea nitrogen, Technicon Laboratory method file N-16b.
70. Singh, H. P., Herbert, M. A., and Gault, M. H.: Effect of some drugs on clinical laboratory values as determined by the Technicon SMA 12/60, Clin. Chem. 18:137, 1972.
71. Skeggs, L. T., Jr.: An automatic method for the determination of carbon dioxide in blood plasma, Am. J. Clin. Pathol. 33:181, 1960.
72. Skeggs, L. T., Jr., and Hochstrasser, H.: Multiple automatic sequential analysis, Clin. Chem. 10:918, 1964.
73. Sobrinho-Simões, M.: A sensitive method for the measurement of serum uric acid using hydroxylamine, J. Lab. Clin. Med. 65:665, 1965.
74. Stern, J., and Lewis, W. H. P.: The colorimetric estimation of calcium in serum with o-cresolphthalein Complexone, Clin. Chim. Acta 2:576, 1957.
75. Stevens, D. L.: Total protein, Technicon Laboratory method file N-14b.
76. Stevens, D. L., and Skeggs, L. T., Jr.: Creatinine, Technicon Laboratory method file N-11b.
77. Stoner, R., and Plestina, J.: Normal values, Channel 12 2:4, 1973.
78. Total protein, Technicon method No. SE4-0014FC4, March, 1974.
79. Urea nitrogen (UN), Technicon method No. SE4-0001FD4, April, 1974.
80. Uric acid, Technicon Laboratory method file N-13b.
81. Uric acid, Technicon method No. SE4-0013FH4, August, 1974.
82. Weichselbaum, T. E.: Accurate and rapid method for the determination of proteins in small amounts of blood serum and plasma, Am. J. Clin. Pathol. 10:40, 1946 (tech. section).
83. White, W. L., Erickson, M. M., and Stevens, S. C.: Practical automation for the clinical laboratory, ed. 2, St. Louis, 1972, The C. V. Mosby Co.
84. Wolf, P. L., Williams, D., Coplon, N., and Coulson, A. S.: Low aspartate transaminase activity in serum of patients undergoing chronic hemodialysis, Clin. Chem. 18:567, 1972.
85. Young, D. S., Thomas, D. W., Friedman, R. B., and Pestaner, L. C.: Effects of drugs on clinical laboratory tests, Clin. Chem. 18:1041, 1972.
86. Zall, D. M., Fisher, D., and Garner, M. O.: Photometric determination of chlorides in water, Anal. Chem. 28:1665, 1956.

4

Glucose and nonprotein nitrogen compounds

Glucose

Carbohydrates play the important role of chief fuel supply for the body processes. The limited storage capacity, approximately 300 g of carbohydrates for a 70-kg individual, will produce an energy supply for about 12 hours. When the storage capacity is not sufficient, the body will compensate by drawing on fat and protein, which will be converted to carbohydrates.

The polysaccharides and disaccharides are enzymatically hydrolyzed to monosaccharides in the gastrointestinal tract. They are then absorbed in the intestine by an active cellular process, or diffusion. The rate of the absorption depends on the different sugars and is affected by the emptying time of stomach, type and amount of food in the intestine, physiologic state of intestinal mucosa, and endocrine activity such as the salt-regulating functions of adrenal cortical hormone and the state of the thyroid activity. Insulin has no effect on this process.

METABOLIC PROCESS

The only sugar found in the blood of a fasting, healthy individual is the hexose, glucose, which is stored in the form of its polymer, glycogen. Glycogen (approximately 70% of the total carbohydrate in the body) is present only as an intracellular component, while normally free glucose is found only in extracellular fluids. After passing through cell membranes, the glucose is immediately phosphorylated by combining with the terminal phosphate of adenosine triphosphate under the influence of an appropriate enzyme. Through a series of enzyme-catalyzed reactions, the glucose may be converted into fat or amino acids after a degradation to two-carbon compounds and stored as glycogen or oxidized to CO_2 and H_2O with the release of energy.

Liver

The liver has a primary role in the regulation of blood sugar levels. The monosaccharides glucose, fructose, and galactose are removed from portal venous blood by liver cells, phosphorylated in the presence of specific enzymes (hexokinase and glucokinase) and are converted to glycogen or metabolized through pyruvate and Krebs' cycle intermediates. The enzyme hexokinase of liver acts maximally when glucose concentrations are lower than those usually occurring in plasma whereas liver glucokinase activity is not maximal until higher than normal blood glucose concentrations are present. As blood glucose levels increase above normal, the rate of glucose phosphorylation by glucokinase increases. The liver and the kidney (to a lesser extent) are the two endogenous sources of blood glucose. It really is not clear how the kidney effects the release of glucose into the circulation, but the liver is capable of releasing or removing glucose from circulation and of synthesizing it from a number of amino acids and fatty acid fragments.

The blood sugar level acts as a stimulus to the liver and determines whether glycogenolysis or glycogenesis will be present. Hypoglycemia is usually associated with a decreased hepatic gluconeogenesis and glucose output. In the presence of normal amounts of insulin, hyperglycemia inhibits the hepatic output of glucose since it accelerates its deposition, with the result of restoring the glucose to a normal state.

Endocrine glands

The blood glucose level is influenced by the effect of the thyroid, adrenal cortex, and anterior pituitary hormones on the liver. Insulin, produced by the pancreatic islet cells, acts mainly on extrahepatic tissue, while glucagon, produced by the alpha cells of the pancreas, elevates blood glucose levels by a glycogenolytic effect on liver glycogen.

Although epinephrine is not involved in glucose regulation, an increased secretion may cause excessive hepatic glycogenolysis and hyperglycemia.

Peripheral tissues

Glucose is utilized by all peripheral tissues for energy. However, only the cardiac and skeletal muscles are capable of storing glycogen in significant amounts. It has been found that within certain levels of glucose concentration and in normal individuals, there exists a direct relationship between that concentration and the utilization of carbohydrates by extrahepatic tissues. The rate of utilization is lower for diabetic persons. The rate of absorption into the muscle will vary for a hypoglycemic individual.

Insulin is the one hormone that will increase the extrahepatic uptake of glucose. Insulin facilitates glucose transfer through cell membranes before its phosphorylation. It initially inhibits glucose release by the liver and, if followed by hypoglycemia and releasing of epinephrine, glycogenolysis results. Therefore when there is decrease in insulin there is an increase in glucose and vice versa. An excess of growth hormone has an antiinsulin effect.

Decreased glucose concentration—hypoglycemia

Hypoglycemia may result from a normal hepatic glucose output with an increased peripheral uptake, a hepatic gluconeogenesis decrease when combined with a normal peripheral utilization of glucose, and a combination of these two factors.

Hepatic destruction resulting from poisoning by arsphenamine, carbon tetrachloride, phosphorus, benzol, or chloroform may result in hypoglycemia, since the main source of glucose is destroyed. However, in individuals with hepatic damage (not destruction) either hypoglycemia or hyperglycemia may be observed. Psychoneurotic individuals have shown evidence of hypoglycemia since nervous tissues require adequate and continuous supplies of glucose for maintenance of normal activity. The function of the abnormality in the neuroses has not been defined but serious and permanent brain damage can result from episodes of hypoglycemia.

Von Gierke's disease, a glycogen storage disease, is characteristic of hypoglycemia since the increased hepatic stores of glycogen are not available for glucose formation.

There is an imbalance between peripheral removal and hepatic output in hypofunction of the thyroid gland, adrenal cortex, or anterior pituitary. These deficiencies cause a rate decrease in peripheral removal and a greater rate decrease in hepatic formation of glucose with lowered glucose levels.

Hyperinsulinism resulting from pancreatic islet cell hyperplasia, carcinoma, or adenoma produces hypoglycemia. In these cases the low glucose levels result from the inability of the hepatic glucose output to keep up with the accelerated rate of insulin production. Insulin increases the peripheral uptake of the glucose, favoring glycogenesis by the liver; however, the counterregulatory effect of hypoglycemia increases the hepatic glucose output.

If a fasting blood sugar (FBS) is low, hyperinsulinism may be indicated. Other tests that may demonstrate hyperinsulinism are the occurrence of hypoglycemia after a 12- to 18-hour fast, or a persistent or increasing depression of the blood glucose concentration after an intravenous glucose tolerance test over a period of 4 to 6 hours.

Increased glucose concentration—hyperglycemia

Hyperglycemia may be caused by a normal hepatic output of glucose with a decreased rate of peripheral removal, an increase in hepatic production and release of glucose with a normal removal rate by peripheral tissues, or the combination of these two factors.

A chronic hepatic dysfunction may cause increased blood glucose values because the relative insensitivity of hepatic cells to various homeostatic mechanisms.

It is unusual to observe an increased glucose level secondary to gastrointestinal dysfunction. Usually the cause of hyperglycemia is a gastrectomy, which causes food to move into the intestine at a fast rate, resulting in the dumping syndrome.

A combination of anoxia and epinephrine secretion is possibly responsible for hyperglycemia in convulsive conditions such as eclampsia, epilepsy, tetany, and intracranial trauma where there may be damage to the fourth ventricle.

General anesthesia mobilizes liver glycogen and raises the glucose level because of adrenal medullary stimulation, acidosis, or anoxia. Anoxia may cause hyperglycemia since liver glycogen is unstable in the presence of a deficient oxygen supply.

An increase in adrenal cortical activity, primary or secondary to pituitary basophilism with increased ACTH production, will increase the glucose, since the hepatic gluconeogenesis increase is greater than the increase in peripheral uptake.

If the rate of peripheral utilization cannot keep pace with the hepatic output, the thyroid hormone may elevate the blood glucose.

Hyperglycemia may be seen in individuals with a pheochromocytoma. This is evident if the blood is drawn at the time when excess epinephrine is discharged from the adrenal medullary tumor.

In diabetes mellitus, persistent hyperglycemia and glycosuria are observed. The glycosuria occurs when the renal tubular reabsorptive capacity for glucose is exceeded. A person with mild diabetes may show normal fasting glucose levels, so a glucose tolerance test or a postprandial blood sugar analysis may be the determining factor of a diabetic diagnosis. The overproduction of glucose by the liver is the primary abnormality in the syndrome. Within limits of any glucose level, the diabetic individual utilizes less carbohydrate than the normal person. However, at the diabetic elevated glucose level, the individual will utilize carbohydrate at a rate comparable to that of a normal subject. The diabetic syndrome can result from an absolute or relative insulin lack caused by lesions in the islets of Langerhans, disorders of endocrine glands, pancreatectomy, or insulin antagonists.

Table 4-1
Conditions accompanied by abnormal glucose values*

Increased glucose	Decreased glucose
Diabetes	**Hyperinsulinism**
Hyperthyroidism	Hypothyroidism
Hyperpituitarism	Hypopituitarism
Nephritis	Hepatic disease
Coronary thrombosis	Addison's disease
Infections	Pernicious vomiting
Pregnancy	Starvation
Uremia	

*Diseases of primary importance to physician are in boldface.

METHODS OF GLUCOSE DETERMINATION

Folin-Wu method

An alkaline copper solution is added to a protein-free filtrate. The glucose present reduces the copper to cuprous oxide. The cuprous oxide in turn reduces a colorless phosphomolybd*ic* acid solution to blue phosphomolyb*ous* acid. The depth of color is measured in a colorimeter. The normal values for this method are 80 to 120 mg/dl of blood.

Benedict method

A special bisulfite copper solution is used that increases the specificity of the test for glucose at the expense of the saccharoids. This solution is added to a protein-free filtrate and the glucose reduces the copper to cuprous oxide. The cuprous oxide in turn reduces a colorless phosphomolybd*ic* acid solution to blue phosphomolybd*ous* acid. The depth of color is measured in a colorimeter. The normal values are 70 to 100 mg/dl of blood.

Somogyi-Nelson method

A Somogyi protein-free filtrate that removes reducing substances other than sugar is made. A copper tartrate reagent is added to the filtrate and the glucose reduces the copper to cuprous oxide. The cuprous oxide then reacts with an arsenomolybdate reagent to give a blue color. The depth of color is measured in a colorimeter. The normal values are 65 to 95 mg/dl of blood.

Glucose oxidase method (modified)

This is an enzymatic method that will respond to glucose only. No other physiologic constituent is measured. Only a Somogyi protein-free filtrate, which is free of metallic ions, can be used. The tungstic acid filtrate cannot be used with this method. The enzyme, glucose oxidase, catalyzes (or promotes) the oxidation of glucose with oxygen to gluconic acid and hydrogen peroxide. The hydrogen peroxide is detected by a chromogenic oxygen acceptor in the presence of peroxidase (horseradish). The normal range is 65 to 95 mg/dl.

o-Toluidine method

Glucose reacts with *o*-toluidine in hot glacial acetic acid to yield a blue-green N-glucosylamine. The absorbance is measured at 625 nm. This method can be applied directly to serum, plasma, cerebrospinal fluid, and urine. Whole blood or moderately hemolyzed serum requires deproteinization.

The reaction is not based on the reducing properties of glucose; therefore, only a few physiologically occurring compounds react with *o*-toluidine to yield substances absorbing at the wavelength used to measure the colored compound formed with glucose. Mannose and galactose, and to a lesser degree lactose and xylose, are potential sources of error but are normally not present in serum or plasma in significant concentrations.

It is the opinion of many clinical chemists that when all factors are considered—such as susceptibility of enzymatic methods to inhibitors, expense of reagents, stability of reagents, and ease performance—the *o*-toluidine method is the choice for the routine clinical chemistry laboratory.[13, 14]

DETERMINATION OF GLUCOSE
Specimen stability

When blood is drawn and permitted to stand uncentrifuged at room temperature, the average decrease in serum glucose is approximately 7% in 1 hour. In separated unhemolyzed serum the glucose is usually stable up to 8 hours at 25° C and up to 72 hours at 4° C. Variable stability, related to bacterial contamination, can be observed over longer storage periods. Glucose can be preserved by adding sodium fluoride to the serum.

o-Toluidine method of Hultman,[10] modified by Dubowski,[3] and Hyvärinen and Nikkilä[11]

Reagents

Tungstic acid reagent	(D-78)
o-Toluidine reagent (stabilized)	(D-79)
Glucose standard, 100 mg/dl	(D-80)

Procedure—serum or plasma

1. Transfer 5.0 ml of *o*-toluidine reagent to tubes of about 10-ml capacity. Label tubes *blank, standard,* and *unknown.*
2. Add 100 μl of glucose standard to the *standard* tube and mix thoroughly.
3. Add 100 μl of serum or plasma to the *unknown* tube and mix thoroughly.
4. Place loosely stoppered or capped tubes in a boiling water bath or a heating block preset at 100° C. Remove all tubes after 10 minutes and place in a cold water bath.
5. When the tubes have cooled, read the absorbance of each tube against the blank at 625 nm.

Calculation

$$\text{mg glucose/dl} = \frac{A_x}{A_s} \times 0.1 \times \frac{100}{0.1}$$

$$= \frac{A_s}{A_x} \times 100$$

Procedure—cerebrospinal fluid (CSF)

The procedure for measuring CSF glucose is the same as for serum or plasma except that 200 μl of spinal fluid is used and 100 μl of water is added to the standard tube.

Calculation

$$\text{mg glucose/dl} = \frac{Ax}{As} \times 0.1 \times \frac{100}{0.2}$$

$$= \frac{Ax}{As} \times 50$$

The normal values for glucose in spinal fluid are approximately two thirds that of the blood glucose or 45 to 75 mg/dl of fluid. Increased values may accompany brain tumors. Decreased values are seen in most types of meningitis. If the analysis cannot be done within the hour, the spinal fluid may be placed in a test tube containing sodium fluoride as a preservative.

Procedure—whole blood

1. Prepare a deproteinized sample by transferring 200 µl of blood sample into a tube containing 1.8 ml of tungstic acid reagent. Mix, let stand for 5 minutes, and centrifuge.
2. Add 1.0 ml of the clear supernatant to a tube containing 5.0 ml of *o*-toluidine reagent, mix, and label *unknown*. Set up the *blank* and *standard* as follows:
   Blank: 5.0 ml *o*-toluidine reagent plus 1.0 ml water
   Standard: 5.0 ml *o*-toluidine reagent plus 0.9 ml water plus 100 µl glucose standard
3. Proceed, starting with step 4 of the serum or plasma method.

Dextrostix method

Dextrostix,* a rapid screening test, is a strip that may be used for measurement of glucose in a drop of whole blood. The test is based on the action of the enzyme glucose oxidase, which catalyzes the oxidation of glucose to gluconic acid and hydrogen peroxide. The hydrogen peroxide in the presence of peroxidase oxidizes the chromogen system, which is related to the amount of glucose present in the blood. *The test is not intended to replace quantitative procedures.*

Procedure

1. Place a large drop of whole blood (sufficient to cover the entire reagent area) on the reagent area of the printed side of the strip. (A thin film of of blood tends to dry and is removed with difficulty during the washing.)
2. Allow the blood to remain on the strip for exactly 60 seconds; time accurately.
3. Quickly wash the blood from the reagent area, using a stream of water from a wash bottle and holding the strip vertically. This requires only 1 or 2 seconds.
4. Promptly (1 or 2 seconds) and decisively match the color with the color chart. Longer periods for matching allow the color of the strip to change giving unsatisfactory results. The system has been devised for whole blood. Do not use serum or plasma as the color shades will not correspond with the color chart.

Reaction

The values on the color chart are 40, 65, 90, 130, 150, 200, and 250 mg of glucose/dl of blood with color shades from gray to blue-purple.

---

*Ames Co., Division of Miles Laboratories, Inc.

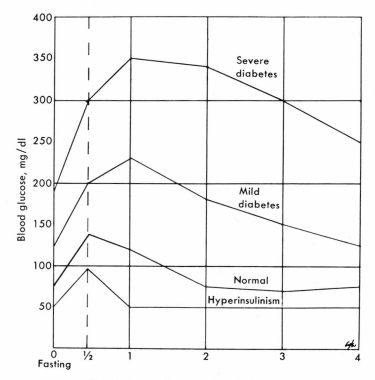

Fig. 4-1. Typical glucose tolerance curves in the Janney-Isaacson test and the intravenous test.

GLUCOSE TOLERANCE

When glucose is given to a normal individual and to a diabetic one, the normal individual removes the glucose at a faster rate than the diabetic. This is the principle of the glucose tolerance test. In the diabetic person, the delay in reaction is due to lack of insulin. Criteria for a normal glucose tolerance test[4] are listed below:

FBS   < 100 mg/dl   1½ hours < 140 mg/dl
1 hour < 160 mg/dl   2 hours < 120 mg/dl

Many methods for the glucose tolerance tests are used. Brief discussions of three methods follow.

Janney-Isaacson test

In the Janney-Isaacson test, fasting blood and urine specimens are taken. A known amount of glucose is given orally. Blood and urine specimens are then taken at ½, 1, 2, 3, and 4 hours after the ingestion of glucose. Glucose levels are determined on all blood and urine specimens. Normal and abnormal results are indicated in Fig. 4-1.

Intravenous test

In the intravenous test, the following factors are taken into consideration: When glucose is taken by mouth, it is absorbed into the bloodstream from the small intestine. In health, the rate of absorption is fairly uniform. In diseases of the stomach and intestine, however, the rate of absorption may be altered. This would affect the results of the glucose tolerance test. The situation may be remedied by giving the glucose intravenously.

The intravenous test is performed as follows: Fasting blood and urine specimens are taken. Glucose is given intravenously by a physician. Blood and urine specimens are again taken at ½, 1, 2, and 3 hours after the injection. Glucose tests are run on all blood and urine specimens. Normal and abnormal results are indicated in Fig. 4-1.

Exton-Rose test

In the Exton-Rose test, fasting blood and urine specimens are taken. A known amount of glucose is given. One half hour later, blood and urine specimens are taken. A known amount of glucose is again given. Blood and urine specimens are again taken at ½ hour. Glucose levels are determined on all blood and urine specimens.

In a normal person, (1) the fasting blood specimen is normal, (2) the ½-hour blood specimen does not exceed the fasting blood specimen by more than 75 mg, (3) the 1-hour blood specimen does not exceed the ½-hour specimen by more than 10 mg, and (4) all urines are negative for glucose.

In the diabetic person, the 1-hour blood specimen exceeds the ½-hour specimen by 10 mg or more, and urines are usually positive for glucose.

In renal glycosuria, the blood follows the normal curve, but the urine specimens are positive for glucose.

In alimentary glycosuria, the blood glucose follows the normal curve, the first and second urine specimens are negative for glucose, but the third specimen is positive for glucose.

Procedures for glucose tolerance tests
Janney-Isaacson test

1. Instructions to patient: The patient is instructed by the physician to maintain a normal carbohydrate diet (at least 300 g of carbohydrate per day) for 3 days prior to the test. The test is performed in the morning, and the patient is told not to eat or drink anything following the evening meal of the previous day.
2. Prepare the glucose solution as follows: Weigh 200 g of glucose (dextrose) into a container and add 500 ml of tap water. Stir to dissolve. Place solution in a refrigerator for at least 1 hour in order to chill and thus make more palatable. Just prior to use, add a little lemon juice for flavoring.
3. Collect about 5 ml of venous blood, place in an oxalated tube containing fluoride, and invert several times to dissolve the oxalate. Label the test tube: Fasting blood sugar. Collect a urine specimen and label: Fasting specimen. Store this blood specimen and all succeeding blood specimens in a refrigerator until you are ready to run the tests.
4. Obtain the patient's weight. Multiply this number by 2 and give the patient this number in milliliters of the glucose solution. This results in 1.75 g of glucose per kilogram of weight. Example: A patient weighs 100 pounds. Multiplying 100 by 2, we get 200. The patient then receives 200 ml of the glucose solution. If a patient weighs over 250 pounds, simply give him all the glucose solution.
5. Immediately after giving the glucose, note and record the time.
6. Obtain blood and urine specimens at exactly ½, 1, 2, and 3 hours after

	Blood (mg/dl)	Urine
Fasting specimen	75	Neg.
½-Hour specimen	135	Neg.
1-Hour specimen	120	Neg.
2-Hour specimen	75	Neg.
3-Hour specimen	70	Neg.

Fig. 4-2. Sample report for glucose tolerance test.

glucose ingestion and label. If the physician suspects hyperinsulinism, 4- and 5-hour blood specimens may be requested to give a more complete picture or curve.

7. Run glucose determinations on all blood specimens. Run *qualitative* glucose tests on the urine specimens.
8. Make the report as indicated in Fig. 4-2. Normal and abnormal curves are given in Fig. 4-1.

Intravenous test

1. Instructions to patient: The patient is instructed by the physician to maintain a normal carbohydrate diet (at least 300 g of carbohydrate per day) for 3 days prior to the test. The test is performed in the morning, and he is told not to eat or drink anything after the evening meal of the previous day.
2. Obtain a fasting blood and urine specimen and label.
3. Obtain the patient's weight. Multiply this by 0.15 to get the dosage in grams (equivalent to 0.3 g/kg). Example: Patient weighs 100 pounds. And 100 times 0.15 equals 15 g.
4. Now the dosage in grams must be multiplied by a glucose concentration factor to get the number of milliliters to inject. If the sterile glucose solution on hand is a 50% solution, the factor is 100/50. If the glucose solution is a 33% solution, the factor is 100/33. Example: Patient is to get 15 g of glucose. The glucose on hand is a 50% solution. Therefore, the patient gets $15 \times 100/50 = 30$ ml of the 50% glucose solution.
5. Using *sterile* technique, the physician injects the proper amount of *sterile* glucose solution *very slowly* into a vein, taking about 5 minutes to complete the injection.
6. Immediately after giving the glucose, note and record the time.
7. Obtain blood and urine specimens at exactly ½, 1, 2, and 3 hours after glucose injection and label.
8. Determine glucose levels of all blood specimens. Run *qualitative* glucose tests on all urine specimens.
9. Make the report as indicated in Fig. 4-2, noting that it was an intravenous test.

Exton-Rose test

1. Instructions to patient: The patient is instructed by the physician to maintain a normal carbohydrate diet (at least 300 g of carbohydrate per day) for 3 days prior to the test. The test is performed in the morning, and the patient is told not to eat or drink anything after the evening meal of the previous day.
2. Weigh 100 g of glucose (dextrose) into a container. Dissolve in 650 ml

of tap water. Divide the solution into two equal portions and place in a refrigerator for at least 30 minutes in order to chill and thus make more palatable. Just prior to use, add a little lemon juice for flavoring.
3. Obtain a fasting blood and urine specimen and label.
4. Have the patient drink the first portion of glucose. Note and record the time. Exactly ½ hour later collect another blood and another urine specimen. Label each of them: ½-hour specimen.
5. Have the patient drink the second portion of glucose. Note and record the time. Exactly ½ hour later collect another blood and another urine specimen. Label each of them: 1-hour specimen.
6. Run glucose determinations on the three blood specimens. Run *qualitative* glucose tests on the three urine specimens.
7. Make the report as indicated by the accompanying chart, simply filling in the blood values and reporting the urine results as positive or negative.

Report of Exton-Rose test

	Blood (mg/dl)	Urine (pos. or neg.)
Fasting specimen	_____	_____
½-hour specimen	_____	_____
1-hour specimen	_____	_____

Nonprotein nitrogen compounds

The nonprotein nitrogen (NPN) of the blood is the part of the nitrogenous substances not precipitated by protein precipitants. It is made up of amino acids, uric acid, creatinine, creatine, ammonia, urea nitrogen, and a fraction of nitrogen consisting of polypeptides and other aggregations of amino acids. The urea nitrogen constitutes about 45% of the total NPN. The nonprotein nitrogenous constituents of the blood represent products of the intermediary metabolism of the ingested and tissue protein.

In general, they are the waste products of metabolism and are removed from the bloodstream by the kidneys. When they accumulate in the blood, they point to a flaw in the filtering system of the kidneys.

CREATININE

Creatinine is a waste product that is removed from the bloodstream by the kidneys. Its normal concentration is 0.6 to 1.3/dl of serum. Increased values may be found in nephritis, urinary obstruction, and intestinal obstruction.

For the determination of creatinine, the method of Folin and Wu is commonly used. In this method, sodium picrate in alkaline solution is added to a serum or plasma filtrate. The ensuing reaction is called the Jaffe reaction. It consists of a reaction between creatinine and sodium picrate to form creatinine picrate. The creatinine picrate is red in color. The depth of color is proportional to the creatinine concentration.

Determination of creatinine[5-7]
Specimen stability

Serum or plasma specimens for creatinine are stable for about a week when stored at 4° C or for at least a month when frozen.

Principle

Creatinine yields a definite color reaction in the presence of picric acid in alkaline solution, due to the formation of a red tautomer of creatinine picrate (Jaffe reaction).

Reagents

Sulfuric acid, $\frac{2}{3}$ N	(D-81)
Sodium tungstate, 10%	(D-82)
Picric acid, saturated	(D-30)
Sodium hydroxide, 10% (2.5 N)	(D-83)
Alkaline picrate solution	(See below)

Procedure

1. With a 2.0-ml Ostwald-Folin pipet, pipet 2 ml of plasma into a 125-ml Erlenmeyer flask.
2. Add 14 ml of distilled water.
3. Add 2 ml of $\frac{2}{3}$ N H_2SO_4. Mix by swirling.
4. Add 2.0 ml of 10% sodium tungstate. Stopper and shake.
5. Filter.

Alkaline picrate solution

The alkaline picrate solution is made up *just before using*. Set up the blank, standards, and unknowns first. Then make up the alkaline picrate solution and add to the tubes.

Determine the amount of solution needed for the test and mix in the following ratio:

    5 volumes of picric acid (saturated)
    1 volume of 2.5 N NaOH
    Mix well and use immediately.

Either in cuvets or in test tubes add the following:

Blank

    5.0 ml of distilled water
    2.5 ml of alkaline picrate solution

Standards

Two standards of different concentrations are set up to check the calibration curve.

Standard 2 (from calibration curve)
    1.0 ml of dilute standard (0.010 mg/ml)
    4.0 ml of distilled water
    2.5 ml of alkaline picrate solution
    Mix.

Standard 4 (from calibration curve)
    2.0 ml of dilute standard (0.010 mg/ml)
    3.0 ml of distilled water
    2.5 ml of alkaline picrate solution
    Mix.

Unknown

    5.0 ml of protein-free filtrate
    2.5 ml of alkaline picrate solution
    Mix.

    Allow the tubes to stand for 20 minutes. Read in a colorimeter or spectrophotometer at 520 nm against the *blank* zero. Read absorbance

Table 4-2
Blood creatinine calibration chart

Tube	Working standard (ml)	H$_2$O (ml)	Alkaline picrate (ml)	A (520 nm)	Creatinine equivalent (mg/dl)
Blank	0.0	5.0	2.5	0.00	0.0
1	0.5	4.5	2.5		1.0
2	1.0	4.0	2.5		2.0
3	1.5	3.5	2.5		3.0
4	2.0	3.0	2.5		4.0
5	2.5	2.5	2.5		5.0
6	3.0	2.0	2.5		6.0
7	3.5	1.5	2.5		7.0

(A). Calculate values from the calibration curve. The standards should coincide with their values on the curve.

Calibration curve

Stock standard (1 ml = 1 mg)

Weigh on analytic balance 1.0000 g of creatinine. If creatinine zinc chloride is used, weigh out exactly 1.6026 g of creatinine zinc chloride. Dissolve in 0.1 N HCl and make up to 1000 ml with 0.1 N HCl.

Working standard (1 ml = 0.01 mg)

Dilute 10 ml of stock creatinine standard to 1000 ml with 0.1 N hydrochloric acid.

Prepare a work sheet and set up the standards as shown in Table 4-2. Allow the color to develop for 20 minutes at room temperature. Read absorbance at 520 nm and record readings in the A column on the work sheet. Plot A readings against the creatinine equivalent (mg/dl) on linear graph paper.

Normal values

Creatinine is the least variable nitrogenous constituent of the blood. In early nephritis, values of 2 to 4 mg/dl are noted, and in chronic hemorrhagic nephritis with uremia 4 to 35 mg/dl may be noted. Creatinine is more readily excreted by the kidneys than urea or uric acid, and an increase of creatinine to 4 or 5 ml/dl or more in the blood is evidence of marked impairment of kidney function. Abnormally high creatinine values are accompanied by increased urea content of blood.

Serum creatinine and creatine[8]

Reagents

Picric acid, 0.04 M	(D-84)
Sodium hydroxide, 0.75 N	(D-85)
Sodium tungstate, 10%	(D-82)
Sulfuric acid, ⅔ N	(D-81)

Creatinine standard:

Stock standard: 1.5 mg/ml in 0.1 N HCl. Stable indefinitely in the refrigerator.

Working standard: 15 μg/ml. Dilute stock standard 1:100 with distilled water. Prepare fresh daily.

Creatine, working standard: 15 μg of creatine as creatinine per milliliter. No stock standard is stable, and working standard is made each day used.

Creatinine procedure

1. Into a 15-ml centrifuge tube pipet 2.0 ml of serum, 3.0 ml of distilled water, 1.0 ml of 10% sodium tungstate, and 2.0 ml of $2/3$ N H_2SO_4. Mix and centrifuge.
2. Into test tubes pipet:

Blank

3.0 ml of distilled water

Standard

1.0 ml of working standard and 2.0 ml of distilled water

Unknown

3.0 ml of protein-free centrifugate

3. To each tube add 1.0 ml of picric acid and 1.0 ml of 0.75 N NaOH. Mix.
4. Allow to stand exactly 20 minutes and then read the absorbance against the reagent blank at 520 nm.

Calculation

$$\frac{\text{A of unknown}}{\text{A of standard}} \times 0.015 \times \frac{100}{0.75} = \text{mg creatinine/dl serum}$$

Creatine procedure

1. Determine creatinine (preformed) as described above.
2. Determine preformed creatinine and creatine as creatinine by pipeting into graduated centrifuge tubes:

Blank

6.0 ml of distilled water

Standard

1.0 ml of working standard and 5.0 ml of distilled water

Unknown

3.0 ml of protein-free centrifugate, 0.25 ml of 0.75 N NaOH, and 2.75 ml of distilled water

3. Add to each tube 1.0 ml of 0.04 M picric acid.
4. Immerse tubes in a boiling water bath so that the water level is above the liquid level in the tubes. Heat until the volume in each tube has decreased below 4.0 ml (requires between 1.5 and 2 hr).
5. Cool tubes to room temperature and add water to 4.0 ml.
6. To each tube add and mix 1.0 ml of 0.75 N NaOH.
7. At exactly 20 minutes read the absorbance against the reagent blank at 520 nm.

Calculations

$$\frac{\text{A of unknown}}{\text{A of standard}} \times 0.015 \times \frac{100}{0.75} = \text{mg total creatinine/dl serum or}$$

preformed creatinine + creatine as creatinine

Total creatinine − preformed creatinine = mg creatine as creatinine/dl serum

Normal values

For adults, the normal creatine/dl of serum is 0.2 to 0.6 mg in men and 0.35 to 0.9 mg in women. It is increased in hyperthyroidism and severe muscular disease.

Urine creatinine

Procedure

1. Suggested urine dilutions

    Dilution A (for volume less than 1000 ml/24 hr)

    Dilute 1.0 ml of urine to 3.0 ml with distilled water.

    Multiply the result by 3.

    Dilution B (for volume above 1000 to 1500 ml/24 hr)

    Dilute 1.0 ml of urine to 2.0 ml with distilled water.

    Multiply the result by 2.

2. Into 100 ml volumetric flasks pipet the following. (NOTE: Upon the addition of the 1.5 ml of 2.5 N NaOH, start timing for 20 minutes.)

    Blank

    20.0 ml of saturated picric acid
    1.5 ml of 2.5 N NaOH

    Standards (two)

    Standard 1

    0.5 ml of creatinine standard (1 mg/ml)
    20.0 ml of saturated picric acid
    1.5 ml of 2.5 N NaOH

    Standard 2

    1.0 ml of creatinine standard (1 mg/ml)
    20.0 ml of saturated picric acid
    1.5 ml of 2.5 N NaOH

    Unknown

    1.0 ml of diluted urine
    20.0 ml of saturated picric acid
    1.5 ml of 2.5 N NaOH

3. Swirl flasks to mix.
4. Stand 20 minutes.
5. Dilute to mark with distilled water.
6. Stopper and mix thoroughly.
7. Read absorbance in spectrophotometer or colorimeter at 520 nm against blank.

The purpose of the standards is to check the reagents and other conditions of the test. If the reading of the standard varies too much from the curve, the reagents should be checked. It may be necesary to construct a new curve.

Calibration curve

With the stock standard (1 ml = 1 mg) make the following dilutions, using 0.1 N HCl:

Dilution A (1 ml = 0.25 mg)

Mix in a test tube 2.5 ml of stock standard and 7.5 ml of 0.1 N HCl.

Dilution B (1 ml = 0.1 mg)

Mix in a test tube 1.0 ml of stock standard and 9.0 ml of 0.1 N HCl.

In 100-ml volumetric flasks, set up the following concentrations according to the work sheet shown in Table 4-3. Allow the color to develop 20 minutes. Dilute to mark with distilled water and mix well. Pour into proper tubes and read absorbance at 520 nm.

Table 4-3
Urine creatinine calibration chart

Tube	Stock standard (1 ml = 1.0 mg)	1:4 dilution (1 ml = 0.25 mg)	1:10 dilution (1 ml = 0.1 mg)	Picric acid (ml)	NaOH (2.5 N)	A (520 nm)	Creatinine equivalent (mg/dl)
Blank	0.00	0.00	0.00	20	1.5	0.00	0.00
1	2.00	0.00	0.00	20	1.5		2.00
2	1.50	0.00	0.00	20	1.5		1.50
3	1.00	0.00	0.00	20	1.5		1.00
4		3.00	0.00	20	1.5		0.75
5		2.00	0.00	20	1.5		0.50
6		1.00	0.00	20	1.5		0.25
7			2.00	20	1.5		0.20
8			1.00	20	1.5		0.10

Plot the absorbance readings on the vertical (ordinate) of linear graph paper, allowing each division of the graph paper to represent 0.005.

Plot the milligrams per milliliter on the horizontal (abscissa), allowing each division to represent 0.01 mg/ml.

The curve does not form a straight line.

Normal values

The daily excretion of creatinine by an adult of medium weight averages about 1.25 g. The average adult normal range for a male is 20 to 26 mg/kg and for females is 14 to 22 mg/kg. The value is nearly constant from day to day for a given individual, being influenced by the diet very little unless there is a heavy meat diet which contains much preformed creatinine.

Comments

The excretion of creatinine is, to a certain extent, a measure of muscular efficiency and of the amount of active muscle tissue in the body. Relative to body weight, therefore, less creatinine is excreted by obese persons.

Creatinine excretion is decreased in disorders associated with muscular atrophy and muscular weakness. It increases with increased tissue catabolism as in fever.

Urine creatine

Creatine occurs in only very small amounts in the urine of normal adults, but it is found in larger amounts in the urine of children. Creatine ingestion in adults has little effect on the urinary excretion. In a fasting state, the amount is markedly increased. The normal adult range is 0 to 200 mg/24 hr. Creatine also appears in the urine after high water ingestion. It is found in many pathologic conditions associated with malnutrition and disintegration of muscular tissue and in fevers. Large amounts are found in diseases of the muscles.

The method of determining urine creatine is based upon the Jaffe reaction. Creatine, if boiled with acid, is transformed into creatinine. By determining the content of creatinine before and after the acid treatment it is possible to calculate the amount of creatine originally present in the urine.

Procedure

1. Pipet 10 ml of diluted urine into a Folin-Wu sugar tube.
2. Add 10 ml of 1 N HCl.
3. Cover top of tube tightly with aluminum foil and put into boiling water bath for 30 minutes. Use timer.
4. Cool in room-temperature water bath and dilute to 25-ml mark with distilled water.
5. Stopper and mix well by inverting tube several times.
6. In 100-ml volumetric flasks set up the following:

Total creatinine

2.5 ml of the above acid-treated urine (2.5 ml = 1.0 ml of diluted urine)
20.0 ml of saturated picric acid
2.0 ml of 2.5 N NaOH
Swirl to mix.

Preformed creatinine

1.0 ml of diluted urine (*not* the acid-treated urine)
20.0 ml of saturated picric acid
1.5 ml of 2.5 N NaOH
Swirl to mix.

Blank

20.0 ml of saturated picric acid
1.5 ml of 2.5 N NaOH
Swirl to mix.

Standards

Standards are the same as for the urine creatinines; urine creatines may be run with urine creatinines.

7. Swirl flasks to mix.
8. Let stand 20 minutes. Use timer.
9. Dilute to mark with distilled water.
10. Stopper and mix thoroughly.
11. Read absorbance at 520 nm against the *blank*.
12. Calculate the two unknowns (*preformed creatinine and total creatinine*) from the urine creatinine standard curve.
13. The results of the preformed and total creatinine test are then multiplied by the dilution of urine.

Calculations of the creatine

Total creatinine − Preformed creatinine = mg creatine/ml

or

(T. creatinine − P. creatinine) × 100 = mg creatine/dl

or

$$\text{mg/dl creatine} \times \frac{\text{24-hour urine vol}}{100} = \text{mg creatine/24 hr}$$

URIC ACID

Uric acid is a waste product that is removed from the bloodstream by the kidneys. Several conditions in which an increase or a decrease may be expected are listed in Table 4-4.

Table 4-4
Conditions accompanied by abnormal uric acid values

When increased		When decreased
Gout	Acute infections	Salicylate therapy
Nephritis	Intestinal obstruction	Atophan therapy
Arthritis	Urinary obstruction	Yellow atrophy of liver
Eclampsia	Metallic poisoning	
Leukemia	Hypertension	
Polycythemia	Following exercise	
Diabetes		

Specimen stability

Uric acid in serum is stable for at least a week when stored at 4° C and for a month when frozen.

Procedures

The more commonly used methods for the determination of uric acid are discussed briefly here.

Folin method

A blood filtrate is treated with urea-cyanide and phosphotungstic acid to form a phosphotungstate complex. The uric acid present reduces the phosphotungst*ate* complex to the blue phosphotungst*ite* complex. The depth of color is measured in a colorimeter and compared with a standard.

Brown method

A blood filtrate is treated with sodium cyanide, urea, and phosphotungstic acid to form a phosphotungstate complex. The uric acid present reduces the phosphotungst*ate* to the blue phosphotungst*ite* complex. The depth of color is measured in a colorimeter and compared with a standard.

Archibald method[1]

The uric acid in serum or plasma reduces phosphotungstate acid in an alkaline solution, thus producing a blue color. This method may also be used for urine without any modifications of the procedure. The phosphotungstate is not only a color-developing agent. It also precipitates the proteins. Plasma or serum is preferable to whole blood because the glutathione and ergothioneine from the red cells cause interference.[12]

Reagents

Uric acid stock standard, 1 mg uric acid/ml	(D-38)
Polyanethol sodium sulfonate (Liquoid)	(D-86)
Glycerin-silicate reagent	(D-87)
Phosphotungstic acid	(D-88)

Sodium hydroxide, 0.5 N (D-29)
Uric acid special reagent (D-89)
Uric acid working standard, 0.005 mg/ml
  Pipet 0.5 ml of stock uric acid standard into a 100-ml volumetric flask and dilute to the mark with water. Prepare a fresh working standard daily.

Procedure

1. Pipet into a 125-ml Erlenmeyer flask 2.0 ml of plasma or serum. With urine make a 1:10 dilution and pipet 2.0 ml into the flask.
2. Add 16 ml of distilled water; swirl flask to mix.
3. Add 0.8 ml of 0.5 N NaOH, swirling the flask for at least 1 minute. For large volumes a mechanical shaker is most helpful.
4. Let the flask stand for 10 minutes.
5. Slowly add 1.2 ml of the phosphotungstic acid solution to the flask, with constant swirling of the flask to ensure a complete precipitation of protein.
6. Let the flask stand for 5 minutes.
7. Filter through Whatman No. 42 filter paper into large test tubes or flasks.
8. Set up the following tubes:

Blank
  5.0 ml of water

Standard
  5.0 ml of working standard

Unknown
  5.0 ml of filtrate

9. Add 2.5 ml of glycerin-silicate reagent to each tube. Set timer for 15 minutes. Begin timing upon the addition of 2.0 ml of uric acid special reagent. Mix immediately. After 15 minutes the color is fully developed.
10. Set zero with blank and read absorbance measurements on a spectrophotometer at 700 nm.

Calculations

$$\frac{\text{Absorbance of unknown}}{\text{Absorbance of standard}} \times 5 = \text{mg uric acid/dl}$$

e.g., Urine—dilution, 1:100; filtrate, 5 ml

$$\frac{\text{Absorbance of unknown}}{\text{Absorbance of standard}} \times 10 \times 5 = \text{mg uric acid/dl in urine}$$

Normal values[8]

	Men	Women
Serum or plasma	2.5 to 7.0 mg/dl	1.5 to 6.0 mg/dl
Urine, 24-hour	250 to 750 mg	250 to 750 mg

Henry method[9]

The uric acid in serum, plasma, or urine reduces an alkaline phosphotungstate solution to a tungsten blue. The alkali used is sodium carbonate. The depth of color is measured photometrically and compared with a standard.

Reagents

Phosphotungstic acid (The directions given in the original publication were in error and were corrected by the authors in a later publication. For this reason the directions for preparing this reagent are given here instead of in Appendix D.)

Weigh 40 g of reagent grade sodium tungstate and transfer to a 1-L round-bottom boiling flask containing about 300 ml of distilled water. Swirl to dissolve. Add 32 ml of 85% o-phosphoric acid and mix. Attach a reflux condenser to the flask and reflux gently for 2 hours. Cool to room temperature and dilute to 1000 ml with distilled water. Weigh 32 g of lithium sulfate monohydrate and add to the above solution. Mix to dissolve. This reagent is stable indefinitely when stored in a refrigerator.

Sodium carbonate, 14% (D-90)

Uric acid stock standard, 1 mg uric acid/ml (D-38)

Working standard, 0.01 mg/ml

Dilute stock standard 1:100 in a volumetric flask.

Procedure—serum

1. Prepare a Folin-Wu protein-free filtrate.
   Avoid: oxalated plasma, which may cause a crystalline precipitate.
2. Set up the following tubes:

   Blanks (2 tubes)
   3.0 ml of distilled water into each tube.

   Standards (3 tubes)
   3.0 ml of working standard into each tube.

   Unknown (1 tube per test)
   3.0 ml of unknown filtrate into its numbered tube.

3. *To all tubes add*
   1.0 ml of 14% Na_2CO_3 and mix.
   1.0 ml of phosphotungstic acid and mix.
4. Allow tubes to stand 15 minutes to develop color.
5. Read against the blank at 660 nm within a 30-minute period. The blank must be colorless.

Calculation

$$\frac{A \text{ of unknown}}{A \text{ of standard}} \times 10 = \text{mg uric acid/dl}$$

The three standards are averaged unless they vary a great deal. In that case, repeat the test. The two blanks are used to check against each other.

Procedure—urine

1. If the urine is cloudy, warm it to about 60° C to dissolve the urates and centrifuge.
2. Dilute 1:10 with distilled water.
3. Proceed with the method given for the serum uric acid. Everything is the same as in the preceding method with the exception of the calculation.

Calculation

$$\frac{A \text{ of unknown}}{A \text{ of standard}} \times 100 = \text{mg uric acid/dl}$$

NOTE: **If the A reads less than 0.2 or more than 0.8, repeat the test, using the proper dilution.**

UREA NITROGEN

Urea is the principal waste product of protein catabolism. Its formula is NH_2CONH_2. Since it is difficult to determine urea itself, and a relatively simple matter to analyze for the nitrogen in urea, it has become customary to determine the urea nitrogen. Most methods depend on the urea being converted to ammonium carbonate:

$$\underset{\text{Urea}}{\underset{NH_2}{\overset{NH_2}{C}}=O} + \underset{\text{Water}}{2H_2O} \xrightarrow[38°]{\text{Urease}} \underset{\text{Ammonium carbonate}}{(NH_4)_2\,CO_3}$$

The normal values for the blood urea nitrogen (BUN) are 6 to 20 mg/dl of blood. The corresponding values for urea may be found by multiplying these figures by the factor 2.14. Several conditions in which an increased or a decreased urea nitrogen value may be expected are given in Table 4-5.

Karr method

A protein-free filtrate is treated with a buffer solution and the enzyme urease. The purpose of the buffer solution is to control the pH of the solution, since the urease reacts better at a certain pH (6.8). Upon incubation, the urease decomposes the urea and liberates ammonia, which forms ammonium carbonate. After incubation, gum ghatti is added to form a protective colloid and keep the solution from becoming cloudy. Nessler's solution is then added. This is an alkaline solution of the double iodide of mercury and potassium ($HgI_2 \cdot 2KI$). It reacts with the ammonium carbonate to form yellow dimercuric ammonium iodide. The depth of color is measured in a colorimeter and compared with a standard.

Van Slyke-Cullen method

In the Van Slyke-Cullen method, oxalated blood is treated with the enzyme urease to form ammonium carbonate. Addition of potassium carbonate liberates the

Table 4-5
Conditions accompanied by abnormal urea nitrogen values

When increased		When decreased
Nephritis	Dehydration	Acute liver destruction
Intestinal obstruction	Malignancy	Amyloidosis
Urinary obstruction	Pneumonia	Pregnancy
Metallic poisoning	Surgical shock	Nephrosis
Cardiac failure	Addison's disease	
Peritonitis	Uremia	

ammonia gas, which is collected in a boric acid solution. This is then titrated with a standard acid solution and the urea nitrogen found by calculation.

Gentzkow method

Oxalated blood is diluted. To this is added the enzyme urease, which decomposes the urea to form ammonium carbonate. Next, the proteins are precipitated with sulfuric acid and sodium tungstate. The resulting filtrate containing the ammonium carbonate is treated with Nessler's solution to form the yellow dimercuric ammonium iodide. The depth of color is measured in a colorimeter and compared with a standard.

Folin-Svedberg method

A protein-free filtrate is made. To this is added a buffer solution to control the pH. Next, the enzyme urease is added to decompose the urea and form ammonium carbonate. The mixture is incubated to enhance the reaction. Sodium borate is added to the solution to liberate the ammonia gas, which is collected in a weak hydrochloric acid solution. Nessler's solution is then added to convert the ammonium ion to the yellow dimercuric ammonium iodide. The depth of color is measured in a colorimeter and compared with a standard.

Berthelot reaction

The urea in serum, plasma, or urine is split, by the action of urease, into ammonia and CO_2. The ammonia is determined photometrically by the phenolhypochlorite reaction of Berthelot, using sodium nitroprusside as a catalyst. Because of the very high dilution used, it is not necessary to precipitate and remove proteins.

Determination of urea nitrogen
Specimen stability

Serum or plasma specimens for urea nitrogen determinations are stable for about a week when stored at 4° C and at least a month when frozen. Urine is particularly susceptible to loss of urea as a result of bacterial decomposition; therefore, in addition to refrigeration, several thymol crystals will help to reduce the loss of urea.

Berthelot reaction[2]

Reagents

Urease solution, buffered*	(D-91)
Phenol color reagent*	(D-92)
Alkali-hypochlorite reagent*	(D-93)

Urea standard
 Transfer 0.429 g of urea to a 1-L volumetric flask and dilute to volume with distilled water. Add a few drops of chloroform as a preservative and store in the refrigerator (1 ml = 0.2 mg of urea N).

Procedure—serum or plasma

1. Set up the following in test tubes (16 × 125):

   Blanks
   0.2 ml of buffered urease solution

---

*These reagents are available commercially from Hyland Laboratories in the form of a kit with the trade name UN-Test.

Standard
> 0.2 ml of buffered urease solution
> 0.02 ml of urea standard

Unknown
> 0.2 ml of buffered urease solution
> 0.02 ml of serum or plasma
>> The 0.02 ml aliquots of standard and unknown are added with TC micropipets. Sahli hemoglobin pipets are satisfactory.

2. Incubate the tubes in a water bath at 37° C for 15 minutes.
3. Then add 1.0 ml of phenol color reagent to each tube and mix.
4. Add 1.0 ml of alkali-hypochlorite reagent and mix promptly again.
5. Incubate tubes at 50° to 60° C for 5 minutes or at 37° C for 20 minutes.
6. Add 8.0 ml of water to all tubes and mix. If mixing must be accomplished by inversion, cover the tops of the tubes with clean Saran Wrap or Parafilm and hold with the thumb while inverting the tube.
7. Read A at 625 nm. When specimens read higher than 0.800 A, dilute with water, read again, and calculate according to dilution.

Calculation

$$\frac{RU}{RS} \times 0.004 \times \frac{100}{0.02} = \text{mg urea N/dl}$$

$$\frac{RU}{RS} \times 20 = \text{mg urea N/dl}$$

Procedure—urine

1. Since urine normally contains some preformed ammonia, it is necessary to remove this ammonia before proceeding with the test.
2. Measure and record the volume of the urine.
3. Add 5 g of Permutit to a 10-ml aliquot of the urine and shake vigorously for 1 minute. Allow to stand for 10 minutes.
4. Filter.
5. Transfer 1.0 ml of filtered urine to a 50-ml volumetric flask and dilute to the mark with water. Stopper and mix.
6. Using an 0.02-ml aliquot of diluted urine, follow the procedure as with serum or plasma.

Calculation

$$\frac{RU}{RS} \times 20 \times 50 = \text{mg urea N/dl urine}$$

$$\frac{RU}{RS} \times 1000 = \text{mg urea N/dl urine}$$

Normal values

The normal values for urea nitrogen are 6 to 17 g/day. Decreased values may be found in acute nephritis, acidosis, and cirrhosis of the liver. Increased values may be seen in febrile conditions.

References

1. Archibald, R. M.: Colorimetric measurement of uric acid, Clin. Chem. **3:**102, 1957.
2. Chaney, A. L., and Marbach, E. P.: Modified reagents for determination of urea and ammonia, Clin. Chem. **8:**131, 1962.
3. Dubowski, K. M.: An o-toluidine method for body-fluid glucose determination, Clin. Chem. **8:**215, 1962.
4. Fajans, S. S., and Conn, J. W.: Approach to prediction of diabetes mellitus by modification of glucose tolerance test with cortisone, Diabetes **3:**296, 1954.
5. Folin, O.: On the determination of creatinine and creatine in urine, J. Biol. Chem. **17:**469, 1914.
6. Folin, O., and Doisy, E. A.: Impure picric acid as a source of error in creatine and creatinine determinations, J. Biol. Chem. **28:**349, 1917.
7. Folin, O., and Wu, H.: A system of blood analysis, J. Biol. Chem. **38:**81, 1919.
8. Henry, R. J.: Clinical chemistry; principles and technics, New York, 1964, Harper & Row, Publishers.
9. Henry, R. J., Sobel, C., and Kim, J.: A modified carbonate-phosphotungstate method for the determination of uric acid and comparison with the spectrophotometric uricase method, Am. J. Clin. Pathol. **28:**152, 645, 1957.
10. Hultman, E.: Rapid specific method for determination of aldosaccharide in body fluids, Nature **183:**108, 1959.
11. Hyvärinen, A., and Nikkilä, E. A.: Specific determination of blood glucose with o-toluidine, Clin. Chem. Acta **5:**42, 1960.
12. Jorgensen, S., and Nielson, A. T.: Uric acid in human blood corpuscles and plasma, Scand. J. Clin. Lab. Invest. **8:**108, 1956.
13. Sudduth, N. C., Wisish, J. R., and Moore, J. L.: Automation of glucose measurement using ortho-toluidine reagent, Am. J. Clin. Pathol. **53:**181, 1970.
14. Wenk, R. E., Creno, R. J., Loock, V., and Henry, J. B.: Automated micro measurement of glucose by means of o-toluidine, Clin. Chem. **15:**1162, 1969.

5

Electrolytes

In good health, the body maintains electric neutrality. This means that there is a balance or equality between cation and anion groups. Sodium, potassium, calcium, and magnesium ions make up the cation group, sometimes called the total base. In the anion group (from inorganic and organic acids) are the electrolytes consisting of chloride, bicarbonate, protein, phosphate, sulfate, bromide, and iodide ions. These two groups in balance are shown in Table 5-1.

Since the complete set of the cation and anion groups is seldom run, a check may be made of the electrolyte balance as suggested in the following paragraphs. The sodium ion has the highest concentration of the cation group while the chloride and bicarbonate ions have the largest concentration of the anion group. Total the bicarbonate and chloride ions and subtract that total from the sodium ions. Normally, this difference averages 10 mEq/L with variation either way of 1 or 2 mEq. However, in a serious illness (pathologic condition) the difference may be appreciably greater than 10 or may be less than 5.

If there is a difference greater than 15 mEq/L, look at the other anions for an elevated value. Here are three of the most common causes of a great difference:

1. Kidney damage or uremia resulting in an elevated BUN. In addition there is a retention of endogenous acids that are either organic (amino or keto acids) or inorganic (phosphate or sulfate).

Table 5-1
Electrolyte composition of human plasma (approximate)

Cation	mEq/L	Anion	mEq/L
Na (sodium)	143.0	Cl (chloride)	104
K (potassium)	4.5	HCO_3 (bicarbonate)	29
Ca (calcium)	5.0	Protein	16
Mg (magnesium)	2.5	HPO_4 (phosphate)	2
		SO_4 (sulfate)	1
		Organic acids	3
Total	155.0	Total	155

2. Diabetic acidosis with a retention of keto acids in the blood. There will be an elevation in the blood sugar, blood acetone, and a positive acetone result with the urine.
3. Various poisonings in which there are unmeasured acids in the bloodstream. An example would be poisoning from methanol where it is converted to formic acid.

A difference of less than 5 mEq/L is not as prevalent as the greater variation. Decreased albumin may be the cause of a close difference. The smaller amount of protein requires less sodium for electric neutrality; therefore there will often be a low sodium accompanying a low albumin. However, look at the electrolyte results and repeat the test showing the most suspicious result first.

The total of the bicarbonate and chloride should *always be less* than the sodium. A quality control specimen is most beneficial in eliminating doubts of a technical error. The electrolytes should always be added and compared with previous work on the patient. Other tests such as urea nitrogen, glucose, and protein are taken into consideration when totaling the electrolyte balance. Records of each abnormal test, with the patient's name and test results, are invaluable tools for the laboratory.

There are times when results will not "check out" even after repetition of all tests and checking of the instruments. A new specimen should be requested because of two main problems that may arise before the blood reaches the laboratory:

1. A wet or contaminated syringe may have been used.
2. Blood may have been drawn during an intravenous infusion. Infusion may cause a great dilution factor or, if the infusion contains electrolytes, the sample will be contaminated.

Since urinary electrolytes are not excreted in a uniform concentration throughout the day, a 24-hour specimen should be collected. Electrolytes in the urine depend on the following factors: (1) acid-base balance of the body, (2) intake of water and various ions, (3) state of body water: whether the patient is dehydrated or has an excess of fluids (edema), and (4) functioning ability of the kidneys.

Refer to Table 5-2 for approximate normal values obtained from normal individuals under an average diet and exercise program.

Table 5-2
Approximate normal values for urine electrolytes

Electrolytes	Range
Sodium	43 to 217 mEq/24 hr
Potassium	26 to 133 mEq/24 hr
Chloride	170 to 254 mEq/24 hr
CO_2	None
Calcium	25 to 200 mEq/24 hr
Phosphorus	0.2 to 0.6 mEq/24 hr

Table 5-3
Conditions accompanied by abnormal sodium values

When increased	When decreased
Nephritis	**Addison's disease**
Pyloric obstruction	**Alkali deficit**
Hypercorticoadrenalism	Dehydration
	Myxedema
	Sprue

SODIUM

Sodium is absorbed into the bloodstream from the small intestine. It plays a major role in controlling the water balance between blood and body cells. The normal concentration of sodium is 136 to 148 mEq/L of serum. A decreased value is of primary concern in Addison's disease—functional failure of the adrenal glands characterized by anemia, digestive disturbances, and bronzelike pigmentation of the skin. Conditions in which increased or decreased sodium values may be expected are given in Table 5-3.

The following pages describe a colorimetric procedure. For methods employing a flame photometer, refer to the instrument manufacturer's manual.

Specimen stability

Serum or heparinized plasma sodium is stable for at least 2 weeks at room temperature or at 4° C.

Sodium by chemical analysis[3]

Although flame photometry is established in most laboratories, there may be instrument problems or no flame photometer available. Therefore the method of Albanese and Lein for a chemical determination is given. This method gives a reasonable correlation with a flame photometer method.

Principle

With this method, the sodium is precipitated as sodium uranyl acetate. It is then dissolved in water and determined photometrically by the intensity of its yellow color.

Reagents

Uranyl zinc acetate reagent	(D-94)
Ethanol, 95% *(do not use absolute ethyl alcohol)*	
Trichloroacetic acid, 10%	(D-95)
Sodium standard, 140 mEq/L	(D-96)

Procedure

Do all tests (unknowns) in duplicate and average results. This will give a closer limit (±3.6%) in agreement with the flame photometer.

1. Pipet 0.5 ml of serum into a 15 × 125 mm test tube.

2. Pipet 2.0 ml of 10% trichloroacetic acid in the tube and mix.
3. Let stand for 5 minutes.
4. Set up the following tubes in duplicate:

Blank

0.5 ml of distilled water.

Standard

0.5 ml of standard (0.32 mg of Na) into 3 test tubes.

Unknown

0.5 ml of supernatant into the test tubes.

5. Add 1.0 ml of uranyl zinc acetate reagent to all tubes and mix.
6. Refrigerate for 1 hour; then centrifuge.
7. Carefully pour off and discard supernatant. Drain tubes thoroughly.
8. Wash down the sides of the tube with 2 ml of 95% ethyl alcohol and resuspend the precipitate in the wash fluid. This method of mixing is sometimes called "spanking the tube."
9. Centrifuge the tubes and drain the supernatant as before.
10. Pipet 5.0 ml of distilled water into each tube and dissolve by spanking the tube. If turbid, centrifuge.
11. Transfer to spectrophotometer tubes and read against the reagent blank at 430 nm.

Calculation

$$\frac{A \text{ of unknown}}{A \text{ of standard}} \times 140 = \text{mEq Na/L}$$

POTASSIUM

Potassium is absorbed into the bloodstream from the small intestine. A large portion enters the tissues of the body. Here it plays a major role in regulating the distribution of water between the tissues and blood.

The concentration of potassium is normally 3.6 to 5.0 mEq/L of serum. Conditions in which increased or decreased potassium values may be expected are given in Table 5-4.

Table 5-4
Conditions accompanied by abnormal potassium values

When increased	When decreased	
Addison's disease	Hyperinsulinism	Hypercorticoadrenalism
Acute infections	Diabetes	Malignant growth
Pneumonia	Hereditary periodic paralysis	Chronic nephritis
Uremia	Overdosage of testosterone	Severe diarrhea
Acute bronchial asthma	Overdosage of desoxycorticosterone	Sprue

Specimen stability

Serum or heparinized plasma potassium is stable for at least 2 weeks at room temperature or at 4° C.

Potassium by chemical analysis[10]

Flame photometry is the preferred method for potassium determinations because of the accuracy and the speed of the instrument. However, if the flame photometer is not available, the method of Lochhead and Purcell can be used for the chemical determination. In the determination of potassium, regardless of the method used, there must be *no* hemolyzed cells in the serum. The specimen must be allowed to clot, centrifuged as soon as possible, and the serum removed from the clot. The use of heparinized plasma minimizes hemolysis and eliminates the waiting time for the clot to retract before centrifuging.

Principle

With the chemical method of analysis given, potassium is precipitated directly from the serum or plasma as potassium sodium cobaltinitrite. The cobalt in the precipitate is determined photometrically. The alkaline solutions of cobalt, in the presence of a trace of an amino acid such as glycine, reduced the Folin-Ciocalteu phenol reagent to a blue color. The intensity of the color is then compared with a standard.

Reagents

Sodium cobaltinitrite reagent	(D-97)
Sodium acetate, half-saturated	(D-98)
Wash solution, saturated with potassium sodium cobaltinitrite	(D-99)
Glycine, 7.5%	(D-100)
Sodium carbonate, 25%	(D-101)
Phenol reagent of Folin-Ciocalteu, stock	(D-102)
Potassium standard, 4.0 mEq/L	(D-103)

Working phenol reagent
   Prepare fresh. Mix 1 volume of the stock phenol reagent with 2 volumes of distilled water.

Procedure

Before starting the procedure, refilter the volume of sodium cobaltinitrite needed for the test and make certain that a centrifuge is available for the immediate centrifugation needed throughout the test.

1. Using 12- or 15-ml conical-tipped graduated centrifuge tubes, set up the following in duplicate (the *blank* is set up later).

   Standard
      0.2 ml of standard.

   Unknown
      0.2 ml of serum or plasma.

2. Pipet 0.2 ml of half-saturated sodium acetate. Mix.
3. While constantly shaking the tubes, slowly add 0.5 ml of sodium cobaltinitrite.
4. Let the tubes stand at room temperature for 45 minutes. Use timer for exact timing.
5. Add 1.0 ml of distilled water. Mix.
6. Centrifuge *immediately* for *exactly* 15 minutes at 3000 rpm.
7. Carefully pour off supernatant and drain the tubes for about 15 minutes.

8. Add 1.0 ml of wash solution without disturbing the precipitate.
9. *Immediately* centrifuge for 15 minutes.
10. Pour off supernatant and drain the tubes for 5 or 10 minutes.
11. Pipet 1.0 ml of 70% ethanol, and mix thoroughly with a glass stirring rod. Add 3.0 ml of 70% ethanol, completely rinsing the precipitate off the stirring rod.
12. *Immediately* centrifuge for 5 minutes.
13. Pour off supernatant and drain the tubes for at least 5 minutes.
    NOTE: The test may be stopped at this point and continued the next day. If it is to be held overnight, add 1 drop of distilled water to each tube to replace any lost by evaporation and stopper the tubes firmly.
14. Label a *blank* tube and add 2.0 ml of distilled water to the blank and all tubes.
15. Place the tubes in a boiling water bath for 15 to 20 minutes, until the precipitate dissolves. If a minute amount of precipitate remains, this will not affect the analysis.
16. While the tubes are still hot, add 1.0 ml of glycine solution.
17. Add 1.0 ml of sodium carbonate solution. Mix.
18. Add 1.0 ml of working phenol reagent. Mix.
19. Place the tubes in a 37° C water bath for 15 minutes.
20. Cool to room temperature.
21. Add distilled water to the 6.0 ml mark. Stopper and mix.
22. Read absorbance against a water blank at 660 nm.

Calculation

$$\frac{A \text{ of unknown} - A \text{ of blank}}{A \text{ of standard} - A \text{ of blank}} \times 4 = \text{mEq K/dl}$$

CHLORIDE

The majority of methods used for the determination of chloride are based on the principle of precipitating the chloride as an insoluble salt. The analysis is done on either the precipitate or the excess precipitant. However, one of the most common methods of determining chloride concentration is the method of Schales and Schales, who in 1941 adapted the mercurimetric method for use in chloride determination. The chloride ion, in a tungstic acid filtrate of plasma or serum, is titrated with a standard solution of mercuric ions, with the formation of the soluble compound mercuric chloride ($HgCl_2$), which does not dissociate to form mercuric ions. The organic indicator 5-diphenylcarbazone yields a purple color when the excess mercury combines with the indicator. The color is stable although rather difficult to determine with urine, pleural fluid, and serum with elevated bilirubin.

Since the titration end point depends on the individual's experience and eyesight, potentiometric methods have gained popularity by being simple, fast, accurate, and versatile. Icteric sera or turbid specimens pose no problem with these methods. In general, potentiometric methods utilize two electrodes, a voltmeter, and a galvanometer. The electrodes are immersed in a dilute nitric acid solution containing the specimen to be analyzed. By adding silver ions the chloride is precipitated as silver chloride. When the chloride has been precipitated, free silver ions change the potential, resulting in a shift in the reading on the galvanometer.

The Buchler-Cotlove Chloridometer has a spool of silver wire that is immersed with the electrodes in a dilute nitric acid solution. The small stirrer, also attached to the electrode assembly, thoroughly mixes the specimen as it is being analyzed. With this instrument, the silver ions are released into the solution at a constant rate and when the chloride in the solution has been used completely, the machine automatically shuts off. Since the silver ions are being added at a constant rate, the amount of chloride in the sample is proportional to the length of time in seconds necessary to complete the titration. A reagent blank and chloride standard are run and the reaction time of the unknown is calculated against the standard. This potentiometric method is very satisfactory because it has greater sensitivity than the titration or colorimetric methods. The operation is simple, rapid, and automatic, eliminating technical errors and time-consuming precipitation of serum, tissue, urine, spinal fluid, or other biologic fluids.

Large quantities of bromide given to a patient will result in a falsely elevated chloride. This may be corrected by converting the bromide mg/dl to bromide mEq/L. The bromide value is then subtracted from the chloride value. If the bromide is calculated as sodium bromide, convert to bromide. Refer to Chapter 15. The calculation of an example is as follows:

$$Cl(mEq/L) - \frac{Bromide\ (mg/dl \times 10)}{80} = True\ chloride\ value$$

$$112 - \frac{120 \times 10}{80} =$$

$$112 - 15 = 97\ mEq\ chloride/L\ (true\ chloride\ value)$$

Chloride is absorbed into the bloodstream from the small intestine.

Note that whole blood has lower values than serum or plasma. Red cells, which make up almost half of whole blood, contain much less chloride than serum or plasma.

The normal concentration of chlorides is 95 to 103 mEq/L of serum or plasma and 119 to 130 mEq/L of spinal fluid.

An increase in the chloride value is of primary concern in nephritis, prostatic obstruction, and eclampsia (convulsions). Conditions in which an increased or a decreased chloride level may be expected are given in Table 5-5.

Table 5-5
Conditions accompanied by abnormal chloride values

When increased	When decreased	
Nephritis	Addison's disease	Ether anesthesia
Prostatic obstruction	Burns	Typhus fever
Eclampsia	Diabetes	Anaphylactic shock
Anemia	Fevers	Uremia
Cardiac conditions	Intestinal obstruction	Vomiting
Hyperventilation	Metallic poisoning	Polycythemia
Hypoproteinemia	Pneumonia	Profuse sweating
Serum sickness	Heat cramps	Fasting
Urinary obstruction	Diarrhea	Hypercorticoadrenalism

Specimen stability

Serum or heparinized plasma chloride is stable for at least 2 weeks at room temperature or at 4° C.

Schales-Schales titration method*
Principle

A protein-free filtrate is made and an indicator added. The filtrate is then titrated with a standard mercuric nitrate solution. This reacts with the chloride to form undissociated mercuric chloride. When all the chloride has been removed, the addition of more mercuric nitrate gives a color reaction with the indicator. The amount of standard mercuric nitrate used in the titration is an index of the chloride content. This is then found by calculation.

Reagents

Diphenylcarbazone indicator	(D-104)
Standard sodium chloride, 10 mEq/L	(D-105)
Standard mercuric nitrate solution	(D-106)

Procedure
1. Prepare a tungstic acid filtrate.
2. Pipet 2.0 ml of the filtrate into a small flask.
3. Add 4 drops of the indicator diphenylcarbazone. Using a microburet graduated in hundredts, titrate with the standard mercuric nitrate solution. The end point is reached when the solution turns a pale violet or light purple color.

Calculation

Where E equals the number of milliliters of mercuric nitrate solution required for 2 ml of standard sodium chloride solution:

$$\text{ml standard mercuric nitrate used} \times \frac{100}{E} = \text{mEq Cl/L}$$

CO_2 CONTENT
Specimen stability

Plasma (heparinized) should be removed from the cells as soon as possible after blood is drawn. The CO_2 is stable for about 24 hours when stored in a tightly stoppered tube at 4° C.

Method using the Van Slyke manometric apparatus†
Principle

CO_2 gas, released from a measured volume of plasma by the addition of acid, is measured manometrically.

Reagents

Caprylic alcohol

---

*Adapted from Schales, O., and Schales, S. S.: Simple and accurate method for determination of chloride in biological fluids, J. Biol. Chem. 140:879, 1941.
†Arthur H. Thomas Co.

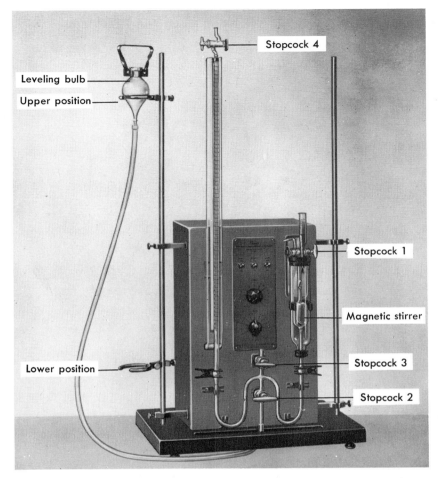

Fig. 5-1. Van Slyke manometric blood gas apparatus. (Courtesy Arthur H. Thomas Co., Philadelphia, Pa.)

Lactic acid, 0.1 N (D-107)
Sodium hydroxide, 5 N (carbonate-free) (D-108)
Mercury

Procedure

1. Before begining analysis for CO_2, remove metal jewelry such as rings and watches. Mercury forms amalgams with most metals, so do not dip coins or trinkets into the mercury. Although there is a temptation for the inexperienced operator to "play" with this element it must be stressed that continual carelessness may result in mercurial poisoning.
2. Expel water from the chamber by placing a vacuum hose into the cup and as the water is suctioned out of the cup, lift the mercury bulb carefully so the remaining water in the chamber and a few drops of mercury are expelled into the waste jar. Close stopcock 1 (Fig. 5-1). Place the leveling bulb in lower position.
3. Add 3 drops of caprylic alcohol to the cup and carefully turn stopcock 1 until the stopcock channel is filled. Caprylic alcohol prevents foaming.

4. Add 3.0 ml of 0.1 N lactic acid to the cup. The lactic acid liberates the carbon dioxide.
5. Using the Van Slyke pipet (or a 1.0 ml graduate measuring pipet but not a blow-out pipet), and 0.5 ml of serum or plasma to the cup. By placing the tip of the pipet into the lactic acid near the bottom of the cup before releasing the serum, contact with air is avoided. For the *blank,* use 3.5 ml of lactic acid instead of 0.5 ml of serum and 3.0 ml of lactic acid.
6. Slowly turn stopcock 1, releasing the serum and lactic acid into the chamber. Allow a drop of the caprylic alcohol to remain behind.
7. Seal the stopcock channel with cercury and lower the bulb to bring the mercury meniscus down to the 50-ml mark. With the bulb in the lower position, agitate the contents in the extraction chamber for 2 minutes.
8. Carefully turn stopcock 2 and let the solution rise (without oscillation) to the 2.0 ml mark.
9. Read the mercury manometer for P_1. The P_1 equals the millimoles of the initial gas pressure.
10. Add 0.3 ml of 5 N NaOH to the cup.
11. Turn stopcock 1 carefully, admitting the alkali to the reaction chamber. No air must enter the chamber.
12. Seal the stopcock with mercury, and by raising and lowering the bulb three times, raise and lower the contents of the chamber. This ensures complete absorption of carbon dioxide by the alkali.

Calculations

The *blank* value represents the amount of carbon dioxide in the reagents. The value usually runs between 1 and 4 mm. The P_1 value equals millimeters of the initial gas pressure. The P_2 value equals millimeters of pressure due to all gases except carbon dioxide:

$P_1 - P_2 =$ mm of carbon dioxide pressure

The millimeter value of CO_2 pressure in the *blank* is subtracted from that of the serum, and the result is multiplied by the appropriate temperature factor (Table 5-6):

$(P_1 - P_2) -$ Blank $\times$ Factor $=$ mEq/L

Care of the Van Slyke apparatus

Since the stopcocks are under pressure, they must be kept well greased. With the Arthur H. Thomas unit, whose method is described, the apparatus may be dismantled for cleaning. The ball-and-socket joint makes dismantling extremely easy. However, the unit may be cleaned without being dismantled. The following method may be used for the old and the new Van Slyke units.

Procedure

1. Attach a vacuum line to stopcock 4 at the top of the manometer column and suction out all the mercury.
2. With stopcock 1 open, and the vacuum line attached to the open stopcock 4, pour acetone into the cup until the complete system is filled.
3. Close stopcock 4 and allow the acetone to remain in the unit for 15 minutes to dissolve the grease particles.
4. Keeping stopcock 2 closed at all times, drain the system by vacuum through stopcock 3.
5. Repeat, filling the system with acetone, and let the actone remain in the system for 15 minutes. After draining the acetone through stopcock 2, the system should be free of grease particles.

Table 5-6
Temperature factors for CO_2 determinations

Temperature (°C)	Factor
18	0.2416
19	0.2404
20	0.2392
21	0.2380
22	0.2366
23	0.2354
24	0.2342
25	0.2330
26	0.2320
27	0.2308
28	0.2298
29	0.2286
30	0.2276
31	0.2266
32	0.2256
33	0.2246
34	0.2236

6. Fill the system with either chromic acid or 10% nitric acid. Let the cleaning solution remain in the system for at least 1 hour before draining.
7. Rinse the system at least three times with distilled water, then twice with acetone.
8. Leaving stopcocks 1 and 4 open, attach the vacuum to stopcock 3 to completely dry the system.
9. Carefully grease the stopcocks and fill the system with clean mercury.

Cleaning mercury

A simple way to clean mercury is to mix the mercury with sugar and then dissolve the sugar by washing with water. Place the dirty mercury in a large mortar. Pour the sugar into the mortar and grind the sugar and mercury with a pestle. The dirty sugar is then rinsed or dissolved off the mercury by running water. The water may be vacuumed off. The procedure is repeated until the sugar remains clean. All moisture is then vacuumed off the mercury, and it is ready to be filtered into the storage bottle. Place a small pinhole in the bottom of the filter paper. The mercury is then filtered to remove small traces of dirt not previously removed and to absorb any remaining moisture.

BLOOD pH, P_{CO_2}, AND P_{O_2}

The pH is a measure of the hydrogen ion concentration. The normal pH values for arterial blood are 7.35 to 7.45. Increased pH values are found in uncompensated alkalosis. Decreased pH values are found in uncompensated acidosis.

Blood pH may be determined electrometrically, colorimetrically, or from calculations by applying the carbon dioxide tension and bicarbonate content of plasma in the Henderson-Hasselbalch equation. (Refer to Chapter 6.) The pH of whole blood and that of plasma are considered to be the same for practical application.

The colorimetric procedures employ indicator color changes that have been established by potentiometric methods and correlated with hydrogen ion concentrations. Unless the methodology is carefully controlled, errors of as much as 1.0 pH may occur. The measurements are affected by temperature, indicator characteristics, color of the sample, salt concentration, and colloidal material in the sample.

In the electrometric method the pH is measured by determining the voltage developed by two electrodes in a solution. In many of the earlier investigations of blood pH a hydrogen electrode was employed. However, a glass electrode was subsequently developed that was more practical for measuring the pH of biologic fluids. The electrode consists of a glass tube closed at the bottom by a thin-walled bulb of soft glass that has high electric conductance. The bulb contains a buffered chloride solution and a silver chloride–coated wire that is connected to a conductor cable which may be attached to a pH amplifier. A theoretical explanation of the response of the glass electrode to hydrogen ion concentration is lacking. A potential is developed at the

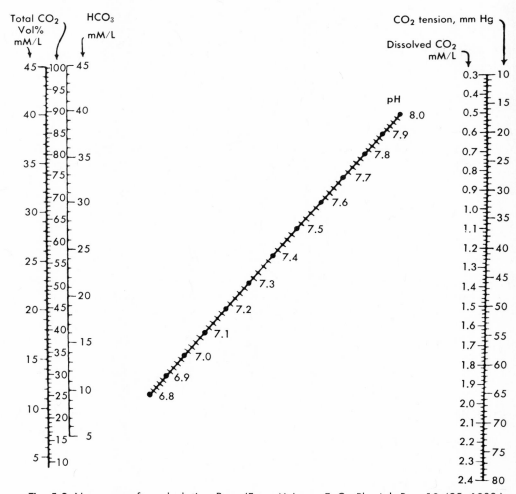

Fig. 5-2. Nomogram for calculating P_{CO_2}. (From McLean, F. C.: Physiol. Rev. **18**:495, 1938.)

glass-solution interface that depends upon the pH of the solution in which the electrode is immersed. For measurement of the potential a second or reference electrode is required—to complete the electric circuit and to furnish a constant potential with which the varying potential of the glass electrode can be compared. As the reference electrode, a calomel electrode, prepared from mercury and mercurous chloride in an aqueous solution of potassium chloride, may be used. It is usually connected to the unknown solution by a saturated potassium chloride bridge.

The pressure of carbon dioxide (P_{CO_2}) in arterial blood may be determined by an indirect calculation of the pressure of CO_2 from a nomogram (Fig. 5-2) based on the Henderson-Hasselbalch equation. This equation is based on three unknowns: the blood pH, the plasma bicarbonate concentration, and the partial pressure of the CO_2 in the blood. If any two are known, the third may be calculated. The nomogram provides a means of obtaining a P_{CO_2} value without long calculations.

Great care must be used in doing a blood pH determination, for an error of 0.02 unit can mean an error of about 1.8 mm Hg in the P_{CO_2} value.

An example in using the nomogram is as follows:

A straight line drawn through the given points on any two scales crosses the other two scales at points indicating simultaneously occurring values. For example, a patient has the following results: an arterial blood pH of 7.40 and a total CO_2 of 28 mEq/L (or millimoles per liter). The P_{CO_2} value is needed. Draw a line from the total CO_2 column reading 28 through the pH column, intersecting the 7.40 point. The line intersecting the mm Hg column gives a value of 45. Therefore, the P_{CO_2} is 45 mm Hg.

Astrup studies

During the great Danish polio epidemic of 1952 simpler and faster methods of determining the acid-base status of patients were needed. At that time Astrup started measuring blood pH routinely; and gradually with Siggaard-Andersen and their associates,[1, 2, 12, 13] a unique system was developed for evaluating acid-base metabolism, which included those blood constituents they considered necessary: the actual pH, carbon dioxide tension, standard bicarbonate, and base excess or deficit. The procedure involves measuring both the actual blood pH and the pH after equilibration of two portions of the blood with carbon dioxide–oxygen gas mixture of two known CO_2 tensions, one higher and one lower than normal. The values obtained are applied to a nomogram to derive other pertinent information.

Terminology

Astrup and associates[5] have defined the terms that they employ in describing acid-base metabolism.

actual pH A measure of the hydrogen ion concentration of anaerobically drawn blood at the patient's partial pressure of CO_2 and temperature. It reflects total acid-base changes caused by variation in the P_{CO_2} and/or base excess (respiratory and nonrespiratory disturbances). A decreased pH is termed total acidosis and an increased pH, total alkalosis.

actual P_{CO_2} (mm Hg) The partial pressure of CO_2 of the CO_2 concentration in a gas phase in the patient's blood drawn anaerobically. As a measure of the respiratory component, it is directly proportional to the CO_2 production and inversely proportional to the respiration. The P_{CO_2} multiplied by 0.03 will give the concentration of carbonic acid in plasma in millimoles per liter. When the P_{CO_2} is decreased, the condition is termed respiratory alkalosis and when increased, respiratory acidosis.

base excess (BE) (mEq/L) The nonrespiratory quantity, a measure of the excess or deficit of base. By definition it is zero for fully oxygenated blood with pH of 7.40 at P_{CO_2} of 40 mm

Hg. When nonvolatile acid accumulates in the blood, the BE value decreases; and when base accumulates or nonvolatile acid is removed, the BE value increases. A negative BE value is termed nonrespiratory acidosis and a positive BE value, nonrespiratory alkalosis.

buffer base (BB) (mEq/L) A nonrespiratory quantity introduced by Singer and Hastings, considered to be the sum of all buffer anions that are available to accept hydrogen ions, mainly bicarbonate and proteinate ions. At a constant base excess it varies with the protein and hemoglobin concentrations. However the compounds that function as buffer bases depend upon the pH of the blood.

standard bicarbonate (mEq/L) The concentration of bicarbonate in plasma from blood that has been completely oxygenated to saturate the hemoglobin and equilibrated with a Pco_2 of 40 mm Hg.

actual bicarbonate (mEq/L) A measure of the bicarbonate of the plasma from anaerobically drawn blood.

total CO_2 (mEq/L) The CO_2 of plasma obtained from the bicarbonate and carbonic acid in anaerobically drawn blood.

acidosis A condition caused by the accumulation in the body of an excess of acid or the loss from the body of base.

alkalosis or basosis A state caused by accumulation in the body of an excess of base or by the loss of acid.

Comments on measurements

A mean normal temperature for man of 37° C is used in all measurements. If the patient's temperature varies by more than 2° C, the measurements should be corrected to actual body temperature; or the water circulation thermostat may be reset if a number of tests are to be run for a different body temperature. The pH of blood decreases by approximately 0.0147 pH unit (0.01 unit for plasma) for each degree C increase in temperature. It is also necessary that the temperature of the electrode and liquid junction be maintained within ± 0.1° C of the body temperature, usually 37° C.

The pH meter must be well grounded; otherwise the instrument will drift.

Blood samples for actual pH must be introduced into the electrode anaerobically so that there will be no loss of CO_2 resulting in a simultaneous rise in pH. There must be no air bubbles in the polyethylene capillary or glass electrode capillary.

Normal values[5]

	Men	Women
Actual pH, whole blood	7.38 to 7.44 (7.40)	7.39 to 7.44 (7.41)
Actual Pco_2 (whole blood), mm Hg	34 to 45 (39.3)	31 to 42 (36.4)
Base excess (whole blood), mEq/L	−2.4 to +2.3 (−0.1)	−3.3 to +1.2 (−1.0)
Standard bicarbonate (plasma), mEq/L	23.0 to 25.4	22.4 to 25.8
Total CO_2 (plasma), mM/L	23.5 to 27.9	22.2 to 27.4

CALCIUM AND PHOSPHORUS

Calcium and phosphorus tend to maintain an equilibrium in the blood. They are usually considered together, since disturbances of one quite often result in a disturbance of the other. A low calcium is often accompanied by a high phosphorus. This does not always hold true, however; during bone formation high levels of both are sometimes noted. This is found when fractures are healing and also explains the higher phosphorus level in children. Rickets is another exception, in that calcium is usually normal while phosphorus may be down to 2 mg/dl. In some cases, however, the results may be reversed, or both calcium and phosphorus levels may be low.

Patients with renal problems will have an elevated phosphorus level due to retention of phosphates, usually excreted in large quantities by normally functioning kidneys. An elevated BUN will generally accompany this situation.

Calcium exists in the body in an ionized form and in a protein-bound form, which is mainly attached to albumin. When a calcium determination is done, the *total* calcium is being determined, regardless of the proportion of the above two forms. The ionized fraction is normally about 4.2 to 5.2 mg/dl, or about 50% to 60% of the total calcium, while the protein-bound calcium is 40% to 50% of the total.

CALCIUM

Calcium is absorbed into the bloodstream from the small intestine. It is used in the formation of bone and the clotting of blood. The normal calcium values are 8.5 to 10.5 mg/dl of serum. In growing children, these values are slightly increased.

Decreased calcium values are of primary concern in tetany—a disease caused by faulty calcium metabolism and characterized by convulsions and muscular twitchings. Increased and decreased calcium values may be found in those conditions given in Table 5-7.

Specimen stability

Serum should be removed from the clot within 1 hour after the blood is drawn. The serum calcium is stable for 1 day at 4° C and for 1 year when frozen. For a urine calcium determination, a 24-hour specimen should be collected in a bottle containing 10 ml of 6 N HCl. This is stable for several days at room temperature.

Clark-Collip modification of Kramer-Tisdall method
Principle

Ammonium oxalate is added to serum to precipitate the calcium as calcium oxalate. This is centrifuged. The precipitate is washed with dilute ammonium hydroxide to remove impurities and excess oxalate. Dilute sulfuric acid is added to convert the calcium oxalate to oxalic acid. After being heated to about 70° to 90° C (to en-

Table 5-7
Conditions accompanied by abnormal calcium values

When increased	When decreased	
Carcinoma	**Tetany**	Osteomalacia
Hyperparathyroidism	Celiac disease	Parathyroidectomy
Hypervitaminosis	Hypoparathyroidism	Pregnancy
Multiple myeloma	Nephritis	Rickets
Polycythemia vera	Nephrosis	Sprue
		Vitamin D deficiency

hance the reaction), the oxalic acid is titrated with standard potassium permanganate. The titration is an oxidation-reduction reaction, the oxalic acid being oxidized and the permanganate being reduced. The calcium content is found by calculation and reported in mg/dl of serum.

Reagents

Sodium oxalate, 0.1 N	(D-109)
Potassium permanganate, 0.1 N	(D-110)
Ammonium oxalate, 4%	(D-111)
Ammonium hydroxide, dilute (0.5%)	(D-112)
Sulfuric acid, 1 N	(D-113)

Sodium oxalate, 0.01 N

Dilute exactly 10 ml of 0.1 N sodium oxalate to 100 ml in a volumetric flask. Keep tightly stoppered.

Potassium permanganate, 0.01 N

Dilute 10 ml of 0.1 N potassium permanganate to 100 ml in a volumetric flask. Prepare and standardize just prior to use.

Procedure

1. *Standardization* of 0.01 N potassium permanganate: Pipet 5.0 ml of 0.01 N sodium oxalate into a 125-ml flask or similar container. Add 10 ml of 1 N (approximate) sulfuric acid. Heat to about 90° C by placing into a beaker containing boiling water. Using a buret, titrate with the approximately 0.01 N potassium permanganate. The titration will require about 5.0 ml so add 4.0 ml rather rapidly and then titrate drop by drop. The end point is reached when a single drop causes the solution to remain a faint pink color.

    Computation of factor. The factor indicates the strength of the potassium permanganate solution in comparison to an exact 0.01 N solution. If the potassium permanganate solution is weaker than 0.01 N, the factor will be less than 1.0; if stronger than 0.01 N, the factor will be more than 1.0. Of course, if the potassium permanganate is exactly 0.01 N, the factor will be 1.0. Find the factor as follows: the number of milliliters of 0.01 N sodium oxalate used in the titration is divided by the number of potassium permanganate required to reach the end point. Example: If 4.8 ml of potassium permanganate were needed, the factor is:

$$\frac{5.0 \text{ (ml of 0.01 N sodium oxalate used)}}{4.8 \text{ (ml of potassium permanganate required)}} = 1.04$$

2. Pipet 2.0 ml of serum into a 15-ml graduated centrifuge tube. Add 2.0 ml of distilled water and 1.0 ml of 4% ammonium oxalate. Mix thoroughly (Vortex mixer).
3. Let stand for 30 minutes in order to allow the ammonium oxalate to precipitate the calcium as calcium oxalate.
4. Centrifuge at 3000 rpm for at least 10 minutes. The precipitated calcium oxalate will then be seen as a small white precipitate in the bottom of the tube.
5. With a slow steady motion, invert the centrifuge tube, thus discarding the supernatant fluid and leaving the precipitate in the bottom of the tube. In this inverted position, place the tube on a piece of filter paper and allow to drain for a few minutes.
6. Remove any impurities the precipitate by washing it *twice* in the following manner: Add 2.0 ml of dilute ammonium hydroxide to the centrifuge tube. Suspend all the precipitate by tapping the bottom of the tube. Wash down

the sides of the tube with another 1.0 ml of the dilute ammonium hydroxide. Centrifuge for 10 minutes. Pour off the supernatant fluid with a slow steady motion and, with the centrifuge tube still in the inverted position, place it on a piece of filter paper to drain.
7. Add 1.0 ml of 1 N (approximate) sulfuric acid. (This converts the calcium oxalate to oxalic acid.) Dissolve the precipitate by tapping the bottom of the tube. Wash down the sides of the tube with another milliliter of the sulfuric acid.
8. Fill a 250-ml beaker about half full with water and bring water to a boil. Place the centrifuge tube in this for 1 minute. (This is done because the following oxidation-reduction reaction proceeds better at a high temperature.)
9. Remove from water bath. Using a buret graduated in hundredths, titrate with the 0.01 N potassium permanganate solution until a faint pink color persists for 1 minute.

Calculations

Take the number of milliliters of 0.01 N potassium permanganate used and the factor for the 0.01 N potassium permanganate and make the calculation according to the following:

$$\text{ml used} \times \text{Factor} \times 0.2 \times \frac{100}{2.0} = \text{mg calcium/dl}$$

where ml used = ml of 0.01 N potassium permanganate used in titration
Factor = Factor for potassium permanganate
0.2 = Amount of calcium equivalent to 1.0 ml of 0.01 N potassium permanganate
100 = Amount of serum reported on (100 ml)
2.0 = Amount of serum used

Example: If the factor for the 0.01 N potassium permanganate is found to be 1.04, and 0.9 ml of the potassium permanganate is used in the titration, the calcium value is found as follows:

$$\text{ml used} \times \text{Factor} \times \frac{\text{Calcium}}{\text{equivalent}} \times \frac{\text{Volume}}{\text{correction}} = \text{mg/dl}$$

$$0.9 \times 1.04 \times 0.2 \times \frac{100}{2} = 9.4 \text{ mg/dl}$$

Bachra-Dauer-Sobel method[4]
Principle

Calcium is determined by titration with disodium ethylenediamine tetraacetate (EDTA), using cal-red* as the indicator. With urine, icteric serum, and hemolyzed serum, the calcium is first obtained as the oxalate. With clear serum (nonicteric or nonhemolyzed) the calcium is titrated directly.

Reagents

Potassium hydroxide, 1.25 N	(D-114)
EDTA solution	(D-115)
Cal-red indicator	(D-116)
Sodium citrate, 0.05 M	(D-117)

---

*Cal-red, or 2-hydroxy-1-(2-hydroxy-4-sulfo-1-naphthylazo)-3-naphthoic acid, may be obtained from Scientific Service Laboratories, Inc.

Calcium standard (1 ml = 100 μg Ca) (D-118)
Caprylic alcohol
Ammonium oxalate, 10% (D-119)
Hydrochloric acid, 1 N (D-120)
Ammonium hydroxide, 5% (D-121)

Procedure

1. Set up the following in 15-ml conical centrifuge tubes:
   Standard (3 tubes)
      0.5 ml of working standard
   Unknown
      0.5 ml of serum plus 1 small drop of caprylic alcohol.
2. Proceed with one tube at a time to obtain a sharp end point. If samples remain in an alkaline medium beyond 10 minutes, the end points are not sharp.
3. Pipet 2.5 ml of 1.25 N potassium hydroxide into the tube and mix.
4. Shake the indicator solution each time before using, to mix. Pipet 0.25 ml of indicator solution into the tube. Mix.
5. Immediately titrate with EDTA solution until the color changes from wine red to blue. The color change may be observed by placing the tube against a bright light.

Calculations

$$\frac{\text{Titration volume of unknown}}{\text{Titration volume of standard}} \times 0.05 \times \frac{100}{0.5} =$$

$$\frac{\text{Titration volume of unknown}}{\text{Titration volume of standard}} \times 10 = \text{mg Ca/dl}$$

or

$$\frac{\text{Titration volume of unknown}}{\text{Titration volume of standard}} \times 5 = \text{mEq/L calcium}$$

ICTERIC AND HEMOLYZED SERUM

There are instances when the only available specimen for analysis is either icteric or hemolyzed. Using the specimens that are either icteric or hemolyzed will give invalid results. The following treatment will enable the specimen to be used.

Procedure

1. Pipet 0.50 ml of serum into a conical tipped centrifuge tube.
2. Add 0.63 ml of distilled water and 0.13 ml of 10% ammonium oxalate. Mix.
3. Incubate in a 56° C water bath for 15 minutes.
4. Centrifuge for 10 minutes at 3000 rpm.
5. Pour off the supernatant and invert the tube on a filter paper or blotter to drain.
6. Dissolve the precipitate in 0.25 ml of 1 N HCl and add 0.25 ml of 0.05 M sodium citrate. Mix.
7. The tube is now ready for the addition of KOH (step 3 of the preceding method).
8. The procedure is carried out in the same way as with the clear serum.

Urine calcium

Calcium on a 24-hour collection of urine may be done with the serum method after the following preparation of the urine sample. If calcium determination only is to be done, the specimen may be collected in a gallon bottle containing 10 ml of glacial acetic acid. If other tests are requested, where an acidified specimen would be undesirable, collect the specimen without the acid. Shake and mix the urine well, taking care that if there is any precipitate in the urine, none is left adhering to the walls of the bottle. Allow the urine to come to room temperature and measure the total volume in a graduated cylinder. The urine must be at room temperature, and the measurement must be accurate. Pipet 10 ml of the well-mixed urine into a test tube and acidify to pH 1 with glacial acetic acid. Use nitrazine paper to check the pH. Proceed as follows:

Procedure
1. Place the tube of acidified urine in a water bath (use a Pyrex beaker with a thermometer and beaker clamp).
2. Heat the specimen to 60° C for 15 minutes, mixing occasionally with a stirring rod.
3. Pipet 0.5 ml of the mixed urine into a conical tipped centrifuge tube.
4. Add 0.1 ml of 10% ammonium oxalate. Mix.
5. Add 1 drop of methyl red indicator (0.1%). Mix.
6. With a pipet or dropper add, dropwise, 5% ammonium hydroxide (NH_4OH) until an orange color is obtained.
7. Place the tube in a boiling water bath for 20 minutes. Use the Pyrex beaker again.
8. Cool the tube to room temperature and centrifuge for 10 minutes at 3000 rpm.
9. Decant the supernatant and invert the tube on filter paper or blotter to drain.
10. Dissolve the precipitate in 0.25 ml of 1 N HCl.
11. Add 0.25 ml of 0.05 M sodium citrate. Mix.
12. Proceed as in step 3 of the serum procedure. The standard is set up just as it was for the serum. (NOTE: If less than 0.5 ml of EDTA is used in the titration of the calcium in the urine, use a larger volume of urine—1.0 ml or more—instead of the 0.5 ml that was to be precipitated.) This is step 3 of the urine method.

Calculations

$$\frac{\text{Titration volume of unknown}}{\text{Titration volume of standard}} \times 10 = \text{mg Ca/dl}$$

$$\frac{\text{Titration volume of unknown}}{\text{Titration volume of standard}} \times 5 = \text{mEq Ca/L}$$

Calculate calcium output per 24 hours with the following:

$$\frac{\text{Titration volume of unknown}}{\text{Titration volume of standard}} \times 0.1 \times \frac{\text{Urine volume}}{(24 \text{ hr})} = \text{mg Ca/24 hr}$$

$$\frac{\text{Titration volume of unknown}}{\text{Titration volume of standard}} \times 0.05 \times \frac{\text{Urine volume}}{(24 \text{ hr})} = \text{mEq Ca/24 hr}$$

Calcium by atomic absorption

Refer to Chapter 8.

Ionized (diffusible) calcium

Determining ionized calcium is not a practical procedure for the routine laboratory. However, the new ion-selective electrodes have made this task somewhat simpler. A very low ionized calcium results in tetany, causing severe muscle spasms and tremors. Therefore, a rapid method for calcium is desirable, as well as a method for determining the ionized calcium content. The following method of estimating ionized (diffusible) serum calcium is given. This differs from the McLean and Hastings[11] nomogram by utilizing the albumin/globulin ratio instead of the total protein in order to give a closer estimation of the ionized calcium value.

An explanation of the K value is as follows. Originally, a nomogram was developed from data obtained by Hopkins and co-workers.[9] Total calcium, ionized (diffusible) calcium, and total serum protein with an A/G ratio of 1.8 were analyzed. The resultant factor (K) was 11.7.

Hanna and co-workers[7] assayed 100 thoroughly screened patients to obtain the K factor used in the following formula.

By using Hanna's findings, the following constant is derived. First, the chemical equation would be:

$$Ca\ proteinate \rightarrow Ca^{++} + Protein$$

Following the law of mass action, "The speed of a chemical reaction is proportional to the active masses of the reacting substances," the following equation is:

$$\frac{\text{Ionized (diffusible) calcium} \times \text{Total protein}}{\text{Nonionized (nondiffusible) calcium}} = K$$

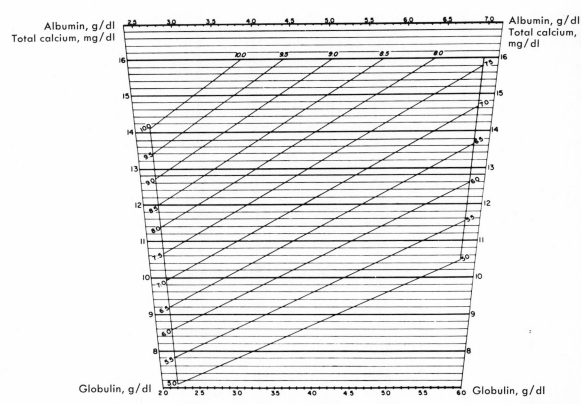

Fig. 5-3. Nomogram for estimating diffusible serum calcium. (From Hanna, E. A., Nicholas, H. O., and Chamberlin, J. A.: Clin. Chem. **10:**235, 1964.)

By assaying over 200 tests in duplicate the following values were obtained for the new K factor: Total calcium, 10.35 mg/dl; ionized calcium, 6.16 mg/dl; albumin, 4.40 g/dl, globulin, 3.60 g/dl, and A/G ratio, 1.22.

The following equation is set up to obtain the K factor:

$$\frac{6.16 \times 8.0}{10.35 - 6.16} = 11.8 \ (\text{K value})$$

Since it is understood how the K value is obtained, the following formula may be used to calculate an ionized calcium:

$$\frac{11.8 \times \text{Total calcium (mg/dl)}}{11.8 + \text{Total protein (g/dl)}} = \text{Ionized calcium (mg/dl)}$$

Using the nomogram

The nomogram shown in Fig. 5-3 is constructed with the K value being 11.8. The following is an example to show how the nomogram works. The patient has an albumin of 4.0 g/dl, a globulin of 3.0 g/dl, and a total calcium of 10 mg/dl. Using the nomogram, draw a line from the top line (albumin, 4.0 g/dl) to the bottom line (globulin, 3.0 g/dl). The total calcium line at 10 mg/dl intersects the protein line, giving an ionized calcium result of 6.3 mg/dl.

PHOSPHORUS

Phosphorus is absorbed into the bloodstream from the small intestine. It is used in forming bone and in regulating the pH of blood. The normal phosphorus values are 3.0 to 4.5 mg/dl of serum, with growing children having slightly higher values. Changes from normal values are sometimes of special importance (Table 5-8).

Specimen stability

Serum inorganic phosphate is stable for about 3 days at 4° C and for 1 year when frozen.

Table 5-8
Conditions accompanied by abnormal phosphorus values

When increased	When decreased
Nephritis	**Rickets**
Healing bone fractures	Hyperparathyroidism
Hyperinsulinism	Myxedema
Hypervitaminosis	Osteomalacia
Hypoparathyroidism	Sprue
Uremia	Idiopathic steatorrhea
Pyloric obstruction	Neurofibromatosis
Starvation	Ether and chloroform anesthesia
Following Pituitrin administration	Following insulin and Adrenalin administration
	Lobar pneumonia

Fiske-SubbaRow method of determination of inorganic phosphate[6]
Principle
Trichloroacetic acid is added to serum to precipitate the proteins. After filtering, a molybdate solution is added to unite with the phosphorus to form phosphomolybdic acid. The phosphomolybdic acid is then reduced by an organic reagent to form a blue color. The intensity of blue color is proportional to the amount of phosphorus present. The depth of color is measured in a colorimeter.

Reagents

Trichloroacetic acid, 10%	(D-95)
Molybdate reagent	(D-122)
Sulfuric acid, 10 N	(D-123)
Reducing reagent (aminonaphtholsulfonic acid)	(D-124)
Phosphorus stock standard, 0.1 mg P/ml	(D-125)
Phosphorus working standard (1 ml = 0.01 mg P)	

Using a 100-ml volumetric flask, dilute 10 ml of stock standard to 100 ml with distilled water.

Procedure
Preparation of protein-free filtrate
1. Pipet 1.0 ml of serum into a 15-ml test tube.
2. Add 9.0 ml of 10% TCA. Stopper and shake.
3. Allow the tube to stand for 15 minutes and then centrifuge or filter through Whatman No. 42 filter paper.

Development of color
1. Set up the following in 15-ml test tubes:

   Blank
   5.0 ml of TCA into tube
   3.6 ml of distilled water

   Standards (three tubes)
   4.5 ml of TCA into each tube
   1.6 ml of distilled water
   2.5 ml of working standard

   Unknown
   5.0 ml of filtrate into the test tube
   3.6 ml of distilled water

2. Add 1.0 ml of molybdate reagent to all tubes. Mix.
3. Add 0.4 ml of cold reducing reagent. Mix thoroughly.
4. Allow the tubes to stand for 15 minutes (use timer) for the full development of color
5. Read against reagent blank at 660 nm.

Calculations

$$\frac{A \text{ of unknown}}{A \text{ of standard}} \times 0.025 \times \frac{100}{0.5} =$$

$$\frac{A \text{ of unknown}}{A \text{ of standard}} \times 5 = \text{mg phosphorus/dl}$$

Urine inorganic phosphate
To determine urine inorganic phosphate, it is necessary to make a 1:5 or a 1:10 dilution with distilled water. This dilution is treated in the same manner as the serum.

The result obtained is then multiplied by the appropriate dilution factor, either 5 or 10.

OSMOLALITY

Measurement of solute (ionic and nonionic) concentration of biologic fluids is important in the care of certain medical and surgical patients. These solutes can be divided into three general categories: (1) electrolytes, (2) organic solutes of small molecular weight (urea, glucose, creatinine), and (3) colloids (chiefly protein).

When a solute is added to a solvent the colligative properties of the solvent change in a linear fashion with the increased amount of solute; the freezing point and vapor pressure decrease, and the osmotic pressure and boiling point increase. With the present commercial instruments, the determination of freezing point lowering is more satisfactory than other colligative properties for measuring solute concentration in biologic fluids. The freezing point of a solution is related to the osmotic concentration of that solution or to the concentration of particles of solute per unit amount of solvent (water). The molal freezing point lowering is the change in freezing point produced by 1 mole of a nonionic solute in 1000 g of solvent. When 1 mole of a nonionic solute is added to 1 kg of water, the freezing point is decreased $1.858°$ C. There is an interaction between the solute and solvent referred to as "activity," which includes, among other factors the degree of dissociation of the solute. The addition of 1 mole of an ionic solute such as sodium chloride lowers the freezing point almost twice as much because of the dissociation. For example, 1 mole of NaCl plus 1 kg of water lowers the freezing point $\phi \times n \times 1.858°$ C, where n is the number of ions from the molecule and ϕ is the osmotic coefficient, or about 0.93 at this concentration. The latter varies with the sodium chloride concentration in solution.

Measurement of freezing point depression is carried out using an osmometer, the operation of which is based upon the comparison of the freezing point of an unknown with the freezing point of a standard solution of known molality. The unit for reporting is milliosmol per kilogram of water or simply milliosmol.

Pure water may be cooled or supercooled to $-40°$ C without freezing. However, the addition of a crystal of ice or a dust particle will cause rapid ice formation and a temperature return to $0°$ C, the normal freezing point. Each gram of water at $0°$ C produces 80 calories of heat (latent heat of fusion). Crystallization of the ice does not occur instantaneously; the more dilute the solution, the more rapidly the equilibrium temperature is reached.

Normal values for osmolality of human serum from 104 blood donors had a mean value and standard deviation of 290 ± 3.8 mOsm/L.[14] A deviation of ± 10 mOsm from normal is considered significant.

For clinical situations the ratio between the serum sodium concentration and osmolality (mEq/mOsm) is often used.[8] For normal patients this should range from 0.43 to 0.50. The ratio will be elevated in hypernatremic patients without a change in other serum metabolites. In hyponatremia in the absence of an increase in other serum solutes, the ratio should be normal. A calculation of the milliosmolal concentration may be made from the serum concentrations of electrolytes, urea, and glucose, by applying the following:

$$\text{mOsm} = 1.86 \ (\text{Na, mEq/L}) + \frac{\text{Glucose (mg/dl)}}{18} + \frac{\text{BUN (mg/dl)}}{2.8}$$

For normal patients the determined osmolality should not exceed the calculated by more than 5 to 8 mOsm/L.

For urine concentration studies the measurement of osmolality is a direct indication of the formation of concentrated urine by the kidney and is more reliable than specific gravity measurements. Osmolality determinations have definite application in following dialysis therapy, the course of uremia, hyperosmolal nonketonic coma in diabetes, and fluid administration in postsurgical patients.

Plasma hypotonicity resulting from low plasma sodium concentration may occur after surgical operations, with congestive cardiac failure and in hepatic failure with ascites. Trauma, acute illness, and many chronic wasting diseases produce accumulation of water and salt, with water in excess of salt resulting in hypotonicity of the body fluids. Patients with advanced liver and kidney disease or severe diabetes mellitus may have an increased plasma osmolality of 20 to 50 mOsm/L from nonelectrolyte solutes such as urea and other related products, glucose, bilirubin, and pathologic constituents.

CONVERSION OF UNITS
Conversion of milligrams/dl to milliequivalents/liter

Milligrams per deciliter (100 ml) is a term that deals with weight, whereas milliequivalents per liter deals with combining power. For example, the element calcium has a specific weight, which may be given as mg/dl. However, it also has combining power, determined by its valence, which may be expressed as mEq/L.

The following abbreviations are usually used:

$$mg = milligram$$
$$ml = milliliter$$
$$mEq \text{ (or meq)} = milliequivalent$$

To convert mg/dl to mEq/L, use the formula:

$$mEq/L = mg/dl \times \frac{10}{\text{Equivalent weight}}$$ where 10 is used to convert 1 dl to 1 L.

Example: Convert 325 mg of sodium/dl to mEq/L. Equivalent weight of sodium is 23. Use the formula:

$$mEq/L = mg/dl \times \frac{10}{\text{Equivalent weight}}$$
$$= 325 \times \frac{10}{23}$$
$$= 141$$

A shortcut in the above formula may be taken by first dividing 10 by 23, which equals 0.435; this is known as the conversion factor for sodium. The formula then becomes:

$$mEq/L = mg/dl \times \text{Conversion factor}$$

Thus, to convert 325 mg of sodium/dl to mEq/L by using the conversion factor of 0.435:

$$mEq/L = mg/dl \times \text{Conversion factor}$$
$$= 325 \times 0.435$$
$$= 141$$

By multiplying the units in column two of Table 5-9, by the factor given, the desired unit (in column four) may be obtained.

Table 5-9
Electrolyte conversion factors

Electrolyte	Unit	Factor	Unit
Na	mg/dl	0.435	mEq/L
Na	mEq/L	2.3	mg/dl
Na	mEq/L	0.0585	g/L as NaCl
K	mg/dl	0.256	mEq/L
K	mEq/L	3.91	mg/dl
K	mEq/L	0.0746	g/L as KCl
Cl	mg/dl	0.282	mEq/L
Cl	mEq/L	3.55	mg/dl
Cl	mEq/L	0.0585	g/L as NaCl
CO_2	mEq/L	0.45	mEq/L
CO_2	vol%	2.24	vol%
Ca	mg/dl	0.5	mEq/L
Ca	mEq/L	2.0	mg/dl
PO_4	mg/dl	0.581	mEq/L
PO_4	mEq/L	1.72	mg/dl

Conversion of volumes percent to millimoles per liter

Gases are generally expressed in millimoles or milliequivalents per liter instead of volumes percent. A mole is the molecular weight expressed in grams and a millimole is one thousandth part of this.

To convert carbon dioxide in volumes percent to millimoles (mM) per liter:

$$mM/L = vol\% \times 0.446$$

Example: Convert 50 vol% of carbon dioxide to millimoles per liter.

$$\begin{aligned} mM/L &= vol\% \times 0.446 \\ &= 50 \times 0.446 \\ &= 22.3 \end{aligned}$$

References

1. Astrup, P., Jorgensen, K., Siggaard-Andersen, O., and Engel, K.: The acid-base metabolism: a new approach, Lancet **1**:1035, 1960.
2. Astrup, P., and Siggaard-Andersen, O.: Micromethods for measuring acid-base values of blood. In Sobotka, H., and Stewart, C. P., editors: Advances in clinical chemistry, vol. 6, New York, 1963, Academic Press, Inc.
3. Albanese, A. A., and Lein, M.: Microcolorimetric determination of sodium in human biologic fluids, J. Lab. Clin. Med. **33**:246, 1948.
4. Bachra, B. N., Dauer, A., and Sobel, A. E.: The complexometric titration of micro and ultramicro quantities of calcium in blood serum, urine and inorganic salt solutions, Clin. Chem. **4**:107, 1958.
5. Current concepts of acid-base measurements, Ann. N. Y. Acad. Sci. **133**:3, 1966.
6. Fiske, C. H., and SubbaRow, Y.: The colorimetric determination of phosphorus, J. Biol. Chem. **66**:375, 1925.
7. Hanna, E. A., Nicholas, H. O., and Chamberlin, J. A.: Nomogram for estimating diffusible serum calcium, Clin. Chem. **10**:235, 1964.
8. Holmes, J. H.: Measurement of osmolality in serum, urine and other biologic fluids by the freezing point determination, pre-workshop manual on urinalysis and renal function stud-

ies, 1962, American Society of Clinical Pathologists, Advanced Instruments, O-2 Holmes.
9. Hopkins, T., Howard, J. E., and Eisenberg, H.: Ultrafiltration studies on calcium and phosphorus in human serum, Bull. Johns Hopkins Hosp. **91**:1, 1952.
10. Lochhead, H. B., and Purcell, M. K.: Rapid determination of serum potassium employing glycine-phenol reagent, Am. J. Clin. Pathol. **21**:877, 1951.
11. McLean, F. C., and Hastings, A. B.: The state of calcium in the fluids of the body, J. Biol. Chem. **108**:285, 1935.
12. Siggaard-Andersen, O.: The acid-base status of the blood, Scand. J. Clin. Lab. Invest. **15** (Supp. 70):1963.
13. Siggaard-Andersen, O., Engel, K., Jorgensen, K., and Astrup, P.: A micro method for determination of pH, carbon dioxide tension, base excess and standard bicarbonate in capillary blood, Scand. J. Clin. Lab. Invest. **12**:172, 1960.
14. Stevens, S. C., Neumayer, F., and Gutch, C. F.: Serum osmolality as a routine test, Nebraska Med. J. **45**:447, 1960.

6

The clinical pathology of hydrogen ion homeostasis

ROBERT C. ROSAN

The lung is the fundamental organ for the defense of the body's hydrogen ion homeostasis (HIH). For every equivalent of a "fixed" acid excreted by the kidney, the lung excretes 100 times more carbonic acid—actually about 10 Eq/day/70 m^2 lung surface or very approximately, 0.5 mEq/breath for a sedentary adult. This rate may be augmented 20 times or more in severe HIH disease or heavy exercise. However, regulation of carbon dioxide transfer is not the only way that the lung participates in HIH; oxygen uptaken itself is related in a quite direct way. The lung regulates oxygen uptake, protecting the bloodstream against too much, as well as too little, oxygen perfusion.

How can the lung protect us against too much oxygen? If the alveolar sac collapsed at the end of every breath, the tension of oxygen within it would vary from 0 mm Hg at the end of the breath to near atmospheric pressure when it filled with fresh air, or about 150 mm Hg $P_{A_{O_2}}$. In newborns, at least, such a $P_{A_{O_2}}$ is high enough to cause retinal arteriospasm and quite likely blindness from the sequelae of vasospastic retinal ischemia. Thus the simple maintenance of alveolar volume helps regulate arterial oxygen tension. On the other hand, if a normal adult breathes *increased* atmospheric tensions of oxygen, he will eliminate portions of his lung from ventilation exchange—he will "rest" a number of breathing units and disturb the ventilation:perfusion ($\dot{V}/\dot{Q}$) regional equilibria. This effect returns the arterial oxygen tension to normal (Table 6-1).

Normally, however, the lung's oxygen transfer system defends HIH by meeting the oxygen demands of tissue, which is, of course, a problem of augmenting oxygen intake when metabolism is increased by work, fever, shivering, and the like. The oxygen burns fats or glucose to carbon dioxide and water, with nearly one CO_2 molecule produced for each O_2 consumed. Since the CO_2 diffuses back into the blood, mostly to be excreted by the same lung involved in O_2 uptake, the system must be quite finely tuned. Even some of the metabolically produced water is volatilized from the lung! However, in general more molecules of oxygen are needed than carbon dioxide is excreted—about 20% more on an ordinary diet. Metabolic business is more than the mere metabolism of glucose to water.

When the alveoli cannot supply or the pulmonary veins or left ventricle cannot deliver oxygen at a sufficient rate for tissue demands, there is a drop in the gradient

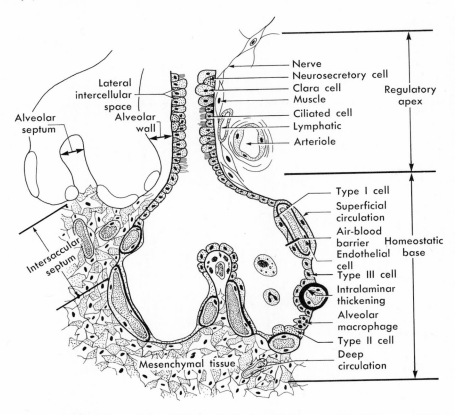

Fig. 6-1. Regenerating injured lung. Many of the features recapitulate the appearance of fetal lung.

of oxygen tension at the vital mitochondrial level. Normally, mitochondria function at a Pm_{O_2} of about 2 to 3 mm Hg. The driving arterial pressure for this is usually around 95 mm Hg Pa_{O_2}. (This partial pressure is then attenuated by the capillary bed and modulated by tissue diffusivity, so that the appropriate mitochondrial environment may be maintained.) However, when the driving pressure and, subsequently, the mitochondrial Pm_{O_2} fall too low, mitochondrial metabolism is depressed. Ultimately—and while some oxygen is still available—glucose is no longer metabolized in the mitochondrial carboxylic acid cycle. Instead, glucose catabolism stops at pyruvate. The majority of pyruvate is converted to lactate by omnipresent isoenzymes of lactic dehydrogenase. In serum, the lactate:pyruvate ratio is about 20:1. At usual cellular pH, some of the lactate is protonated and becomes lactic acid. The lactate is rather freely diffusible over cell membranes and readily enters the bloodstream. Normally, the liver clears lactate from the blood and converts it ultimately to nonionizable glycogen or neutral fat deposits or metabolizes it to carbon dioxide and water. However, when the liver's capacity is exceeded, lactic acidemia occurs—there is excess lactic acid in the blood. Thus the two main origins of lactic acidemia are overproduction and underconsumption. In general, increases in the oxygenation of peripheral tissue (usually from increased lung action) effect the most rapid reductions in blood lactic acid. This mechanism enables the liver to "catch up." Liver clearance itself is slower. Lactic acid is a classic organic "weak"

Table 6-1
Useful physiologic approximations of partial oxygen tension

Po_2	Remarks
2	Ordinary mitochondrial oxygen tension
10	Ninety percent of oxygen released from adult hemoglobin
15	Umbilical artery Pao_2 of fetus near term
25	Approximate P_{50} value for adult hemoglobin; also approximate umbilical vein Pvo_2 of fetus near term
35	Fetal lung Pao_2; also, at this level, about 70% oxygen—roughly two thirds—is released from adult hemoglobin in health
40	Usual resting normal adult venous Pvo_2
60	Less than a 10% improvement in saturation of adult hemoglobin when Pao_2 above this level
70	Pulmonary arteriolospasm induced by alveolar Pao_2 below this level
75	Below this level, cyanosis clinically visible in patients with normal hematocrits
80	Adult hemoglobin about 95% saturated
95	Pao_2 in health at rest for normal adult
100	Alveolar mean Pao_2 in healthy adult
140	Retrolental fibroplasia Pao_2 above this level
150	Highest Po_2 achieved in the water-saturated upper respiratory passages
160	Approximate dry air Po_2 at sea level

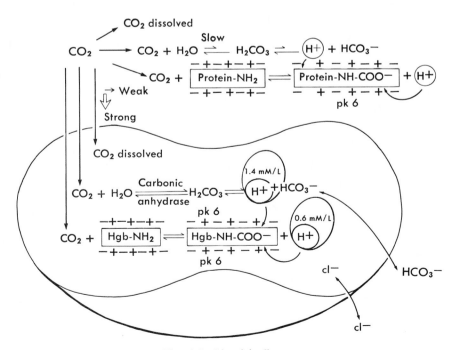

Fig. 6-2. Blood buffers.

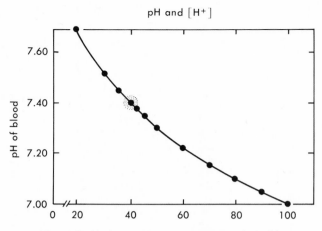

Fig. 6-3. Hydrogen ion concentration (nEq/L).

Table 6-2
Hydrogen ions

	pH	H⁺
Normal	7.40 ≈ 40 mEq/L	
Alkaline	7.45 ≈ 35	
	7.50 ≈ 30	
Acid	7.35 ≈ 45	
	7.30 ≈ 50	
	7.25 ≈ 55	
	7.20 ≈ 60	
	7.15 ≈ 65	

acid, but in terms of biologic homeostasis it is a fairly strong acid, only a little less ionized than acetic acid. However, somewhat misleading is the fact that blood lactate is usually reported as lactic acid, without regard to the blood hydrogen ion concentration or ionization constant of lactic acid.

One may consider that the protons of excess lactic acid are soaked up stepwise; first, by intracellular buffer systems; second, in the buffer systems of the extracellular extravascular space; and third, in the blood. Then, when the buffering capacity of the major blood buffer, hemoglobin, and the secondary protein ampholyte buffers have all been exceeded, the lung again intervenes in HIH and compensates for lactic acidosis. By a remarkable regulatory system, it accelerates the elimination of CO_2 while simultaneously delivering sufficient oxygen to assist aerobic metabolism keep the lactic acid production within control. The excess (H^+) from normal (HCO_3^-) production is reduced when CO_2 is blown off, while the retained CO_2, dissolved in the blood, combines with excess lactic acid hydrogen ion to form carbonic acid, which ionizes very poorly. Indeed, the whole backbone of pulmonary HIH is the stoichiometry of carbonic acid formation in the blood and the excretion of CO_2 from the lung. The process is continuous and integrated in reality, not stepwise as above (Figs. 6-2 and 6-3; Table 6-2).

HOW HEMOGLOBIN BUFFERS [H+]

In 5 L blood volume, 121 mEq = 24.2 mEq/L × 5 L. CO_2 is converted to H_2CO_3, which in turn produces [H+] as follows:

$$121 \text{ mEq } CO_2 + H_2O \longrightarrow 7 \text{ mEq } [H^+]$$
$$+ \text{ Hgb} \longrightarrow 3 \text{ mEq } [H^+]$$
$$\text{Total} = 10 \text{ mEq } [H^+]$$

However, that same blood volume contains 45 mM Hgb (when Hgb = 15 g/dl or 9 mM/L).

And: 1 M Hgb → 0.22 Eq [H+] a buffer capacity.

So: 45 mM × 0.22 mEq = 10 mEq [H+] buffer.

On an ordinary North American middle class diet, the respiratory quotient (CO_2 produced : O_2 utilized) approaches 0.7 to 0.8, and of the order of 70 mEq H+ is produced daily by a 70-kg person—which is about 1 mEq/kg and also 1 mEq/m² pulmonary gas exchange surface. This quantity may be easily exchanged by a few dozen breaths, so great is the reserve of the body. The transportation of carbon dioxide in the red cell and plasma is an important part of this story. Let us consider blood, freed of its protein components, and thus very much like the sea from which all life has risen. It is salty.

Bicarbonate, which exists in the blood and is the central nonprotein buffer of the body, must therefore be considered the salt of a weak acid. As a salt, it is almost completely ionized, but when it reacts with hydrogen ion to form carbonic acid, that product is quite poorly ionized. Thus bicarbonate is typical of the buffers formed by the strong salts of weak acids. Its major difference is that the acid itself, carbonic acid, is easily decomposed to volatile products which can be vented from the lung.

Let us see how this works out:

$$K [H_2CO_3] = [H^+] + [HCO_3^-]$$
$$K \propto [BHCO_3] = [B^+] + [HCO_3^-] \text{ and } \propto \to 1$$

and simplifying:

$$K_1 [H_2CO_3] = [H^+] \times [HCO_3^-]$$

or:

$$[H^+] = K_1 \frac{[H_2CO_3]}{[HCO_3^-]}$$

When carbon dioxide dissolves in extracellular water:

$$CO_2 + H_2O \longleftrightarrow H_2CO_3$$

Thus:

$$CO_2 \propto H_2CO_3$$

But:

$$H_2CO_3 \longleftrightarrow H^+ + HCO_3^-$$

And the dissociation constant:

$$K^1_{A_{H_2CO_3}} = \frac{[H^+] \times [HCO_3^-]}{[H_2CO_3]}$$

Or:

$$K_A = \frac{[H^+] \times [HCO_3^-]}{(CO_2)}$$

Therefore, on rearranging terms:

$$[H^+] = K_A \frac{[CO_2]}{[HCO_3^-]}$$

But:

$$\log_{10}(-H^+) = pH \text{ by definition of pH}$$

And:

$$\log_{10}(-K_A) = pK_1 \text{ by definition of } pK_1$$

Hence:

$$pH = pK_1 - \log \frac{[CO_2]}{[HCO_3^-]}$$

Or:

$$pH = pK_1 + \log \frac{[HCO_3^-]}{[CO_2]}$$

Experimentally, in normal blood, $pK_{1H_2CO_3} = 6.1$

So:

$$pH = 6.1 + \log \frac{[HCO_3^-]}{[CO_2]}$$

According to standard gas law:

1 mole gas = 22.26 L at standard pressure and temperature

Or:

1 ml CO_2 = 0.04 mM CO_2 STP

But, solubility coefficient for CO_2 in water ≈ 0.75

Thus:

$$pH = 6.1 + \log \frac{HCO_3^-}{0.04 \times 0.75 \, (P_{CO_2})}$$

$$pH = 6.1 + \log \frac{HCO_3^-}{0.03 \, (P_{CO_2})}$$

$$pH \approx 6.1 + \log \left(\frac{20}{1}\right)$$

$$pH = 7.4$$

However, it is difficult to measure accurately HCO_3^- in the clinical laboratory, but it is easy to measure total CO_2.

$$T_{CO_2} = HCO_3^- + H_2CO_3$$

Or:

$$HCO_3^- = T_{CO_2} - H_2CO_3$$

And as seen above, $H_2CO_3 = 0.03\, P_{CO_2}$

And as also seen above, $pH = pK_1 + \dfrac{HCO_3^-}{0.03\, P_{CO_2}}$

So:

$$pH = 6.1 + \log \dfrac{T_{CO_2} - 0.03\, P_{CO_2}}{0.03\, P_{CO_2}}$$

Therefore, when T_{CO_2} and P_{CO_2} are known, H can be calculated. Or, when pH, P_{CO_2} and T_{CO_2} are known, $HCO_3^- = T_{CO_2} - 0.03\, P_{CO_2}$ can be calculated.

Given the terms P_{CO_2}, T_{CO_2}, and pH, if we know two of these terms, we can calculate the third.

Note that over the biologic range $Ta_{CO_2} \propto Pa_{CO_2}$, and $Tv_{CO_2} \propto Pv_{CO_2}$.

What we have just outlined is a verbal picture of normal reaction to metabolic acidosis, called "compensation." Unfortunately, the term metabolic acidosis is often used to cover a host of defects in HIH that range from failure of the kidney tubule to take back metallic cations of the glomerular filtrate, through excess production of amino acids in hereditary enzymatic errors, and up through the example of lactic acidosis already given. We prefer not to use the term metabolic acidosis when metabolism is not the issue, for example, in cases of "renal" acidosis, but the incorrect usage is so widespread that the only defense against this misconception is to keep one's mind sharpened to the problem. Metabolic acidosis, as used herein, refers strictly to a disturbance of HIH which arises at the level of intermediary metabolism. Indeed, we prefer the specific term hypoxic metabolic acidosis (HMA) to describe what happens when a lack of oxygen leads to mitochondrial failure and excess protons in the blood. In true metabolic acidosis the role of the pulmonary oxygen function in defeating lactic acidemia is generally underemphasized (Table 6-3).

The lung regulates HIH through two functional mechanisms: the regulation of oxygen supply to the bloodstream to provide sufficient oxygen for aerobic catabolism

Table 6-3
Changes in blood buffer pathology

	(HCO_3^-)	(H^+)	Pa_{CO_2}	State	Common conditions
Ventilatory (respiratory)	+	+	+	Acidemia	Alveolar hypoventilation Increased dead space
	−	−	−	Alkalemia	Mechanical ventilation Central neural lesions
Metabolic	−	+	−	Acidemia	Cellular hypoxia Renal tubular failure Loss of extracellular cation
	+	−	+	Alkalemia	Loss of extracellular cation Na^+ therapy Anion depletion

and thus to prevent lactic acidosis (HMA), and the regulation of carbon dioxide egress from the bloodstream to optimize the carbonic acid buffer system of the blood and thus to prevent "respiratory"—really ventilatory—conditions of defective HIH. In order to discuss these functions as they apply to mechanical activity of the lung, we shall make some oversimplifications.

LUNG FUNCTION

For purposes of oxygen diffusion into the blood, it would be ideal if the lung could be held open to some fixed volume. In that way, there would be a steady diffusion of oxygen down the trachea, through the bronchi, and into the ultimate alveoli, and the gradients of oxygen would be constant at every level. The partial pressure of oxygen which would drive oxygen across the capillary wall and into the blood would always be the same, and thus there would be no variation in the amount of oxygen available to the capillary bed, and so no variation in the partial tension of oxygen which ultimately drives oxygen into the tissue, the cell, and the mitochondrion. Since oxygen diffuses quite well in air, to hold the lung open at constant volume would provide a continuous percolation of oxygen from the mouth to the lung capillary. Indeed, as we shall see, the lung can come quite close to this ideal.

On the other hand, for the purpose of carbon dioxide excretion, it would be most efficient if the lung had a minimal internal volume which held very little CO_2, for this gas diffuses slowly in air but is rapidly dissolved by water, such as is found in the fluid blanket that lines the pulmonary tree or in the cells that line its spaces. Thus, in a simplified way, we may say that oxygen dynamics are managed by the alveoli or "live" space, and carbon dioxide dynamics are managed by the air tubes, or "dead" space. Now let us examine the alveoli further.

Air-blood barrier

Since oxygen diffuses comparatively slowly in fluid, since the lung lining cells are mostly fluid, and since cells have no major mechanisms to transport oxygen other than by diffusion, it is necessary for the cellular barrier between the alveolar lumen and the capillary lumen to be very thin. Indeed, a specialized thin organoid, the air-blood barrier (ABB), has arisen to carry out this function. Thus, the ABB is the fundamental organoid structure of the lung. Embryologically, it begins to appear in appreciable numbers in the midpoint of gestation. It can be shown that its usual or "mean harmonic" thickness of 0.5 μm, practically the same in fetus and adult, is precisely that needed to permit normal oxygen diffusion into the lung capillary bed during air breathing at a physiologic rate. Quite likely, no other part of the alveolar sac or capillary network is of significant effect in the transfer of oxygen. For practical purposes, lung capillaries without air-blood barriers cannot oxygenate blood. This is of great importance in the prematurely born and in the repair of injured lung, for in both instances the number of air-blood barriers per alveolar sac seems relatively reduced, while the total number of capillaries in the lung is still fairly high. Capillaries that transport blood toward the left atrium but that do not oxygenate their blood must necessarily add to the venous admixture, and they are physiologic handicaps. Only a true air-blood barrier can transport oxygen (Fig. 6-1).

Haldane and Bohr effects

THE HALDANE EFFECT

Loss of oxygen from hemoglobin, or reduction, causes more proton acceptors of the hemoglobin molecule to be exposed because reduction alters the tertiary

Table 6-4
Haldane effect

Haldane effect	Arterial		Venous		A-V difference	
	pH	mEq/L [H$^+$]	pH	mEq/L [H$^+$]	pH	mEq/L [H$^+$]
Absent	7.40	40	7.32	48	0.08	8
Present	7.40	40	7.37	43	0.03	3

molecular shape of hemoglobin. Conversely, a rise in the number of free protons that can interact with hemoglobin drives oxygen from hemoglobin. Or, simply, protons reduce hemoglobin and increase the buffer capacity for protons; oxygen oxidizes hemoglobin and decreases the buffer capacity for protons.

BOHR EFFECT

Acidemia decreases and alkalosis increases the avidity of hemoglobin for oxygen at any given P_{O_2}, but particularly at the P_{O_2} levels encountered in the systemic capillary bed (P_{CO_2} 80 → 35 mm Hg). Thus, the production of acid—CO_2 and lactate—by tissue metabolism is a feedback regulator that drives oxygen release. This works because, at any given oxygen saturation along the length of the capillary, increasing acidosis increases the P_{CO_2}, which actually drives the oxygen into the tissue. These increases in P_{CO_2} and the driving force of acidosis are both modulated by the same mechanism, a rearrangement of molecular hemoglobin in the presence of increased H$^+$ which decreases the forces that bind oxygen, thereby releasing O_2 to the plasma.

The Bohr effect in the lung is quite different. Here, the principle-fixed parameter is alveolar oxygen tension ($P_{A_{O_2}} \approx 100$ mm Hg), and ordinarily it varies very little in health. Even severe degrees of exertional acidosis do not significantly depress oxygen saturation in lung capillaries. For example, a very large production of CO_2 ($P_{V_{CO_2}}$ 40 → 80 mm Hg) results in negligible Bohr mechanics (1% to 2% decrease in oxygen saturation). This graphically demonstrates the complex shape of the Bohr curve, a shape achieved partly by the stepwise cooperation of each hemoglobin tetramer with its neighbor in the modulation of the oxygen binding site affinity. Another example would be lung performance at high altitude, where $P_{A_{O_2}}$ is still fixed, but at a much lower value than at sea level. For any given usual $P_{V_{CO_2}}$, the difference in oxygen affinity and hence oxygen saturation by hemoglobin at sea level and at 1 mile (1.6 km) is negligible. Thus hemoglobin saturation with oxygen is maintained, but since $P_{A_{O_2}}$ has dropped, the dissolved oxygen of the plasma has also dropped. However, only a negligibly small fraction of total blood oxygen is carried by the plasma—about 1.5%. The Bohr effect gives the human animal the possibility for intense muscular effort over great geographic areas.

The lung functions in such a way that the needs of blood pumped into the alveolar capillary bed by the heart are met by adjustments of alveolar gas exchange, that is, the lung matches ventilation to perfusion. In a single breath, a certain gas-exchange volume—the "tidal volume"—is inhaled and diffuses peripherally toward the alveoli, and then in exhalation the carbon dioxide diffuses outward. At rest,

Table 6-5
Approximate partition of carbon dioxide in normal adult blood*

CO_2 fraction	Percent of total	Arterial (mEq/L)	Venous (mEq/L)	Difference (mEq/L)
$(H_2O + CO_2)$	9	1.2	1.4	0.2
$(Hgb + CO_2)$	27	1.1	1.7	0.6
$[HCO_3]$	64	19.7	21.1	1.4
$[TCO_2]$	100	22.0	24.2	2.2

*The clinical difference between the arterial and venous carbon dioxide partitions approximates the usual standard error of the CO_2 analysis, when clinical blood gas machines are in ordinary use. Clinically, venous blood can be used for any common type of CO_2 analysis except the measurement of alveolar ventilation.

normal tidal volumes are roughly about 10% of all the air in the lung. In normal individuals, within their individual alveoli, the alveolar gas tensions, P_{AO_2} and P_{ACO_2}, are practically equal to their arterial counterparts, Pa_{O_2} and Pa_{CO_2}. In the case of oxygen, this equality is accounted for by the peculiarities of the oxyhemoglobin dissociation curve. In the case of carbon dioxide, it is accounted for by the linear relationship of P_{CO_2} to T_{CO_2}. In any case, true gas exchange occurs only at the ABB in the alveolar sac. Thus for ventilation to be functional, the atmospheric tension of oxygen must always exceed the alveolar tension, or oxygen will not be driven inward. Once in the alveolus and regardless of the absolute P_{AO_2}, and as long as normal hemoglobin (in erythrocytes) can perfuse the alveolar capillary, the P_{AO_2} will continue to drive. Thus, the P_{AO_2} of a continuously perfused but nonventilated alveolus will drive until it eventually approaches zero.

Lung volume

All the tidal volume that is inhaled does not interact in gas exchange. Only about two thirds of tidal volume—roughly two thirds of 10% of total lung volume—actually reaches the alveoli. Like a train that is too long for the station, some of the tidal volume is still in the conducting tubes when inhalation stops and exhalation starts. Conversely, some of the exhaled gas is still in the conducting tubes when one inhales, and this old "stale" air is actually driven back into the alveoli ahead of the new fresh air. Thus the oxygen is diluted by old air on the way to the ABB. The useless one third of tidal volume is for the most part still in the conducting tubes, and so it is a measure of "dead" (that is, nonexchanging) lung space. When the lung "compensates" in systemic acidosis or alkalosis, it does so primarily by changes in alveolar ventilation clinically; this is usually observed as a change in tidal volume (Fig. 6-7; Table 6-6).

Alveolar ventilation

The concept of alveolar ventilation (which is measured as a carbon dioxide function) and dead space (which is often most particularly significant to oxygen transfer) can be confused by the best of physiologists. The fundamental question about a lung "space" is whether oxygen in that space can diffuse into the red cell of

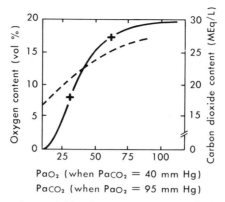

Fig. 6-4. Comparison of oxyhemoglobin dissociation curves of oxygen and carbon dioxide in vivo.

Over most of the physiologic range, every change in $PaCO_2$ produces a nearly equivalent change in CO_2 content. That is, the relationship between TCO_2 and $PaCO_2$ is approximately linear. The same approximation holds for $PvCO_2$ (for ranges higher than 20 mm Hg), $TvCO_2 \propto PvCO_2$.

However, oxygen dissociation is quite different. Within or above the usual exercise range ($PaO_2 = 60 \rightarrow 90$ mm Hg), small changes in PaO_2 produce **no** significant change (less than 10%) in arterial oxygen content, because in these ranges the avidity of hemoglobin for oxygen is quite high and largely independent of PaO_2. However, when the patient has developed severe clinical hypoxemia ($PaO_2 < 60$ mm Hg) this effect is lost and indeed reversed in large measure. Small decreases in PaO_2 in patients already hypoxemic lead to disproportionately large reductions in oxygen content.

But we may also state the converse—when arterial blood becomes desaturated, it takes only a small change in PaO_2 to unload a great deal of oxygen to the tissues. In fact, the usual venous PO_2 is about 35 mm Hg, and at this PvO_2 only a small driving pressure or oxygen gradient is needed to get the oxygen out of the blood. Over the usual range $PaO_2\ 95 \rightarrow 35$ mm Hg, roughly one third of all the hemoglobin is available to tissue. If the hypoxemic patient has to work harder to get oxygen into his blood, still he oxygenates his tissues with the same or greater ease than do normal persons, for his venous PO_2 is lower. When PvO_2 falls $50 \rightarrow 10$ mm Hg, two thirds of the hemoglobin is available to tissue. Thus, hemoglobin has really two dissociation curves—one for transport from the alveolus to the erythrocyte and one for delivery to the tissues which is remarkably responsive to PaO_2 in the systemic capillaries.

On the other hand, whatever the TCO_2, and whether in the lung or in the systemic capillary, the greater the PCO_2, the greater the TCO_2, and vice versa. Since the atmospheric PCO_2 is practically negligible compared to the $PaCO_2$ and PA_{CO_2}, there is a continual driving pressure to rid the body of CO_2. It can be seen, therefore, by simple inspection that hemoglobin is not quite central to the transport problem of CO_2. The reason that CO_2 exchange requires hemoglobin is that aqueous solutions of CO_2 produce hydronium ions which must be neutralized, and not that there is an intense regulatory effect of hemoglobin upon CO_2 transport itself, as is the case with oxygen.

A simple experiment to demonstrate the above is to hyperventilate for a short time, without exercise. The alveoli, and subsequently the blood, will wash out their CO_2 and continue to do so even though the $PvCO_2$ falls markedly. On the other hand, even the most violent exercise, such as ski racing, may be carried out at high altitude where the PA_{CO_2} is quite low, and yet the arterial O_2 content remains near normal. Thus, when the lung suffers from a disease in which some alveoli are poorly ventilated, overventilation of those alveoli which remain can reduce $PaCO_2$ to normal, but cannot elevate PaO_2 to normal. No physiologic process can increase hemoglobin oxygen affinity at usual PA_{O_2} tensions.

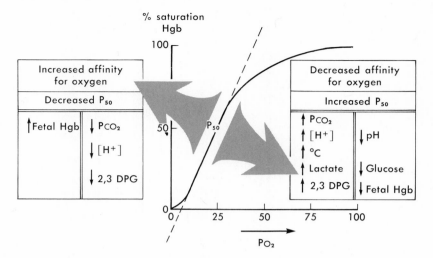

Fig. 6-5. The P_{50} concept. P_{50} is the Po_2 developed at 50% saturation of hemoglobin. It is a useful concept when applied to the linear portion of the hemoglobin dissociation curve in the region of Po_2 that characterizes the systemic capillary bed (and therefore the pulmonary arterial bed). It could be totally misleading if applied to the pulmonary capillary bed.

An increase in P_{50} means that an increased Po_2 is generated without a corresponding change in hemoglobin saturation. This in turn means that the oxygen affinity of the hemoglobin has been decreased, generally by intramolecular forces. Some situations that decrease affinity are shown on the graph. Decreased oxygen affinity is very useful to raise the driving pressure in the capillary plasma and extracellular fluid in order to drive oxygen from the red cell to the tissue cell. Increased affinity is of no ordinary physiologic importance in the range of Po_2 values where the P_{50} is meaningful, namely, the systemic capillary bed.

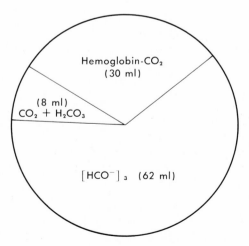

Fig. 6-6. The carbon dioxide "compartments" as volume. The total burden or arteriovenous difference of carbon dioxide is about 11 mEq CO_2 in the blood volume of the normal adult (2.2 mEq/L × 5 L). At any one instant, about 1 dl of blood normally perfuses the alveolar capillaries, at rest. Thus, about 0.2 mEq or 5 ml CO_2 is presented to the lung and ready for exchange at any one instant.

The clinical pathology of hydrogen ion homeostasis 153

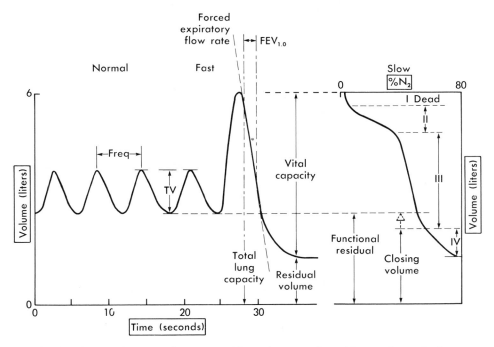

Fig. 6-7. Lung volumes and spaces and how they are plotted from spirometric data.

Table 6-6
The approximate rule of 1200

Tidal volume $\approx$ 600 ml
Residual volume $\approx$ 1200 ml
Functional residual capacity $\approx$ 2400 ml
Inspiratory capacity $\approx$ 3600 ml
Vital capacity $\approx$ 4800 ml
Total lung capacity $\approx$ 600 ml
RV/TLC $\approx$ 0.2
IC/VC $\approx$ 0.75
TV/TLC $\approx$ 0.1
TV/FRC $\approx$ 0.25

an alveolar capillary during the brief time that the red cell is moving by. If it can, that is "live" space; if not, it is dead space. Clearly, no oxygen transfer occurs in the trachea, bronchi, and smaller conducting tubes; none of these possesses air-blood barriers. The anatomy being self-evident, we call that anatomic dead space. Not so clearly, the following are also dead space: alveoli with relatively or absolutely decreased blood supply, alveoli with normal blood supply but an air-blood barrier that is too thick; aveolar sacs that are too large relative to the distance that oxygen molecules must diffuse; alveolar sacs that are open and perfused but whose tubular airway is blocked.

Thus, "live" space is the volume of gas that enters the alveolus and whose oxygen is actually exchanged, and dead space is everything that is left. What one usually measures as dead space in the routine pulmonary function laboratory is a hodge-podge convention—the volume of (inspired) gas which can enter (or exit) the total lung tree (thus excluding blocked airways) in a single breath but which is not exchanged. Or, simply, dead space is the volume of gas in a single inspiration that cannot oxygenate hemoglobin. Thus, the fraction of minute alveolar ventilation that we might identify as "minute dead space" varies from one physical position to another, from health to sickness, from fever to hypothermia, from exercise to sleep, all depending to some extent on alveolar perfusion, that is, whether the blood supply matches the ventilation. Indeed, even when alveolar ventilation is normal, if the diffusing capacity is impaired, or when there is pathologically non-uniform ventilation, lung function may still be disturbed and systemic blood unsaturated.

The important concept to grasp is that the term alveolar ventilation refers only to the volumetric exchange of atmospheric air across the ABB and *not* merely to an exchange of gas. For example, a patient with quite normal lungs intubated with a long endotracheal tube will gradually build up $P_{A_{CO_2}}$ because of the increased dead space. Although his lungs may be mechanically normal and perfusion and alveolar anatomy quite adequate, we must say of him—by definition—that he has poor alveolar ventilation. Indeed, $P_{A_{CO_2}}$ and, therefore, by inference, Pa_{CO_2} are direct measures of alveolar ventilation for many clinical purposes. For example, the deep Kussmaul breathing of diabetic acidosis is an attempt to increase alveolar ventilation through an increase in tidal volume. Obviously, true increases in alveolar ventilation also have an implicit time function.

The *minute alveolar ventilation,* which integrates for time, is a most important clinical measure of lung function. All other things being equal, rapid shallow breathing tends to decrease and deep slow regular breathing tends to increase minute ventilation. This is true whether the patient breathes for himself, or is put on a mechanical ventilator. It is sometimes useful to think of minute alveolar ventilation as a term that refers to the venting of CO_2. To measure it, we assume that all expired CO_2 comes only from perfused alveoli (and ignore the 0.03% CO_2 of inspired air); and we assume further that $P_{A_{CO_2}} = Pa_{CO_2}$, which is true when lung circulation is normal and pulmonary venous CO_2 cannot be shunted into pulmonary arteries.

THE ALVEOLAR GAS EQUATION

At the end of expiration, let:

$\dot{V}$ = volume moved/unit time

$\dot{V}_{CO_2}$ = CO_2 volume in alveolus

$\dot{V}_A$ = total alveolar volume

A = alveolar

Then:

$$\dot{V}_{CO_2} \propto \dot{V}_{alveolar} \times \frac{\%CO_{2A}}{100}$$

Rearranging terms:

$$\dot{V}_A \propto \frac{\dot{V}_{CO_2}}{\%CO_{2A}}$$

But:

$$\%CO_{2A} \propto P_{A_{CO_2}}$$

And:

$$P_{A_{CO_2}} \propto P_{a_{CO_2}}$$

So:

$$\dot{V}_A \propto \frac{\dot{V}_{CO_2}}{P_{a_{CO_2}}}$$

Thus, alveolar *hyper*ventilation is a conceptually simple diagnosis. No amount of increased air exchange can raise the normal Pa_{O_2}, which is limited by the molecular mechanics of hemoglobin. However, the body has no defense against excess venting of CO_2, whose transport enjoys no such molecular complexity. All other things being normal, increases in ventilation must always blow off carbon dioxide and reduce the Pa_{CO_2}. Rapid removal of carbon dioxide diminishes hydrogen ion production in red cells and ultimately results in alkalemia and disturbances of the chloride shift. This is not a "respiratory" but rather a ventilatory alkalosis, but clinical usage has established respiration as the term applied to breathing rather than to mitochondrial metabolism.

In clinical parlance, alveolar hypoventilation almost always refers to a time function such as minute alveolar ventilation. Hypoventilation, by definition, must lead to hypoxemia—arterial unsaturation—for alveolar ventilation refers to the volume of exchangeable fresh air that enters the alveoli. Hypoventilation is an exchange that is too small because not enough oxygen molecules are present. So, if the number of molecules or the oxygen tension of the inspired gas is increased, the hypoventilation—and the arterial oxygen tension—will improve. The most straightforward way to diagnose alveolar hypoventilation is to put the patient on supplemental oxygen. If the Pa_{O_2} improves, alveolar hypoventilation was present. Of course, supplemental oxygen does not directly help the patient to blow off carbon dioxide, which also builds up in alveolar hypoventilation. So the second part of the diagnosis is that Pa_{O_2} improves with no corresponding change in Pa_{CO_2} when extra Fi_{O_2} is supplied.

At this point, however, it is well to remember that improved tissue oxygenation will diminish lactic acidosis and tend to improve the metabolic component of acidosis in the patient who hypoventilates. Conceivably, therefore, the Pa_{CO_2} might even rise!

Thus, the diagnosis of hypoventilation with supplemental oxygen should be made in a few breaths. Additionally, supplemental concentrations of oxygen, particularly if over 60%, have complex acute and chronic effects upon the lung. Acutely, there is dilation of pulmonary arterioles. Later, one may actually see induced hypoventilation and pulmonary hypoperfusion at the alveolar ABB level.

Pulmonary hemodynamics

One of the most remarkable features of the lung is its reserve vascular capacity—the distensible, compliant, voluminous microcirculatory bed. For example, the pressure needed to drive blood through the pulmonary circulation is quite low—roughly 10% of that needed for the systemic blood. What happens when hard-working tissues call for more pulmonary gas exchange? Most remarkably, the lung increases its flow not through an increase in the driving pressure of the right heart but primarily through a decrease in the already low pulmonary resistance. Partly this occurs because the individual vessels are so distensible and partly for the overlapping reason that at rest a significant proportion of the microcirculatory bed is quite poorly perfused. Most important, this resting bed is parallel to the well-perfused part.

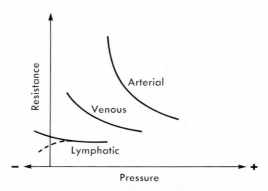

Fig. 6-8. Idealized diagram of lung vascular flow.

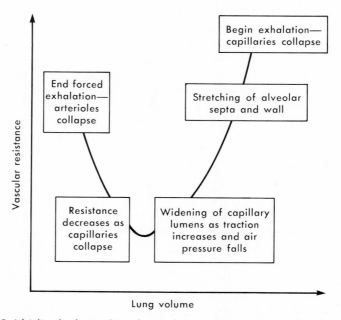

Fig. 6-9. Idealized relationship of some lung vascular and ventilatory factors.

Dilation (or "recruitment") of the parallel resting bed lowers resistance of the whole system, even though all capillaries of the pulmonary bed may happen to have similar individual resistances. Not only is this remarkable compliance of great use during slow changes of pulmonary adaptation to different work loads; it is also of value in damping the relatively pulsatile stroke-by-stroke right ventricular output. Along with the musculoelastic arteries and elastic arterioles, this system damps the 25/8 mm Hg systolic/diastolic pressure waves (and their reflections) to a mean 15 mm Hg. So compliant is this bed that it will normally maintain a fairly constant pressure until flow has been augmented about 300%. It is only at this large figure that pressure rises linearly with flow (Figs. 6-8 and 6-9).

Typically, in precapillary pulmonary arterioles, mean pressure falls to about 12 mm, and in postcapillary venules to 8 mm Hg. Thus the lung capillary bed is perfused at rest by a pressure of 4 mm Hg (72 mm H_2O) or a column of blood the length of an ordinary cigarette! This remarkably low pressure suffices to put 6 L/min of blood through the lung at rest. How do we know the flow and resistance of the pulmonary bed? Flow may be calculated from ordinary gas dynamics.

For example, the common anesthetic gas, nitrous oxide, is extremely soluble in blood, so much so that if a person breathes this gas in a closed box or plethysmograph the pressure around the body will measurably drop during a single breath. Indeed, the pressure of the plethysmograph will drop in a manner that accurately reflects quite small pulsatile changes in the pulmonary microcirculatory bed, normal or abnormal. From that pressure drop and knowledge of simple gas dynamics, and with the assumption that only the limited supply of blood prevents still further gas uptake, one may calculate flow. In a somewhat similar manner, oxygen can be used as the test gas too, if one knows the precapillary and postcapillary gas tensions (Pv_{O_2} and Pa_{O_2}). The measurement of flow from this oxygen dilution technique is derived from the generalized Fick principle of flow measurements from dilution data.

How do the vessels stay open against air pressure? In the systemic circuit, the microcirculatory bed is surrounded by tissue which presses upon vessel walls and tends to close the vessels. In the lung, the mass of surrounding tissue bed is generally negligible, it is infraalveolar air pressure that pushes upon the capillary walls. If air pressure substantially exceeds capillary pressure, the capillaries tend to collapse, whereas if capillary pressure substantially exceeds air pressure, intravascular fluid will be pushed into the alveolar space to form pulmonary edema (Fig. 6-10).

In addition, the alveoli themselves obviously possess an interface between the thin layer of fluid that lines them and the air that fills them. An air-water interface has a very high surface tension, as any child who has floated a needle on water knows. Imagine a globule of water on a surface; the surface forces are directed inward, which shrinks and rounds up the globule. In an alveolus, the globule is inside out but still the forces are directed inward, which tends to shrink the alveolus and "suck" the plasma out of nearby capillaries. Thus, alveolar pressure and surface tension, and of course the oncotic pressure of the extravascular extracellular fluid, are three important factors modulating the patency of pulmonary capillaries and opposing intravascular pressure.

There is a fourth, and quite important, factor, a hydraulic one which takes note of the fact that the lung is a large three-dimensional organ. A column of blood the height of the lung, with zero hydrostatic pressure at the tip, exerts a pressure of 23 mm Hg at the bottom—nearly twice mean pulmonary artery pressure! In actuality, the driving pressure of right ventricular output is added to the hydrostatic

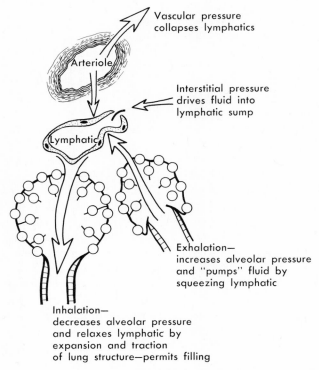

Fig. 6-10. The lympathic as a lung sump.

pressure of the blood column. Clearly, at the top of the column—the lung apex—hydrostatic pressure is negligible and the intravascular hydraulic pressure originates from cardiac action. It turns out to be just sufficient to oppose intraalveolar air pressure and keep the capillaries open in health.

In the midportion of the erect normal person's lung, hydraulic and cardiac forces combine. However, the small pressure drop of about 4 mm Hg along the capillary beds is sufficient to lower the pressure in the veins to less than that in the alveolar spaces. Thus pressure of ventilated gases tends to collapse (drain) the venous bed, and cardiac-hydraulic pressure tends to dilate (fill) it. Evidence suggests that the venous bed, particularly its venular portion, is somewhat independent of the true capillary bed. Indeed, the thin musculoelastic tissue of the larger veins throughout the entire lung compares favorably in mass with the similar tissue on the pulmonary arterial side.

At the base of the lungs, the considerable sum of cardiac-hydraulic forces is to some extent transmitted through the capillary bed and onto the venous side, so that intraalveolar air pressure becomes a negligible factor. The pulmonary veins in this area are difficult to collapse. Clearly, from the preceding outline, alveolar ventilation of the upper lobe bronchial segments may become precarious in disease. Indeed, poor blood supply may be implicated in the classic location of reinfection tuberculosis there, in addition to the crooked path that air must take to reach that area. On the other hand, the lower lobe is the usual site for early pulmonary edema and for extreme venous hypertrophy in cases of a too-low left ventricular minute volume (as in mitral stenosis). Naturally, the effects of hydrostatic pressure are different

Table 6-7
Comparison of extremes of normal regional ventilation*

Property	Apex	Base
V̇:Q̇	+	−
pH	+	−
P_AO_2	+	−
P_ACO_2	−	+
Air exchange	=	+
Blood perfusion	−	++

*In the erect position, normally the ratio V̇:Q̇ goes down from apex to base, but ventilation tends to be more constant than perfusion.

when the patient is prone or supine than when he is standing. Frequent changes in position are important to normal lung action. Continuous lying in bed on one's back, particularly with weak chest muscles and poor excursions of the thorax, may even be associated with pulmonary edema. This occurs in the dependent parts of the lung even though cardiac action be normal (Table 6-7).

Lobular regulation

The mechanism by which the lobules regulate their individual perfusion and by which they normally integrate one with another is not understood. It is known, for example, that lobular arteriolar territories may overlap, but the extent and the means for modulation of this are not known. While the arteriolar muscle mass is low, it is not negligible, particularly when one considers the weak precapillary driving forces. It is demonstrable that alveolar hypoxia induces pulmonary arteriolospasm and that the effect is independent of central neural mechanisms—it occurs in excised lung. It does not occur because of hypoxemia—low Pa_{O_2}—but is induced rather precipitously by fairly marked alveolar hypoxia—$P_{A_{O_2}} < 70$ mm Hg. The most likely mechanism appears to be mediation of the neurosecretory cells which are found in the bronchiolar mucosa, with their tips facing the lumen and their bases near the vascular bed. These cells are known to secrete catecholamines when hypoxic, but not hypoxemic, conditions are employed. The arteriolospasm may progress to a remarkable collapse of the capillary bed, but bronchial circulation and a few collaterals prevent cellular ischemia. It appears possible that irritation of this catecholamine-secreting system has something to do with inappropriate "regional" (lobular?) ventilation:perfusion ratios in chronic bronchitis. It is also known that in massive arterial blockade such as by embolus, alveolar atelectasis or collapse of the hypoxemic lobules occurs more or less in accord with the depression of perfusion, and there is also considerable regional arteriolospasm. Why this coupling does not obtain in chronic bronchitis is an enigma.

Ventilation:perfusion ratio

The most misunderstood of all pathophysiologic disturbances of lung function is probably the most common one that is of clinical importance: ventilation:perfu-

Table 6-8
Clinical conditions of Pa_{O_2} and Pa_{CO_2}*

Pa_{CO_2}	↓Pa_{O_2} hypoxemia	Normal Pa_{O_2}	↑Pa_{O_2} oxygen intoxication
↑Pa_{CO_2} hypoventilation	Alveolar hypoventilation Respiratory acidosis Metabolic alkalosis	Increased oxygen tension ↑Pa_{O_2} ↑ Fi_{O_2} "Mixed" metabolic respiratory acidosis	Hypothermia Positive end-expiratory pressure
Normal Pa_{CO_2}	$\dot{V}:\dot{Q}$ inequity Bronchitis	Acute metabolic acidosis or alkalosis	Oxygen therapy
↓Pa_{CO_2} hyperventilation	Heart failure "Interstitial" pneumonitis $\dot{V}:\dot{Q}$ inequity "Stiff lung" Fever Pulmonary edema Shock	Normal Nonpulmonary hypoxia "Primary" hyperventilation	Oxygen therapy Metabolic acidosis "Pure" or "mixed" respiratory alkalosis

*After Weinberg.

sion inequality (VQI). While many lesions may account for this problem, the most common has to do with failure of individual lobules to match ventilation and perfusion. This is not self-evident. Another possible cause might be, for example, the failure of oxygen to diffuse from the center of the individual lobule to its air-blood barriers and then into the erythrocyte within the time that particular erythrocyte spends at the ABB. However, for all ordinary infralobular distances (that is, less than 2 mm), the diffusion time in alveolar gas is less than about 0.02 second. This is less than a third of the entire time that the red cell is available in transit (0.75 second). Even under conditions of markedly augmented cardiac output, such as severe exercise, the transit time does not diminish much beyond 0.25 second, and so the local morphology remains quite favorable. Of course, in bullous emphysema and marked centrolobular dilation, the conditions are different. But these are not as common as plain VQI (Table 6-8).

It is instructive to consider a technique used to measure V:Q ratios, for it involves a fundamental concept of gas exchange. The method is to choose a gas to which the ABB is impermeable, fill the lung with it, and then measure its gas tension in exhaled air. If the respiratory gases themselves, O_2 and CO_2, dilute the test gas evenly in all alveoli—that is, if input to and the uptake of oxygen from the alveoli is everywhere equal—then within each alveolus the dilution or partial pressure of the test gas will always be equal. If the partial pressures of the O_2 and CO_2 within the alveolus do not change during a breath, then the dilution of the test gas in the alveolus will not change with time, as long as it is inhaled.

Actually, the alveolar concentrations of respiratory gases do not change significantly during ventilation, for steady state dynamics are the basis of continuous perfusion of oxygen into red cells and continuous removal of carbon dioxide from blood. But what would happen if the lung were filled with the diluted test gas and then the subject no longer breathed that gas but switched to air? The test gas would be evenly diluted by time, as it were, and would appear in even increments with each exhaled breath or portion of a breath. However, if the gas were distributed unevenly, these dynamics could not occur. If the gas had been unevenly distributed during inspiration and unevenly blown off during expiration, some lobules would still dump high concentrations of gas while the effluent from others would be blown off normally. Or simply put, at the end of a given time, the partial tension of test gas in the expired air would be high.

Why does the test work? Uneven distribution is usually unevenly blown off. Also, the lung cannot effectively compensate for poor oxygen uptake in one lobule by supranormal oxygen uptake in another because the oxyhemoglobin curve does not permit that. Another possible compensatory mechanism, increased perfusion of normal remaining lobules, which would require increased cardiac output, often is not significant in this VQI test. At rest, of course, the principal volume into which the test gas is diluted is the functional residual capacity; most often, however, variances in this lung volume are not a problem. Finally, the common gas nitrogen is eminently suitable for many of the clinical analyses of VQI for it is universal and easily measured. Of course, such measurements are sensible only if one knows the patient's minute volume and understands how normal anatomic dead space must be differentiated from underventilated alveoli.

Clearly, consideration of total ventilation and total perfusion is less important than whether the ventilation:perfusion ratio is uniform throughout the lung. Though the totals be normal, still VQI may be abnormal. But as we have seen, any significant degree of VQI will usually result in unsaturation of pulmonary venous blood, that is, a stream of unsaturated blood from defective alveoli mixing with more normal blood from normal alveoli. Thus, VQI is usually a type of alveolar hypoventilation characterized by its regional distribution and by a lowered Pao_2. It seems most likely—although quite difficult to prove—that the major types of regional VQI have a lobular pathogenesis.

Conversely, it does not appear from the pathologist's point of view that VQI most usually refers to lesions of a single alveolar unit within a lobule. Instead, most evidence suggests that the basis of VQI has to do with the distributing tubular airways themselves, which are ipso facto assigned to lobular units or collections of lobular units.

This observation leads to the idea that chronic bronchiolar-bronchial (BB) diseases must be quite important in the pathogenesis of VQI. Such diseases might be expected to show, in the first approximation, a significant degree of nonuni-

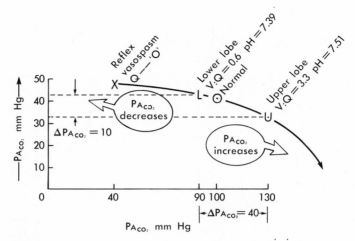

Fig. 6-11. Regulation of normal regional alveolar ventilation. $\dot{V}:\dot{Q}$ ratios related to $P_{A_{O_2}}$ and $P_{A_{CO_2}}$.

formity of the BB tree. However, pathologists rarely evaluate BB systems in terms of intricate comparisons between and among supposedly similar airways. This practice may indeeed account for the pathologic paradigm that one cannot predict physiologic function from histologic inspection, a dogma that seems more of a monument to our currently invincible ignorance than a statement of eternal verity.

The relationships of some dynamic and static properties of the lung are shown in Fig. 6-7. Functional residual capacity (FRC)—but not closing volume—is determined from a quick maximal forced expiratory flow. Closing volume—but not functional residual capacity—is determined from the slow washout of an inert gas. Functional residual capacity is the least volume that the collective alveoli exhibit, their smallest state of inflation. It is from this position that the maximal forces of lung expansion are generated. The alveoli are prevented from further shrinkage, and the force of recoil is generated largely by the surfactant lining. Thus compliance and FRC are linked. (A good example of this is hyaline membrane disease.)

Consider a decrease in FRC. Atelectasis is the term applied to pathologic contracture of the real FRC. Since an atelectatic lobule holds much less gas than the ABB surface area could exchange, atelectasis is an example of alveolar hypoventilation. Thus venous admixture increases and $P_{a_{O_2}}$ decreases in atelectasis. However, most often reflex activity—probably entirely within the lung—markedly reduces the blood flow to the atelectatic lobule and at least partially "restores" the $\dot{V}:\dot{Q}$ ratio locally, that is, matches poor ventilation with poor perfusion. This tends to reduce the venous admixture. One may think this reflex cannot be invoked for an entire lung. But actually, it probably does occur in the whole lung in the deep-diving water mammals, such as whales, and quite possibly in human "shock" lung and allied conditions. Before birth, it is the normal mechanism for suppressing pulmonary flow.

What about increases in FRC? The most common cause of marked increase occurs in the chronic case of emphysema. The residual volume—the volume that cannot be exchanged no matter how much work is done to move it—grows. Since the expansion of the chest is ultimately limited, a large residual volume must mean that some of the pulmonary parenchyma is destroyed. The destruction means attenuation of the ABB total surface as capillary bed and alveolar septa disappear. Generalized loss of the capillary bed results in diffuse alveolar hypoventilation based on generalized insufficient diffusion of oxygen—the P_{AO_2} can still drive, however, and indeed the patient increases his minute volume to make sure P_{AO_2} remains high. These patients are often referred to as "pink puffers." Another factor in severe cases is that the mean free path of intraalveolar oxygen increases as the alveoli simplify into cystlike structures, or even bullae. It begins to take longer for oxygen to reach the red cell than for the red cell to pass through the lung. Clearly, in these cases there must be by definition measurable increases in dead space. Thus the P_{ACO_2} and $Paco_2$ increase. Not so clearly, lung compliance typically increases, that is, the lung is easier to inflate because it has lost some of its structure and because its individual "balloons" (alveoli) are larger.

If, however, the emphysema has a markedly patchy or lobular distribution, the situation may be different. Indeed, in such cases, it is usually one part of the individual lobule, the central part, which becomes weakened and overexpanded. Centrolobular emphysema is thus patchy over the whole lung and within the individual lobule. The weak parts are overly compliant yet are strategically located astride the gas path for the entire lobule. Hence, during expiration, as lung volume contracts and all the alveoli build up the pressure needed to cause the expired breath to flow outward, the tube collapses. The alveoli nestled about the weak centrolobular paths cause the gas path to collapse too early in expiration. Not only does this cause the closing volume of the lobule to increase, but it interferes with gas exchange. As the P_{AO_2} drops, lung reflexes switch off the microcirculatory bed, and perfusion drops even more as the unoxygenated blood is reflexly shunted into the pool of venous admixture. The Pao_2, but not the $Paco_2$, drops rather spectacularly. With time, the shunt vessels which are now exposed to the right ventricular pressure wave become more muscular and elastic, for they are without the assistance of the low compliance microcirculatory bed to damp the wave. The right heart strains harder against the arteriolar muscle, and the arteriolar muscle hypertrophies further. New shunts develop. The subsequent right ventricular hypertrophy is handicapped by hypoxemia in coronary vessels. All of this is the classic picture of the "blue bloater."

BLOOD GAS AND pH MEASUREMENTS (CO_2, O_2, pH—COP)

Blood gas and pH primary measurements (COP) are at present made almost entirely by electrodes. In principle, a wide variety of other devices might be applied, but the clinical technology of COP has been quite conservative. One of the most promising alternatives is the small mass spectrometer, typically with a quadrapole rather than a magnetic analyzer. Limited purpose mass spectrometers are already in use for the analysis of the breathed component of blood gases, but cost, commercial competition of the well-established electrode techniques, and the reluctance of clinical technology as a discipline have conspired to keep these machines out of the clinical chemistry COP business. Unlike electrodes, mass analyzers have the potential capability of both long-term online use and switching applications so that several different patients might be monitored simultaneously from one instrument. Another possibility is that since COP and the analysis of breathed gases obviously must be in equilibrium, in principle one could determine the blood gases noninvasively from calculation of the difference between inhaled and exhaled components of blood gases—the uptake of gas by blood. Since there are already noninvasive techniques for the online estimation of hemoglobin saturation, the new possibilities are the more attractive. Unhappily, state-of-the-art mass spectrometry does not show promise of any striking cost reductions by manufacturers, whereas the justification of their high cost could only be their application to online COP measurements. Since there are as yet no continuously serviceable intravascular endpieces, this noninvasive methodology has not received much attention.

Analyses

Blood gas measurements by glass (pH, CO_2) and Clark polarographic (Po_2) electrodes are still the backbone of all clinical COP. Although the use of indwelling umbilical artery catheters in newborns and arterial loop prostheses in children and adults has led to some experimentation with continuous online COP electrode techniques, the clinical problems are larger than the instrumental ones and continuous recording COP instruments are not in routine hospital use. Some progress has indeed been made in the techniques of continuous recording with conventional electrodes arranged so that their tips are specially protected and mechanically cleansed (flushed) in situ. However, these techniques do not show promise of widespread adoption now. Nevertheless, there can be absolutely no doubt about the bright future for online COP analyses operating through laboratory computers and, ultimately perhaps, through feedback loops that automatically adjust the patient's assisted ventilation, oxygen supply, and parenteral infusions.

At present, most COP measurements are singleton analyses from punctures of, or catheters in, intermediate sized muscular arteries such as the radial or umbilical arteries. Small needle punctures (25 to 27 gauge) are not difficult in the radial arteries of adults. In the nursery, it is almost exclusively the need for COP data that justifies the indwelling arterial catheter. In this case, the catheter must be in the aorta. This is unfortunate, since there is a small but definite morbidity—sometimes serious—which accompanies such practices, most usually related to embolism of vital aortic branches such as those to the splanchnic, renal, spinal, or lower limb tissues. The use of venous samples when Po_2 data are not needed is acceptable and even desirable. There is excellent correlation with arterial data. A beginning has been made for skin surface COP analyses.

It is commonly assumed that venous blood is useless for CO_2 and pH analyses simply because it is venous. In clinical actuality, arteriovenous differences are usually fixed. The technique of phlebotomy by which the sample is drawn has little to do with the problem. Venous obstruction, via tourniquet, to the point of hypoxic pain shifts pH by 0.04 unit or less, and one assumes that competent personnel do not perform such painful phlebotomies. On the other hand, muscular activity such as fist-pumping can markedly increase—even quadruple—local lactate accumulation and thus metabolically shift pH. For all these reasons, pH studies by phlebotomy should include only the initial portion of blood drawn, even though this necessitates changes of syringe and is inconvenient for the phlebotomist.

A widespread arbitrary attitude has arisen with respect to the supposed deficiencies of capillary Po_2 values which some authorities believe should be redressed. It is often said that capillary Po_2 is of no value in the typical cyanotic patient (Pao_2 less than 70 mm Hg). This idea is not supported by any reliable data. What *is* true, however, is that at Pao_2 values *over* 80 mm Hg, capillary tensions depart significantly from arterial oxygen tensions on comparative linear regression plots. However, there is little nominal clinical interest in the Po_2 over 80 mm Hg, except in the special instances of retrolental fibroplasia and cardiac catheterization or other diagnostic procedures not related to intensive care. To protect against retrolental fibroplasia, a hyperoxic toxicity of the eyes of newborns, it is necessary to know when the Pao_2 is over 90 mm Hg or the capillary Po_2 over 70 mm Hg. While the generally cited "magic" number for eye toxicity is 140 mm Hg, there is absolutely no physiologic need for neonatal Pao_2 to exceed 70 to 80 mm Hg. That clinical limit may be closely guarded if patients whose capillary Po_2 exceeds 70 mm Hg are given supplemental oxygen very conservatively and with great caution, if at all. There are no convincing clinical data to show what other value, if any, there is in accurate (as opposed to precise) knowledge of the Pao_2 of patients in intensive care when the values lie in the near normal ranges, 80 to 100 mm Hg. On the other hand, logical arguments can be made that since capillary Po_2 represents the gas picture at the local tissue level more accurately than Pao_2, it is a more relevant measurement.

It is thus most unfortunate that several new models of COP machines require larger blood volumes than are conveniently gathered by capillary technique. Some of these machines are doubly plagued, since their working parts are opaque and the operator has a poor view of the loading chamber and the electrode. There is no inherent reason why manufacturers cannot standardize on smaller volumes. Blood gases alone—not to mention other clinical chemical procedures—may consume more than 1% of a small premature infant's blood volume every day during a common intensive care schedule of four daily COP analyses when one of the new blood gas machines is used.

Quality control

Quality control is an unsolved problem in blood gas analysis. Wild errors are difficult to eliminate, even in the most careful laboratories. In one well-known study, when tonometrically equilibrated blood macrosamples were carefully examined in a regular hospital setting at four different accurately established Pco_2 levels by two different operators, pooled coefficients of variance ranged from approximately 7% to 10%. In other words, 1 out of 20 analyses departed from the true value by more than 20% and another 6 by nearly 10%. But these are pooled data; in one series, a coefficient of variance of about 12.7% was recorded, equivalent to a ±25% for a third of the tests done by an experienced reliable technologist on a familiar machine! The largest error was 16 mm Hg Pco_2. The differences in pH were accordingly also large when the two laboratory sites were compared to each other with respect to the "true" value, for differences from "true" value approximated as much as 0.053 pH units, and interlaboratory differences were even higher. These difficulties do not begin to account for the even greater difficulty of knowing what the "true" value is, since each individual blood or pool has its own Henderson-Hasselbalch pK, as well as other individual variations in hematocrit, 2,3-diphosphoglycerate, lactic acid, and so on. Many of these difficulties are not removed by tonometry, and some are even increased. Commercial pH/CO_2 standards, such as Versatol, are valuable and should always be used in all available ranges. However, they do not ensure that quality control is adequate at present. (One very useful clinical check, often ignored, is the estimation of lactic acid. At Pao_2 values approaching 31 mm Hg, all patients probably have lactic acidemia.)

In general, the study also showed that aqueous buffers are much less useful than generally supposed in the standardization of blood gas machines. While considerable recent progress has been made in the field of noninvasive continuous oximetry through the use of computer-program corrected, multispectral, on line, temperature-controlled instruments, it appears that, for all their vagaries, blood gas machines will continue to be the methodologic means of choice for clinical COP studies. A more widespread knowledge of their inherently high variances and of the intrinsic problem of standardization in a technique where there is no recognized internal standard product available is needed on the part of clinical laboratory personnel.

In addition to lactate analysis, four indirect quality control checks are of value in coping with COP quality control. The first check is of the carbon dioxide, typically in a Natelson Microgasometer. In this instrument, the equilibrium $H_2CO_3 \rightleftharpoons H^+ + HCO_3^-$ is driven toward the left by means of an excess of weak nonvolatile acid (lactic). The equilibrium $H_2O + CO_2 \leftrightarrow H_2CO_3$ is in turn driven to the left by sharp reduction of the ambient pressure—a vacuum is created. The measured total carbon dioxide content, T_{CO_2}, is then estimated by manometry. Alternatively, and more conveniently in automated clinical chemistry, the released CO_2 can be reconverted to acid and estimated by analysis of the spectrophotometric change in a pH-sensitive dye. While the clinical T_{CO_2} estimation is universally carried out on separated plasma, and there are significant differences in the HCO_3^- buffer behavior of plasma compared with whole blood, the measurement is nonetheless a valuable check. In general, when the difference between the derived T_{CO_2} blood value (from a nomogram plot based on the glass electrode) and the measured T_{CO_2} (from plasma) exceeds 5 mm Hg (an arbitrary limit in our laboratory), it seems advisable to repeat the blood gas estimation.

A second type of quality control may be achieved through the so-called delta check. In this method, estimates of the patient's last preceding COP values are compared with the current ones. In a general way, however, COP delta checks are of quite limited value among patients during intensive care, just the population among whom the quality control need is the greatest. The reason is that modern parenteral and ventilation therapy and remarkable physiologic function of the un-aided lung conspire to permit such rapid clinical shifts of COP parameters that the delta check loses its value as a statistical tool. One would be forced to check the majority of analyses in many dynamic situations where the critically ill are involved. Of course, the laboratory quality of blood measurements is such that replicate analyses are desirable in any case, whenever possible. However, in those events, the whole philosophy of the delta check is lost.

A third method of quality control, for pH and P_{CO_2}, is the estimation of total base deficit from the combined molarities of sodium plus potassium, plus an arbitrary constant of 7 mEq/L to account for the sum of all other cations. When the total $Na^+ + K^+ + 7$) departs from the calculated base deficit derived from the nomogram by a factor greater greater than 10 mEq/L, we arbitrarily repeat the blood gases if possible.

A fourth type of quality control, for P_{O_2}, is the use of oximetry. Here, however, corrections are complex because of the sigmoid oxyhemoglobin dissociation curve. The availability of multifactorial ear oximeters and microprocessor technology may change this difficulty.

All these methods of quality control assume that the usual standardization of the blood gas machines has been carried out by tonometry and by the use of aqueous standards. However, as previously discussed, these methods do not usually eliminate many of the sources of variance that make blood gas analysis so capricious a laboratory methodology. Nor do they account for what are undoubtedly wide physiologic variances in the "normal" assumed constants that are implicit in basic COP equations. It must be noted that all blood buffer measurements obviously depend on the amount of red cell hemoglobin present and functional. There is no excuse for a manufacturer to retail a blood gas machine which does not estimate hemoglobin directly in the machine, thus producing results not directly corrected for this variable.

All parts of a blood gas machine, in addition to the COP itself, are temperature-sensitive. These variations bear no simple relationship to one another. In general, electronic equipment should not be operated in high humidities, particularly when the ambient temperature is also high. Furthermore, it is in the nature of temperature regulation equipment that as room temperature approaches incubator bath temperature, the latter becomes more difficult to regulate. Usually, the only temperatures known are those of the tonometer, the patient, and the electrode chamber. In a general way, lowering the temperature of blood lowers the P_{CO_2}, P_{O_2}, and hydrogen ion concentration, while raising it has the opposite effect. In principle, corrections should be made; in practice, the machine variables themselves are not linear with temperature change and it is inexpedient to change the electrode water bath temperature to match the patient. A much more important value is the hemoglobin in red cells; this is indeed the main blood buffer, and all nomograms either account for its possible range or assume a given fixed value. Of course, even these notations do not include variances in the molecular type of hemoglobin present, the diphosphoglycerate content of the red cells, or a great many other important variables. Some blood gas machines now do estimate the hemoglobin and internally correct for the estimated value. It seems quite likely that, with the advent of the single-chip microprocessor computer module, more of the variables involved in COP methodology will be preprogrammed. Lacking that, however, it is still possible to draw up programs for desk-top microcomputers that will handle a good deal more known and readily available information than even the most complex nomogram can handle.

As noted, increases in temperature of blood increases the pH—decreases the H^+ ion concentration—at a rate of about 0.015 pH units per degree C. However, blood cooled in vitro and recycled to body temperature does not return to its former pH unless it is completely mixed and kept in a sealed container. The effect of sedimentation, or poor mixing, is explained by the fact that the temperature effect on the buffer action of whole blood differs from that of plasma. On the other hand, use of an unsealed container promotes loss of CO_2 to the air, where the ambient CO_2 tension is invariably lower than that of blood. When CO_2 is lost in this manner, the Henderson-Hasselbalch equilibrium shifts in the direction of decreased H^+—increased pH—and the blood appears to be more alkaline than it was in vivo, by an unpredictable amount usually associated with a loss of the magnitude of 5 mEq/L CO_2. Unfortunately for the laboratory, the equilibrium of the loss itself is an exponential one, that is, the majority of the CO_2 lost upon exposure to air disappears in the first few minutes. Temperature corrections involve both linear effects for $pK'_{H_2CO_3}$ and exponential effects for blood gas concentrations. In this light, laboratory corrections tend to be empirical. The only excuse for cooling blood is to prevent further metabolism and hypoxemic metabolic acidosis of the sample. It is far better practice to have the blood gas machine reasonably and rapidly accessible to the bedside. Indeed, we would not find any administrative excuse for cooling COP blood samples a valid one.

Many of the laboratory mistakes that cause preventable variance in results are simple to correct. Incubation baths and their thermostats should function flawlessly. Electrodes must be intact; those that drift are useless. All grounded components should go to a *common* ground. Electrodes should never be allowed to dry out. Protein films are particularly heinous laboratory villains. The electrode tips should be checked with a magnifying glass under oblique light. The reference electrode path must be clean, liquid- and bubble-free, and show obvious KCl crystals. Only temperature-compensated buffers traceable to National Bureau of Standards and similarly certified gas mixtures should be used.

A pH electrode contains a fixed reservoir of relatively great hydrogen ion concentration behind a membrane which is selectively but not exclusively permeable to H^+. When it is dipped into blood, a transmembrane potential difference exists between the pressure (voltage) that drives H^+ through the membrane and into the blood and that which attempts to push H^+ out of the blood and through the membrane. The electrons, of course, flow in the opposite direction to their respective H^+. The potential difference can be measured because it does force a flow of electrons. However, there are many almost insurmountable problems with such a simple circuit. Most of these are diminished if the voltage generated is compared to the voltage generated by a reference electrode immersed in the same blood. This could be done by comparing the two voltages in two separate circuits, which is indeed done in certain exquisite researches. Instead, it is quite practical to connect the two electrodes together in one circuit and measure the sum (or difference) of their potentials at a single point. However, the very minute current that flows in such a circuit must be measured with a precision of a few nanoamps (0.001 pH unit = 50 nanovolts at a membrane impedance [resistance] of 50 femtohms—$50 \times 10^{12} \, \Omega$).

The meter that actually measures this circuit therefore must not draw current itself. Modern solid state circuits, based in part on operational amplifiers, are suitable for the need. It is common, but not inherently necessary, that the direct current thus detected is "chopped" or converted to an alternating current before measurement, because this helps to reduce thermal electronic drift in the amplifier. However, modern complex but cheap integrated circuits largely eliminate the need for such conversion, which was based more on the economics of amplifier manufacturer than on inherent electric problems. Actually, it can be appreciated that every junction between dissimilar materials in the circuit, such as between electrode plug and the receptacle in the amplifier, generates a potential. The usual failure, however, is in the KCl bridge that connects the reference and test electrodes. Such a system should be reproducible to +0.002 pH unit and accurate to 0.01 pH unit according to standard aqueous buffers. It is, of course, very sensitive to temperature fluctuations that change the electronic forces at the numerous junctions in unpredictable ways. While temperature should be controllable to 0.2° C, it should never vary during a determination by more than a small fraction of that. Adequate shielding is essential; often, the electrostatic potential on a dry synthetic fabric garment worn by the operator will cause the amplifier to develop oscillating electronic poltergeists.

Electrodes

A Pco_2 electrode is one in which there is again an impermeable membrane, but this time impermeable to ions and selectively permeable to CO_2. In essence, the CO_2 is passed into a standardized bicarbonate buffer system. The pH of the interior bicarbonate system then generates an electromotive force in opposition to the buffer system of the blood. Thus, the Henderson-Hasselbalch equation is used to measure itself, as it were! The measurement is carried out very much like the pH measurement previously discussed. Since the diffusion characteristics of the CO_2 are membrane-limited, choice and integrity of this plastic component are among the most important determinants of the utility of the method.

Oxygen electrode measurements are more complex and clinically less reliable, or at least cannot be easily checked by an independent means. In principle, two electrodes are connected through a constant voltage source. The electrons are permitted to leak off one electrode and through a buffer solution to the other electrode. The electrons are actually produced at a silver–silver chloride anode as that system ionizes into solution. At the other electrode, the electrons selectively but not exclusively ionize the free oxygen present. The system is ingeniously arranged so that if oxygen is not ionized, those electrons tend not to flow. How is this done? The two electrodes are connected by buffered AgCl-KCl. At the cathode oxygen is ionized from aqueous solution, whereas at the anode electrons are pushed into solution. In the water, which is always slightly ionized, the KCl is "oxidized" to KOH by $O^=$; the Cl^- that is liberated rejoins the silver–silver chloride electrode to regenerate the latter system. A background or polarizing current flows continuously between the two electrodes from the constant voltage source. However, the superimposed signal or test current is proportional to the number of electrons in the oxygen circuit. The two circuits are easily separated, e.g., by arbitrarily imposing a convenient frequency on the polarizing current, since the signal is essentially direct current. Clark's contribution was the invention of a semipermeable membrane which separated the two electrodes and simultaneously provides for a natural constant gate on the oxygen diffusion process.

In service, many of the problems with oxygen electrodes are associated with contaminated membranes, or less frequently broken leaking membranes. The buffer system is itself banal. Unfortunately, the electrolysis produces powerful oxidants such as Cl^- and peroxides, which tend to alter the membrane and poison the platinum detector electrode. Oxygen electrodes of this type are inherently unstable after a matter of minutes to hours, and even the most carefully designed and meticulously manufactured ones are difficult to use in chronic installations. For a variety of reasons, voltage errors increase as a function of the area of the electrode. Therefore, very small (about 25 μm diameter) platinum electrodes are used which in turn are associated with the generation of miniscule currents, yet they are more susceptible to damage. To amplify these currents, systems similar to those used for pH and P_{CO_2} measurements are required. A sensitivity of 20 pA/mm Hg P_{O_2} is usual.

Since the F_{IO_2} of air is invariably higher than the P_{O_2} of normal blood (150 > 95 mm Hg), exposure of normal or hypoxic blood samples to air raises the P_{O_2}. If the exposure happens to take place in an environment of increased F_{IO_2}, as is common in intensive care oxygen devices, the error is greater. On the other hand, if the blood P_{O_2} exceeds 150 mm Hg, as sometimes happens during poorly regulated ventilation therapy, the blood will lose oxygen to the air. In the worst case, that of a newborn infant, the conditions that may lead to retrolental fibroplasia will not be detected when this sort of carelessness occurs. Occasionally, the sample may contain undetected bubbles, which is a good reason to use a transparent electrode chamber. Bubbles passing the membrane cause a characteristic erratic oscillation quite similar to that caused by static electricity in the operator's garments.

HYDROGEN ION HOMEOSTASIS

Each day, the typical adult body produces CO_2 at a rate approximating 150 mEq/kg of body weight (or still more roughly, 0.15 mEq/kg/min). A most important 60 mEq, or 40 minute production of the body total per day, is needed to cover about 1 mEq/kg/day secretion of nonvolatile "fixed" acid in the kidney, in order to buffer the protons with bicarbonate. The "fixed" acid H^+ is exchanged within the proximal renal tubular cell for the Na^+ of the glomerular infiltrate and migrates into the lumen. The infratubular H^+ is then covered by HCO_3^-, which immobilizes the protons in innocuous H_2CO_3. The carbonic acid decays to CO_2 and H_2O; the CO_2

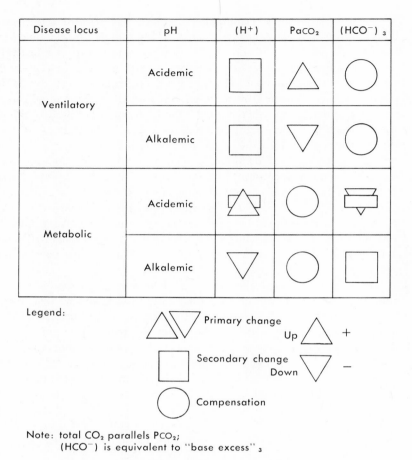

Fig. 6-12. Plasma pH adjustments in disease.

diffuses rapidly back into the kidney, where it is reconverted to H_2CO_3 and then into HCO_3^- and H^+. The final result is that, while the CO_2 diffuses back into the kidney, the original "fixed" acid H^+ has been converted into water.

$$\underbrace{\text{"Fixed" acid} \quad H^+ + HCO_3^-}_{\text{Tubular lumen}} \longrightarrow H_2CO_3 \longrightarrow \underset{\text{To kidney cells}}{CO_2 \downarrow} + \underset{\text{To urine}}{H_2O \downarrow}$$

In the distal tubular cell, additional H^+ is produced from the ionization of carbonic acid and is exported into the lumen where it exchanges for Na^+ and K^+. The ionization, of course, involves HCO_3^- production, which is diverted back to the bloodstream to be covered by the resorbed "fixed" cations, Na^+ and K^+. It is readily seen that the bicarbonate buffer, HCO_3^-, is essential to renal tubular performance. It is also readily seen that HCO_3^- really does function as a buffer and as a base in renal homeostasis. Thus, the statement that bicarbonate measures the kidney's ability to chronically adjust buffer base and Pa_{CO_2} measures the lung's ability to acutely adjust hydrogen ion makes some sense (Tables 6-9 to 6-10).

Table 6-9
Classification of changes in blood buffer components

	pH		Base excess (mEq/L)		T_{CO_2}* (mm/L)		Pa_{CO_2}	
Normal	7.35	7.45	−2.5	+2.5	24	30	35	45
Moderate	7.25	7.50	−8	+8	20	35	30	60
Marked	7.10	7.60	−15	+15	10	45	20	75
Severe	< 7.10	> 7.10	< 15	> 15	< 10	> 45	< 20	> 75

*Measured in plasma.

Table 6-10
Common mechanisms in "metabolic" buffer diseases

Acidosis		Alkalosis	
Anion gap	Anion gap normal	Urine Cl⁻ < 15 mEq/L	Urine Cl⁻ > 15 mEq/L

Renal tubular defect
  Chronic nephritis
  Congenital renal tubular defect
Nonrenal defect
  Organic acidosis
  Diabetic acidosis
Combined renal and nonrenal defect
Intoxication
  (e.g., methanol)
	Renal tubular defect
	  Failure of proximal tubular HCO_3^- resorption
	  Failure of distal tubular acid secretion
	Nonrenal defect
	  HCO_3^- loss from nonrenal site
	  Anion intoxication (e.g., NH_4Cl)
		Chloride depletion
		Gastrointestinal loss of chloride
		Excess use of diuretic drug
			Increased distal tubular Na^+ resorption
			Adrenal hyperplasia
			Primary aldosteronism
			Congenital renal tubular defect
			Chronic K^+ depletion
			Mechanism not known

BUFFER BASE CONCEPT

"Buffer base" is the term applied to the total of all acceptors of hydrogen ion (proton) in the whole blood. Normally, this unwieldly collection is simplified to the sum of all bicarbonate in 'true' plasma plus all hemoglobin in red cells:

$$H_2CO_3 \rightleftharpoons H^+ + HCO_3^-$$
$$\underline{H \cdot Hgb \rightleftharpoons H^+ + Hgb^-}$$
$$S \rightleftharpoons H^+ + B^-$$

For example, in metabolic acidosis, H⁺ increases. To keep S constant (roughly 50 mEq/L), then B⁻ must decrease. But B⁻ is the sum of Hgb⁻ and HCO_3^-. Only the bicarbonate change can occur, since hemoglobin is fixed, so HCO_3^- diminishes. This happens quickly at the lung level when CO_2 is blown off and slowly at the renal level when HCO_3^- ties up H⁺.

Ventilatory homeostasis

In pure acute ventilatory (respiratory) acidosis, a condition always associated with alveolar hypoventilation or increased dead space or both, CO_2 is excreted by the lung in less than normal amounts. Carbon dioxide retention in the lung (that is, increased $P_{A_{CO_2}}$) leads directly and proportionately to hypercarbemia (increased Pa_{CO_2}), which in turn leads to increased plasma H_2CO_3 by mass action:

$$CO_2 + H_2O \longrightarrow H_2CO_3 \xleftarrow{\quad > \quad} H^+ + HCO_3^-$$

In normal adults, for every 1 mm Hg Pa_{CO_2} increase, H^+ increases in the blood only by 0.8 mEq/L, while the plasma HCO_3^- increases by 3.5 mEq/L. Clearly, another increment of H^+ is also produced at this time but is sequestered in red cells, extravascular extracellular fluid, and ultimately the tissue cells themselves; of this whole increment, about a third is actually buffered by hemoglobin.

In pure acute ventilatory (respiratory) alkalosis, a condition always associated with alveolar hyperventilation—whether induced by drugs, cerebral injuries, ventilation therapy, hyperthyroidism, reflexes of interstitial pneumonitis, or whatever—pulmonary CO_2 excretion is increased beyond normal. This leads to a reduction in blood H^+ and HCO_3^-, with the magnitude of the H^+ alteration remarkably similar to that in acute ventilatory acidosis: 0.8 mEq/L H^+ disappears for every 1 mm Hg Pa_{CO_2} diminution. Now, the hydrogen ions are released by buffer in hemoglobin and the rest of the body in the ratio 1:2. Thus the compensatory mechanisms for acute pure ventilatory acidosis and alkalosis are two forms of the same phenomenon.

When pure ventilatory (respiratory) acidosis is chronic and fully compensated rather than acute, the kidney forces an upward adjustment of HCO_3^- dynamics and an increase in H^+ renal excretion. More sodium ion is resorbed by the proximal tubule, which permits more hydrogen ion to be exported, which forms more urinary bicarbonate. However, the bicarbonate of the blood also is increased in chronic ventilatory acidosis because of CO_2 retention. The resorbed urinary sodium ion tends to cover this bicarbonate and thus no longer "drags" chloride out of the urinary infiltrate of the tubule; hence, the chloride is excreted. Indeed, increased urinary secretion of chloride is a hallmark of chronic respiratory acidosis. This chronic condition is marked by a more complete defense of body pH, perhaps 3 times more efficient than in the case of acute ventilatory acidosis. About 0.24 to 0.3 mEq/L H^+ now appears in blood for each 1 mm Hg Pa_{CO_2} increase, compared with the earlier acute figure of about 0.8 mEq/L H^+.

In chronically compensated pure respiratory alkalosis, the defense is still better, indeed about twice as efficient as in the case of chronic acidosis. The reasons for this are not clear yet but may in part involve a greater production of organic "fixed" acid (such as lactic acid) as well as a more efficient compensation by the kidney. As one might suspect, the renal defense is almost precisely opposite to that in the case of chronically compensated pure ventilatory acidosis. Characteristically, more urinary sodium is excreted, less urinary bicarbonate forms, less hydrogen ion is exported by the blood to the urine, and the blood ratio $H^+:HCO_3^-$ becomes larger. As small a decrease as 0.17 mEq H^+ will appear in the blood for each 1 mm Hg Pa_{CO_2} diminution.

Metabolic homeostasis

Metabolic acidosis should be divided into those types associated with tissue hypoxia—such as due to oxygen lack, glucose lack, or other causes of mitochondrial failure—and those types associated with anomalous excretion of cation. In all mitochondrial (cellular) hypoxia, the production of organic acid tends to increase by quite a large amount. This is far larger than the normal renal "fixed" acid excretion load. The magnitude of this change is conventionally measured by means of the so-called anion gap:

$$G = (Cl^- + HCO_3^-) - Na^+$$

$$G = 8 \text{ mEq/L} \pm 4 \text{ in health}$$

Mitochondrial hypoxia also can be checked by means of serum lactate determinations. Indeed, every laboratory that deals with intensive care should be equipped to perform lactic acid analysis, because the elevation of this moiety is a hallmark of hypoxic metabolic acidosis, which may forecast deterioration of a previously compensated pH.

On the other hand, if the acidosis is caused by renal incompetence (acid retention) or loss of bicarbonate (vomiting of pancreatic juice), for example, according to the equation the anion gap will be small. This brings up the interesting point that since metabolic acidosis and alkalosis are slow to develop compared to the speed with which the lung can react, all metabolic pH derangements are almost immediately compensated to some extent by pulmonary action. Practically simultaneous with the onset of metabolic acidosis, alveolar ventilation increases. To the contrary, primary pulmonary acidosis and alkalosis are but slowly improved by the kidney.

When the lung blows off CO_2, the net effect is to convert ionized H^+ of metabolic acidosis to nonionized (H) of H_2O:

$$*H^+ + HCO_3^- \rightleftarrows H_2CO_3 \rightleftarrows CO_2 \uparrow + *HOH$$

The water load produced is still very small compared to the water burden of the entire body, and indeed hyperventilation increases the "invisible" water loss from the alveolar air and sometimes, by virtue of increased muscular effort, from the sweat. The decline in Pa_{CO_2} is a direct consequence of reduced P_{ACO_2}, and it is nearly equimolar and directly parallel to the decline in plasma HCO_3^-. By a day or so after this compensation the normal kidney has a fully mounted defense of its own. In the proximal tubule, this comes about through increased sodium ion resorption coupled to increased hydrogen ion secretion. In the distal tubule, the dynamics are more complex. Here, the acid load is buffered by the formation of ammonia:

$$H^+ + NH_3 \longrightarrow NH_4^+$$

Because the kidney and the lung can act in concert in the compensation of metabolic acidosis, substantial defense is achieved. In fact, the defense is limited only by the muscular effort of breathing. That is, at a certain high level of attempted alveolar hyperventilation, the muscular activity itself produces so much lactic acid that it aggravates rather than compensates the metabolic acidosis. This is actually observed when, for example, a diabetic individual with emphysema goes into hyperglycemic ketoacidosis. Normally, however, compensation extends down to 25 mm Hg Pa_{CO_2}. Even at this level, a fall in plasma HCO_3^- is still accompanied by a slightly larger fall in Pa_{CO_2}. Failure to achieve this 1:1.1 ratio may be interpreted as evidence that the metabolic acidosis is complicated by ventilatory (respiratory) acidosis and/or by unusual organic acid production.

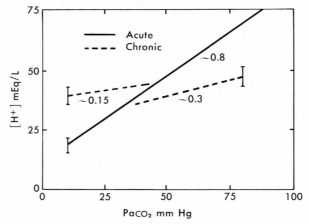

Fig. 6-13. Compensation slopes for acute and chronic respiratory acidosis. The changes in H+ with changes in Pa_{CO_2} are illustrated for "pure" respiratory acidosis and alkalosis, acute and chronic. The simplified Henderson equation for comparison is:

$$H^+ = \frac{(K)\, Pa_{CO_2}}{HCO_3^-} = \frac{24 \times Pa_{CO_2}}{HCO_3^-}$$

$$H^+ = 40 \text{ (in health)}$$

The slopes in Fig. 6-13 reflect the role of other buffers, principally hemoglobin, in acute "respiratory" acidosis and renal function in the chronic form. Note that there is only one curve for the acute form, which probably indicates a single continuous compensatory mechanism. By determination of H^+/Pa_{CO_2} ratios or, alternatively, Pa_{CO_2}/HCO_3^- ratios, it is possible to show whether the reaction falls in the expected range of slopes from charts. This helps determine the nature of the underlying defect and its compensation.

Metabolic alkalosis, compensated or not, does not usually cause substantial increases of Pa_{CO_2}. The degree of hypoventilation necessary for the lung to achieve such compensation would generally result in hypoxemia. The respiratory quotient (CO_2 out : O_2 in) is usually about 0.7 to 0.8, that is, nearly equimolar exchange of these gases. Thus alveolar hypoventilation has a marked effect on oxygen uptake. Clinically, this fact is often obscured by both the sigmoid characteristics of the oxyhemoglobin dissociation curve and the thoughtless referral to the aphorism that alveolar ventilation is measured by Pa_{CO_2}. Commonly, only the most severe forms of metabolic alkalosis receive pulmonary compensation. Even in such cases, there is almost always less than 60 mm Hg Pa_{CO_2}, which probably defines the physiologic limit of compensatory alveolar hypoventilation in the adult. In the newborn, the situation is much less clear, although some additional possibility for compensation is probably afforded by the increased oxygen affinity of fetal hemoglobin.

"Metabolic" alkalosis cannot arise from "metabolic" disease in the sense that, for example, diabetes, shock, and general causes of mitochondrial deterioration are "metabolic." Instead, "metabolic" alkalosis is almost always an "ionic" alkalosis associated with some condition that destroys salt (sodium chloride) homeostasis. Either too much chloride is excreted by nonurinary routes or too much sodium is resorbed by the kidney. Usually conditions in which too much chloride is excreted (such as persistent vomiting) can be repaired with chloride therapy, but those in which excessive sodium is resorbed (such as adrenal hyperplasia) are not repaired by additional chloride. Naturally, if excess chloride is lost outside the kidney, urinary chloride is low—typically, under 15 mEq/L with normal urine flow. On the other hand, in the metabolic alkalosis of high sodium resorption by the kidney, there often is simultaneously high chloride excretion in the urine, typically over 15 mEq/L urine.

While we have previously considered that the proximal renal tubule takes up Na^+ from the lumen in exchange for H^+ secreted into the lumen, in fact most Na^+ is not really exchanged for anything; it must be accompanied by anion, which is often Cl^-. Not enough HCO_3^- can ever be present in the proximal tubule for this huge task. If there is a deficiency of Cl^-, then some Na^+ is not resorbed at the proximal tubule and goes on to the distal tubule. Here again is an opportunity for exchange, this time with either H^+ or K^+. Exchange with H^+ is again limited by the kinetics; not enough H^+ comes into the lumen to provide sufficient HCO_3^- to cover the Na^+. Instead, the exchange to some extent occurs with K^+ and this may ultimately lead in metabolic alkalosis to a body depletion of K^+, the major intracellular cation. Clearly, if enough exogenous chloride is provided to the glomerular filtrate, the whole series of events may be reversed. When the kidneys are intrinsically normal and the urine has a low chloride content, the anatomy of the extracellular fluid must be reconstructed.

On the other hand, chloride-resistant "metabolic" alkalosis usually implies that the kidneys are defective, either because they are the target for abnormal extrinsic hormone activity (as in hyperaldosteronism) or because they are intrinsically abnormal (as in Bartter's syndrome). The urine has a chloride content over 15 mEq/L in the typical case. The pathogenesis assumes that a normal salt load is presented in the kidney, but the (distal) nephron is too avid in the exchange of Na^+ for K^+ and H^+. The increased Na^+ resorption pulls increased HCO_3^- and water loads back into the body. Blood volume expands along with the entire extracellular space, which augments the glomerular filtration load and increases the presentation of Na^+ to the distal tubule. There can be little improvement if such a patient receives parenteral sodium chloride.

SUMMARY

How can we summarize the problem of COP measurements? The vascular compartment, which is what one evaluates in COP estimations, is but a fraction of the body's mass. Indeed, the total extracellular water content descends from a value as high as 35% at birth to as low as 20% in the lean adult, while the intracellular water remains practically a constant portion of the actual mass of cells (exclusive of fat). If one assumes extracellular water to be a "mean" of 25% of total body mass, it is usually divided so that one fourth is blood plasma and three fourths is extravascular but still extracellular fluid. Another 45% of total body mass is intracellular water and the remaining 30% is "solid" tissue (exclusive of fat).

Although it is widely realized that intracellular contents—potassium, for example—are poorly mirrored by their concentration in plasma, there is a universal propensity to equate acidemia—what we measure—with acidosis and alkalemia with alkalosis. All classification schema of buffer base pathology perpetuate this error. Unhappily, the laboratory estimation of COP is but a static reflection of the intracellular dynamics of some indeterminate time in the past. What we "correct" with fluid therapy is largely the present intravascular space, with a crude surmise as to the needs of the cells themselves. In particular, intracellular and extracellular homeostasis need not approximate nor even parallel each other during the acute crises that characterize the intensive care unit, exactly the time and place that our need for intracellular information is greatest. Many a laboratory error in COP estimation relates better to the physician's ignorance of these facts than to the inherent problems of COP quality control. Many another error is due to our inability to deal with all the variables that modulate COP measurements and are within the very sample we evaluate.

In practice, derangements of buffer defenses are often not "pure." In the newborn nursery, hypoxic metabolic acidosis invariably complicates ventilatory (respiratory) acidosis; indeed, the first may often cause the second. In diabetic ketoacidosis, dehydration and intracellular potassium deficits markedly affect renal defenses while alveolar ventilation is maximally exercised by Kussmaul breathing patterns. The patient with failing kidneys is often at increased risk for pneumonia. Pulmonary hypertension may lead to heart failure and poor renal perfusion. Renal hypertension leads to renal failure but also to strokes and destruction of the brain-lung neural loops. Finally, the application of modern assisted ventilation therapy in the intensive care unit may be so efficient as to effect a "flip-flop" of metabolic or respiratory acidosis to respiratory alkalosis before anyone realizes what has happened.

Physicians sometimes forget that COP data are only indications of statistical likelihood. The classic case is the confusion between chronic compensated ventilatory (respiratory) acidosis and ventilatory acidosis accompanied by primary metabolic alkalosis. A clinical history, hemogram, and urinalysis (including volume and chloride analysis) still have something to contribute. Furthermore, the current quality control problem in COP estimations must be accepted. For instance, if seven critically ill patients each get one COP study per day and one set is wrong, the results may be catastrophic. If one patient in intensive care gets seven COP studies in one day and one set is wrong, we can cope with that. The COP game is still one of skill and art and may remain so for several years.

References

1. Bear, R. A., and Gribick, M.: Assessing acid-base imbalances through laboratory parameters, Hosp. Pract. 9:157-165, 1974. (Readable referenced summary of confidence limits methodology applied to hydrogen ion homeostasis.)
2. Cumming, G., Semple, S. J., and Arnott, W. M.: Disorders of the respiratory system, Oxford, 1973, Blackwell Scientific Publications. (Best comprehensive text on relationship between physiological lung function and hydrogen ion homeostasis in health and disease, with interpretations.)
3. Fleischer, W. R., and Gambino, S. R.: Blood pH, Po_2 and oxygen saturation, Chicago, 1972, American Society for Clinical Pathology. (Basic reference for standard clinical laboratory blood gas methods, with interpretations by Sigaard-Andersen and Weisberg base-excess methodology.)
4. Hause, M. E.: Acid-base imbalance. In White, W. L., Erickson, M. M., and Stevens, S. C.: Chemistry for medical technologists, ed. 3, St. Louis, 1970, The C. V. Mosby Co., pp. 222-235. (Details of electrodes and blood gas machines with technology pointers and interpretations by Sigaard-Andersen methodology.)
5. Rosan, R. C.: Hyaline membrane disease and a related spectrum of neonatal pneumopathies. In Rosenberg, H. S., and Bolande, R. P., editors: Perspectives of pediatric pathology, vol. 2, Chicago, 1975, Year Book Medical Publishers, Inc., pp. 15-60. (Pathologist's view of injury and regeneration of the infant lung, with a recent advance in conceptual model of lung.)
6. West, J. B.: Respiratory physiology—the essentials, Baltimore, 1974, The Williams & Wilkins Co. (Short concise paperback that covers physiological lung function in health and disease, with interpretations by Sigaard-Andersen base-excess methodology.)

7

Hepatobiliary system

The liver, biliary tract, and pancreas are often considered together because of their similar symptoms, interrelated functions, and anatomic location. The liver is not only one of the largest organs but it is considered the crossroad of the body. Within the liver, the portal and systemic circulations join to drain through a common venous outflow. The liver supports the intermediary metabolism of all foodstuffs; it is the major nucleus of synthetic, catabolic, and detoxifying activities in the body; it is crucial in the excretion of heme pigments; and through its Kupffer cells it participates in the immune response.

Since the liver has a tremendous reserve capacity, hepatic disease does not become apparent until widespread damage is in evidence. Focal lesions may remain silent although they may produce considerable hepatic damage. Diffuse liver diseases gradually deplete the reserve resulting in liver damage, jaundice, and in some cases liver failure.

EVALUATION OF LIVER DISEASE

Laboratory analysis is essential in the evaluation of the patient suspected of having liver disease. The routine battery of tests used reflect three aspects of hepatic activity: response to inflammation, synthetic function, and clearance function, including biochemical alterations ("detoxification") and bile function.

Inflammation and necrosis

The liver is a rich source of transaminases, and any damage to the hepatic tissue triggers a release of transaminase into the circulation, giving a good indication of hepatic inflammation and necrosis. Transaminase will clear from the serum within 3 days, so a serial measurement will monitor the progress of the cell damage.

If the transaminase increases over 1,000 units and maintains this elevation for several days, this usually indicates an acute, diffuse hepatocellular disease as opposed to extrahepatic obstruction. There will be a serum transaminase increase when liver cell necrosis develops from any cause. When the analysis results are in a nondiagnostic range (less than 40.0 units), flocculation tests, although considered obsolete, are sometimes helpful. The flocculation test is not specific for liver disease but will become positive with an inbalance (elevation) of albumin and gamma globulin. Since the gamma globulin is usually elevated when the disease is active, the analysis is effective in assessing hepatic inflammation within the group of chronic active liver diseases.

Synthetic function

Because the liver has extensive metabolic activities, the hepatic synthetic function can be evaluated by measuring several serum components. The level of a component in the serum reflects both production and destruction of the substance. Therefore, if there is a constant rate of destruction (catabolism), the serum concentration reflects the production change (Production–Destruction = Rate of component change). Tests most utilized in the evaluation of hepatic synthetic function are total protein, A/G ratio, protein electrophoresis, and prothrombin. The static serum level reflects the balance between the synthetic and catabolic rates. A low serum albumin concentration does not necessarily indicate reduced synthesis but may result from increased catabolism of albumin. Therefore, an increased rate of albumin loss from the body may cause a lowered serum albumin concentration even though the synthetic rate remains normal. An example of this is a protein-losing enteropathy. A decreased serum albumin concentration may also result from an increased plasma volume or a large extravascular albumin pool (for example, ascites formation). If the preceding factors are absent and if nutrition is adequate, an albumin concentration below 3.0 g/dl indicates reduced synthesis.

Individuals with severe hepatic disease or obstructive jaundice may have coagulation defects that result in long prothrombin times.

The prothrombin time reflects several clotting factors in blood synthesized by the liver, especially the vitamin K dependent factors.

Since vitamin K is fat soluble and requires bile salts for absorption, it is given to the patient and a "pro-time" done. For an individual with obstructive jaundice, the prolonged pro-time usually returns to normal with ingestion of vitamin K, while in the individual with hepatocellular disease the pro-time remains prolonged.

If a prothrombin time greater than 3 minutes above the control does not correct following parenteral administration of vitamin K, a hepatic parenchymal disease, not a vitamin K deficiency, is indicated. Serial monitoring with the prothrombin test gives a good assessment of the hepatic disease course and its severity.

Clearance function

The liver functions as a clearing station in bile formation and drug or chemical detoxification. The clearance function is evaluated by measurement of endogenously produced compounds such as bile acids, bilirubin, and administered compounds such as BSP. The single analysis is performed more frequently but kinetic analyses of bilirubin and BSP give more information of the hepatic function.

BILIRUBIN

Serum bilirubin is found in either conjugated or free form. The van den Bergh reaction is the conventional method for measurement of the direct and indirect reacting fractions of the two forms. Conjugated hyperbilirubinemia reflects a defect in the hepatic transport of bilirubin into bile canaliculi, resulting from extrahepatic biliary obstruction or hepatic disease. However, two metabolic disorders, Rotor's and Dubin-Johnson syndromes, have conjugated hyperbilirubinemia in the absence of structural hepatic alterations. Unconjugated hyperbilirubinemia may be a result of impaired hepatic clearance of bilirubin from the plasma or an excess production of bilirubin.

Present analytic methods determine the total bilirubin and the direct fraction on the assumption that the difference between the total and the direct fraction represents the indirect (unconjugated) bilirubin.

Radioimmunoassay techniques are showing promise in the investigation of hepatic disorders, but generally, the Malloy-Evelyn modification of the van den Bergh procedure is used in the laboratory.

There are precautions to be taken in the collection and handling of the specimen. The blood is preferably collected fasting or in the postabsorptive state to avoid lipemia, which interferes with analysis. Hemolysis is also to be avoided since both lipemia and hemolysis produce falsely low results. Serum should be kept in the refrigerator or a darkened area prior to analysis because of deteriorating effects of light on bilirubin.

Because of the complexity of the liver, no single test can provide an overall picture of its function. Most disorders affect the liver in some way because of the liver's circulatory system, hepatic cells, biliary passages, and reticuloendothelial tissue. The tests are based on the liver's capacity in synthesizing a product when the precursor prothrombin has been furnished, enzymes in serum whose levels are a reflection of hepatic damage, plasma disappearance rate of administered BSP dye, and albumin production and bilirubin removal, which depend on the liver.

ENZYMES

The enzymes tests may be utilized in monitoring their elevation in certain clinical situations.

Transaminase

The SGOT and SGPT levels rise with the loss of integrity of the cell membrane. They are sensitive in indicating hepatobiliary disease. However, as previously stated, their main purpose is pointing to acute parenchymal injury, where the SGOT:SGPT ratio is less than 1. The SGPT is more specific for hepatobiliary disease. The SGOT level will elevate with myocardial or skeletal muscle injury and may show a slight increase following drug administration of morphine-like narcotics or sympathomimetic drugs to patients with biliary tract disease or those who have had a cholecystectomy.

Alkaline phosphatase

Along with the transaminases, the alkaline phosphatase test is widely used for monitoring liver disease.

Alkaline phosphatase is one of four phosphomonoesterases found in serum; the others are acid phosphatase, 5'-nucleotidase, and glucose-6-phosphatase. Only the last two are substrate-specific. Normal serum levels are derived from bone, the hepatobiliary system, and the intestinal tract. The serum alkaline phosphatase will be increased in conditions associated with increased osteoblastic activity or disorders of the liver and bile ducts. Increases tend to be most marked in cholestatic disorders and space-occupying lesions, where levels are often 3 or more times normal. Lower elevations are seen in acute parenchymal injury and chronic parenchymal disease.

5'-Nucleotidase

The serum 5'-nucleotidase analysis has essentially the same usefulness as that for the serum leucine aminopeptidase. The highest levels are found in inflammatory and obstructive lesions of the biliary tree. Increased levels are not unusual in normal pregnancy and are probably derived from the placenta.

Leucine aminopeptidase

This enzyme is increased in almost all instances where there is an elevated alkaline phosphatase and generally parallels its activity. It is useful only in helping to distinguish increased levels of alkaline phosphatase from osseous and hepatobiliary sources since it is not increased by osteoblastic activity.

Lactic dehydrogenase (LDH)

The total LDH activity is less sensitive and less specific than the transaminases. The determination of LDH isoenzymes is a sensitive and specific indicator of hepatic necrosis. The hepatic isoenzyme (LDH_5) is the most cathodic and therefore is referred to as the slow fraction. However, there is little or no LDH_5 in serum, so a significant increase may occur without being reflected in total LDH activity. Such increases are found with hepatic cell injury of any cause and provide the same information as the transaminases.

Isocitric dehydrogenase (ICD)

Isocitric dehydrogenase is present in the liver, heart, muscles, various tumors, red cells, and platelets. Without the presence of hemolysis, an elevation of ICD indicates liver cell injury. This rise is accredited to the stability of the hepatic form over other forms. The test has no advantage over the SGPT analysis.

OCCULT BLOOD IN STOOL

In the case of obstructive liver disease associated with jaundice, the presence of blood in the gastrointestinal tract suggests neoplasm.

Liver function tests

Many tests may be utilized, since no single test is sufficient for clinical evaluation of the liver. There are older tests that are now being revived and performed by radioimmunoassay. Some analyses used for liver function evaluation are:

Bilirubin

Serum, conjugated and unconjugated	Little value in diagnosis of liver disease; used in evaluation of hemolytic disease
Urine bilirubin	
Urine urobilinogen	
Fecal urobilinogen	

Protein

Albumin and globulin (A/G ratio)	Hepatic and nonhepatic
Electrophoresis	Hepatic and nonhepatic
Flocculation and turbidometric test	Becoming obsolete but will detect hepatic and nonhepatic disease
Serum mucoprotein level	Not widely used
Serum haptoglobin level	May have value
Blood ammonia level	Treatment of hepatic coma
BUN	Insensitive to liver damage

Dye excretion test

BSP (sulfobromophthalein)	Used for hepatic function and hepatic blood flow studies
Rose bengal	The old method suffered from methodology interference by icteric serum; the RIA method aids in differentiation between obstructive jaundice and parenchymal liver disease

Enzyme levels

Transaminases and alkaline phosphatase	Widely used
Cholinesterase	Seldom used

Lipid metabolism

Serum or plasma cholesterol	Limited but useful
Cholesterol esters	Little value
Bile acids	Research limitations

Detoxification and synthesis

Hippuric acid	Once widely used, now abandoned
Prothrombin time and response to vitamin K administration	Useful

BLOOD AMMONIA

A major source of blood ammonia is from the gastrointestinal tract, with a minor amount coming from the kidney. Ammonia released by bacteria and ingested as ammonium salts is absorbed into the portal vein. The liver usually removes most of the ammonia from the portal vein blood, converting it to urea, with only a small amount escaping to the hepatic vein to be transported to the systemic circulation.

Elevated ammonia level in a hepatic disorder appears to be dependent on an impaired parenchymal function and shunting of portal blood past the liver.

Laboratory examinations for proteins

The blood proteins are made up partially of antibodies, enzymes, and hormones. These substances have varied functions. As a group, one of their major functions is to maintain the water balance between the blood and tissues.

Under normal conditions, the water balance is at equilibrium. However, if blood proteins are lost, the density of the blood decreases, and water flows toward the tissues. Swelling, or edema, results. This is found in nephrosis, a disease of the kidneys characterized by appearance of albumin in the urine.

The proteins measured most frequently in the clinical laboratory are albumin, globulin, and total protein. These substances are the most abundant solid material in blood, and their values are given in grams per deciliter (g/dl) of serum. The normal values are given below.

Albumin	4.0 to 5.0 g/dl of serum
Globulin	2.0 to 3.0 g/dl of serum
Total protein	6.0 to 8.0 g/dl of serum

Some significant diseases in which abnormal protein values may be expected are given in Table 7-1.

Some of the more commonly used methods for the determination of proteins are discussed briefly here.

Kingsley biuret reaction

1. The albumin is determined as follows. A sodium sulfate solution is added to a portion of serum. This precipitates the globulin, which is removed by filtering or by centrifuging. (If the centrifuge is used, ether is added to aid in the separation.) A biuret reagent is added to the filtrate containing the albumin and a reddish purple complex is formed (the biuret reaction). The depth of color is measured in a colorimeter and compared with a standard.

Table 7-1
Conditions accompanied by abnormal protein values

Disease	Albumin	Globulin	Total protein
Nephrosis	Decreased	Increased	Decreased
Multiple myeloma	Normal	Normal	Increased
Infectious hepatitis	Decreased	Increased	Decreased
Cirrhosis of liver	Decreased	Normal	Decreased

2. The total protein is determined as follows. Another portion of serum is diluted with sodium chloride solution. The same biuret reagent is added. This forms a reddish purple complex. The depth of color is measured in a colorimeter and compared with a standard.

3. The globulin is determined as follows: Globulin = Total protein — Albumin.

Micro-Kjeldahl method

1. The total protein is found as follows. (The test calls for two separate determinations and a calculation.)

In the first determination, the total nitrogen is determined on serum by a micro-Kjeldahl method. This entails adding a digestion mixture, heating to decompose nitrogenous substances, adding Nessler's solution to produce a yellow-orange color, and then measuring the depth of color in a colorimeter.

In the second determination, the nonprotein nitrogen is determined on a protein-free filtrate by the micro-Kjeldahl method. This again calls for digestion and nesslerizing.

In the calculation, the total protein is found from the formula: Total protein = (Total nitrogen – Nonprotein nitrogen) × 6.25, where the 6.25 is a factor converting nitrogen to protein.

Example: Total nitrogen value is found to be 1000 mg/dl and nonprotein nitrogen is found to be 30 mg/dl. Therefore, total protein = (1000 – 30) × 6.25 = 6062.5 mg/dl, or 6.1 g/dl.

2. The albumin is determined as follows. The globulin is removed from the serum by precipitating with sodium sulfate and filtering. The filtrate contains the albumin. This is digested and nesslerized to determine the total nitrogen. The albumin is found from the formula: Albumin = (Total nitrogen – Nonprotein nitrogen) × 6.25, where 6.25 is a factor converting nitrogen to protein. The nonprotein nitrogen value used above is derived from the total protein determination.

Example: Total nitrogen value is found to be 700 mg/dl, and nonprotein nitrogen is 30 mg/dl. Therefore: Albumin = (700 – 30) × 6.25 = 4187.5 mg/dl, or 4.2 g/dl.

3. The globulin is found as follows: Globulin = Total protein – Albumin.

Example: Using the above figures for total protein and albumin, Globulin = 6.1 – 4.2 = 1.9 g/dl.

DETERMINATION OF TOTAL PROTEIN AND ALBUMIN

Specimen stability. Serum proteins, including albumin, are stable for at least 1 week when stored at 4° C.

Total protein method of Weichselbaum[22]

Principle. Protein produces a violet color in the presence of cupric ions in an alkaline solution, reaching its maximum color within 15 minutes. The color then is stable for several hours.

Reagents

Biuret reagent	(D-126)
Albumin standard, stock solution, 10 g/dl	(D-127)
Albumin standard, working solutions, 2.0, 3.0, 4.0, 5.0, 6.0, and 8.0 g/dl	(D-128)

Procedure

1. Add 100 µl of the working standards and the unknown specimens to tubes containing 5.0 ml of biuret reagent.
2. Prepare a reagent blank by adding 100 µl of water to 5.0 ml of biuret reagent.
3. Mix well and incubate the tubes at 37° C for 15 minutes.
4. Cool to room temperature and measure absorbance at 550 nm. Sample blanks are required only in cases of hyperlipemia and hyperbilirubinemia and are prepared by adding 100 µl of the sample to 5.0 ml of a reagent prepared as the biuret reagent, except that the copper sulfate is omitted.

Calculation

The total protein concentration may be obtained from an absorbance-concentration plot of the standards, or if the instrument response is linear, only a single standard (6.0 g/dl) is required and the total protein concentration may be calculated.

$$\frac{A_s}{A_u} \times 6 = \text{g protein/dl}$$

Albumin, bromcresol green (BCG) method[5]

Principle. Bromcresol green dye binds specifically to albumin, resulting in an increase in absorbance at 628 nm. The addition of 4.0 ml of 30% Brij 35/L provides a low blank level, prevents turbidity, increases sensitivity, and is required for linearity. Concentratons of Brij greater or less than the amount specified result in decreased sensitivity and loss of linearity.

Reagents

Succinate buffer, 0.10 M, pH 4.0	(D-129)
Bromcresol green stock solution, 0.6 mM	(D-130)
Bromcresol green working solution	(D-131)
Brij 35, 30%*	
Albumin standard, stock solution, 10 g/dl	(D-127)
Albumin standard, working solutions, 2.0, 3.0, 4.0, 5.0, 6.0, and 8.0 g/dl	(D-128)

Procedure

1. Add 25 µl of the working standards and the unknown specimens to tubes containing 5.0 ml of the working BCG solution.

---

*Available from Technicon Instruments Corp.

2. Mix and allow to stand at room temperature for 10 minutes.
3. Measure the absorbance at 628 nm using the working BCG solution as the reference. If a serum sample is lipemic, a serum blank is prepared by adding 25 μl of the sample of the sample to 5.0 ml of 0.1 M succinate buffer. Its absorbance, with water as a reference, is subtracted from the absorbance of the unknown.

Calculation

The albumin concentration of an unknown serum may be obtained from an absorbance-concentration plot of the standards; or if the instrument response is linear, only a single standard solution (4.0 g/dl) is required and the serum albumin concentration may be calculated.

$$\frac{A_s}{A_u} \times 4 = \text{g albumin/dl}$$

ELECTROPHORESIS

The application of electrophoresis to the separation of serum proteins and the methodology are given in Chapter 16.

THYMOL TURBIDITY

When the serum of a normal person is added to a buffered thymol solution, no appreciable precipitation occurs. However, when the serum of a patient with liver disease is added to a buffered thymol solution, a definite precipitation is produced. Consequently, the procedure can be used as a test of liver function.

The precipitation (turbidity) reaction is not completely understood. It is activated with an increase in the gamma globulin and is inhibited with a quantitative or qualitative decrease or change in the albumin.

The method of Maclagan, modified by Shank and Hoagland, is commonly used. In this method, serum is obtained. To this is added a thymol solution which has been buffered to a pH of 7.55. After 30 minutes, any turbidity produced is measured in a colorimeter and compared with a standard.

The normal values for the thymol turbidity test are 0 to 5 S.H. units. These values are exceeded in the following diseases: infectious hepatitis, cirrhosis of the liver, cancer of the liver, diabetes, nephrosis, Weil's disease, and parenchymatous liver disease.

Specimen stability. Serum specimens for thymol turbidity determinations are usually stable for about 5 days at 4° C.

Shank-Hoagland modification of Maclagan method[13, 21]

To obtain a visual comparison of turbidity, the original Maclagan method used Kingsbury's gelatin standards. These permanent standards were also used for protein determinations in urine with sulfosalicylic acid.[10]

Shank and Hoagland adapted the method for a photometer or spectrophotometer by using $BaSO_4$ suspensions for standardization and reading the turbidities.

If the pH of the thymol-barbital buffer used is changed from a pH 7.80 to a pH 7.55, the sensitivity of the test is greater for detecting hepatocellular dysfunction, without an introduction of false positives.

Temperature control is important because the turbidity of the test decreases with an increase in the temperature. The pH of the thymol buffer also decreases with a temperature increase, affecting the sensitivity of the test. A maximum error of 10% may be found if the temperature deviates beyond ± 3° C; therefore the safe range is between 22° and 28° C. If necessary, use a controlled-temperature water bath at 25° C.[21]

Principle. Serum from patients with hepatitis or any disease that causes an increase in beta globulin will produce definite turbidity when mixed with a thymol solution in barbiturate buffer. Lipemia will also cause turbidity. The specimen should be obtained while the patient is in a fasting state.

Reagent

Thymol-barbital buffer, pH 7.55 (D-132)

Procedure
1. Pipet 6.0 ml of thymol-barbital buffer into two 19 × 150 mm cuvet.
2. Pipet 0.1 ml of water into the *blank* cuvet containing the thymol buffer. Mix.
3. Pipet 0.1 ml of serum into the *unknown* cuvet containing the thymol buffer. Mix.
4. Let the tubes stand for 30 minutes at a temperature of 25° ± 3° C. Use a controlled-temperature water bath if necessary.
5. Set the blank to 0 A and read the unknown absorbance at 650 nm.

Calculations

The calculations may be done by reading from a calibration curve or by using a constant (K), which is averaged from three independent batches of barium chloride reagent.

NOTE: Since this method uses a 0.0962 *normal* solution of barium chloride, the results are recorded in Shank-Hoagland units. If the method used a 0.0962 *molar* solution of barium chloride, the results would be recorded in Maclagan units. Two Shank-Hoagland units equal one Maclagan unit.

The emphasis on molarity and normality in relation to the barium chloride solution is made for the following reason. In 1946, Shank and Hoagland[21] published the method of mixing barium chloride with dilute sulfuric acid for the barium sulfate suspensions. A misprint using the symbol N (normality) instead of M (molarity) occurred, resulting in the values for the unknowns being twice the results of the original method and the standards being only half the value expected. Although the authors corrected the error in reprints that were requested, many established this now-popular method in their laboratories, using the 0.0962 normality value. The Commission on Liver Disease of the Armed Forces Epidemiological Board has recommended that Shank-Hoagland units be adopted.

Calibration curve

Make the following reagents:
1. Barium chloride standards, 0.0962 N
   a. Weigh 1.173 g of barium chloride ($BaCl_2 \cdot 2H_2O$) and place into a 100-ml volumetric flask.
   b. Dissolve and dilute to mark with water (distilled).
2. Sulfuric acid, 0.2 N
   a. In a volumetric flask, dilute 6.0 ml of concentrated sulfuric acid to 1 L with distilled water. CAUTION: Add the acid to the water and then dilute to 1 L.
   b. Standardize to 0.2 N.
3. Stock barium sulfate
   a. Pipet 5.0 ml of the barium chloride standard into a 100-ml volumetric flask. Chill to 10° C.
   b. Dilute to mark with cold (10° C) 0.2 N sulfuric acid.
   c. Mix well.

Table 7-2
Chart for thymol turbidity calibration

Units	BaSO$_4$ (ml)	H$_2$SO$_4$ (ml)	A (average of three readings)
0	0	10.00	
5	1.35	8.65	
10	2.70	7.30	
15	4.05	5.95	
20	5.40	4.60	
25	6.75	3.25	
30	8.10	1.90	
35	9.45	0.55	

Set up the tubes as indicated in Table 7-2 and read each tube 3 times against the zero standard at a wavelength of 660 nm. Average each set of readings and draw the curve.
Obtain the constant (K):

$$K = \frac{\text{Units}}{A}$$

Obtain an average K value from the readings and use this to calculate the thymol turbidity units of serum specimens:

$A \times K =$ Thymol turbidity units

Normal values

Shank-Hoagland units (buffer pH 7.55)	0 to 5 S.H. units
Shank-Hoagland units (buffer pH 7.80)	0 to 4 S.H. units

CEPHALIN—CHOLESTEROL FLOCCULATION TEST[2, 9, 11]

A cephalin-cholesterol emulsion is flocculated slightly or not at all by serum with normal proteins. Sera with altered proteins, as found in parenchymatous liver disease and certain other diseases, cause flocculation. Flocculation is caused by the precipitation of globulin and cholesterol. Cephalin acts as an emulsifying agent. Albumin inhibits flocculation. A positive test results from an increase in globulin and/or decrease in albumin or a change in its antiflocculation properties.

1. Specimen: 0.4 ml unhemolyzed serum. Test should be set up on fresh serum since the stability of the test is unpredictable. Frozen serum should not be used.
2. Precautions: False positives may result from:
   a. Exposure to light during the incubation of the test
   b. Dirty glassware (particularly heavy metal or strong acid contamination)
   c. Serum contamination with bacteria

Reagents

Saline, 0.9%	(D-17)
Cephalin-cholesterol stock ether solution*	(D-133)
Cephalin-cholesterol emulsion	(D-134)

*Difco Bacto-Cephalin Cholesterol Antigen.

Procedure

1. Set up each test in duplicate. Add 0.2 ml serum in a 12-ml tube (unscratched, conical-tipped centrifuge tube) containing 4.0 ml saline. Mix gently.
2. Set up an emulsion control with each run, consisting of 4.0 ml saline. If a 4+ positive control serum is available, the test should be included.
3. Add 1.0 ml cephalin-cholesterol emulsion to each of the tubes. Cover each well with parafilm and mix gently and thoroughly.
4. Place tubes in dark at 25° C ± 3° C.
5. Read the reaction at 24 hours and again at 48 hours as follows:

Negative	No flocculation
1+	Minimal flocculation
2+	Definite flocculation
3+	Considerable flocculation but definite residual turbidity in the supernate
4+	Complete flocculation with clear supernate

If more than minimal flocculation occurs in the "emulsion control" or the 4+ positive control does not display a 4+, the test should be repeated with fresh emulsion and fresh serum specimens.

Comments. A 3+ or 4+ flocculation in 48 hours is abnormal and positive. Flocculation may occur as a result of decrease in serum albumin which stabilizes the emulsion. The low serum albumin may result from hepatic disease, malnutrition, nephrosis, or malabsorption. Alpha globulin may stabilize the emulsion. A decrease in alpha globulin will cause a flocculation, as will an increase in gamma globulin.

Positive reactions occur in sera heated at 56° C for 30 minutes or if the reaction mixture is exposed to light during the test period.

The clinical value of the test is in hepatocellular damage with decreased serum albumin and increased gamma globulin.

FIBRINOGEN

Fibrinogen is used in the clotting of blood. The normal values range from 200 to 400 mg/dl of plasma. These values are increased in infections and decreased in cirrhosis of the liver, phosphorus poisoning, and typhoid fever.

Fibrinogen, quantitative[12]

Principle. This method is similar to that of Reiner and Cheung[18] with the exception that fibrinogen is removed from the plasma by precipitating with calcium chloride instead of thrombin.

Reagents

Saline, 0.9%	(D-17)
Calcium chloride, 2.5%	(D-135)
Sodium hydroxide, 0.2 N	(D-136)
Biuret reagent	(D-126)
Protein standard, 3 mg/ml	(D-137)

Procedure

1. Use either a 3204 QBD Vacutainer tube (EDTA) or a regular syringe and a test tube containing EDTA to collect 5.0 ml of venous blood.
2. Centrifuge for 15 minutes at 3000 rpm.

3. Add 25 ml of saline to an Erlenmeyer flask.
4. Pipet 1.0 ml of EDTA plasma into the flask. Swirl to mix.
5. Add 1.0 ml of 2.5% $CaCl_2$; thoroughly mix.
6. Let stand for 1 hour. (A firm fibrin clot will form.)
7. With a glass stirring rod, pick up the fibrin clot by twisting the clot around the glass rod, squeezing out all of the saline plasma solution. (NOTE: Small parts of the clot may have separated from the main clot. These may be found by placing the flask against a dark background.)
8. Place all of the clot in a 125-ml Erlenmeyer flask.
9. Add about 15 ml of saline and swirl the flask to wash the clot.
10. Carefully pour off the saline, taking care not to lose any of the clot.
11. Do three saline washings.
12. After the last washing, carefully twist the clot around a clean glass stirring rod and gently press against the side of the flask to eliminate excess saline.
13. Transfer the clot to a 15-ml centrifuge tube. (Place the stirring rod against the lip of the test tube and gently pull the rod back away from the clot, which will slip off quite easily.)
14. Add 2.0 ml of 0.2 N NaOH.
15. Place a marble on the tube and heat in a boiling water bath until the fibrin clot dissolves.
16. Set the tube in a room temperature water bath to cool.
17. Add 2.0 ml of biuret reagent.
18. Stopper and mix. Let stand for 30 minutes.
19. Set up a blank and standard as follows:

Blank

2.0 ml of 0.2 NaOH
2.0 ml of biuret reagent. Mix.

Standard

1.0 ml of protein standard, 3 mg/ml
1.0 ml of 0.2 N NaOH
2.0 ml of biuret reagent. Mix.

20. Read the standard and test against the blank at 550 nm

Calculation

$$\frac{A \text{ of unknown}}{A \text{ of standard}} \times 300 = \text{mg fibrinogen/dl}$$

Normal values

Men	237.6 to 348.4 mg fibrinogen/dl (293.0 ± 55.4)
Women	255.1 to 356.1 mg fibrinogen/dl (305.6 ± 50.6)

BLOOD AMMONIA[4, 19]

Blood is alkalinized with a mixture of potassium carbonate and potassium bicarbonate that converts the ammonium ions to ammonia gas. Ammonia diffuses from the blood-reaction mixture, is collected in sulfuric acid on a receiving rod, and is measured colorimetrically.

Specimen collection

Venous blood is drawn with a minimum of stasis in a heparinized Vacutainer tube and analyzed within 30 minutes of the time of collection. Longer delay will result in variable increases in ammonia concentration.

Equipment

Diffusion bottles, 50 ml capacity, 42 × 73 mm with 13-mm inlet*
Receiving bottles, 10 ml capacity, 25 × 53 mm with 13-mm inlet*
Rubber stoppers to fit the above bottles*
Receiving rods, 6 × 80 mm, glass rod with a constriction 12 mm from one end.†

The short end below the constriction is round so that when dipped into the acid it retains an acid film over its surface. Rod is fitted into the bored rubber stopper. This combination, when fitted into the 50-ml bottle, will allow the rod to extend halfway into the bottle and at the same time will seal it.

Stainless steel mixing rods, aproximately 8 mm in diameter and 40 mm in length*

These rods blend the reaction mixture and increase the diffusion surface.

Rotator*

This device is a vertically mounted wheel on which are placer 18 spring clips to hold the diffusion bottles in a horizontal position 6 to 12 cm from the center of the wheel. The wheel is rotated at 50 rpm.

Spoon to deliver approximately 3.0 g of alkali mixture

Reagents

Alkali mixture

Mix 2 parts, by weight, potassium carbonate ($K_2CO_3 \cdot 1\frac{1}{2}H_2O$) and 1 part potassium bicarbonate ($KHCO_3$). Store in sealed containers.

Sulfuric acid, 1 N

Mix 1 volume concentrated H_2SO_4 and 35 volumes deionized water.

Nessler's reagent of Vanselow

Dissolve 45.5 HgI_2 and 34.9 g KI in as little water as possible (approximately 150 ml). Add 112 g of KOH in 140 ml water and dilute to 1 L. Allow to stand for 3 days. Dilute 1:20 before use with deionized water.

Ammonium sulfate standard

Stock: Dissolve 1.415 g ammonium sulfate in water and dilute to 1 L to give 0.30 mg NH_3-N/ml.

Working standard: Prepare on the day of use by diluting the stock standard 1:100 (3 µg NH_3-N/ml).

Procedure

1. Into each diffusion bottle place 3 g of alkali mixture and 2 steel mixing rods. Arrange bottles in a horizontal position serially and rotate gently by hand to align the steel rods.
2. Add to diffusion bottles in duplicate:

Blank

2.0 ml of deionized water

Unknown

2.0 ml of blood

Standard

2.0 ml of working standard

3. Cap each bottle with a solid rubber stopper immediately after the pipetting and place it on the rotator. When pipetting has been completed, rotate the bottles for 1 minute to facilitate mixing of the blood and alkali reagent.

---

*Obtainable from Wheaton Glass Co.
†Obtainable from Macalaster Bicknell of Conn., Inc.

4. Remove bottles from the rotator in the order of analysis, replace the rubber stoppers with glass receiving rods that have been dipped into 1 N sulfuric acid up to the constriction, and return the bottles to the rotator.
5. Rotate the bottles a 50 rpm for 20 minutes. The diffused ammonia is received by the acid on the glass rod.
6. Prepare receiving bottles by adding 3.0 ml of the diluted Nessler's solution.
7. At the end of the diffusion period carefully remove the receiving rods from the bottles and immerse them directly into the Nessler's solution in the receiving bottles. Color development is complete in 5 minutes.
8. Measure the absorbance of the yellow-orange solution at 395 nm in a spectrophotometer, using the Nessler reagent as the reference.

Calculation

Absorbance readings are corrected for the reagent blank, and calculations are made from the standard as follows:

$$\frac{A \text{ of unknown}}{A \text{ of standard}} \times 3 \times 100 = \mu g \ NH_3\text{-}N/dl \ blood$$

Normal values

The normal fasting ammonia nitrogen (NH_3-N) concentration of venous blood is between 60 and 150 μg/dl, with a mean of 102 ± 23 (the SD) μg/dl.

LIPID METABOLISM[16]

Lipid metabolism in the liver includes functions related to promotion of the digestion and absorption of fats, storage of fats, oxidation and conjugation of fatty acids, and metabolism and excretion of cholesterol. Serum lipid levels may be altered in disease of the liver and biliary tract.

The biologically important lipids include the neutral fats (triglycerides), the sterols, the phospholipids and related compounds, and vitamins A, E, and K. The triglycerides are composed of three fatty acids bound to glycerol. Naturally occurring fatty acids contain an even number of carbon atoms. They may be saturated (no double bonds) or unsaturated (dehydrogenated, with various numbers of double bonds). The sterols include the various steroid hormones and cholesterol. The phospholipids are constituents of the cells, especially in the nervous system.

Triglycerides, free fatty acids, phospholipids, and sterols all are bound to the albumin, alpha-globulin, and beta-globulin fractions of the plasma proteins. The lipid-protein complexes containing the most lipid and the least protein are the least dense, and those containing the least lipid and the most protein are the most dense. On the basis of their density, the complexes are divided into very low-density (VLDL), low-density (LDL), and high-density lipoproteins (HDL). In the ultracentrifuge the speed with which the various beta-lipoprotein fractions rise to the top of the plasma can be measured. This speed is expressed in Svedberg units of flotation (S_f units). Triglycerides are bound to very low-density beta-lipoproteins (VLDL), cholesterol to low-density beta-lipoproteins (LDL), phospholipids to alpha-lipoproteins, and free fatty acids to albumin.

Short-chain fatty acids liberated by hydrolysis of triglycerides in the intestine are absorbed into the portal blood. Long-chain fatty acids (more than 10 carbon atoms long) are reesterified to form triglycerides in the intestinal mucosa. They are then complexed with lipoproteins, cholesterol esters, and phospholipids to form chylomicrons.

Chylomicrons are the least dense of the plasma lipoproteins, forming particles sufficiently large (0.1 to 0.5 μm in diameter) to be seen in a darkfield microscope.

The intestinal cells esterify all long chain fatty acids absorbed through the intestinal wall and secrete the resulting chylomicrons and VLDL into the extracellular spaces between the intestinal cells. From there they enter lymph channels to be collected into the thoracic duct.

Much of the lipid metabolism of the body was formerly thought to be performed by the liver. Since the discovery that most tissues have the ability to oxidize fatty acids completely and that adipose tissue is extremely active metabolically, the former emphasis on the role of the liver in metabolism has been modified. However, the concept of a central and unique role for the liver in lipid metabolism is still an important one. It facilitates the digestion and absorption of lipids by the production of bile, which contains cholesterol and bile salts synthesized within the liver. Also, the liver has active enzyme systems for synthesizing and oxidizing fatty acids, for synthesizing triglycerides, phospholipids, cholesterol, and plasma lipoproteins, and for converting fatty acids to ketone bodies (ketogenesis).

The laboratory test most frequently performed is that for cholesterol levels in the serum. However, neutral fat and phospholipid concentrations may show marked changes. When there is damage to the parenchymal cells, lowered lipid concentrations are noted. At the onset of jaundice in viral hepatitis a decrease in serum cholesterol occurs, largely resulting from a decrease in the esterified fraction. If liver damage is severe, esterified cholesterol concentrations may be below detectable limits. In cirrhosis of the Laennec type, serum lipids and cholesterol levels are usually low.

Biliary obstruction, cholangiolitic hepatitis, and toxic hepatitis (from drugs and chemicals) are associated with elevated concentrations of serum lipids.

Elevated levels of triglycerides in the plasma are associated primarily with the cardiovascular diseases. Furthermore, there is evidence that abnormally high values in the well individual, if allowed to persist, may predict the development of pathology.

The triglyceride level varies independently of the amounts of cholesterol and phospholipids present and may become elevated in certain other disease states such as idiopathic hyperlipemias, uncontrolled diabetes mellitus, chronic hemorrhage, hypothyroidism, glycogen storage disease, and nephrotic syndrome. However, the triglyceride level alone is not diagnostic of these conditions. Low triglyceride levels are of no significance.

An early method of determining triglycerides was "by difference." Total lipid, phospholipids, cholesterol, and cholesterol esters were determined gravimetrically or by a method permitting calculation of results in gravimetric terms, then: Triglycerides = Total lipid − Phospholipid − Cholesterol esters.

The disadvantages of this approach are: (1) the factors used for phospholipid and cholesterol esters (25 and 1.67, respectively) are averages that do not apply exactly to any one sample; (2) gravimetric determinations of total lipids include lipids other than triglycerides, cholesterol, cholesterol esters, and phospholipid; thus, in some conditions spuriously high values for triglycerides can be obtained; and (3) normally, the triglycerides are only about one tenth to one fourth of the total lipids, and with low levels of triglycerides the precision of this method is extremely poor.

Other methods determine the glycerol released from triglycerides. These methods involve extraction, removal of phospholipids and other chromogens, saponification or transesterification to free the glycerol, and the colorimetric or fluorometric determination of glycerol.

More recently, a completely enzymatic method has become available from several clinical reagent supplies (such as CalBiochem).

This method involves the following reactions:

$$\text{Triglycerides} \xrightarrow{\text{Enzymatic hydrolysis}} \text{Glycerol + Free fatty acids}$$

$$\text{Glycerol + Adenosine triphosphate} \xrightarrow{\text{Glycerol kinase}} \alpha\text{-Glycerol phosphate + Adenosine diphosphate}$$

$$\text{Adenosine diphosphate + Phosphoenolpyruvate} \xrightarrow{\text{Pyruvate kinase}} \text{Adenosine triphosphate + Pyruvate}$$

$$\text{Pyruvate + Nicotinamide-adenine dinucleotide (reduced) (colored at 340 nm)} \xrightarrow{\text{Lactic dehydrogenase}} \text{Lactate + Nicotinamide-adenine dinucleotide (oxidized) (colorless at 340 nm)}$$

Since this method requires the use of several purified enzyme preparations, the detailed method is not given here. It is recommended that these enzyme preparations be purchased from a reliable company and that their directions be followed for the determination of triglycerides.

The normal triglyceride distribution as suggested by Fredrickson and associates[7] is as follows:

Age (years)	Triglycerides (mg/dl)
0-29	10-140
30-39	10-150
40-49	10-160
50-59	10-190

Total cholesterol

Cholesterol is a fatlike substance that is found in blood, bile, and brain tissue. It serves as a precursor of bile acids and various steroid hormones.

Cholesterol is a very important member of a group of substances called sterols. Sterols are complex hydroaromatic alcohols distributed throughout all living matter. After extensive research, some evidence has shown that the animal sterol (cholesterol) is a mixture of two or more sterols that differ in respect to their biologic and physiochemical properties.

The animal sterols are probably absorbed from the intestinal tract along with neutral fats and other lipids. They form esters with the fatty acids and thus act as a vehicle for their carriage into the bloodstream. Some cholesterol is formed in the disintegration of red blood cells, and some is synthesized by the body, but just "where" it is synthesized is unknown.

The cholesterol is more or less evenly divided between the plasma and the erythrocytes. In the blood cells and the tissue cells the cholesterol is almost all free, while in the serum about 60% to 70% is present as cholesterol esters of fatty acids. The serum and plasma studies have more clinical value than those of whole blood.

Table 7-3
Conditions accompanied by abnormal total cholesterol values

When increased		When decreased
Obstructive jaundice	Lipemia	Pernicious anemia
Nephrosis	Celiac disease	Hyperthyroidism
Diabetes	Leukemia	Severe infections
Hypothyroidism	Multiple sclerosis	Epilepsy
Xanthomatosis	Pregnancy	Gaucher's disease
Eclampsia	Aplastic anemia	Inanition
		Intestinal obstruction

Normally, the serum cholesterol is relatively constant. However, disorders such as pernicious anemia may lower the cholesterol level to as low as 50 mg/dl. Hemolytic jaundice produces a low cholesterol level, while obstructive jaundice maintains an elevated level of cholesterol.

Total cholesterol consists of free cholesterol and cholesterol esters. In discussions, when the word "cholesterol" is used by itself, it implies total cholesterol.

The normal values for total cholesterol are 150 to 340 mg/dl of serum. Increased or decreased values may be found in those conditions listed in Table 7-3. Abnormal values are of primary interest in the disorders given in boldface.

The normal value for cholesterol esters is 70% to 78% of the total cholesterol. Decreased values are found in diseases of the liver.

DETERMINATION OF TOTAL SERUM CHOLESTEROL

Specimen stability. Serum cholesterol is stable for 7 days at room temperature and for at least a year when frozen.

Method of Abell, Levy, Brodie, and Kendall[1, 6]

Principle. The cholesterol esters in plasma or serum are saponified by incubation with alcoholic KOH. Free cholesterol is extracted into petroleum ether and an aliquot of the extract is evaporated to dryness. A modified Liebermann-Burchard reagent is added to the dried extract and the cholesterol is determined photometrically at 620 nm.

Reagents

Absolute ethanol
Hexane, spectroanalyzed grade
Acetic acid, A.R.
Sulfuric acid, A.R. (Store tightly stoppered.)
Acetic anhydride, A.R.
Potassium hydroxide, 33%
   Dissolve 10 g reagent grade KOH in 20 ml water.
Alcoholic KOH solution (Prepare immediately before use.)
   Add 6.0 ml of 33% KOH to 94 ml absolute ethanol.
Liebermann-Burchard reagent, modified
   Cool 30 volumes of acetic anhydride to 5° C. Slowly add 1 vloume of

chilled concentrated sulfuric acid with mixing. Chill mixture to 5° C and add 3 volumes of glacial acetic acid. Add sufficient anhydrous sodium sulfate to the mixture to give a concentration of 2% (w/v). This reagent is stable for at least 2 weeks at room temperature.

Cholesterol standard, 200 mg/dl

Dissolve 200 mg of purified cholesterol in about 50 ml of warm absolute ethanol. Cool to room temperature and dilute to 100 ml with ethanol.

Procedure

1. Set up the following in 40-ml glass-stoppered centrifuge tubes and add reagents in the order listed.

	Reagent blank	Standard	Unknown serum
Standard solution (ml)		1.0	
Serum (ml)			0.5
Ethanol (ml)	5.0	4.0	
Alcoholic KOH (ml)	0.5	0.5	5.0

2. Stopper the tubes, shake vigorously, and place in a 37° to 40° C water bath for 55 minutes. Cool the tubes to room temperature.
3. Add 10 ml of hexane to each tube, stopper, and shake the tubes vigorously for 1 full minute. Add 5.0 ml of water to each tube and centrifuge at 2000 rpm for 5 minutes or until the emulsions break and the two phases are clearly separated.
4. Transfer 4-ml aliquots from each hexane layer (upper phase) to clean, dry tubes. Place the tubes in a fume hood and evaporate the solvent in a gentle stream of air or nitrogen.
5. Place all tubes in a 25° C water bath for 5 minutes.
6. Add 6.0 ml of the modified Liebermann-Burchard reagent to the reagent blank tube. Use this tube to zero the spectrophotometer.
7. At timed intervals of 1 minute, add 6.0 ml of the reagent to each of the remaining tubes and mix.
8. Allow all of the tubes to remain in the 25° C water bath for exactly 30 minutes and then read the absorbance at 620 nm. (Remove and read each tube at 1-minute intervals.)

NOTE: The linearity of this method may vary with the type of spectrophotometer used. This should be checked with a series of standards of varying concentrations.

Calculations

$$\frac{A_{unkn}}{A_{std}} \times 2 \times \frac{100}{0.5} = \text{mg cholesterol/dl serum}$$

Normal values[17]

Age (years)	Total cholesterol (mg/dl)
20-29	144-275
30-39	165-295
40-49	170-315
50+	177-340*

---

*The range here may be higher for females than for males.

Bile pigment metabolism

Jaundice, caused by an abnormal accumulation of bilirubin in the tissues of the body, is a physical sign of liver damage or bile duct damage. Bilirubin is formed from hemoglobin by the reticuloendothelial cells throughout the body. It is transported in the blood (linked to protein, mostly albumin) to the liver, removed by the parenchymal cells, and converted by conjugation to a water-soluble form. Most of the bilirubin is conjugated with glucuronic acid. Conjugated bilirubin, but not the free, is excreted from the liver cells into the bile canaliculi, becoming one of the constituents of the bile that flows into the duodenum. The conjugated bilirubin is reduced by bacteria in the intestine to urobilinogen. Some of the urobilinogen is absorbed from the intestinal tract, returned to the liver, and re-excreted into the bile as urobilinogen. There is usually a small amount that is not removed by the liver cells, which passes into general circulation and is excreted by the kidneys as urine urobilinogen.

Most of the urobilinogen is excreted in the feces although some is oxidized to urobilin by bacteria in the colon. The stool contains urobilinogen and urobilin.

An increase in bilirubin concentration in the blood may occur from an increase in the rate of destruction of red cells (as in hemolytic anemia, transfusion reactions, massive infarcts of lungs or heart), which produces more bile pigment than the normal liver can excrete; or a damaged liver may fail to excrete bilirubin formed in normal amounts. Jaundice will also result if obstruction of the excretory ducts of the liver prevents the excretion of bilirubin. There is a genetic defect that limits the enzyme activity in glucuronide conjugation and causes hyperbilirubinemia.

BILIRUBIN

Bilirubin is a bile pigment that is normally present in blood. The normal values are 0.5 to 1.4 mg/dl of serum. Increased values are found in the following diseases: obstructive jaundice, hemolytic jaundice, hepatic jaundice, infective hepatitis, and pernicious anemia.

Bilirubin is usually determined now by the method of Malloy and Evelyn. This method is essentially the same as the outdated procedure of van den Bergh.

In the van den Bergh method, both a qualitative test and a quantitative test for bilirubin were made. These tests were based on a color reaction. The absence of color development was normal for the qualitative test. If color did develop, however, it was then measured quantitatively.

It was formerly thought that the time interval of color development had a definite correlation with the origin of the bilirubin present. If the reaction took place immediately, the bilirubin came from bile (obstructive jaundice). If the reaction was delayed, the bilirubin came from hemolyzed red cells (hemolytic jaundice). Recent investigations, however, have ruled out this theory.

The qualitative test of van den Bergh was known as the direct reaction, and the quantitative test was referred to as the indirect reaction.

Like the van den Bergh, the method of Malloy and Evelyn also calls for a direct and an indirect reaction.

In the direct reaction, serum that is free from hemolysis is obtained. To this is added Ehrlich's diazo reagent, which reacts with the bilirubin to form a pink to reddish purple color. The depth of color is measured in a colorimeter.

In the indirect reaction, the tube containing the serum and the diazo reagent is used. To this is added methanol, which releases free bilirubin from its glucuronide, permitting it to react with the diazo reagent. The depth of color is then measured in a colorimeter.

DETERMINATION OF TOTAL AND DIRECT BILIRUBIN

Specimen stability. Serum should be protected from light until analysis is performed. Bilirubin is light-sensitive, and if serum is allowed to stand in a well-lighted area there may be as much as a 50% drop in bilirubin concentration within 1 hour. Serum specimens are stable for up to a week when stored in the dark at 4° C and for 3 months when frozen.

Malloy-Evelyn modification for determination of direct and total bilirubin[14]

Principle. In 1883, Ehrlich described the diazotizing reaction of bilirubin with a mixture of sulfanilic acid, hydrochloric acid, and sodium nitrite, resulting in the final violet compound called "azobilirubin."

Thirty years later, in 1913, van den Bergh and Snapper quantitated this reaction. This bilirubin determination is known as the van den Bergh procedure using Ehrlich's diazo reagent.

Malloy and Evelyn in 1937 published their adaptation of the above procedure to the photoelectric cell colorimeter.[14]

The serum is treated with Ehrlich's reagent, and the bilirubin in the serum reacts with the reagent, forming a red-violet (or purple) compound known as "azobilirubin." The intensity of the color is proportional to the bilirubin present in the serum and is measured photometrically and compared with a standard.

Reagents

Methanol, absolute, A.R. grade	
Diazo blank solution	(D-138)
Sulfanilic acid, solution A, 0.1%	(D-139)
Sodium nitrite stock solution B, 5%	(D-140)
Sodium nitrite working solution B, 0.5%	(D-141)
Working diazo reagent	(Prepare just before using.)

Mix 0.3 ml of working solution B with 10 ml of solution A.

Precautions

1. Dilute the serum sample with water just before analysis. This is done to minimize the effect of the proteins slowly precipitating from the serum. Since "salt" interferes with the diazo reaction, do not use saline as a diluent.
2. The reagents must be added with exact measurements and in the sequence given in the method, or a precipitate will form, invalidating the test.
3. Bilirubin is sensitive to light (artificial and sunlight). If the serum is allowed to stand in a well-lighted area, there may be as much as a 50% drop in bilirubin concentration within 1 hour. Keep the serum in *total darkness* until analysis.

Procedure

1. Set up two colorimeter tubes for each test. Label one *blank* and the other *unknown*. Pipet into *each* tube:
   a. 0.4 ml of serum
   b. 3.6 ml of distilled water
2. To the blanks, add 1.0 ml of diazo blank solution. Mix.
3. To the unknowns, add 1.0 ml of diazo reagent. Mix.
4. *Exactly* 1 minute later (use timer), read the unknown (tube 2) against its blank (tube 1) at 540 nm wavelength. If only the total bilirubin is requested, omit this reading and proceed to the next step.

5. Pipet 5 ml of fresh methanol to each tube and mix by inverting gently. Avoid bubbles. Old methanol has a tendency to cause decreased results.[15]
6. Let the tubes stand at room temperature for 15 minutes and read the unknowns against their blanks. This is the total bilirubin.
7. CAUTION: If the solutions are transferred to small cuvets, bubbles may cling to the sides of the cuvets. Clear or dislodge the bubbles by tapping the cuvet gently.

NOTE: This method may be converted to the micromethod by scaling down 10 or 20 times. Since the final volume is $\frac{1}{10}$ or $\frac{1}{20}$ the volume in the macroprocedure, it is necessary to use an instrument adapted for microcuvets.

Calculation

Example: The bilirubin standard has a value of 5 mg/dl. Since 0.4 ml of the serum standard is used in the analysis, there is 0.02 mg of bilirubin in the 0.4 ml of serum.

A_1 = Absorbance of sample at 1 minute
A_s = Absorbance of standard at 15 minutes
A_t = Absorbance of sample at 15 minutes

$$\frac{A_1}{2A_s} \times 0.02 \times \frac{100}{0.4} =$$

$$\frac{A_1}{A_s} \times 2.5 = \text{mg bilirubin/dl at 1 minute or } \textit{direct bilirubin}$$

$$\frac{A_t}{A_s} \times 0.02 \times \frac{100}{0.4} =$$

$$\frac{A_t}{A_s} \times 5 = \text{mg } \textit{total bilirubin/dl}$$

$Total\ bilirubin - Direct\ bilirubin = Indirect\ bilirubin$

Normal values

Direct levels	Less than 0.4 mg/dl
Total levels	Less than 1.4 mg/dl

RECOMMENDED STANDARDIZATION[3, 8]

The need for a reliable bilirubin standard has become most evident in recent years. There is a more knowledgeable understanding of the severe jaundice caused by Rh and ABO incompatibility between a mother and her infant. Exchange transfusions are used to keep the bilirubin below 20 mg/dl in full-term infants and below 18 mg/dl in premature infants. Determining the need for the exchanges is very dependent on the bilirubin result.

There has been great variation from one laboratory to another. In order to standardize the method and obtain a uniform bilirubin standard, a subcommittee of the American Academy of Pediatrics* has made a study of various types of standards in use. The standard recommended was a crystalline bilirubin, defined in terms of its solubility and color intensity in chloroform. The color intensity must be read under strictly controlled conditions of temperature, light, weight, volume, wavelength, absorbance, and light path.

---
*The Joint Bilirubin Committee is composed of members from the American Academy of Pediatrics, the College of American Pathologists, the American Association of Clinical Chemists, and the National Institutes of Health.

Obviously, most laboratories do not have all the ideal conditions along with the proper spectrophotometer for measuring the color intensity (molar absorptivity). Various manufacturers* have bilirubin standards in serum that meet the requirements suggested by the commitee. These are more stable, uniform, and accurate than the standards most hospital laboratories are able to prepare. The controls are made under rigid quality control conditions, are assayed, and are then lyophilized. Lyophilizing is the process of quickly freezing the serum under unusually low temperatures and then using a high vacuum to dehydrate the frozen mass quickly. The product is stable under refrigeration. When reconstituted, the serum should be kept in the dark and used within 2 hours.

A calibration curve may be made with the lyophilized serum, or standards may be included with each group of bilirubin determinations.

DETOXIFYING MECHANISM

Conjugation is a function of the liver essential for the excretion of bilirubin, bile acids, and steroid metabolites. The same mechanism is used to convert foreign substances and toxic constituents to nontoxic products. The most frequently used test of this protective function of the liver is based upon a conjugation reaction whereby detoxication of benzoic acid with glycine results in the formation of hippuric acid that is excreted in the urine. The test is also a measure of the metabolic function of the liver, since the ability to synthesize the hippuric acid depends not only on the conjugating enzyme system of the liver but also on the availability of glycine stores in the liver.

EXCRETORY CAPACITY—BROMSULPHALEIN (BSP, SULFOBROMOPHTHALEIN) TEST

The ability of the liver to remove from the circulation and eliminate from the body certain dyes has been used as the basis for liver function tests. The Bromsulphalein test is a sensitive indicator of the functioning ability of the liver cells.

When Bromsulphalein, an organic dye, is injected into the bloodstream, it is removed by the liver and excreted into the bile. A healthy liver removes the dye at a faster rate than a diseased liver. Consequently, the rate of removal can be used as a test of liver function.

A commonly used method for the Bromsulphalein test is the procedure of Rosenthal and White, introduced in 1925. In this method, the patient is given 2 mg of dye per kilogram of body weight. The sterile solution of dye is injected into the patient's vein by the physician, with special care being taken to see that a tourniquet is *not* used while the injection is being made.

Two samples of blood are taken, one 5 minutes after the injection and the other 30 minutes after the injection. Serum is obtained. Sodium hydroxide is added to bring out the color of the dye. The concentration of dye is then found either from a colorimeter reading or from a comparison with standards representing known quantities of Bromsulphalein.

With a healthy liver, the 5-minute sample of serum contains 20% to 50% of the dye and the 30-minute sample contains no dye. With a diseased liver, the 30-minute sample contains from 5% to 100% of the dye.

In 1939, MacDonald increased the amount to 5 mg of dye per kilogram of body weight. This measure is used almost exclusively because it increases the sensitivity of the test in detecting liver impairment. With the larger amount of dye injected,

---

*Dade; Hyland Laboratories; and General Diagnostics Division, Warner-Chilcott.

Hepatobiliary system

usually only one blood specimen is taken, 45 minutes later. Normally, less than 5% of the dye is retained 45 minutes after the injection.

In 1957, Seligson and co-workers introduced the use of an alkaline buffer to develop the color and an acid buffer for the blank. This was found to minimize errors caused by icterus, hemolysis, and lipemia. They also showed that free BSP has a different absorptivity and absorption peak than when it is bound to albumin as in serum. This was corrected by adding large amounts of an anion, p-toluenesulfonate, to the alkaline buffer.

Procedure for Bromsulphalein (BSP) test

The general considerations that follow pertain to the procedure with Bromsulphalein.

1. Specimen: serum.
2. Preservation of serum: may be stored overnight in a refrigerator.
3. Precautions:
    a. If the patient has obstructive jaundice, this test should not be run, because the obstruction prevents the removal of dye and thus renders the test inaccurate.
    b. In some patients, the injected dye may cause chills, faintness, or headache. On rare occasions there may be a severe reaction. For this reason, the dye should be injected into the patient only by a physician.
4. To calculate the amount of BSP dye needed, weigh the patient and determine the amount of dye to be injected (5 mg/kg) by the following formula:

$$\frac{\text{Weight in pounds}}{22} = \text{ml of dye needed}$$

5. Table 7-4 shows the amount of dye needed according to the patient's weight in pounds.

Table 7-4
BSP dye dosages

Patient weight (pounds)	BSP dye (ml,* at 5 mg/kg)
60	2.7
70	3.2
80	3.6
90	4.1
100	4.5
110	5.0
120	5.5
130	5.9
140	6.4
150	6.8
160	7.3
170	7.7
180	8.2
190	8.6
200	9.1
210	9.5
220	10.0

*The amount of dye is given to the nearest 0.1 ml.

6. Exactly 45 minutes after the injection of the dye, obtain a blood specimen from the arm not injected. Allow blood to clot, centrifuge, and remove the serum.

Seligson-Marino-Dodson method for sulfobromophthalein (BSP)[20]

Principle. Serum is diluted with an alkaline buffer and the absorbance is measured at 580 nm. Acid buffer is then added and the absorbance measured. The difference of absorbance represents the BSP present.*

Reagents

Alkaline buffer, pH 10.6 to 10.7	(D-142)
Acid reagent, 2 M NaH_2PO_4	(D-143)
BSP standard, 5 mg/dl†	(D-144)

Procedure

1. Prepare the following tubes:
   Standard (50% retention standard)
   1.0 ml of standard into a colorimeter or test tube.
   7.0 ml of alkaline buffer.
   Mix.
   Unknown (serum)
   1.0 ml of serum into a colorimeter or test tube.
   7.0 ml of alkaline buffer.
   Mix.
2. Read the absorbance of both tubes against a water blank at a wavelength of 580 nm or a filter of similar wavelength.
3. To the *unknown* tube:
   Add 0.2 ml of acid reagent.
   Mix.
   Read the absorbance against a water blank at 580 nm.

Calculation

$$\frac{\text{Absorbance of alkaline buffer} - \text{Absorbance of acid buffer}}{\text{Absorbance of standard}} \times 50 = \%BSP$$

Make the following corrections for body weight:
110 to 149 lb. Add 1% retention
150 to 169 lb. No correction
170 to 189 lb. Subtract 1% retention
190 to 279 lb. Subtract 3% retention

The color of the Seligson method is stable for at least 24 hours and obeys Beer's law up to 100% retention.

Normal values

For dosage 5 mg/kg of BSP:

Time	Retention
30 minutes	Less than 10%
45 minutes	Less than 6%
60 minutes	Less than 3%

BSP is eliminated faster in infants and children than in adults.

---

*Available from Hynson, Westcott & Dunning, Inc., or from Dade.
†Diluted standard is stable for 1 week.

References

1. Abell, L. L., Levy, B. B., Brodie, B. B., and Kendall, F. E.: A simplified method for the estimation of total cholesterol in serum, J. Biol. Chem. **195:** 357, 1952.
2. Bunch, L. D.: A rapid cephalin-cholesterol flocculation test, Am. J. Clin. Pathol. **28:**111, 1957.
3. Committee report: Recommendation on a uniform bilirubin standard, Clin. Chem. **8:**405, 1962.
4. Conn, H. O.: Blood ammonia. In Meites, S., editor: Standard methods of clinical chemistry, vol. 5, New York, 1965, Academic Press, Inc.
5. Doumas, B. T., Watson, W. A., and Biggs, H. G.: Albumin standards and the measurement of serum albumin with bromcresol green, Clin. Chim. Acta **31:**87, 1971.
6. Dryer, R. L.: The lipids. In Tietz, N. W., editor: Fundamentals of clinical chemistry, Philadelphia, 1970, W. B. Saunders Co.
7. Fredrickson, D. S., Levy, R. I., and Lees, R. S.: Fat transport in lipoproteins—an integrated approach to mechanisms and disorders, N. Engl. J. Med. **276:**148, 1967.
8. Gambino, S. R.: Bilirubin measurement in the newborn: need for a common reference standard in serum, Hosp. Top., Feb., 1964.
9. Hanger, F. M.: The flocculation of cephalin-cholesterol emulsions by pathological sera, Trans. Assoc. Am. Physicians **53:**148, 1938.
10. Kingsbury, F. B., Clark, C. P., Williams, G., and Post, A. L.: The rapid determination of albumin in urine, J. Lab. Clin. Med. **2:**981, 1926.
11. Knowlton, M.: Cephalin-cholesterol flocculation test. In Seligson, D., editor: Standard methods of clinical chemistry, vol. 2, New York, 1958, Academic Press, Inc.
12. Lackland, H.: Quantitative fibrinogen, Lab. Digest **28:**3, 1965.
13. Maclagan, N. F.: The thymol turbidity test as an indicator of liver dysfunction, Br. J. Exp. Pathol. **25:**234, 1944.
14. Malloy, H. T., and Evelyn, K. A.: The determination of bilirubin with the photoelectric colorimeter, J. Biol. Chem. **119:**481, 1937.
15. Michaelsson, M.: Bilirubin determination in serum and urine, Scand. J. Clin. Lab. Invest. **13** (Supp. 56):1, 1961.
16. Orten, J. M., and Neuhaus, O. W.: Human biochemistry, ed. 9, St. Louis, 1975, The C. V. Mosby Co.
17. Reed, A. H., Cannon, D. C., Winkelman, J. W., Bhasin, Y. P., Henry, R. J., and Pileggi, V. J.: Estimation of normal ranges from a controlled sample survey. 1. Sex- and age-related influence on the SMA 12/60 screening group of tests, Clin. Chem. **18:**57, 1972.
18. Reiner, M., and Cheung, H. L.: A practical method for the determination of fibrinogen, Clin. Chem. **5:**414, 1959.
19. Seligson, D., and Hirbara, K.: The measurement of ammonia in whole blood, erythrocytes and plasma, J. Lab. Clin. Med. **49:**962, 1957.
20. Seligson, D., Marino, J., and Dodson, E.: The determination of sulfobromophthalein in serum, Clin. Chem. **3:**638, 1957.
21. Shank, R. E., and Hoagland, C. L.: A modified method for the quantitative determination of the thymol turbidity reaction of serum, J. Biol. Chem. **162:**133, 1946.
22. Weichselbaum, T. E.: An accurate and rapid method for the determination of proteins in small amounts of blood serum and plasma, Am. J. Clin. Pathol. **16** (tech. sec. 10):40, 1946.

8

Atomic absorption spectroscopy

Atomic absorption spectroscopy is based on the fact that atoms of an element in the ground or unexcited state will absorb light of the same wavelength that they emit in the excited state. When atoms of a metallic element are raised from the ground state to a higher energy level, or excited state, they absorb energy from a source such as heat or light. Since the excited state is extremely unstable, the atoms quickly return to the ground state. In doing so, they release the absorbed energy in the form of light. The wavelengths of that light or the resonance lines are characteristic for each element. No two elements have identical resonance lines. Atomic absorption spectroscopy involves the measurement of light absorbed by atoms in the ground state. Flame photometry or flame emission spectroscopy employs the measurement of light emitted by atoms in the excited state. Most elements have several resonance lines but usually one line is considerably stronger than the others. The wavelength of this line is generally the one selected for measurement. Occasionally, however, it may be necessary to select another resonance line, either to reduce sensitivity or because a resonance line of an interfering element is very close to the one of interest.

If a beam of monochromatic light, wavelength 3247 Å, is passed through a flame and monochromator to a detector and a treated serum specimen is aspirated into the flame, only the copper atoms in the flame absorb light from the beam. The amount of light absorbed is proportional to the amount of copper in the serum specimen.

The sensitivity of atomic absorption techniques has stimulated investigation into the role of trace metals in biologic processes. A summary of these investigations was made by Christian.[3] Schroeder and Nason[8] also have published an excellent review article on trace elements.

HOLLOW CATHODE LAMPS

The source of the monochromatic light is usually a hollow cathode lamp. This consists of a sealed glass tube or envelope containing a rod-shaped anode and a cup-shaped cathode that is about 1 cm in diameter and 1 cm deep. The cathode cup contains the metal of interest. Surrounding the electrodes within the glass envelope is a filler gas, usually neon or argon, at a pressure of about 3 mm Hg. When current is applied to the electrodes, the filler gas within the cathode becomes partially ionized and causes a cloud of metallic ions to sputter off the cathode. These metallic atoms collide with the gas ions and with each other, causing the metal atoms to enter the excited state in which they emit radiation characteristic of the component metal.

The newest hollow cathode lamps have cathodes shielded to contain the glow within the cathode. Glow outside the cathode is not only wasted, so far as absorption measurements are concerned, but it also shortens the life of the lamp. A restrictive shield is used for volatile or molten metals and an open shield is used for metals difficult to sputter.

Multi-element lamps are now available, some containing as many as seven elements.

SENSITIVITY AND DETECTION LIMITS

Sensitivity refers to the concentration of an element in aqueous solution that will produce an absorption of 1%. Generally, it is expressed as µg/ml/1%. *Detection limit* is defined as the concentration of an element in aqueous solution that gives a signal twice the size of the background variability.

INTERFERENCES

Although it may be claimed that certain elements can be determined without interferences, there are certain effects related to the flame that must be considered.

Matrix interferences

A significant difference between the surface tension or viscosity of the sample and that of the standards will cause them to be aspirated or atomized differently. High acid concentration can alter the absorption signal. Standards should have the same acid content as the samples. A total solids content of greater than 2% in aqueous solution may cause interference. For these solutions it may be more accurate to use the *method of additions* to minimize bulk interference effects.

Background absorption

A high salt content in a sample will cause an erroneously high analysis due to the light-scattering effect of the salt. Light is bounced off or reflected from the small particles of salt that still survive in the flame. Since this is light that does not reach the detector, it is measured as absorption. To correct for this light-scattering effect the apparent absorption should be measured at a nearby wavelength that is known to be nonabsorbing for the metal of interest. The background absorption is then subtracted from the analytic line absorption before the metal concentration of the sample is calculated.[9]

Acetylene cylinders

Erratic and inconsistent results are obtained when the pressure of an acetylene tank falls below 75 psi. This is especially true in work with metals that require a very rich acetylene flame. It is a problem caused by chemical interference from acetone vapors. Acetylene is dissolved in acetone in the tank. As the tank pressure decreases, the rate at which the acetone comes out of the tank increases.[6, 7]

NOTE: It is dangerous to leave acetylene trapped in a line. When shutting the instrument down for any extended period of time, close the main valve at the tank and bleed all lines. No pressure should be registered on any of the acetylene gauges.

STANDARDS

Stock standards may be purchased from several sources. Harleco supplies atomic absorption standards with a concentration of 1000 µg/ml. For optimal accuracy, dilute working standards should be prepared just prior to use. Very dilute standards will change with time, as a result of plating of the metal to the sides of the container. The presence of 1% HNO_3 will, in many solutions, retard this plating and lengthen the life of the standard.

PERKIN-ELMER ATOMIC ABSORPTION SPECTROPHOTOMETER, MODEL 303

The discussion to follow concerning instrument parts, settings, and methods refers specifically to the Perkin-Elmer instrument. The general information given applies to all models of Perkin-Elmer AA instruments. For specific instrument settings for models other than the 303, see the appropriate instrument manual.

Burner heads

There are at least five different burner heads available for this instrument. The one most frequently used in clinical work is the three-slot (Boling) burner head. The wide cross section of the flame permits more precise and more sensitive determinations, particularly for metals whose compounds do not dissociate readily in the flame—for example, calcium, chromium, tin. It will also burn concentrated solutions without clogging. Both the fuel and air flows should be adjusted to a reading of 10 on the flowmeter. Use a standard solution and adjust the fuel for maximum absorption when running calcium, magnesium, chromium, strontium, or tin. If a very lean (clear blue) flame is used, the heat will warp the slots unless water or some other solution is aspirated continuously.[1]

The short path burner head may be used when less sensitivity is desired.

Organic solvents

Organic solvent extractions generally enhance sensitivity and will improve detection limits by a factor of about four. Many solvents have been used but the one found most satisfactory is methyl isobutyl ketone. The fuel-to-air ratio should be adjusted to the settings recommended for the analysis. Then the fuel flow should be reduced gradually until the flame is as blue as possible but does not lift off the burner. The nebulizer also must be readjusted for organic solvents. Select an element with a wavelength greater than 250 nm that requires a blue flame (using air, not N_2O)—for example, copper. Loosen the capillary locknut by turning it counterclockwise. While aspirating a standard in the solvent, turn the capillary counterclockwise until bubbling ("blowback") occurs in the solvent at the sample tube intake. Then slowly turn the capillary knob clockwise until maximum absorption is indicated on the null meter. Adjust the null meter needle back to midscale if necessary. When organic solvents are to be burned for more than 1 hour, it is necessary to replace the silicone O-ring in the burner end cap assembly with a Corkprene O-ring. The silicone O-ring must be replaced when aqueous solutions are burned again.[1]

METHOD OF ADDITIONS

When a sample matrix has a very high solids content whose effect on absorption is hard to duplicate with aqueous standards, the method of additions should be used.

Take three equal aliquots of the sample. Add a known volume of diluent to the first aliquot. Add the same volume of known standards to each of the other two aliquots so that each will have a different known concentration of the metal. Determine the percent absorption on each of the three aliquots. Convert percent absorption readings to absorbance and plot against concentration. For example:

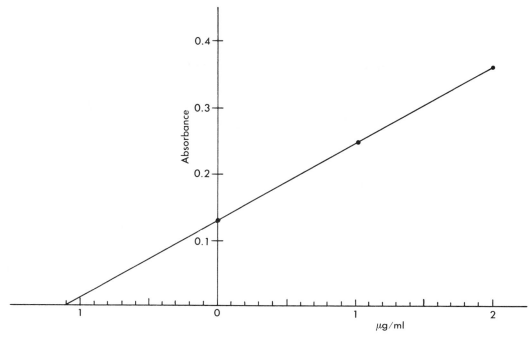

Fig. 8-1. Method of additions.

If 1 μg/ml of the metal was added to the second aliquot and 2 μg/ml to the third, plot the absorbance of the first aliquot at zero concentration, the second at 1 μg/ml concentration, and the third at 2 μg/ml. Draw a line through the three points and extend it back until it intersects the concentration axis. This point will give the concentration of the metal in the original sample (Fig. 8-1).

METHODS
Calcium

Reagents

Lanthanum stock solution, 5% in 25% (v/v) HCl	**(D-145)**
Lanthanum working solution, 0.5% in 2.5% (v/v) HCl	**(D-146)**
Calcium stock standard, 50 mg Ca/dl	**(D-42)**

Intermediate standards

Pipet the following volumes of calcium stock standard into 100-ml volumetric flasks. Add 1 drop of concentrated HCl to each flask. Dilute to volume with water and mix thoroughly.

ml Ca stock	mg Ca/dl
5	2.5
10	5.0
15	7.5
20	10.0
25	12.5
30	15.0

Preparation of working standards and samples

If possible, use an automatic dilutor for these dilutions. Set it to aspirate 0.25 ml of sample (or intermediate standard) and to dispense the sample along

Table 8-1
Instrument settings

Element	Source Lamp ma*	Wavelength Å	Dial	Range	Slit	Burner head	Acetylene pressure Tank	Flow†	Air pressure Regulator	Flow†	Scale expansion
Calcium	H-C‡	4227	211	Visible	4	3-slot	8	10	30	10	1X
Magnesium	H-C	2852	285	UV	5	Standard§	8	9‖	30	8	1X
Copper	H-C	3247	325	UV	4	3-slot	8	10	30	10	5X
Zinc	H-C	2138	214	UV	5¶	3-slot	8	10	30	10	1X
Lead	H-C	2833	283	UV	4	3-slot	8	10	30	10	5-10X

*Use current given on lamp. Space was left for user to fill in.
†All flowmeter readings should be made at the center of the ball.
‡Hollow cathode.
§Use either standard or short path burner.
¶Set to 4 when using a multi-element lamp containing copper.
‖Use a standard solution of 0.5 μg Mg/ml and adjust the acetylene flow to obtain maximum absorption.

with 4.75 ml of lanthanum working solution. Prepare working standards from the intermediate standards in the same manner. Using an automatic dilutor will ensure uniformity in the sample and standard dilutions.

Urine specimens should be collected in a jar containing 15 ml (for a 24-hour collection) of glacial acetic acid. If a specimen contains any sediment, a portion of the urine should be centrifuged. The clear portion may then be diluted in the same manner as the serum.

Procedure

1. Install the calcium lamp in the lamp compartment.
2. Turn the power switch on and adjust the source to the current given on the lamp.
3. Set the instrument to the proper wavelength, range, and slit (Table 8-1).
4. Install the 3-slot burner head.
5. Turn on compressed air supply and adjust the burner regulator to 30 psi and the flowmeter to 10.
6. Open the valve on the tank of acetylene and adjust the pressure coming from the tank to 8 psi. Adjust the fuel flowmeter to 10.
7. Be sure the U in the burner drain tube is filled with liquid and that the end of the tube is below the surface of a liquid. Light the flame.
8. Allow the flame to aspirate water for about 5 minutes; then adjust absorption to zero while aspirating lanthanum working solution.
9. Determine percent absorption on diluted standards and samples.
10. Convert percent absorption to absorbance (Table 8-2) and plot the absorbance of the standards versus concentration on linear graph paper.

A very detailed referee method for the determination of calcium in serum has been published by Cali and co-workers.[2]

Table 8-2
Conversion of percent absorption (R) to absorbance

R	.0	.1	.2	.3	.4	.5	.6	.7	.8	.9
0.0	.0000	.0004	.0009	.0013	.0017	.0022	.0026	.0031	.0035	.0039
1.0	.0044	.0048	.0052	.0057	.0061	.0066	.0070	.0074	.0079	.0083
2.0	.0088	.0092	.0097	.0101	.0106	.0110	.0114	.0119	.0123	.0128
3.0	.0132	.0137	.0141	.0146	.0150	.0155	.0159	.0164	.0168	.0173
4.0	.0177	.0182	.0186	.0191	.0195	.0200	.0205	.0209	.0214	.0218
5.0	.0223	.0227	.0232	.0236	.0241	.0246	.0250	.0255	.0259	.0264
6.0	.0269	.0272	.0278	.0283	.0287	.0292	.0297	.0301	.0306	.0311
7.0	.0315	.0320	.0325	.0329	.0334	.0339	.0343	.0348	.0353	.0357
8.0	.0362	.0367	.0372	.0376	.0381	.0386	.0391	.0395	.0400	.0405
9.0	.0410	.0414	.0419	.0424	.0429	.0434	.0438	.0443	.0448	.0453
10.0	.0458	.0462	.0467	.0472	.0477	.0482	.0487	.0491	.0496	.0501
11.0	.0506	.0511	.0516	.0521	.0526	.0531	.0535	.0540	.0545	.0550
12.0	.0555	.0560	.0565	.0570	.0575	.0580	.0585	.0590	.0595	.0600
13.0	.0605	.0610	.0615	.0620	.0625	.0630	.0635	.0640	.0645	.0650
14.0	.0655	.0660	.0665	.0670	.0675	.0680	.0685	.0691	.0696	.0701
15.0	.0706	.0711	.0716	.0721	.0726	.0731	.0737	.0742	.0747	.0752
16.0	.0757	.0762	.0768	.0773	.0778	.0783	.0788	.0794	.0799	.0804
17.0	.0809	.0814	.0820	.0825	.0830	.0835	.0841	.0846	.0851	.0857
18.0	.0862	.0867	.0872	.0878	.0883	.0888	.0894	.0899	.0904	.0910
19.0	.0915	.0821	.0926	.0931	.0937	.0942	.0947	.0953	.0958	.0964
20.0	.0969	.0975	.0980	.0985	.0991	.0996	.1002	.1007	.1013	.1018
21.0	.1024	.1029	.1035	.1040	.1046	.1051	.1057	.1062	.1068	.1073
22.0	.1079	.1085	.1090	.1096	.1101	.1107	.1113	.1118	.1124	.1129
23.0	.1135	.1141	.1146	.1152	.1158	.1163	.1169	.1175	.1180	.1186
24.0	.1192	.1198	.1203	.1209	.1215	.1221	.1226	.1232	.1238	.1244
25.0	.1249	.1255	.1261	.1267	.1273	.1278	.1284	.1290	.1296	.1302
26.0	.1308	.1314	.1319	.1325	.1331	.1337	.1343	.1349	.1355	.1361
27.0	.1367	.1373	.1379	.1385	.1391	.1397	.1403	.1409	.1415	.1421
28.0	.1427	.1433	.1439	.1445	.1451	.1457	.1463	.1469	.1475	.1481
29.0	.1487	.1494	.1500	.1506	.1512	.1518	.1524	.1530	.1537	.1543
30.0	.1549	.1555	.1561	.1568	.1574	.1580	.1586	.1593	.1599	.1605
31.0	.1612	.1618	.1624	.1630	.1637	.1643	.1649	.1656	.1662	.1669
32.0	.1675	.1681	.1688	.1694	.1701	.1707	.1713	.1720	.1726	.1733
33.0	.1739	.1746	.1752	.1759	.1765	.1772	.1778	.1785	.1791	.1798
34.0	.1805	.1811	.1818	.1824	.1831	.1838	.1844	.1851	.1858	.1864
35.0	.1871	.1878	.1884	.1891	.1898	.1904	.1911	.1918	.1925	.1931
36.0	.1938	.1945	.1952	.1959	.1965	.1972	.1979	.1986	.1993	.2000
37.0	.2007	.2013	.2020	.2027	.2034	.2041	.2048	.2055	.2062	.2069
38.0	.2076	.2083	.2090	.2097	.2104	.2111	.2118	.2125	.2132	.2140
39.0	.2147	.2154	.2161	.2168	.2175	.2182	.2190	.2197	.2204	.2211
40.0	.2218	.2226	.2233	.2240	.2248	.2255	.2262	.2269	.2277	.2284
41.0	.2291	.2299	.2306	.2314	.2321	.2328	.2336	.2343	.2351	.2358
42.0	.2366	.2373	.2381	.2388	.2396	.2403	.2411	.2418	.2426	.2434
43.0	.2441	.2449	.2457	.2464	.2472	.2480	.2487	.2495	.2503	.2510
44.0	.2518	.2526	.2534	.2541	.2549	.2557	.2565	.2573	.2581	.2588
45.0	.2596	.2604	.2612	.2620	.2628	.2636	.2644	.2652	.2660	.2668
46.0	.2676	.2684	.2692	.2700	.2708	.2716	.2725	.2733	.2741	.2749
47.0	.2757	.2765	.2774	.2782	.2790	.2798	.2807	.2815	.2823	.2832

Table 8-2
Conversion of percent absorption (R) to absorbance—cont'd

R	.0	.1	.2	.3	.4	.5	.6	.7	.8	.9
48.0	.2840	.2848	.2857	.2865	.2874	.2882	.2890	.2899	.2907	.2916
49.0	.2924	.2933	.2941	.2950	.2958	.2967	.2976	.2984	.2993	.3002
50.0	.3010	.3019	.3028	.3036	.3045	.3054	.3063	.3072	.3080	.3089
51.0	.3098	.3107	.3116	.3125	.3134	.3143	.3152	.3161	.3170	.3179
52.0	.3188	.3197	.3206	.3215	.3224	.3233	.3242	.3251	.3261	.3270
53.0	.3279	.3288	.3298	.3307	.3316	.3325	.3335	.3344	.3354	.3363
54.0	.3372	.3382	.3391	.3401	.3410	.3420	.3429	.3439	.3449	.3458
55.0	.3468	.3478	.3487	.3497	.3507	.3516	.3526	.3536	.3546	.3556
56.0	.3565	.3575	.3585	.3595	.3605	.3615	.3625	.3635	.3645	.3655
57.0	.3665	.3675	.3686	.3696	.3706	.3716	.3726	.3737	.3747	.3757
58.0	.3768	.3778	.3788	.3799	.3809	.3820	.3830	.3840	.3851	.3862
59.0	.3872	.3883	.3893	.3904	.3915	.3925	.3936	.3947	.3958	.3969
60.0	.3979	.3990	.4001	.4012	.4023	.4034	.4045	.4056	.4067	.4078
61.0	.4089	.4101	.4112	.4123	.4134	.4145	.4157	.4168	.4179	.4191
62.0	.4202	.4214	.4225	.4237	.4248	.4260	.4271	.4283	.4295	.4306
63.0	.4318	.4330	.4342	.4353	.4365	.4377	.4389	.4401	.4413	.4425
64.0	.4437	.4449	.4461	.4473	.4485	.4498	.4510	.4522	.4535	.4547
65.0	.4559	.4572	.4584	.4597	.4609	.4622	.4634	.4647	.4660	.4672
66.0	.4685	.4698	.4711	.4724	.4737	.4750	.4763	.4776	.4789	.4802
67.0	.4815	.4828	.4841	.4855	.4868	.4881	.4895	.4908	.4921	.4935
68.0	.4948	.4962	.4976	.4989	.5003	.5017	.5031	.5045	.5058	.5072
69.0	.5086	.5100	.5114	.5129	.5143	.5157	.5171	.5186	.5200	.5214
70.0	.5229	.5243	.5258	.5272	.5287	.5302	.5317	.5331	.5346	.5361
71.0	.5376	.5391	.5406	.5421	.5436	.5452	.5467	.5482	.5498	.5513
72.0	.5528	.5544	.5560	.5575	.5591	.5607	.5622	.5638	.5654	.5670
73.0	.5686	.5702	.5719	.5735	.5751	.5768	.5784	.5800	.5817	.5834
74.0	.5850	.5867	.5884	.5901	.5918	.5935	.5952	.5969	.5986	.6003
75.0	.6021	.6038	.6055	.6073	.6091	.6108	.6126	.6144	.6162	.6180
76.0	.6198	.6216	.6234	.6253	.6271	.6289	.6308	.6326	.6345	.6364
77.0	.6383	.6402	.6421	.6440	.6459	.6478	.6498	.6517	.6536	.6556
78.0	.6576	.6596	.6615	.6635	.6655	.6676	.6696	.6716	.6737	.6757
79.0	.6778	.6799	.6819	.6840	.6861	.6882	.6904	.6925	.6946	.6968
80.0	.6990	.7011	.7033	.7055	.7077	.7100	.7122	.7144	.7167	.7190
81.0	.7212	.7235	.7258	.7282	.7305	.7328	.7352	.7375	.7399	.7423
82.0	.7447	.7471	.7496	.7520	.7545	.7570	.7595	.7620	.7645	.7670
83.0	.7696	.7721	.7747	.7773	.7799	.7825	.7852	.7878	.7905	.7932
84.0	.7959	.7986	.8013	.8041	.8069	.8097	.8125	.8153	.8182	.8210
85.0	.8239	.8268	.8297	.8327	.8356	.8386	.8416	.8447	.8477	.8508
86.0	.8539	.8570	.8601	.8633	.8665	.8697	.8729	.8761	.8794	.8827
87.0	.8861	.8894	.8928	.8962	.8996	.9031	.9066	.9101	.9136	.9172
88.0	.9208	.9245	.9281	.9318	.9355	.9393	.9431	.9469	.9508	.9547
89.0	.9586	.9626	.9666	.9706	.9747	.9788	.9830	.9872	.9914	.9957

$$A = -\log\left(1 - \frac{R}{100}\right)$$

A = absorbance
R = % absorption

Magnesium in serum and urine

Reagent

    Magnesium stock standard, 100 mEq/L   (D-147)

Intermediate standards

Pipet 1.0, 2.0, 3.0, and 4.0 ml volumes of stock magnesium standard into 100-ml volumetric flasks and dilute to volume with distilled water. Mix. These standards equal 1, 2, 3, and 4 mEq Mg/L.

Working standards

Dilute 1.0 ml of each of the intermediate standards to 50 ml with distilled water. Mix.

Sample preparation

Dilute 0.5 ml of serum or urine to 25 ml with distilled water and mix. Urine dilutions may need to be adjusted according to the magnesium concentration.

If an automatic dilutor is available, set it to aspirate 0.1 ml of sample and to deliver the sample plus 4.9 ml of distilled water. Dilute the intermediate standards in the same manner.

Procedure

1. Install the magnesium lamp in the lamp compartment.
2. Turn the power switch on and adjust the source to the current given on the lamp.
3. Set the instrument to the proper wavelength range, and slit (Table 8-1).
4. Install the standard burner head.
5. Turn on compressed air supply and adjust the burner regulator to 30 psi and the flowmeter to 8.
6. Open the valve on the acetylene tank and adjust the pressure coming from the tank to 8 psi. Adjust the fuel flowmeter to 9.
7. Be sure the U in the burner drain tube is filled with water and that the end of the tube is below the surface of the water. Light the flame.
8. Allow the flame to aspirate water for about 5 minutes; then adjust absorption to zero.
9. Determine percent absorption on diluted standards and samples.
10. Convert absorption to absorbance (Table 8-2) and plot the absorbance of the standards versus concentrations on linear graph paper.

Automated calcium and magnesium

Calcium and magnesium can be automated by using a Technicon sampler and pump to dilute and deliver the specimen to the burner of the atomic absorption spectrophotometer. A manifold diagram is shown (Fig. 8-2) that may be used for either calcium or magnesium.[6] Lanthanum working solution (D-146) containing 0.5 ml Brij 35/L is used as the diluent in both procedures. The intermediate standards used in the manual procedures are poured into the sample cups and run along with the serum and urine specimens. Samples may be run at a rate of 50/hr with a 2:1 sample-wash ratio. An automatic null recorder readout and a 10 mv potentiometric recorder are required.

Copper in serum and urine

Reagents

Copper stock standard, 1000 µg/ml	(Purchase prepared standard.)	
Copper dilute stock standard, 100 µg/ml	(Dilute stock 10 to 100.)	
Sodium diethyldithiocarbamate, 1%	(D-148)	
Sodium ethylenediamine tetraacetate (EDTA), 1%	(D-149)	

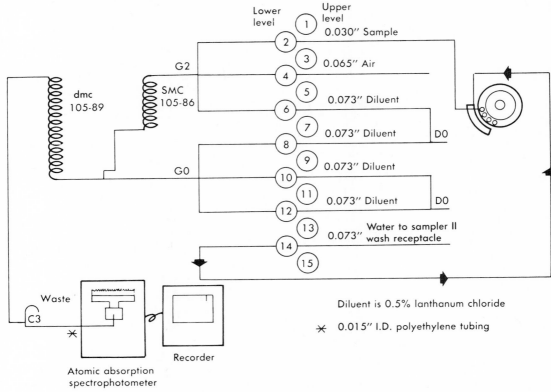

Fig. 8-2. Calcium or magnesium by atomic absorption.

Methyl isobutyl ketone (MIBK)
Sulfuric acid, concentrated
Nitric acid, concentrated

Working standards

Dilute 0.5, 1.0, 1.5, and 2.0 ml volumes of dilute stock copper standard, 100 μg/ml, to 100 ml with water. These standards are equivalent to 50, 100, 150, and 200 μg/dl. For serum standards, dilute 1.0 ml of each working standard with 1.0 ml of water. For urine standards, extract 1.0 ml of each working standard as described beginning with step 3 in sample preparation—urine.

Sample preparation—serum[1]

1. Dilute 1.0 ml of each unknown serum with 1.0 ml of water and mix.
2. Aspirate diluted standards and unknowns directly into the flame after setting the zero while aspirating water. The wavelength setting is 325 nm.
3. Compare the absorbance of the unknowns against the absorbance of the standards.

Sample preparation—urine

1. Transfer 25 ml of urine to a porcelain evaporating dish and add 2.0 ml of concentrated sulfuric acid. Boil gently. Add concentrated nitric acid repeatedly until the digestants are completely clear and colorless.
2. Transfer to a 25ml volumetric flask. Rinse the evaporating dish with a small amount of water and transfer the washings to the volumetric flask. Dilute to volume and mix.

3. Transfer a 10-ml aliquot to a 60-ml separatory funnel. Add 1.0 ml of 1% sodium diethyldithiocarbamate, 7.0 ml of 1% EDTA and 3.0 ml of MIBK to the funnel. Stopper and shake the funnel for 10 minutes.
4. Discard the lower aqueous layer and centrifuge the upper layer.
5. Adjust the flame and nebulizer as directed under Organic solvents. Determine the absorption on the extract and compare with the standards.
6. The calculated copper content is divided by 10 to obtain μg Cu/dl.

Zinc in plasma and urine[4]

Reagents

Zinc stock standard, 1000 μg/ml (Purchase prepared standard.)
Zinc dilute stock standard, 10 μg/ml (Dilute stock to 100.)
Dextran, 6%*
Dextran, 3% (1 part 6% dextran + 1 part water)

Working standards

Dilute 5.0, 10, 15, and 20 ml volumes of dilute zinc standard, 10 μg/ml, to 100 ml with water and mix. Mix 2.0 ml of each standard with 2.0 ml of 3% dextran.

Sample preparation—plasma

1. Hackley and associates[4] found that serum zinc values were about 15% higher than plasma zinc. They recommend collecting blood specimens in tubes containing 0.1 ml of a 30% sodium citrate solution.
2. Mix 2.0 ml of plasma with 2.0 ml of water.
3. Determine percent absorption of standards and unknowns, using the 3-slot burner head.

Sample preparation—urine

1. Urine specimens should be acidified with a quantity of glacial acetic acid equal to about 1% of their volume.
2. Aspirate the undiluted urine directly into the flame along with undiluted working standards and determine percent absorption.

Lead in blood[5] and urine

Reagents

Triton X-100, 5% (v/v)
Ammonium pyrrolidine dithiocarbamate (APDC), 2% (w/v) in 5% Triton X-100
   This solution, when stored in an amber bottle in a refrigerator, is stable for about 1 month.
Methyl isobutyl ketone (MIBK), water-saturated
Lead stock standard, 1000 μg/ml (Purchase prepared standard.)
Lead dilute stock standard, 50 μg/ml (Dilute stock 5 to 100.)

Working standards—whole blood

To each of six tubes containing 1.0 ml of deionized water, add 0, 0, 25, 50, 75, and 100 μl of lead dilute stock standard, respectively. This provides two blanks and four standards equivalent to 25, 50, 75, and 100 μg Pb/dl. To each of these tubes is added 5.0 ml of pooled whole blood. The blanks and standards are then extracted in the same manner as the unknown samples.

---

*Abbott Laboratories.

Procedure—whole blood
1. To a tube containing 1.0 ml of deionized water, add 5.0 ml of the unknown whole blood specimen.
2. Add 1.0 ml of APDC solution to the unknown, blanks, and standard tubes and mix.
3. Add 3.0 ml of MIBK, stopper, and shake vigorously about 60 times.
4. Centrifuge for 8 minutes at about 2000 rpm.
5. Using a 3-slot burner, adjust the flame and nebulizer in the manner described under organic solvents. Set the zero while aspirating water-saturated MIBK.
6. Aspirate the organic phase from each of the blanks, standards, and unknown tubes.

Calculations

Subtract the average absorbance of the two blanks from the absorbance of each of the standards. This corrects for any lead present in the pooled blood. Prepare a calibration curve by plotting the corrected absorbance readings versus the corresponding standard concentrations.

Working standards—urine

Originally, urines were extracted in separatory funnels using 50 ml of urine and 50 ml of water containing the amounts of standards listed below. Since the manipulations of six or more separatory funnels can present problems, the method was modified to permit the extractions to be made in 50-ml screw-capped tubes. By reducing the volumes of urine and water to 30 ml, maintaining the amounts of standards added, and multiplying calculated results by $50/30$ or $5/3$, equivalent results are obtained.

To each of five 50-ml screw-capped tubes, add 30 ml of deionized water that has been adjusted to pH 2.2 to 2.8 with 6 N HCl. Add 0, 25, 50, 100, and 300 μl of lead dilute stock standard to the five tubes. These standards are equivalent to 0 (blank), 2.5, 5.0, 10, and 30 μg Pb/dl, respectively.

Procedure—urine
1. Transfer 30 ml of urine to a 50-ml screw-capped tube and adjust the pH to 2.2 to 2.8.
2. Add 1.0 ml of APDC to the blank, standards and unknowns. Mix.
3. Add 3.0 ml of MIBK, cap, and mix by inversion for 2 minutes.
4. Centrifuge at about 2000 rpm for 8 minutes.
5. Transfer organic phase (and emulsion layer, if present) to 10 × 100-mm tubes using transfer pipets.
6. Stopper and centrifuge for about 5 minutes.
7. Using a 3-slot burner, adjust the flame and nebulizer in the manner described under organic solvents. Set the zero while aspirating water-saturated MIBK.
8. Aspirate the organic phase from the blank, standard, and unknown tubes.

Calculations

Subtract the absorbance of the blank from that of the standards and unknown. Plot the corrected readings of the standards versus their respective concentrations and determine the lead concentrations of the unknowns directly from the graph. Multiply results by $5/3$.

Interpretation of biologic lead levels

	Blood lead (μg/dl)	
	Nonoccupational	Occupational
Normal	10-50	10-70
Equivocal; may be considered as evidence for or against; recheck in 2 to 4 weeks	50-70	70-80
Evidence for lead intoxication; recheck in 1 to 4 weeks	80+	90+

	Urine lead (μg/L)	
	Nonoccupational	Occupational
Normal	10-80	10-150
Equivocal; recheck in 2 to 4 weeks	80-150	150-200
Evidence for lead intoxication; recheck in 1 to 4 weeks	150+	200+

References

1. Analytical methods for atomic absorption spectroscopy, Norwalk, Conn., 1973, The Perkin-Elmer Corp.
2. Cali, J. P., Mandel, J., Moore, L., and Young, D. S.: A referee method for the determination of calcium in serum, National Bureau of Standards Special Publication 260-36, 1972.
3. Christian, G. D.: Medicine, trace elements, and atomic absorption spectroscopy, Anal. Chem. **41:**24A, 1969.
4. Hackley, B. M., Smith, J. C., and Halsted, J. A.: A simplified method for plasma zinc determination by atomic absorption spectrophotometry, Clin. Chem. **14:**1, 1968.
5. Hessel, D. W.: A simple and rapid quantitative determination of lead in blood, Atomic Absorption Newsletter **7:**55, 1968.
6. Manning, D. C.: The effect of acetone vapor in the air-acetylene flame, Atomic Absorption Newsletter **7:**44, 1968.
7. Manning, D. C., and Chabot, H.: The effect of acetylene containing acetone vapor in atomic absorption analysis, Atomic Absorption Newsletter **7:**94, 1968.
8. Schroeder, H. A., and Nason, A. P.: Trace-element analysis in clinical chemistry, Clin. Chem. **17:**461, 1971.
9. Slavin, W.: Atomic absorption spectroscopy, New York, 1968, Interscience Publishers, p. 72.

9

Enzymes

INTRODUCTION
What is an enzyme?

A phenomenon that intrigues biochemists is how the living cell performs with such speed and precision several thousand chemical reactions that otherwise occur with immeasurable slowness at the same temperature and pressure. It has long been known that the answer is to be found in the catalysts created by the living cell. A catalyst is defined as a substance that promotes a chemical reaction but generally not its extent; it is not consumed during the course of the reaction and only a very small amount is needed. The biologic catalysts that mediate all the chemical reactions necessary for life are called enzymes. All known enzymes are globular proteins, although many require metal ions (Mg^{++}, Mn^{++}, Ca^{++}, Cu^{++}, Co^{++}, Zn^{++}, and others) and organic molecules, or cofactors, to operate.

Enzymatic activation

Before a chemical reaction can occur, whether catalyzed or uncatalyzed, the molecules involved in the reaction must acquire a certain energy content, the so-called activation energy. In uncatalyzed reactions, an increased reaction temperature is usually employed for this purpose. As heat is absorbed, the internal energy of the molecules increases and they react more readily. In enzyme-catalyzed reactions, high unphysiologic temperatures are not necessary, since these catalysts, in some still unknown way, reduce the amount of energy required for activation.

The differences in activation energy between uncatalyzed and catalyzed reactions can best be illustrated by an example. In the absence of any catalyst, the energy of activation for the decomposition of hydrogen peroxide to water and hydrogen is 18,000 calories per mole of H_2O_2. The presence of colloidal platinum lowers this to 11,700 calories. A further reduction to 2000 calories per mole occurs when the reaction is catalyzed by the enzyme catalase.

Enzyme-substrate complex

The combination of an enzyme with its substrate is a basic step in the catalytic process. The enzyme-substrate complex formed dissociates on completion of the reaction to yield the end products and free enzyme. The enzyme now is ready to bind with another substrate molecule.

$$E + S \rightleftharpoons ES \rightarrow P + E$$

Enzymes usually are quite specific in their action. A very small change in the conformation of a molecule may mean that the enzyme will no longer accept it as a substrate. This led Emil Fischer, in 1894, to propose his famous "lock and key" theory, which states that the surface of an enzyme is shaped in such a way that its substrate fits onto it like a key into a lock, and that only slightly altered molecules may be unable to fit onto the enzyme and therefore will not be acted upon by it.

Factors affecting enzyme activity

It is important to understand basic enzyme kinetics in enzyme analysis in order to understand the basic enzymatic mechanism and to select a method for enzyme analysis. The conditions selected to measure the activity of an enzyme would not be the same as those selected to measure the concentration of its substrate. Perhaps the most important of the many factors affecting the rate at which enzymatic reactions proceed are substrate concentration, inhibitors, temperature, and pH.

SUBSTRATE CONCENTRATION

Generally, when the substrate is present in excess, the reaction rate is maximal and zero-order kinetics are said to be in operation. In zero-order kinetics, the rate of reaction is constant; that is, the same amount of substrate is converted by a fixed amount of enzyme during each unit of time.

The reaction rate, therefore, is independent of the substrate concentration and a straight line is obtained when the product concentration is plotted with respect to time. If the substrate concentration is inadequate to saturate the enzyme, then the rate of reaction is less than maximal. It decreases steadily and is proportional to the substrate concentration existing at any one particular moment. The reaction then is considered to exhibit first-order kinetics. Although the substrate concentration changes continually, the fraction converted remains the same. Therefore, the graphic representation is a curve.

Most enzymatic reactions are actually a mixture of both types of reaction with the type changing from zero order to first order as the substrate concentration diminishes. In testing for zero-order kinetics, the reaction is allowed to proceed for twice the recommended time. The amount of product formed at the end of the double time interval should be twice the amount obtained at the end of the recommended time.

Michaelis constant

The amount of substrate required for saturation differs from enzyme to enzyme and, for any given enzyme, it also differs from one substrate to another. A measure of the affinity an enzyme has for a particular substrate is given by the Michaelis constant (K_m).

Michaelis developed a set of mathematical expressions to calculate activity in terms of reaction rate from laboratory data. The Michaelis constant, K_m, is defined as the substrate concentration at half the maximum velocity.

In the equation:

$$E + S \underset{k_2}{\overset{k_1}{\rightleftharpoons}} ES \overset{k_3}{\longrightarrow} E + P \tag{1}$$

k_1, k_2, and k_3 represent the rate constants of the reactions. The Michaelis constant is defined as:

$$K_m = \frac{k_2 + k_3}{k_1} = [S]_{\frac{V_{max}}{2}} \tag{2}$$

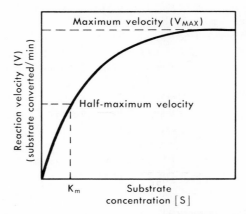

Fig. 9-1. Relationship between reaction velocity, substrate concentration, and K_m.

The breakdown of ES into product and free enzyme occurs at a rate determined by the concentration of ES. Since direct measurement of the amount of ES (and often E) is not possible, the velocity of the reaction is determined instead, and substitutions are made in equation 2 and the terms rearranged. Using this relationship, the equation can be expressed in the following form:

$$v = \frac{V_{max}\,[S]}{K_m + [S]} \tag{3}$$

or

$$K_m = \frac{[S]\,(V_{max} - 1)}{v} \tag{4}$$

where V = observed velocity at a given substrate concentration [S]
V_{max} = maximum velocity when all the enzyme is saturated by substrate
K_m = substrate concentration required for the reaction to reach half maximal velocity ($v = \frac{1}{2} V_{max}$)

This becomes apparent when numerical values are substituted in equation 4:

$$K_m = [S]\left(\frac{2}{1} - 1\right) \tag{5}$$

The relationship between reaction velocity, substrate concentration, and K_m can be seen in Fig. 9-1.

Michaelis constants have been determined for most of the commonly used enzymes. The size of K_m tells several things about a particular enzyme. A small K_m indicates that the enzyme requires only a small amount of substrate to become saturated. Therefore, V_{max} is reached at relatively low substrate concentrations. A large K_m indicates the need for high substrate concentrations to achieve maximum reaction velocity. The substrate with the lowest K_m on which the enzyme acts as a catalyst is generally assumed to be the enzyme's natural substrate, though this is not true for all enzymes.

EFFECTS OF INHIBITORS

Enzyme inhibitors are substances that alter the catalytic action of an enzyme and consequently slow down or, in some cases, stop catalysis. Three common types

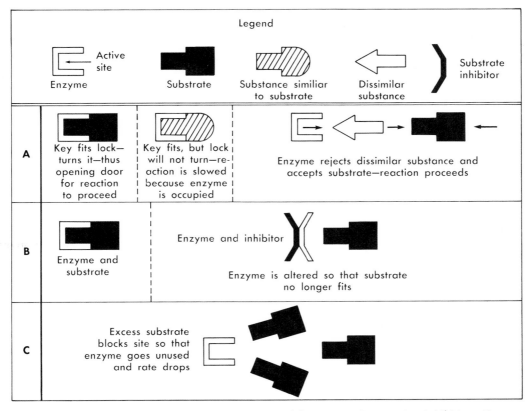

Fig. 9-2. Lock and key theory of enzyme inhibition. **A,** Competitive inhibition; **B,** noncompetitive inhibition; **C,** substrate inhibition.

of enzyme inhibition are competitive, noncompetitive, and substrate inhibition. Most theories concerning the mechanisms of inhibition are based on the existence of the enzyme-substrate complex.

Competitive inhibition occurs when a substance resembling the substrate is present in the reaction mixture along with the enzyme and substrate. According to the "lock and key" theory, a particular location on the enzyme surface, known as the active site, has a strong affinity for the substrate. The substrate molecule is held in such a way that its conversion to the reaction products is most favorable. Consider the enzyme active site as the lock and the substrate as the key (Fig. 9-2). The key is inserted in the lock, turned, the door is opened, and the reaction proceeds. However, when an inhibitor that resembles the substrate is present, it will compete with the substrate for the position in the enzyme lock. When the inhibitor wins and gains the lock position, it is unable to turn and open the door. Hence, the reaction is slowed down because some of the available active sites are occupied by the inhibitor. If a dissimilar substance is present that does not fit the site, the enzyme rejects it and accepts the substrate, and the reaction proceeds normally.

Noncompetitive inhibitors are substances that bind to the enzyme at a location other than the active site and alter the active site in such a way that it cannot accept the substrate.

Substrate inhibition may occur when excessive amounts of substrate are present.

Additional amounts of substrate added to the reaction mixture after the maximum velocity has been reached may actually decrease the reaction rate. It is theorized that because there are so many substrate molecules competing for the active sites on the enzyme surfaces, they block the sites and prevent substrate molecules from occupying them. This causes the reaction rate to drop.

EFFECTS OF TEMPERATURE

Like most other chemical reactions, the rate of an enzyme-catalyzed reaction increases with an increase in temperature. A 10° C rise in temperature will increase the activity of most enzymes by 50% to 100%. Variations in reaction temperature as small as 1° or 2° may cause changes of 10% to 20% in the results.

The effects of temperature on the activity of an enzyme present themselves as two forces acting simultaneously but in opposite directions. With increasing temperature the activity increases and at the same time denaturation of the enzyme protein (inactivation) is accelerated. The higher the temperature, the greater the inactivation.

The temperature at which denaturation becomes a decisive factor in activity measurements varies from enzyme to enzyme. While for many enzymes heat inactivation is negligible below 30° C and begins between 30° and 40° C, some enzymes can be heated to 60° C without appreciable loss of activity. Therefore, measurements of enzyme activities should be performed at a temperature that does not exceed the recommended temperature in the method used.

EFFECTS OF pH

Enzymes are affected considerably by changes in pH. The most favorable pH, the point where the enzyme is most active, is considered the optimum pH. Extremely high or low pH values result in complete loss of activity for most enzymes. The optimum pH varies greatly from one enzyme to another. For example, the optimum pH for pepsin activity is 1.5 to 1.6, whereas for pancreatic lipase it is 8.0.

Enzyme nomenclature

Through the years, as more and more enzyme and their properties were identified, difficulties in terminology arose from instances where the same enzyme had been given different names or, conversely, the same name applied to different enzymes. In 1964, the Commission on Enzymes of the International Union of Biochemists adopted a numerical system designed to simplify identification of the growing population of enzymes.

In the system devised by the Commission, enzymes are divided into six major categories according to the type of reaction they catalyze:

1. Oxidoreductases
2. Transferases
3. Hydrolases
4. Lyases
5. Isomerases
6. Ligases

This grouping together with the respective substrates provides the basis for the numbering and the naming of individual enzymes.

The numbering system for an enzyme consists of four figures separated by periods. The first figure refers to the major category cited above. The second and third figures give additional information about the substrates and coenzymes that are a part of the reaction. The fourth figure is simply the serial number for a specific enzyme.

In addition to the numbers, two types of names have been given to each enzyme: systematic and trivial. The systematic name attempts to identify the substrate precisely and to describe the action of the enzyme as exactly as possible. The trivial name is sufficiently short for general use but is not necessarily descriptive of the reaction it catalyzes. In a great many cases, it is the name already in current use. In the 1972 revision,[9] the Commission has changed the term "trivial name" to "recommended name."

Enzyme units

Enzyme concentrations are reported in terms of activity units. A number of different enzyme units have come into usage, largely through having their origin in a certain investigator's technique. Some results have been expressed as Bodansky, Karmen, Somogyi, Wroblewski, and other units. Such variation in reporting made the comparison of results very difficult.

In an effort to standardize units, the Enzyme Commission,[25] in 1961, proposed that the unit of enzyme activity be defined as that amount of enzyme that will catalyze the transformation of 1 micromole (μM) of substrate per minute under optimal conditions of pH, temperature, and substrate concentration, and that this unit be called the International Unit, U. It was further recommended that the concentration be expressed in terms of U/ml or mU/ml, whichever gives the more convenient numerical value. The symbol IU, rather than U, is commonly used to differentiate the International Unit from other units of activity.

In 1972, the Commission[9] revised the definition of an enzyme unit. It is now recommended that the unit in which enzymatic activity is expressed be the amount of activity that converts 1 mole* of substrate per second. The new unit of enzymatic activity is named the *katal* (symbol: kat). An activity of 1 katal, corresponding to the conversion of 1 mole of substrate per second, will very often be too great for practical use. In most cases, activities will be expressed in microkatals (μkat), nanokatals (nkat), or picokatals (pkat), corresponding to reaction rates of micromoles, nanomoles, or picomoles per second, respectively.

Enzyme activities expressed in the former enzyme units (IU) may be converted to katals with the following formula:

$$1 \text{ IU} = 1 \ \mu\text{M/min} = \frac{1}{60} \mu\text{M/sec} = \frac{1}{60} \mu\text{kat} = 16.67 \text{ nkat}$$

The changeover to the new units is expected to be a slow process.

Choice of methods

Whenever practicable, the preferred methods for determining enzyme activity are the kinetic or time-rate methods. Most of these methods are based on the fact that the reduced nicotinamide adenine dinucleotides, NADH and NADHP, absorb light with a peak at 340 nm, while the oxidized forms, NAD$^+$ and NADP$^+$, show no absorption between 300 and 400 nm. Any reaction in which either NAD$^+$ or NADP$^+$ is reduced or NADH or NADHP oxidized can be measured by recording the increase or decrease, respectively, of the absorbance at 340 nm.

Generally, the reagents required for these methods have limited stability and are subject to contamination by activators or inhibitors. Therefore, it is recommended that the reagents be purchased from a reliable reagent manufacturer and that their directions be followed in performing the enzyme analyses. For this reason, no kinetic enzyme methods are given in this chapter.

---

*The mole has been defined as the amount of substance of a system that contains as many elementary entities as there are carbon atoms in 0.012 kg of the nuclide ^{12}C.

Table 9-1
Conditions accompanied by abnormal amylase values

When increased	When decreased
Acute pancreatitis	Cirrhosis of liver
Carcinoma of head of pancreas	Carcinoma of liver
Duodenal ulcer	Abscess of liver
Perforation of gastric ulcer	Hepatitis
Hyperthyroidism	Acute alcoholism
Mumps	Toxemias of pregnancy
Acute injury of spleen	
Renal disease with impaired renal function	
High intestinal obstruction	

Methods
AMYLASE (α-1,4-GLUCAN, 4-GLUCANOHYDROLASE), EC 3.2.1.1

Amylase is a hydrolytic enzyme that catalyzes the splitting of polysaccharides with an α-1,4 glucosidic link, such as starch and glycogen. Many tissues, including liver, kidney, and muscle, contain amylase, but those of greatest importance are the α-amylases of the pancreas and salivary glands. Pancreatic amylase is synthesized in the endoplasmic reticulum of the acinar cell and then aggregated as active enzyme in the zymogen granules.

Pancreatic enzymes are secreted into the many small ducts of the pancreas and these in turn empty into the main duct, the duct of Wirsung, which runs the entire length of the pancreas before emptying into the duodenum.

The amylase found in normal serum may have its origin, at least in part, in the liver, because its concentration is not affected by pancreatectomy. Urinary amylase is derived from plasma.

Amylase has a molecular weight of 45,000 and a pH optimum of 7. The enzyme is activated by Ca^{++} and therefore can be classified as a metaloenzyme. Monovalent anions also are required for maximum activity. Chloride ions are most important (not less than 0.01 M), Br^- and I^- having less effect. Fluoride inhibits amylase activity, as does oxalate, citrate, polyphosphate, or EDTA, each of which binds Ca^{++}.

Some of the conditions in which abnormal amylase activities are found are listed in Table 9-1.

SACCHAROGENIC METHOD

The basic saccharogenic method, which involves the measurement of reducing sugars hydrolyzed from polysaccharides, was introduced by Kjeldahl[14] in 1880 and modified by Somogyi[32] in 1938. Somogyi's modification, which depends only on the serum in the reaction mixture to maintain the pH, has been and continues to be widely used.

Henry and Chiamori[12] modified the method to include buffering of the reaction mixture as well as the use of optimal substrate concentration. The reducing substances (expressed as glucose) are determined in a protein-free filtrate of the assay reaction mixture and a blank. The unit of activity is that amount of enzyme which catalyzes the liberation of 1 mg of glucose per milliliter of specimen.

AMYLOCLASTIC METHOD

The amyloclastic method differs from the saccharogenic method in that the rate of disappearance of substrate is determined. In 1908, Wohlgemuth[36] introduced the basic assay that is still in use. Iodine is added to reaction mixtures of buffered starch and serially diluted samples. The starch-iodine color is a function of the extent of starch hydrolysis. The difficulties inherent in this procedure include interference from icteric or lipemic sera, hemolyzed blood, and any other reducing agents.

CHROMOGENIC METHODS

In recent years synthetic chromogenic amylase substrates have been developed. These substrates are derivatives of polysaccharides containing chromogens or dyes as part of their structure. The hydrolytic action of amylase releases the dye from the large carbohydrate substrate molecule. After precipitation of the remaining substrate, the chromogen in the supernate is measured colorimetrically.

One of the first dye-bound substrates was prepared by Rinderknecht and associates,[26] who used the dye, Remazolbrilliant blue.

Another method uses a dyed amylopectin substrate.[2, 4]

Klein, Foreman, and Searcy[16] introduced a chromogenic method using as substrate Cibachron Blue F3GA dyed-amylose, a compound of unknown structure. It is presumed that the treatment of amylose with Cibachron Blue (a monochlorotriazine dye) results in formation of a covalent bond preferentially at the primary (C6) hydroxyl of the glucosyl portions of the linear $1 \rightarrow 4$ carbohydrate polymer, producing an ether.

Determination of amylase activity
Specimen stability

Amylase is quite stable. Negligible activity is lost at room temperature in the course of a week or at refrigerator temperatures over a 2-month period.

KLEIN-FOREMAN-SEARCY METHOD[16]
Principle

The α-amylase activity is determined by the enzymatic cleavage of the dye-polysaccharide compound, Cibachron Blue F3GA–amylose. The soluble dye products released from this substrate are directly proportional to the amylase activity of the serum or urine. The absorbance of the dye products is measured at 625 nm.

Reagents*

Cibachron Blue F3GA-amylose substrate
Phosphate buffer, 0.1 M, pH 4.3
Dye standard, 0.2 mg dye/ml

Procedure

1. **Place 1 substrate tablet (200 mg) into a test tube containing 1.9 ml of distilled water. The tablet will disintegrate rapidly. Mix the resultant slurry thoroughly by vortexing and warm to 37° C.**

---

*Available as test kit, Amylochrome, from Roche Diagnostics, Division of Hoffmann-LaRoche, Inc.

2. Add 0.1 ml of serum or urine and mix again by vortexing. Incubate this mixture for exactly 15 minutes at 37° C. If more than one specimen is to be assayed allow 1 minute between sample additions.
3. Prepare a reagent blank by placing 1 substrate tablet in 2.0 ml of distilled water and treating as a test sample. The resulting blank solution is stable for 2 weeks at room temperature.
4. At the end of the 15-minute incubation period, add 8.0 ml of the acid phosphate buffer to each tube. Mix thoroughly and centrifuge.
5. Transfer the clear supernate to a cuvet and read absorbance at 625 nm against the reagent blank. Determine the amylase activity from a calibration curve.

Calibration curve
1. Pipet 1.0, 2.0, and 3.0 ml of dye stock standard into 10-ml volumetric flasks, dilute the contents of each flask to the 10-ml mark with phosphate buffer, and mix well.
2. Read the absorbance of each standard at 625 nm against phosphate buffer.
3. Plot the absorbance of the standards against their equivalent concentrations of 200, 400, and 600 dye units/dl on linear graph paper.

Normal values
Serum 45 to 200 dye units/dl
Urine 0.66 to 5.43 dye units/min[15]

Comments

In the case of an extremely turbid, hemolyzed, or discolored specimen, a serum blank is necessary. The blank is assayed as above except that the sample is added after the phosphate buffer in step 4. Subtract the blank value from the test value.

Urine for amylase determination should be collected over a precisely timed period, for example, 2 hours. To determine units per minute, perform assay as described above and use the following formula:

$$\frac{\text{dye units/ml} \times \text{volume of 2-hr urine}}{\text{Time (min)}} = \text{Dye units/min}$$

LIPASE (GLYCEROL ESTER HYDROLASE), EC 3.1.1.3

Lipases are a group of enzymes that hydrolyze the glycerol esters of long-chain fatty acids. It is believed that the products of the reaction are 1 mole of β-monoglyceride and 2 moles of fatty acids per mole of substrate. A small amount of the ester is hydrolyzed completely to glycerol. The enzyme preferentially attacks the end ester bonds of the glycerol ester.

Lipase activity is inhibited by heavy metal ions, quinine, many aldehydes, eserine, and diisopropylfluorophosphate but not by fluoride ions.

The most important source of lipase is the pancreas, although some lipase is secreted by the gastric mucosa and intestinal cells.

In general, serum lipase values parallel amylase values. At the onset of acute pancreatitis, the lipase level increases more rapidly and remains elevated longer (up to 7 to 10 days) than does the amylase value. Elevations are observed in obstruction of the pancreatic duct, in about 50% of the cases of carcinoma of the pancreas, and occasionally in patients with duodenal ulcers, intestinal obstruction, and renal diseases (retention of lipase).

In 1932, Cherry and Crandall[6] introduced a lipase method that used a 50%

emulsion of olive oil with 5% (w/v) gum acacia in a phosphate buffer of pH 7.0 as substrate. After a 24-hour reaction period at 37° C, the fatty acids formed were titrated with 0.05 N NaOH to a phenolphthalein end-point. A unit of lipase activity was defined as the quantity of enzyme which liberated acid equivalent to 1.0 ml of 0.05 N NaOH. In the ensuing years, various modifications have appeared involving substrate emulsifiers, incubation times, indicators, pH optima, and types of buffers.

In recent years, two methods have been reported that avoid the difficulty of titration against a buffer. Both employ modified Cherry-Crandall substrates and remove the liberated fatty acids from the reaction mixture by solvent extraction after a 30-minute incubation period.

Massion and Seligson[17] measured photometrically the color change produced in a balanced methyl red solution by the extracted fatty acid. The color is not linear and requires several standards for an accurate calibration curve.

The method of Dirstine, Sobel, and Henry[8] is similar in that the fatty acids produced are extracted with a solvent and then reacted to form a compound that can be measured photometrically. The fatty acids are complexed with copper (copper nitrate) to form a soluble, colored, copper soap that can be quantitated accurately in extremely small concentrations.

The usefulness of this method is limited by the necessity for three extraction and wash steps that substantially increase the time required for performance. The advantages of sensitivity and short incubation time are outweighed by the disadvantages of solvent extractions and centrifugations. Therefore the simpler titrimetric procedure is preferred for the routine clinical laboratory.

Determination of serum lipase
Specimen stability

Lipase in serum is stable for a week at room temperature and for several weeks at 4° C. Serum should be free of hemolysis, since hemoglobin inhibits lipase activity.

TIETZ-FIERECK METHOD[35]
Principle

An aliquot of serum is incubated at 37° C with a stabilized olive oil emulsion and buffer of pH 8.0 for 3 hours. The olive oil is hydrolyzed by lipase to fatty acids, diglycerides, and to a small extent monoglycerides and glycerol. The fatty acids liberated are titrated with 0.05 N NaOH either electrometrically to a pH of 10.5 or visually to a light blue color with thymolphthalein as the indicator. The number of milliliters of 0.05 N NaOH used to reach the end-point is equivalent to the (conventional Cherry-Crandall) lipase units per milliliter of specimen. International Units per liter of specimen may be obtained by multiplying the conventional units by the factor of 278.*

Reagents†

Olive oil emulsion	(D-150)
Buffer, stock, 0.8 M TRIS	(D-151)
Buffer, working, 0.2 M, pH 8.0	(D-152)

---

*One International Unit of lipase activity is that amount that will catalyze the hydrolysis of $\frac{1}{3}$ μM of substrate ($= 1$ μEq) per minute under the conditions of the test. Since decimal figures are inconvenient, the International Unit in this case is expressed in enzyme activity per liter of specimen rather than per 1 ml of specimen.

†All reagents may be obtained from Sigma Chemical Co.

Sodium hydroxide, 0.05 N (D-153)
Thymolphthalein indicator, 1% (D-154)
Ethanol, 95%

Procedure

1. Into each of two test tubes labeled *blank* and *test,* pipet 2.5 ml of water and 10 ml of olive oil.
2. Using a 1-ml volumetric pipet, add exactly 1.0 ml of working buffer to each tube.
3. Into the tube labeled *test,* add 1.0 ml of serum. Stopper both tubes and shake vigorously.
4. Incubate the tubes for 3 hours at 37° C. The amount of liberated fatty acids may be determined by electrometric titration (5a) or by visual titration (5b).
5a. *Electrometric end-point determination.* Transfer the contents of the test and blank tubes to 1-oz medicine glasses. Wash both tubes with 3 ml of 95% ethanol and add washings to their respective glasses. Add 1 ml of serum to the blank and mix the contents of both glasses. Slowly titrate the contents of each glass to pH 10.5 with 0.05 N NaOH. A magnetic stirrer is recommended for efficient mixing.
5b. *Visual end-point determination.* Transfer the contents of the test and blank tubes to 50-ml Erlenmeyer flasks. Wash both tubes with 3 ml of 95% ethanol and add washings to their respective flasks. Add 1.0 ml of serum to the blank and 4 drops of thymolphthalein indicator to both flasks. Mix thoroughly. Titrate the contents of both flasks to a light but distinct blue color with 0.05 N NaOH. Test and blank should have the same color intensity.

Calculations

Subtract the blank value from the test value.
Example: Test   5.35 ml 0.05 N NaOH
         Blank 4.45 ml 0.05 N NaOH
                0.90 ml 0.05 N NaOH

In the example, the unknown serum contains 0.9 Cherry-Crandall unit of lipase activity per milliliter. To convert to International Units per liter, multiply this value by 278.

Normal range

0.1 to 1.0 Cherry-Crandall unit.

ALKALINE AND ACID PHOSPHATASE[22, 37]

The two phosphatases routinely analyzed are called alkaline and acid phosphatases according to the optimum pH at which they are measured.

The optimum pH of alkaline phosphatase is 9.7. Alkaline phosphatase is found in serum, bone, kidney, liver, mammary gland, intestine, lung, spleen, leukocytes, seminiferous tubules, and adrenal cortex. However, it is found chiefly in bone and the liver. Since this enzyme plays an important part in bone formation, there is an increase of activity in children. At birth, the alkaline phosphatase activity is low, rising rapidly in the first month of life, remaining highly elevated the first 2 years, and then decreasing to the normal child's range of activity.

Serum alkaline phosphatase increases in bone disease. There have been various theories regarding how this increased action takes place. One theory is that in the absence of normal bone synthesis, the bone's capacity for cellular activity is greater. With this greater activity there is an increased formation of phosphatase. Other explanations are given. The increased enzyme action may be one result of an overproduction of the enzyme in the bone attempting to compensate for the lesion, or there may be a forced discharge of the enzyme from the injured bone tissue. Diseases associated with this increase in the alkaline phosphatase are Paget's disease, rickets, bone atrophy, osteomalacia, osteoporosis, and bone malignancy. Moderately increased values have been found in hyperparathyroidism.

Alkaline phosphatase is also excreted into the bile by the normal liver. The serum values are usually increased in jaundice due to mechanical obstruction. Most liver diseases cause an elevation of results.

A decrease in the enzyme reaction, giving consistently low results, is found in individuals with hypophosphatasia (a hereditary bone disease), anemia, infectious hepatitis, or cretinism and in children who have an unusually early development.

The optimum pH of the acid phosphatase is 5.0. This enzyme was first discovered in male urine. Further investigation has shown the prostate gland to be extremely rich in this enzyme, thus leading to extensive investigation of tumors of the prostate. Normal blood plasma or serum contains small amounts of acid phosphatase. This may have its origin in the liver, spleen, bones, kidney, or prostate. It is present in women and children also, although the analysis for acid phosphatase is usually requested for diagnosis of prostate carcinoma.

Gutman and co-workers[10] (1936) first demonstrated that serum acid phosphatase levels increased significantly in many cases of metastatic prostatic carcinoma but seldom increased in the absence of metastases. However, there are three possibilities of obtaining normal results when metastasis has occurred:

1. Prostatic cells may produce very little enzyme because of very low androgen (hormone) level or because of anaplasia.
2. Cancer may have affected the lymph and blood channels so that the excessive amount of the enzyme cannot get into the bloodstream.
3. Castration, administration of estrogens, intensive irradiation, or radical prostatectomy inhibits the activity of the enzyme.

Women may also show elevations of acid phosphatase if they suffer from disease of any one of several organs (liver, spleen, kidney, or bone). There is an increase of the enzyme activity with carcinoma of the breast with extensive skeletal metastasis.

Since the red cells contain a high level of acid phosphatase, it is imperative that serum used for the acid phosphatase analysis contain no hemolysis.

Alkaline phosphatase, EC 3.1.3.1

The normal values for alkaline phosphatase depend upon the method of determination. The values for three of the commonly used methods are listed here:

	Adults	*Children*
Bodansky units	1.5 to 4.5	5 to 14
King-Armstrong units	4 to 13.0	15.0 to 30.0
Bessey-Lowry-Brock units	0.8 to 2.3	2.8 to 6.7

Several conditions in which an increased or decreased value may be expected are listed in Table 9-2. Abnormal values are of primary interest in those disorders given in boldface.

For the determination of alkaline phosphatase, the three methods discussed next are commonly used.

Table 9-2
Conditions accompanied by abnormal alkaline phosphatase values

When increased		When decreased
Rickets	Neurofibromatosis	Scurvy
Hyperparathyroidism	Myositis ossificans	Hypoparathyroidism
Carcinoma of bone	Obstructive jaundice	Chronic nephritis
Paget's disease	Idiopathic steatorrhea	Celiac disease
Osteomalacia	Occlusion of pancreatic duct	Osteolytic sarcoma
Multiple myeloma	Abscess of liver	

BODANSKY METHOD

Serum is added to sodium glycerophosphate that has been buffered to pH 8.6 in order to ensure optimum conditions of pH and electrolyte content. The mixture is incubated at 37° C for exactly 1 hour. This allows the phosphatase enzyme in the serum to decompose the sodium glycerophosphate and liberate phosphorus.

A phosphorus determination is then run on this incubated serum and also on nonincubated serum. The phosphorus determination consists of adding trichloroacetic acid to both specimens to precipitate the proteins, adding molybdic acid to form a phosphate complex, adding a reducing agent to reduce the complex, and finally measuring the depth of color in a colorimeter.

The alkaline phosphatase units are then found by subtracting the phosphorus value of the nonincubated specimen from the phosphorus value of the incubated specimen. Thus:

$$\text{Alkaline phosphatase units} = \text{Phosphorus value of incubated specimen} - \text{Phosphorus value of nonincubated specimen}$$

Example:

$$\text{Alkaline phosphatase units} = 6 \text{ mg/dl} - 3 \text{ mg/dl} = 3$$

These units are usually reported as Bodansky units, a Bodansky unit being that amount of phosphatase which will liberate 1 mg of phosphorus/dl of serum under the conditions described above.

KING-ARMSTRONG METHOD

Serum is added to disodium phenylphosphate that has been buffered to pH 9.0. The mixture is incubated at 37.5° C for 30 minutes to allow the phosphatase to decompose the disodium phenylphosphate and liberate phenol. A blank containing all the above ingredients is prepared. This is not incubated.

A phenol reagent is then added to the test and the blank. The reagent forms a color complex with the liberated phenol. The depth of color is compared with a standard and measured in a colorimeter. This determines the amount of phenol present.

The nonincubated specimen (blank) subtracted from the incubated specimen (test) gives the units of alkaline phosphatase.

Alkaline phosphatase units = Phenol value of incubated specimen −
Phenol value of nonincubated specimen

The report is given in King-Armstrong units, one unit being the milligrams of phenol liberated by 100 ml of serum under the above conditions.

BESSEY-LOWRY-BROCK METHOD

Serum is added to p-nitrophenylphosphate that has been buffered to pH 10.3 to ensure optimum conditions for reaction. The mixture is incubated at 37° C for exactly 30 minutes. This allows the phosphatase to decompose the p-nitrophenylphosphate and liberate p-nitrophenol. The p-nitrophenol color is measured. Hydrochloric acid is added to decolorize the p-nitrophenol. The color due to serum is then measured. These units are reported as millimoles (mM) of substrate split per hour per liter of serum:

Alkaline phosphatase units or mM = p-Nitrophenol value of incubated serum −
Decolorized value of the same specimen

Acid phosphatase, EC 3.1.3.2

The normal values for acid phosphatase are 0 to 1 Bodansky unit or 0 to 2 King-Armstrong units. Increased values are found in carcinoma of the prostate, hyperparathyroidism, and acute granulocytic leukemia.

For the determination of acid phosphatase, the following three methods are commonly used.

BODANSKY METHOD

This is exactly the same procedure as the Bodansky alkaline phosphatase procedure described above except that the buffer solution is an acid buffer of pH 5.0 rather than the alkaline buffer of pH 8.6. The acid phosphatase units are found similarly:

Acid phosphatase units = Phosphorus value of incubated specimen −
Phosphorus value of nonincubated specimen

KING-ARMSTRONG METHOD

This is exactly the same procedure as the alkaline phosphatase procedure of King and Armstrong, which has been described above, except that the buffer solution is an acid buffer of pH 5.0 instead of the alkaline buffer of pH 9.0. The acid phosphatase units are found similarly:

Acid phosphatase units = Phenol value of incubated specimen −
Phenol value of nonincubated specimen

BESSEY-LOWRY-BROCK METHOD, MODIFIED

This is the same procedure as for alkaline phosphatase except that an acid buffered substrate, pH 4.8, is used and serum blanks are not incubated. The p-nitrophenol liberated during the incubation period is measured. The nonincubated serum blanks are also measured. These units are reported either as Sigma units or as mM (millimoles):

Acid phosphatase units or mM = p-Nitrophenol value of incubated serum −
Nonincubated serum

Determination of alkaline phosphatase
Specimen stability

Alkaline phosphatase in serum is stable for at least 8 hours at room temperature, for a week at 4° C, and for a month when frozen. Serum alkaline phosphatase activity may increase when specimens are kept in warm room temperatures. Therefore, it is best to separate serum from cells and refrigerate it as soon as possible after blood is drawn.

BESSEY-LOWRY-BROCK METHOD[5]
Principle

p-Nitrophenyl phosphate is colorless in solution; but upon hydrolysis the phosphate group liberates *p*-nitrophenol, which is highly colored in an alkaline solution. The rate of hydrolysis of the *p*-nitrophenyl phosphate is proportional to the concentration of the enzyme present in the serum. The reaction may be shown as follows:

$$p\text{-Nitrophenyl phosphate} + H_2O \xrightarrow{\text{Phosphatase}} p\text{-Nitrophenol} + H_3PO_4$$

(colorless in solution)   (colorless in acid but highly colored—yellow—in alkaline solution)

*Reagents**

Alkaline buffer, 0.1 M glycine, pH 10.5	(D-155)
p-Nitrophenyl phosphate stock substrate, 0.4%	(D-156)
Alkaline buffered substrate, pH 10.3 to 10.4	(D-157)
Sodium hydroxide, 0.02 N	(D-158)
Hydrochloric acid, concentrated	

Procedure

1. Use either 16 × 150 mm test tubes or spectrophotometer tubes for the test.
2. Into the *reagent blank* and *test* tubes, pipet 1.0 ml of alkaline buffered substrate.
3. Place the tubes (blank and tests) into a 37° C controlled water bath for 5 minutes, for the tubes and reagent to come to water bath temperature.
4. Set a timer for 30 minutes. Upon addition of the water to the reagent blank the timer is started, and every ½ minute thereafter the serum is added to its tube in the water bath.
5. Pipet 0.1 ml of distilled water to the blank and 0.1 ml of serum to each tube at ½-minute intervals.
6. Allow the tubes to incubate until *exactly* 30 minutes after the water was added to the blank.
7. Obtain an empty test tube rack in which to place the tubes as they are taken from the water bath.
8. Have in preparation a fast-flowing 10 ml serologic pipet (blowout) with a flask of 0.02 N NaOH. This must be added at timed intervals, also. The addition of the sodium hydroxide will stop the enzyme reaction.
9. *Exactly* 30 minutes by the timer, remove the blank from the water bath and add 10 ml of 0.02 N NaOH. Continue at ½-minute intervals until the reaction has been stopped on all the tests.
10. Stopper the tubes with clean stoppers and mix by inverting.

---

*May be purchased from Sigma Chemical Co.

11. Transfer the blank and tests to spectrophotometer tubes if these were not used during the incubation.
12. Read and record the absorbance (A) of the tests against the reagent blank at 410 nm.
13. Add 0.1 ml (2 drops) of concentrated HCl to each tube and mix by shaking tube. This removes the color due to the p-nitrophenol.
14. Reread each tube against the reagent blank at 410 nm.
15. Subtract the second reading from the first reading. The value for this corrected absorbance is then read from a curve or a table made from the curve.

NOTE: When a value greater than 10 mM units is obtained, repeat the test with a smaller volume of serum or a shorter incubation time. Adjust the calculations accordingly: for example, use a 15-minute incubation and multiply the result by 2 for the correct values, or dilute the serum with saline and multiply by the dilution made. One millimolar (mM) unit of phosphatase activity liberates 1 millimole of p-nitrophenol/L of serum/hr (1 mM = 0.1391 g). If you are using the Sigma units, these are the same as the millimolar units.

Sources of error

1. Incorrect temperature of the water bath. The temperature must not vary more than ± 1° C. With an increase in temperature of 10° C the enzyme reaction will increase 2 to 4 times the initial value. The temperature limit of the animal enzyme is 40° to 50° C. Temperatures above this will cause a decrease in enzyme activity, and at 60° C the enzyme is destroyed. Plant enzymes can stand a higher temperature than the animal enzyme. The animal enzymes are less stable and have a tendency to become denatured when heated in the presence of water.
2. Incorrect timing of the incubation period. The timing must be exact.
3. A substrate that is either too old or has an altered pH.
4. Incorrect pipeting.
5. Improper lighting. Light may either increase the enzyme rate or inhibit the rate of reaction. Ultraviolet rays or radium dispersion will inhibit enzyme reaction. Red or blue light will increase the reaction of the digestion enzymes such as pepsin, trypsin, and amylase.
6. Glassware not chemically clean. If the glassware has been acid-washed and not thoroughly rinsed, the enzyme activity will be inhibited.
7. Violent shaking of the enzyme solutions, which will cause a denaturation of the protein enzyme, resulting in an inhibiting effect.
8. Separate storage, if possible, for glassware used for enzyme studies. Heavy metals such as mercury inhibit enzyme activity.

Determination of total acid phosphatase and prostatic acid phosphatase[5]
Specimen stability

Acid phosphatase is unstable and serum must be separated from the cells and refrigerated immediately. The serum must not be hemolyzed, since the red cells contain high levels of acid phosphatase. Since prostatic acid phosphatase may lose up to 50% of its activity within an hour at room temperature, 0.01 ml of 20% acetic acid/ml of serum is added to stabilize the enzyme.

TOTAL ACID PHOSPHATASE METHOD

The total acid phosphatase test uses the same stock substrate as the alkaline phosphatase.

*Reagents**

Acid buffer, 0.09 M citric acid, pH 4.8	(D-159)
p-Nitrophenyl phosphate stock substrate, 0.4%	(D-156)
Acid buffered substrate, pH 4.8 to 4.9	(D-160)
Sodium hydroxide, 0.1 N	(D-161)
Acetic acid, 20% (v/v)	(D-162)

Procedure

1. Using either 16 × 150 mm test tubes or spectrophotometer tubes, set up the following test with a serum blank for each serum to be analyzed.
2. The serum blanks are not incubated with the tests and may be read while the serum tests are incubating.
3. Into the *reagent blank* and *test* tubes, pipet 0.5 ml of acid buffer substrate.
4. Place the tubes (blank and tests) into a 37° C controlled water bath for 5 minutes, for the substrate to come to bath temperature.
5. Set the timer for 30 minutes. (NOTE: When placing the serum into the serum *test* tubes for incubation, it may also be added to the serum blank tube within the ½-minute interval between the tubes.)
6. Start timing at the addition of 0.2 ml of water into the *reagent blank*. Mix *gently* and replace the tube into the water bath.
7. At ½-minute intervals, pipet 0.2 ml of serum into the serum test tubes. Mix *gently* and replace into the water bath.
8. Incubate for 30 minutes.
9. *Serum blanks:* To the tubes containing 0.2 ml of serum, add 6.0 ml of 0.1 N NaOH. Stopper and mix by inversion or mix laterally by shaking the tube gently.
    a. Obtain the absorbance readings by reading the tests against a distilled water bank at wavelength of 410 nm.
    b. Refer to the calibration chart for the acid phosphatase units.
10. *After incubating the serum tests for 30 minutes,* pipet 5.0 ml (use a fast-flowing serologic pipet) of 0.1 N NaOH into each tube at ½-minute intervals to stop the enzyme activity. The color is stable for several hours.
11. Obtain an absorbance reading by reading the tests against the *reagent blank* at a wavelength of 410 nm.
12. Determine the acid phosphatase units from the calibration chart.
13. Subtract the "unit-results" of the serum blanks from the "unit-results" of the serum tests. This is the *corrected total acid phosphatase* of the serum sample.

NOTE: When a value greater than 2.8 mM units is obtained, repeat the test with a smaller volume of serum or a shorter incubation time. Adjust the calculations accordingly: for example, use a 15-minute incubation period and multiply the result by 2 for the correct value, or dilute the serum with saline and multiply by the dilution made.

Normal values

Male total acid phosphatase	0.13 to 0.63 mM unit
Female total acid phosphatase	0.01 to 0.56 mM unit

NOTE: One millimolar (mM) unit of phosphatase activity liberates 1 mM of p-nitrophenol per liter of serum per hour.

---

*May be purchased from Sigma Chemical Co.

PROSTATIC ACID PHOSPHATASE METHOD

Since the enzyme to be measured is highly unstable, refrigerate the blood immediately after drawing. After the clot has retracted (allow 30 minutes) centrifuge and keep the serum at 0° to 5° C, or add 0.01 ml of 20% acetic acid per milliliter of serum to keep it stabilized.

*Reagents**

Tartrate acid buffer, pH 4.8	(D-163)
Acid buffer, 0.09 M citric acid, pH 4.8	(D-159)
p-Nitrophenyl phosphate stock substrate, 0.4%	(D-156)
Acetic acid, 20% (v/v)	(D-162)

Procedure

1. Using either 16 × 150 mm test tubes or spectrophotometer tubes, prepare two tubes as follows:
   Tube 1
       0.5 ml of stock substrate
       0.5 ml of tartrate acid buffer
   Tube 2
       0.5 ml of stock substrate
       0.5 ml of acid buffer
2. Place the tubes into the 37° C water bath for 5 minutes.
3. Set timer at 30 minutes. Start timing with the addition of 0.2 ml of serum into each tube. Mix *gently* and replace tube into water bath.
4. After the 30-minute incubation, add 5.0 ml of 0.1 N NaOH to each tube. Mix either by stoppering the tube and inverting or by lateral shaking of the tube.
5. Now, since the reaction as been stopped by the addition of NaOH, the tubes are read against a distilled water blank at 410 nm.
6. From the calibration chart, obtain the acid phosphatase units for tubes 1 and 2.
7. Subtract the "unit-results" of tube 1 from the "unit-results" of tube 2. This is the prostatic acid prosphatase of the serum.

Sources of error

The usual sources of error given earlier pertain to all enzymes, but two factors should be mentioned that frequently cause inaccurate results and thus misrepresent the correlation between the acid phosphatase and the prostatic acid phosphatase.
1. Care may not have been taken with the acid phosphatase determination. This enzyme is *unstable,* and correct precautions must be taken.
2. Several authors[13, 23, 29] have reported elevated total acid phosphatase and prostatic acid phosphatase after the physician has done a simple rectal palpation of the prostate. The blood should be drawn for the phosphatase analysis before the examination, or a time lapse of 24 to 48 hours be allowed before drawing the blood for analysis.

CALIBRATION CURVE FOR ALKALINE AND ACID PHOSPHATASE

The calibration may be done at room temperature. There is no incubation period in the calibration method.

---

*May be purchased from Sigma Chemical Co.

Table 9-3
Phosphatase calibration

Tube	Working standard (ml)	0.02 N NaOH (ml)	1 Column for scale reading % T or A	2 Equivalent to mM units, serum phosphatase	
				Alkaline	Acid
1	1.0	10.0		1.0	0.28
2	2.0	9.0		2.0	0.56
3	4.0	7.0		4.0	1.12
4	6.0	5.0		6.0	1.67
5	8.0	3.0		8.0	2.23
6	10.0	1.0		10.0	2.80

Reagents

p-Nitrophenol, 10 mM/L* (D-164)
Sodium hydroxide, 0.02 N (D-158)

Working standard: Discard after use—stable 1 day
  a. Into a 100-ml volumetric flask, pipet 0.5 ml of p-nitrophenol standard solution.
  b. Dilute to 100-ml mark with 0.02 N NaOH.
  c. Stopper and mix thoroughly.

Procedure

1. Pipet the following volumes into the tubes. Refer to Table 9-3.
2. Using a wavelength of 410 nm, read the absorbance of the six tubes and record results in column 1. These are read against a reference blank containing 0.02 N NaOH.
3. Prepare the alkaline phosphatase curve by plotting the results in column 1 against the mM units of alkaline phosphatase in column 2.
4. Prepare the acid phosphate curve by plotting the results in column 1 against the mM units of acid phosphatase in column 2.

TRANSAMINASE

In 1940, an enzyme preparation was obtained from pig heart muscle and pigeon breast muscle. It was noted that in this preparation properties of both aspartic aminopherase and glutamic aminopherase were exhibited, thereby suggesting that only one enzyme existed. It had been noted previously that the differences were related primarily to activity rather than specificity. Therefore, the name suggested for the enzyme was *transaminase*. By utilizing tissue extracts, it was shown that the transaminase is largely limited to two reactions that can be demonstrated.[31]

REACTION 1—ASPARTATE AMINOTRANSFERASE, EC 2.6.1.1

Aspartate aminotransferase (glutamic oxaloacetic transaminase—GOT) catalyzes the transfer of an amino group from aspartic acid to α-ketoglutaric acid to form

---

*May be purchased from Sigma Chemical Co.

glutamic and oxaloacetic acids:

$l(+)$ Glutamic acid + Oxaloacetic acid $\rightleftharpoons$ α-Ketoglutaric acid + $l(+)$ Aspartic acid

REACTION 2—ALANINE AMINOTRANSFERASE, EC 2.6.1.2

Alanine aminotransferase (glutamic pyruvic transaminase—GPT) catalyzes the transfer of an amino group from alanine acid to α-ketoglutaric acid to form glutamic and pyruvic acids:

$l(+)$ Glutamic acid + Pyruvic acid $\rightleftharpoons$ α-Ketoglutaric acid + $l(+)$ Alanine

The transaminase enzymes are found in the tissues of many organs. There are large amounts of the enzymes present in the heart muscle, skeletal muscle, brain, liver, and kidneys, in descending order of concentration. Deterioration of the organs causes a release of abnormal amounts of the transaminase enzymes into the blood. An elevated serum aspartate aminotransferase may be found in cases of myocardial infarction, various liver diseases, and occasionally renal diseases.

Normally, the ratio of GOT/GPT is 1.0 or over, except in viral or infectious hepatitis, wherein the ratio is less than 1.0. Then the GPT value is higher than the GOT value, and both values are increased.[7, 30]

Determination of GOT and GPT[24]
Specimen stability

Serum GOT and GPT activities are stable for 1 week at 4° C and for 1 month when frozen.

REITMAN-FRANKEL METHOD

The procedures for the GOT and GPT utilize similar conditions with the exceptions that alanine instead of aspartic acid is used in the GPT and only a 30-minute incubation period is necessary. Since the two tests may be run simultaneously, either of the following incubation methods may be chosen:
1. The GOT and GPT may be started together. Then, at the end of the 30-minute incubation period, the GPT tubes are removed from the water bath and the hydrazine (color reagent) is added immediately. This stops the reaction and starts the color development. The tubes may sit at room temperature until the GOT tubes have finished incubating or they may be completed.
2. The GPT may be started 30 minutes after the GOT.

The following method is written so that both tests may be done either simultaneously or individually.

The instrument is calibrated and tests are read at 505 nm.

*Reagents**

Phosphate buffer, 0.1 M, pH 7.4	(D-165)
GOT substrate, α-ketoglutarate, 2 mM/L; dl-aspartate, 200 mM/L	(D-166)
GPT substrate, α-ketoglutarate, 2 mM/L; dl-alanine, 200 mM/L	(D-167)
2,4-Dinitrophenylhydrazine, 1 mM/L	(D-169)
Sodium hydroxide, 0.4 N	(D-169)

---

*May be purchased from Sigma Chemical Co.

Procedure

1. Including a reagent blank for the GOT and for the GPT, pipet 1.0 ml of the proper substrate into each tube.
2. Place the tubes in a 37° C water bath for 5 minutes.
3. Set the timer for 60 minutes. Start timing when 0.2 ml of distilled water is added to the GOT or GPT reagent blank, whichever group is started first. Mix by swirling and replace tube in water bath.
4. At ½-minute intervals, pipet 0.2 ml of serum into each tube. Mix by swirling. Replace tube in water bath.
5. After incubating *exactly* 30 minutes, remove the GPT tubes from the water bath at ½-minute intervals and *immediately* add 1.0 ml of color reagent to stop the reaction and start the color development.
6. After incubating *exactly* 60 minutes, remove the GOT tubes from the water bath at ½-minute intervals and *immediately* add 1.0 ml of color reagent to stop the reaction and start the color development.
7. After a minimum of 20 minutes of standing with the color reagent (hydrazine), add 10 ml of 0.4 N NaOH. Stopper and mix by inverting the tubes. The final color develops in alkaline solution.
8. After standing 5 minutes, read at a wavelength of 505 nm against the water blank. The blank tube should read very close to the same value on each run.
9. Calculate results from the calibration curve. If a value exceeds the upper limit of the curve, dilute serum with saline and multiply by the dilution made. For example, with a 1:10 dilution, multiply by 10.

Normal range

GOT	8 to 40 units
Doubtful	40 to 50 units
GPT	5 to 35 units
Doubtful	35 to 45 units

CALIBRATION CURVE

The standards for the calibration curve are *not* incubated. The curve may be plotted either with absorbance versus the activity, on linear graph paper, or with the percent transmittance versus the activity, on semilog paper. The units of the GPT and GOT are shown in the chart at 37° and 40° C. The method was presented for a 37° C water bath, since the other enzyme tests in this chapter are calibrated for the use of a 37° C water bath.

Reagents

Pyruvate standard, 2 mM/L (D-170)

The remaining reagents are those used in the methods for the GOT and GPT.

Procedure

1. Pipet the following volumes into the tubes. Refer to Table 9-4.
2. Add 1.0 ml of color reagent to each tube. Mix.
3. Pipet 10 ml of 0.4 N NaOH to each tube. Stopper and invert to mix.
4. Wait 5 minutes;; then read at 505 nm against a distilled water blank.
5. Record the readings of the five tubes in column 1.
6. Using the alkaline phosphatase as an example, plot the readings from column 1 against those of column 2 or column 3.

LACTATE DEHYDROGENASE (LDH), EC 1.1.1.27

Lactate dehydrogenase is an enzyme that is found in almost all tissues. For this reason, a number of pathologic conditions can cause an elevation of LDH activity. LDH activity is of particular interest as an aid in the diagnosis of myocardial infarction, hepatitis, and cirrhosis.

Table 9-4
Transaminase calibration

Tube	Pyruvate (ml)	Sub-strate (ml)	H₂O (ml)	1 Absorbance % T	2 GOT		3 GPT	
					37° C	40° C	37° C	40° C
1	0	1.0	0.2		0	0	0	0
2	0.1	0.9	0.2		24	21	28	25
3	0.2	0.8	0.2		61	54	57	51
4	0.3	0.7	0.2		114	102	97	87
5	0.4	0.6	0.2		190	169	—	—

Determination of lactate dehydrogenase
Specimen stability

Serum LDH activity is stable for 1 week at 4° C and for 1 month when frozen.

BABSON-PHILLIPS METHOD[3]
Principle

LDH catalyzes the oxidation of lactate to pyruvate. In this reaction NAD is reduced to NADH which is coupled to the reduction of a tetrazolium salt, 2-*p*-iodophenyl-3-*p*-nitrophenyl-5-phenyl tetrazolium chloride (INT), with phenazine methosulfate (PMS) serving as an intermediate electron carrier. The bright red formazan produced is measured at 520 nm.

$$\text{Lactate} + \text{NAD} \xrightleftharpoons{\text{LDH}} \text{NADH} + \text{Pyruvate}$$

$$\text{NADH} + \text{PMS} \longrightarrow \text{NAD} + \text{Reduced PMS}$$

$$\text{Reduced PMS} + \text{INT} \longrightarrow \text{PMS} + \text{Reduced INT (red color)}$$

Reagents

LDH color reagent	**(D-171)**
LDH buffer reagent	**(D-172)**
LDH substrate, 0.1 M *l*(+) lactate	**(D-173)**
LDH control reagent	**(D-174)**
Hydrochloric acid, 0.1 N	**(D-32)**

Working standards

Commercial lyophilized enzyme standards such as Versatol-E (V-E) and Versatol-E-N (V-E-N) are used to prepare a standard curve.
Example:

Tube	V-E-N (ml)	V-E (ml)	Units	Example
1	0.5	—	Given value	50
2	1.5	0.5	Values (1* + 3*) ÷ 2	88
3	0.5	0.5	Values (1* + 5*) ÷ 2	125
4	0.5	1.5	Values (3* + 5*) ÷ 2	163
5	—	0.5	Given value	200

*Tube number.

These standards are set up and read in exactly the same manner as the unknown serum specimens.

Procedure

1. Pipet 0.1 ml of serum and 0.2 ml of buffer reagent into each of two tubes, one labeled *test* and the other, *control*.
2. Add 0.5 ml of substrate to the "test" and 0.5 ml of control reagent to the control. Mix and place in a 37° C water bath for 5 minutes to warm.
3. At convenient but precisely timed intervals, pipet 0.2 ml of color reagent into each tube, mix, and return immediately to the water bath. Incubate for exactly 5 minutes.
4. Add 5 ml of 0.1 N HCl to each tube in the same sequence and time intervals as in step 3. Mix well.
5. Read absorbance at 520 nm against distilled water within 20 minutes.

Calculations

Subtract the absorbance of the control from that of the test for standards and unknown sera. Plot the corrected absorbances of the standards versus LDH activity on linear graph paper. Determine the LDH activity of the unknowns from the standard curve, using the corrected absorbances.

Chemical fractionation of lactate dehydrogenase

Elevation in total LDH activity can be caused by a variety of pathologic conditions. Fractionation of the total activity into proportional amounts of isoenzyme activity has been demonstrated by a variety of techniques. However, it is important to remember how the isoenzyme activities make up the total LDH activities. In any procedure, the proportion of the total LDH activity is at best a measurement of the grouped activities of all five isoenzymes. The proportion of the total activity that is detected by the isoenzyme separation technique is a formation of the following three parameters:

1. Inherent chemical characteristics of each individual isoenzyme as related to temperature, pH, substrate concentration, buffer, coenzyme concentration, and so forth
2. Total reaction conditions for the assay
3. Relative amount of each isoenzyme in the total LDH activity

Combinations of the three parameters listed will regulate the proportional amount of isoenzyme determined.

The chemical inhibition of LDH heart and liver tissue isoenzymes has been studied by Babson.[1] Chemical inhibition has also been used to assess LDH isoenzymes in serum. This procedure separates the slow or hepatic type isoenzymes from the fast or cardiac type isoenzymes. To inhibit the activity of the electrophoretically fast migrating isoenzymes, 2 M lactate is used as a substrate. The slow migrating LDH enzymes are similarly inhibited when the substrate consists of 0.02 M lactate and 2 M urea. The simultaneous use of these two substrates provides an *isoenzyme index* (I.I.), which can reflect an elevation of fast, slow, or intermediate LDH isoenzymes.

The chemical inhibition method indicates groupings of predominating isoenzymes that do correlate to the fast, slow, and intermediate electrophoretically migrating isoenzymes. The isoenzyme index may be divided into three groups of LDH isoenzyme information. First, an isoenzyme index of less than 0.80 indicates a predominance of the faster moving isoenzymes. Second, an isoenzyme index of greater than 1.20 indicates predominance of the slower moving isoenzymes.

Finally, there may be a combination of both the fast and the slow moving isoenzymes, *or* a predominance of the intermediate moving isoenzymes. In either case, an isoenzyme index of 0.80 to 1.20 may be expected. Therefore, when the total serum LDH is elevated, an isoenzyme index in this middle range must be suspect for one of these last two possibilities. An abnormal total LDH and an intermediate type isoenzyme index can reflect simultaneous cardiac and hepatic tissue damage, or diseases that produce isomorphic LDH_3 elevations.

The isoenzymes are divided into three groups and their occurrences are:

Group I (I.I. below 0.80). Tissues belonging to this group are distinguished by a high percentage of LDH_1 and LDH_2 and consequently a low percentage of LDH_4 and LDH_5. This fraction occurrence is seen in heart muscle but also is found in kidneys, brain, and erythrocytes.

Group II (I.I. above 1.20). Tissues in this group possess high percentages of LDH_4 and LDH_5, whereas LDH_1 shows the smallest percentage. This isoenzyme group is found in liver tissue, skeletal muscle, skin, and leukocytes.

Group III (I.I. 0.80 to 1.20). The vast majority of tissues belong to this group. As its distinctive feature, this group shows maximum activity in the middle fractions—LDH_2, LDH_3, and LDH_4. Tissues giving these middle elevations are lungs, spleen, ureter, uterus, pancreas, prostate, ovary, gallbladder, appendix, connective tissue, duodenum, thyroid gland, and lymph gland.

Reagents

Reagents listed for total lactate dehydrogenase
LDH color reagent	(D-171)
LDH buffer reagent	(D-172)
Hydrochloric acid, 0.1 N	(D-32)
$l(+)$ Lactate, 2 M	(D-175)
$l(+)$ Lactate, 0.02 M to 2 M urea	(D-176)

Procedure
1. Into each of two tubes place 0.2 ml of LDH buffer reagent.
2. Add 0.5 ml of the heart-inhibiting substrate to one tube ("H" tube) and 0.5 ml of the liver-inhibiting substrate to the other ("L" tube).
3. Add 0.1 ml of serum to each tube, mix, and incubate 10 minutes at 37° C to provide the inhibitory effects.
4. Following this, add 0.2 ml of color reagent—that is, the NAD, the intermediate electron carrier, and the tetrazolium salt—to each tube and continue the incubation for exactly 5 minutes.
5. Terminate the reaction by adding 5.0 ml of 0.1 N HCl.
6. Read the absorbance of each tube at 520 nm with a distilled water blank.

Calculation

$$\text{Isoenzyme index} = \frac{A \text{ of ``H'' tube}}{A \text{ of ``L'' tube}}$$

Interpretation

I.I.	LDH isoenzyme fractions
I. Below 0.80	Elevation of fast fractions, LDH_1 and LDH_2—"cardiac"
II. Above 1.20	Elevation of slow fractions, LDH_4 and LDH_5—"hepatic"
III. 0.80 to 1.20	a. Both fast and slow fraction elevated
	b. Intermediate fraction elevated LDH_3

CREATINE KINASE (CREATINE PHOSPHOKINASE, CPK), EC 2.7.3.2

Because of the tissue specificity of creatine phosphokinase, there has been an increased interest in the determination of this enzyme as an aid in the diagnosis of acute myocardial infarction and primary muscle disease. High concentrations are found only in heart muscle, skeletal muscle, and brain. Only negligible activity is found in liver, pancreas, kidney, lung, or erythrocytes. Serum CPK activity elevates earlier after myocardial infarctions than either LDH or GOT.

Determination of creatine kinase
NUTTAL-WEDIN METHOD[19]
Principle

In the method given, the forward CPK reaction is coupled to the pyruvate kinase reaction. The pyruvate formed is measured colorimetrically (at 440 nm) through the formation of the hydrazone with 2,4-dinitrophenylhydrazine.

$$ATP + Creatine \xrightleftharpoons{CPK} ADP + Creatine\ phosphate$$
$$ADP + Phosphoenolpyruvate \xrightleftharpoons{PK} Pyruvate + ATP$$

Interferences are eliminated by the use of a serum blank.

Serum CPK activity is heat-labile. When serum is heated to 56° C for 15 minutes, there is a total loss of activity. Specimens will lose an average of 45% of their original activity over a period of 4 hours at room temperature. The addition of a sulfhydryl agent to the reaction mixture, however, will fully reactivate the enzyme. This has been found to be true of specimens kept up to 5 days at refrigerator temperatures. Several sulfhydryl agents have been used, such as cysteine and glutathione, but mercaptoethanol proved to be much more stable. The mercaptoethanol must be added separately and just before incubation since it is somewhat unstable at the pH of the reaction mixture.[19]

Hemoglobin concentration up to 1.25 g/dl and lipemia have no effect on the measurement of CPK activity with the use of a serum blank.

Reagents

Sodium phosphate, dibasic, 0.1 M	(D-177)
Potassium phosphate, monobasic, 0.1 M	(D-178)
Phosphate buffer, pH 7.4	(D-179)
Pyruvic acid standard, 0.5 μM/ml	(D-180)
CPK blank substrate	(D-181)
Creatine substrate	(D-182)
2,4-Dinitrophenylhydrazine, 1 mM	(D-168)
Sodium hydroxide, 0.4 N	(D-169)
Mercaptoethanol, 0.25 M	(D-183)

Standard curve

1. To prepare a standard curve, set up the following tubes:

Tube	Standard (ml)	PO₄ buffer (ml)	μM/ml	IU
1	0.0	1.0	0.1	0
2	0.2	0.8	0.2	33
3	0.4	0.6	0.3	67
4	0.6	0.4	0.4	100
5	0.8	0.2	0.5	133
6	1.0	0.0	0.6	167

2. Add 1.0 ml of 2,4-dinitrophenylhydrazine to each tube, mix, and allow to stand at room temperature for 20 minutes.
3. Add 10 ml of 0.4 N NaOH and allow to stand for 5 minutes.
4. Read at 440 nm using tube 1 as the reagent blank.
5. Plot absorbance against concentration on linear graph paper.

Procedure

1. Set up two tubes for each unknown serum to be tested. Label one tube *test* and the other *blank*.
2. Pipet 0.1 ml of 0.25 M mercaptoethanol into the bottom of each tube.
3. Pipet 1.0 ml of creatine substrate into each test tube and 1.0 ml of CPK blank substrate into each blank tube and one tube to be used as a reagent blank.
4. Place all tubes in a 37° C water bath for 5 minutes to warm.
5. At timed intervals, add 0.1 ml of serum to each of the two substrates and mix. Exactly 30 minutes later, add 1.0 ml of 2,4-dinitrophenylhydrazine to stop the reaction.
6. Mix and allow to stand at room temperature for 20 minutes.
7. Add 10 ml of 0.4 N NaOH, mix, and allow to stand for 5 minutes.
8. Read each tube against the reagent blank at 440 nm.
9. Subtract the blank A from the test A and determine the CPK activity from the standard curve.
10. Results are expressed in International Units (IU) per liter of serum; that is, micromoles of pyruvate (creatine phosphate) produced per liter of serum per minute at 37° C.

$$\frac{\mu M \text{ pyruvate in tube}}{30 \text{ min (incubation)}} \times 10{,}000^* = \mu M/L/\min \text{ (IU) CPK activity}$$

Normal values

Nuttal and Wedin found that normal values ranged from about 6 to 25 IU for females and from 6 to 43 IU for males.

γ-GLUTAMYL TRANSPEPTIDASE (γ-GLUTAMYLTRANSFERASE), EC 2.3.2.2

γ-Glutamyl transpeptidase (GGT) is an enzyme that catalyzes the transfer of the γ-glutamyl group from peptides containing this group to other peptides and to L-amino acids.

The use of the enzyme for diagnostic purposes has only recently begun to attract attention in this country. The enzyme first came to the attention of clinical chemists through the British journal, *Nature*. Hanes, Hird, and Isherwood,[11] in 1950, described an enzyme obtained from sheep kidney that exhibited a transpeptidase function.

For the next 10 years, most of the early work on clinical applications of the enzyme was carried on in Poland and published in Polish. In 1961, the results were summarized and commented on in the English language,[34] setting the stage for broader international development of the basic ideas. Orlowski and co-workers reported that while the highest enzyme activity was demonstrated in the kidney, great changes in serum GGT activity were found in certain diseases of the liver.

In 1963, a Boston research group[28] confirmed the value of GGT as a screening test for diseases of the liver, common duct, or pancreas in the anicteric patient.

---

*0.1 ml of serum converted to 1 L.

Researchers used a variety of substrates for determining GGT activity until Orlowski and Meister,[21] in 1963, reported on the use of γ-glutamyl-*p*-nitroanilide as a substrate for the rapid determination of GGT activity. The colorless substrate, which exhibits a maximum absorbance at 315 nm and no absorbance at 410 nm, is enzymatically converted to *p*-nitroaniline, which exhibits high absorbance in the range of 350 to 420 nm. This substrate is now generally accepted as being the best one known to date.

Many workers have contributed to determining the site of origin of the enzyme appearing in the serum. The main consensus of opinion is that the serum GGT comes from the hepatobiliary system, although the enzyme or an isoenzyme is found in a number of other tissues.

Naftalin and co-workers[18] found that in the human subject the enzyme is produced along the whole tract from liver parenchyma to common bile duct and also in the pancreatic acini and small ductules. No GGT activity was found in human heart muscle[18] or in skeletal muscle.[27]

In 1969, Szasz[33] published a kinetic method for the determination of γ-glutamyl transpeptidase that uses γ-glutamyl-*p*-nitroanilide as the substrate and glycylglycine as the acceptor of the γ-glutamyl moiety.

He found the optimum substrate concentration to be 4 mM and calculated the Michaelis constant of GGT in human serum to be $K_m = 0.96$ mM. The Michaelis constant of human serum GGT for glycylglycerine, for the range 5 to 50 mM glycylglycerine, is $K_m = 6.65$ mM.

Szasz found that the presence or absence of Mg^{++} did not alter GGT activity; however, since the presence of $MgCl_2$ seemed to promote the solubility of the substrate, the buffer substrate mixtures routinely contained $MgCl_2$ at a concentration of 10 mM

The relationship between enzyme concentration (that is, amount in serum) and enzyme activity was reported to remain linear within the range up to 134 mU/ml.

Szasz detected no significant loss of enzyme activity in four sera stored at −20°, 4°, and 20° C for 1 week. Normal values were given as 3.2 to 13.5 mU/ml for females and 4.5 to 24.8 mU/ml for males.

At least four companies are now marketing kits for GGT assays based on the method of Szasz.*

---

*Boehringer Mannheim Corp., Dade Reagents, Sigma Chemical Co., Worthington Biochemical Corp.

References

1. Babson, A. L.: The chemical differentiation of tissue lactic dehydrogenase, Clin. Chim. Acta **16**:121, 1967.
2. Babson, A. L., Kleinman, N. M., and Megraw, R. E.: A new substrate for serum amylase determination, Clin. Chem. **14**:802, 1968.
3. Babson, A. L., and Phillips, G. E.: A rapid colorimeter assay for serum lactic dehydrogenase, Clin. Chim. Acta **12**:210, 1965.
4. Babson, A. L., Tenney, S. A., and Megraw, R. E.: New amylase substrate and assay procedure, Clin. Chem. **16**:39, 1970.
5. Bessey, O. A., Lowry, O. H., and Brock, M. J.: Method for rapid determination of alkaline phosphatase with 5 cubic millimeters of serum, J. Biol. Chem. **164**:321, 1946.
6. Cherry, I. S., and Crandall, L. A., Jr.: The specificity of pancreatic lipase: its appearance in the blood after pancreatic injury, Am. J. Physiol. **100**:266, 1932.
7. DeRitis, F., Coltorti, M., and Giusti, G.: An enzyme test for the diagnosis of viral hepatitis; the transaminase serum activities, Clin. Chim. Acta **2**:70, 1957.
8. Dirstine, P. H., Sobel, C., and Henry, R. J.: A new rapid method for the determination of serum lipase, Clin. Chem. **14**:1097, 1968.

9. Enzyme Nomenclature, Recommendations (1972) of the Commission on Biochemical Nomenclature on the Nomenclature and Classification of Enzymes together with their Units and the Symbols of Enzyme Kinetics, New York, 1973, American Elsevier Publishing Co., Inc.
10. Gutman, A. B., Tyson, T. L., and Gutman, E. B.: Serum calcium, inorganic phosphorus and phosphatase activity in hyperparathyroidism, Paget's disease, multiple myeloma, and neoplastic disease of bones, Arch. Intern. Med. **57:**379, 1936.
11. Hanes, C. S., Hird, F. J. T., and Isherwood, F. A.: Synthesis of peptides in enzymic reactions involving glutathione, Nature **166:**228, 1950.
12. Henry, R. J., and Chiamori, W.: Study of the saccharogenic method for the determination of serum and urine amylase, Clin. Chem. **6:**434, 1960.
13. Hock, E., and Tessier, R. N.: Elevation of serum acid phosphatase following prostatic massage, J. Urol. **62:**488, 1949.
14. Kjeldahl, J.: Dinglers Polytech. J. **235:**370, 1880.
15. Klein, B., and Foreman, J. A.: Urinary amylase excretion as measured by Amylochrome assay, Clin. Chem. **19:**1226, 1973.
16. Klein, B., Foreman, J. A., and Searcy, R. L.: A new chromogenic substrate for determination of serum amylase activity, Clin. Chem. **15:**784, 1969.
17. Massion, C. G., and Seligson, D.: Serum lipase: a rapid photometric method, Am. J. Clin. Pathol. **48:**307, 1967.
18. Naftalin, L., Child, V. J., and Morley, D. A.: Observations on the site of origin of serum γ-glutamyl transpeptidase, Clin. Chim. Acta **26:**297, 1969.
19. Nuttal, F. Q., and Wedin, D. S.: A simple rapid colorimetric method for determination of creatine kinase activity, J. Lab. Clin. Med. **68:**324, 1966.
20. Orlowski, M.: The role of gamma-glutamyl transpeptidase in the internal diseases clinic, Arch. Immunol. Ther. Exp. **11:**1, 1963.
21. Orlowski, M., and Meister, A.: γ-Glutamyl-p-nitroanilide: a new convenient substrate for determination and study of L- and D- γ-glutamyltranspeptidase activities, Biochem. Biophys. Acta **73:**679, 1963.
22. Orten, J. M., and Neuhaus, O. W.: Human biochemistry, ed. 9, St. Louis, 1975, The C. V. Mosby Co.
23. Ozar, M. B., Isaac, C. A., and Valk, W. L.: Methods for elimination of errors in serum acid phosphatase determinations, J. Urol. **74:**150, 1955.
24. Reitman, S., and Frankel, S.: A colorimetric method for the determination of serum glutamic oxalacetic and glutamic pyruvic transaminases, Am. J. Clin. Pathol. **28:**36, 1957.
25. Report of the Commission on Enzymes of the International Union of Biochemistry, New York, 1961, Pergamon Press.
26. Rinderknecht, H., Wilding, P., and Haverback, B. J.: A new method for the determination of α-amylase, Experientia **23:**805, 1967.
27. Rosalki, S. B., and Thomson, W. H. S.: Serum gamma-glutamyl transpeptidase in muscle disease, Clin. Chim. Acta **33:**264, 1971.
28. Rutenburg, A. M., Goldbarg, J. A., and Pineda, E. P.: Serum γ-glutamyl transpeptidase activity in hepatobiliary pancreatic disease, Gastroenterology **45:**43, 1963.
29. Sigma Technical Bulletin 104, St. Louis, 1974, Sigma Chemical Co.
30. Sigma Technical Bulletin 505, St. Louis, 1967, Sigma Chemical Co.
31. Sobotka, H., and Steward, C. P.: Advances in clinical chemistry, vol. 1, New York, 1958, Academic Press, Inc.
32. Somogyi, M.: Micromethods for estimation of diastase, J. Biol. Chem. **125:**399, 1938.
33. Szasz, G.: A kinetic photometric method for serum γ-glutamyl transpeptidase, Clin. Chem. **15:**124, 1969.
34. Szczelik, E., Orlowski, M., and Szewczuk, A.: Serum γ-glutamyl transpeptidase activity in liver disease, Gastroenterology **41:**353, 1961.
35. Tietz, N. W., and Fiereck, E. A.: A specific method for serum lipase determination, Clin. Chim. Acta **13:**352, 1966.
36. Wohlgemuth, J.: Über eine neue Methode zur quantitativen Bestimmung des diastatischen Ferments, Biochem. Z. **9:**1, 1908.
37. Woodard, H. Q.: The clinical significance of serum acid phosphatase, Am. J. Med. **27:**902, 1959.

10

Renal function

The English physician and clinician Richard Bright first introduced urine examinations as a routine procedure for the physicians in Guy's Hospital (1827). Bright is known universally now for his description of chronic nephritis, Bright's disease. Physicians soon developed the analysis of urine to a fine art, and before the turn of the century many procedures were published on the macroscopic and microscopic examination of urine. Thomas Addis brought the microscopic study of the urinary sediment to a peak of perfection by the introduction of quantitative methods of study, which he believed were necessary for an accurate evaluation of renal disease.

In the 1930s physicians began to realize that too much faith had been placed on the reading of the urinary sediment during the earlier years of medicine. When Henry Christian, an authority on renal disease, stated that it was impossible to determine during life the anatomic changes in the kidney that would be found after the patient's death, the importance of the urine examination seemed to decline in popularity. With the development of new blood tests and mechanical equipment for registering various body functions, urine analysis became the "stepchild" of the laboratory.

Kidney function tests have been developed to aid the physician in determining the anatomic changes in the kidney during life and the importance of urinary examination is again recognized. Other body malfunctions are also determined with the help of urinary studies.

Discussion of the kidneys and the analysis of urine is divided into the following three sections: (1) renal function tests, (2) renal insufficiency and dialysis, and (3) routine urine examination.

Renal function tests

The success of dialysis and kidney transplants in saving lives of patients who formerly would have died from uremia has increased the laboratory testing of blood and biologic fluids as a guide in diagnosis and prognosis because of the chemical abnormalities characteristic of renal insufficiency. Hemodialysis and peritoneal dialysis will remove from or add to solutes in the blood and tissue fluids as required for survival, either relieving the kidneys until they are able to resume normal functioning or substituting for them by use of frequent dialysis. The awareness of the general public concerning availability of these lifesaving devices and the use of artificial kidneys in the home and the hospital have required an educational program for patients and their families as well as for hospital personnel associated with renal programs.

The two kidneys serve to regulate the composition of the blood and maintain the internal composition of the body compatible with life. They are located retroperi-

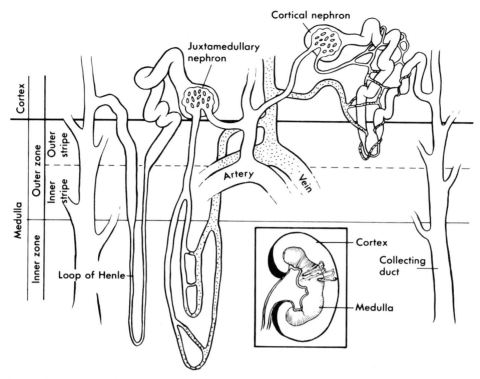

Fig. 10-1. Diagram of cortical and juxtamedullary nephrons with a cross section of the kidney.

toneally in the back of the abdominal cavity, one on each side of the vertebral column at approximately the L-3 level. The kidneys are major excretory organs for eliminating from the blood normal metabolic end products, excess or unwanted substances, and chemical poisons. Some of these may cause injury to the tissues and alter the normal elimination procedure.

ANATOMY OF THE KIDNEYS (Fig. 10-1)

Each kidney measures about $12 \times 6.5 \times 4$ cm and weighs between 115 and 170 g. Externally it is shaped like a bean with a depression at its inner edge called the hilus where the renal artery from the aorta enters the organ and the renal vein leaves it to join with the vena cava. Nerve fibers enter along the artery. The hilus is also partially taken up by the pelvis, a funnel-shaped structure that is formed by fusion of the calyces and tapers to form the ureter, which then courses downward to empty into the bladder.

If a kidney is cut longitudinally, two distinct zones may be seen on the cut surface, the outer or cortex and the inner or medulla. The cortex contains glomeruli, which are the filtering units of the kidney. The medulla of each kidney contains about 1 million microscopic tubules that converge into a number of pyramidal areas, which then empty into the calyceal system and then into the renal pelvis.

Each tubule originates from a malpighian corpuscle, the outer wall of which is known as Bowman's capsule, while the capillary network within is called the glomerulus. The filtrate within the capsule is plasma water extruded by hydrostatic pressure from the thin-walled capillary network of the glomerulus; it empties into the tubule which evolves as a series of convolutions (proximal or first convoluted tubule) in the cortex. The tubule then straightens and dips downward into the medulla as a thin descending segment, making a U turn known as Henle's loop, to rise again into the cortex where the wall structure changes and the tubule enlarges. It again becomes convoluted (distal or second convoluted tubule) near its own glomerular unit before it straightens out to join with others into the collecting ducts. Many ducts combine, forming channels (papillary ducts of Bellini) to carry the urine through the pyramids to the kidney pelvis. By this arrangement a large surface area of the tubules comes in contact with a meshwork of very small blood vessels. Each glomerulus, with its particular tubule, is an essentially independent functional unit and is referred to as a *nephron*.

The kidney itself synthesizes small amounts of hippuric acid and ammonia. Otherwise, all known urinary constituents are brought to it by the blood and are excreted or reabsorbed unchanged. Each kidney is supplied with blood from the abdominal aorta through the renal artery that enters at the hilus. Normal renal blood flow is about 25% of total circulation. The renal artery divides into several branches, and when these reach the border between the cortex and medulla they divide laterally to form the arch or arcuate arteries. From the latter, interlobular arteries enter the cortex, forming at intervals small afferent arteries and then arterioles, which form the capillary network of the individual glomeruli. There are some 50 capillary loops formed that coil and intertwine. The capillaries reunite to pass out of the glomerulus and form the efferent arteriole. This efferent arteriole divides into another network of capillaries, surrounding the tubule in the cortex and the loop of Henle in the medulla. They eventually unite to form the interlobular and medullary veins, from which the blood flows into the arcuate veins between the cortex and medulla. The arcuate veins form interlobar veins that merge into the renal vein, which then leaves the kidney at the hilus and empties into the inferior vena cava.

The blood supply to the kidneys is about 25% of the cardiac output, or approximately 1160 ml/min in men and 940 ml/min for women. It is under pressure that is exerted by contraction of the heart, and over 10% of the fluid is removed as glomerular filtrate. Glomerular filtration rate is about 127 ml/min and 118 ml/min for males and females, respectively.

PHYSIOLOGY OF THE KIDNEYS

In the formation of urine, kidneys perform a series of functions. They excrete end products of protein metabolism (urea, uric acid, creatinine) and about one half of the water eliminated from the body, maintain normal concentration of essential blood constituents by excreting those not needed and reabsorbing the ones required, and help preserve normal acid-base balance of body fluids.

By physical forces such as ultrafiltration and diffusion fluid is removed from the blood in the glomerular capillaries into the filtration chamber of the capsules. The filtrate has essentially the same composition as protein-free plasma, although a small amount of protein may pass through the glomeruli and be reabsorbed in part by the tubules. Filtration depends on the ability of the hydrostatic pressure of the blood to overcome the colloid osmotic pressure of the blood proteins and the tissue tension in the capillaries. There is decreased filtration when the blood pressure falls or a lowering of cardiac output results from shock, surgery, dehydration, or cardiac decompensation, and also when renal damage causes a reduction in the total number of

Table 10-1
Comparison of average amounts of biochemical constituents in glomerular filtrate and excreted urine

Constituent	In 180 L of filtrate	In 24-hour urine collection
Fluid volume (L)	180	1.5
Sodium (g)	580	4.5
Potassium (g)	30	2.5
Chloride (g)	650	7.0
Calcium, diffusible (g)	9	0.2
Magnesium, diffusible (g)	4	0.08
Urea (g)	60	30
Glucose (g)	170	Below detectable limits

functioning glomeruli. When glomeruli are damaged, as in glomerulonephritis, plasma albumin leaks into the filtrate.

In the normal adult between 120 and 130 ml of glomerular filtrate is produced per minute. If this filtrate were excreted unchanged, death would occur rapidly. Approximately 97% of the water, and—selectively—other constituents such as electrolytes, glucose, and amino acids are reabsorbed during the passage of the filtrate down the tubule. Assuming 125 ml of filtrate is produced per minute (180 L/day) and that it is principally protein-free plasma, the major constituents removed per day from the blood may be approximated as shown in Table 10-1. These figures illustrate the enormous reabsorptive action of the kidneys, whereby most of the constituents filtered from the blood are promptly returned to it by absorption at the tubules.

As the filtrate moves from the glomerulus downward through the tubule, selective reabsorption by tubular cells of glucose, amino acids, sodium, and potassium, but not waste materials, occurs by active cellular transport. About 97% of the filtered water is passively reabsorbed, resulting in a urine with a high concentration of waste materials. Reabsorption of organic compounds such as glucose and amino acids is essentially complete in the proximal tubule. It is the function of the kidneys to conserve these but not to regulate their concentration in the blood. Thus needed constituents are restored to the blood by passage through the cells lining the tubules.

Glucose is transferred by means of a transport system that utilizes the coupling of glucose with phosphate. When there is an excess of glucose in the glomerular filtrate, the capacity of the tubular transport system is exceeded and glycosuria will result. About 85% of the sodium is actively reabsorbed; with it the associated chloride anion is passively reabsorbed, as is water, the latter as a result of the osmotic gradient produced by the sodium movement. The osmotic concentration of the fluid entering the descending limb of Henle's loop is the same as that of the original glomerular filtrate but it progressively increases to the bend in the depths of the medulla. In the ascending limb of the loop of Henle, sodium is actively reabsorbed or transported out of the tubular lumen (and with it chloride, passively). Although the loop is impermeable to water the filtrate is dilute by the time it reaches the distal convoluted tubule because of solute loss. In the presence of antidiuretic hormone (ADH), water moves passively from the urine as it passes through the distal tubule and collecting ducts in response to concentration gradients, yielding a urine with concentration similar to that of the interstitial tissue fluid at the tip of the papilla. In the absence of ADH, the distal tubule is impermeable to water, which will therefore be retained although sodium reabsorption may continue, depending on need of the body to produce a dilute urine. One third to one half of the filtered load of urea is passively reabsorbed along the length of the tubule, mainly by diffusion in response to osmotic gradients.

Certain tubular cells are capable of removing additional waste materials from the surrounding peritubular fluid and capillaries and transferring it to the tubular lumen. By this action, the tubules supplement glomerular filtration in removal of waste materials (urea, creatinine) from the blood.

CHEMICAL TESTS OF RENAL FUNCTION[22]

The basic activities of the kidneys that are measurable by chemical means are glomerular filtration, tubular reabsorption and excretion, and renal blood flow. When glomeruli or tubules are damaged or destroyed, they are not replaced. Any action that results in alteration or destruction of kidney tissue may be critical because of accumulation, in blood and other body fluids, of materials normally eliminated by the kidneys.

Chemical tests of renal function are used to determine the nature and extent of renal dysfunction. However, they give no indication as to whether the damage is anatomic or functional. The latter may result from shock, congestive heart failure, severe dehydration, or other conditions in which the normal blood supply to the kidneys is altered. These tests provide a good index of renal function if the influence of various extrarenal factors is considered—such as the decrease of normal renal function measurements with increasing age, severe muscular exercise, and marked reduction in dietary protein.

Glomerular filtration

The ability of the glomeruli to filter soluble constituents from the blood can be measured by clearance tests. These are based upon the rate at which kidneys remove a substance from the plasma. Clearance is an expression of the volume of blood plasma containing the amount of a substance excreted into the urine in 1 minute. By measuring the volume of urine excreted per minute (V) and the concentration of a constitutent in blood (B) and in urine (U) samples obtained about the same time, the clearance of that substance may be calculated: UV/B.

An ideal test substance for glomerular filtration should be able to move freely across the glomerular membrane, maintaining the same composition in the filtrate as in the plasma, and be neither reabsorbed nor secreted by the tubules. Inulin and mannitol are two compounds that meet these requirements. However, these are not used in routine clinical studies because the procedure is tedious and time-consuming and because catheterization of the patient is required.

Two clearance tests generally employed are those for urea and creatinine. While the excretion of these endogenous substances is influenced by tubular processes, their clearance rates as determined by standard procedures give an approximation of the glomerular filtration that is sufficient for clinical purposes as an indication of renal function. They do not yield an accurate measure of glomerular filtration rate.

INULIN CLEARANCE[1, 7, 19]

Inulin, a vegetable starch (fructose polysaccharide), is not metabolized in the body but is excreted into the urine by glomerular filtration and neither reabsorbed nor excreted by the tubules. Inulin clearance is a measure of glomerular filtration and is used as the reference test.

Collection of specimens

1. The patient should be fasting, but during the 2-hour period prior to the test he should drink about a liter of water. Insert a urethral catheter to provide continuous drainage of the bladder. Before starting the infusion obtain a blood sample and some urine to use as blanks.
2. Administer intravenously a priming dose of 50 mg of inulin per kilogram of body weight and follow immediately with a sustaining solution that contains 70 ml of 10% inulin diluted to 500 ml with saline, given at about 4 ml/min, or the equivalent of the expected rate of renal clearance of inulin. Where there is impaired renal function, decrease the amount for maintenance accordingly.
3. After an equilibrium period of 30 to 45 minutes see that the bladder is emptied and washed with sterile saline. Collect three accurately timed urine specimens over an interval of approximately 20 minutes. At the end of each collection period inject 20 ml of sterile saline and 20 ml of air into the bladder to wash out all of the urine. Combine the washing with the specimen. Obtain blood samples in heparinized tubes about the midpoint of each urine collection period for plasma inulin determinations.

Reagents[11]

(All chemicals are reagent grade.)
Trichloroacetic acid, 10% (D-95)
Hydrochloric acid, concentrated
Indole-3-acetic reagent, 0.5% in 96% ethyl alcohol
   Store in the refrigerator.
Inulin standard, 0.05 mg/ml
   Prepare a paste of 12.5 mg of inulin in several drops of water and add 50 to 75 ml of hot (90° C) distilled water. Transfer to a 250-ml volumetric flask quantitatively, add sufficient benzoic acid to saturate, and dilute to volume.

Procedure[11]

1. Place 1.0 ml of plasma into a centrifuge tube. Slowly add 4.0 ml of 10% trichloroacetic acid. Allow to stand 10 minutes and filter.
2. Into duplicate test tubes pipet plasma filtrate, 1.0 ml; urine diluted 1:100 with water, 1.0 ml; standard, 1.0 ml; and water (reagent blank), 1.0 ml.
3. To each tube add 0.2 ml of indole-3-acetic acid reagent and 8.0 ml of concentrated hydrochloric acid.
4. Incubate in a 37° C water bath for 75 minutes.
5. Cool to room temperature and read absorbance at 520 nm against water. The color intensity increases with standing. Read as rapidly as possible in the order in which the tests were set up.

Calculations

$$\text{Plasma:} \quad \frac{A \text{ unknown} - A \text{ plasma blank}}{A \text{ standard} - A \text{ reagent blank}} \times 0.05 \times \frac{100}{0.2} = \text{mg inulin/dl plasma}$$

$$\text{Urine:} \quad \frac{A \text{ unknown} - A \text{ urine blank}}{A \text{ standard} - A \text{ reagent blank}} \times 0.05 \times \frac{100}{0.01} = \text{mg inulin/dl urine}$$

248 *Chemistry for the clinical laboratory*

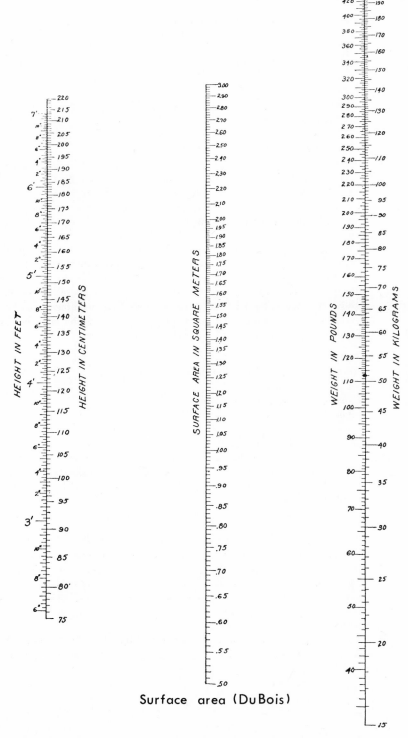

Fig. 10-2. Chart for estimating surface area of body. (From Boothby, W. M., and Sandiford, R. B.: Boston Med. Surg. J. **185**:337, 1921.)

Clearance corrected to average body surface area:

$$C_{in} = \frac{UV}{P} \times \frac{1.73}{S}$$

where U = mg inulin/ml urine
        V = ml urine excreted/min
        P = mg inulin/ml plasma
        S = body surface area in square meters (Fig. 10-2)
     1.73 = average body surface area in square meters

Normal values

The range of normal values for inulin clearance found by different investigators varies. Calculated per 1.73 m² of body surface, it is approximately[20]:

    Male   125 ± 15 ml/min
    Female   110 ± 15 ml/min

Comments. By this method inulin is hydrolyzed to fructose and the products of hydrolysis react with the indole-3-acetic acid to form a purple color. Interference from glucose is less than with other methods, and the color intensity follows Beer's law.

A number of procedures for the determination of inulin have been proposed based on the reactions of the products formed by hydrolysis. Addition of diphenylamine gives a blue color,[2, 10, 15] resorcinol a red color,[7, 18] vanillin a red color, and indole-3-acetic acid a purple color.[11] Each method has limitations and the largest source of error appears to be the glucose present in the specimens. Many modifications have been suggested to eliminate the glucose or to correct for it.

Inulin clearance is the most accurate measure available of glomerular filtration. However, physicians are reluctant to permit the use of procedures involving catheterization of a patient because of the risk of urinary infection. This is especially true in the presence of some renal diseases.

UREA CLEARANCE

Urea, the most abundant end product of protein metabolism, is distributed throughout the intracellular and extracellular body fluids in equal concentrations. Since most of it is excreted from the body by the kidneys, the clearance of this waste product is a measure of kidney function, principally the effectiveness of glomerular filtration.

The concentration of urea in the glomerular filtrate is the same as in the plasma from which it was filtered. About one half is reabsorbed into the blood as the filtrate passes through the tubule. With the exception of production rate, the most important factor influencing the level of urea in blood is glomerular filtration rate (GFR). An increased blood urea level indicates a breakdown in the excretory functions of the kidney.

The clearance of urea measures the volume of plasma required to supply the quantity of urea excreted in a urine specimen in 1 minute. Each minute, the kidney clears or removes urea from the determined volume of plasma. Actually the kidneys never remove all of the urea from the blood and the level in the renal vein is lower than in the renal artery, but the concept is a practical means of assessing renal efficiency. Since the rate of excretion depends on urinary flow or volume, two tests have been devised: maximum urea clearance, where the flow rate of the urine is high, and standard urea clearance, where the urine flow is less than 2 ml/min.

Moller-McIntosh-Van Slyke method[14]

Because exercise interferes with the test, the patient is instructed not to exercise before or during the test. He is requested to eat a light breakfast, which may include water but not coffee or tea, and should include little or no protein. The specimens are then collected in the manner given below. Two urine specimens are obtained so that one may serve as a check on the other. In the collection of urine it is essential that the patient completely empty the bladder and that the exact time interval between collections be measured.

Collection of specimens
1. One hour after breakfast, give the patient a glass of water to drink. Ask him to empty the bladder completely and to discard the urine. Record the exact time in which the urine is voided.
2. Exactly 1 hour later, ask the patient to empty the bladder completely and to save the urine. Record the time at which this urine is collected and label the urine: 1-hour specimen. (If the patient cannot void at the prescribed time, the exact period of time that elapses should be recorded on the specimen. For example, if the patient voids 8 minutes after the prescribed time, the specimen should be labeled: 1-hour 8-minute specimen.)
3. Make a venipuncture and collect blood for a blood plasma urea nitrogen determination.
4. Give the patient another glass of water.
5. Exactly 1 hour after the collection of the first urine specimen, tell the patient to empty the bladder completely and to save the urine. Label: 2-hour specimen.

Procedure
1. Measure the volume of the 1-hour specimen of urine.
2. Calculate the milliliters passed per minute (ml/min) and record the results. Example: urine is collected at the prescribed 1-hour or 60-minute interval, the urine specimen measures 120 ml, and 120 divided by 60 equals 2 ml/min.
3. Measure the volume of the 2-hour specimen of urine.
4. Calculate the ml/min.
5. Analyze the blood and two urine specimens for urea nitrogen concentration.

Calculations
1. The following data, obtained above, must be available: 1-hour urine specimen, ml/min and urea nitrogen value; 2-hour urine specimen, ml/min and urea nitrogen value; and plasma sample, urea nitrogen value.
2. If the patient is a child or an unusually small adult, refer to Note A, p. 251, before continuing.
3. The urea clearance of the 1-hour urine specimen and that of the 2-hour urine specimen are calculated separately. The two should not differ by more than 15%.
4. If the volume of urine excreted is more than 2 ml/min, use formula A. If the volume of urine excreted is 2 ml or less/min, use formula B.

Formula A:

$$\text{Urea clearance (in percent of normal)} = \frac{\text{Urine urea nitrogen (mg/dl)}}{\text{Plasma urea nitrogen (mg/dl)}} \times V \times \frac{100}{75}$$

where V = ml of urine per minute
      75 = average normal value

Table 10-2
Square roots of small numbers

V	$\sqrt{V}$	V	$\sqrt{V}$	V	$\sqrt{V}$	V	$\sqrt{V}$
0.2	0.45	0.7	0.84	1.2	1.1	1.7	1.3
0.3	0.55	0.8	0.89	1.3	1.14	1.8	1.34
0.4	0.63	0.9	0.95	1.4	1.18	1.9	1.38
0.5	0.71	1.0	1.00	1.5	1.23	2.0	1.42
0.6	0.78	1.1	1.05	1.6	1.27	2.1	1.45

Formula B:

$$\text{Urea clearance (in percent of normal)} = \frac{\text{Urine urea nitrogen (mg/dl)}}{\text{Plasma urea nitrogen (mg/dl)}} \times \sqrt{V} \times \frac{100}{54}$$

where V = ml of urine per minute
54 = average normal value

5. If formula B is used, the $\sqrt{V}$ will be needed. For convenience, the square roots of various small numbers are given in Table 10-2. If V = 0.2, the $\sqrt{V}$ = 0.45, etc.
6. Sample calculations here are based on the following data obtained from a urine specimen and the plasma specimen:
   Volume of urine, 138 ml
   Rate of excretion (ml/min), 2.3
   Urine urea nitrogen, 244 mg/dl
   Plasma urea nitrogen, 12 mg/dl

Since the volume of urine excreted per minute is more than 2 ml, formula A must be used:

$$\text{Urea clearance (in percent of normal)} = \frac{\text{Urine urea nitrogen (mg/dl)}}{\text{Plasma urea nitrogen (mg/dl)}} \times V \times \frac{100}{75}$$

$$= \frac{244}{12} \times 2.3 \times \frac{100}{75}$$
$$= 62.3\%$$

Note A:

1. If the patient is a child or an unusually small adult, a correction for size is essential.[13] In such cases, the milliliters excreted per minute must be multiplied by a size correction factor. This is derived in the following manner.
2. Obtain the patient's height and weight. Refer to Fig. 10-2 and place a ruler at the height and weight. Read the surface area from the middle column.
3. Example: Patient is 4 feet tall and weighs 65 pounds. Placing a ruler at these points gives a surface area of 0.97 m².
4. To obtain the size correction factor, divide the surface area into 1.73. (The 1.73 is the surface area of the average man.) Example:

$$\text{Patient's surface area} = 0.97 \text{ m}^2$$

$$\text{Size correction factor} = \frac{1.73}{\text{Surface area}}$$

$$= \frac{1.73}{0.97}$$

$$= 1.78$$

5. Now multiply this size correction factor by the milliliters of urine passed per minute. Use this corrected value for the milliliters passed per minute and continue with the calculations.

Normal values

Maximum average	75 ml/min (64 to 99)
Minimum average	54 ml/min (40 to 68)

Comment. The most common sources of error are incomplete emptying of the bladder and inaccurate timing.

Clinical findings. The clearance rate is lower than the filtration rate because of tubular reabsorption. The amount reabsorbed depends upon the flow rate of the urine in the tubules and the concentration gradients for urea between the urine and blood. When the urine flow is rapid and large, more of the filtered urea is excreted in the urine and the clearance is higher.

When kidney damage occurs, the remaining functioning nephrons increase their capacity to maximum. As GFR decreases, the blood urea concentration rises gradually although it remains within normal limits until the GFR is approximately 40% of normal. Additional loss of filtration capacity causes a rapid increase in blood urea. Thus in chronic renal failure a measurement of blood urea is not a sensitive indicator of renal function, since marked kidney damage can occur without elevating blood urea above the normal range. However, when the urea clearance is decreased, there is renal impairment.

With refinement in other techniques, urea clearance is less often requested because of the inherent variability. Most clinicians depend on clearance of creatinine, which is described next.

CREATININE CLEARANCE[8, 21]

The blood level of creatinine, formed from muscle creatine and phosphocreatine, depends on skeletal muscle mass and is relatively constant in any individual from day to day. In the same manner as urea, most of the creatinine is excreted in the kidney by glomerular filtration. Serum creatinine concentration rises when filtration rate falls. Protein variations in the diet have a slight but minimal effect. A 24-hour urine specimen is preferred although shorter periods, 2 or 6 hours, may be used.

Collection of specimens

1. At the beginning of the test ask the patient to empty the bladder completely and to discard the urine. Record the exact time at which the urine was voided.
2. Instruct the patient to collect all urine voided for the next 24 hours in a single container in which 15 ml of toluene has been placed as a preservative. At the end of the 24-hour period ask the patient to void and add this urine to the total collection.
3. Near the end of the urine collection draw a fasting blood sample for serum creatinine determination.
4. Record the height and weight of the patient for calculating surface area of the body.

Procedure

1. Measure the volume of the 24-hour urine specimen.
2. Determine the creatinine content of the urine and serum.

$$C_{cr} = \frac{UV}{P} \times \frac{1.73}{S}$$

where C_{cr} = endogenous creatinine clearance corrected to average body surface area
U = mg creatinine/ml urine
V = ml urine excreted/min
P = mg creatinine/ml serum
S = body surface area in square meters (Fig. 10-2)
1.73 = average body surface area in square meters

Normal values

Male	105 ml/min (72 to 141)
Female	95 ml/min (74 to 130)

Comments. The method generally employed for determining creatinine, the Jaffé reaction, whereby an orange-yellow color is formed with an alkaline picrate solution, is not specific. Normally some 10% of the reaction is due to noncreatinine chromogen. However, the results are considered acceptable for routine clinical purposes.

Clinical findings. Creatinine clearance is used as an estimate of glomerular filtration. It appears to be the most practical test available for clinical evaluation of renal function and in general is comparable to inulin clearance tests. In normal persons the ratio of C_{cr} to C_{in} is about 1.1 to 1.0. It is useful in following the progress of patients with renal insufficiency.

In advanced renal disease endogenous creatinine clearance is significantly higher than the filtration rate. Apparently creatinine is partly excreted by the renal tubules in such a situation. At this point the ratio of C_{cr} to C_{in} may be 1.4 or 1.5 to 1.0. However, this does not occur until there has been extensive renal damage and therefore the error in measurement is not large enough to be misleading.

Tubular functions

Without the reabsorptive activity of the tubules, the high rate of filtration by the glomeruli would rapidly eliminate all water and water soluble constituents from the body. By reabsorption and secretion the tubules maintain the volume, osmolality, and electrolytes of body fluids within normal limits. While they act to conserve water and essential solutes in a manner dependent on the dietary intake and storage requirements, they excrete waste products and other substances not needed by the body.

Tubular reabsorption is designated as being either active or passive. Active transport refers to the transfer of a constituent such as glucose from a lower to a higher concentration, as from a low level in urine to a much higher level in blood. Energy is expended in the process. Reabsorption is passive when the substance diffuses down a gradient of chemical activity or electric potential, that is, from a high concentration in the urine to a lower one in the plasma, for example, urea.

Another term used in relation to the active absorption of certain substances is threshold. Some constituents of blood are not normally found in urine until their concentration exceeds a certain amount or threshold level. A constituent such as glucose that is completely reabsorbed from the tubular fluid at normal blood levels is described as having a high threshold. The capacity of the tubules to reabsorb glucose is not saturated until the blood level exceeds approximately 180 mg/dl. Below that level all glucose in glomerular filtrate is actively reabsorbed. At saturation of the tubular capacity for reabsorption of glucose, the rate of reabsorption becomes constant and is designated as the tubular maximum for glucose (T_{mG}) or maximal reabsorptive capacity for glucose. Other T_m-limited substances are phosphate, sulfate, amino acids, uric acid, proteins, and certain organic acids. If a given mechanism transports more than one substance, competition for the carrier may occur.

The hypertonic urine at the tip of Henle's loop becomes hypotonic in the ascending limb. The volume and concentration of urine are largely controlled by antidiuretic hormone (ADH). ADH is secreted into blood by the neurohypophysis in response to blood osmolality shifts. The concentration or dilution of urine occurs in the distal convolution and the collecting tubules. As the tubular fluid enters the distal tubule, it is hypotonic, becoming isotonic by the time it leaves. Further reabsorption of water continues during passage through the collecting ducts under the influence of ADH. In the absence of antidiuretic hormone, the urine remains hypotonic during its passage from the ascending limb of Henle to the renal papilla.

CONCENTRATION TESTS[3]

Regulation of the concentration of urine is one of the principal functions of the renal tubules and is often diminished in renal disease. Damaged kidneys have fewer nephrons to excrete a larger amount of solute. The formation of hypertonic urine by the distal tubule and collecting duct is fundamental to the concentrating mechanism. Therefore the ability to concentrate may be diminished before retention of nitrogenous waste products occurs in certain diseases such as polycystic kidneys, pyelonephritis, and hydronephrosis, since these conditions tend to damage the medulla earlier than the cortex of the kidney.

Measurements of concentrating ability are carried out by determining either osmolality or specific gravity of urine under standardized or known conditions. Anatomic or environmental conditions can cause urine flow to fluctuate from 0.25 to 25 ml/min. However, the osmolality tends to be inversely proportional to the volume when the tubules are functioning normally. The determination of milliosmolal concentration, calculated from freezing point depression, is a direct measure of the total excretion of solutes and of the concentration. The results can be compared with those of serum. If higher than serum, active concentrating ability is indicated.

If the urine is free of glucose and protein, measurement of specific gravity is sufficiently accurate for most clinical purposes. Since urea and salts are the principal solutes in urine, normal dietary intake must be maintained when the test is performed. An appropriately sealed refractometer may be used to determine specific gravity if the volume of urine is too small for the standard urinometer.

Mosenthal test

The Mosenthal test measures the specific gravity of urine at various intervals under specified conditions.

Collection of specimens
1. On the day of the test, as well as the day before, the patient is to have a full regular diet and is to drink at least a pint of fluids at each meal, but he is not to eat or drink between meals.
2. For the morning of the test, he is instructed to completely empty the bladder before breakfast at 8 AM and to discard the urine. The test is usually started at 8 AM.

3. He is then to completely empty the bladder into separate labeled containers at exactly 10 AM, 12 noon, 2 PM, 4 PM, 6 PM, and 8 PM.
4. Collect in a single container all the urine voided between 8 PM and 8 AM.

Procedure

1. Measure the volume and specific gravity of each urine specimen.
2. Report the volume and specific gravity of each specimen. See sample report:

Specimen	10 AM	12 noon	2 PM	4 PM	6 PM	8 PM	8 PM to 8 AM
Volume (ml)	150	156	170	165	200	110	300
Specific gravity	1.017	1.020	1.014	1.022	1.012	1.010	1.018

Interpretation

Normal values are indicated by a fluctuation in the specific gravity of at least 7 points; that is, there is at least a 7-point difference between the lowest and highest specific gravity. In kidney disease the specific gravity fluctuates only a few points and the volume of the night urine (8 PM to 8 AM) is usually over 600 ml. The test is useful for detecting slight degrees of renal dysfunction. Ability to concentrate is frequently abnormal in chronic glomerulonephritis, but it may be normal in acute glomerulonephritis if filtration rate is not impaired. Polyuria or nocturia interferes with the accuracy of the test.

Fishberg test[9]

The Fishberg test measures the specific gravity of urine at various intervals under specified conditions.

Collection of specimens

1. The patient is told to eat an evening meal consisting of a high-protein diet and not more than 200 ml of fluid.
2. After the evening meal, no food or fluid is to be taken until the completion of the test on the following morning.
3. The patient is further instructed to void at will during the night. In the morning, he should empty the bladder into separate containers at 8 AM, 9 AM, and 10 AM.

Procedure

1. Measure the volume and specific gravity of each urine specimen.
2. Report the volume and specific gravity of each specimen. See sample report:

Specimen	8 AM	9 AM	10 AM
Volume (ml)	210	190	150
Specific gravity	1.020	1.026	1.025

Interpretation

If kidneys are healthy, at least one of the specimens should have a specific gravity over 1.024. With diseased kidneys all the specific gravities are below 1.020. Achievement of a high specific gravity depends on both the availability of excretable solute and a maximal stimulus to converse water. The test is a sensitive indicator of early renal dysfunction. Patients with advanced renal disease may show a fixed urinary specific gravity value between 1.010 and 1.012, the same as an ultrafiltrate of plasma. In edema and congestive heart failure low urinary specific gravity levels may be found, simulating renal impairment.

DILUTION TESTS

In renal disease the ability of the kidneys to dilute urine is impaired. With normal renal function, following a water load, most of the water ingested will be excreted within 4 hours, and the specific gravity of at least one hourly urine specimen will be below 1.003. In renal impairment a low specific gravity is not reached and the urine volume remains low. However, since the ability to concentrate is lost before the ability to dilute, this type of test is of little value in assessing early change in renal function.

TUBULAR EXCRETORY FUNCTION

Substances such a p-aminohippuric acid, Hippuran, and Diodrast are actively excreted by the tubule cells and may be employed to measure tubular secretory function. The limit or maximum capacity of the tubules to reabsorb or secrete these compounds is termed the tubular excretory maximum (T_m). Para-aminohippuric acid (PAH) is most frequently used in renal function studies. At very high blood levels of PAH the transport system will operate at maximum capacity or saturation. The plasma concentration must be above saturation to force the tubule cells to excrete at their maximum capacity over a sufficient period of time to conduct clearance studies. The high constant blood level can be achieved only by a sustained infusion of a solution of PAH. By determining the maximum amount of PAH that can be excreted per minute by the tubule cells and the percentage of PAH removed from the blood during passage through the kidneys the tubular excretory maximum can be calculated. It is an index of functional tubular tissue, since the excretion of PAH diminishes as the mass of tubular cells is reduced. In normal renal function, the maximal tubular excretory capacity for PAH (T_{mPAH}) corrected to 1.73 m², the average surface body area, is [20]:

Men	79.8 ± 16.7 mg/min
Women	77.2 ± 10.8 mg/min

The test is not used as a clinical method, since it is time-consuming, difficulty is encountered in maintaining saturation of the plasma with the test compound, and application is limited to one transport mechanism.

Renal blood flow

The kidneys are dependent on the blood supply for transport to and from them of biologic constituents that are selectively either eliminated from the body into the urine or reabsorbed into the blood. Specific ions or molecules are actively transported in the exchange between the tubular fluid and the blood, while other solutes and water passively diffuse between the lumen and blood according to osmotic or electric potential gradients.

The total renal blood flow is that volume which enters the renal arteries each minute (approximately 1200 ml). The *effective* renal blood flow, or renal plasma flow, is that amount which perfuses the *functional* renal tissue each minute. The latter may be measured by determining PAH clearance and the effective renal blood flow calculated by applying the hematocrit value. At low plasma levels about 90% to 93% of PAH or Diodrast is removed in one passage through the kidneys. Agreement is lacking among investigators concerning the disposition of the remaining 7% to 10%.[4] Suggested explanations have been (1) that it is in the blood flowing through inactive or nonfunctioning tissue or (2) that there is a complex formed between PAH and glucose which has a lower clearance rate than PAH. Variation in effective renal plasma flow is great even when corrected to standard body surface area. One standard deviation for any population is nearly 20% of the total mean value.

Phenolsulfonphthalein (PSP), a dye that is readily excreted by the tubules, may be used also to determine renal blood flow. The test lacks sensitivity in that there can be 40% renal impairment before detection by use of the phenol red, but the ease with which the test can be performed has made it popular.

PAH CLEARANCE[7]

Since PAH is not normally present in blood, sufficient amounts must be administered to maintain a constant blood level during the clearance test. Below 5 mg/dl the PAH clearance of plasma is not dependent on the concentration. Some investigators[12] prefer a screening procedure that employs a single intramuscular injection of 8 ml of 20% PAH-Na solution and 4 ml of an anesthetic containing epinephrine. The patient is asked to drink several glasses of water and the clearance determination is started after 30 to 50 minutes. Two timed clearance periods of 45 to 60 minutes are used where urine and blood specimens are obtained in the same manner as for other clearance tests.

A more exact PAH clearance may be carried out by using the procedure as described for inulin clearance. Frequently the two clearances are performed simultaneously by including the proper amounts of both chemicals in the priming and infusion solutions.

Collection of specimens

The preparation of the patient, administration of PAH, and collection of urine and blood samples are the same as for inulin clearance. A priming solution of 8 mg of the sodium salt of PAH per kilogram of body weight is administered and followed by a sustaining infusion containing 8 ml of a 20% PAH solution diluted to 500 ml with saline administered at a rate of 4 ml/min.

Procedure

The procedure is the same as that employed for the determination of sulfonamides, except the standard solution is prepared from *p*-aminohippuric acid. For the analyses plasma and urine diluted 1:100 are used. The calculations are the same as for inulin clearance.

Normal values

The normal PAH clearance at low plasma levels corrected to 1.73 m^2 of body surface as a measure of renal plasma flow is as follows:

    Male   700 ± 135 ml plasma cleared/min
    Female   600 ± 100 ml plasma cleared/min

The effective renal blood flow may be calculated from the plasma clearance by applying the patient's hematocrit.

    Male   1275 ± 245 ml blood cleared/min
    Female   1090 ± 180 ml blood cleared/min

Clinical findings. A decreased effective renal blood flow will be noted when (1) the cardiac output is reduced or arterial blood pressure is lowered as in congestive heart failure or shock; (2) the renal vascular system is damaged as in glomerulonephritis, intercapillary glomerulosclerosis, or renal arteriosclerosis; (3) the mass of functioning kidney tissue is decreased as in polycystic disease, pyelonephritis, renal hypoplasia, or malignancy; or (4) there is constriction of the afferent glomerular arterioles in conditions such as essential hypertension.

An increased renal blood flow may be found following the administration of pyrogenic substances.

PHENOLSULFONPHTHALEIN (PSP) EXCRETION TEST[5, 6, 17]

Phenolsulfonphthalein is a nontoxic dye that binds reversibly to the albumin fraction of plasma. When administered most of it is excreted by the renal tubules and the rate of excretion depends principally on the renal plasma flow. Therefore as used in low concentration in a test procedure it provides an indication of renal blood flow.

Collection of specimens

The nurse and physician usually have the task of instructing the patient, injecting the dye, and collecting the urine specimens. This consists of the following:

1. The patient is told not to take any stimulants (coffee, tea, or tobacco) for 2 hours prior to the test and also during the test.
2. The patient voids and then is given several glasses of water to drink to promote urine flow. The test is started about 30 minutes later.
3. Exactly 1.0 ml (6 mg) of sterile phenolsulfonphthalein is injected intravenously and the time noted and recorded.
4. Exactly 15, 30, 60, and 120 minutes after injection urine is collected. The time of collecting each sample is recorded.

Reagents

Sodium hydroxide, 10% (2.5 N)   (D-83)
   For method B
PSP stock standard, 120 mg/500 ml   (D-184)
   Working standard, 1.2 mg/500 ml

Pipet 5.0 ml of stock standard into a 500-ml volumetric flask, add 1.0 ml of 10% NaOH, and dilute to the mark with distilled water. Mix thoroughly.

Procedure

1. Measure and record the volume of each specimen.
2. Add 10% sodium hydroxide dropwise to each specimen until the maximum red color is brought out. About 5.0 ml is usually required. The dye is colorless in acid urine and red in alkaline urine. Check the pH. It should be higher than 9.0.
3. Dilute the urines according to volumes, as follows:
   a. If the volume is less than 50 ml, dilute to 250 ml with water.
   b. If the volume is between 50 and 450 ml, dilute to 500 ml with water.
   c. If the volume is above 450 ml, dilute to 1000 ml with water.
4. Mix each specimen.
5. Determine the percent of dye in each specimen by one of the two following methods:

   Method A: Compare with Dunning standards representing known percentages of the dye.

   Method B: Compare in a photoelectric colorimeter or spectrophotometer. Filter about 10 ml of the urine. Pipet 6.0 ml of filtered urine into a cuvet and 6.0 ml of diluted working standard into another cuvet. Read absorbance (A) against a water blank at 520 nm.

Calculation

$$\frac{A \text{ of unknown}}{A \text{ of standard}} \times 1.2 = \text{mg PSP in specimen}$$

NOTE: **If diluted urine volume is 250 ml, divide by 2. If diluted urine volume is 1000 ml, multiply by 2. The sum of the mg PSP in the specimens is the total PSP excreted. Since 6 mg is usually injected, calculate the percent excreted as follows:**

$$\frac{\text{Total mg PSP}}{6} \times 100 = \text{Percent excreted}$$

Normal values

Time (min)	Percent PSP excreted Average	Range
15	35	28-51
30	17	13-24
60	12	9-17
120	6	3-10

The 15-minute value is the most important for evaluating renal function since some damaged kidneys are able to excrete normal amounts from the blood in 2 hours. It is important that the patient be loaded with water before starting the test to minimize the errors in emptying the bladder.

Clinical findings. Excretion is decreased in congestive heart failure, chronic nephritis, cirrhosis with normal kidneys, cystitis, pyelonephritis, prostatic obstruction, and edema. In marked renal damage the excretion of PSP tends to vary directly with urine volume. Mild nephritis, essential hypertension, and nephrosclerosis cause some delay in excretion. In chronic renal insufficiency the excretion decreases as the disease progresses. Excretion of PSP may be normal in acute glomerulonephritis even when glomerular filtration is reduced and plasma urea is elevated.

Bile or blood in the urine will interfere and must be removed before analysis for the PSP content is made. Bromsulphalein will also interfere, so a period of 24 hours should be allowed between the two tests.

Blood chemistry in renal disease[16, 23]

Measurements of the concentration of various chemical constituents in blood are used to assess renal function. However, because of their large reserve capacity the kidneys can maintain normal blood levels when only one half of one kidney is functioning normally.

In the initial stages of renal insufficiency the involvement may be located primarily in either the glomeruli or the tubules or both. As a disease progresses, destruction of nephron units occurs until finally the quantity of functioning kidney mass is inadequate to maintain normal body chemistry. In chronic renal failure the nephrons are destroyed without the possibility of reversal, but blood tests do not reflect the condition until extensive kidney damage has occurred with the accompanying retention of nitrogenous metabolic products.

In acute renal failure, depending on the cause, the nephrons may be only temporarily injured and the process can be reversed if proper therapy is initiated before permanent damage occurs. In these cases the changes in blood constituents do provide an indication of the degree of renal dysfunction.

UREA

Factors other than renal function affect the urea concentration in blood. A patient on a restricted-protein diet or one with severe liver disease may have markedly reduced kidney function, yet a normal blood urea level. Since urea is synthesized in the liver, a malfunctioning liver will produce less urea and the level in the blood may be normal even in the presence of renal damage. A high-protein diet or rapid cell destruction, as in intestinal hemorrhage, may increase the blood urea above the normal concentration range.

Blood urea concentration does not start to increase until glomerular filtration has fallen below 50%. However, the test is useful in serial determinations as a guide during treatment of renal failure, and a sudden rise or elevated value should be considered as indicating probable impairment of renal function.

CREATININE

The concentration of creatinine in serum is a more sensitive index of renal function than is urea. It depends essentially on muscle mass and is largely independent of protein intake. In a situation where there is severe muscle wasting, the blood level may be exceptionally low. For each individual the muscle mass is relatively constant and therefore the serum creatinine concentration varies inversely with glomerular filtration. A markedly elevated value is indicative of severe renal insufficiency.

POTASSIUM

In chronic renal insufficiency, elevated serum potassium levels are not present until the final stage of the disease. Except when oliguria or anuria occurs, the kidneys are able to maintain a normal level of urinary potassium by tubular secretion. Hyperkalemia may be found in patients with decreased urine output when there is an increased supply of potassium such as occurs in tissue breakdown or exogenous potassium administration. In acute renal failure, an elevated serum potassium is a most important chemical finding because of the ion's toxic effect on the heart.

Potassium depletion may be the result of, or a cause of, renal disease. It may occur postoperatively when losses arising from surgical trauma are not replaced, from administration of diuretics or adrenal steroids with inadequate intake, and from renal losses. Hypokalemia can produce tubular damage with impairment of the ability of the kidneys to concentrate.

SODIUM

As a result of decreasing ability of the kidneys to concentrate, the serum sodium concentration is often reduced in progressive renal failure. This appears to result from the inability of the functioning tubules, because of the increased osmotic load, to bring about the necessary reabsorption of sodium from the tubular fluid. In the final stage of chronic renal failure or when the urine flow is very low, serum sodium is increased and the patient is then maintained on a restricted low-sodium diet.

BICARBONATE

A characteristic finding in renal disease is a lowered plasma bicarbonate, or metabolic acidosis, resulting from the accumulation of acid waste products. Normally the kidneys eliminate acid by combining it either with ammonia or with buffers, principally sodium dihydrogen phosphate, to form titratable acid. The phosphate buffer is obtained from inorganic phosphate in glomerular filtrate, derived from serum. In chronic renal insufficiency, the ability of the kidneys to provide the required ammonia and phosphate buffer decreases as the disease progresses.

CALCIUM AND PHOSPHATE

In chronic renal failure the plasma calcium concentration is frequently below normal. However, factors other than serum phosphate concentration appear to influence the level, which may remain normal even with increased serum phosphate or may be decreased with little change in phosphate concentration.

In the later stages of chronic renal disease, there is often marked elevation of serum phosphate. However, this does not occur until there is marked decrease of renal function. The metabolic bone disease, renal osteodystrophy, which is associated with azotemia may cause malacia, sclerosis, or cystic changes in bone; but there is no consistent calcium or phosphate pattern characteristic of the condition.

MAGNESIUM

Hypermagnesemia is frequently observed in advanced chronic renal failure. However, normal or low values have been found for some patients. It has been suggested that the calcium level may influence the magnesium concentration. The fact that the ratio of bound to ionized serum magnesium is increased in uremia may be important.

PROTEINS

Hypoproteinemia is characteristic of patients with chronic renal insufficiency. It may result from loss of large quantities of protein in the urine, dietary restriction of protein to control azotemia, and defects in protein metabolism.

RENIN-ANGIOTENSIN SYSTEM

Renin is a proteolytic enzyme arising in the juxtaglomerulosa apparatus of the kidney. It acts on an $alpha_2$ globulin (angiotensinogen) synthesized in the liver, releasing a decapeptide, angiotensin I, which forms angiotensin II by the action of converting enzyme during passage through the lungs. The half-life of the angiotensin II is 1 to 2 minutes; it is rapidly destroyed by the action of angiotensinase found in many tissues.

The renin-angiotensin system influences the fluid volume of the body by regulating aldosterone secretion from the adrenal cortex. Release of renin from the kidneys is stimulated by low plasma volume or serum sodium. Aldosterone secretion is regulated by the renin-angiotensin system in a feedback control. A decrease in extracellular fluid volume or intraarterial vascular volume will decrease renal arterial pressure and increase renin secretion, which is responsible for increased angiotensin II that causes increased rate of secretion of aldosterone. The aldosterone causes sodium and water retention, expanding fluid volume and thus eliminating the stimulus that initiated increased renin secretion. Standing and constriction of the thoracic inferior vena cava decrease intraarterial vascular volume.

Hypertension that is not relieved by restoration of normal body water, exchangeable sodium, and antihypertensive drugs may respond to bilateral nephrectomy. Increased plasma renin activity levels will distinguish this group of patients. Renin activity measurements of blood from left and right renal veins are helpful in diagnosing unilateral renal disease.

Plasma renin activity is based on measurement of angiotensin I by radioimmunoassay using a specific antibody system. The plasma samples are incubated with reagents which block conversion to angiotensin II. The generated angiotensin I competes with that labeled with ^{125}I for binding sites of the antibody. Variables that influence renin activity (posture, salt intake) must be controlled.

Normal levels of plasma renin activity are:

Normal salt diet	Range (ng/ml/hr)
Supine	0.5 to 1.6
Upright	1.9 to 3.6

Low salt diet	
Supine	2.2 to 4.4
Upright	4.0 to 8.1
After diuretic and 4 hours upright	6.7 to 15.1

Increased plasma renin activity levels are found with low salt diets, diuretic therapy, renal artery stenosis, malignant hypertension, nephrotic syndrome, renal ischemia, cirrhosis of the liver, decompensated congestive heart failure, valvular heart disease, estrogen therapy or pregnancy, and untreated Addison's disease. Decreased renin activity levels occur with increased salt intake, salt-retaining steroid therapy, antidiuretic hormone therapy, and supine posture. A suppressed renin state is characterized by failure to stimulate renin activity by low sodium intake, diuretic therapy, and upright posture.

Renal insufficiency and dialysis
C. F. GUTCH, M.D.

The clinical syndrome of renal insufficiency is the result of a decrease in renal function due to injury or disease, so that metabolic wastes are not adequately removed from the body by urinary excretion. Complicated disturbances of water and electrolyte balance accompany the retention of catabolic products.

Renal insufficiency may be either acute or chronic. Acute renal failure develops rapidly—within hours or a few days—and is usually manifested by oliguria (24-hour urine output less than 400 ml). Often it is the result of acute tubular necrosis, which is a potentially reversible lesion, if the patient can be kept alive for the 2 to 3 weeks required for recovery to begin. The mechanism of acute tubular necrosis is unclear, but decreased perfusion—due to renal vasoconstriction, or shunting of blood away from the kidneys—seems to play a major role. Shock or severe hypoxia, pancreatitis, head injury, severe dehydration, carbon monoxide intoxication, trauma, and septicemia are frequent precipitating factors. Other causes include crush syndrome (with production of hemoglobinemia or myoglobulinemia), sickle cell crisis, hemolysis after transurethral resection, mismatched transfusion, severe burns, and hemorrhagic fever.

Acute renal failure (not always limited to tubular lesions only) may also result from specific nephrotoxic substances—mercury, bismuth, phosphorus; some antibiotics; organic compounds such as carbon tetrachloride; and mushroom poisoning.

Rapid onset of oliguria, or even anuria (urine output less than 100 ml/24 hr), is often a feature of acute fulminating glomerulonephritis, necrotizing arteriolar disease, or the consumption coagulopathies. Such lesions, unlike acute tubular necrosis, are often not reversible, although sufficient return of function to maintain life may be regained. Renal function patterns, urine sediment, and renal biopsy serve to permit differentiation.

Renal failure from acute tubular necrosis has an overall mortality rate of 50% with conservative treatment. Early and adequate dialysis has improved the outlook for patients whose oliguria follows hypotension or hemolysis; it has also greatly reduced their morbidity. The mortality of persons with renal failure as a result of extensive trauma remains high even with dialysis, although such patients die of sepsis or other complications rather than uremia.

Chronic renal insufficiency may follow acute or subacute glomerulonephritis, which progresses relentlessly to loss of functional renal mass. However, not infrequently the patient may be unaware of the underlying condition for many years. Symptoms may not become apparent until there is marked elevation of serum urea and creatinine, by which time the functional kidney mass is down to 25% to 50% of normal. Insidious chronic renal insufficiency of this sort may follow an unrecognized glomerulonephritis of many years' duration.

Pyelonephritis, the other major cause of chronic renal failure, is the result of direct bacterial invasion of kidney parenchyma. It is often indolent and unrecognized until renal failure supervenes.

Polycystic disease, a congenital anomaly in which functional renal tissue is gradually compressed by fluid-filled cysts, often remains undetected until the fourth or fifth decade of life, when kidney failure begins to manifest itself.

Diabetic glomerulosclerosis is not infrequently an end-stage complication of long-standing diabetes, particularly the diabetes that has its onset in childhood or adolescence. Persons with gout may develop a nephropathy as a result of urate deposits in renal parenchyma. People with primary hypertension may develop arteriolar nephrosclerosis, in which damage to renal arterioles causes loss of function due to ischemia.

Obstruction of the lower urinary tract (such as prostatic enlargement) or other anatomic defect may lead to damming back of urine, resulting in striking distention of the renal pelvis and calyces (hydronephrosis) and decreased renal function due to the pressure. There are many other, less common disorders that can cause insidious loss of renal function to the point of chronic renal failure.

The body can adjust to rather severe biochemical disturbances, particularly if these have been gradual in development. Persons can maintain an active life with less than 10% of normal renal function.

DIALYSIS

Dialysis is defined as the separation of solutes and/or crystalloids by differential diffusion across a porous (semipermeable) membrane between two solutions. When there is a concentration gradient between the two solutions, involving one or more substances, differential diffusion occurs. Natural or artificial membranes may be used as the semipermeable screen in clinical application of dialysis.

In the reversible renal failure of acute tubular necrosis, dialysis therapy can remove metabolic wastes and minimize biochemical abnormalities until function of the patient's own kidneys returns. Occasionally the nature of the basic renal lesion is not apparent in a patient with acute renal failure, and dialysis may be undertaken while appropriate diagnostic procedures are initiated.

Additionally, some patients with permanent renal insufficiency of such severity as to be incompatible with life can be maintained by regular dialysis treatment for prolonged periods. Several hundred such people are now alive after 3, 4, or more years of regular maintenance hemodialysis, even though they have little or no renal function of their own. Such patients have numerous biochemical abnormalities, many of which are even now poorly understood. Their lives are not normal but can be useful and satisfying.

Hemodialysis

The first hemodialysis apparatus, or artificial kidney, was designed by Abel, Rowntree, and Turner in 1913.[24] Cylindrical tubes of collodion were used as the semipermeable membrane. These inventors did not have adequate anticoagulants to keep blood from clotting and they had difficulty controlling the electrolyte concentration of the bath, so that their apparatus could not be used clinically. Thirty years later (1943) Kolff developed and used the first practical artificial kidney, employing a length of cellophane tubing wound in a spiral manner on a rotating drum. The rotating type of kidney was described in 1943 by Kolff and Berk,[34] and detailed plans were published by Kolff in 1947.[33] Several other types of hemodialyzers were developed about the same time. In 1947, Alwall[25] (Sweden) developed an efficient apparatus, and in the United States Skeggs and Leonards[41] devised a layer type dialyzer utilizing cellophane sheets placed between grooved rubber plates. In 1955 Kolff[35] perfected the twin-coil dialyzer, which was then mass produced by Travenol Laboratories and was the prototype of present coil dialyzers. The Kiil dialyzer, a simplified layer design, was modified in 1960 by Scribner's group in Seattle.[39] The Kiil and the Kolff coil are currently the two most widely used dialyzers.

Initially, hemodialysis was considered suitable only for patients with reversible renal failure. Since only one or two dialyses were usually performed, access to blood vessels was obtained by cannulating an artery and a vein in an arm or leg with metal, glass, or plastic cannulas. Scribner and co-workers[40] realized that if repeated dialysis was to be done for patients, permanent and easier means of access to the blood vessels was needed. They devised indwelling cannulas of Teflon and Silastic, with an external connecting loop outside the skin. Creation of this external arteriovenous fistula prevented clotting when not in use, yet the arterial and venous tubes could be attached to the artificial kidney whenever needed.

In recent years intensive research has gone on to develop smaller and more economical hemodialysis devices. Cuprophane, a copper-treated cellulose membrane, is thinner than cellophane and more permeable. It is in common use in both Kiil and coil dialyzers. Other types of membranes are being investigated, including some formed from collagen from animal sources. A hollow cellulose-fiber kidney—somewhat reminiscent of the original apparatus of Abel, Rowntree, and Turner but using very tiny capillary tubes and having a small blood volume—has been developed and used clinically.

DIALYSATE SOLUTIONS

Hemodialysis involves the passage of blood along one side of a semipermeable membrane while a special bath solution flows along the other side. The membrane has pores of 25 to 30 Å diameter, so that red cells, leukocytes, and serum proteins do not pass through; but solutes—including the metabolic wastes such as urea, creatinine, and uric acid, and other undesired products—diffuse through. On the bath side of the membrane is a solution of electrolytes prepared to approximate the concentration of plasma water. Thus desired electrolytes are maintained in equilibrium on both sides of the membrane, while wastes and unwanted ions such as sulfate and phosphate pass from the blood to the area of lesser concentration, the bath.

The several varieties of dialysis fluid delivery systems may be classified into three general categories:
1. Recirculating system. In recirculating systems a quantity of dialyzing fluid (usually 100 L) is continually recirculated through the dialysis apparatus. This bath is changed every 2 or 3 hours to maintain a gradient of waste product from blood to bath and to control bacterial growth in the bath.
2. Single-pass system. For the single-pass systems, fresh dialysis fluid flows through the dialyzer only once, passing on to the waste collector.
3. Recirculating single-pass system. The recirculating single-pass system combines the other two. A small recirculating volume (8 to 12 L) is continually replenished with fresh dialysis fluid at a slower rate. The volume used varies with each system and with the duration of dialysis.

Fluid for dialysis may be prepared by the single-batch method, whereby a given volume is made up and checked for proper electrolyte concentration immediately before use. The tanks for such systems are usually of 100, 200, 300, or 385 L capacity. For acutely ill patients, fluid prepared with tap water is usually satisfactory. However, when patients on chronic dialysis are dialyzed repeatedly, the chemicals in the raw water (usually sodium, calcium, magnesium, carbonate, sulfate, and sometimes fluoride) may assume significance. Filtering, softening, or deionization may be necessary to ensure uniformity.

The dialysis bath may be prepared using reagent grade chemicals. A common formula for the Kolff 100-L unit is as follows:

Component	(g/100 L)	Na^+	K^+	Ca^{++}	Mg^{++}	Cl^-	HCO_3^-
		(mEq/L)					
NaCl	570	97				97	
$NaHCO_3$	300	36					36
KCl	30		4			4	
$CaCl_2$	28			5		5	
$MgCl_2$	7.5				1.5	1.5	
TOTAL		133	4	5	1.5	107.5	36

Dextrose or invert sugar is added to a concentration of 200 to 400 mg/dl to provide appropriate osmolality. Lactic acid is used to adjust the pH to 7.4, and 10% carbon dioxide with 90% oxygen is bubbled through the fluid continually to maintain the carbonate concentration.

When the quantity prepared is other than 100 L, appropriate adjustments must be made in the amount of each chemical used. The calcium and magnesium chloride should be dissolved in a separate container of warm water and added to the solution of sodium chloride and bicarbonate after these are dissolved, the lactic acid added, and the tank nearly full, to avert precipitation of the calcium and magnesium.

Most dialysis centers now employ commercially prepared concentrates that, appropriately diluted, yield the desired concentration of electrolytes. Sodium acetate is used in these rather than bicarbonate to achieve stability. This may slow correction of a patient's acidosis slightly since the acetate must be metabolized by the body to bicarbonate. Also in prolonged dialysis, there may be a tendency to metabolic alkalosis at the end.

Commercial concentrates may be supplied in bulk (50 gallons) or in premeasured containers for specific machines. There are a number of formulations available with various concentrations of sodium, calcium, potassium, and magnesium to meet varying needs of chronic dialysis patients. It is extremely important to double check all concentrate containers and to verify that the correct dilution is obtained.

Verification of the dialysis bath is of critical importance to the well-being of the patient. Several deaths have resulted from bath errors. Bath checks should include (1) pH, (2) test for solute concentration, either osmolality by freezing point depression or refractivity on a high-grade refractometer, and (3) accurate titration for total chloride content since sodium and chloride are the major ions present. The ionic content may alternatively be registered by an electric conductivity probe. Conductivity (reciprocal of resistance) is an accurate index of ion content of the solution if the instrument is temperature compensated. All tests should be recorded as a permanent record and instruments and methodology verified against known standards at frequent intervals.

Many large dialysis centers, and some home dialysis equipment, use an online dialysis fluid–mixing system. In these units, concentrate is metered and continually mixed with water, usually by a set of interconnected proportioning pumps. Thus a fresh supply of appropriately mixed fluid is constantly provided. These systems most commonly are used with layer type dialyzers where the duration of the run and quantity of dialysate used are greater than in coil systems. Such central proportioning equipment has built-in filters, heaters, one or more conductivity controls, and temperature controls. These are designed to operate within very narrow limits and to be "fail-safe" in that flow of dialysis fluid is stopped and alarm indicators activated if there is any deviation from preset limits.

DIALYSIS EQUIPMENT AND PROCEDURES

The setup of two types of dialyzers will be described as examples of procedures for extracorporeal hemodialysis therapy.

Twin-coil dialyzer (Fig. 10-3)

The coil and the arterial and venous tubing sets are separately packaged and are factory sterilized. Arterial blood enters the coil at the inner core and leaves at the peripheral. Because of the internal resistance in the coil, a pump is necessary to propel the blood through the unit.

Assembly (Fig. 10-4)
1. Soak the coil for 5 minutes in warm (37° C) water or dialysate.
2. Fasten the coil (with inflatable cuff, if required) in the canister.
3. Using sterile technique, connect the arterial set to the inlet side of the coil.
4. Position the pumping segments in the blood pump and clamp.

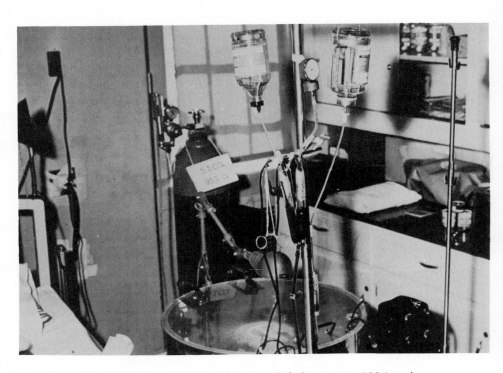

Fig. 10-3. Hemodialysis with twin-coil dialyzer using 100-L tank.

5. Suspend from a standard one or more liter bottles of heparinized saline and attach the infusion tubing to the priming tube inlet.
6. Attach the venous set to the outlets of the coil.
7. Suspend a bubble trap chamber from an IV standard and attach one side-arm tubing to the positive pressure monitor.
8. Prepare the dialyzing fluid in the tank at the required temperature (39° C) with a continuous bubbling of the CO_2-O_2 gas mixture (10% to 90%) if using a bicarbonate bath. Verify the dialysate bath as described previously and start the recirculating pump.

Testing

1. Slowly pump sterile, heparinized saline from the infusion set through the coil at a rate not to exceed 75 ml/min until the saline reaches the bubble trap on the outlet side. Rinse the coil with 500 to 1000 ml of saline.

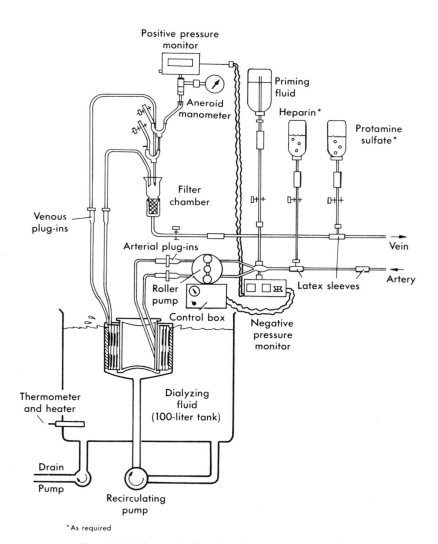

Fig. 10-4. Schematic of twin-coil system in operation.

2. Cross-clamp the outlet tubing below the bubble trap.
3. When the manometer pressure reaches 200, stop the blood pump.
4. Pressure will drop slowly, 5 to 10 mm/min. A precipitous drop indicates a leak or inadequate pump occlusion.

Connecting patient to dialyzer

1. Adjust the bubble trap pressure to 70 to 100 mm Hg.
2. Fill the proximal arterial line with saline from the infusion set.
3. Attach the arterial line to the patient's arterial cannula.
   a. If transfusion is to be used for priming, attach heparinized blood to side arm of arterial set and start pumping. When blood fills the venous line, clamp, stop the pump, and attach the venous line to the venous cannula. Remove the clamps from both lines and start the blood pump slowly.
   b. If the patient is to be bled for priming, attach the arterial line to the arterial cannula and start the blood pump very slowly. When blood reaches the venous line, attach to the venous cannula.
   c. For saline priming, attach both arterial and venous lines to appropriate cannulae and start the blood pump slowly.
4. As soon as blood enters the coil, watch the canister for evidence of blood leakage.
5. Attach the heparin infusion line to the atrerial line or give heparin intermittently as determined by clotting times.
6. During the first 15 minutes carefully monitor the patient's blood pressure. The blood pumping rate may be gradually increased to 200 to 300 ml/min. Adjust the level of the blood in the bubble trap and the pressure on the manometer by a screw-clamp below the bubble trap. With increased pressure more ultrafiltration results.

Coil dialyzers require a high rate of bath flow over the membrane and this is accomplished by a pump delivering approximately 30 L/min. In the simplest type of coil dialyzer the bath is made up in a 100-L tank and recirculated through the coil. The bath is changed at intervals of 2 to 3 hours to maintain the diffusion gradient from blood to bath. To replace the dialyzing fluid during the dialysis of a patient, the blood pump is kept running while the bath fluid is removed from the tank with the exhaust pump. The recirculating pump is stopped before all of the fluid is out of the tank and is not restarted until the new batch of dialyzing fluid is prepared and in the tank. Each new batch of dialyzing fluid must be verified. If it is not isotonic before being pumped through the coil, hemolysis will occur in the cellulose tubes of the coil.

In the recirculating single-pass (RSP) coil dialyzer the coil is immersed in a small (8-L) compartment with a recirculating pump. Fresh dialysate from a separate tank is continually added at a slower rate, usually 500 ml/min, while an equal quantity of used dialysate runs off through an overflow.

Kiil dialyzer (Fig. 10-5)

The modified Kiil dialyzer consists of three rigid plastic boards that fit together in sandwich fashion with two sheets of membrane (cellophane or Cuprophane) between each layer of the boards. Layer type dialyzers have less resistance than the coil dialyzers and the patient's arterial pressure is usually sufficient to provide adequate blood flow. Limited thickness of the blood film, uniform distribution, and avoidance of laminar flow are critical factors in dialyzer efficiency. The Kiil and other layer type dialyzers employ the single pass of dialysate at flow rates of 300 to 1500 ml/min.

Renal function 269

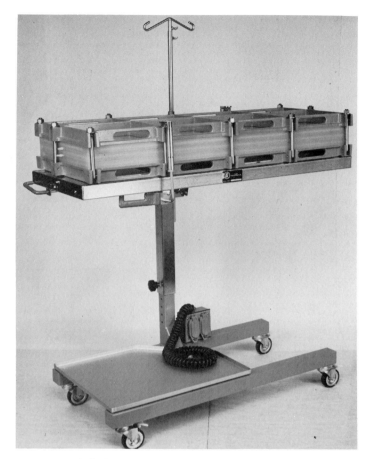

Fig. 10-5. Kiil dialyzer. (Courtesy Cobe Laboratories, Inc., Denver.)

Dialyzer assembly

Membrane preparation
1. Fill the special membrane pan with hot water and carefully lay the membranes on the water.
2. Soak the membranes for at least 15 minutes.

Stacking and clamping
1. Center the bottom dialyzer board on the bottom clamp with grooves upward.
2. Remove one wet membrane from the pan, stretch gently, and place carefully onto the board.
3. Place a blood port in the port housing at each end of the board.
4. Layer a second membrane over the first.
5. Position the middle board.
6. Repeat the procedure with a second set of membranes and blood ports.
7. Position the top board with grooves downward.
8. Place the top dialyzer clamp on the assembled boards and clamp the dialyzer, torquing the bolts to a designated pressure.

Pressure testing
1. Clamp off all tubes on one end of dialyzer.
2. Test each blood compartment separately.

3. With manometer attached, pump each blood compartment to 260 mm Hg air pressure. After allowing 2 minutes for the membranes to stretch, the rate of fall of pressure should not be greater than 6 mm/min.
4. Test the dialysate compartments in a similar fashion.

Sterilizing
1. With dialyzer in a vertical position, fill both blood compartments from below, to displace air, with 4% formalin solution.
2. Fill dialysate compartments is a similar fashion and clamp all tubes.
3. Minimum sterilization time is 2 hours; maximum, 10 days.

Preparation for dialysis
1. Attach the dialysate hoses to a source of warm water and flush the compartments for 30 minutes at a flow rate of 500 ml/min.
2. Attach blood tubing sets and, after step 1 is complete, allow 800 to 1000 ml of sterile heparinized saline to flow by gravity through the blood compartments.
3. Near the end of the saline rinse, check the fluid with a Clinitest tablet to determine whether all formalin has been removed.

Connecting patient to dialyzer
1. Check the dialysate supply and set the flow to *on*.
2. Connect the arterial line to the patient's arterial cannula.
3. When the venous line is filled with blood, connect to the venous cannula.
4. Check the dialysate outflow for evidence of blood (leak).
5. Set the heparin infusion pump to deliver to the arterial line.
6. Attach arterial pressure monitor tubing to arterial line drip chamber.
7. Adjust negative pressure at dialysate outflow to govern ultrafiltration.

CLINICAL APPLICATION

Acute patients are dialyzed according to the severity of their biochemical disturbances, with frequent monitoring of serum values. Chronic patients are dialyzed more routinely, usually three times a week for 5 to 10 hours at a time, depending on the equipment.

Peritoneal dialysis

Peritoneal dialysis accomplishes essentially the same things as does hemodialysis.[31] The lining of the peritoneal cavity, although a living membrane, acts in many ways like the inert semipermeable membrane of the artificial kidney. The clearance of some metabolic products of large molecular size such as creatinine and uric acid is somewhat greater than by hemodialysis.

The first human use of peritoneal dialysis for removal of substances ordinarily excreted by the kidney was reported by Ganter[29] in Germany in 1923. Subsequent clinical use was not encouraging, although animal work showed that transperitoneal to blood transfer of solutes, and vice versa, did occur and that uremic animals showed improvement when treated by peritoneal dialysis. As with hemodialysis, increasing knowledge of body fluid and electrolyte balance and of methods for accurate measurement has made the procedure more physiologic and safe. The advent of effective antibiotics reduced the risk of peritonitis. By 1946 a group of investigators[28] in Boston described a technique of continuous peritoneal lavage with a balanced electrolyte solution which was effective in treating uremia. Technique problems consisted of "streaming" of the fluid, resulting in contact with only a fraction of the surface area of the peritoneum, and walling off by omentum. Grollman and co-workers[32] in 1951 described a technique of intermittent peritoneal dialysis that was simple, efficient, and economical in the use of fluid. This was essentially the same technique used today. However, clinical acceptance was slow until 1959 when, with their co-workers, Doolan[27] and Maxwell[37] separately reported effective results from intermittent peritoneal dialysis.

The procedure has been made generally available by the commercial preparation of prepackaged sterile fluids of appropriate electrolyte concentration, pH, and osmolality. Such fluids contain sodium, calcium, magnesium, and chloride in concentrations approximating that of normal plasma. Lactate or acetate is included which, when metabolized, provides a suitable source of bicarbonate. Dextrose is furnished, usually in concentration of 1.5 or 7 g/dl, to provide the desired osmolality. The standard solution with 1.5 g dextrose/dl has a measured osmolality of approximately 350 mOsm/L, while the 7% dextrose solution exceeds 600 mOsm/L (measured by freezing point, the osmolality is lower than the theoretical value). Glucose molecules diffuse more slowly across the peritoneal membrane than do those of water and electrolytes. Therefore the hypertonic dextrose solution serves to remove fluid from the blood and prevent or relieve overhydration. Potassium may be added to dialysis fluid; if hyperkalemia is a problem, it is omitted. Heparin may be added to reduce fibrin deposition in the catheter, and antibiotics may be included at the discretion of the physician.

PATIENT PREPARATION

In the usual case, the patient's abdomen is prepared as for a paracentesis. A small (1 cm) skin incision is made in the midline below the umbilicus, and a trocar inserted until the tip is felt to pass into the peritoneal cavity. A plastic or silicone rubber catheter is placed through the trocar, which is then withdrawn. Bleeding is usually minimal. The catheter may be secured with a snug purse-string skin suture or simply secured with tape. The fluid administration set is connected to the catheter and to the bottles of fluid.

DIALYSIS FLUID

Commercially prepared sterile dialysis solutions in 1-L bottles are used. The fluid should be warmed to approximately body temperature before infusion. The customary volume of fluid for each exchange for an adult patient is 2 L. Smaller volumes are used for children or small infants (initial volumes of 300 to 700 ml can usually be employed with youngsters under 1 year of age).

PATIENT DIALYSIS

Utilizing a high IV standard, 2 L of fluid can be instilled within 7 to 10 minutes. The fluid is left in the peritoneal cavity for 45 to 60 minutes and then drained by gravity through the closed tubing system into a container on the floor.

Urea concentration in the dialysis fluid will reach 65% to 80% of that in serum in 45 to 60 minutes, while approximately 30% to 50% of the dextrose will have been absorbed. The usual period for drainage of the fluid is 10 minutes, or until the *brisk flow* tapers off. If the patient initially has no ascitic fluid present, it is common to get incomplete return of the first two or three exchanges of dialysis fluid. Barry and co-workers[26] pointed out that it is often necessary to establish a pool of fluid in the abdomen, following which drainage is usually quite free, with return of 2 L or more in 10 minutes. Continued slow or erratic drainage may be due to obstruction of the catheter by omentum or to plugging with fibrin clot and may require manipulation or repositioning of the catheter. Such difficulties are less frequent with Silastic catheters[30] than the more rigid, small plastic ones.

The exchanges are repeated continually for such period of time as the condition of the patient indicates. Five to six exchanges of 2 L each will remove the normal 24-hour accumulation of metabolic wastes of an average size adult who is not catabolic. If the patient is severely azotemic initially, many more exchanges obviously will be required. A careful running record of the volumes infused and removed, as well as medications and vital signs, is essential for monitoring the patient's course.

The use of 7% dextrose solutions for peritoneal dialysis is only rarely indicated, the prime indication being found in the severely waterlogged patient with incipient pulmonary edema. Such hyperosmotic solutions attract vast quantities of fluid rapidly and can result in hypovolemia and shock. When used, the 7% solution should not be left in the abdomen longer than half an hour, and rarely should more than two or three such exchanges be used. On the other hand, severely azotemic persons (those with a BUN of 150 mg/dl) may have a serum osmolality sufficiently high that there is an inadequate gradient for removal of water from the body if only 1.5% solutions are used. Use of a bottle each of 1.5% and 7% dextrose solutions (yielding an effective dextrose concentration of 4.25%) will almost always result in the desired rate of fluid removal.

CLINICAL APPLICATION

Peritoneal dialysis removes metabolic wastes more slowly than does hemodialysis—a 12- to 15-hour peritoneal dialysis will be approximately equal to a 4-hour hemodialysis, using the commonly employed coil equipment, or 6 to 8 hours of Kiil dialysis without a blood pump. This is often actually advantageous to the stable patient, since the water and solute shifts between the various bodily compartments are more gradual and less likely to be accompanied by adverse symptoms attributable to the dialysis procedure.

Peritoneal dialysis in diabetic individuals is difficult and generally unsatisfactory because of inability to metabolize the large load of glucose. The acidosis of azotemia contributes to insulin resistance. Dialysis solutions in which sorbitol is the osmotic agent are available and have been suggested for use with diabetic persons as well as other patients. However, this sugar is slowly metabolized and can, over the course of a prolonged dialysis, diffuse throughout extracellular fluid, resulting in a hyperosmolar syndrome.

Several types of apparatus have been designed to automate peritoneal dialysis. Employing a system of pumps, solenoid valves, and automatic timers, such equipment greatly reduces the amount of nursing time needed for conducting the exchanges. On-site preparation of dialysis fluid, using flash sterilization or Millipore filtration, is a feature of such apparatus as those designed by Tenchoff and associates[42] and by McDonald.[38]

TRANSPLANTATION

For the person with end-stage renal failure, the prospect of receiving a transplanted kidney, which might function normally, is a goal that dialysis therapy cannot approach. However, the immunologic defenses of the body are such that homograft tissues (grafts between different individuals of the same species) are rejected by these defense mechanisms. Rejection of a kidney homograft may occur within the first few days or weeks or months later. It is manifested by edema, cellular infiltration of the parenchyma, proliferation of vascular endothelium, and fibrinoid necrosis of arterial and arteriolar walls. The result is loss of the transplanted kidney. Only when the transplant takes place between identical twins is rejection avoided.

Homograft reaction may be suppressed by irradiation and by some drugs, but in humans it is never completely overcome. The amount of irradiation necessary for immunosuppression is dangerously close to the lethal dose. Numerous drugs have been investigated as immunosuppressive agents. Of these, azathioprine (Imuran) and prednisone have proved most useful. More recently the use of antilymphocyte globulin (ALG), prepared from serum of horses that have been injected with human lymphoid tissue, has made possible the use of smaller and less toxic amounts of the other suppressive agents.

Tissue-matching techniques have been devised and are used to determine the best "match" between recipient and possible donor. Donor and recipient sera and leukocytes are cross-matched and are matched against a panel of cells and sera in a manner somewhat akin to red blood cell typing and matching.

Upward of 3,000 renal homografts have been done throughout the world. Using current immunosuppressive chemotherapy and antilymphocyte globulin, and with careful tissue typing technique, persons who receive a transplanted kidney from a related living donor now have an approximately 90% chance of having that kidney survive and function at least 2 years. It is projected that the 5-year survival may approach the 90% figure also. Data for cadaveric transplantation (in which the donor kidney is taken from a person after death) are less complete but, with the best methods of matching and suppressive therapy, may approach a 2-year survival rate of 70%.

DIALYSIS FOR ACUTE POISONING OR DRUG INTOXICATION

Dialysis is an effective means of removing many poisons from the body. However, dialysis is never a substitute for appropriate and intensive supportive therapy. Once emergency care has been established (adequate airway, circulatory support, gastric lavage, etc.), the question of indications for dialysis should be raised. Substances that are completely or largely excreted by the kidney are theoretically dialyzable.

If possible, the intoxicating substance or substances should be identified and quantitative levels in body fluids determined.

The decision to dialyze must be based on the clinical picture and not solely on serum levels. If the substance cannot be verified or if quantitation in the laboratory may require considerable time but there is evidence from the history that the drug ingested is one that meets the criteria for dialysis, the procedure should be instituted. When one or more of the substances ingested is dialyzable, the duration of coma can be reduced and morbidity and mortality decreased.

A patient may be considered a candidate for dialysis when he is poisoned with an agent (agents) that meet the criteria for dialysis and one or more of the following clinical indications are present:
 1. Hypotension not responsive to adjusting circulatory volume
 2. Apnea or marked respiratory depression
 3. Significant hyperthermia
 4. Severe shift in acid-base balance
 5. Significant electrolyte shift
 6. Hyperosmolality

According to Maher and Schreiner,[36] standards for deciding on the applicability of dialysis for poisoning are as follows:

1. The substance should diffuse through a dialyzing membrane at a reasonable rate.

2. A significant quantity of the poison should be in serum water or in body compartments which rapidly equilibrate with serum water.

3. The severity of clinical intoxication should be directly related to the blood concentration and to duration of exposure to the poison.

4. The amount of the poison that will be removed by dialysis must add significantly to the quantity that will be removed by normal mechanisms (metabolism, conjugation, eliminating by bowel and kidney) under the physiologic circumstances at the time.

It should be appreciated that renal excretion of substances which meet these criteria can often be greatly accelerated by forced diuresis (high fluid intake, mannitol, and/or pharmacologic diuretic). When dialysis is deemed appropriate, the use of a high-efficiency apparatus (in terms of quantity of solute removed per unit time) is indicated. A coil dialyzer is the preferred hemodialysis unit. Indeed two coils may be connected in series, or a coil and a Kiil dialyzer operated in series, and so forth. Peritoneal dialysis is appreciably slower in removing almost all poisons than is hemodialysis. However, a peritoneal dialysis can be started and under way in less than 30 minutes after the decision to dialyze is made, whereas in most institutions 2 or 3 or more hours usually elapse before a hemodialysis can be started. Maher reported use of all three modalities—forced diuresis, peritoneal dialysis, and hemodialysis—simultaneously in one case of poisoning!

Transactions of the American Society of Artificial Internal Organs each year includes an updating of available information on dialyzable poisons. This information should be on hand at all major poison control centers.

In general, ingestion of such agents as methyl alcohol, ethylene glycol and propylene glycol, sulfonamides, sodium chlorate, acetone, carbon tetrachloride, heavy metals in soluble compounds, and mushrooms proved toxic *(Amanita phalloides)* should call for immediate dialysis. Early removal may prevent the severe complications of these agents, such as blindness, liver necrosis, or renal failure.

In most other intoxications, the indications are relative and depend on the clinical situation. Agents falling into this category include barbiturates, salicylates, chloral hydrate, amphetamines, antibiotics, boric acid, thiocyanate, meprobamate, isoniazid, and others. Glutethimide (Doriden) is a special problem because of its partitioning in body fat; dialysis is generally reserved for the most seriously ill patient who appears to be worsening in spite of all other supportive measures.

Dialysis is not of benefit for excess intake of the phenothiazines, chlordiazepoxide, diazepam, Imipramine, other thymoleptics, dextropropoxyphene (Darvon), or the anticholinergic medications.

Routine urine examination

A routine examination of the urine consists of the tests listed in Fig. 10-6. These tests are discussed on the following pages.

URINE COLLECTION

For routine qualitative clinical studies a random voided urine specimen is suitable. The first urine voided in the morning is usually preferred because of the concentration of the constituents and, therefore, the tendency to reveal abnormalities. However, glycosuria and albuminuria are less likely to be revealed by this sample.

```
Name: Mr. John Doe                                    6-4-70
Color . . . . . . . . . . . . . . . . . . . . . . . . . . straw
Transparency . . . . . . . . . . . . . . . . . . . . cloudy
Reaction . . . . . . . . . . . . . . . . . . . . . . . . acid
Specific gravity . . . . . . . . . . . . . . . . . . . 1.020
Albumin . . . . . . . . . . . . . . . . . . . . . . . . . 3+
Sugar . . . . . . . . . . . . . . . . . . . . . . . . . . . negative
Acetone . . . . . . . . . . . . . . . . . . . . . . . . . negative
Microscopic examination. . . . . . . . . Centrifuged specimen shows 4 to 6
                                          granular casts per high dry field
```

Fig. 10-6. Sample urine analysis report.

The specimen should be delivered to the laboratory and examined as soon as possible. Bacterial growth causes rapid decomposition and alteration of urinary constituents. When a delay is unavoidable, the specimen should be refrigerated or a few drops of toluene added as a preservative. For preservation of the structure of cells and casts in urine, 40% formalin may be added. The microscopic examination to be valid should be performed when the urine is fresh. Hydrochloric acid is a suitable preservative when inorganic constituents other than chlorides are being investigated.

When quantitative measurements such as electrolytes, proteins, electrophoresis, or hormones are requested, a 24-hour collection is necessary. The patient should be instructed to empty his bladder at a specific time (for example, 7 AM) and discard this urine. He should save all urine voided up to and including the 7 AM specimen on the following day. Each specimen as voided should be placed in a container having the preservative required for the test procedure being conducted.

Catheterization is rarely used for collection of urine specimens since it is difficult to prevent contamination of the bladder and the upper urinary tract during passage of a catheter. This may result in a genitourinary tract infection and pyelonephritis.

The quantities of preservatives that may be used for routine qualitative examinations per 100 ml of urine are 6 to 8 drops of formalin, 0.2 g of boric acid, and 0.1 g of thymol.

For quantitative chemical analysis sufficient toluene to form a layer on top of the specimen is satisfactory for most examinations. Preservation by acidification with sulfuric or hydrochloric acids (1 ml of concentrated acid/dl of urine) may be used for a majority of the inorganic constituents and nitrogen. When special tests are requested, such as hormone assays, refer to the recommended procedures described with each method.

The bottles for the collections should be clean and dry before the preservatives are added.

COLOR

The color of the urine often indicates urinary abnormalities. Some of these colors and their usual causes are given in Table 10-3.

Procedure
1. **Mix urine by inverting or swirling the container.**
2. **Note and record the color.**
3. **If the color is dark amber, the urine should be shaken to produce foam. If the foam is yellow, a test for bile should be run.**

Table 10-3
Usual causes of urinary colors

Color	Usual cause
Straw to amber	Urochrome, a pigment found in normal urine
Colorless	Reduced concentration
Silvery sheen or milky appearance	Pus, bacteria, or epithelial cells
Smoky brown	Blood
Black	Melanin
Port wine	Porphyrins
Yellow foam	Bile or medications
Orange, green, blue, or red	Medications

Table 10-4
Rapid urine tests using tape, dipsticks, and tablets

Test	Method
Multitest	
Protein, glucose	Uristix
Protein, glucose, pH	Combistix
Protein, glucose, pH, occult blood	Hema-Combistix
Protein, glucose, pH, occult blood, ketones	Labstix
Acetoacetic acid, acetone	Acetest, Ketostix
Single tests	
Glucose	Clinistix, Tes-Tape
Sugar	Clinitest, Galatest
Protein	Albustix, Albutest, Bumintest
Bilirubin	Ictotest
Occult blood	Hemastix, Occultest
Salicylate	Phenistix
Phenylpyruvic acid	Phenistix

RAPID TESTS OF URINE

Due to time-consuming and demanding techniques required with many of the chemical tests for urine analyses, the modern tablet, dipstick, and tape test for urinalysis are used extensively (Table 10-4). These tests are sensitive and must be used with recognition of their limitations. Each tape, dipstick, or tablet is designed for a special purpose, and if each is used according to the manufacturer's specifications and instructions, accurate results may be obtained.

The strips are made with selected cellulose of standard porosity. The tips are impregnated with chemicals that will react with various substances in the urine, producing colors that can be measured against standard color charts. Some color reactions are timed. This requirement is given in the instructions and *must* be followed exactly!

Precautions
1. When not in use, the tablets or strips must be kept in their bottles, tightly stoppered to keep out the moisture.
2. Bottles that are so marked should be placed in refrigeration.
3. If the strips (dipsticks) are kept too long in the urine or in the urine stream, the chemicals impregnated in the cellulose layer will dissolve and inaccurate results will be obtained.
4. Directions must be carefully followed when there is more than one reaction on a strip. For example, on Combistix the sections indicating pH, protein, and glucose are separated by water-impermeable barriers of plastic.

TRANSPARENCY

The transparency of urine may be classified as clear or cloudy. The degree of transparency depends on the amount of suspended materials: the more suspended materials, the greater the cloudiness. In normal urine, the main causes of cloudiness are crystals and epithelial cells. In pathologic urine, the principal causes of cloudiness are pus, blood, and bacteria.

If the urine is clear, it may inform the physician that the patient is not concentrating the urine, that is, removing metabolic waste materials. On the whole, however, this test has little significance, for both normal and pathologic urine may be either clear or cloudy.

Procedure
1. Mix urine.
2. Note degree of transparency.
3. Report as clear or cloudy. Also report any abnormal contents such as clumps or mucus or pieces of tissue.

REACTION

Freshly voided urine may be either acid or alkaline, the normal pH range being 4.0 to 8.0. If allowed to stand, acid urine becomes alkaline because urea decomposes and liberates ammonia. Increased acidity of urine is found in fevers and diabetes. Increased alkalinity is seen in alkali therapy and in retention of urine.

Labstix and Combistix, in the pH portion of the strip, are impregnated with a combination of the acid-base indicators, bromthymol blue and methyl red. Distinctive colors are produced at different levels within the urinary pH range of 5 to 9.

DISORDERS AND THE URINE pH[43]

Tests for pH of urine should be done a short time after the specimen has been voided. As the urine stands, it will become alkaline, due to the decomposition of urea into ammonia. However, when fresh urine is persistently acid or alkaline, this may be due to one of the following disorders:

A. Consistently acid urine
  1. Tuberculosis of the kidney
  2. Pyrexia
  3. Methyl alcohol poisoning due to the formation of formic acid
  4. Metabolic disorders
     a. Phenylketonuria
     b. Alkaptonuria

5. Metabolic acidosis due to severe diarrhea, starvation, diabetic ketosis, uremia, and certain types of poisoning
6. Respiratory acidosis due to conditions in which there is CO_2 retention

B. Consistently alkaline urine
1. Certain genitourinary infections
2. Metabolic alkalosis due to overdosage of sodium bicarbonate, potassium citrate, or other alkalies
3. Respiratory alkalosis due to hyperventilation or cardiac failure
4. Bronchial carcinoma (in some cases)
5. Cushing's syndrome or hyperaldosteronism
6. The use of certain diuretics, such as Diamox and chlorothiazide, which are carbonic anhydrase inhibitors

SPECIFIC GRAVITY

Specific gravity is a term used to compare the weight of a liquid with the weight of water. For example, sulfuric acid weighs 1.840 g/ml, whereas water weighs 1.000 g/ml. Since sulfuric acid weighs 1.840 times as much as water, the specific gravity is 1.840.

Because of its dissolved substances, urine weighs more than water, the normal specific gravity being 1.008 to 1.030. Increased values are found in acute nephritis, fevers, and diabetes mellitus (because of the dissolved sugar). Decreased values are found in chronic nephritis and diabetes insipidus (because of the copious volume).

Measuring the specific gravity of the urine is the most convenient way of measuring the concentrating and diluting function of the kidney, meaning the ability of the kidney to produce urine of a specific gravity greater or less than 1.010.

Procedure
1. Mix the urine by inverting or swirling the container.
2. Pour into a urinometer cylinder.
3. Place the urinometer in the urine and spin it as you would a top.
4. When it comes to rest, record the reading. When the reading is taken, the the urinometer must *not* be touching the sides of the cylinder.
5. If there is not sufficient urine to fill the cylinder, some laboratories report "qns" (quantity nonsufficient). The temperature-compensated hand refractometer will measure the specific gravity of a minute volume (1 to 2 drops) of urine.

In addition to a refractometer, an osmometer may be employed to measure the concentration of dissolved solutes in urine. Both methods are especially useful when very small amounts are available. The refractive index of a solution depends on the concentration of the dissolved substances, a relationship that approaches linearity for a pure solution. The osmolality measures the number of molecules and ions in a solution and is usually performed by a freezing point depression method. The normal range for an adult with adequate fluid intake is 500 to 850 mOsm/kg of water. Using Long's coefficient, determine the total solids per liter by multiplying the last two figures of the specific gravity of urine by 2.66.

PROTEIN IN URINE

A small amount of protein is filtered by normal glomeruli and is found in normal urine. However, when excretion exceeds 150 mg/24 hr, it is abnormal. In normal individuals, proteinuria may result from injections of adrenaline, exercise, or renal ischemia caused by dehydration, surgery, hemorrhage, or salt depletion. Continuous proteinuria is indicative of kidney disease.

Urinary protein in most renal diseases is largely albumin. In proliferative or membranous glomerulonephritis there is an increase in proteins other than albumin. In myeloma, macroglobulinemia, and primary amyloidosis, the excretion of globulin may be greater than that of albumin since there is an increase of abnormal non-albumin proteins in the serum. Proteins of low molecular weight are filtered by the glomeruli but are usually reabsorbed from the proximal tubules unless they are damaged. A marked increase in glomerular permeability (generalized glomerular disease) causes excessive proteinuria.

Some renal diseases that are characterized by having increased amounts of protein in the urine are polycystic disease, pyelonephritis, hypercalcemic or hypokalemic nephropathy, stones, renal insufficiency due to dehydration, and congenital malformations.

Orthostatic proteinuria is often found in young adults as a result of prolonged standing in the erect position or strenuous exercise. For these individuals, urine obtained in the morning before arising is free of protein. To obtain a proper urine specimen for examination, the individual should empty his bladder at midnight without arising in order to remove any urine formed during the previous evening while he was in an upright position.

Several of the rapid screening tests for urinary protein (Labstix, Combistix, Albustix, Albutest) utilize the principle of protein error of a pH indicator. Bromphenol blue is yellow at pH 3 but in the presence of protein the color is a greenish blue. For the paper strips, the proper pH in the protein portion is maintained with citrate buffer. The protein reagent area is sensitive to albumin and globulin but less sensitive to hemoglobin and Bence Jones protein. The insensitivity to myeloma proteins is a disadvantage. A highly buffered alkaline urine will give a positive reaction; therefore the pH must be checked for each specimen.

For the tablet form, salicylic acid–buffered cellulose is used with the bromphenol blue indicator. When a drop of urine is placed on the tablet and followed by 2 drops of water, a green color on the surface of the tablet is the result of adsorbed protein.

The majority of the other tests for urinary protein depend on protein precipitation with agents such as heat and acetic acid, nitric acid, sulfosalicylic acid, and trichloroacetic acid. In acidified solutions, albumin and globulin are coagulated by heat in the presence of inorganic salt. Proteoses are also precipitated. Phosphate salts, if present, precipitate out and form a white cloud when urine is heated. The addition of an acid, however, dissolves the salts.

Sulfosalicylic and trichloroacetic acids precipitate protein in urine with a turbidity that may be measured by a spectrophotometer, colorimeter, nephelometer, or visually and compared with known standards. Bumintest tablets may be used to prepare a 5% sulfosalicylic acid solution and the results may be graded from trace to 4+. Substances that are also precipitated by sulfosalicylic acid are urinary proteose, polypeptides, tolbutamide metabolite, and x-ray contrast media.

Qualitative tests for protein

If it is necessary to prepare any of the reagents used in the following tests, take the number that follows the reagent and refer to Appendix D.

Purdy test

Reagents
Sodium chloride, saturated (D-185)
Acetic acid, 50% (D-186)

Procedure

1. Pour 5 ml of urine into a large Pyrex test tube.
2. Add 1 ml of the saturated solution of sodium chloride. (This prevents the precipitation of mucin, if it should be present.)
3. Add 5 to 10 drops of the 50% acetic acid. Mix.
4. Using a test tube holder, heat the *upper* portion over a flame. If a white precipitate forms, albumin is present.
5. Report the test as negative, 1+, 2+, 3+, or 4+, depending on the amount of precipitate.

Sulfosalicylic acid test

Reagent

Sulfosalicylic acid, 20% (D-187)

Procedure

1. Pour 5 ml of urine into a test tube.
2. Add 1 drop, per milliliter of urine, of the 20% sulfosalicylic acid. Mix. If albumin is present, a white cloud will form.
3. Report the test as negative, 1+, 2+, 3+, or 4+, depending on the amount of precipitate.

Roberts' test

Reagent

Roberts' reagent (D-188)

Procedure

1. Follow either method A or method B:
   Method A: Pipet a few milliliters of Roberts' reagent into a test tube. Tilt the tube to about a 30-degree angle. Using a medicine dropper, overlay with clear urine by allowing the urine to flow *slowly* down the sides of the test tube.
   Method B: Pour about 4 ml of urine into a test tube. Using a 1-ml serologic pipet, pipet out 1 ml of the Roberts' reagent. Place the tip of the pipet in the bottom of the test tube containing the urine. Allow the pipet to drain slowly.
2. If albumin is present, within a few minutes a white ring will form at the junction between the two liquids.
3. Report the test as negative, 1+, 2+, 3+, or 4+, depending on the amount of precipitate.

Heller test

Reagent

Concentrated nitric acid

Procedure

1. Place 3 ml of concentrated nitric acid in a small test tube.
2. Tilt the tube to about a 30-degree angle. Using a medicine dropper, overlay with urine by allowing the urine to flow *slowly* down the sides of the tube.
3. If albumin is present, within a few minutes a white ring will form at the junction between the two liquids.
4. Report the test as negative, 1+, 2+, 3+, or 4+, depending on the amount of precipitate.

Quantitative tests for protein in urine

A quantitative protein determination may be helpful in following the course of a kidney disease.

For the quantitative determination of protein in urine, the following methods may be used: (1) Esbach's method, (2) Shevky-Stafford rapid approximate sedimentation method, and (3) sulfosalicylic acid turbidity method. The first two methods call for the precipitation of the protein by an acid. The amount of precipitation is then measured in a special tube. The last method also calls for the precipitation of the protein by an acid, but the amount of precipitation is then compared with a set of standards.

PROCEDURES FOR QUANTITATIVE PROTEIN
Esbach's method

Reagents

Hydrochloric acid, 5%	(D-189)
Esbach's reagent	(D-190)

Procedure

1. Acidify the urine with the 5% hydrochloric acid to pH 5.
2. Filter.
3. Fill the Esbach tube to "U" mark with the filtered urine.
4. Add Esbach's reagent to the "R" mark.
5. Close the tube with a rubber stopper and invert several times to mix.
6. Place in an upright position for 24 hours.
7. Read off the height of the coagulation. This will be the number of grams per 1000 ml of urine. To obtain percentage, divide reading by 10.

Shevky-Stafford[44] rapid approximate sedimentation method

The following procedure deviates in a few details from the original description. The first variation is in the dilution of the urine. Nephritic urines are usually diluted tenfold. However, urines yielding very small amounts of protein should be diluted much less or not at all for accurate results. Occasionally a urine will be found with more than 2.8% protein. If a 1:10 dilution is not adequate, use a 1:20 dilution.

Reagent

Tsuchiya's reagent	(D-191)

Procedure

1. Pipet 4.0 ml of diluted urine into a special graduated centrifuge tube.* Since the 4-ml mark on the tube may serve as a measurement, the serologic pipet is used for convenience, to avoid spilling the urine and to allow better control for the proper measurement to the 4-ml mark.
2. Add Tsuchiya's reagent until the tube is filled to the 6.5-ml mark.
3. Stopper and invert slowly three times to mix.
4. Let stand for 10 minutes.
5. Centrifuge for 10 minutes (use timer) at 1800 rpm.
6. Read the volume of precipitate on the scale.

Calculation

Milliliters of precipitate read $\times$ 7.2 $\times$ Urine dilution = Grams of protein/Liter of urine

---

*Special equipment needed: The graduated centrifuge tube is a special calibrated tube that may be obtained from Phipps & Bird, Inc.

Sulfosalicylic acid turbidity method

Reagents

Sulfosalicylic acid, 3%	(D-192)
Hydrochloric acid, 1.25%	
Albumin standard, stock solution, 10 g/dl	(D-127)
Protein standards, working, 25, 50, and 100 mg/dl	(D-193)

Procedure

1. Filter or centrifuge an aliquot of urine specimen, single specimen or 24-hour collection.
2. Into test tubes pipet:
   Unknown: 2 ml of urine
   Unknown blank: 2 ml of urine
   Standards: 2 ml of 25 mg albumin/dl
   2 ml of 50 mg albumin/dl
   2 ml of 100 mg albumin/dl
   2 ml of 0.85% NaCl (blank)
3. To tubes containing unknown and standards add 8 ml of sulfosalicylic acid solution. Mix by inversion and allow to stand for 5 minutes.
4. To tubes containing blanks for unknowns add 8 ml diluted HCl. Mix by inversion and allow to stand for 5 minutes.
5. Measure absorbance at 500 nm against reagent blank for standards and unknown blanks for unknowns.
6. Read values from the standard curve plotted on semilog paper. Report as mg/dl urine or g/24-hr volume of urine.

Normal values

2 to 8 mg/dl or up to 150 mg/24-hr specimen

Comment. In addition to albumin and globulin, sulfosalicylic acid will precipitate urinary proteoses, polypeptides, and Bence Jones protein.

Proteoses

Proteoses are intermediate products in the digestion of proteins. Under normal conditions, proteoses are not found in the urine. They may be present, however, in cancer, osteomalacia, pneumonia, diphtheria, and atrophy of the kidneys.

Reagents

Acetic acid, 5%	(D-194)
Trichloroacetic acid, 10%	(D-95)

Procedure

1. Acidify the urine with 5% acetic acid, using nitrazine paper as an indicator. Mix.
2. Pour 10 ml of the acidified urine into a small beaker and bring to a boil.
3. Filter the urine while it is still hot in order to remove any albumin or globulin that may be present.
4. Allow the filtrate to cool.
5. Add several drops of 10% TCA to the filtrate. If proteoses are present, a white cloud will form.

Bence Jones protein

This protein, discovered by Bence Jones, has the following peculiar characteristics: it precipitates out at a temperature of about 50° C, partially or completely disappears at 100° C, and then reappears upon cooling to room temperature. Its presence in the urine is considered a significant diagnostic aid in multiple myeloma, a malignant tumor of the bone marrow.

Reagent

    Acetic acid, 5% (D-194)

Procedure
1. Pour about 10 ml of the urine into a test tube.
2. Place the thermometer and the test tube containing the urine in a beaker of water.
3. Heat gently and watch both the thermometer and the urine.
4. Bence Jones protein is present if *all* the three following conditions are met: (a) At about 50° to 60° C, a white cloud will appear, which turns into a a white precipitate and usually clings to the sides of the tube. (b) Upon the addition of a few drops of 5% acetic acid and further heating to the boiling point, the precipitate will partially or completely disappear. (c) If the urine is filtered while hot, using a heated funnel and test tube, the white cloud will reappear as the solution cools to room temperature. (NOTE: If a heated funnel and tube are not used, a false positive may be read.)

GLUCOSE OR SUGAR (REDUCING SUBSTANCES)

Glucose and other carbohydrates are normally present in urine in quantities too small to be measured by the usual test procedures. The presence of measurable quantities of urinary glucose (glycosuria) may occur as a temporary condition following general anesthesia and administration of certain drugs, in pregnancy, after shock, or due to emotional disturbances. However, persistent glycosuria may be diagnostic of diabetes mellitus. Other sugars that may appear in the urine under certain conditions are lactose, fructose, pentose, and galactose.

The following discussion covers some methods that may be used to test for glucose.

In Benedict's method and the Clinitest method, glucose reduces copper. The reduced copper then forms a colored compound, cuprous oxide. As the particle size of cuprous oxide increases, the color of the precipitate varies from green to yellow to orange.

The Clinitest tablet includes all of the reagents and the heating mechanism for reduction of the copper by sugar or other reducing substances. Sodium carbonate and a small amount of citric acid act as an effervescent couple to facilitate rapid solution of the tablet. The sodium carbonate yields CO_2, which forms a protective shield over the heated solution. This prevents oxidation of the reducing sugar in the hot alkaline solution by atmospheric oxygen. If oxidation does occur (the solution being shaken and the shield broken) during the reaction, the readings will be low or negative even though the specimen may contain 0.25% glucose. Sodium hydroxide produces the alkaline medium for the reaction, and the necessary heat is obtained from the heat of solution of the sodium hydroxide and from reaction between the sodium hydroxide and citric acid. The heating process takes place as the tablet dissolves, keeping a suitable temperature, not underheating, overheating, or caramelizing the specimen.

Clinistix, Combistix, and Labstix are strips impregnated with the enzyme glucose oxidase, peroxidase, and *o*-tolidine that produce a color reaction when moistened with a solution containing glucose. In the presence of oxygen and glucose oxidase, glucose forms gluconic acid and hydrogen peroxide. The hydrogen peroxide in the presence of peroxidase oxidizes the colorless dye, *o*-tolidine, to its blue oxidized form.

The test is specific for glucose and sensitive. Therapeutic doses of ascorbic acid or certain antibiotics may inhibit the color development of enzymatic tests.

In the Galatest method, glucose reduces bismuth oxide to a gray or black precipitate of metallic bismuth.

For the Tes-Tape method; the color change depends on an enzyme system reaction that is specific for glucose. The color varies from light green to deep blue.

Benedict method
Qualitative test

Reagent

Benedict's qualitative reagent (D-195)

Procedure

1. Place 8 drops of urine in a test tube. (A few milliliters of urine may be poured into a test tube and the urine discarded by inverting the tube. Then quickly place the tube in an upright position. The amount of urine prevented from draining is about 8 drops.)
2. Add 5 ml of Benedict's qualitative reagent.
3. Place tube in a boiling water bath for 5 minutes. If glucose is present, it will reduce the copper in the Benedict's solution. The color will vary from green to yellow to orange, depending on the amount of glucose present.
4. Report the test as negative, 1+, 2+, 3+, or 4+.

Quantitative test

A quantitative urinary sugar test may aid the physician in determining the severity of a diabetic case and also in following its course during treatment.

For the quantitative determination of sugar in urine, the method of Benedict is commonly used. In this test, the following takes place. Urine, containing glucose, is slowly added to 5 ml of a copper solution of known strength. The amount of urine required to cause complete reduction of the copper is recorded. This figure is then used in a formula to calculate the percent of glucose present.

Reagents

Sodium carbonate
Benedict's quantitative reagent (D-196)

Procedure (Benedict)

1. Use a 250-ml beaker and half fill with water. Heat to boiling.
2. Pipet *exactly* 5.0 ml of Benedict's *quantitative* reagent into a large Pyrex test tube. Add 1 to 2 g of sodium carbonate. Mix.
3. Place the tube in the boiling water and allow the solution to come to a boil.
4. Use a 1-ml serologic pipet having 100 graduations or a buret and pipet out exactly 1.0 ml of urine. Add it drop by drop to the boiling Benedict's solution until the blue color completely disappears. Take the pipet or buret reading and make the calculation as follows:

Calculations

The Benedict's solution is made so that 1.0 ml of a 1% glucose solution will exactly neutralize the 5.0 ml of Benedict's solution. To check the strength of the Benedict's solution, proceed as follows. Using the analytic balance, weigh out 1.000 g of glucose. Transfer to a 100-ml volumetric flask and add distilled water to the 100-ml mark. Mix. Use this 1% glucose solution in place of the urine and continue as in the procedure. It should require 1.0 ml. If it does not, take the number of milliliters required and substitute that figure in the formula for the 1.0 shown here (milliliters required for neutralization). To find the percentage of glucose in the urine, use this formula:

$$\frac{1.0 \text{ (ml required for neutralization)}}{\text{ml used in titration}} \times 1\% = \% \text{ glucose in g/dl}$$

Example: If 0.5 ml of urine is used in the titration:

$$\frac{\text{ml required for neutralization}}{\text{ml used in titration}} \times 1\% = \% \text{ glucose}$$

$$\frac{1.0}{0.5} \times 1\% = 2\% \text{ glucose}$$

Clinitest method

Reagent

Clinitest tablets*

Procedure

1. Place 5 drops of urine in a test tube.
2. Add 10 drops of distilled water.
3. Add 1 Clinitest tablet. If there is no glucose present, the solution will be blue. If glucose is present, however, the color will vary from green to yellow to orange, depending on the amount of glucose present.
4. Report the test as negative, 1+, 2+, 3+, or 4+.

Galatest method

Reagent

Galatest powder†

Procedure

1. Place a small mound (about half the size of a dime) of the Galatest powder on a piece of white paper. NOTE: Keep the vial tightly closed when not in use.
2. Add a small drop of urine. If glucose is present, the powder will change color. The color will vary from gray to black. The blacker the color, the greater the concentration of glucose.
3. Report the test as negative, 1+, 2+, 3+, or 4+.

Tes-Tape method

Reagent

Tes-Tape‡

Procedure

1. Tear off a piece of Tes-Tape and dip one end into the urine.
2. When moist, remove from urine and continue holding tape for 1 minute.
3. Compare with color chart that comes with Tes-Tape.
4. Report the test as negative, 1+, 2+, 3+, or 4+.

---

*May be purchased from your local supply house or from Ames Co.
†May be purchased from your local supply house or from Denver Chemical Manufacturing Co., Inc.
‡May be purchased from your local supply house or from Eli Lilly & Co.

CARBOHYDRATES

This section considers tests for lactose, levulose, galactose, and pentose, a fermentation test for glucose, and a phenylhydrazine test.

Lactose

Shortly before and shortly after the birth of a baby, a woman's urine may give false positive tests for glucose. This is often caused by the presence of lactose, which reacts similarly to glucose. The two sugars may be distinguished by a test introduced by Max Rubner.

Rubner test

Reagents

Lead acetate
Ammonium hydroxide, concentrated

Procedure

1. Mix the urine and measure 10 ml into a large test tube.
2. Add 3 g of lead acetate, shake, and filter.
3. Bring the filtrate to a boil, add 2 ml of concentrated ammonium hydroxide, and boil again. If lactose is present, the solution turns brick red and a red precipitate forms. If glucose is present, the solution also turns red but the precipitate is yellow.

Levulose

Levulose is often referred to as fructose. Levulose reacts similarly to glucose but may be distinguished from it by the following test of Seliwanoff.

Seliwanoff test

Reagents

Ethyl alcohol
Seliwanoff's reagent
Place 0.05 g of resorcinol in a 125-ml Erlenmeyer flask. Add 60 ml of distilled water and 30 ml of concentrated hydrochloric acid. Mix.

Procedure

1. Mix the urine and place 5 drops in a small test tube.
2. Add 5 ml of Seliwanoff's reagent and boil for not more than 30 seconds.
3. Centrifuge, pour off the fluid, and add 3 ml of ethyl alcohol. Mix. If levulose is present, a bright red color will be produced.

Galactose

Galactose may be found in the urine of nursing infants having disturbances of the digestive tract. It may be distinguished from other sugars (with the exception of lactose) by the following mucic test.

Mucic test

Reagent

Nitric acid, concentrated

Procedure

1. Mix the urine and pour 50 ml into a small beaker.
2. Add 10 ml of concentrated nitric acid and boil until the volume is reduced to about 10 ml.

3. Add 10 ml of water and let stand overnight. If galactose (or lactose) is present, a fine white precipitate will form.

Pentose

Pentose may be found in the urine when the diet is rich in fruit juices, grapes, prunes, and plums. It has no pathologic significance.

Tauber test

Reagent

    Tauber's reagent
        Dissolve 1 g of benzidine in 25 ml of glacial acetic acid. Mix.

Procedure
1. Mix the urine and pipet 0.5 ml into a test tube.
2. Add 2.5 ml of Tauber's reagent and boil vigorously for 2 minutes.
3. Cool by immersing tube in cold water.
4. Add 5.0 ml of distilled water and mix. If pentose is present, a pink to red color is produced. Disregard any yellowish brown color.

Glucose

Fermentation test

The fermentation test may be used to distinguish glucose from lactose and pentose. When glucose and yeast are mixed, fermentation occurs and carbon dioxide is given off. This, however, does not occur with lactose and pentose.

Reagents

    Fresh yeast
    Glucose, 5%

Procedure
1. Place a small test tube in an inverted position in each of three large test tubes (see Fig. 10-7, which shows either positive or negative result for tube 3).
2. Label these tubes as follows: *Positive control, Negative control,* and *Test.*
3. *Test:* Boil the urine to be tested. Cool. Pour 25 ml of the urine into a small beaker and add one third of a cake of fresh yeast. Make an emulsion by mixing thoroughly. Pour into the tube labeled *Test.* Stopper and slowly invert so that the small tube becomes *completely* filled.
4. *Positive control:* Pour 25 ml of 5% glucose into a small beaker and add one third of a cake of fresh yeast. Make an emulsion and pour into the tube labeled *Positive control.* Stopper and slowly invert to that the small tubes becomes *completely* filled.
5. *Negative control:* Make an emulsion with 25 ml of water and the remaining third of the yeast cake. Pour this into the tube labeled *Negative control.* Stopper and slowly invert so that the small tube becomes *completely* filled.
6. Plug the three tubes with cotton and place in a 37° C incubator for 2 hours.
7. Remove the tubes from the incubator. The tube labeled *Positive control* should contain gas in the small tube (Fig. 10-7), whereas the tube labeled *Negative control* should not contain gas in the small tube. If this is not so, repeat the test with fresh yeast. If this is so, look at the tube marked *Test.* If this tube contains gas in the small tube, the test is positive for glucose. However, if this tube does not contain gas, the test is negative for glucose.

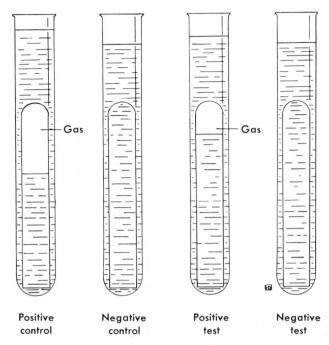

Fig. 10-7. Reading the results in a fermentation test.

Phenylhydrazine test
Kowarsky test as modified by Blumel

The phenylhydrazine test helps to distinguish glucose from other sugars occurring in urine. When sugars are treated with osazone, they form characteristic crystalline structures.

Reagents
Phenylhydrazine
Acetic acid, glacial
Sodium chloride, saturated

Procedure
1. Place 5 drops of pure phenylhydrazine in a large test tube.
2. Add 10 drops of glacial acetic acid and 1 ml of a saturated solution of sodium chloride. Mix.
3. Add 4 ml of urine and 4 ml of distilled water. Mix.
4. Add 4 glass beads to prevent bumping and boil until the volume is lowered to about 2 or 3 ml. Allow to cool for 30 minutes.
5. Using a medicine dropper, place a sample of the sediment on a glass slide and examine with the low-power (16-mm) objective of the microscope. If glucose is present, yellow needlelike crystals (usually arranged in clusters) may be seen. These crystals are phenylglucosazone.

ACETONE

When glucose is not oxidized, as in diabetes, it is believed that the body endeavors to compensate by oxidizing fats. This is referred to as abnormal fat catabolism.

During abnormal fat catabolism, diacetic acid is produced. When diacetic acid decomposes, it liberates acetone and β-oxybutyric acid:

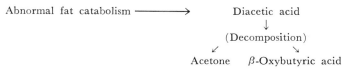

Acetone in the urine is significant, for it often indicates acidosis—a decrease in the alkali reserve. This could mean approaching death.

Acetone may be found in the following conditions: prior to diabetic coma, after anesthesia, in toxemias of pregnancy, in fevers, and in gastrointestinal disorders.

In discussions of acetone, textbooks use various terminology, and it is well to remember the following terms: (1) the presence of acetone in the urine is frequently referred to as ketosis; (2) diacetic acid is often called acetoacetic acid; (3) the three substances—diacetic acid, acetone, and β-oxybutyric acid—are often referred to as acetone or ketone bodies.

In the presence of acetone, sodium nitroprusside forms a complex compound that is purple to red in color. This is the basis for the following tests: Acetest, Ketostix, acetone test (Denco), Lange test, Rothera test, and Rantzman test.

Acetest is a tablet containing nitroprusside, phosphate buffers, glycine, and lactose. A purple color develops when a drop of urine containing acetone is placed on the surface. Ketostix and the ketone portion of Labstix contain a stable alkaline phosphate-buffered nitroprusside reagent that includes glycine. When the strips are dipped into urine containing ketone bodies, the characteristic purple color appears. BSP dye and large amounts of phenylpyruvate in urine will give a false positive reaction.

In the presence of acetone, salicylaldehyde also forms a complex compound that is purple to red in color. This is the basis of the Frommer test.

If the acetone test is positive, many laboratories run a test for diacetic acid. In diabetes, it is believed that the presence of diacetic acid is an additional indication of approaching coma.

Procedure for acetone in urine
Acetest

Reagent

Acetest tablets*

Procedure

1. Place an Acetest tablet on a piece of white paper.
2. Place a drop of urine on the tablet. If acetone is present, a lavender to purple color develops within 30 seconds. Compare with color chart that comes with the reagent.
3. Report the test as positive or negative.

DIACETIC ACID

As was previously mentioned, diacetic acid results from abnormal fat catabolism, which may occur in severe diabetes. Its presence in the urine may forewarn the physician of approaching coma.

---

*May be purchased from your local supply house or from Ames Co.

Either Gerhardt's or Lindemann's test is used to test for diacetic acid. In Gerhardt's test, a reaction between diacetic acid and ferric chloride produces a red color. In Lindemann's test, a reaction between diacetic acid and iodine decolorizes the iodine.

Procedure for diacetic acid in urine
Gerhardt test

Reagents

Concentrated nitric acid
Ferric chloride, 10% (D-197)

Procedure

1. Pour 5 ml of *fresh* urine into a test tube.
2. Add 10% ferric chloride drop by drop until no further precipitation is produced.
3. Centrifuge.
4. Transfer the supernatant fluid to a test tube. Add a few more drops of 10% ferric chloride to this supernatant fluid. If no color is produced, the test is negative. If a red color is produced, however, it is due either to diacetic acid or to a drug. To decide which, proceed as follows.
5. Pour 5 ml of *fresh* urine into a beaker. Add 5 ml of distilled water and 2 drops of concentrated nitric acid.
6. Boil down to 5 ml. Cool.
7. Add a few drops of the 10% ferric chloride. If *no* red color is produced, the test is *positive* for diacetic acid. The boiling drives off or oxidizes the diacetic acid but not the drug.

β-OXYBUTYRIC ACID

β-Oxybutyric acid is a reduction product of diacetic acid. When found in the urine, it is usually accompanied by both diacetic acid and acetone and, like these, its presence indicates the abnormal catabolism of fats.

Hart method

Reagents

Acetic acid, glacial
Hydrogen peroxide, 3%
Ammonium hydroxide, concentrated
Sodium nitroprusside, saturated (freshly prepared) (D-198)

Procedure

1. Mix urine. Pour 20 ml into a small beaker. With a marking pencil, mark the level to which the urine comes (the 20-ml level). Add 20 ml of distilled water and a few drops of glacial acetic acid. Drive off any acetone or diacetic acid by boiling down to the 20-ml level.
2. Add 10 ml of distilled water. Mix. Label two large test tubes as 1 and 2. Pour 10 ml of the mixture into each tube. Add 1 ml of 3% hydrogen peroxide to tube 1. (This oxidizes any β-oxybutyric acid present to diacetic acid and acetone.) Add none to tube 2. Warm gently and allow to cool.
3. To each tube add 10 drops of glacial acetic acid and 10 drops of a freshly prepared concentrated solution of sodium nitroprusside. Mix both tubes by inversion.
4. Using a medicine dropper, overlay both tubes with concentrated ammonium hydroxide. If tube 1 shows a reddish purple ring at the junction of the two

Table 10-5
Some significant sediments and their usual causes

Sediment	Usual causes
Pus	Inflammation of urethra, bladder, vagina, or kidneys
Red cells	Bleeding from bladder, uterus, or kidneys
Casts	Nephritis
Bacteria	Bacterial infection of urethra, bladder, vagina, or kidneys
Trichomonas	Trichomonas infestation of vagina

liquids, the test is positive for β-oxybutyric acid. Tube 2 should show no ring.

MICROSCOPIC EXAMINATION

The microscopic examination consists of obtaining, examining, and reporting the urinary sediments. The sediments, which are examined with a microscope, are magnified either 100 or 450 times.

Knowing the type and quantity of sediments aids the physician in diagnosing diseases of the urethra, bladder, and kidneys. Some significant sediments and their usual causes are given in Table 10-5. The pus cells mentioned in the table are often referred to as white blood cells or WBC. The red cells are often reported as red blood cells or RBC. The casts are solidified sediments that have been forced out of urinary tubules. *Trichomonas* is a parasite that is sometimes found in the urine of women. With the exception of *Trichomonas,* the sediments are illustrated in Fig. 10-8.

The manner of performing and reporting the microscopic examination varies from laboratory to laboratory. For example, some laboratories obtain the sediment by centrifuging the urine; other laboratories do not centrifuge the urine. Some laboratories report the number of sediments per low-power field (magnification 100); other laboratories report the number of sediments per high dry field (magnification 450). Some laboratories report all sediments seen; other laboratories report only the more significant sediments.

In performing the microscopic examination, guard against the following sources of error:

1. Failure to see casts. If the light is too bright, casts are very difficult to see. Therefore, always lower the source of light.
2. Failure to distinguish red cells from pus cells. If in doubt, place a drop of 5% acetic acid on the sediment. If the cells are red cells, they will hemolyze. If the cells are pus cells, it is possible to see their nuclei.
3. Failure to distinguish red cells from fat globules. The red cells are uniform in size, whereas fat globules usually come in various sizes.

A suggested procedure for the microscopic examination is given below.

Procedure for microscopic examination
1. Mix the urine specimen.
2. Place 5 ml of urine into a test tube. Centrifuge at high speed for 5 minutes and decant the supernatant fluid.

292 *Chemistry for the clinical laboratory*

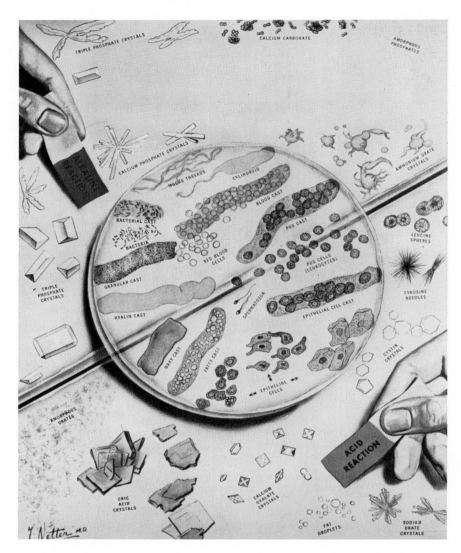

Fig. 10-8. Urinary sediments. (From Sharp & Dohme Seminar, November, 1940, Sharp & Dohme, West Point, Pa.)

3. The grayish white precipitate (urinary sediment) may be loosened by gently shaking or tapping the bottom of the tube.
4. Place the sediment on a glass slide.
5. When the sediment is on the slide, smear it over the surface of the slide with the lip of the test tube. This "spreading out" will prevent the high dry objective of the microscope from touching the sediment and eliminates the use of coverslips. However, coverslips are preferable.
6. Place the slide on the stage of a microscope and examine sediment under low power and high dry power.
7. Identify the contents, using Figs. 10-8 and 10-9 as guides.
8. Make the report according to the practice of your particular laboratory.
9. Sample microscopic report: Centrifuged specimen shows 10 to 12 RBC and 0 to 2 granular casts per high dry field. Bacteria present.

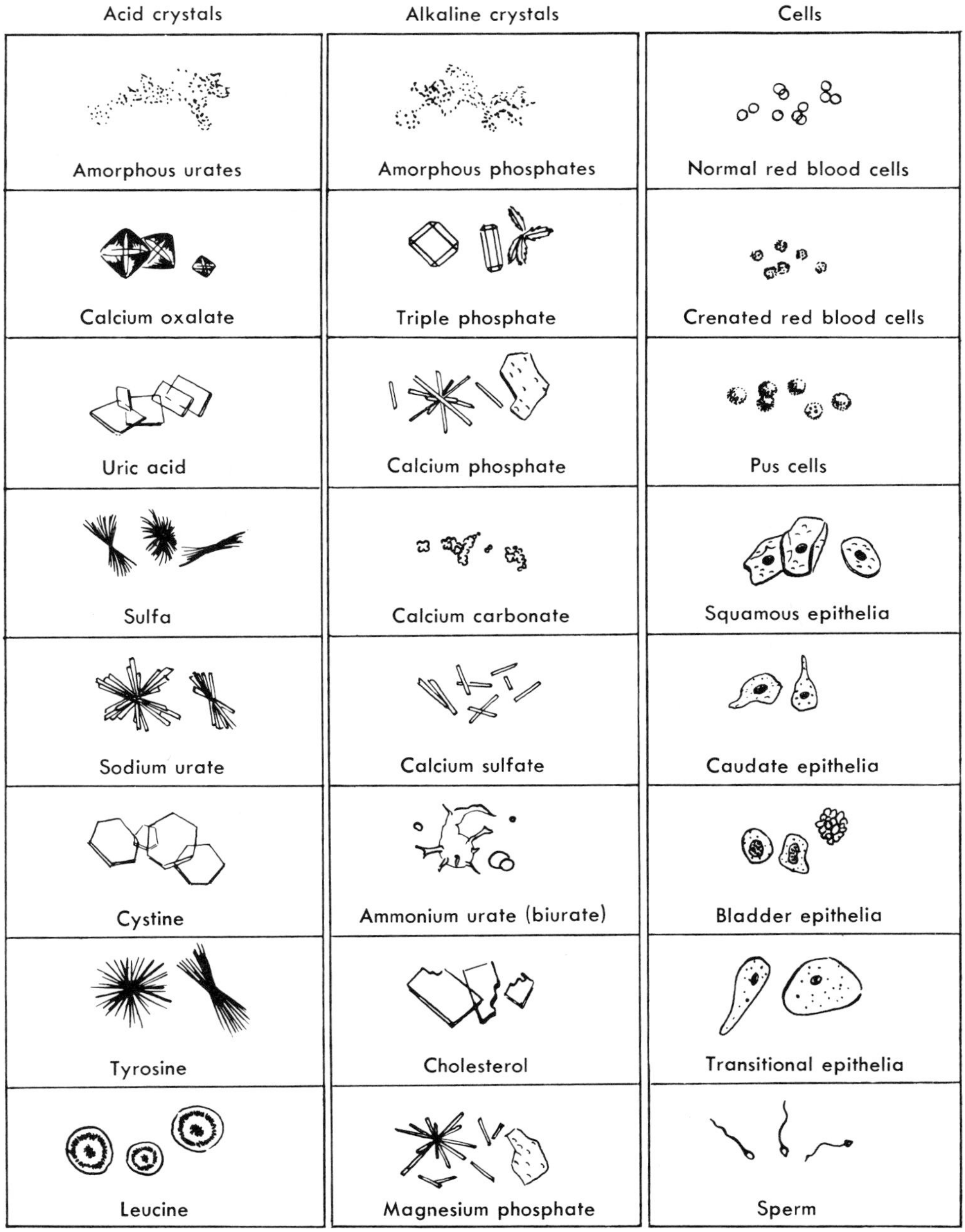

Renal function 293

Acid crystals	Alkaline crystals	Cells
Amorphous urates	Amorphous phosphates	Normal red blood cells
Calcium oxalate	Triple phosphate	Crenated red blood cells
Uric acid	Calcium phosphate	Pus cells
Sulfa	Calcium carbonate	Squamous epithelia
Sodium urate	Calcium sulfate	Caudate epithelia
Cystine	Ammonium urate (biurate)	Bladder epithelia
Tyrosine	Cholesterol	Transitional epithelia
Leucine	Magnesium phosphate	Sperm

Continued.

Fig. 10-9. Sediments found in urine. (Drawings by Robert Pribbenow and Samuel Taylor.)

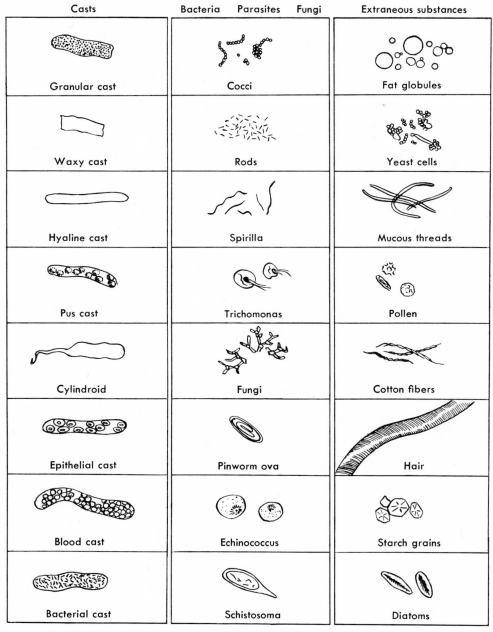

Fig. 10-9, cont'd. Sediments found in urine. (Drawings by Robert Pribbenow and Samuel Taylor.)

BILIRUBIN (BILE)

Bilirubin is a red bile pigment. It is found in the urine in infectious hepatitis, obstructive jaundice, and many other liver diseases. In liver disease, bilirubin may be detected in the urine even before the blood level is elevated.

The majority of tests for bilirubin depend on the oxidation of bilirubin to various colored compounds. Among these compounds are biliverdin, which is green, bilicyanin which is blue, and choletelin, which is yellow.

Ictotest is a test system consisting of an asbestos-cellulose mat and a reagent tablet containing sulfosalicylic acid with a stabilized diazonium salt, p-nitrobenzene diazonium p-toluene sulfonate (bilazo). Bilirubin in urine reacts with the reagent to form a blue or purple color.

Procedure for bilirubin in urine
Ictotest

Reagent

Ictotest reagent tablets*

Procedure

1. Place 5 drops of urine on a special mat that comes with the reagent.
2. Put 1 Ictotest tablet in the center of the moistened area.
3. Place 2 drops of water on the tablet so that the water just flows over the sides. If a bluish purple color appears on the mat around the tablet, the test is positive for bilirubin.
4. Report the test as positive or negative.

UROBILINOGEN

Urobilinogen is a colorless compound derived from bilirubin. Small traces are normally present in urine. Large amounts are found in liver disease, lead poisoning, hemolytic anemia, and other diseases of a hemolytic nature.

Urobilinogen is formed in the intestine by the bacterial action on bilirubin. This is reabsorbed in the small intestine and excreted partly by the kidney and partly by the liver. With hemolytic anemia, where there is an increased destruction of the red blood cells, and in parenchymal liver disease, the urinary urobilinogen is increased. With complete obstructive jaundice, there may be no urobilinogen in the urine, since bilirubin cannot reach the intestine to be changed to urobilinogen.

For the determination of urobilinogen in urine, methods commonly used are: (1) Ehrlich qualitative test, (2) Ehrlich semiquantitative test, and (3) Schwartz-Watson test. In all these methods, a reaction takes place between urobilinogen and Ehrlich's reagent (p-dimethylaminobenzaldehyde) to produce a cherry red color. The extent of the reaction may be compared with standards.

Procedure for urobilogen in urine
Schwartz-Watson test

Reagents

Ehrlich's reagent	(D-199)
Sodium acetate, saturated	(D-200)
Sodium acetate, crystals	
Chloroform	

---

*May be purchased from your local supply house or from Ames Co.

Procedure

1. Pipet 1 ml of fresh urine into a test tube.
2. Add 1 ml of Ehrlich's reagent. Mix.
3. Add 4 ml of supersaturated aqueous solution of sodium acetate. Mix.
4. Test with Congo red paper. Paper reaction:
    a. If red, continue to step 5.
    b. If not red, add sodium acetate crystals (shake to dissolve) until a positive result of red is obtained.
5. Add 3 ml of chloroform.
6. Mix carefully by inverting twice and releasing the stopper; then *shake vigorously*. (NOTE: If the tube is not shaken vigorously, the test will produce a false positive reaction.)
7. Results
    a. *Negative*—the aqueous layer is not colored.
    b. *Positive*—the aqueous layer forms a purple-red aldehyde reaction.
    c. NOTE: *Porphobilinogen* will be indicated by a red-brown aldehyde layer.
8. Results with ultraviolet light: *Uroporphyrin* and *coproporphyrin* are fluorescent in ultraviolet light.

Diazo substances

Diazo substances are unknown substances that give a color reaction with Ehrlich's diazo reagent. They may appear in the urine in febrile conditions, measles, typhoid fever, and tuberculosis.

Reagents

Ehrlich's diazo reagent	(D-201)
Ammonium hydroxide	

Procedure

1. Pipet 2 ml of urine into a small test tube and add 2 ml of Ehrlich's diazo reagent. Shake well.
2. Using a medicine dropper, overlay with 2 ml of full-strength (28%) ammonium hydroxide. If a red ring appears at the junction of the two liquids, the test is positive for diazo substances. Disregard a yellow or orange color.

UROBILIN

When urine stands, urobilinogen is oxidized to a brown pigment known as urobilin. Small amounts of urobilin are normally found in urine. Large amounts may be seen in diseases of the liver and in diseases of a hemolytic nature.

To determine urobilin in urine, the method of Schlesinger is commonly used. In this method, calcium chloride is added to the urine to precipitate bile. The urine is then filtered. Lugol's solution is added to the filtrate to convert urobilinogen to urobilin. A solution of zinc acetate is then added. This reacts with the urobilin to produce a green fluorescence.

Procedure for urobilin in urine

Reagents

Calcium chloride, 10%	(D-202)
Lugol's solution	(D-203)
Zinc acetate, saturated alcoholic solution	(D-204)

Procedure (Schlesinger)
1. Mix the urine and pour 12 ml into a small flask. (Measure with a graduated cylinder.)
2. Add 3 ml of 10% calcium chloride.
3. Mix and filter to remove any bile present.
4. Take 10 ml of the filtrate and add a few drops of Lugol's solution.
5. Mix and add 10 ml of a saturated alcoholic solution of zinc acetate.
6. Mix and filter. Let filtrate stand about an hour.
7. Examine in sunlight, concentrating the sunlight on the solution with a hand lens. The presence of a green fluorescence is positive for urobilin.

PORPHYRINS

The porphyrins are pigments or complex organic compounds that possess a basic structure of four pyrrole rings linked by methene bridges. Small amounts may be found in normal urine. Increased amounts are found in porphyria, a metabolic disease in which porphyrins are retained in pathologic quantities in the blood and other tissues, and in feces and urine. Urine containing increased amounts of porphyrins usually turns dark red upon standing.

The two predominant porphyrins excreted are coproporphyrin and uroporphyrin (refer to Chapter 16, p. 514).

Reagents
    Acetic acid, glacial
    Ethyl ether
    Hydrochloric acid, 5% (v/v)   (D-189)

Procedure
1. Mix the urine, pour 25 ml into a separatory funnel, and add 10 ml of glacial acetic acid.
2. Extract the mixture twice with 50-ml portions of ether, saving the *waste fluids* from the extractions, since they will be used later in the procedure.
3. Combine the two ether extractions and wash with 10 ml of 5% hydrochloric acid, saving the *washings*.
4. Examine the *washings* under ultraviolet light. If there is a strong fluorescence, the test is positive for coproporphyrin.
5. Examine the urinary *waste fluids* (which resulted from the above ether extractions) under ultraviolet light. If there is a strong red fluorescence, the test is positive for uroporphyrin.

PORPHOBILINOGEN

Porphobilinogen is a precursor of uroporphyrin. It is found in the urine in acute intermittent porphyria. Clinically the latter may be characterized by periodic attacks of intense abdominal colic with nausea and vomiting, constipation, neurotic behavior, and neuromuscular disturbances.

Reagents
    Ehrlich's reagent, modified
        Dissolve 0.7 g of *p*-dimethylaminobenzaldehyde in a mixture of 100 ml of distilled water and 150 ml of concentrated hydrochloric acid. Mix well and store in a brown bottle.
    Hydrochloric acid, concentrated
    Sodium acetate, saturated   (D-200)
    Chloroform
    Butanol

Procedure

1. Obtain a fresh sample of urine and acidify to pH 4 or 5 with concentrated hydrochloric acid, using nitrazine paper as an indicator.
2. Pipet 3 ml of the urine into a large test tube and add 3 ml of the modified Ehrlich's reagent. Mix by inversion.
3. Add 6 ml of saturated sodium acetate and mix. If urobilinogen, porphobilinogen, or other Ehrlich's reactive substances are present, there will be a pink or red color.
4. If a positive reaction is noted in step 3, pipet 5 ml of that solution into a test tube, add 2 ml of chloroform, and mix. If color is extracted into the chloroform layer, reextract the aqueous phase for complete extraction of color. Color produced by urobilinogen will be removed by chloroform; that produced by porphobilinogen will not be extracted.
5. If the pink color is not extracted by chloroform, use another 5-ml aliquot from step 3 and extract with 2 ml butanol. Color produced by urobilinogen will be extracted by butanol but that produced by porphobilinogen will not be extracted. Other Ehrlich's reactive compounds may be extracted.
6. Report as negative; positive for urobilinogen; positive for porphobilinogen; positive for Ehrlich's reactive compounds. (Other compounds that react with Ehrlich's reagent are sulfonamides, procaine, and 5-hydroxyindole acetic acid.)

HIPPURIC ACID SYNTHESIS

One liver function (hippuric acid synthesis) test which is dependent on renal function is based on the principle that a healthy liver will convert benzoic acid to hippuric acid. The latter is then excreted in the urine. When the liver is not functioning properly (intrinsic hepatic disease), however, this process of conversion and excretion is decreased.

Quick's method

Reagents

Sodium benzoate	
Oil of peppermint	
Ammonium sulfate	
Glacial acetic acid	
Hydrochloric acid, concentrated	
Congo red indicator	(Table C-3)
Phenolphthalein indicator	(D-9)
Sodium hydroxide, 0.2 N	(D-136)

Procedure

1. The patient is instructed to eat a light breakfast. One hour after breakfast, he is told to empty the bladder and discard the urine. He is then given a dose of 6 g of sodium benzoate dissolved in 30 ml of water and flavored with a little oil of peppermint. This is followed with a half glass of water.
2. Collect all urine voided within the next 4 hours. Ask the patient to empty the bladder at the end of the 4-hour period and add this to the collection.
3. Measure the volume of urine. If the volume is 150 ml or less, it does not have to be boiled down. However, if the volume is more than 150 ml., acidify with a few drops of glacial acetic acid and boil down to 150 ml.
4. Add 5 g of ammonium sulfate for each 10 ml of urine. Stir to dissolve. Filter.
5. Add concentrated hydrochloric acid to the filtrate until the indicator Congo

red turns blue (pH 3.0). Usually about 2 ml is needed. Stir for several minutes to aid crystallization. Place in the refrigerator for 1 hour.
6. Using a Buchner filter and suction, filter off the crystalline hippuric acid. Wash the container with ice-cold distilled water and pour the washings into the Buchner filter. Wash the precipitate several times with ice-cold distilled water.
7. Carefully transfer the precipitate and filter paper to a beaker. Add enough distilled water to cover and heat gently to dissolve the crystals. Add a few drops of phenolphthalein indicator and titrate to a permanent pink color with 0.2 N sodium hydroxide. Record the milliliters of sodium hydroxide used.

Calculations

One milliliter of the 0.2 N sodium hydroxide is equivalent to 0.0358 g of hippuric acid. Therefore:

$$\text{ml of NaOH used} \times 0.0358 = \text{g of hippuric acid}$$

Some hippuric acid, however, still remains in solution. To correct for this, take your answer obtained above and add 0.1 g for each 100 ml of urine used. If you wish to express the hippuric acid in terms of benzoic acid, multiply your corrected answer by 0.68.

Normal values

The normal values are 3 g or more of benzoic acid in 4 hours or, if sodium benzoic acid is given intravenously, the normal value is at least 1 g of benzoic acid in 1 hour.

MELANIN

The word melanin is Greek for black. When present in the urine, melanin usually occurs as a colorless compound. Upon standing, however, it oxidizes to a black pigment that is readily discernible. Melanin is not present in normal urine but may be found in the urine of patients with melanotic tumors.

For the determination of melanin, the following methods are commonly used: (1) ferric chloride test and (2) bromine test. Both methods call for the oxidation of the colorless compound (melanogen) to the colored compound (melanin).

Procedures for melanin in urine
Ferric chloride test

Reagent

Ferric chloride, 10% (D-197)

Procedure

1. Mix the urine and pour about 8 ml into a small flask.
2. Add about 6 drops of 10% ferric chloride. Mix.
3. Allow to stand 30 minutes. If the solution turns black, melanin is present.

Bromine test

Reagent

Bromine water (D-205)

Procedure

1. Mix the urine and pour about 25 ml into a small flask.
2. Add about 25 ml of the bromine reagent. Mix.
3. Allow to stand 30 minutes. If the solution turns black, melanin is present.

OCCULT BLOOD

When blood occurs in the urine, the condition is referred to as hemoglobinuria. It may be found in the following disorders, all of which are accompanied by excessive red cell destruction: transfusion reactions, severe burns, and various chemical poisonings.

For the detection of occult blood, the following tests are commonly used: (1) benzidine test, (2) o-tolidine test, (3) guaiac test, and (4) Occultest. The principle of these tests is: the hemoglobin of the blood reacts with peroxide to liberate oxygen and the liberated oxygen then reacts with an organic reagent to produce a colored compound.

Occultest is a tablet containing o-tolidine, strontium peroxide, calcium acetate, tartaric acid, and sodium bicarbonate. The interaction of hydrogen peroxide with o-tolidine catalyzed by the peroxidative activity of blood produces a characteristic blue color. Hemastix and the occult blood portion of Labstix contain o-tolidine and cumene hydroperoxide, which when dipped into urine containing occult blood develop the blue color of oxidized o-tolidine. Large quantities of ascorbic acid may inhibit the sensitivity of this test.

Procedures for occult blood in urine
o-Tolidine test

Reagents

Hydrogen peroxide, 3%
o-Tolidine reagent (D-206)

Procedure

1. Place 1 ml of urine in a test tube.
2. Add 1 ml of the o-tolidine reagent.
3. Add about 10 drops of 3% hydrogen peroxide. Mix. A green to blue color is a positive reaction.

Occultest

Reagent

Occultest tablets*

Procedure

1. Place a piece of filter paper (which comes with the reagent bottle) on a clean surface.
2. Place an Occultest tablet on the filter paper.
3. Place a large drop of urine on the tablet so that the urine just flows over the the sides. If a blue color appears on the filter paper around the tablet, the test is positive for occult blood.

CALCIUM

Calcium is normally present in the urine. Increased amounts are found in hyperparathyroidism. Decreased quantities are observed in hypoparathyroidism.

Quantitative test

The urine of the average person contains 50 to 400 mg of calcium/day. Increased values may be found in hyperparathyroidism, whereas decreased values are seen in rickets. Refer to atomic absorption calcium method, p. 206.

---

*May be purchased through your local medical supply house or ordered directly from Ames Co.

Wang method

Reagents

Trichloroacetic acid, 20%	(D-207)
Sodium acetate, 20%	(D-208)
Ammonium hydroxide, 10% (w/v)	(D-209)
Acetic acid, 10% (v/v)	(D-210)
Sulfuric acid, 2 N	(D-211)
Ammonium oxalate, 0.1 M	(D-212)

Potassium permanganate, 0.01 N (D-110 diluted 1:10)
Wash solution
  Place 80 ml of distilled water in a large flask; add 80 ml of ether and 80 ml of alcohol; then add 5.0 ml of ammonium hydroxide. Mix.

Procedure

1. Collect a 24-hour specimen of urine. Measure and record the volume. Mix. Pour 100 ml into a flask. Add 25 ml of the 20% trichloroacetic acid. Add 2 g of acid-washed charcoal. Shake well and set aside for 15 minutes.
2. Filter. Pipet 5.0 ml of the filtrate into a 15-ml centrifuge tube. Add 1.0 ml of the 20% sodium acetate solution. Mix. Add 1.0 ml of the 0.1 M ammonium oxalate solution. Mix again.
3. Adjust the acidity to pH 5 as follows. Find the approximate pH with nitrazine paper. If above pH 5, add the 10% acetic acid dropwise until the solution is green to bromcresol green or pink to methyl red. If below pH 5, add the 10% ammonium hydroxide dropwise until the solution is green to bromcresol green or pink to methyl red. Mix again and place in the refrigerator overnight.
4. The next morning, centrifuge for 15 minutes at high speed. Discard the supernatant fluid and leave the precipitate in the bottom of the tube. Keep the tube in an inverted position, place on a piece of filter paper, and let drain for a few minutes.
5. Wash the precipitate *twice* as follows. Add 2.0 ml of the wash solution (preparation given above). Mix well. Wash down the tube with another 2.0 ml of the wash solution. Centrifuge for 15 minutes.
6. Invert the tube, discard the supernatant fluid, and leave the precipitate in the bottom of the tube. Keep the tube in an inverted position, place on a piece of filter paper, and let drain for a few minutes.
7. Place the tube (right side up) in an oven at 85° to 100° C for 1 hour. This removes organic solvents.
8. Add 2.0 ml of the 2 N (approximate) sulfuric acid. Mix to dissolve the precipitate. Place in a boiling water bath for 2 minutes. Remove from water bath. Titrate while hot (over 70° C) with the 0.01 N potassium permanganate. The end point is a faint pink color that persists for at least 1 minute. Record the milliliters of permanganate used.

Calculations

One milliliter of the 0.01 N potassium permanganate is equivalent to 0.2 mg of calcium. Therefore:

$$\frac{\text{mg of calcium}}{\text{(24-hour specimen)}} = \frac{\text{ml of urine (24-hour specimen)}}{4} \times 0.2 \times \text{ml of KMnO}_4 \text{ used} \times \text{Factor}$$

  4 = milliliters of urine used in the test (5 ml of a 4 to 1 mixture of urine and 20% trichloroacetic acid).
  Factor = number telling how strong or how weak the potassium permanga-

nate solution is in comparison to an exact 0.01 N $KMnO_4$ solution. The factor is determined in preparing the solution.

Sample calculation: If the 24-hour specimen measures 1000 ml, the number of milliliters of 0.01 N potassium permanganate used is 2, and the factor is 1.0, the calculation is performed as follows:

$$\begin{matrix}\text{mg of calcium}\\ \text{(24-hour specimen)}\end{matrix} = \frac{1000}{4} \times 0.2 \times 2 \times 1.0 = 100 \text{ mg}$$

SULFATES

The normal values for total urinary sulfates in a 24-hour period are 0.6 to 1.2 g. Increased values are found in acute fevers. Decreased values are seen in conditions associated with decreased metabolic activity.

Folin method

Reagents

Hydrochloric acid, concentrated
Barium chloride, 5%   (D-213)

Procedure

1. Collect a 24-hour specimen of urine. Mix. Pour 50 ml into a 250-ml flask. Add 4 ml of concentrated hydrochloric acid. Mix. Place a watchglass or small beaker over the mouth of the flask and boil gently for ½ hour.
2. Cool the flask in cold water. Add sufficient distilled water to make the total volume equal 150 ml. Mix. Without stirring the solution, add dropwise 10 ml of a 5% solution of barium chloride. This precipitates the sulfur as barium sulfate. Set aside for 1 hour.
3. Handling the crucible with tongs, weigh a Gooch crucible on the analytic balance. Record the weight. Shake the solution and then filter it through the weighed Gooch crucible. Wash the precipitate with 200 ml of cold distilled water. Dry in an oven. Ignite, cool, and weigh.

Calculations

1. To obtain the weight of the barium sulfate, subtract the weight of the crucible before filtering from the weight of the crucible after ignition.
2. To find the weight of sulfur in this amount of barium sulfate, use the following formula:

$$\text{Weight of sulfur} = \text{Weight of BaSO}_4 \times \frac{32.064}{233.402}$$

32.064 = atomic weight of sulfur
233.402 = molecular weight of barium sulfate

3. To find the amount of sulfur in the 24-hour specimen, use this equation:

$$\begin{matrix}\text{Weight of sulfur}\\ \text{in 24-hour specimen}\end{matrix} = \begin{matrix}\text{Weight of sulfur}\\ \text{in sample}\end{matrix} \times \frac{\text{ml of 24-hour specimen}}{\text{ml used in test}}$$

Sample calculations: A 24-hour specimen measured 1000 ml and 50 ml was used in the test. The weight of the Gooch crucible before filtering was 5.000 g. The weight of the Gooch crucible after ignition was 5.364 g.

a. The weight of the barium sulfate is found by subtracting the weight of the Gooch crucible before filtering from the weight after ignition. The weight of the barium sulfate is therefore 5.364 g minus 5.000 g which is 0.364 g.

b. The weight of sulfur in 0.364 g of barium sulfate is found by using the formula:

$$\text{Weight of sulfur} = \text{Weight of BaSO}_4 \times \frac{32.064}{233.402}$$

$$= 0.364 \times \frac{32.064}{233.402}$$

$$= 0.05 \text{ g}$$

This is the weight of sulfur in the 50 ml of urine that was used in the test.

c. To find the weight in the 1000 ml of the 24-hour specimen:

$$\frac{\text{Weight of sulfur}}{\text{in 24-hour specimen}} = \frac{\text{Weight of sulfur}}{\text{in sample}} \times \frac{\text{ml of 24-hour specimen}}{\text{ml used in test}}$$

$$= 0.05 \times \frac{1000}{50}$$

$$= 1.0 \text{ g}$$

AMYLASE

The normal values for urinary amylase (or diastase) depend upon the method of determination. Greatly increased values are found in acute pancreatitis. See p. 221 for procedure.

CHLORIDE

The average person excretes 10 to 15 g/day of chloride (expressed as sodium chloride). Decreased values may be found in acute infections, severe diarrhea, nephritis with edema, and dehydration conditions. See p. 121 for procedure.

CALCULI

Calculi, the Latin word for stones, are formed by the crystallization of salts. Their formation is often started by an infection in the urinary system and is enhanced by sedentary habits and the presence of urinary salts that are difficult to dissolve, two of the chief offenders being calcium oxalate and calcium phosphate. In mild cases, the stones may be passed in the urine; in severe cases, surgery is essential.

Procedure

To perform the various tests, follow the procedures given in Table 10-6.

Reagents

Ammonium molybdate solution
Dissolve 3.5 g of ammonium molybdate in 75 ml of water. Add 25 ml of concentrated nitric acid and mix.

Hydrochloric acid, 10% (v/v)
Add 10 ml of concentrated hydrochloric acid to 90 ml of water. Mix.

Sodium cyanide, 5%
Weigh 25 g of sodium cyanide and transfer to a 500-ml volumetric flask. Add 1 ml of concentrated ammonium hydroxide. Dissolve and dilute to volume with water.

Sodium nitroprusside, 5%
Weigh 5 g of sodium nitroprusside (nitroferricyanide) and transfer to a 100-ml volumetric flask. Dissolve and dilute to volume with water. Mix. Store in a brown bottle.

Table 10-6
Analysis of urinary calculi

Stone	Physical characteristics	Procedure	Positive results
Urates or uric acid	1. Multiple, smooth, round, without luster 2. Single stones may have bumpy surface, miniature craters 3. Crushed stones are yellow	1. Pulverize stone 2. Place in small evaporating dish 3. Add several drops of concentrated nitric acid 4. Heat gently and evaporate to dryness 5. Add 2–3 drops of concentrated NH_4OH	Deep yellow, orange-red, or crimson, turning purple with NH_4OH, or bluish violet
Phosphates	1. Rather large compact balls 2. Also appear as large friable masses 3. Color: clay or chalky 4. Note: White color, porous, corallike formations are usually calcium oxalate, not phosphate	1. Pulverize stone 2. Use microscope slide instead of evaporating dish 3. Add 4–5 drops of ammonium molybdate solution 4. Warm slide over flame	Distinct yellow precipitate forms; this is $(NH_4)_3PO_4 \cdot 12MnO_3$
Carbonate	Same appearance as phosphate	1. Pulverize stone 2. Use a dark background for easier visibility 3. Place a larger quantity of stone on a slide 4. Add 8–10 drops of 10% HCl	Foaming effervescence
Oxalates	1. Smooth luster or 2. Irregular, buff-colored stone with elaborate crystalline structure 3. Occasionally the structure may appear loose and porous	1. Pulverize stone 2. Mix with an equal portion of resorcinol 3. Add 3 drops of concentrated H_2SO_4	Dark blue-green color
Cystine	1. Pale yellow or white granules 2. Appearance similar to calcium oxalate	1. Pulverize stone 2. Add 1 drop of NH_4OH and 1 drop of NaCN 3. Wait 5 minutes 4. Add 2–3 drops of sodium nitroprusside	Upon standing, a beet red color may change to orange-red

PHENYLKETONURIA[43]

Phenylketonuria is an inherited disorder of phenylalanine metabolism that will cause severe mental retardation. This disorder usually appears between the second and sixth week of the infant's life. The urines of all infants are now usually tested for phenylpyruvic acid weekly up to the age of 6 months, if they have convulsions, severe diaper rashes, or eczema. Normal children are checked either biweekly or monthly during this period of 6 months. Phenylketonuria is diagnosed by the excess phenylpyruvic acid found in urine or serum. Urine will also contain phenylacetic and phenyllactic acid.

By early treatment with diets low in phenylalanine, mental defects may be avoided or corrected. If undetected, abnormal metabolism of the amino acid (phenylalanine) will impair the normal development of the brain.

Qualitative methods

Ferric chloride method

Reagents

Sulfuric acid, 1 N	(D-113)
Ferric chloride, 5%	(D-214)

Procedure

1. Place 5 ml of fresh urine into a test tube and acidify with 2 drops of diluted sulfuric acid.
2. Add 5 drops of 5% ferric chloride solution. Mix.
3. Positive result: solution turns to a bluish green.

NOTE: The ferric chloride method has been used also for determination of acetoacetic acid. Positive result: solution turns to red. Other substances may interfere with the test.

Phenistix test

The reagent strip reacts rapidly with the phenylketones in urine, the color of the strip changing from gray to a distinctive blue-gray. The strip is either dipped into a fresh specimen of urine or pressed against the wet diaper. Within 30 seconds the color is developed on the strip and it is then compared with the color scale whose reference points are zero and 15, 40, and 100 mg/dl of phenylketone.

Reagents on the strip

Ferric ammonium sulfate
Supplies the ferric ions, which react with phenylpyruvic acid in a properly acidified medium

Cyclohexylsulfamic acid
Provides the optimum acidity for the above reaction

Magnesium sulfate
The magnesium ions are incorporated to prevent the urine phosphates from interfering with the color reaction.

Procedure

1. Dip the test strip in fresh urine and remove immediately, or press against a wet diaper to moisten.
2. At the end of 30 seconds compare the color against the reference points on the color chart.

NOTE: The only other substance giving a typical blue-gray color of phenylpyruvic acid in a 30-second interval is β-imidazolepyruvic acid. This is excreted in the urine due to the deficiency of the enzyme histidine α-deaminase, a rare disorder of children. Substances not affecting Phenistix are as follows:

Phenylacetate	Thymol
Phenyllactate	Toluene
Phenylalanine	Baby powders
Pyruvate	Baby oils
Lactate	Antiseptics
Tyrosine	Homogentisic acid
Tryptophan	Acetoacetic acid

URIC ACID

The urine of the average person contains about 250 to 750 mg/day of uric acid. Increased amounts are found after an attack of gout. (See p. 109.)

UREA NITROGEN

The normal urine urea nitrogen is 6 to 17 g/24 hr. Decreased values may be found in acute nephritis, acidosis, and cirrhosis of the liver. Increased values may be seen in febrile conditions. (See p. 113.)

CREATININE

The normal range for 24-hour urine creatinine is 20 to 26 mg/kg for males and 14 to 22 mg/kg for females. Creatinine excretion is decreased in disorders associated with muscular atrophy and muscular weakness. It is increased with increased tissue catabolism, as in fever. The concentration is not dependent on diet or diuresis. (See pp. 50 and 105.)

CREATINE

The 24-hour urine specimen of the average person contains up to 200 mg of creatine. Increased values are found in fever, malnutrition, and pregnancy. (See p. 106.)

References

Renal function

1. Alving, A. S., and Miller, B. F.: A practical method for the measurement of glomerular filtration (inulin clearance), Arch. Intern. Med. **66**:306, 1940.
2. Alving, A. S., Rubin, J., and Miller, B. F.: A direct colorimetric method for the determination of inulin in blood and urine, J. Biol. Chem. **127**:609, 1939.
3. Alving, A. S., and Van Slyke, D. D.: The significance of concentration and dilution tests in Bright's disease, J. Clin. Invest. **13**:969, 1934.
4. Brodwall, E. K.: Renal extraction of PAH in normal individuals, Scand. J. Clin. Lab. Invest. **16**:1, 1964.
5. Chapman, E. M.: Further experience with the fractional 'phthalein test, N. Eng. J. Med. **214**:16, 1936.
6. Chapman, E. M., and Halstead, J. A.: The fractional phenolsulphonphthalein test in Bright's disease, Am. J. Med. Sci. **186**:223, 1933.
7. Dick, A., and Davies, C. E.: Measurement of the glomerular filtration rate and the effective renal plasma flow using sodium thiosulphate and p-aminohippuric acid, J. Clin. Pathol. **2**:67, 1949.
8. Enger, E., and Blegen, E. M.: The relationship between endogenous creatinine clearance and serum creatinine in renal failure, Scand. J. Clin. Lab. Invest. **16**:273, 1964.
9. Fishberg, A. M.: Hypertension and nephritis, ed. 5, Philadelphia, 1954, Lea & Febiger.
10. Harrison, H. E.: A modification of the diphenylamine method for determination of inulin, Proc. Soc. Exp. Biol. Med. **49**:111, 1942.
11. Heyrovsky, A.: A new method for the determination of inulin in plasma and urine, Clin. Chim. Acta **1**:470, 1956.
12. Josephson, B., and Ek, J.: The assessment of the tubular function of the kidneys. In Sobotka, H., and Stewart, C. P., editors: Advances in clinical chemistry, vol. 1, New York, 1958, Academic Press, Inc.
13. McIntosh, J. F., Moller, E., and Van Slyke,

D. D.: Studies of urea excretion. III. Influence of body size on urea output, J. Clin. Invest. **6:**467, 1929.
14. Moller, E., McIntosh, J. G., and Van Slyke, D. D.: Studies of urea excretion. II. Relationship between urine volume and the rate of urea excretion by normal adults, J. Clin. Invest. **6:**427, 1928.
15. Rolf, D., Switchen, A., and White, H. L.: A modified diphenylamine procedure for fructose or inulin determination, Proc. Soc. Exp. Biol. Med. **72:**351, 1949.
16. Rosenheim, M. L., and Ross, E. J.: Chronic renal failure. In Black, D. A. K., editor: Renal disease, Philadelphia, 1963, F. A. Davis Co.
17. Rowntree, L. G., and Geraghty, J. T.: An experimental and clinical study of phenolsulphonphthalein in relation to renal function in health and disease, Arch. Intern. Med. **9:**284, 1912.
18. Schreiner, G. E.: Determination of inulin by means of resorcinol, Proc. Soc. Exp. Biol. Med. **74:**117, 1950.
19. Smith, H. W.: The kidney, New York, 1951, Oxford University Press.
20. Smith, H. W.: Principles of renal physiology, New York, 1956, Oxford University Press.
21. Tobias, G. J., McLaughlin, R. F., Jr., and Hopper, J., Jr.: Endogenous creatinine clearance, a valuable clinical test of glomerular filtration and prognostic guide in chronic renal disease, N. Engl. J. Med. **266:**317, 1962.
22. Van Slyke, D. D.: Renal function tests, N. Y. J. Med. **41:**825, 1941.
23. Wills, M. R.: Biochemical consequences of chronic renal failure: a review, J. Clin. Pathol. **21:**541, 1968.

Renal insufficiency and dialysis

24. Abel, J. J., Rowntree, L. G., and Turner, B. B.: On the removal of different substances from the circulating blood of living animals by dialysis, J. Pharmacol. Exp. Ther. **5:**275, 1914.
25. Alwall, N.: On the artificial kidney. I. Apparatus for dialysis of blood in vivo, Acta Med. Scand. **128:**317, 1947.
26. Barry, K. G., Shambaugh, G. E., Goler, D., and Matthews, F. E.: A new flexible cannula and seal to provide prolonged access to the peritoneal cavity for dialysis, Trans. Am. Soc. Artif. Intern. Organs **9:**105, 1963.
27. Doolan, P. D., Murphy, W. P., Jr., Wiggins, R. A., Carter, N. W., Cooper, W. C., Watten, R. H., and Alpen, E. L.: An evaluation of intermittent peritoneal lavage, Am. J. Med. **26:**831, 1959.
28. Fine, J., Frank, H. A., and Seligman, A. M.: The treatment of acute renal failure by peritoneal irrigation, Ann. Surg. **124:**857, 1946.

29. Ganter, G.: Ueber die Beseitigung giftiger Stoffe aus dem Blute durch Dialyse, München Med. Wochenschr. **70:**1478, 1923.
30. Gutch, C. F.: Peritoneal dialysis, Trans. Am. Soc. Artif. Intern. Organs **10:**406, 1964.
31. Gutch, C. F., Stevens, S. C., and Watkins, F. L.: Periodic peritoneal dialysis in chronic renal insufficiency, Ann. Intern. Med. **60:**289, 1964.
32. Grollman, A., Turner, L. B., and McLean, J. A.: Intermittent peritoneal lavage in nephrectomized dogs and its application to the human being, Arch. Intern. Med. **87:**379, 1951.
33. Kolff, W. J.: New ways of treating uremia; the artificial kidney, peritoneal lavage, intestinal lavage, London, 1947, J. & A. Churchill, Ltd.
34. Kolff, W. J., and Berk, H. T. J.: Artificial kidney, dialyzer with great area, Geneesk. Geds. **5:**21, 1943; Acta Med. Scand. **117:**121, 1944.
35. Kolff, W. J., and Watschinger, B.: Further development of a coil kidney; disposable artificial kidney, J. Lab. Clin. Med. **47:**969, 1956.
36. Maher, J. F., and Schreiner, G. E.: Current status of dialysis of poisons and drugs, Trans. Am. Soc. Artif. Intern. Organs **15:**461, 1965.
37. Maxwell, M. H., Rockney, R. E., and Kleeman, C. R.: Peritoneal dialysis. I. Technique and applications, J.A.M.A. **94:**917, 1959.
38. McDonald, H. P.: An automatic peritoneal dialysis machine for hospital or home peritoneal dialysis: preliminary report, Trans. Am. Soc. Artif. Intern. Organs **15:**108, 1969.
39. Scribner, B. H., Caner, J. E. Z., Buri, R., and Quinton, W.: Technique of continuous hemodialysis, Trans. Am. Soc. Artif. Intern. Organs **6:**88, 1960.
40. Quinton, W., Dillard, D., Cole, J. J., and Scribner, B. H.: Eight months' experience with Silastic-Teflon bypass cannulas, Trans. Am. Soc. Artif. Intern. Organs **8:**236, 1962.
41. Skeggs, L. T., and Leonards, T. R.: Studies of an artificial kidney. I. Preliminary results with a new type of continuous dialyzer, Science **108:**212, 1948.
42. Tenchoff, H., Shilipeter, G., Van Paasschen, W. H., and Swanson, E.: A home peritoneal dialysate delivery system. Trans. Am. Soc. Artif. Intern. Organs **15:**103, 1969.

Routine urine examination

43. Kark, R., Lawrence, J., Pollak, V., Pirani, C., Muehrcke, R., and Silva, H.: A primer of urinalysis, ed. 2, New York, 1963, Harper & Row, Publishers.
44. Peters, J. P., and Van Slyke, D. D.: Quantitative clinical chemistry, vol. 2, Baltimore, 1956, The Williams & Wilkins Co.

11

Gastrointestinal system

The gastrointestinal tract is the fibromuscular tube that originates in the mouth and ends with the anus. Through this portal foodstuffs, fluids, and orally administered substances enter the body. These may be digested and/or absorbed.

DIGESTION AND ABSORPTION

Digestion is generally defined as the process of breaking down food so that it can be absorbed through the lining of the intestinal tract. The human digestive system performs three basic functions: ingestion, digestion, and absorption.

Ingestion is the basic mechanics of taking food into the mouth, grinding and breaking the material into smaller bits, and passing the material on by swallowing. Assisting the masticated material on its way are secretions of the parotid, submaxillary, sublingual, and buccal glands that produce about 1.5 L of saliva each day. There are two types of saliva: a very fluid material that aids in diluting and dissolving dry food and a more viscous fluid that contains mucus for lubricating food as it moves down the esophagus. The saliva acts also as a lubricant in the mouth, preventing the tongue from adhering to the roof and aiding in proper speech.

Saliva contains the digestive enzyme ptylalin (salivary α-amylase) which attacks starch molecules. Ptyalin functions well at the pH of the saliva (6.5 to 6.8) but is inactive at the acid condition of the stomach. However, food acts as a buffer in the stomach allowing ptyalin action until the acid of the stomach has worked through the mass of food.

Muscle contraction (peristalsis) in the wall of the esophagus and throughout the gastrointestinal tract moves the food (bolus) through the system. The two muscle layers taking part in the peristaltic action are an inner layer, which acts as a constrictor by contracting the muscle, and an outer longitudinal layer, which is used for propulsion of the material. The consistency of the food materials determines the rate of travel to the stomach. Solid masses move at the slowest rate, leaving smaller particles to be picked up by the next peristaltic wave. A homogenized mass requires about 6 seconds to reach the stomach from the mouth.

The temperature of liquids has a distinctive effect upon the esophageal peristalsis. Drinking a liquid at a low temperature between 0.5° to 3° C will drastically cut back or even stop the peristaltic action. However, the esophageal sphincter muscle will contract and relax with each swallow even though there is little to no peristaltic activity above the sphincter. This often results in an accumulation of liquid above the sphincter. The relaxation is delayed with a followup of an exaggerated contraction. Drinking liquids of a higher temperature increases the esophageal process accordingly, and there is less dramatic effects of contraction and relaxation during peristalsis.

Digestion takes place in the stomach, which is a distensible thick-walled muscular sac having a maximum capacity of approximately 2.5 L for the adult human. The stomach wall is lined with thick mucosa arranged in rugal or pleatlike folds. The cells in this mucosa fall within three main types: goblet or mucus-secreting cells, which act as a protective coating for the mucosa and as a lubricating agent for the food particles; chief (peptic) cells, which produce the protein-splitting enzyme pepsin; and parietal (oxyntic) cells, which produce hydrochloric acid, glycoprotein, or mucoprotein (parietal cells). Rennin, a milk-clotting enzyme (found in young mammals and possibly in infants), involves a partial hydrolysis of the milk protein (caseinogen) to a soluble casein. In the presence of Ca^{++}, the soluble casein forms a coagulum of calcium caseinate that is more digestible. The fat-splitting enzyme lipase shows little activity in the acid reaction of the stomach contents. However, it is not destroyed and can become active in a more suitable pH of the duodenal contents. Infants may have some activity.

After the contracting action of the stomach, the food material has the consistency of a cream and the digestion is well advanced. The proteins of the food have been partly broken down to peptides (molecular weight of about 1000), and starches have been broken down to smaller units by the ptyalin or amylase. However, the simple carbohydrates (mono- and disaccharides) and fats are unchanged. The digested food from the stomach passes through the pylorus in regular spurts into the duodenum as chyme. This requires from 1 to 4 hours, depending on the volume and type of food intake. The sequence in which food types leave the stomach is carbohydrates, proteins, fats. Hunger pangs result from the stomach contracting when empty.

The duodenum is the first segment (about 20 cm) of the small intestine. While the small intestine and most of the large intestine are mobile because of their mesentery connection to the dorsal wall, the duodenum is tightly secured in the abdominal cavity by ligaments that connect it to the liver, stomach, and dorsal wall. The major part of digestion and absorption occurs in the duodenum and jejunum, while the undigested and unabsorbed residues go from the ileum to the colon. In the duodenum the chyme is mixed with digestive juices from the pancreas and with bile from the gallbladder. The principal digestive enzymes from the pancreas are trypsin and chymotrypsin, which complete the breakdown of proteins to peptides; peptidases, which convert peptides to amino acids; lipase, which breaks down fats; and amylase, which finishes the conversion of starch to maltose. The alkaline (pH approximately 8.5) pancreatic secretions aid in adjusting the pH of the chyme to provide an optimum pH for action of the enzymes. About 500 ml of pancreatic juice is produced daily.

The liver is important to digestion, since it contributes bile to the digestive juice. The secretion excreted from the liver contains water, bile salts, bile pigments, inorganic salts, and a mixture of lipid materials including fats (3 g/L neutral fat), cholesterol (0.6 to 1.7 g/L unesterified cholesterol), and lecithin (1 to 5 g/L).

Bile is formed in all parts of the liver and is emptied into interlobular bile ducts which form larger ducts until two main ducts unite in the portal fossa and form the hepatic duct. The hepatic duct passes downward and joins the duct from the gallbladder, the cystic duct. The hepatic and cystic ducts form the common bile duct that passes downward and enters the duodenum below the pylorus. Bile is stored in the gallbladder until it is needed and there it is concentrated by removal of water and salts. Although the bile is secreted constantly, it passes to the duodenum only after food has been eaten. The muscular wall of the gallbladder contracts, passing the bile down the duct. Being alkaline, bile aids in the neutralization of the chyme.

The bile contributes to digestion by the action of the bile salts, since it is not a digestive enzyme. The bile salts act as emulsifying agent so that fat droplets within the chyme are formed during the peristaltic motion of the intestine. With the fat broken into droplets greater surface area is provided for action of the lipase (fat-splitting enzyme) in effectively digesting the fat. When the bile duct is obstructed, the absence of bile salts in the intestine impairs digestion of the fat and causes much fat to be lost in the feces. The bile salts are reabsorbed from the lower part of the intestine and returned to the liver.

Absorption of products of digestion is the main activity in the small intestine. However, there are present some amylase, lipase, and enzymes that break down disaccharides to monosaccharides. Food absorption is a complex process involving in part the diffusion of substances from the intestine through the lining cells into the blood and lymph capillaries.

Most of the food is absorbed in the small intestine and the residue is moved into the large intestine. Since most of the water is absorbed, the waste material takes on a solid form. During this period bacterial activity occurs. By fermentation and putrefaction various gases (carbon dioxide, hydrogen, nitrogen, methane, hydrogen sulfide) and acids (acetic, butyric, lactic) are produced. The large intestine contains a significant amount of ammonia, presumed to be a product of the putrefaction of nitrogenous substances by bacteria. Bacteria decomposing lecithin may produce choline and related toxic amines such as neurine.

From 12 to 24 hours is required for the waste to pass through the colon and rectum. The resulting product, called feces, contains large quantities of bacteria (approximately half of the mass), heavy metals, bile pigments and other substances secreted by the body, and indigestible remnants of food.

GASTRIC SECRETION

Cells of the gastric glands secrete daily over 3 L of gastric juice, the composition of which varies with the physiologic condition of the individual. In health, the major constituents of gastric secretion consist of (1) hydrochloric acid, secreted by the parietal cells of the glands, which aids digestion by providing acidity required for production of pepsin and which hydrolyzes polypeptides and disaccharides; (2) pepsinogens (digestive enzymes), from the zymogen granule–containing chief or peptic cells of the stomach, which produces three pepsins (I, II, and III) that hydrolyze the bonds between aromatic amino acids and a second amino acid to yield polypeptides of different sizes; and (3) mucus (mucopolysaccharides and mucoproteins), secreted by glands in the pyloric region and the neck cells of the glands in the rest of the stomach, which appears to function in a protective manner. A mucoprotein secreted by the glands of the body and fundus of stomach, intrinsic factor, is required for the action of vitamin B_{12}. Other substances present in gastric secretion are electrolytes (sodium, potassium, calcium, chloride, bicarbonate, phosphate), the digestive enzymes rennin and lipase, nondigestive enzymes (lactic dehydrogenase, GOT, GPT, isocitric dehydrogenase, leucine aminopeptidase, β-glucuronidase, alkaline phosphatase, ribonuclease), and small amounts of serum albumin and gamma globulin.

In disease, the gastric contents may contain blood, bacteria, and lactic acid. Although various tests have been used for gastric analysis, the measurement most frequently utilized is that for the hydrochloric acid, since it has continually proved to be simple and sensitive to gastrointestinal disease. Measurement of the pH reflects the ability of the stomach to secrete hydrochloric acid.

Regulation of gastric secretion

Gastric secretion is produced in response to psychic factors (sight, taste, smell) mediated by the vagus nerve, to gastrin, and to some products of digestion in the intestine. The small intestine acts as an inhibitor of gastric secretion. The major part of digestion and absorption occurs in the small intestine which includes the duodenum, jejunum, and ileum.

GASTROINTESTINAL HORMONES

Several of the gastrointestinal hormones have been isolated and synthesized. These include gastrin, secretin, and cholecystokinin-pancreaozymin. They are polypeptides produced in the mucosal endocrine cells of the stomach and small intestine. They promote stimulation, inhibition, and regulation of the motor and secretory activities of the stomach, small intestine, liver, biliary tract, and pancreas.

Gastrin

Gastrin is produced by cells in the lateral walls of the glands in the antral portion of the gastric mucosa. Its release is stimulated by insulin-induced hypoglycemia, antral distention, and protein and polypeptides in the stomach (after eating). The release is inhibited by overdistention of the antrum and by acid in the antrum (indicating a negative feedback control mechanism with gastrin and hydrochloric acid functioning as a closed loop).

Approximately equal amounts of two gastrins are secreted, the structures being the same except that gastrin II contains a —SO_3H group on the tyrosine residue in the 12 position and gastrin I lacks this group. Both forms are equally active. A biologically active "big gastrin," which consists of gastrin bound to an alpha globulin, has been found in blood. Gastrin is the most potent known stimulant of gastric acid secretion. It also stimulates gastric mucosa to secrete pepsin and intrinsic factor, pancreas and liver to secrete water and bicarbonate, and small intestinal mucosa to release secretin.

Circulating gastrin is measured in serum by radioimmunoassay techniques. Normal fasting serum concentrations are less than 155 pg/ml with average values about 70 pg/ml. The values tend to be lowest during early morning hours and highest during the day.

Serum gastrin concentration is increased in patients with pernicious anemia and Zollinger-Ellison syndrome (ulcerogenic islet cell tumor). In patients with Zollinger-Ellison tumors the elevated fasting gastrin does not increase after eating but markedly increases after intravenous infusion of calcium. In normal individuals and patients with peptic ulcer disease, feeding a test meal causes an increase in serum gastrin concentrations. In duodenal ulcer, results are not uniform; they may be either normal or slightly increased. In response to eating, however, serum gastrin levels of persons with duodenal ulcer are higher than for normal individuals. Intragastric accumulation of acid to a pH of 1.5 or lower inhibits gastrin release. An inverse relationship exists between gastric acid secretion and serum gastrin concentration. There is no general trend of serum gastrin elevation in gastric ulcer patients.

Secretin

Secretin is secreted by cells in the glands of the mucosa of the upper portion of the small intestine. In structure it resembles glucagon. Acid is the only known stimulant to release of secretin. Secretin promotes secretion of bicarbonate and water in the intrahepatic and extrahepatic biliary ducts and of pepsin by the stomach. It inhibits gastric acid secretion and, smooth muscle contraction in the stomach, small and large intestines, and lower esophageal sphincter. Secretin increases cardiac output and blood flow to the pancreas and the small intestine and decreases hepatic arterial blood flow.

A radioimmunoassay method has been reported, but absolute values for serum secretin are not available. Increased levels have been reported following continuous intraduodenal infusion of hydrochloric acid and following a meal.

Cholecystokinin-pancreozymin

Cholecystokinin-pancreozymin (CCK-PZ) stimulates gallbladder contraction and pancreatic enzyme secretion. Its functions are similar to those of gastrin and secretin, but it can also inhibit gastric activity.

CCK-PZ is secreted from the mucosa of the upper small intestine in response to intraluminal acid, amino acids, and fatty acids. In patients with celiac sprue there is decreased secretion of the hormone.

INTRINSIC FACTOR

Intrinsic factor, a mucoprotein secreted by the gastric mucosa, is necessary for cyanocobalamin (vitamin B_{12}) absorption from the small intestine. Cyanocobalamin is a complex cobalt-containing vitamin required for normal erythropoiesis. When insufficient absorption of vitamin B_{12} occurs, it causes an anemia characterized by a number of megaloblasts in the blood. In megaloblastic anemia, failure to absorb the needed quantity of vitamin B_{12} may be caused by a dietary deficiency, conditions such as sprue and celiac disease, and malabsorption secondary to intestinal disorders (hepatic or biliary obstruction and pancreatic lesions). In the absence of liver disease and chronic leukemia, the serum level of vitamin B_{12} is a reflection of the status of the vitamin. There is an increase in serum B_{12} levels when there is hepatocellular damage and in acute and chronic liver disease. In pernicious anemia there is intrinsic factor deficiency resulting from idiopathic atrophy of the gastric mucosa.

Folic acid deficiency is more often found than vitamin B_{12} deficiency in conditions causing malabsorption. Deficiency of folic acid or vitamin B_{12} is suspected when blood or bone marrow shows characteristic patterns of megaloblastic anemia. In megalobastic anemia of infants and in pregnancy, the cause is usually folate deficiency. Serum folate levels aid in determining etiologic diagnosis of megaloblastic anemia and for evaluation of malnutrition.

Radioimmunoassay techniques are used for measuring serum vitamin B_{12} and folate levels. The normal range for serum vitamin B_{12} is 200 to 1000 pg/ml and for serum folate 4 to 16 ng/ml.

GASTRIC ANALYSIS

Measurement of gastric acid is useful in diagnosis of Zollinger-Ellison syndrome and suspected carcinoma or atrophic gastritis and for indication of completeness of vagotomy. In determining rates of gastric acid production, the two procedures in use are: (1) the basal secretion, obtained the morning after an overnight fast in an unstimulated stomach, which measures vagal and hormonal factors acting on the gastric mucosa, and (2) maximal acid output for a patient previously given a parenteral antihistaminic drug and then an amount of parenteral histamine or betazole (Histalog) in an effort to obtain maximal stimulation of parietal cells.

Basal gastric secretion

Following an overnight fast and with an unstimulated stomach, the patient is intubated and the first aspirate discarded. Continuous aspiration is begun and four 15-minute samples are obtained over a period of an hour. For each sample the volume (ml), pH, and titratable acidity are measured. The total acid output for the 1-hour period is expressed in milliequivalents.

The basal gastric acid output for normal males varies between 1.3 and 4.0 mEq/hr with slightly lower values for females. The volume for fasting specimens is about 30 to 70 ml/hr with a pH between 1.6 and 1.8. Basal gastric acid concentrations are usually lower than normal for gastric carcinoma and benign gastric ulcer and are increased for duodenal ulcer or jejunal ulcer following partial gastrectomy with gastrojejunostomy. In Zollinger-Ellison syndrome a high acid output is found.

Maximum acid output

If the augmented histamine test follows the basal secretion study, an antihistaminic drug to control side effects of histamine is administered 30 minutes prior to completion of the collections for the basal secretion test. When the latter test is completed, histamine acid phosphate is administered subcutaneously and gastric contents are collected as four 15-minute samples for a period of 1 hour. When Histalog is used, it is not necessary to administer an antihistaminic drug and eight 15-minute samples are obtained.

Normally, the maximum rate of acid secretion is reached within 15 minutes after histamine administration, maintained for about 30 minutes, and returned to baseline by the end of an hour. The normal range of maximal acid output is wide, although most values are between 20 and 25 mEq/hr. Hypersecretion may be found in cases of duodenal ulcer and is characteristic of Zollinger-Ellison syndrome. Anacidity is usually noted in adults with pernicious anemia and in some cases of gastric carcinoma.

Response to Histalog is similar to that of histamine except that there is a longer (30 to 75 minutes) latency period for peak secretion and the period of duration is extended (45 to 90 minutes).

Titratable acidity

Titratable acidity is determined by titration of an aliquot (5 ml) of the gastric secretion with 0.1 N NaOH to a pH of 7.0 or 7.4 using a pH meter or to a colorimetric end-point with phenol red indicator (color change, yellow to red; pH 6.8 to 8.4).

$$\text{Titratable acidity (mEq/L)} = \text{ml NaOH} \times 20$$

$$\text{Acid output (mEq)} = \frac{\text{Titratable acidity} \times \text{Specimen volume (ml)}}{1000}$$

Microscopic examination

Make a smear of the fasting specimen. Allow to dry and stain with Gram's stain. Examine for Boas-Oppler bacilli and sarcinae. Discussion follows.

Boas-Oppler bacilli

Boas-Oppler organisms are large gram-positive rods (Fig. 11-1). They are found in cancer of the stomach but must be present in large numbers to be of any significance. The presence of the bacilli indicates gastric stagnation.

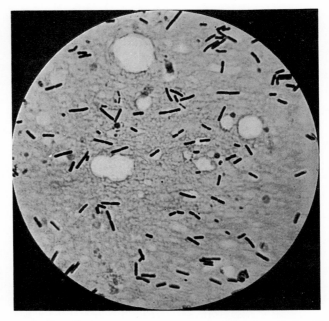

Fig. 11-1. Boas-Oppler bacilli. (From Levinson, S. A., and MacFate, R. P.: Clinical laboratory diagnosis, ed. 5, Philadelphia, 1956, Lea & Febiger.)

Sarcinae

Sarcinae are large gram-positive cocci arranged in groups that look like bales of cotton. They may be found in ulcers but must be present in large numbers to be of any significance. Their presence also indicates gastric stagnation.

Bile

Small amounts of bile may be normally present in the gastric contents. Large amounts give the stomach secretions a yellow or green color. This usually indicates an obstruction at the beginning of the small intestine.

Procedure
1. Pour 5 ml of concentrated nitric acid into a small test tube.
2. Tilt the tube to about a 30-degree angle and *carefully* overlay the acid with the unfiltered gastric contents.

Reaction
If bile is present, a display of colors—yellow, green, red—will form at the junction of the two liquids.

DUODENAL ANALYSIS

The normal duodenal contents are colorless. Upon stimulation of the gallbladder, however, bile flows into the duodenum and gives the contents a brown, yellow, or green color. If bile fails to flow, the presence of gallstones may be suspected.

The samples of duodenal contents are usually obtained by the physician and brought to the laboratory. The method of obtaining the specimens is similar to the procedure in a gastric analysis except that the tube is allowed to pass into the duodenum. The gallbladder is then stimulated with a solution of magnesium sulfate.

An analysis of the duodenal contents consists of tests for amylase, trypsin, and bile and a microscopic examination. The procedures follow.

Amylase

The amylase of the duodenal contents is referred to as amylopsin or pancreatic amylase. The normal values depend upon the method of determination. For the procedure of Myers and Fine, which follows, the normal values are 5 to 200 units. Methods for determining serum amylase may be used after proper dilution of the duodenal contents. Decreased values are found in chronic pancreatitis, pancreatic insufficiency, and fibrocystic disease of the pancreas.

Reagents

Sodium carbonate, 1%	(D-215)
Starch, 1%	(D-216)
Gram's iodine solution	(D-217)

Procedure (Myers-Fine)

1. Test the duodenal contents with nitrazine paper. If acid, add the 1% sodium carbonate dropwise until it is alkaline.
2. Label six large test tubes 1 to 6 for a serial dilution of the duodenal contents.
3. Place 1.0 ml of duodenal fluid in tubes 1 and 2. Pipet 1.0 ml of distilled water into all tubes except tube 1. Mix the contents of tube 2 by drawing them into a 1-ml pipet and ejecting them several times. Transfer 1.0 ml of the contents of tube 2 to tube 3. Mix the contents of tube 3 and transfer 1.0 ml to tube 4. Continue this process through the remaining tubes, discarding 1.0 ml from tube 6. The number of milliliters of duodenal fluid now in each tube is given below:

Tube	1	2	3	4	5	6
ml	1.0	0.5	0.25	0.125	0.062	0.031

4. Add 5.0 ml of the 1% starch solution to each tube. Mix. Incubate at 37° C for exactly 30 minutes.
5. Add 10 ml of cold water and 2 drops of Gram's iodine to each tube. Mix. Record the last tube in which there is a complete disappearance of blue color.

Calculation

By definition, the amylase activity equals the number of milliliters of starch solution that will be digested by 1.0 ml of duodenal contents under the conditions of the above test. Therefore:

$$\text{Amylase activity} = \frac{1.00}{\text{ml used}} \times 5$$

where 5 = number of milliliters of starch solution digested
ml = number of milliliters of duodenal contents in the last tube in which there is a complete disappearance of blue color

Example: If tube 3 is the last tube in which there is a complete disappearance of blue color, the calculation is made as follows. As shown in step 3, tube 3 contains 0.25 ml of the duodenal contents. Therefore:

$$\text{Amylase activity} = \frac{1.00}{0.25} \times 5$$
$$= 20$$

Trypsin

Trypsin is an enzyme produced by the pancreas and used to digest protein. It is normally present in the duodenal fluid. The normal values depend on the method of determination. For the procedure of Gross, which follows, the average tryptic activity is 2.5. Decreased values are found in chronic pancreatitis, pancreatic insufficiency, and fibrocystic disease of the pancreas.

Reagent

Acetic acid, 1% (D-218)

Procedure (Gross)

1. Prepare a casein solution by dissolving 1 g of pure casein in 1 L of 0.1% sodium carbonate. Add 1 ml of chloroform to prevent bacterial decomposition.
2. Heat 100 ml casein solution to 40° C and pipet 10 ml into each of ten test tubes.
3. Centrifuge the duodenal fluid and add supernatant in following amounts:

Tube	1	2	3	4	5	6	7	8	9	10
ml	0.1	0.2	0.3	0.4	0.5	0.6	0.7	0.8	0.9	1.0

4. Mix the contents of each tube and place in a 37° C incubator for exactly 15 minutes.
5. Remove the tubes from the incubator and add 3 drops of 1% acetic acid to each tube, noting whether the addition of the acid causes an increased turbidity. Record the first tube which shows *no* increased turbidity upon the addition of the acid. This is the first tube in which the casein has been completely digested by the trypsin. Calculate the tryptic activity as follows.

Calculation

By definition, the tryptic activity is expressed as the effect of 1.0 ml of duodenal fluid on 10 ml of a 0.1% casein solution under the conditions of the above test. Thus:

$$\text{Tryptic activity} = \frac{1}{\text{ml used for digestion}}$$

Example: Tube 4 was the first tube which showed no increased turbidity upon addition of the 1% acetic acid. As noted in step 4, tube 4 contains 0.4 ml of duodenal fluid. Therefore:

$$\text{Tryptic activity} = \frac{1}{\text{ml used for digestion}}$$
$$= \frac{1}{0.4}$$
$$= 2.5$$

Bile

Bilirubin is the bile pigment normally present in duodenal fluid. Its absence often indicates an obstruction in the hepatic duct or common bile duct.

Procedure

1. Bile gives the duodenal fluid a yellow or green color. Note the color and report a 1+, 2+, 3+, or 4+, depending on the depth of color.

Microscopic examination

A microscopic examination of the duodenal fluid may give early warning of inflammation in the duodenum or biliary tract and also of the presence of parasites.

Procedure

1. Centrifuge a fresh sample of duodenal fluid and place the sediment on a glass slide.
2. Look for an abnormal amount of pus cells, which indicates an inflammatory condition. A few cells are normally present.
3. Also look for the cysts of amebae, the eggs of hookworms, and parasites such as *Giardia lamblia* and *Strongyloides stercoralis*.

FECAL ANALYSIS

Feces are waste products left in the wake of intestinal digestion and absorption. The average normal output of feces is 200 g/24 hr. In disease, deviations from the normal composition may occur. A few qualitative tests follow.

Occult blood

Occult blood may be found in the stool of a person having ulcers or cancer. The patient should be on a meat-free diet for 4 days prior to the test to eliminate a false positive reaction. The following tests may be used to test for occult blood.

Reagents for occult blood tests

Benzidine, saturated, in glacial acetic acid   (D-219)
Hydrogen peroxide, 3%   (May be purchased)
Glacial acetic acid   (May be purchased)
Ethyl alcohol, 95%
o-Tolidine, 4%   (Make fresh monthly)
  Using a volumetric flask, dissolve 4 g of o-tolidine in 100 ml of glacial acetic acid. Keep in brown bottle and refrigerate.
Hematest tablets

Benzidine test

Procedure

1. Place a piece of feces about the size of a pea on a spot plate.
2. Add a few drops of a saturated solution of benzidine in glacial acetic acid and mix with the feces.
3. Add a few drops of 3% hydrogen peroxide. Mix.

Reaction

A blue to green color represents a positive reaction. Disregard any other color. Report the test as positive or negative.

Guaiac test

Additional reagent needed

Guaiac reagent   (Prepare just prior to use)
  Into a test tube, place 2 ml of 95% ethyl alcohol and 0.2 g of powdered guaiac. Add 2 ml of 3% hydrogen peroxide. Mix.

Procedure

1. Place a piece of feces about the size of a pea on a spot plate.
2. Add 2 drops of glacial acetic acid and mix with the feces.
3. Add a few drops of the guaiac reagent. Mix.

Reaction

A blue to green color is a positive reaction. Disregard any other colors. Report the test as positive or negative.

o-Tolidine test

Procedure

1. Place a piece of feces the size of a pea on a spot plate.
2. Add a few drops of 3% hydrogen peroxide and mix with the feces.
3. Add a few drops of o-tolidine reagent. Mix.

Reaction

A blue to green color is a positive reaction. Disregard any other colors. Report the test as positive or negative.

Hematest*

Hematest, a tablet for detecting occult blood in stool, contains calcium acetate, tartaric acid, and strontium peroxide (which produce hydrogen peroxide when moistened) and o-tolidine as the chromogen.

Procedure

1. Place a small streak of solid feces on the special filter paper square supplied with the product. Do not use an emulsion of stool.
2. Place Hematest tablet on streak of stool.
3. Add 1 drop of water on tablet, wait 5 to 10 seconds, and flow a second drop on tablet allowing it to run down the sides onto the filter paper.

Reaction

Negative—If within 2 minutes color does not appear on the moistened filter paper square around the tablet, the test is considered negative.

Positive—A blue color appearing on the filter paper around the tablet within 2 minutes indicates a positive reaction. The amount of color and the time of its appearance provide a rough indication of the amount of blood in the stool.

Bile pigment

Bile pigments are not normally present in feces but may be found in diarrhea and obstructive jaundice.

Gmelin test

Reagents

Barium chloride, 10% (D-220)
Nitric acid, concentrated

Procedure

1. Using an evaporating dish or mortar, mix 7 ml of 10% barium chloride with a piece of feces about the size of a marble.
2. After an emulsion is made, filter the specimen.
3. When the funnel is completely drained, add a few drops of concentrated nitric acid to the filter paper.

Reaction

A display of colors, one of which must be green, is a positive reaction. Report the test as positive or negative.

---

*Ames Co.

Urobilin

Urobilin is normally present in feces. However, it is usually absent in obstructive jaundice.

Schmidt qualitative test

Reagent

Mercuric chloride, 10%   (D-221)

Procedure

Place a piece of feces in an evaporating dish or mortar and mix thoroughly with 2 ml of 10% mercuric chloride and let the mixture stand overnight.

Reaction

If urobilin is present, a deep red color will be evident after 8 hours.

Fecal lipids[1,2]

Normally the lipids of the feces have little to do with the lipids of the diet. They resemble the lipids of the blood and are undoubtedly secreted by the small intestine. Abnormal increase in the lipid is due to the food lipids, not those secreted by the liver or intestinal mucosa. Such increases may result from blockage of the bile ducts or pancreatic ducts, failure of the pancreas to secrete pancreatic juice, or malabsorption. Malabsorption results when there is increased motility of the upper intestine and the food rushes through too rapidly. When the feces contain large amounts of fat, fatty acids, and soaps, the condition is called steatorrhea.

The term "malabsorption syndrome" is commonly used to describe a condition wherein absorption of the main nutrients is defective. This usually results in the passage of bulky, pale, fluid or semifluid, offensive, fatty stools. There are five main groups of disorders in which a malabsorption syndrome may develop. One group results from surgical operations on the gastrointestinal tract; a second is associated with constitutional diseases, such as diabetes; a third is due to granulomatous or neoplastic disease of the intestines; a fourth group is attributable to faulty preparation of food for absorption in the intestinal lumen; and a fifth group, which may be termed the "sprue group," includes such clinical conditions as celiac disease, idiopathic steatorrhea (nontropical sprue), and tropical sprue.

The estimation of fecal fat output is an important aid in diagnosing a malabsorption syndrome.

DETERMINATION OF FAT IN FECES
van de Kamer modified method[3,4]

An aliquot of a 72-hour collection of feces is saponified with concentrated potassium hydroxide in ethyl alcohol. This solution contains the soaps that were originally present in the stool plus those derived from the neutral fats and fatty acids. Hydrochloric acid is added to the solution to liberate the fatty acids. They are then extracted with petroleum ether. Ethyl alcohol is added to an aliquot of the petroleum ether layer, and the fatty acids are titrated with a standard solution of potassium hydroxide. Thymol blue is used as an indicator.

Preparation of patient and collection of feces

1. The patient should be on a known fat intake of at least 50 g/day of fat. If lower quantities are consumed, the fat in the feces originating from bacteria, intestinal cells, or bile (a total of 1 to 2 g is excreted every 24 hours) will influence the results.

2. The patient is given a carmine red dye capsule at the beginning of a 72-hour period and another at the end of 72 hours.
3. The collection of the stool should begin when the first red dye marker appears. All stool is then collected in the preweighed jar (2 or 4 quarts) until the second red dye marker appears. This may take more or less than 72 hours depending on intestinal motility.
4. Refrigerate the jar during the collection.

Reagents

Ethyl alcohol, 95%, containing 0.4% capryl alcohol	(D-222)
Ethyl alcohol, 95%, neutral to thymol blue	
Potassium hydroxide, KOH, 33%	(D-223)
Hydrochloric acid, HCl, 25% (sp. gr. 1.13)	(D-224)
Petroleum ether, boiling point 90° to 100° C	
Sodium hydroxide, 1 N	(D-225)
Thymol blue, 0.2% in 50% ethyl alcohol	(D-226)

Special equipment

1. Large mixer* to homogenize the entire 72-hour stool collection
2. Infrared lamp heater
3. Round-bottom centrifuge bottles, 250-ml capacity
4. Centrifuge carrier to hold the 250-ml bottles
5. Mason-type glass jars, 2- and 4-quart capacity
6. Microburet, 1.0-ml capacity

Procedure

1. Weigh the jar containing the stool collection and record the weight.

    Weight of stool = Total weight − Weight of empty jar

2. If the stool is formed (not watery), add enough water to equal about two-thirds the weight of the stool. Weigh the jar and record:

    Weight of blended stool = Total weight − Weight of empty jar

3. Blend until the stool and water are thoroughly homogenized.
4. Weigh two 250-ml round-bottom centrifuge bottles for duplicate analysis.
5. Pipet about 10 g of the homogenized feces into each 250-ml centrifuge bottle. Reweigh the bottles:

    Weight of blended sample = Total weight − Weight of empty bottle

6. To each bottle add:
    a. 10 ml of 33% KOH
    b. 40 ml of 95% ethyl alcohol containing 0.4% capryl alcohol
    c. 10 to 12 glass beads
7. Attach an air condenser to each bottle and boil for 30 minutes over an infrared lamp. (A 25-ml volumetric pipet with the mouthpiece inserted through a cork serves well as a condenser.)
8. Cool in a room temperature water bath.
9. Add 17 ml of 25% HCl and cool again in the water bath.
10. Add 50 ml of petroleum ether (boiling point: 90° to 100° C).
11. Stopper the bottles and shake vigorously for 1 minute.
12. Centrifuge at moderate speed for 5 minutes or allow the bottles to stand until the layers separate. (The concentration of the ethanol is such that,

---

*May be obtained from Ivan Sorvall, Inc. This mixer has an adapter to fit mason-type collection jars. The total specimen may be homogenized in its collection jar.

after the mixture has been shaken, the petroleum ether and the acid ethanol layers separate rapidly.)
13. Transfer (use volumetric pipet) 25 ml of the petroleum ether layer to a 50-ml beaker or Erlenmeyer flask.
14. Add 3 or 4 glass beads (or small piece of filter paper) and evaporate the petroleum ether. Do not evaporate to dryness or char the residue. When the ether layer is gone, the odor will no longer be evident.
15. Add 10 ml of neutral 95% ethyl alcohol.
16. Titrate the fatty acids with 1.00 N NaOH from a microburet, using thymol blue as an indicator (4 drops).

Calculation

The calculations are based upon the assumption of an average molecular weight of 265 for fatty acids in stool:

A = Milliliters of 1.00 N NaOH used in titration
B = Total weight (in grams) of blended stool
S = Grams of blended feces taken for analysis

$$\frac{A \times 265 \times 1.04 \times 2 \times B}{1000 \times S} =$$

$$\frac{A \times 0.551 \times B}{S} = \text{Grams of fat in 72-hour stool}$$

The correction factor 1.04 must be used, since the petroleum ether layer increases 1% in volume when shaken with alcoholic hydrochloric acid and 3% of the amount of fatty acids remains in solution in the acid alcohol layer.

Precautions

1. Use either glass beads or a small piece of filter paper to prevent bumping (irregular boiling). Do not use pumice, which absorbs fatty acids.
2. It is not necessary to evaporate the petroleum ether quantitatively, since small amounts do not affect the titration.
3. The extraction is complete after 1 minute. Excessive shaking, beyond a 1-minute period, does not extract more fat.
4. The molecular weight of the fatty acids depends on the kind of fat consumed. For general purposes (both in cases of normal fat excretion and in steatorrhea) 265 was chosen, with palmitic acid being the main component.
5. Weigh and label the jars before sending any to the patient.
6. Weighings must be accurate. If accurate technique is used, the method has an accuracy of ± 2%.

Normal values

Normal values are expressed as percent of fat retained or "% retention." For normal, healthy adults, a fat retention of 95% to 98% is found. For example, if a patient has ingested 150 g of fat during the 3-day stool collection period, then a total stool fat of 3 to 7.5 g would be considered normal.

References

1. Frazer, A. C.: The malabsorption syndrome, with special reference to the effects of wheat gluten. In Sobotka, H., and Stewart, C. P., editors: Advances in clinical chemistry, vol. 5, New York, 1962, Academic Press, Inc.
2. Orten, J. M., and Neuhaus, O. W.: Human biochemistry, ed. 9, St. Louis, 1975, The C. V. Mosby Co.
3. van de Kamer, J. H.: Total fatty acids in stool. In Seligson, D., editor: Standard methods in clinical chemistry, vol. 2, New York, 1958, Academic Press, Inc.
4. van de Kamer, J. H., ten Bokkel Huinink, H., and Weijers, H. A.: Rapid method for the determination of fat in feces, J. Biol. Chem. **177:** 347, 1949.

12

Cerebrospinal fluid

Cerebrospinal fluid is a clear colorless fluid having a specific gravity of 1.004 to 1.008. The fluid, being chiefly a dialysate from the blood, maintains a pH of 7.35 to 7.40. The protein content is extremely low, with no fibrinogen; the glucose is approximately two thirds that of the blood sugar; and the chloride is about 25% higher than the plasma chloride. The total volume of cerebrospinal fluid is between 120 and 150 ml.

The brain, which is the largest and most complex mass of nervous tissue in the body, may be regarded as the elaborate end of the cerebrospinal axis, occupying the entire cranial cavity. A narrow central canal that extends the length of the spinal cord is continued anteriorly into the brain where it widens out to form the ventricles. There are five ventricles, one of which is not a true ventricle, since it has no connection to the ventricular system. The two lateral ventricles are located one in each of the cerebral hemispheres, under the corpus callosum that connects the two hemispheres. The cavity of the lateral ventricles is large and may become overdistended with cerebrospinal fluid in certain pathologic conditions. The third ventricle, located behind the lateral ventricles, is connected with them by small openings and with the fourth ventricle by a slender canal, the cerebral aqueduct. The fourth ventricle is located in front of the cerebellum, behind the pons varolii and medulla oblongata.

The spinal cord and brain are covered by three layers of connective tissue, known as meninges. One of these, the dura mater, is the outermost tough membrane of fibrous connective tissue. The cranial dura mater is arranged in two layers, the outer layer adhering to the bones of the skull and the inner one covering the brain. Except where they separate to form sinuses for the passage of venous blood, these layers are closely connected. The spinal dura mater is a loosely applied membrane fastened to the bony neural arches of the vertebrae. The cranial and spinal dura mater form one complete membrane.

The pia mater, the layer that is innermost and closely applied to the brain and cord, is a vascular covering and nutritive membrane. In the brain it is attached to the cerebral cortex, passing into various fissures and protruding into the ventricles. In the spinal column this delicate membrane (containing many small blood vessels) is attached to the surface of the spinal cord.

Between the dura mater and the pia mater is the arachnoid mater, a serous membrane consisting of a delicate network of fibrous tissue that contains many blood vessels. The network of the arachnoid is filled with spinal fluid. A bloody spinal fluid received in the laboratory may be caused by subarachnoid hemorrhage (beneath the arachnoid membrane, within the network of blood vessels and spinal fluid). Inflammation of the coverings is known as meningitis.

Cerebrospinal fluid is formed in the choroid plexuses, tufts of capillary blood vessels in the cerebral ventricles, and moves through the ventricular system and out of three small openings in the roof of the fourth ventricle into the subarachnoid cavity. It circulates over the cerebral hemispheres and downward over the spinal cord, carrying nutriment to the nerve cells and removing wastes. Spinal fluid fills the central canal, ventricles, and space between the arachnoid and pia mater. The return from the subarachnoid space is through arachnoid villi of the dural sinuses to the venous blood.

The spinal fluid has three functions:
1. To protect the brain from injury by acting as a fluid buffer.
2. To act as a medium of exchange for the transfer of dialyzable material between the bloodstream and the tissues of the brain and the spinal cord.
3. To equalize pressure between the brain and the spinal cord. The pressure usually remains constant, supporting the soft bulk of the central nervous system. However, since this pressure is closely related to the blood pressure, coughing, sneezing, or any stress that causes a rise in venous pressure will also increase the cerebrospinal fluid pressure.

COLLECTION OF CEREBROSPINAL FLUID

Spinal fluid is obtained by making a spinal puncture. This procedure is performed by a physician. The patient should lie on one side with his hips and knees drawn to arch the back. A local anesthetic is usually applied. The tap for fluid is made by the insertion of a needle about 10 cm. long, having a bore of 1 to 1.5 mm and provided with a stylet, into the small of the back between the third and fourth lumbar vertebrae.

Blood may be picked up on or in the spinal needle during passage through the tissue. Therefore, the first few drops may be discarded or placed in a tube for microbiologic examinations. The usual practice is to collect fluid in three sterile small screw-top tubes, labeled in order of collection. The first tube is delivered to the microbiology laboratory for a smear and culture; the second, to the hematology laboratory for a cell count; and the third, to the chemistry laboratory for chemical examinations. Any fluid remaining in the last tube is forwarded to the serology laboratory for specified tests.

PHYSICAL APPEARANCE OF SPINAL FLUID WITH VARIOUS DISEASES

Xanthochromia. The spinal fluid is a clear yellow and shows rapid coagulation. There are no blood pigments present, but there are large amounts of globulin. This disorder results from a tumor pressing on the spinal cord and affecting the flow of spinal fluid. A pocket of fluid collects and stagnates. Xanthochromia may appear after a cerebral hemorrhage.

Acute meningitic infection. The fluid may appear slightly to very turbid until there is an appearance of almost pure pus.

Tuberculous meningitis. The fluid is clear, although upon standing for 12 to 24 hours a pellicle will form. With many cases of acute meningitis, tuberculous meningitis, and suppurative meningitis a pellicle of fibrin and blood cells forms in the fluid when the specimen is allowed to stand.

Suppurative meningitis. Pellicle formation is extremely rapid. It is difficult to work with this type of material because of its speedy coagulation.

Spinal fluid contains glucose, protein, and chloride. In disease, the concentration of these materials may vary, and white cells and bacteria may also be present.

The composition of the fluid is usually altered in meningitis, poliomyelitis, encephalitis, and latent syphilis.

MICROSCOPIC EXAMINATION

This includes an examination for microorganisms and a count of the different types of cells. In disease, the spinal fluid may contain the following organisms: staphylococci, streptococci, meningococci, tubercle bacilli, and influenza bacilli.

These organisms may be identified by a smear and culture. The procedures follow.

Smear

Procedure
1. Centrifuge the spinal fluid at high speed for 5 minutes and discard the supernatant fluid.
2. Using a bacteriologic loop, smear some of the sediment over a glass slide.
3. Warm the bottom of the slide to "fix" the bacteria.
4. Stain with Gram's stain, using the usual gentian violet, Gram's iodine, acetone, and safranin. (If examination for tubercle bacilli is requested, make another smear and stain for AFB.)
5. Identify the organisms, using a textbook on microbiology as a guide.

Culture

The culture should be made while the spinal fluid is fresh. If this is not possible, store the fluid in the 37° C incubator. This is permissible, since spinal fluid itself is a good culture medium.

In many hospitals, the routine procedure for a spinal culture begins with streaking the fluid onto the following media: (1) blood agar plate or slant and (2) chocolate agar plate or slant. (These media are usually stored in a refrigerator and should be warmed to room temperature before use.)

Procedure
1. The sediment that was used to prepare the smear may be used to make the culture. If this sediment is not available, centrifuge a sample of the spinal fluid at high speed for 5 minutes. Discard the supernatant fluid. Use the sediment for the culture.
2. Warm blood agar plate gently over a flame.
3. Streak the plate or slant with the spinal fluid sediment and place in a 37° C incubator.
4. Warm chocolate agar plate or slant gently over a flame.
5. Streak the plate or slant with the spinal fluid sediment and place the plate or slant in the CO_2 jar. Light the candle. Place the lid on the jar so that it is tightly secured. The candle, of course, will go out. This creates an increased CO_2 tension, which aids in the growth of meningococci, place the CO_2 jar in the 37° C incubator.
6. Twenty-four hours later, examine the plates or slants for growth, make slides, and identify the organisms as directed in textbooks on microbiology.

Cell count

The spinal fluid cell count consists of the total cell count and the differential cell count. Since the cells disintegrate rapidly, the counts should be made while the fluid is fresh—preferably within an hour following the withdrawal of fluid.

Spinal fluid often contains contagious material. Consequently, take particular care in handling the specimens. In making transfers of the fluid, it is a wise policy to use medicine droppers.

Total cell count

The cells found in spinal fluid are usually white cells (leukocytes). Red cells may be introduced by the puncture and may cause a slight increase in the count. The normal values for the total cell count are 0 to 10 cells/mm^3. Increased values, such as 20 to 1000 cells, may be found in meningitis, encephalitis, poliomyelitis, and latent syphilis.

Procedure
1. Mix the spinal fluid thoroughly. Draw the spinal fluid counting solution to the 0.5 mark and the spinal fluid to mark 11 in a white cell counting pipet, making a dilution of the spinal fluid of 19:20. Mix thoroughly. Transfer a small portion (2 or 3 drops) to a counting chamber, which is generally used for counting blood cells.
2. Using either the 4-mm or the 16-mm objective, count the cells in the *entire* ruled area. This will be the nine large squares as seen with the low-power (16-mm) objective of the microscope.
3. After the count is made, the counting chamber should be soaked in alcohol to prevent contagion.

Calculation

The total area of the counting chamber consists of nine large squares or 9 mm^2. Since the counting chamber is 0.1 mm deep, this will be a volume of 0.9 mm^3 ($9 \times 0.1 = 0.9$). The report, however, is to be given as the number in 1.0 mm^3. Therefore, multiply the number counted by 1.0/0.9 (or 10/9).

Example: The number of cells counted in the entire ruled area was 9. Multiplying this by 10/9:

$$9 \times 10/9 = 10 \text{ cells/mm}^3$$

Spinal fluid counting solution (Tuerk solution)
1. Pipet 1.0 ml of glacial acetic acid into a volumetric flask and dilute to 100 ml with distilled water.
2. Add a few drops of alcoholic gentian violet to color.
3. Stopper and mix thoroughly.
4. Filter several times, until clear.

CHEMICAL EXAMINATION

The chemical constituents of the cerebrospinal fluid that have clinical significance are total protein, glucose, and chloride. However, changes in these may be the result of metabolic alterations and not necessarily a reflection of intracranial disturbances.

Protein

The protein fractions of spinal fluid are similar to those of blood serum. Fibrinogen is normally absent. In many pathologic states associated with an increase in permeability of the blood-brain barrier, such as bacterial meningitis, protein in the spinal fluid is increased. Current practice generally requires a quantitative determination of the total protein. However, since the qualitative tests for globulin will indicate the presence of abnormal increases which are always associated with pathologic changes, they are included.

GLOBULIN

One of the traditional qualitative globulin tests is the Pandy test.

Reagents

    Pandy's reagent   (D-227)

Pandy test

1. Pipet 1 ml of Pandy's reagent into a small test tube.
2. Using a medicine dropper, add a few drops of the spinal fluid.
3. If an abnormal amount of globulin is present, the solution will become *markedly* turbid.
4. Report the test as positive or negative, disregarding a faint turbidity, which may be found in normal spinal fluid.

TOTAL PROTEIN

The normal values for total protein in spinal fluid are 15 to 45 mg/dl. Increased values are found in meningitis and latent syphilis.

If red cells are present in a sample of cerebrospinal fluid, the fluid must be centrifuged and decanted before an analysis is done. Cells will elevate the values. Elevated proteins are found in most types of meningeal inflammation, latent syphilis, brain tumor, subarachnoid hemorrhage, and sometimes contamination of blood. The blood contamination results from trauma during the withdrawal of the fluid.

Meuleman's method[1]

Principle. Trichloroacetic acid is added to the spinal fluid and the protein is precipitated as a fine suspension. The turbidity of the suspension is measured in a colorimeter and compared with a standard.

Reagents

    Trichloroacetic acid, 3%   (D-229)
    Working standard

        Prepare fresh daily. Transfer 0.5 ml of serum with a known protein concentration to a 100-ml volumetric flask. Dilute to volume with saline, stopper, and mix thoroughly.

        To determine the protein concentration of this solution in mg/dl, multiply the serum protein concentration in grams per dl by 5. For example, control values of a serum standard (6.6 g/dl) × 5 = 33 mg/dl working standard solution.

Procedure

1. Set up the following tubes:

    Blank

        5.0 ml of 3% trichloroacetic acid

    Standard

        1.0 ml of working standard
        4.0 ml of 3% trichloroacetic acid

    Unknown

        1.0 ml of spinal fluid
        4.0 ml of 3% trichloroacetic acid

2. Mix and let stand for 10 minutes.
3. Remix and read absorbance (A) at 450 nm (or read % T and convert to A).

Calculations

$$\frac{\text{A of unknown}}{\text{A of standard}} \times \text{Concentration of standard} = \text{mg protein/dl}$$

Glucose

The normal values for glucose in spinal fluid are approximately two thirds that of the blood glucose, or 40 to 80 mg/dl of fluid. Increased values may accompany brain tumors. Decreased values are seen in most types of meningitis.

The determination should be made within an hour after the withdrawal of spinal fluid, because glucose decomposes upon standing. Refer to Chapter 4 for the quantitative analysis for glucose.

Chloride

The chloride is expressed in terms of sodium chloride. The normal values are 119 to 130 mEq/L of sodium chloride/dl of spinal fluid. Decreased values are found in acute meningitis, particularly in tuberculous meningitis. The test is seldom used.

Refer to Chapter 5 for the quantitative analysis for chloride.

Enzymes

A number of different enzymes have been measured in cerebrospinal fluid. The values are increased in a variety of pathologic states but the activity appears nonspecific.

Reference

1. Meuleman, O.: Determination of total protein in spinal fluid with sulfosalicylic and trichloroacetic acid, Clin. Chim. Acta 5:757, 1960.

13

Radioassay

NANCY VanDILLEN

The availability of radioisotopes in sufficient quantities for use in research and industrial processes has been possible only since World War II. The last few years have brought a snowball effect in the possible applications of radioisotopes. Laboratory methods using radioisotopes and the required instrumentation have rapidly flooded the clinical field, while proper training or understanding of the radioisotope concepts has lagged behind.

BASIC THEORY

In the fifth century B.C., two philosophers, Leucippus and Democritus, arrived at a theory of the atom not too far from the modern view: "All matter is composed of invisible, indestructible, and indivisible particles separated by voids."

John Dalton (1766-1844) developed the atomic theory. During the years 1803 to 1808, Dalton constructed the following principles:

1. Each element is composed of many invisible particles called atoms.
2. Atoms of each element are identical in mass and size.
3. Atoms of different elements have different masses.
4. In a chemical reaction atoms separate or combine in a simple integral ratio.

Dalton's four principles are still applicable. However, not all the atoms of any one element are alike in mass, size, and structure. There are atoms that differ slightly in the nucleus structure. These offweight atoms are called isotope atoms.

The atom, which is no longer considered indivisible, consists of a positively charged nucleus surrounded by negatively charged electrons. These orbital electrons, each carrying a negative charge, neutralize the atom in the various energy level arrangements. The outermost electrons are labeled valence electrons because they participate in chemical change. By gaining, losing, or sharing these electrons, atoms of one element may combine with those of another element to form molecules. The number of outer electrons determines the atom's valence or combining power.

The number of protons in the nucleus gives the atomic number of the atom. When an atom is in the neutral or uncombined state, before it takes part in a reaction, the protons are balanced by an equal number of electrons. In this state, the atom as a whole may be called a neutral, balanced structure.

This balance is upset when an atom gives up electrons. The valence is positive when the atom has an excess of positive charges and negative when the number of negative charges is greater.

The identity of an atom is determined by the charge of the nucleus. As shown in

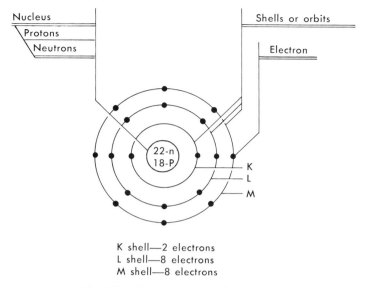

K shell—2 electrons
L shell—8 electrons
M shell—8 electrons

Fig. 13-1. Basic structure of the atom.

Fig. 13-1, an atom is made up of protons (positively charged particles) and electrons (negatively charged particles). The positively charged nucleus consists of negative neutrons and excess protons. The charge is equal to the number of protons in the nucleus and is known as the atomic number (Z). The number of neutrons is called the neutron number (N), and the mass number (A) is equal to the total of protons and neutrons (A = Z + N). Certain combinations of the nucleons (protons and neutrons) possess stability while others are relatively unstable.

Particles

The atom is made up primarily of three particles: protons, neutrons, and electrons. The proton (Greek, meaning first) was used as the elementary unit of positive charge observed in positive-ray experiments and is identical with the nucleus of the hydrogen atom.

Existence of the neutron was postulated by Rutherford (1920) and was observed but not identified by Bothe and Becker (1930) when beryllium and other light elements were bombarded by alpha particles. Chadwick (1932) finally identified the neutron.

The electron was initially (1891) used as a unit of electricity. About 6 years later, the value of the charge-to-mass ratio (e/m) for the electron was published. Studies of the charge carried by an ion in gases have been made by various investigators. J. S. Townsend concluded from his observations of electrified gas liberated by the electrolysis of oxygen that each particle carried a charge of about 3×10^{-10} esu. Measurements of the charge have been made by J. J. Thomson (3.4×10^{-10} esu), H. A. Wilson (3.1×10^{-10} esu), and Millikan and Bergeman (4.06×10^{-10} esu or 1.60203×10^{-19} coulombs) using the well-known oil drop method. The results of the various measurements lead to the conclusion that the charge carried by the alpha particle (9.3×10^{-10} esu) is between 2e and 3e. It is generally viewed that the charge e carried by the hydrogen atom is the fundamental unit of electricity; therefore, the charge carried by the alpha particle is an integral multiple of e and may be either 2e or 3e.

Radioactivity

Radioactivity is a spontaneous process characteristic of atoms and unstable nuclei whereby the nucleus releases energy either as a particle with kinetic energy or as electromagnetic energy. With the release of this energy, the nucleus may become stable or may be still in an unstable condition, requiring another energy release before acquiring stability.

Nuclear stability is governed by various combinations of protons and neutrons. The neutron/proton (n/p) ratio is stable when the ratio is approximately 1:1. Some examples of this are ^{16}O, ^{12}C, ^{2}H (deuterium), ^{4}He, ^{6}Li, and ^{40}Ca. Above calcium, the nuclei are in an unstable condition because as the atomic number increases, the negative coulombic forces increase at a rate greater than that of the positive nuclear forces. Extra neutrons are needed to increase the average distance between protons within the nucleus, thus reducing the coulombic force. If the nuclide is too rich in neutrons, instability will result. Stability may be obtained by conversion of a neutron into a proton within the nucleus ($n \rightarrow p + e^- + \nu$). Also, an excess of protons will produce an unstable nucleus. Stability can now be obtained by an increase in the n/p ratio by converting a proton into a neutron within the nucleus. With these two possible reactions, positron emission ($p \rightarrow n + e^+ + \nu$) and electron capture ($p + e^- \rightarrow n + \nu$), the neutrino ($\nu$) is emitted to conserve energy and momentum.

Radioactivity was observed as early as 1867 when fogging of silver chloride emulsions by uranium salts was reported. The blackening of the emulsions also occurred when they were separated from the uranium salts by thin sheets of paper. At the time, the cause of the blackening or radioactivity was attributed to luminescence effect. It was not until Madame Curie studied radiation emitted by uranium that the phenomenon was correctly explained. The fogging of the photographic plates was then known to be caused by a radiation penetration similar to x-rays of Roentgen. Madame Curie (1898) also observed that radiation evolved from two new elements, polonium and radium.

Radioactive decay

Between 1899 and 1902 Rutherford, Owens, and Soddy discovered that some elements lost their radioactive properties in a consistent rate peculiar to that element. This phenomenon was defined as radioactive decay. Later, the term radioactive disintegration was proposed by Rutherford and Soddy, who believed that the source of some radioactive elements lay in samples of other radioactive elements that produced daughter radioactive atoms, electrons, alpha particles, and gamma radiation. Disintegration may be explained as a spontaneous, radioactive transformation of one species of nucleus into a nucleus of a different type. This is usually accompanied by the emission of radiation.

A limited number of types of transformation have been observed in the decay of natural or artificially produced radioactive nuclides. The only particles emitted are alphas, negative electrons, and positive electrons. The processes may be accompanied by gamma rays, or the gamma rays may be the only radiation emitted by the nucleus. Various mechanisms of radioactive decay are alpha decay, beta decay, positron emission, electron capture, isomeric transition, and spontaneous fission.

Alpha decay

Alpha decay is usually found in isotopes of the heavy elements. Isotope (Greek, meaning equal place) denotes a different mass of the same element. Alpha radiation consists of a flow of alpha particles. These particles are identical to helium nuclei having two protons and two neutrons without two orbital electrons. As the alpha particle loses energy it attracts electrons and eventually becomes a neutral helium atom. The alpha particle's initial energy is great but is stopped in a few centimeters of air or in a few microns of water or soft tissue. ^{226}Ra and ^{222}Rn are typical alpha emitters. With few exceptions, nuclei which decay by alpha emission have an atomic number greater than 82. Alpha decay is seldom used in clinical laboratory work.

Beta decay

Beta decay is an isobaric transition leading to the emission of an electron-antineutrino pair. The electron and the antineutrino pair are emitted as reaction products when the neutron (in nucleus) makes a spontaneous transformation into a proton. When the isobaric transition occurs the proton number (Z) of the nucleus increases by 1, but the mass number (A) does not change.

The emitted antineutrino has no measurable rest mass and no charge and produces no radiobiologic effects because of its evasive nature. However, it shares the disintegration energy with the emitted electron; therefore, the spectrum of the emitted betas is continuous instead of the alpha emission line spectrum.

The beta radiation is a flow of either positive or negative electrons. The electron particles are identical, with the exception that the negative electron (negatron) carried a -1 charge while the positive electron (positron) carried a $+1$ charge. One is the antiparticle of the other. When emitted from radioactive nuclei, they are called beta particles and the symbols are β^- and β^+. Otherwise, they are represented as e^- and e^+. Examples of such radionuclides are ^{3}H, ^{14}C, ^{32}P, ^{35}S, and ^{90}Sr.

Positron emission

Positron decay is one way by which an unstable atom can convert a nucleon from a proton to a neutron. The transformation is accompanied by the production of a positron and a neutrino with emission of these two particles. Positron emission occurs if the n/p ratio is not great enough for the stability of the nucleus. The proton mass is exceeded by total neutron and positron masses, therefore a proton decay is energetically unfeasible. Assuming the neutrino rest mass is 0, a calculation is shown illustrating the imbalance of the n/p ratio and the energy needed to convert the proton into a neutron and positron:

$$\begin{aligned} M_n &= 1.0086654 \text{ amu} = 939.550 \text{ Mev} \\ M_e &= 0.0005486 \phantom{\text{ amu}} = 0.511 \\ \hline \text{Total} &= 1.0092140 \phantom{\text{ amu}} = 940.061 \\ M_p &= 1.0072766 \phantom{\text{ amu}} = 938.256 \\ \hline M &= 0.0019374 \text{ amu} = 1.805 \text{ Mev energy needed} \end{aligned}$$

It is a remote possibility that energy would be supplied to a free proton. If, however, the proton is within the nucleus, energy may be given from other nucleons. As with beta decay, the disintegration energy that the nucleus loses in transition goes into generating the two particles and giving them both kinetic energy. The energy spectrum of the emitted positrons is continuous and of a shape similar to that of the beta decay spectrum. The daughter nucleus may be at ground state subsequent to positron emission of the parent nuclide (^{11}C, ^{15}O, ^{13}N, ^{18}F) or it may be in an excited isomeric state with subsequent transitions producing gamma radiation (^{22}Na).

Electron capture

Electron capture is a second process whereby an unstable nucleus can convert a proton to a neutron. If an orbital electron passes through its unstable nucleus, it may be captured by the nucleus and combined with a proton, thus producing a neutron with a neutrino. Because a K electron is involved in about 90% of the captures, it is often called the K capture. The remaining percents are approximately 10% to 12% L and 1% M electrons. The result is the formation of a daughter product in which the atomic number is reduced by 1, the neutron increased by 1, and the mass remains the same.

Electron capture occurs when insufficient energy is available for beta decay. It appears as a competitive process in each case of beta decay. The result of this transformation is the formation of a daughter product in which there is a vacancy in the electron configuration. When an electron is lost in the K, L, or M shell, the existing vacancy is filled by electrons from higher energy levels. Rearrangement of the orbital electrons results in the emission of x-rays characteristic of the daughter nuclide and/or the Auger (ăw-zhā) electrons. If the decay leaves the daughter in an excited isomeric level, then gamma rays or conversion electrons will be emitted in the subsequent transition to the ground level of the nucleus.

The Auger effect involves the transition of an orbital electron from an excited state to a lower energy level producing an x-ray. The x-ray cannot escape from the atom without colliding into another orbital electron, giving it all its energy. This second electron (Auger electron) is now ejected from the atom with a kinetic energy equal to that of the x-ray, less the binding energy of the electron.

Isomeric transition

In some instances the excited isomeric state of a daughter nucleus is long-lived with the transition to ground level delayed beyond nanoseconds into minutes or hours. Since the ground and excited levels have identical nuclei (except for energy content), they are called nuclear isomers, and the transition between them is known as an isomeric transition. This transition takes place by gamma ray emission or by the competing process (internal conversion) where the available excitation energy is transferred to an orbital electron causing its ejection from the atom (conversion electron). This delayed, highly excited level is called the metastable level and is recognized by the addition of m to the mass number, for example, ^{99m}Tc and ^{60m}Co.

Spontaneous fission

Very heavy nuclides, such as ^{254}Cf, have a mode of radioactivity decay in the form of spontaneous nuclear fission, whereby the nucleus is split into heavy fragments resulting in the release of energetic neutrons and gamma radiation.

The radioactive decay by negatron emission occurs when the n/p ratio is extremely high; a neutron may be emitted, although this is not found in the routinely used isotopes.

Spontaneous fission is initiated when the heavy nucleus is struck by a neutron. The neutron is absorbed and the activated nucleus splits into pieces. The nuclides most frequently produced are those in the range of the atomic number 42 (molybdenum) and 56 (barium). However, the range may be from 30 to 64. The nuclei are rarely split into equal parts. Over 200 product nuclides and eighty direct products have been noted to derive from over forty modes of fissioned nuclei splitting.

Fission of some nuclei (^{233}U, ^{235}U, ^{239}Pm, ^{241}Am) is induced by fast or slow neutrons while other nuclei with more stability (^{232}Th, ^{238}U) require fast neutrons. The greater the stability of the nucleus, the greater the energy needed for the attacking particle to initiate fission. Fission has been initiated by deuterons, protons, and gamma rays.

Photon interaction with matter

Radiation acts upon matter by transferring energy. The x-rays and gamma rays have the ability to penetrate materials that are opaque to less energetic radiation. The mechanics of absorption of visible, ultraviolet, and infrared radiation is the transfer of photon energy to the electronic, vibrational, and rotational energy states of the material doing the absorbing. However, with the x-ray and gamma ray photons' greater energies, the transfer removes only a small fraction of its energy from the photon, and the absorption process yields ions and radicals which eventually can have a greater biologic effect. Energy absorption of the x-rays and gamma ray photon requires many energy transfers which lead to their notable penetrating power. In interacting with matter a number of different effects occur, among which are photoelectric absorption, Compton scattering, and pair production. The radiation does not act like waves but rather as particles called photons or quanta.

In photoelectric absorption, the incident photon dislodges an orbital electron, transferring its energy to that electron. Two electrically charged particles, a positive charged atom and a negative electron, are created. The photon ceases to exist. The electron arrangement produced results in the emission of characteristic x-rays and Auger electrons, the products of a competing process.

In Compton scattering, the incident photon gives up part of its energy in dislodging and transferring energy to a loosely bound (or free) electron and then glances off at a reduced energy. The dislodged electron (Compton electron) recoils with a kinetic energy essentially equal to the energy degradation of the photon.

Pair production occurs in a strong electric field near a nucleus where the incident photon, having an energy in excess of 1.02 Mev, undergoes an energy-to-mass transition in which the energy $h\nu$ of the photon is converted to mass. A positron-electron pair of particles is produced. The positron disappears in a collision with an electron.

Rate of radioactive decay

There is no way to accelerate or retard the decay of the radioisotope. The radioactivity is undergoing a consistent, continuous change in an orderly process, with the rate of decay proportional to the number of radioactive atoms present. The equation for this reaction according to the exponential law is:

$$N = N_o e^{-\lambda t}$$

where N = number of atoms remaining at time t
N_o = original number of atoms present when t = 0
$e^{-\lambda t}$ = decay factor
e = base of natural logarithms—2.71828
$-$ = indicative of a decrease in total number of atoms
λ = decay constant for a particular radionuclide
t = time

The time derivative of the above equation which describes radioactive decay is:

$$dN/dt = \lambda N$$

where dN/dt = activity of radioactive material (rate of radioactive decay)

A method for determining a decay constant for a radioactive element (whose rate is neither too short nor long) is to prepare a solution of the radioactive element and measure the activity on successive days. The activity (counts per minute) is plotted against the time (days). The half-life and decay constant are calculated.

Half-life

The time required for the disintegration of half the atoms in a radioactive sample is called the half-life ($t\frac{1}{2}$). This is one of the important characteristics of any radionuclide. About 290 radionuclides have a half-life of 1 to 24 hours, 240 have a half-life of 1 day to 1 year, and 83 have a half-life greater than 1 year. The half-life and decay constant bear a fixed relationship, so if one is known, the other can be calculated.

The following equation computes the time (t) required for half the nuclei to disintegrate. Solve equation by a natural logarithms:

$$N_o \tfrac{1}{2} = N_o e^{-\lambda t \frac{1}{2}}$$

or

$$\ln \tfrac{1}{2} = -\lambda T \tfrac{1}{2} = 2.30 \log 2$$

$$t \tfrac{1}{2} = \frac{0.693}{\lambda}$$

where N = original number of atoms present
N_t = number of atoms remaining at time (t)
λ = the decay constant for a particular radionuclide or fraction transformed per unit of time
e = base of natural logarithms—2.71828
T = half period
N = $N_o \tfrac{1}{2}$ at end of half period
t = $t\tfrac{1}{2}$

During the first half period, one half the total atoms will decay, and the process continues. Half will be lost again in the second period, and so on until after seven half periods there remains less than 1% of the original atoms. However, when a change of decay is obvious within a few hours, instrument efficiency might be the problem. Check the system with a standard.

There are abundant listings of half-life and decay constants for various isotopes. Some of those used more often in the clinical field are in Appendix A. The use of the equation and the decay constant gives a rapid calculation for the radioactivity remaining. A decay curve can be plotted on semilog paper. Since the expressions of the formula appear as a straight line, a two point curve can be drawn. If the curves are made on linear graph paper, multiple points are required.

Mean-life

The mean-life of a radioisotope differs from the half-life in that the former is a quantity based on statistical analysis. The mean-life may be determined by dividing the sum of the times of existence of all atoms by the initial number of atoms.

$$t = \frac{1}{\lambda} = \frac{1}{0.693/t\frac{1}{2}} = 1.44 \; t\tfrac{1}{2}$$

Mean-life is useful for computing the dose caused by infinite exposure, because the product of the mean-life and the initially observed disintegration rate is equal to the total number of atoms that will eventually disintegrate.

Biologic half-life

Physical and mean half-lives of radioisotopes represent only part of the facts in biologic systems that have their own way of eliminating materials. The rate of removal can be expressed as a biologic half-life (turnover/unit of time) and is proportional to the quantity present. If radioactive material with a physical disintegration probability λ_p is incorporated in a compound that is eliminated with a biologic excretion probability λ_b, the effective disappearance probability can be expressed by $\lambda_{eff} = \lambda_p + \lambda_b$.

The effective half-life is as follows: The T_{eff} is less than either T_b or T_p.

$$T_{eff} = \frac{T_b \times T_p}{T_b + T_p}$$

CLINICAL APPLICATIONS

Although research in endocrinology owes much of its success to the development of biologic assays, the new techniques of radioassay have opened a new era of clinical laboratory applications. There is still a definite need for bioassay analysis, but it is cumbersome for the diagnostic laboratory to maintain and control.

In 1959 Rosalyn S. Yalow and Solomon A. Berson introduced a radioimmunoassay (RIA) procedure for the determination of insulin in human plasma. This pioneering work has brought forth an explosive proliferation of tests performed by this method of analysis.

One of the major reasons for the excitement generated by radioassay is its sensitivity. Laboratories are now able to quantitate such minute amounts as one picogram (1×10^{-12} g, or 0.000000000001 g), thereby producing significant data for the physician. There are currently at least eight categories of substances which can now be quantitated by RIA: hormones, drugs, toxins, viral proteins, allergens, vitamins, enzymes, and globulins. Possibilities appear endless—there is a deluge of test kits flooding the market, multitudes of papers are being presented and published, and the reassurance is given that the methodology is quick, economical, and easy to do. Now is the time to stop to evaluate this new tool. Radioassay methods need precision techniques; they are expensive to utilize for small volume work; many kits leave much to be desired in their reagents and directions. Many problem areas that should have been resolved by the manufacturer before releasing the method are left to the laboratory. Caution must be used also in selecting the hardware used for counting the samples. As with all equipment, the counter capability should suit the method. Precautions, evaluations, and quality control must temper the excessive production of new methodologies and instrumentation.

Radioassay is the term used to include several types of displacement analyses, such as radioimmunoassay, immunoradiometric assay, radioenzymatic assay, radioreceptor assay, and competive protein binding assay. A fine distinction exists in these methodologies, but all meet certain criteria:
1. Two forms of ligands, one from patient or standard and the other the identical ligand but labeled with a radioisotope
2. A receptor, such as an antibody, to provide the binding sites for the ligands
3. A method to separate the bound ligand from the unbound
4. Capability to count the radioactivity in the reaction tube

Radioimmunoassay

Radioimmunoassay (RIA) is one type of displacement analysis based on the assumption that if two forms of an antigen, labeled and unlabeled, are present in excess with a limited amount of antibody, they will compete for binding sites on the antibody according to the law of mass action. That is, the more unlabeled antigen present (from patient or standard) the more binding sites it will occupy, thus leaving fewer available for the labeled antigen. This competition is directly proportional to the amount of antigen present. Then by separating the bound antigen from the unbound and counting one or the other or both, we can accurately relate the quantity present by constructing a standard curve and interpolating the counts of the unknowns.

For a radioimmunoassay to be acceptable, three basic reactions must be assumed:
1. The antibody reacts equally with both antigens.
2. Both antigens, labeled and unlabeled, react identically.
3. The separation of antibody-bound antigen from free antigen is complete.

Precisely, this technique is called competitive equilibrium radioimmunoassay. It requires the labeled antigen to have a very high specific activity.

Specimen preparation

The specificity of the binding protein determines the extent of specimen preparation. This is noted in the various methodologies. Often a series of steps involving deproteinization, solvent extraction, or chromatography (column, paper, or thin-layer) must be performed. In certain assays in which the binding sites have a greater specificity or certain antibodies have high affinity constants and specificity, little or no specimen preparation is necessary.

Radioactive tracers

The ligand containing a radioactive label must be in pure form. When iodination is possible, external labeling may be accomplished with either ^{131}I or ^{125}I, which gives high specific activity. When no iodination is possible (steroid hormones), an internal label using either ^{3}H (tritium) or ^{14}C must be used. The specificity of this method is usually less than that of the radioiodine labeled compounds, however.

Antigen

The antigen used for radioactive labeling and analysis must be pure for the assay to be specific. A particular antigen will induce a formation of multiple (heterogeneous) antibodies. There have been no reports of an antigen having specific (homologous) antibodies. Depending on the equilibrium constants (K) the antigen will combine with the multiple antibodies in varying degrees. The heterogenic condition of the antigen may lead to multiple forms of gastrin,[34] parathyroid hormone,[4] and insulin.[17, 21] A further problem with antigen is that two different antigens from the same species may cross-react, or the same antigens in two different species may show evidence of cross-reaction. However, cross-reaction does not take place if antibodies are specific only for the antigens from the animal or species in which the antibodies are formed. This lack of cross-reaction is called species specificity. Specificity is not the same for every antigen and consideration must be made for the specific antigen assayed.

Other factors influencing the antigen-antibody reaction are the composition of the incubating buffer, temperature, pH, anticoagulants, excessive bilirubin, hemolysis, and urea. The importance of the effect of these factors cannot be stressed too often. For example, hemolysis will cause the destruction of insulin[7] as well as affect the counting of activity. If a specimen to be assayed for insulin is left at room temperature overnight, as much as 25% reactivity may be lost.[16]

Antibody

The next requirement for RIA is a suitable antiserum. The antibodies are the gamma globulins or immunoglobulins of serum proteins. Most immunoglobulins belong to the IgG class, with the remaining ones being IgA, IgD, IgE, and IgM. According to Porter and Edelman (Nobel Prize, 1972) the antigen-binding sites appear to be located on the H and L chains of the IgG molecule. A single antibody may have a variable number and location of binding sites. Development of better technology in antisera purification is improving the problems of multiple formation.

The major criterion for establishing an antiserum is the power or energy of interaction between the antigen and antibody. Antiserum is usually purchased commercially, but it can be prepared by mixing a purified antigen with Freund's adjuvant (mineral oil, wax, and killed bacilli) to enhance and prolong the antigenic response. This is injected into an animal over a period of several months. (Small molecules are first attached as haptens to large carriers, such as inert particles, before injection.) The animal is bled, and the specificity titer and binding affinity of the antibody are determined. Optimal titer is then determined by measuring the ability of trace levels of labeled antigen to be bound by antiserum at differing dilutions. The proper dilution will bind approximately half the radiolabeled antigen. The antiserum producing the sharpest slope provides the most sensitive assay. Before freezing for storage, the antisera may be diluted with ethacridine lactate (Rivanol), which precipitates proteins other than IgG. The three variables of the antiserum which are determined for each new batch are titer, specificity, and binding affinity.

Incubation

The incubation period is the time needed for the radiolabeled antigen and the unlabeled antigen to compete for binding sites on the antibody. The time and temperature required for equilibrium to be established vary, according to the specific antigen being measured. In some RIA procedures it is recommended that incubation not be carried to equilibrium. With a long incubation the antigen may be altered by the lengthy exposure to high concentrations of plasma proteins. Sensitivity may be enhanced by delaying the addition of the labeled antigen. Incubating at low temperature can decrease the incubation damage caused by free radicals, oxidants, or proteolytic enzymes. Another protection may be the addition of inhibitors, such as mercaptoethanol, aprotinin, or ε-aminocaproic acid. Increasing the precipitation with aprotinin (Traysol) has been successful with glucagon but very poor for ACTH. When damage may result, the loss is corrected by incubating a control (damage control) without antiserum. Mercaptoethanol inhibits the oxidation of labeled hormones containing methionine during storage. It will also inhibit the antigen-antibody reaction when the concentration exceeds 5 g/L. Benzamidine is less expensive, although as efficient, as aprotinin for inhibiting proteolytic enzymes affecting glucagon assays.

Separation of bound from free antigen

There must be a suitable method for a complete and rapid separation of the bound antigen from the free antigen so that both or either phase can be accurately measured. This method must be simple, reproducible, and economical. Some methods that have been used are electrophoresis, chromatoelectrophoresis, precipitation of the antigen-antibody complex by salts, solvents, or a second antibody, bonding of the antigen or antibody to a solid phase or carrier (Bentonite), and adsorption of the free antigen on coated charcoal. Binding may be to the inner surface of a polystyrene or polypropylene tube (special form of solid phase bonding). In the double antibody technique a second antibody, an anti-IgG from an animal such as a goat or a sheep, is used to precipitate the initial antigen-antibody complex. Filtration or centrifugation is used to complete the separation.

A critique of various separation techniques follows.

Electrophoresis. This technique involves the differential migration in an electric field of the bound and free antigen on cellulose acetate. This has been used for assay of growth hormone.

Chromatoelectrophoresis. In this, the free antigen binds to the cathode, and as the generated heat by electrical current increases water evaporation, the antigen-antibody complex migrates by hydrodynamic flow. This has been used for ACTH, TSH, PTH, HGH, glucagon, and insulin.

Chemical precipitation. This procedure uses precipitating agents such as saturated ammonium sulfate to precipitate the antigen-antibody complex. Although there is an advantage of immediate separation and one incubation, the chemical precipitation may bring down other proteins, thus interfering with a specific separation of bound from free antigen.

Gel filtration on cross-linked dextrans. This method is far too unwieldy for volume work, since the necessary columns require too much space. It has been used for insulin and HGH.

Bonding to solid phase. Either antigen or antibody is bonded to an adsorbent which separates or precipitates either bound or free fractions. This technique makes use of antibodies covalently bonded or fixed to insoluble polymers or covalently cross-linked to one another. Antigen or antibody may also be adsorbed to plastic. A solid adsorbent will adsorb the free antigen and then be either centrifuged or filtered. Among the adsorbents commonly used are charcoal, zirconyl phosphate gel, Fuller's earth, and talc. These must be able to adsorb quantitatively the desired material while at the same time not adsorb the material to be separated as the soluble fraction. Although large amounts of antibodies are required, there is the advantage that all three components of the assay—antibody, labeled antigen, and the means of separating bound from free phases—are in one unit. Bonding is simple, rapid, reproducible, and economical.

Coated charcoal. Charcoal demonstrates excellent characteristics in its adsorbing capability. Dextran-coated charcoal as described by Herbert and associates[18] is used effectively to separate low molecular weight antigens from antigen-antibody complexes. The molecular weight may vary from 10,000 to 250,000, according to the size of unbound antigen being adsorbed. The synthetic glucose polymer forms a coating which allows only smaller molecular weight particles (free antigen, not the antigen-antibody complex) to be adsorbed by the charcoal. The complex is of sufficiently large molecular size to prevent its passing through the dextran coating and being subsequently adsorbed onto the charcoal. Charcoal coated with hemoglobin has also been used. This method of separation is rapid, successful, economical, and widely used.

Double antibody technique. In most cases the antigen-antibody complex represents a relatively large molecule, but it is not sufficiently large or insoluble enough to form a precipitate. Precipitation can often be facilitated by increasing the size of the complex. This is done by adding a second antibody which complexes with the first antibody. Usually the second antibody, known as the precipitating antibody, is raised to the globulins of the animal source of the first antibody. Since more precipitating antiserum is required, a larger animal such as a sheep or goat is used.

This double antibody technique makes use of a preliminary incubation step in which the first antibody and antigen form a complex, followed by a second incubation in which the complex will react with the anti-sheep antibody to form the double antibody. This yields molecules large enough to be filtered or centrifuged. Double antibody techniques demand a lengthy time for equilibration, so rapid methods have been utilized with a resultant sacrifice of specificity and/or sensitivity. The "shortcut" utilizes a chelating agent, such as EDTA, or heat to inactivate serum complement and to equalize the serum content of standards and samples at the time of precipitation.

Solid phase. In this method the antibodies are immobilized on a solid support such as plastic or resin. One of the more effective applications of this technique utilizes the inside walls of plastic test tubes as the solid support. After incubation the free antigen, which has not reacted with the antibody, can be decanted and the support washed. After washing, the solid support is counted to determine the amount of label incorporated in the complex. This procedure is advantageous to the user who wishes to minimize personnel time. Because of higher manufacturing costs, products incorporating solid phase techniques will usually be higher priced.

Calculation

After the free antigen is separated from the bound, the radioactivity of either or both fractions is measured by counting the samples. The results are compared to the standard curve, which is run under similar conditions, and plotted as a graph, relating the test concentration (antigen) to the free-to-bound ratio, or the percentage bound counts to total counts, or percentage free counts to total counts.

The plot of the percentage bound to total counts on cartesian coordinate paper gives a sharp sigmoid curve. Linearity is more apparent by using semilogarithmic paper. The log scale is on the X axis. The percentage count of the bound antigen is inversely related to the concentration.

Radioassay reliability
Sensitivity and specificity[13, 14]

While radioassay methods are sensitive enough to measure amounts as minute as one nanogram (billionth of a gram) or one picogram (trillionth of a gram), its specificity is not the best. Some of the reasons are as follows: drugs may cross-react; the affinity of the antigen for antibody may be changed by concentration of other constituents in biologic material such as bilirubin or urea; an antigen may induce more than one antibody which may not be specific for the antigen of interest; temperature and pH may change the affinity of antigen for antibody; precursors, fragments, and metabolites will often react with antibody; different molecules with similar antigenic sites may react with the antibody.

Accuracy

The accuracy of radioassay technique is almost impossible to assess because of the very factor that gives it its unique position. That is, it is the only methodology available with any degree of routine practicality that is capable of detecting and quantitating certain physiologic constituents. The older bioassays were used for some of these measurements (angiotensin and urinary FSH), but it is recognized that the biologically active part of the molecule is not always the antigenic part, therefore the two methods do not always correspond. The accuracy of radioassays is attested to by the fact that the results so obtained match the clinical picture.

Precision (reproducibility)

As with all analyses, precision depends on quality techniques, equipment, and reagents. When the necessary care and precautions are taken, precision better than

±5% and ±2 coefficients of variation may be expected. With specialized analysts dedicated to this type of analysis, precision is of course better.

Radioassay exhibits a dissociation between immunoactivity and biological activity due to the difference in their binding sites on the molecule. Tissue and plasma carrier protein binding sites correlate more readily with biologic activity. In measuring the biologic specificity two methods of inner quality control are possible: measure the unknown at different dilutions, and measure a second unknown whose secretion depends on the activity of the first. If the results are in agreement, the analysis is considered valid.

Other radioassay techniques

The umbrella term radioassay is used to describe several assay procedures now in use. Most of these techniques are spinoffs from radioimmunoassay (RIA), with only discrete differences, some technical, some semantic.

Competitive equilibrium radioimmunoassay

This topic is discussed on pp. 335-336.

Sequential saturation analysis

In this procedure the addition of the tag is delayed until the unlabeled antigen and the antibody have come to equilibrium. By this method, which may increase the sensitivity of the assay, an excess of antibody sites is added to the unlabeled antigen so that essentially all the unlabeled antigen is bound. At this stage of testing, which is reversible, an excess of labeled antigen is added to react with the antibody sites that are not occupied by unlabeled antigen.

Radioreceptor assay

This is a general term applied to all analyses based on the competition for specific binding sites in a macromolecule by smaller molecules called ligands. Such proteins with specific binding properties are called receptors. If the binding sites in the receptor are first saturated by a radioactive ligand, the addition of an excess of nonradioactive ligand will result in competition for binding sites and in displacement of the radioactivity from the receptor-ligand complex. Following the law of mass action, the degree of displacement will be proportional to the amount of nonradioactive ligand added. Requirements for this analysis are a specific binding protein, a radioactive form of ligand, a technique for either purifying or extracting the sample, and a technique for separating the bound from the unbound radioactivity.

Immunoradiometric assay (IRA)

The variation in this method is that the antibody, not the antigen, is labeled. The advantages of this over RIA are the ease of labeling antibody in comparison to antigen, the stability of antibody over antigen, and low blank values. More of the unlabeled antigen is allowed to react than in RIA procedures.

Competitive protein binding analysis (CPBA)

CPBA is similar to RIA except that, instead of antibody, a different type protein with the specific capability for binding the ligand is used. The protein is usually a purified circulating globulin, such as transcortin (cortisol)[3] or thyroxin-binding globulin (thyroxin).[10] In general, this method is neither as specific nor as sensitive as RIA.

The difference between RIA and CPB methods lies with the receptors. In RIA the binding protein is an antibody and the ligand is either an antigen or hapten. In CPB the receptor is either a carrier plasma protein or target tissue and the ligand is nonantigenic. It is the receptor that primarily determines the sensitivity and specificity of the assay.

CPB methods are sensitive and reproducible in the nanogram and picogram ranges. Greater sensitivity is obtained when a limited number of binding sites is available and the amount of tracer ligand is small. Following the law of mass action for reversible noncovalent reactions, the optimal binding sites are determined by the affinity constant between the protein and ligand. The affinity constants are defined in liters/mole or the dilution at which the binding protein will bind 50% of tracer ligand.

Liquid scintillation counting*

Liquid scintillation counting is a method of detecting radioactivity by means of a solution of fluors and a photomultiplier tube. The scintillation solution converts the energy of the primary particle emitted by the radioactive sample to light, and the phototube responds to this light energy by producing a charge pulse which can be amplified and counted by a scaling circuit.

Some radioactive isotopes decay by the emission of a beta particle or positron. The energy of the decay is divided between the beta particle and the antineutrino in such a way that a frequency distribution of the energy may be plotted.

The particle energies range from less than 1 ev to maximum energy (E_{max}). As noted in Fig. 13-2, the peak of the distribution curve is $\frac{1}{3}$ E_{max}. The beta energy distribution is similar with the exception that each isotope has its own E_{max}. Four major beta emitters used in the biomedical field are ^{14}C, ^{3}H, ^{32}P, ^{35}S.

As the nucleus disintegrates, the emitted beta particle collides with other atoms, losing energy with each collision. By detecting the energy loss, the beta particle can be detected and its energy measured.

The liquid scintillation counting (LSC) spectrometer detects and measures the energy of the emitted particle. Although the spectrometer is most often used to count

*See references 1, 6, 8, 9, 12, 19, and 20.

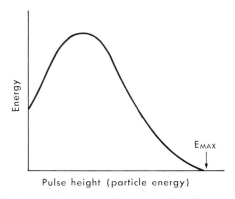

Fig. 13-2. Energy frequency distribution for the beta emitter.

low beta emitters, it is also utilized for counting other isotopes. As with any radiation detection method, the counting efficiency is greatest when the maximum number of emitted particles reaches the detector and interacts in a specific manner. Efficiency is lost when the isotopically emitted particles escape from the detector or are stopped by some contaminant which escapes detection.

The beta-emitting isotope is prepared for counting by either dissolving or suspending it in a solution called a scintillation cocktail. A typical cocktail consists of an organic solvent, such as toluene or p-dioxane, in which is dissolved a combination of fluorescent substances. These fluors include primary fluors, such as p-terphenyl and PPO (2,5-diphenyloxazole), and secondary fluors, such as dimethyl POPOP (2,2'-p-phenylenebis-[4-methyl-5-phenyloxazole]) and TPB (1,1,4,4-tetraphenyl-1,3-butadiene).

The emitted beta particle collides with the toluene molecules, thereby losing its energy. This energy loss excites the toluene molecule, which quickly returns to an unexcited state, transferring the energy to the PPO (primary scintillator), which becomes excited. This energy loss is in the form of ultraviolet light. The detector of the liquid scintillation counter is not sensitive to ultraviolet light; therefore a secondary fluor (POPOP) is added to absorb this light and reemit it as light of a much longer wavelength (visible light). These photons of light emitted by POPOP occur as rapid flashes when the molecule is excited by the energy transfer from the beta particle and are then seen by the photomultiplier tube. The beta particle's scintillation in the cocktail is proportional to the particle's energy.

The photomultiplier tube (PM tube) output is amplified and used as an input to a pulse height analyzer. With the amplified voltage proportional to beta energy (by photomultiplier tube), the pulse height analyzer electronically discriminates for or against the voltage pulse. There are two discriminators on the analyzer; the lower discriminator can be adjusted to reject all pulses smaller than selected, and the upper discriminator rejects pulses greater than the preselected limit. The two discriminators create a counting window. The window setting can be adjusted for a frequency distribution by starting at low voltages and counting the number of events along the energy spectrum (Fig. 13-3). The number of events is plotted on the ordinate and the pulse height (particle energy) on the abscissa.

The heights of pulses produced by beta particles are directly proportional to the energies of the particles.

Quenching

Basically, there are two kinds of liquid scintillation samples:

1. Samples of unknown or varying quenching—these samples require an analyzer

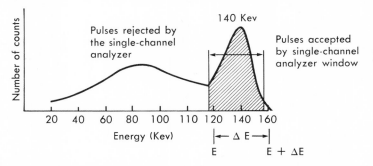

Fig. 13-3. Counting window for frequency distribution.

setup procedure that measures quenching in each sample automatically.
2. Samples of known or constant quenching—for these samples, individual quench measurements are unnecessary. All samples of a given isotope are counted at the same or a known efficiency.

Quench can be defined as any process that interferes with the transmission and conversion of the original beta-energy to the resulting amplified electric pulse. Since the sample being counted might be either a solid or layered on a solid material such as a filter paper or glass disk, the beta particle, rather than interacting with the cocktail, interacts with the solid material surrounding it. This is called self-absorption. The particle's energy may be partially or completely lost to detection. Therefore, the particle *must* totally interact with the cocktail in order for the output pulse to be proportional to the particle's energy.

Chemicals other than phosphors present in the cocktail or colored contaminants (such as hemoglobin) may absorb the ultraviolet or visible light produced by the phosphors, and the scintillation would be partially lost to detection. Also, inert materials suspended in the cocktail, such as debris, might absorb the energy of the beta emission.

These processes result in a loss of the original energy of the beta particle to substances other than the cocktail. Since the resulting scintillation, although reduced, is still proportional to the original energy of the beta particle, the energy spectrum is shifted (relative to the unquenched spectrum) to the left proportionately for all isotopes. In the same cocktail all beta particles will be quenched the same amount relative to each other regardless of their source.

Measuring quench. To determine disintegrations per minute (dpm) from the observed counts, the relative counting efficiency must be related to quench.

Internal standardization method. The sample count is first determined. A known amount of standard isotope is added to each sample and the samples are recounted. The efficiency of each sample is calculated by the following equation.

$$\text{eff} = \frac{\text{cpm (standard + sample)} - \text{cpm (sample)}}{\text{dpm standard}}$$

With the efficiency known, the dpm of the sample is:

$$\text{dpm (sample)} = \frac{\text{cpm (sample)}}{\text{eff}}$$

The sample must be counted twice, however, and self-absorption is not measured.

Channel ratio method. The spectral shift is related to quench. One channel is set to count a relatively unquenched sample as efficiently as possible and a second channel is set to accumulate counts at the lower energy region of the spectrum. As the spectrum shifts to the left because of quench the lower channel accumulates counts while the upper channel loses counts and the ratio of one channel to the other changes. If a series of standards of known dpm are counted and quench is progressively increased, a quench curve can be developed relating efficiency as a function of the channel ratio. When an unknown is counted, the efficiency is determined from the channel ratio and the dpm calculated as in equation:

$$\text{dpm (sample)} = \frac{\text{cpm (sample)}}{\text{eff}}$$

Since this method uses the sample's own count, only one count is required to determine efficiency and all parameters influencing quench are measured. A disadvantage is with very low count rates where counting errors may be large; however, this can be overcome if significant counts are accumulated by increasing the length of time the sample is counted.

Double labeling

Double labeled samples are samples in which two different beta-emitting isotopes are present. In counting samples of this type, both isotopes are counted simultaneously during a single counting period on separate analyzer channels. Since beta-emitting isotopes emit a continuous range of beta energies up to a maximum energy value, the spectra of any two isotopes will overlap to some extent. Therefore, it is not possible to choose two analyzer channels such that pulses from only one isotope are counted in each channel.

However, if a major fraction of the spectrum of the high energy isotope appears above the maximum energy of the low energy isotope, a parameter on the high energy isotope's channel may be chosen that will lower the counting efficiency for the low energy isotope to an insignificant level. Either increase the attenuation or raise the lower level discriminator setting on the high energy channel or both. The increase of either parameter reduces the counting efficiency of the high-energy isotope in its channel. This counting arrangement is workable only if the reduced counting efficiency for the high-energy isotope is great enough to permit reliable counting for its expected activity within a reasonable counting time.

dpm calculations

Disintegrations per minute leads to simplified calculations in double labeled counting, because the efficiency for the low energy isotope in the high-energy channel is essentially zero. Counts in the high-energy channel result only from the high-energy isotope, permitting direct calculation (without correction for the low-energy isotope's count) of the high-energy isotope's disintegration rate.

Efficiency measurements

If the energy separation between two isotopes is not sufficient, pulses from both isotopes will appear in each counting channel. Therefore, four efficiencies must be determined, one for each isotope in each of two channels. All four must be known in order to calculate the disintegration rates of the two isotopes. The two efficiencies are not independent in a given channel. If the channel is chosen to give a desired efficiency for one isotope, the value of efficiency for the other will be fixed. Under this condition, the appropriate efficiency may be chosen for one isotope in each channel.

Automatic external standardization (AES)

The external standard counting modes are intended for counting single labeled or double labeled samples of unknown and varying quenching to provide automatic quench measurement. This method incorporates features of both internal standardization and channel ratio. The sample is first counted and then recounted (usually 1 minute or less) with a source of constant output automatically placed close to the sample (some counters count the sample plus standard first). The Compton electrons produced in the cocktail are counted in two channels and a channel ratio determined:

$$\text{AES (ratio)} = \frac{\text{cpm (sample + standard)} - \text{cpm (sample channel B)}}{\text{cpm (sample + standard)} - \text{cpm (sample channel A)}}$$

If a series of quench standards are prepared, the efficiency can be related to the AES ratio.

Sample efficiency can be determined from the AES ratio and the dpm calculated as follows:

$$\text{dpm (sample)} = \frac{\text{cpm (sample)}}{\text{eff}}$$

A high count rate of an external standard will give good statistical counting while retaining channel ratio sensitivity. Only the cocktail quench is measured and other factors such as self-absorption or loss due to sample artifacts may not be measured. This method may not be successful with multiphase scintillation cocktails since the sample may not be completely soluble in a multiphase solution.

Contamination monitoring[11]

Beta emitters such as tritium cannot be detected by routine monitoring instruments, therefore the liquid scintillation counter is also used for frequent periodic laboratory monitoring. If beta emitters are used, routinely monitor the laboratory to ensure safety conditions. Although external radiation from these sources is not dangerous, these isotopes, if present on laboratory surfaces, can be ingested or disseminated.

The swipe method of monitoring is generally used. Using numerous 25-mm wet Millipore filter disks, each laboratory surface—floors (especially around areas where radioactivity is being used), centrifuges, columns, incubators, sinks, and any other equipment where possible radioactivity might be found—is wiped. The filter paper is then placed in the cocktail and counted. Since debris picked up by the paper is inevitable, quench is likely to be significant. As a result, any count significantly above background (10 cpm above) indicates a contaminated surface. Surfaces found to be contaminated should be immediately decontaminated by washing thoroughly and then retested. A preparation specifically made for radioactive decontamination is recommended.

A record should be kept of the routine monitoring. A schematic diagram of the laboratory showing the floor plan and position of equipment is useful in indicating areas which are or were contaminated. All such areas should be shown along with counts after decontamination. If any questions should arise in the future, there will be adequate documentation.

Gamma ray spectrometry

In comparison to the beta systems, gamma spectrometry is found more frequently in the clinical laboratory. Gamma isotopes are used extensively in vivo and in vitro by the clinical field. The gamma system differs from a liquid system in that the gamma has a solid detector made of a sodium iodide crystal (NaI) that has been activated with thallium. The crystal is sealed to the photomultiplier tube with a transparent silicone grease and the assembly is hermetically sealed and placed into the aluminum can.

Scintillations are produced in the crystal by gamma rays resulting from nuclear disintegrations in the sample. The light output is proportional to the gamma ray energy lost in the crystal, and the output signal of the photomultiplier is proportional to the light output. Therefore, an electric signal from the photomultiplier is a measure of the energy of the gamma ray produced by a disintegration in the sample.

The gamma radiations from the isotope occur at specific energy levels. Nuclear and optic interactions with the sample, the sample container, the sodium iodide crystal, the photomultiplier tube, and the surrounding shielding distort this monoenergetic line to a photopeak having a measurable spread.

Signals from the photomultiplier tube are coupled to the analyzer. The analyzer tests the amplitude (proportional to energy) of each signal and gives one output pulse for each signal that falls between a lower energy limit and an upper energy limit: an energy "window." The pulses from each analyzer channel are counted in a scaler channel to accumulate the total counts in each desired energy range. The accumulated counts may then be expressed in counts per time interval, normally counts per minute (cpm).

A radioactive source is used for the external standardization of gamma counters. Calibrated gamma standards are available: ^{137}Cs, ^{226}Ra, ^{224}Am, or ^{133}Ba. These tubes are usually placed at the end of the sample run. The ionizing radiation causes an electron release resulting in the formation of Compton electrons of known energy and quantity.

RADIATION HANDLING AND HAZARDS*

The safe handling of radioactive isotopes requires that the user maintain at all times a controlled environment such that exposure to personnel is eliminated completely or reduced to acceptable levels.

Controlled areas

A controlled area is one in which the boundaries are specifically designated for radioactive usage. These areas may become contaminated, and appropriate measures must be taken to monitor for radioactive contamination. Areas of contamination must be immediately cleaned up. Only authorized personnel should use radioactive isotopes within a controlled area that has appropriate warning signs posted. There should be control of area traffic and occupancy and of working conditions within the area to protect users and others from the radioactivity. If radioactivity is used in a laboratory where nonradioactive work is also performed, a clear division between controlled areas and other areas must be made. Controlled areas must *never* have carpeting or other absorbent materials on floors, and the floors must have a heavy coat of wax or other strippable material. Table tops must be a nonporous surface without cracks, pockets, or material-collecting areas.

Cold laboratory

This is a standard laboratory where no special facilities are available for handling radioisotopes. Special protective clothing is required and the controlled area must be clearly designated.

Low level laboratory

In addition to the conditions described for a cold laboratory, isotopes must be handled with gloves for alpha and soft beta emitters and with tongs for hard beta- and gamma-emitting sources. Pipeting by mouth is prohibited. All work should be done with protective clothing and in a restricted area of the laboratory over absorbent paper and in a fume hood. Appropriate disposal should be carried out. Counting rooms are considered at this level. Personnel monitor devices are necessary. Appropri-

---

*See references 2, 4, 11, 23, 26, 27, and 33.

ate contamination checks of the work area are required. No smoking, eating, application of cosmetics, or drinking should take place in the area. Floors must be coated with at least a heavy coat of wax.

Intermediate level laboratory

This laboratory requires all the conditions described for a low level laboratory. In addition, shielding of gamma emitters and hard beta emitters is required. A radiochemical hood or glove box with appropriate filters of exhaust air may be necessary for some isotopes. Strippable plastic floors, walls, and work surfaces are highly recommended. Protective clothing should not be commercially laundered. Remote control devices are required.

High level laboratory

In addition to those conditions described for the intermediate level laboratory, a separate room with restricted access is required. Special consideration must be given to walls, floors, and ceilings to provide adequate shielding to adjacent environments. The room should be under negative ambient pressure and exhaust must be appropriately filtered. Unauthorized personnel are not permitted.

Signs

Appropriate signs must be posted where radioactivity is or may be present. This includes controlled areas, storage areas, counting rooms, and all vessels containing radioactivity. Signs must be conspicuous and contain all information necessary to warn individuals of the hazard and must include the yellow-magenta three-bladed radiation symbol. Radiation signs must not be used for any purposes other than where radioactivity is stored or used or where a radiation hazard exists. The types of signs most often used are:

Caution radiation area. This is used in radioactivity type areas where exposure to radiation can occur.

Caution radioactive material. Sign is used in any and all rooms where radioisotopes are used or stored.

Caution radioactive material. This warning is used as a label on any and all vessels containing radioisotope. The type, quantity, and date of assay of material must be clearly specified on the label.

Airborne radioactivity area. This is required where radioactivity may be present in the air, such as iodine or radium (radon).

High radiation area. This sign is required if doses of radiation in excess of 100 mrem/hr can exist. Audible and visible alarms are also required for these areas.

Isotope hazard

The amount of isotope that can be used in any particular laboratory level will depend on the amount and type of emission of the isotope. Typical shielding materials for radionuclides of various emissions are shown in Table 13-1.

Some radioisotopes such as iodine or radium (radon decay product) can be volatile and therefore can be ingested by breathing fumes of the material. Such radioisotopes must be confined to properly ventilated areas such as a fume hood and may not be used without such protective ventilation. Radioisotopes must not be used unless the individual is protected with appropriate clothing, such as a laboratory coat or leaded clothing, depending on the amount and type of isotope being used. Care should be taken to avoid direct contact if the isotopes are spilled or handled in a sloppy man-

Table 13-1
Shielding materials for radionuclides

Radiation	Shielding material		Additional clothing
	Permanent	Temporary	
Alpha	Unnecessary	Unnecessary	Unnecessary
Beta	Lead, copper, iron, aluminum, concrete, wood	Iron, aluminum, plastics, wood, glass, water	Leather, rubber, plastic, cloth
Gamma, x-rays	Lead, iron, copper, lead glass, heavy aggregate concrete, aluminum, ordinary concrete, plate glass, wood, water, paraffin	Lead, iron, lead glass, aluminum, concrete blocks, wood, water	Lead fabrics (but not for hard gamma)

ner. Individuals entering the controlled area may come into contact with the contaminated area, therefore the controlled area must be monitored and contamination of physical surfaces eliminated as it occurs.

Alpha emitters

Usually no shielding is required for pure alpha emitters because the range of the alpha particle is very short and will be completely absorbed by the container holding the radioisotope. The hazard from alpha emitters results from their direct contact with the skin or ingestion into the body.

Beta emitters

The penetrating power of beta particles depends on the energy of the particle. Generally speaking, pure beta emitters with energies less than 200 Kev can be considered similar to alpha radiation and require no additional shielding other than the vessels which contain them. Again, the hazard is from ingestion or contact. Beta radiation greater than 200 Kev requires shielding. However, if the isotope concentrations are 50 μCi or more, care must be taken to see that x-rays generated by the decay do not add significantly to the resulting dose. Beta radiation from an external source will have its primary effect on the skin.

Gamma emitters

The exposure dose from gamma irradiation decreases inversely with the square of the distance from the source. Therefore, the further away from the source, the less exposure received. For example, the body is a type of shielding material, and the dose of gamma irradiation would decrease inversely with the square of the distance as it penetrates. It is probable that the individual receives some exposure from all gamma radiation emitters regardless of the type of absorbing material. Certain levels of gamma irradiation are acceptable, and adequate shielding is necessary to reduce the dose for any particular amount of isotope to the acceptable level. When using gamma emitters, lead, iron, copper, leaded glass, heavy aggregated concrete, aluminum, plate glass, wood, and water are appropriate shielding materials depending on the energy of the isotope and the amount being used. Since gamma emitters may also emit beta particles or electrons, shielding for beta particles should be considered. Adequate shielding should be determined by dosimetry. Advice can be obtained from the Radiation Safety Office.

RADIATION MONITORING[22]

Monitoring of an area or individual is governed by the kind and amount of isotope used. The effectiveness of this control procedure depends on the monitoring system.

Personnel monitors may be required if activities greater than 1 mCi are used. Film badges, pocket ionization chambers, or thermoluminescent dosimeters are effective for gamma and high-energy beta radiation. However, low-energy betas (sulfur, carbon, tritium) or alpha particles are not detectable with this equipment. Film badges should be worn by personnel working with gamma emitters or in the area of high gamma or x-radiation field.

The film emulsion usually consists of silver bromide and silver iodine and is capable of recording an approximate accumulation radiation dose from gamma rays, x-rays, charged particles (β^-, β^+, e^-, α^{++}, etc.), and neutrons. Film badge service can be obtained from commercial firms.

Electroscope ionization chambers operate on a saturation voltage (all paired ions formed are collected) and are usually calibrated so that there is a relationship between the ionization produced in a known mass of air to the unit of radiation exposure (Roentgen). The pocket chambers are available for x-ray and neutron fields. Although they give an on-the-spot indication of radiation exposure, they should be used in conjunction with a film badge or other personnel monitoring device. When daily operation is required, pocket ionization chambers are the most practical.

Pocket dosimeters may be used if there is occasional use or with low activity. These are dose-measuring units consisting of an ionization chamber, condenser, charging device, and electrometer. Precision is poor but they are convenient and portable.

Ring badges contain small film packs attached to a plastic ring to be worn on the hand. Although a higher exposure level is permitted to the hands, they can get more than 15 times the dose indicated on the film badge worn on the coat. Ring badges are used where the hand dose may exceed 25% of the relevant limit (75 R/yr).

When an individual has the possibility of receiving a radioactive dose exceeding 25% of values recommended for maximal permissible dose (MPD), the law requires monitoring devices and accurate records. Such monitoring will be done by film badges worn on the chest or outside a lead apron. The laboratory area will be marked as "Radiation Area."

If the nature of radiation or elevated level of exposure demands constant monitoring, personnel dosimeters of the ionization type are worn with the film badge and dosimeter readings are recorded daily.

The radiation safety officer may find a significant increase in a specific nuclide and request bioassay procedures such as urinalysis or breath analysis. Individuals routinely handling 100 mCi of ^{3}H or 20 mCi of ^{14}C or who are involved in a spill of this magnitude must have such an analysis. Personnel handling ^{125}I or ^{131}I would be subject to thyroid counts at the discretion of the radiation safety officer.

Permissible levels of radiation[25, 28]

The average dose due to background radiation is estimated to be about 100 mrem/yr. There are valid reasons for concern in the increase of environmental radiation levels caused by manmade sources. The use of ionizing radiation has been shown to be beneficial with promise of limiting radiation doses.

The National Council of Radiation Protection (NCRP) has established what are called maximum permissible doses (MPD) for occupational conditions (controlled areas) under the supervision of a radiation safety officer.

Accumulated dose

The permissible dose is based on an accumulated total dose received by a person over the entire lifetime. The whole body MPD recommended is 5 rem/yr for radiation workers and 0.5 rem for nonoccupational personnel. The critical organs considered in this recommendation are gonads, lens of the eye, and red bone marrow.

Radiation workers should not receive more than 3 rem in any 13-week period. The long-term accumulation of the whole body dose equivalent is not to exceed 5 rem multiplied by the person's age in years (N) beyond 18. Subtract 18 from the age and multiply by 5.

$$\text{Total dose} = 5(N-18)$$

A person 23 years old may be exposed to or accumulate a total of $5(23-18) = 25$ rem. This long-term formula applies to critical organs; the rem is doubled for the skin. The skin of the whole body should not receive more than 15 rem/yr.

Weekly dose

The permissible weekly whole-body dose is 0.3 rem, or the 13-week dose of 3 rem when the weekly limit is exceeded.

Emergency dose

An accidental dose of 25 rem to the whole body, occurring only once in the lifetime, has no effect on the radiation tolerance of the individual.

Medical dose

There is no effect on the radiation tolerance status of the individual from medical and dental procedures.

During the entire gestation period, the MPD equivalent to the fetus from occupational exposure of the expectant mother should not exceed 0.5 rem.

SUMMARY OF PROCEDURES FOR HANDLING RADIOISOTOPES

1. Use isotopes only in designated areas (controlled areas) in each laboratory.
2. All equipment, supplies, and working areas must be designated by appropriate approved labels.
3. Laboratory surfaces (floors, walls, and benches) should be coated with a strippable nonporous material (high level laboratory) or a heavy coat of wax (intermediate and low level laboratories).
4. High standards of cleanliness and good housekeeping should be maintained throughout the laboratory.
5. Disposable gloves must be worn when handling radioisotopes.
6. No radioactive material may be pipeted by mouth.
7. Eating, smoking, or the application of cosmetics is not permitted in con-

trolled areas where unsealed radioactive materials are being handled or used.
8. Personal belongings should not be brought into the laboratory.
9. All manipulations must be performed over absorbent paper with absorbent side up and preferably in a tray designed to contain the spill.
10. Individuals working with radioactive material are not permitted to touch objects outside the controlled area while working with radioactivity.
11. No materials may be removed from the radiation control area into the other parts of any laboratory unless appropriate measures are taken to ensure that the material will not come in contact with any other object, person, or apparatus except those specifically designated to handle radioactivity.
12. The amount of radioactive material to be used should be no greater than that needed to carry out the given operation to reduce the seriousness of spills and to reduce exposure to personnel.
13. Whatever animal work is going to be performed using radioactive material, the plan for handling the animals, the plan for the injection of radioactive materials into the animals, the treatment of the animals during their lifetime, and the disposal of all food, water, litter, cages, and the animals themselves must be planned. Areas housing animals containing radioactive material must be properly designated, and it is the responsibility of the user to see that these animals are cared for. No personnel handling animals have the authority to touch or handle any animals or supporting equipment containing radioactive material unless they are approved. The area containing the animals must be monitored frequently to ensure that the animal facilities outside the cages are not contaminated. Records must be kept of results. Animals and supporting equipment must be appropriately disposed of or decontaminated. Monitor every 3 days and after every experiment.
14. Where possible, use only disposable labware for manipulations.
15. Gamma emitters in quantities greater than 50 μCi and 200 Kev energy must be handled behind appropriate shielding.
16. Dispose of all radioactive materials in barrels obtained from the Radiation Safety Office.
17. Dispose of all radioactive animals in refrigerated containers appropriately labeled or bring to Radiation Safety Office.
18. Do not dispose of any radioactive material in the sanitary sewer system.
19. Reduce all liquid waste to solid by pouring over animal litter or sawdust, then put in barrel. Alternatively, pour liquids in plastic container for disposal. Do not mix liquid and solid wastes. Do not mix animals or their wastes with other wastes.
20. Decontaminations with suitable detergent of nondisposable labware may be done with running tap water in an appropriately designated sink after first rinsing once or twice to reduce residual radioactivity. Initial rinses should be disposed of in radioactive wastes.
21. It is the responsibility of any person who spills radioactive materials to notify the Radiation Safety Office and to ensure that the material is effectively cleaned up. Appropriate monitoring of the area and documentation that these materials are appropriately cleaned up are required.
22. Liquid scintillation counting fluid and all contents must be discarded in special double walled containers obtained from the Radiation Safety Office. Vials can be discarded in the regular trash or decontaminated for reuse.
23. All radioactive isotopes are stored under lock and key.

24. All vessels containing radioactive material should be covered when not in use in a vessel which, when tipped over, will not spill or break.
25. All containers containing radioactive materials, including sealed sources and standard sources, should be labeled with radiation warning tape. The isotope, amount, and date should be indicated.
26. Before starting work with radioactive materials, each person shall make known any previous work with radioactive materials or radiation sources and any exposure over the maximum permissible dose.

These many regulations are enforceable by law, and all licensing requirements are set forth in the *Federal Register* (vol. 39, no. 146) under authority of the National Regulatory Commission (formerly the Atomic Energy Commission) of the Food and Drug Administration in the Department of Health, Education, and Welfare.

QUALITY CONTROL AND DATA REDUCTION

Specificity, recovery, reproducibility, and stability are monitored by comparison of dose response curves, or assaying aliquots of serum/plasma pools or commercial controls. The data may be calculated as analysis of variance (anova) for standard deviation (SD) or coefficient of variation (CV).

Most measurements used in monitoring involve counting the nuclear activity such as disintegration of the radioactive atoms, radiation scattering, and production of induced reactions. Since radioactivity decay is a random process, there is a consistent change in the activity of a specific sample and a fluctuation in the decay rate of that sample.

These events follow the statistical law (Poisson). The Poisson distribution describes most of the counting observation. The stability of the counter is a principal factor that may dominate the fundamental statistical error, if its deficiency is greater than 1%. Counter efficiency is therefore involved and counted with the disintegration rate, radiation flux through the counter, and the counting rate. The disintegration rate, $-dN/dt$, is an average value; the actual rate fluctuates in accordance with Poisson statistics. The observed fluctuation depends, however, on the observed counts, not on the number of disintegrations during the time interval concerned.

The chi-square test of goodness of fit determines the probability that repetition of the observations shows greater deviations from the assumed distribution than the observations observed in the first trial. The data should be subdivided into at least five classifications, each containing at least five events. In the counting application, the expected value is the average count observed.

In measuring the radioactivity average, the arithmetic mean is the best value. Accuracy demands a precise and unbiased condition. The precision is influenced primarily by indeterminate errors as found in random decay. Measurement is made by subtracting the deviation of individual samples from their sample mean. The indeterminate errors are beyond the analyst's control. The bias is influenced by determinate errors that introduce a constant error into the data. The measurement is obtained by subtracting the deviation of the sample mean from the true mean. The determinate error is caused by backscattering, radiation self-absorption, temperature changes, incomplete precipitation, or equipment malfunction. Summation probability (cumulative normal frequency distribution) may be used to calculate error of magnitude to expect rather than attempting to calculate the probability of obtaining a specific count. These are a few basic examples of applying statistical methods to the field of radioassay.

It is just as bad to get a set of data that is too consistent as it is to get one that is not consistent enough. This may indicate that spurious pulses of a rather uniform rate are mixed in with the desired pulses. Too much inconsistency usually means instability of some component such as the high-voltage supply or the counter tube itself. It can also mean simply an unlucky run. This possibility should be kept in mind if the apparatus performs satisfactorily after one apparent malfunction.

Statistical methods have developed into a high degree of sophistication and the field of radioassay lends itself to computerization. Although control programs are mandatory for satisfactory assays, manual manipulations can overwhelm the purpose of the test. Minicomputers or programmable calculators present a relief by offering greater speed and increased accuracy with greater reproducibility.

Online application usually interfaces a dedicated system to the counters. The advantage of this application depends on the workload of the counters and the need of the calculator in the other laboratory area.

Offline application permits a more universal use of the system. The input may be through programmable magnetic cards, card reader, teletype, or paper tape. This again depends on the specific need of the laboratory and its investment potential.

Regardless of the type of application chosen, it is of utmost importance to understand the theory and laboratory methods which involve the mathematical manipulations needed to transform raw data into analytic results. This will be the guideline in selecting compatible programs (software) and equipment (hardware) for the laboratory.

GLOSSARY

accuracy Extent to which the value obtained in an assay corresponds to true value.
adjuvant Substance used to coat antigen so that when latter is injected into animals it will not be readily degraded.
agents Substances that react with specific receptors in radioassays; take place of antigens.
alpha particle (alpha ray) Positively charged helium nucleus.
amplifier Component of counter which increases size of pulse without distortion.
antibody Substance produced in animal as result of stimulation with antigen. Commonly symbolized Ab.
antibody coated tube method Solid phase radioimmunoassay in which the reaction tube is coated with antibody.
antigen Anything which, when injected into an animal, will produce an antibody to react with it. Commonly symbolized Ag.
antigen-antibody complex Compound produced by union of antigen and antibody. Commonly symbolized AgAb.
antisera Sera containing antibody against a specific antigen.
avidity Affinity of antibody for antigen.
beta emitters Radioactive substances which, upon disintegration, emit beta particles.
beta particle (beta ray) A positively or negatively charged electron.
binder The specific binding agent, such as antibody or binding protein.
bioassay Procedure whereby potency of a compound is determined by its effect on animals, isolated tissues, or microorganisms as compared with a standard preparation.
bound antigen Antigen in the antigen-antibody complex.
chromatoelectrophoresis Original electrophoretic procedure used to separate bound from free isotope in radioimmunoassay by using certain grades of chromatographic paper and allowing a buffer to flow over the point of application of the incubation mixture on a strip of such paper.
cocktail Scintillation fluid containing organic solvent, fluor, and solubilizer if necessary.

competitive protein binding assay Assays similar to radioimmunoassays but which substitute protein binders as receptors instead of antibodies. Commonly symbolized CPB.
components Separate substances used in radioassay, that is, antigens, antibodies, tracers, separating substances.
Compton effect Radiation effects resulting when gamma rays interact with matter and transfer part of their energy to the ejected electron. The remaining energy appears as a new or "daughter" gamma ray which is scattered at a different angle.
cpm Counts per minute.
crystal Component of counter in which scintillations are produced by nuclear disintegrations.
Curie Unit of measurement of radioactivity intensity. Equivalent to 3.7×10^{10} disintegrations per second or 2.22×10^{12} dpm.
dextran-coated charcoal Component used to separate bound from free isotope in some radioassay procedures (adsorbs free isotope).
displacement analysis Displacement of bound radioactive antigen by patient antigen in radioimmunoassay and competitive binding assay.
dose-response curve Curve in which concentration of standard is plotted versus percentage bound or percentage free isotope.
double antibody technique A method of separating free from the bound isotope in radioimmunoassays in which the antigen-antibody complex is reacted with an antibody against the first antibody causing precipitation of the antigen-antibody complex.
dpm Disintegrations per minute.
efficiency Extent to which counts per minute correspond to disintegrations per minute. Efficiency = (cpm × 100)/dpm.
electron An elementary particle of negative charge found outside the nucleus of an atom. Cathode rays and beta rays are electrons.
e-v Electron-volt; unit of energy.
external standard channels ratio Method for correcting quench process in scintillation cocktail.
fluor A substance which when added to scintillation cocktail shifts the wavelength of the emitted radiation to a region which is optimal for detection by the photomultiplier tube.
free antigen Antigen not bound into the antigen-antibody complex.
gamma counter Scintillation counter which uses solid sodium iodide crystal for detecting gamma radiation.
gamma emitters Radioactive substances which, upon disintegration, emit gamma rays.
gamma rays Electromagnetic radiation originating from nucleus.
half-life The period of time required for an isotope to emit half its energy. The half-life for the most commonly used laboratory isotopes is: ^{125}I, 60 days; ^{131}I, 8 days; ^{3}H (tritium), 12½ years; ^{57}Co, 270 days; ^{14}C, 5570 years.
hapten Nonantigenic substance which, coupled to a carrier protein, acts as an antigen when injected into an animal.
immunology The study of antigens and their reactions with antibodies.
immunoradiometric assay (IRA) An analytic technique similar to RIA except that the antibody, not the antigen, is tagged.
immunosorbents Antigens or antibodies coupled to insoluble polymers to form stable conjugates which retain immunologic binding capacity.
isobaric Two or more atoms having equal atomic weights but different atomic numbers.
isotope One of a group of nuclides of same element having the same number of protons in the nucleus but differing in the number of neutrons.
isotopic Of or relating to an isotope.
Kev Kilo (thousand) electron volts.
labeled tracer Antigen tagged with a radioactive element.
ligand The substance to be bound, or tested.
liquid scintillation counter A counter which measures beta radiation produced when a beta-emitting isotope is dissolved in a liquid scintillation fluid.

Mev Million electron volts.

microcurie (μCi) One one-thousandth of a millicurie.

millicurie (mCi) One one-thousandth of a curie.

nanogram 1 times 10^{-9} g.

neutrino An uncharged elementary particle with a mass nearing zero.

neutron An uncharged particle in the nucleus of an atom with a mass nearly the same as that of the proton.

nucleon An elementary particle in the atomic nucleus, especially a proton or a neutron.

nuclide An individual atom of given atomic number, mass number, and energy content, which exists for a measurable length of time and has a distinct nuclear structure.

percent bound (%B) Percent of isotopic material bound to antibody or receptor.

photoelectric effect Radiation effect resulting when a gamma ray strikes an electron and transfers all its energy to that electron, thus ejecting it from its atom.

photomultiplier tube (**PMT**) A photo tube of exceptionally high sensitivity which not only changes the light signal into an electrical signal but amplifies the signal within the body of the tube.

photon A quantum of electromagnetic radiation.

picogram 1×10^{-12} g.

precision Extent to which an assay is reproducible or repeatable.

proton A nuclear particle with a positive charge equal to and opposite that of an electron, the number of protons being the atomic number of that atom.

pulse-height analyzer (**PHA**) Adjustable threshold device used to set energy window limits of counter.

quench Anything in the scintillation cocktail which prevents transmission of photons from sample to the photocathode of the photomultiplier tube.

rad Radiation absorbed dose. One rad equals the absorption of 100 ergs of radiation energy per gram of matter.

radiation Propagation of electromagnetic radiation (visible light, infrared, x-ray, etc.) through space as well as corpuscular radiation such as alpha particles, beta particles, and electrons.

radioactivity The emission of alpha, beta, or gamma rays in elements that undergo spontaneous atomic disintegration.

radioassay Competitive binding assays involving receptors and labeled agents.

radioimmunoassay Assays in which labeled antigens compete with antigens in patient's biologic fluid for sites on antibodies.

radioisotope Isotope exhibiting radioactivity, usually one created artificially from a nonradioactive element.

radioligand Substance tagged with radioisotope.

receptor Substance that acts in place of antibody in combining with agent (antigen) in radioassays.

rem (**roentgen equivalent man**) Dose of an ionizing radiation producing the same biologic effect as one unit of absorbed dosage of x-rays.

roentgen Unit of exposure dose of gamma or x-radiation. One roentgen is an exposure dose such that the associated corpuscular emission per 0.001293 g of air produces, in air, ions carrying 1 esu (electrostatic unit) of electricity of either sign.

saturation analysis Analyses using antibodies and carrier proteins found in serum as principal saturable reagents. Sometimes used to designate radioimmunoassay and competitive binding assays.

scaler Series of decade counters to measure irradiations.

scintillation spectrometer Scintillation counter designed to permit the measurement of the energy distribution of radiation.

sensitivity Smallest amount of unlabeled antigen that can be distinguished from no antigen.

separation Dividing the bound antigen from the free.

solid crystal counter Scintillation counter used to detect gamma radiation. The detector consists of a thallium doped sodium iodide crystal.

solid phase radioimmunoassay Radioimmunoassay in which antibody or antigen is coupled to insoluble polymers.

specific activity The relative activity of radioisotope per unit of mass (counts per minute per milligram).

specificity Degree of uniqueness with which antibodies bind substance being assayed. Extent of freedom from interference by substances other than the one intended to be measured.

substrate Unknown to be measured.

tagged compound Compound labeled with isotope as marker.

titer Highest effective working dilution of the antiserum.

x-ray Same type of electromagnetic radiation as gamma rays, but originating outside nucleus.

References

1. Baillie, L. A.: Determination of liquid scintillation counting efficiency by pulse height shift, Int. J. Appl. Rad. Isotopes **8**:1, 1960.
2. Basic radiation protection criteria, National Council on Radiation Protection and Measurements, Report No. 39, Washington, D.C.
3. Beitins, I. Z., Shaw, M. H., Kowarski, A., and Migeon, C. J.: Comparison of competitive protein binding radioassay of cortisol to double isotope dilution and Porter-Silber methods, Steroids **15**:765, 1970.
4. Bell, C. G., and Hayes, F. N.: A manual of radioactive procedures, National Bureau of Standards Handbook 80, Washington, D.C., U.S. Government Printing Office.
5. Berson, S. A., and Yalow, R. S.: Immunochemical heterogeneity of parathyroid hormone in plasma, J. Clin. Endocrinol. Metab. **28**:1037, 1968.
6. Bransone, E. D.: The current status of liquid scintillation counting, New York, 1970, Grune & Stratton, Inc.
7. Brodal, B. P.: The influence of haemolysis on the radioimmunoassay of insulin, Scand. J. Clin. Lab. Invest. **28**:287, 1971.
8. Bruno, G. A., and Christian, J. E.: Correction for quenching associated with liquid scintillation counting, Anal. Chem. **33**:650, 1961.
9. Bush, E. T.: Liquid scintillation counting of doubly-labeled samples, Anal. Chem. **36**(6): 1082, 1964.
10. Cassidy, C. E., Benotti, J., and Peno, S.: Clinical evaluation of the determination of thyroxine iodine, J. Clin. Endocrinol. Metab. **28**:420, 1968.
11. Control and removal of radioactive contamination in laboratories, National Bureau of Standards Handbook 48, Washington, D.C.
12. Davidson, J. D., and Peigelson, P.: Practical aspects of internal-sample liquid scintillation counting, Int. J. Appl. Rad. Isotopes **2**:1, 1957.
13. Ekins, R. P.: Problems of sensitivity with special reference to optimal conditions: concentrations of tracer and antiserum, time and temperature of incubation, volume of incubation, etc. In Margoulies, M., editor: Proceedings of International Symposium on Proteins and Polypeptide Hormones, Amsterdam, 1969, Excerpta Medica, p. 672.
14. Ekins, R., and Newman, B.: Theoretical aspects of saturation analysis, Acta Endocrinol. **64**(Suppl. 147):11, 1970.
15. Feldman, H., and Rodbard, D.: Mathematical theory of radioimmunoassay. In Odell, W. D., and Daughaday, W. H., editors: Principles of competitive protein-binding assays, Philadelphia, 1971, J. B. Lippincott Co., p. 158.
16. Fraser, R., Lowy, C., and Rubenstein, A. H.: Radioimmunoassay on urine samples. In Margoulies, M., editor: Proceedings of International Symposium on Proteins and Polypeptide Hormones, Amsterdam, 1969, Excerpta Medica, p. 14.
17. Goldsmith, S. J., Yalow, R. S., and Berson, A.: Significance of human insulin Sephadex fractions, Diabetes **18**:834, 1969.
18. Herbert, V., Lau, K. S., Gottlieb, C. W., and Bleicher, S. J.: Coated charcoal immunoassay of insulin, J. Clin. Endocrinol. Metab. **25**: 1375, 1965; see also Leyendecker, G., Wardlaw, S., and Nocke, W.: Gamma globulin protection of radioimmunoassay and competitive protein binding saturation analysis of steroids, J. Clin. Endocrinol. Metab. **34**:430, 1972.
19. Kallmann, H., and Furst, M.: The basic processes occurring in the liquid scintillator. In Bell, C. G., and Hayes, F. N.: A manual of radioactive procedures, National Bureau of Standards Handbook 80, Washington, D.C., U.S. Government Printing Office, pp. 3-22.
20. Kobayshi, Y.: Liquid scintillation counting—some practical considerations, Waltham, Mass., Tracer Lab, Inc.
21. Lazarus, N. R., Tanese, T., and Recant, L.: Proinsulin and insulin synthesis and release by human insulinoma, Diabetes **18**:340, 1969.
22. Morgan, G. W.: Facilities and equipment for isotopes program, Hospitals, Mar., 1955.

23. Morgan, G. W.: Surveying and monitoring of radiation from radioisotopes, Nucleonics **4** (3):24-37, 1949.
24. National Institute of Health Radiation Safety Guide, DHEW Publication No. NIH73-18, Washington, D.C., U. S. Government Printing Office.
25. Permissible dose from external sources of ionizing radiation, National Bureau of Standards Handbook 59, Washington, D.C., 1954, U. S. Government Printing Office.
26. Quimby, E. H.: Disposal of radioactive waste from hospital laboratories, Lab. Invest. **2**:49, 1953.
27. Recommendations for disposal of carbon 14 wastes, National Bureau of Standards Handbook 53, Washington, D.C., U. S. Government Printing Office.
28. Regulation of radiation exposure by legislative means, National Bureau of Standards Handbook 61, Washington, D.C., 1955, U. S. Government Printing Office.
29. Rubinstein, A. H., Cho, S., and Steiner, D. F.: Evidence for proinsulin in human urine and serum, Lancet **1**:1353, 1968.
30. Rutherford, E., Chadwick, J., and Ellis, C. D.: Radiations from radioactive substances, Cambridge, 1930, Cambridge University Press.
31. Safe handling of radioactive isotopes, National Bureau of Standards Handbook 42, Washington, D.C., U. S. Government Printing Office.
32. Safe handling of radioactive material, National Bureau of Standards Handbook 92, Washington, D.C., U. S. Government Printing Office.
33. Shapiro, J.: Radiation protection, Cambridge, Mass., 1972, Harvard University Press.
34. Yalow, R. S., and Berson, S. A.: Further studies on the nature of immunoreactive gastrin in human plasma, Gastroenterology **60**: 203, 1971.

General references

Chase, G. D., and Rabinowitz, J. L.: Principles of radioisotope methodology, ed. 3, Minneapolis, 1967, Burgess Publishing Co.

Quimby, E. H., Feitelberg, S., and Silver, S.: Radioactive isotopes in clinical practice, Philadelphia, 1958, Lea & Febiger.

Skelley, D. S., Brown, L. P., and Besch, P. K.: Radioimmunoassay, Clin. Chem. **19**:146, 1973.

14

Hormones

The classic definition of hormones describes them as specific substances or messengers secreted into the bloodstream or tissue fluid by glands of the endocrine system or by specialized cells and carried to other parts of the body where they act upon target tissue or cells responsive to them. In addition to these, there are hormones that are produced by one cell type in a multicellular tissue which act upon other cell types with the same tissue, and others that are formed from precursors in the circulation and in tissues. Cells (platelets, erythrocytes, leukocytes) in the blood also respond to hormone concentrations, as contrasted to fixed tissue cells.

Glands that have been traditionally recognized as a part of the endocrine system are pituitary (anterior and posterior), thyroid, parathyroid, adrenal (medulla and cortex), pancreas (islets of Langerhans), and gonads (testes, ovaries, placenta). Other organs and tissues such as the thymus, pineal organ, kidney, gastrointestinal tract, and skin also have endocrine functions.

Endocrine glands are vital to life processes including reproduction, growth, structural and biochemical differentiation, and adaptation to environment and to nutritional states. The activity of each gland is not that of a discrete unit but of an integrated system where dysfunction of one of the glands may cause changes in the function or response of other glands.

HORMONE ACTION AND CONTROL

Hormones may be grouped chemically into three classes: amines, steroids, and peptides (or proteins). Two general theories have been proposed to explain their mode of action. Steroid hormones appear to act by entering the cell and combining with specific nuclear receptors producing a conformational change in the structure. The receptor-hormone complex is transported to the nucleus, where it modifies the process of gene transcription which leads to RNA formation and protein synthesis. It has been proposed that amine, peptide, and protein hormones interact with a stereospecific site on the surface of the target cell.[36] This causes an increase (or decrease) in the activity of the enzyme, adenylate cyclase, which results in an increase (or decrease) in the concentration of cyclic AMP (adenosine-3′,5′-monophosphate) from ATP (adenosine triphosphate). Sutherland and co-workers[64, 65] suggested the second messenger concept whereby the first messenger, hormone, after reaching the target tissues, passes on the message by modifying the intracellular concentration of a second messenger, cAMP, which is responsible for the observed changes in cell function caused by the hormone. Prostaglandins also appear to act as intracellular hormone mediators or second messengers.

The second messenger or single action theory does not explain all of the effects of some hormones, for example insulin. Also, it makes no provision for a feedback of information from cell interior to cell surface or for additional second messengers such as the calcium ion. That all peptide and amine hormones act only at the surface of the target cell has yet to be established.

The control of secretion of hormones depends on various physiologic factors. Feedback control, especially negative feedback system, in various stages of complexity is one of the most important principles of control. For example, a decrease in ionized serum calcium concentration increases parathyroid hormone (PTH) synthesis by the parathyroid gland. The PTH secreted from the gland acts on bone to release calcium ions, on the kidney to prevent renal loss of calcium and promote loss of calcium-binding phosphate ions, and on the intestines to increase calcium absorption. When the serum level of ionized calcium becomes normal, the secretion of PTH decreases to the basal level.

In addition to negative feedback control mechanisms, endocrine glands may be influenced by positive feedback, positive and negative feed-forward loops, diurnal variation and biologic rhythms, external and internal environment (stress, nutrition, temperature, pain, fever, infection, low blood glucose), genetic factors, and pharmacologic doses of hormones. Examples of these will be noted in the discussion of various hormones.

TRANSPORT AND METABOLISM

Transportation of some hormones in the bloodstream is largely by combination with plasma proteins (cortisol bound to transcortin, estrogens bound to albumin, testosterone bound to a sex hormone–binding globulin, thyroxine to thyroxine-binding globulin). The protein-bound hormones are present in equilibrium with small amounts in the free form. It is the free form which is considered to be the metabolically active portion with respect to the target tissue.

The half-life of a hormone in plasma may vary from a few minutes to several days. A physiologically detectable response of a particular target tissue may be immediate or delayed for several hours, or a given hormone may provoke an immediate response in one tissue and a delayed response in another. Hormones are continually lost by metabolic inactivation or by excretion. Some undergo structural alteration to biologically less active or inactive compounds in the liver. These include androgens, estrogens, progesterone, corticosteroids, thyroid hormone, insulin, and epinephrine. Steroid hormones and their metabolites are conjugated in the liver to form compounds more readily soluble in water and hence are more readily excreted in the urine. Some hormones may undergo metabolic inactivation in cells of target tissue (thyrotropin by the thyroid gland) and others may be transformed in target cells to substances of different biologic activity.

All hormone-producing glands must continually provide a characteristic quantity to maintain homeostasis. Hormone deficits, excesses, or altered secretions may be associated with endocrine disorders. Laboratory measurements are performed to determine the concentration of hormones or their metabolites in body fluids.

Pituitary gland (hypophysis)

The pituitary gland is a small mass weighing about 0.5 g and located in a cavity of the sphenoid bone, below the base of the brain. It is divided into anterior and posterior lobes, which are functionally unrelated. The pars intermedia is essentially nonexistent in the human pituitary.

ANTERIOR PITUITARY (ADENOHYPOPHYSIS)
Hypothalamic control

The hypothalamus, a pathway for stimuli originating either in higher brain centers or in the central nervous system, releases neurohormones which exert specific effects on the anterior pituitary.[58] It communicates with the anterior pituitary through the portal system. The anterior pituitary lobe derives its blood supply from two sources: branches of the superior hypophyseal artery and a system of portal veins. The portal system arises from a network of blood vessels formed in the median eminence from branches of the superior hypophyseal artery. These capillaries are in contact with the nerve fibers from the hypothalamus and act as the means of transport for the neurohumors. From this network blood is collected into hypophyseal portal veins, carried down the anterior surface of the pituitary stalk into the sinusoidal capillaries, and brought into contact with the hormone-secreting cells of the anterior lobe. The small neural peptides act to control the secretory activity of the anterior pituitary by stimulation or inhibition of hormone synthesis and release. Those transmitters or releasing/inhibiting factors are as follows:

Adrenocorticotropic-releasing factor (ACTH-RF)
Luteinizing hormone–releasing factor (LH-RF)
Thyroid-stimulating hormone–releasing factor (TSH-RF)
Follicle-stimulating hormone–releasing factor (FSH-RF)
Growth hormone–releasing and inhibiting factors (GH-RF, GH-IF)
Prolactin-inhibiting and releasing factors (P-RF, P-IF)
Melanocyte-stimulating hormone–releasing and inhibiting factors (MSH-RF, MSH-IF)

There is evidence that LH-RF and FSH-RF may be the same molecule but modulated by sex glands since the two do not increase and decrease simultaneously during the reproductive cycle.

The posterior lobe receives its blood from branches of the inferior hypophyseal arteries. It acts largely as a storage site for the hormones vasopressin and oxytocin, which appear to have the hypothalamus as their site of origin.

The structures of thyrotropin-releasing factor (TRH) and luteinizing hormone–releasing factor (LRH) are known. They have been synthesized and used in experimental investigations. A single injection of TRH will cause a rapid increase of plasma TSH and an increase in triiodothyronine level. A prolonged infusion will increase the plasma levels of both triiodothyronine and thyroxine. LRH has proved useful in studies of pituitary failure and of causes of anovulation.

Feedback control

The activity of the anterior pituitary gland is influenced by the concentration of circulating peripheral hormone through feedback mechanisms. The tropic hormones (ACTH, TSH, LH, FSH) produce hormones in target glands which feed back to the pituitary and hypothalamus to maintain the appropriate secretion of the tropic hormone. For example, the anterior pituitary elaborates the thyroid-stimulating hormone (TSH) that causes the thyroid gland to produce and secrete thyroxine. An excess of thyroxine in the blood inhibits the production of TSH and establishes a balance by feedback control. This control of TSH secretion appears to occur at the pituitary level.

For adrenocorticotropic hormone (ACTH), there appears to be control at both the pituitary and the hypothalamus. The pituitary is stimulated to release ACTH, which acts on the adrenal cortex to produce and secrete steroids into the bloodstream. When cortisol in the circulating blood reaches a certain level, it inhibits the release of ACTH from the pituitary. At the level of the hypothalamus, ACTH secretion is controlled by the corticotropin-releasing hormone which is affected by various neurogenic stimuli such as circadian rhythm, anxiety, and hypoglycemia. If exogenous ACTH is administered, the adrenal will increase the output of steroids and the pituitary will be inhibited. However, the control mechanism is not effective when there are disturbances in the feedback regulation such as are found with ACTH accumulation for certain types of adrenal overfunction associated with adrenal hyperplasia (Cushing's syndrome).

Some pituitary hormones apparently may inhibit their own secretion by acting directly on either the hypothalamus or the pituitary gland (autoregulation or short feedback loop). For the female, the hypothalamus appears to have an endogenous rhythm of hormone release which influences the control of the reproductive cycle.

Seven recognized hormones secreted by the anterior pituitary are growth hormone (somatotropin), thyrotropin, corticotropin, melanotropin, follicle-stimulating hormone, luteinizing hormone, and prolactin.

Human growth hormone (HGH)

Human growth hormone or somatotropin is a single polypeptide chain that exercises a generalized growth effect on all tissues and organs. It stimulates bone and cartilage formation and influences intermediary metabolism of protein (enhances amino acid transport and incorporation into proteins), carbohydrates (diabetogenic action), and lipids (stimulates mobilization of fat from adipose tissue).

HGH is species specific, the result of physiochemical variations in molecular structure. The half-life in plasma is about 20 to 25 minutes. It is secreted discontinuously by the pituitary and only in response to various stimuli. There is no evidence of binding to a specific carrier protein.

Measurement of plasma HGH is by radioimmunoassay. In normal individuals, the level is influenced by sex, activity, stress, and metabolism. After the first few weeks of life when the levels are extremely high (15 to 40 ng/ml), the basal level is between 1 and 5 ng/ml. In response to stimuli such as exercise, during deep sleep, and several hours after a meal, HGH will increase temporarily. The tendency for an increased plasma level after exercise is greater in women than in men.

The physiologic effect of decreased or increased HGH secretion depends on the age of the individual. An increased elaboration may be manifested as gigantism in a child and as acromegaly (enlarged hands, feet, frontal bone, and jaw) in the adult. Since other organs of the body are influenced by the pituitary, excess HGH concentrations usually produce changes such as enlargement of the thyroid, altered sexual function, and diabetes. Dwarfism results from the absence of growth hormone in a child. In the adult, decreased secretion in growth hormone produces less obvious changes in metabolism.

To differentiate between low normal and hypopituitary levels of plasma HGH, stimulation tests are used which cause the normal pituitary to increase secretion of HGH, namely, insulin-induced hypoglycemia, arginine infusion, and L-dopa administration.

For insulin-induced hypoglycemia, 0.05 to 0.1 unit of insulin/kg of body weight is injected intravenously. Blood specimens are obtained at intervals over a period of 2 hours and analyzed for glucose and HGH. In normal persons, the blood glucose decreases to 50% of the basal level and a marked increase in plasma HGH occurs 45 to 60 minutes after insulin injection. Any rise in HGH secretion is interpreted as a normal functioning pituitary.

The arginine infusion test usually consists of infusing 30 g of arginine monohydrochloride in distilled water over a period of 30 minutes. Plasma HGH response comparable to that obtained by insulin-induced hypoglycemia is found for women, with a lesser increase for men.

Based on a patient's weight, L-dopa is administered orally in a fasting state and blood samples collected 40, 60, and 90 minutes after ingestion. Patients remain recumbent during the procedure. A plasma level of HGH greater than 6 ng/ml on any sample is considered a normal response.[53] This test is useful in outpatient screening for hyposomatotropin.

As previously noted, growth hormone plasma concentrations in normal individuals vary greatly with different stimuli, including activity. Therefore, for baseline levels of HGH which are elevated in patients with acromegaly, it is necessary to show a lack of suppressibility of the HGH levels with a standard dose of glucose. Following 100 g of glucose given orally, a normal person will have a plasma HGH level within an hour of less than 1 ng/ml. Patients with acromegaly will have little, if any, suppression of the fasting plasma growth hormone concentration.

Growth hormone appears to produce its effect on bone by stimulating the growth of cartilage. In the formation of cartilage, uptake of sulfate occurs in the synthesis of chondroitin sulfate. The incorporation of the sulfate depends on a growth hormone–dependent polypeptide, somatomedin or sulfation factor, with insulin-like properties. Somatomedin appears to influence cartilage growth in other ways and may perhaps be the mediator of other effects of growth hormone.

Animal experiments have indicated the liver as one site for the synthesis of somatomedin. It has been measured in serum using radioimmunoassay. There is a decrease in patients with pituitary dwarfism and an increase in patients following administration of growth hormone as well as in some patients with acromegaly.

Thyroid-stimulating hormone (TSH)

Thyrotropin or TSH is a glycoprotein that stimulates various metabolic activities of the cells of the thyroid gland. Secretion by the anterior pituitary is influenced by a thyrotropin-releasing factor from the hypothalamus and by the level of circulating thyroid hormone.

TSH is responsible for a number of reactions occurring within the thyroid gland, including those necessary for hormone synthesis and secretion. It activates adenyl cyclase in the thyroid cell membrane, resulting in increased intracellular cyclic AMP which acts as the second messenger in the cell. TSH is made up of two subunits (alpha and beta), with the specificity apparently conferred by the beta unit. The feedback control of its secretion by thyroxine and triiodothyronine is mainly in the pituitary. Biologic half-life is about 60 minutes and it is degraded for the most part in the kidney and to a lesser extent in the liver.

Measurement of the concentration of TSH in plasma is performed by radioimmunoassay, and the normal level has been found to be less than 10 μU/ml. A marked increase occurs in primary hypothyroidism, but in hypothyroidism secondary to hypopituitarism TSH is undetectable. In hyperthyroidism TSH is not elevated.

In patients with thyrotoxicosis, there is present a long-acting thyroid stimulating substance (LATS), an immunoglobulin, so called because upon being administered intravenously it is cleared from the blood more slowly than TSH. It is measured by bioassay and usually none is detected in normal serum.

Adrenocorticotropic hormone (ACTH)

Corticotropin or ACTH is a polypeptide responsible for regulating corticosteroid secretion by the adrenal cortex; in experimental situations it influences carbohydrate and lipid metabolism. It is secreted intermittently throughout the day, peaking generally early in the morning, but it depends on the sleep pattern. Regulation of the secretion is by negative feedback of cortisol, diurnal rhythm, and stress. The hormone attaches to a membrane receptor on the surface of the adrenocortical cell, stimulating the action of adenyl cyclase to form cyclic AMP. It does not appear to be altered in the process of exerting its effect on adrenal tissue. The biologic half-life is about 10 minutes.

The plasma concentration of ACTH determined by radioimmunoassay has been found to be less than 100 pg/ml for normal individuals, the level being highest about 6 AM and lowest about 10 PM. Increased blood concentrations are produced by stress and low cortisol output. ACTH is increased in Cushing's syndrome because of the pituitary disease and in nonendocrine tumors releasing ACTH.

Melanocyte-stimulating hormone (MSH)

Melanotropin or MSH was so named because it darkens the skin of certain animals. In man, it increases pigmentation by promoting melanin synthesis.

There are two melanocyte-stimulating peptides, alpha MSH and beta MSH. Nearly all of that found in the human pituitary is the beta MSH. It has been suggested that since the secretion of MSH and ACTH appears to occur at the same time in certain adrenal disorders, they may be from the same cell. The feedback control by cortisol is the same for both hormones.

Measurements for each form of MSH by radioimmunoassay have shown alpha MSH to be below detectable limits and beta MSH to have a concentration of about 0.1 ng/ml in normal plasma. Increased amounts of beta MSH have been found for patients with untreated Addison's disease, in Cushing's syndrome, and with ectopic ACTH-producing tumors. MSH can cause darkening of the skin of humans when corticoids are absent, such as in Addison's disease.

Gonadotropic hormones (follicle-stimulating hormone and luteinizing hormone)

There are two gonadotropic hormones secreted by the anterior pituitary, follicle-stimulating hormone (FSH) and luteinizing hormone (LH). They are glycoproteins, each having two peptide chains, alpha and beta, with the specificity in the beta chain. If both hormones are present, FSH and LH appear to have enhanced activity of each of their effects. They are not sex-specific.

FSH stimulates development of the ovarian follicle in the female and development of testicular tubules and the maintenance and differentiation of spermatozoa in the male. In the female, LH promotes luteinization or formation of the corpus luteum, stimulates estrogen secretion, and causes ovulation. In the male, LH (also termed interstitial cell–stimulating hormone [ICSH]), stimulates the development of the Leydig cells of the testis and causes them to secrete testosterone.

The mechanisms by which gonadotropins produce their effects are complex and can only be partially explained by the second messenger theory. LH appears to bind to a specific receptor on the surface of the responsive cell (corpus luteum) where it activates adenyl cyclase to form cyclic AMP. The cyclic AMP initiates the cellular processes which result in hydroxylation of cholesterol prior to its conversion to pregnenolone and finally to progesterone in the biosynthetic pathway. In a similar fashion, LH acts on the Leydig cells of the testes, but the final product is testosterone. Less

Table 14-1
Normal range of FSH and LH in urine and serum

	Follicle-stimulating hormone		Luteinizing hormone	
	Urine (IU/24 hr)	Serum (mIU/ml)	Urine (IU/24 hr)	Serum (mIU/ml)
Men	5- 25	5- 25	5- 18	6- 30
Women				
Follicular phase	5- 20	5- 20	2- 25	2- 30
Peak of middle cycle	15- 60	12- 30	30- 95	40-200
Luteal phase	5- 15	5- 15	2- 20	0- 20
Postmenopausal	50-100	40-200	40-110	35-120

is known about the action of FSH except that it does stimulate testicular cyclic AMP.

The sex hormones feed back through the hypothalamus, and possibly the pituitary, to influence the subsequent secretion of gonadotropins. The biologic half-life of FSH is approximately 170 minutes and that of LH about 60 minutes.

Both FSH and LH in serum and urine are measured using radioimmunoassay (Table 14-1). They are present in serum at all ages. Urinary concentrations are comparable to those of serum with the exception that very little is found before puberty. For women with normal menstrual cycles, serum FSH increases rapidly on the day of menses, stabilizes, decreases 2 days before the LH peak when there is a concomitant FSH rise of less degree and duration, and finally decreases to its minimum during the latter half of the luteal phase of the cycle. For LH, the concentrations rise on the first day of menses and plateau, peak at midcycle with considerable fluctuation throughout the luteal phase, and gradually decrease to minimum levels on the day preceding the first day of the next menstrual cycle. In men and nonovulating women, the gonadotropin levels are comparable while for women during menopause there is a marked increase.

For evaluation of hypothalamic-pituitary function, serial daily measurements of serum FSH and LH are useful, but there are hour-to-hour fluctuations which are especially pronounced for LH. The greatest variation has been observed during the midcycle surge and the least during the late follicular phase.

Gonadotropins are increased in primary hypogonadism and ovarian failure. They are decreased in hypofunction of the hypothalamus and pituitary, estrogen administration, anorexia nervosa, and estrogen- or androgen-secreting neoplasms of the adrenal gland, ovary, or testis. Assay of the gonadotropins provide a means of differentiating male hypogonadotropic eunuchoidism caused by testicular insufficiency from hypogonadism resulting from pituitary failure. In a condition termed "fertile eunuch" there is LH deficiency. In gonadal agenesis of children, FSH will increase before puberty with a delay of the increase of LH. Patients with polycystic ovarian disease have been found to have plasma LH higher than normal women during the follicular phase while the concentration of FSH is consistently lower.

Prolactin

Prolactin initiates and maintains lactation by acting directly on the mammary gland. However, it depends on the presence of other hormones to achieve its functions. The biologic half-life is 20 to 30 minutes.

Using radioimmunoassay, prolactin has been measured in blood serum of normal adults and children. The normal range is up to 25 ng/ml, and there is a circadian rhythm.

In pregnancy, the serum levels increase progressively, paralleling placental lactogen, and return to nonpregnant levels 4 to 6 weeks postpartum. Prolactin is elevated in serum from most patients showing amenorrhea and galactorrhea. There are a number of stimuli that will cause a temporary rise: surgery, suckling and breast manipulation, exercise, psychic stress, insulin-induced hypoglycemia, and synthetic thyrotropin-releasing hormone administration. It has been implicated as having a role in carcinogenesis and growth of human breast cancer. However, literature is contradictory at this time.

NEUROHYPOPHYSIS (POSTERIOR PITUITARY)

The neurohypophysis is part of a neurosecretory system that includes neurons of the supraoptic and paraventricular nuclei of the hypothalamus and the neurohypophyseal tract which transports axons from these nuclei to terminations in the median eminence, pituitary stalk, and posterior pituitary.

The posterior pituitary secretes two hormones, vasopressin and oxytocin. These are produced in the ganglion cells of the hypothalamic nuclei and are carried to the posterior pituitary, which serves mainly as a reservoir of the hormones. Their structures have been determined (nonapeptides) and the compounds synthesized. Each shares some of the physiologic effect of the other and various stimuli will cause their release into the circulation.

Vasopressin (antidiuretic hormone)

Vasopressin or antidiuretic hormone (ADH) has as its major physiologic function increasing water reabsorption from the distal renal tubules and collecting ducts. When ADH is lacking, urine hypotonic to plasma causes an increased loss of water. Vasopressin probably acts by binding to its renal target cell by disulfide bonds and causing an increase in cAMP, which in turn mediates the effect of ADH on water and sodium reabsorption.

ADH secretion is regulated by the effective osmotic pressure of the plasma, extracellular fluid volume, and central nervous system activity. Its biologic half-life is between 10 and 20 minutes.

Concentration of ADH in plasma may be measured by radioimmunoassay. For healthy adults in recumbent position, the normal range is less than 1.5 pg/ml. The level varies with the state of hydration and with exercise.

A deficiency of ADH produces diabetes insipidus. It may be the result of congenital or acquired lesions in the neurohypophyseal system. Characteristic symptoms are the passage of large quantities of dilute urine (polyuria) and drinking of large amounts of fluid (polydipsia). Abnormally elevated levels of ADH have been reported for some pathologic conditions.

Oxytocin

Oxytocin promotes secretion of milk from the lactating breast (milk-ejecting effect) and stimulates contraction of the smooth muscle of the uterus. Preparations containing oxytocin have been used to induce or speed up labor.

Oxytocin has a biologic half-life of 1 to 4 minutes, is metabolized mainly by the kidney, and is rapidly excreted in the urine. Measurement in the plasma is made by radioimmunoassay. Normal males and nonpregnant women have concentrations in plasma of less than 1.5 pg/ml.

Thyroid

The thyroid consists of two lobes and a connecting isthmus. It is located at the sides of the trachea below the larynx and weighs about 20 g. Normal thyroid function depends on synthesis and secretion of active thyroid hormones which influence various metabolic processes and growth.

An adequate supply of iodine is necessary for synthesis of thyroid hormones. It is derived from two sources, ingested iodine absorbed from the gastrointestinal tract as iodide and iodide from deiodination of thyroid hormones. The iodide is removed selectively from the circulation into the thyroid cell and follicular lumen at a clearance rate of about 35 to 40 ml of plasma per minute; approximately 50% of the iodide from the blood moving through the gland is removed. The iodide is oxidized by iodide peroxidase to a form capable of iodinating tyrosyl residues in thyroglobulin, a glycoprotein that is synthesized within the follicular epithelium and stored within the follicles as colloid. The organic iodination which occurs at or near the surface of the thyroid follicle cells results in formation of monoiodotyrosine (MIT) and diiodotyrosine (DIT) residues. These iodotyrosines undergo a coupling reaction to yield a variety of iodothyronines including 3,5,3′-triiodothyronine (T_3) and 3,5,3′,5′-tetraiodothyronine (T_4) or thyroxine. Storage of the hormones occurs in the colloid of the gland located in the follicles of the lobules. The colloid consists mainly of thyroglobulin, other proteins, and water. The epithelial follicular cells ingest the thyroglobulin colloid by pinocytosis and in these cells the thyroid hormones are released by hydrolysis of the thyroglobulin with proteases and peptidases. Degradation to the amino acids and secretion by the gland are controlled by the thyrotopic hormone (TSH) from the anterior pituitary.

Thyroid hormones exist in the gland as units of the peptide chain of thyroglobulin but are circulated in the blood, noncohesively bound to serum proteins. The active iodothyronines are the amino acids thyroxine and triiodothyronine. Thyroxine accounts for more than 90% of the hormonal iodine circulating in the blood (4.5 to

Thyroxine (T_4)

Triiodothyronine (T_3)

11.0 μg/dl), and about 60% is transported in combination with a glycoprotein, thyroxine-binding globulin (TBG), that has an electrophoretic mobility between the alpha$_1$ and alpha$_2$ globulins. Approximately 30% is bound to prealbumin (TBPA)

and 10% to albumin, with 0.9 to 2.5 ng/dl remaining free in circulation. Normally about one third of the available sites on TBG are occupied by T_3 and T_4; the remainder are unoccupied. The T_3 is less firmly bound to plasma protein (TBG and albumin, not TBPA) and is much more active physiologically than thyroxine. However, the concentration of T_3 in blood (75 to 200 ng/dl) is so low that it does not significantly affect the iodine level. Biologically its effect is 4 to 5 times that of thyroxine. The free hormone causes the characteristic action in the target cells. The biologic half-life of plasma T_4 is 7 days and that of plasma T_3, 1½ to 3 days. Many peripheral tissues take up T_4 and T_3 and also deiodinate them.

Thyroid hormones increase oxygen consumption (calorigenic effect) in many tissues and increase general metabolic rate. They are necessary for skeletal growth and development and alter protein, carbohydrate, and fat metabolism.

No one of the thyroid function tests that aid in assessing the activity of the thyroid gland gives the complete information needed in clinical evaluations. Since the level of circulating hormones is related to the activity of the gland, procedures for determining the serum levels are used in studying the functioning and physiology of the organ. The characteristics of the hormones that provide a basis for methodology are the concentration of iodine in the molecule, the equilibrium that exists in vivo and in vitro between the free and the protein-bound hormone, and the specific protein binding of the hormones.

Protein-bound iodine and butanol-extractable iodine

The protein-bound iodine (PBI) or serum-precipitable iodine has been used extensively as a measure of serum thyroxine. Red blood cells contain about 25% of the amount of protein-bound iodine found in serum. When proteins of serum are precipitated with the usual precipitating reagents, the hormones are quantitatively coprecipitated since approximately 99% of the iodothyronines are bound to serum proteins. Precipitation is performed to eliminate contaminants, but nonhormonal iodine such as contrast materials used in x-ray examinations and drugs containing iodine do interfere and give falsely high values. The serum may also be treated with an ion exchange resin to remove nonbound iodine. The precipitate or treated serum is digested and oxidized with acid (wet ash method)[72] or is dried, oxidized, and ashed in a muffle (dry ash method) to convert the organic iodide to iodate for the iodine analysis. The determination of iodine is based on the catalytic effect of iodide on a combination of arsenious acid and ceric ammonium sulfate.

In an effort to reduce the error introduced into the PBI procedure by nonhormonal iodine, a butanol-extractable iodine (BEI) method was developed. The test is based upon the solubility of T_3 and T_4 in butanol and removal from the organic extract of inorganic iodide and the monoiodotyrosine and diiodotyrosine by use of alkali. Since the method is time consuming and organic iodine compounds interfere, it has little advantage over the PBI procedure.

Thyroxine by column

Interference by exogenous organic iodine compounds may be reduced but not eliminated by employing a thyroxine-by-column method. The procedure utilizes an anion exchange resin that removes T_4, T_3, other iodoamino acids when present, and iodides from an acidified (pH 3.8) solution of the specimen. The acidified wash solutions free the column of residual protein, some organic iodine compounds used in diagnosis and therapy, and mono- and diiodotyrosines if present. Elution of T_3 and T_4 is accomplished by using a 50% acetic acid solution (pH 1.4). Inorganic iodine remains on the column. Three fractions are collected and the iodine content of each fraction is determined. Several procedures, with and without incineration, have been proposed for determining the iodine content of the eluate. There is not definite separation of T_4 from iodinated contaminants such as radiographic contrast media. The following have been found to interfere: diiodohydroxyquin (Floraquin), iothiouracil (Itrumil), bunamiodyl (Orabilex), iodoalphionic acid (Priodax), iopanoic acid (Telepaque), and iophenoxic acid (Teridax).

Thyroxine by competitive protein-binding (Murphy-Pattee)

Murphy and Pattee[44] developed a more specific method for thyroxine assay using the specific binding properties of thyroxine-binding globulin. It is based on the equilibrium between free and bound thyroxine in plasma. Since the measurement is that of total thyroxine, neither inorganic nor organic iodine contaminants interfere.

A number of modifications of the Murphy-Pattee technique have been published. The T_4 is extracted from the serum either with organic solvents or by adsorption. A standard amount of TBG in a barbital buffer containing ^{125}I-thyroxine is added to the extracted T_4. The radioactive T_4 is displaced from the TBG in amounts proportional to the patient's serum T_4 in the extract. The free radioactive T_4 is separated from the bound and either the free or bound ^{125}I-thyroxine measured. In serum diluted with a barbital buffer solution, T_4 is preferentially bound by TBG. Normal values by this technique vary between 4.5 and 11 μg/dl. If values are desired in terms of T_4-iodine, multiply the measured T_4 concentration by 0.653, the proportion of the iodine in T_4. Ethanol extraction of thyroxine from patient sera varies in efficiency (usually 75% to 93%). Some laboratories correct for recovery and some do not.

As long as the binding capacity of the transport proteins is normal, measurement of the concentration of T_4 in the serum reflects T_4 secretion. When the TBG in blood plasma is increased, as in pregnancy and after administration of estrogens or antiovulatory drugs, the serum T_4 is usually elevated although the patient appears euthyroid. The TBG capacity may be decreased following administration of androgens and for patients with a nephrotic syndrome, thus a decreased T_4 level is found without hypothyroidism. Low values are obtained when patients are receiving diphenylhydantoin (Dilantin) since it competes with thyroxine for TBG-binding sites at the thyroxine levels. Tracer doses of ^{131}I may also invalidate thyroxine values by competitive protein-binding for several weeks. When variations of the TBG outside of the normal range occur, a free T_4 should be measured to reflect the thyroid status of the patient.

Thyroid uptake of ^{131}I

The radioactive iodide uptake (RAIU) method is a measure of the rate at which the thyroid gland removes iodine from the circulation, relative to the efficiency of the kidney. It is not necessarily a measure of hormone release and it depends on blood flow, iodine availability, and normal biosynthesis. A tracer dose of ^{131}I is administered orally and the radioactivity measured over the thyroid after 3, 6, and 24 hours. The result is related to the dose administered. Uptake is usually increased in hyperthyroidism for the shorter time periods but may be normal at 24 hours. In hypothyroidism the 24-hour uptake is decreased but the 3- and 6-hour uptake may be normal. The plasma protein-bound radioiodine 24 hours after administration of the dose may also be determined as an indication of hyperthyroidism and hypothyroidism. The method is limited, since antithyroid and iodine-containing drugs and organic iodine x-ray contrast media interfere.

T_3 uptake test

T_3 uptake tests measure the unsaturated binding capacity of serum proteins, mostly TBG, for labeled T_3 and thereby indirectly measure serum thyroxine levels. A continuous equilibrium exists in vivo and in vitro between free and protein-bound thyroid hormones. Free T_4 will easily replace bound T_3, but the reverse reaction is insignificant. Serum is added to a buffer solution of radioiodine-labeled T_3 and unoccupied TBG binding sites bind the labeled T_3. The unbound labeled T_3 is removed by an adsorbent. T_3 is bound to the TBG molecule at places not occupied by T_4.

If the concentration of TBG in the serum is normal, the number of available binding sites for labeled T_3 binding is inversely proportional to the number of binding sites occupied by T_4. When there is a high level of endogenous thyroid hormone, the binding sites in the plasma are relatively saturated, leaving fewer sites available for the labeled T_3 and more ^{125}I-labeled T_3 in a free state to be adsorbed. When the endogenous hormone level is decreased, as in hypothyroidism, fewer binding sites are occupied by T_4, leaving more sites than normal available for labeled T_3 and less ^{125}I-labeled T_3 in a free state to be adsorbed. Since the test measures the unoccupied binding sites, the result is affected not only by the level of thyroxine but also by variations in the concentration of TBG. The procedure is rapid and not affected by inorganic iodine.

In conditions where the TBG is being produced in increased amounts, such as pregnancy and estrogen therapy, T_3 uptake will be decreased.

The T_3 uptake is increased when TBG production is decreased in conditions such as nephrosis and liver disease, for patients receiving ACTH or corticosteroid therapy where fewer unbound sites are available, and by the action of drugs that release thyroxine from its binding sites such as salicylates, diphenylhydantoin, and phenylbutazone. The test is being replaced by free thyroxine assay.

Free thyroxine

The physiologic action of thyroxine depends on the concentration of the unbound hormone, and a measurement of this fraction is useful in the assessment of the activity of the thyroid gland. Levels of free and bound thyroxine in the circulation depend on the quantity of thyroxine and the availability of thyroxine-binding globulin. Because of the low concentration of free T_4, a direct measurement is not made. However, by adding radioiodine-labeled T_4, which equilibrates in a manner similar to endogenous T_4, to a serum sample and dialyzing, the concentration of the free T_4 can be assessed by measuring the fraction of labeled T_4 which is dialyzed across the semipermeable membrane under established conditions. An alternate procedure involves the measurement of total thyroxine and the latent binding capacity simultaneously to give the TBG percent saturation. The percent saturation of the unknown serum is compared with the percent saturation of standards having a known free thyroxine value. Free thyroxine is controlled by the relationship between circulating T_4 bound to TBG and the unsaturated binding capacity. Since methods of analysis vary, the normal range will differ somewhat in different laboratories. However, it will be approximately 0.9 to 2.5 ng/dl.

The test is important in assessing thyroid status in hypoproteinemia, in pregnancy, and in patients using antiovulatory birth control compounds or receiving drugs such as diphenylhydantoin (Dilantin) where total thyroxine levels may provide misleading information. In pregnancy and when estrogens are administered, free thyroxine remains normal in the euthyroid state. Clark and Horn[11] proposed a free thyroxine index obtained from the product of the total serum thyroxine and the T_3 uptake. In euthyroid states with pregnancy or with estrogen therapy, the index is within the normal range.

Triiodothyronine (T_3)

Triiodothyronine is physiologically more active than thyroxine (T_4). It is derived from thyroid secretion and from monodeiodination of T_4 in peripheral tissue. Sensitive radioimmunoassay procedures for measuring T_3 are now available. Triiodothyronine covalently bonded to a suitable protein (bovine serum albumin) to form an antigen is injected into rabbits to produce highly specific antibodies. Antibody specificity depends on the site on the T_3 molecule where the protein is bound to form the antigen. When using whole serum, T_3 must be displaced from and prevented from rebinding with the TBG. For this purpose ANS (8-anilino-1-naphthalene-sulfonic acid) appears to be the best compound for inhibiting the T_3 binding on active sites of TBG. T_3 in serum and radioactive T_3 are mixed with the antibody; after equilibration has occurred, the free T_3 is separated from the antibody-bound T_3 by a suitable adsorbent. Normal values range between 75 and 200 ng/dl.

The usefulness of serum T_3 assays is being observed as more data are being made available.[10, 69] In T_3 toxicosis, the serum T_3 is elevated while the T_4 remains normal. Both are elevated in Graves' disease, but the proportion of T_3 relative to T_4 is greater. Response to pituitary TSH and hypothalamic TRH appear to be more rapid for T_3 than T_4. When there is a decreased thyroid reserve, the T_4 may be decreased but the patient may show a euthyroid status by preferentially secreting T_3.

Hypothyroidism and hyperthyroidism

In Table 14-2 a summary of normal values is given for a variety of laboratory tests which reflect various aspects of thyroid function.

Hypothyroidism, caused by a deficiency of thyroid hormone, is associated with a number of clinical symptoms depending on the age of the patient and the extent of the deficiency. It is found in cretinism, myxedema, and some forms of goiter. Cretinism, a congenital defect caused by hypothyrodism in infancy, is characterized by retardation in mental, physical, and sexual development. In myxedema, a condition

Table 14-2
Thyroid function tests—summary of normal values

Method	Normal range
Protein-bound iodine	4.0 to 8.0 µg/dl serum
Butanol-extractable iodine	3.2 to 6.2 µg/dl serum
Thyroxine	4.5 to 11.0 µg/dl serum
By column, as iodine	2.9 to 6.4 µg/dl serum
As iodine	2.9 to 7.2 µg/dl serum
Free	0.9 to 2.5 ng/dl serum
Triiodothyronine	75 to 200 ng/dl serum
T_3 uptake	10% to 14.6%
Thyroid-stimulating hormone	Up to 10 µU/ml serum

resulting from atrophy or removal of the thyroid gland during adulthood, low metabolic rate, loss of hair, apathy, and thickening of the skin resembling edema are observed. Some forms of goiter are associated with hypothyroidism. When a defect of thyroid physiology occurs in which there is an inadequate output of thyroid hormone, the thyroid gland enlarges. Thyroid cells increase in size and number to compensate in part for the deficient activity of the cells. If the defect is relatively slight, hyperfunction may produce normal hormone levels and the enlarged thyroid gland is classified as a simple goiter. Iodine is required for elaboration of the thyroid hormone; when foods and drinking water are deficient in iodine, the cells cannot synthesize the hormone. With decreased thyroxine secretion, the subject is lethargic, passive, and sensitive to cold.

Hyperthyroidism, caused by overproduction of thyroid hormone, is characterized clinically by hypermetabolism associated with cardiovascular and neuromuscular changes. Some physical characteristics of the disease are increased metabolic rate and heart rate, excessive sweating, heat intolerance, gastrointestinal upsets, protrusion of eyeballs, insomnia, and increased appetite accompanied by weight loss and nervousness. Diffuse enlargement of the thyroid gland occurs in exophthalmic goiter (Graves' disease). In toxic adenoma, the enlargement is uneven as a result of the tumors. In various forms of Graves' disease a gamma globulin produced by the lymphocytes of the patient and termed the long-acting thyroid stimulator (LATS) is found in significant concentration. The mechanism of action of LATS on the thyroid appears similar to that of TSH. However, its appearance alone does not provide sufficient explanation for all aspects of Graves' disease.

Calcitonin (thyrocalcitonin)

The thyroid gland also secretes thyrocalcitonin, a polypeptide hormone that lowers serum calcium. It is formed in thyroid parafollicular cells and in some cases in parathyroid and thymus tissue. The structure is known and it has been synthesized. Radioimmunoassay methods of measurement in plasma have been developed which indicate a circulating concentration ranging from 0.02 to 0.4 ng/ml for the normal adult.

The most important physiologic function of calcitonin is to lower serum calcium when it increases above the normal range. Calcitonin secretion is proportional to the blood calcium when the total calcium exceeds 9 mg/dl; below this level it is undetectable in normal plasma. It appears to act by inhibiting bone resorption and calcium release and has a half-life of 4 to 12 minutes. Patients with medullary carcinoma of the thyroid have continued hypersecretion of calcitonin.

Parathyroid glands

The parathyroid glands consist usually of two pairs of discrete small masses embedded in thyroid tissue, two on each side of the neck (the location and number of individual parathyroids may vary). They secrete the parathyroid hormone (PTH), a polypeptide, which influences calcium and phosphate metabolism. It is synthesized and secreted continuously, not stored in the tissue.

PTH increases calcium absorption from the intestines, regulates excretion in the urine, and stimulates release from the bone. When there is a drain on the blood calcium, the parathyroid glands are stimulated; if sufficient dietary calcium is not supplied, bones will be demineralized. The primary function of the PTH is to maintain proper concentration of ionized calcium in the blood, the physiologically active fraction of the calcium which is approximately half of the total circulating calcium.

Calcium is essential to many biologic processes and its concentration in cellular and extracellular fluids in maintained relatively constant by a system of controls. Three of the most important controls are hormonal; they are parathyroid hormone, calcitonin, and 1,25-dihydroxycholecalciferol (vitamin D). PTH and the active metabolite of vitamin D ($1,25[OH]_2D_3$) act together to maintain the normal serum calcium concentration and increase mobilization of magnesium from bone. The most important function of calcitonin appears to be that of lowering serum calcium concentration when it exceeds the normal limit. The three hormones act on bones, kidneys, and intestines in maintaining mineral homeostasis.

Actions of parathyroid hormone on its target tissues, bone and kidneys, involve activation of adenyl cyclase and then increasing cyclic AMP concentration which mediates the physiologic response. In the normal state, the circulating ionized calcium levels comprise a negative feedback mechanism to regulate the secretion of the parathyroid glands. A decrease in serum calcium stimulates increased secretion of PTH and an increase in serum calcium decreases secretion of PTH. By controlling its clearance by the kidneys, PTH regulates in large part serum phosphate concentrations. Levels of serum calcium and phosphate tend to vary inversely.

PTH in blood serum in measured by radioimmunoassay. Purified human PTH is difficult to obtain; therefore, antibodies are prepared against either bovine or porcine PTH as the antigen. The hormone level is assayed in conjunction with the serum calcium concentration. There is an early morning increase in PTH, which is present in urine in very small amounts.

There is an increase in parathyroid hormone production in primary hyperparathyroidism caused by a defect in the normal feedback mechanism of plasma calcium concentration as found in patients with adenoma or hyperplasia of the parathyroids. Serum calcium is elevated, although sometimes only slightly. The serum PTH must be related to the serum calcium level in order to discriminate between normal individuals and patients with primary hyperparathyroidism. PTH assays are also used to aid in locating the site of suspected parathyroid adenoma by obtaining blood samples for analysis from various areas on both sides of the neck.

Secondary hyperparathyroidism is characterized by changes in mineral metabolism associated with compensatory enlargement or tumors of the gland. Such conditions occur in chronic renal failure and with patients requiring long-term hemodialysis.

Upon injection of the hormone into the body or in hyperparathyroidism there is increased excretion of phosphorus in the urine. The PTH decreases tubular reabsorption of phosphorus (TRP). Measurement of TRP is useful as an aid in early diagnosis of primary hyperparathyroidism.

Hypoparathyroidism or deficiency of PTH is usually the result of parathyroidectomy. There is a decrease of calcium ion levels in blood, and tetanic convulsions eventually appear because of changes in electrolyte balance in muscular and nervous tissue. Thus by means of its control of blood calcium the hormone regulates nerve and muscle irritability. Laboratory data characteristic of hypoparathyroidism are decreased serum and urinary calcium, elevated serum phosphate and tubular reabsorption of phosphorus, and low or undetectable PTH.

Islets of Langerhans of the pancreas

The endocrine tissue of the pancreas constitutes approximately 1% of the gland and consists of about 2 million islets of Langerhans scattered throughout the pancreas. Insulin-secreting beta cells and glucagon-secreting alpha cells are located in the islets. Both hormones are important in regulating metabolism throughout the body. Excess insulin will lower blood glucose (hypoglycemia) and insufficient secretion of insulin will cause diabetes mellitus. Glucagon increases blood glucose by stimulating hepatic glycogenolysis. An excess will cause diabetes whereas too little glucagon can produce hypoglycemia.

Insulin

Insulin is a polypeptide with two chains of amino acids joined by disulfide bonds. It is synthesized as a single polypeptide chain, proinsulin, in the endoplasmic reticulum of the beta cells and transported to the Golgi complex where it is packaged into beta granules. Proinsulin is the biosynthetic precursor of insulin. The exact site of conversion of proinsulin to insulin has not been established. The conversion occurs by proteolytic cleavage and leaves a small amount of unconverted proinsulin, a C- (connecting) peptide, and insulin in the granules. Insulin is complexed with zinc and stored. In response to stimuli (glucose) the beta granules move to the plasma membrane of the beta cell and the walls of the granules fuse with the membrane and rupture. The released insulin must then cross the basement membrane of the beta cell and the vascular basement membrane and finally the endothelial cells to reach the bloodstream. In circulation, insulin is degraded by cleavage of the disulfide bonds. It has a half-life of 10 to 25 minutes, with a large amount being bound in the liver and kidneys.

The major control of insulin secretion is exerted by a feedback effect of the blood glucose on the pancreas. The action of glucose on insulin secretion is biphasic; there is an acute release followed by a slowly developing prolonged response. The major function of insulin is the control of carbohydrate metabolism whereby blood glucose concentrations are maintained at a normal level. When the supply of insulin is deficient or lacking, the peripheral tissues decrease glucose use, there is increased hepatic gluconeogenesis, blood glucose levels increase above the threshold, glycosuria occurs, and an osmotic diuresis results. Because of this increased urinary flow or output, dehydration and hemoconcentration may become evident, followed by shock, hypotension, diminished renal flow, anuria, and death. Insulin deficiency also causes a mobilization of depot fat into the blood as free fatty acids and a decrease in protein synthesis. Ketone bodies (acetone, acetoacetic acid, and β-hydroxybutyric acid) appear in the blood since synthesis of fatty acids stops at this stage in the liver cells and in the adipose tissue.

When the concentration of glucose in the blood moving through the pancreas attains a certain level, insulin production is increased. Other agents that will stimulate insulin secretion are amino acids, secretin, fructose, glucagon, sulfonylureas, pancreozymin, and "gut glucagon," a material released by the intestinal tract, which crossreacts immunologically with glucagon. Inhibitors of insulin release are epinephrine, norepinephrine, and the diazoxide and thiazide drugs. Insulin lowers blood glucose by increasing the oxidation of carbohydrates and the deposition of glycogen in muscles and liver.

Serum and urine insulin levels are determined by radioimmunoassay procedures. The normal fasting serum level ranges from 10 to 30 μU/ml. After oral ingestion of glucose, blood insulin in normal adults increases rapidly to a maximum in less than an hour and then decreases to the fasting level in 2 to 3 hours. Following a carbohydrate load, some prediabetic patients will have serum insulin levels below normal and those with maturity onset diabetes will frequently have elevated serum insulin levels in the fasting state and in response to the glucose. Measurement of serum insulin levels during a glucose tolerance test gives additional information about the carbohydrate metabolism of the patient.[35]

Obese patients have elevated fasting serum insulin values and secrete several times as much insulin as normal after glucose ingestion. In functional or reactive hypoglycemia, insulin release corresponds to rising postprandial blood glucose, but in patients with mild diabetes insulin release is delayed. It may or may not be increased in cases of insulinoma since the tumors frequently produce insulin intermittently. To screen for insulinomas, fasting or administration of tolbutamide may be used to produce hypoglycemia, and then the serum is assayed for insulin in conjunction with blood glucose. Elevated serum insulin levels have been found in acromegaly. When hypoglycemia is associated with nonpancreatic tumors, the serum insulin concentration is usually within the normal range.

The symptoms of diabetes mellitus can be relieved by injection of insulin. The hormone cannot be given orally because it is inactivated by digestion with proteolytic enzymes. Several types of insulin are available for clinical use, including regular insulin, crystalline insulin, protamine zinc insulin, NPH insulin, globin insulin, and lente insulin. The chief difference is the time of action.

Glucagon

Glucagon is a polypeptide hormone secreted by the alpha cells of the islets of Langerhans. It is released in response to insulin-induced hypoglycemia or fasting. Orally or intravenously administered glucose will suppress glucagon secretion.

Glucagon stimulates glucose production from glycogen in the liver by increasing cyclic AMP and activating hepatic phosphorylase. It also stimulates increased release of insulin from the pancreatic beta cells independent of the increases in blood glucose.

Serum concentration of glucagon is measured by radioimmunoassay and the fasting level is approximately 100 pg/ml. Increased levels have been found for patients with diabetes of genetic origin and familial multiple endocrine adenomatosis. Decreased levels have been reported for some patients with severe chronic pancreatitis.

Adrenal gland

The superior pole of each kidney is covered by an adrenal gland in the manner of a cap that weighs between 5 and 7 g. The glands are composed of two separate parts: about 80% consists of the outer shell or cortex and the remainder is an inner core or medulla. They are attached to and separated from the kidneys by a layer of fat. The two parts arise from different types of tissue and have unrelated functions.

ADRENAL CORTEX

Secretions of the adrenal cortex are indispensable for life; without mineralocorticoid and glucocorticoid replacement therapy following adrenalectomy death will result. The cortex produces steroid hormones that control water and electrolyte balance (mineralocorticoids), influence the metabolism of carbohydrates and proteins (glucocorticoids), and exert some effect on reproductive function (sex hormones). Histologically, the cortex consists of three separate layers or zones. The zona glomerulosa, the outer and thinnest layer, produces mineralocorticoids and is influenced very little by adrenocorticotropin. The zona fasciculata, the central and thickest layer, and the zona reticularis, the inner layer, are responsible for the production of glucocorticoids and androgens. Activity, function, and growth of the two inner zones are controlled by corticotropin from the anterior pituitary gland (hypothalamus-pituitary-adrenal axis), whereas the control of the outer zone is by the renin-angiotensin-aldosterone system.

Structurally, the adrenocortical steroids are of two types: those that have a keto

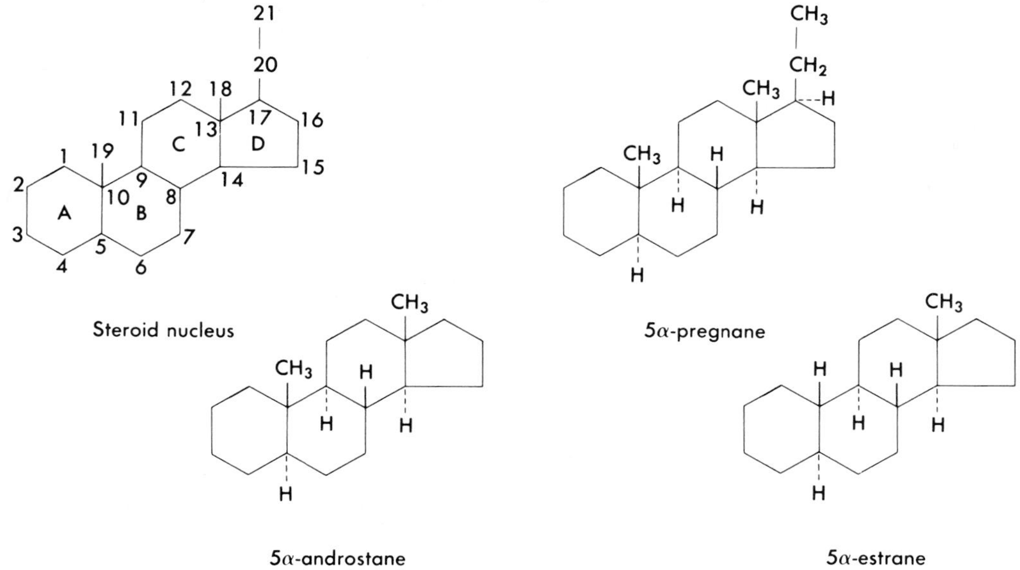

Fig. 14-1. Basic nucleus of steroid groups.

group at position 17 and contain nineteen carbon atoms (17-ketosteroids) and those which have a two-carbon side chain at position 17 and contain twenty-one carbon atoms. Those of the latter group which have a hydroxyl group at the 17 position are known as 17-hydroxycorticosteroids.

Steroid hormones: structure, nomenclature, and biosynthesis

All steroid hormones have in common a cyclopentanoperhydrophenanthrene ring structure. The four rings are identified by the letters A, B, C, and D and the carbon atoms by numbers 1 to 21 (Fig. 14-1). The natural steroids, with the exception of the estrogens, have two angular methyl groups that are attached to carbon atoms 10 and 13. Except where a carbon atom joins two rings or is attached to another substituent, each carbon is saturated by two hydrogen atoms. Substituted groups may be found on carbon atoms 3, 10, 11, 13, 16, and 17. The angular methyl groups occupy positions 18 and 19. The position of a substituted group on the nucleus is indicated by the number of the carbon atom to which it is attached.

Nearly fifty steroids have been isolated from the adrenal cortex but cortisol, corticosterone, and aldosterone and their metabolites produce the major activity attributed to the gland. Steroid hormones may be classified as adrenocortical steroids, progesterones, androgens, and estrogens. They may also be grouped (Table 14-3) according to the number of carbon atoms in the compounds: corticosteroids, including progesterones, have twenty-one carbon atoms; androgens, nineteen carbon atoms; and estrogens, eighteen carbon atoms. The parent hydrocarbon of each—respectively, pregnane, androstane, estrane—is used as the basis for systematic naming (Fig. 14-1). Terminology for steroids has created confusion because several different names may be used to designate the same compound. For example, cortisol is referred to as hydrocortisone, compound F (Kendall), 17-hydroxycorticosterone, or compound M (Reichstein) and has the chemical name 4-pregnen-11β,17α,21-triol-3,20-dione.

Table 14-3
Major steroid hormones and their metabolites

Trivial name	Systematic name
C-18 compounds	
Estrone	1,3,5(10)estratrien-3-ol-17-one
Estradiol-17β	1,3,5(10)estratrien-3,17β-diol
Estriol	1,3,5(10)estratrien-3,16α,17β-triol
C-19 compounds	
Androsterone	5α-androstan-3α-ol-17-one
Dehydroepiandrosterone	5-androsten-3β-ol-17-one
Etiocholanolone	5β-androstan-3α-ol-17-one
11-hydroxyetiocholanolone	5β-androstan-3α,11β-diol-17-one
11-hydroxyandrosterone	5α-androstan-3α,11β-diol-17-one
11-ketoetiocholanolone	5β-androstan-3α-ol-11,17-dione
11-ketoandrosterone	5α-androstan-3α-ol-11,17-dione
Androstenedione	5α-androstan-3,17-dione
Testosterone	4-androsten-17β-ol-3-one
Epitestosterone	4-androsten-17α-ol-3-one
C-21 compounds	
Cortisol	4-pregnen-11β,17α,21-triol-3,20-dione
Cortisone	4-pregnen-17α,21-diol-3,11,20-trione
Corticosterone	4-pregnen-11β,21-diol-3,20-dione
11-deoxycortisol	4-pregnen-17α,21-dione-3,20-dione
21-deoxycortisol	4-pregnen-11β,17α-diol-3,20-dione
Progesterone	4-pregnen-3,20-dione
Aldosterone	4-pregnen-18al-11β,21-diol-3,20-dione
Tetrahydrocortisol	5β-pregnan-3α,11β,17α,21-tetrol-20-one
Tetrahydrocortisone	5β-pregnan-3α,17α,21-triol-11,20-dione
Tetrahydrocorticosterone	5β-pregnan-3α,11β,21-triol-20-one
Tetrahydro S	5β-pregnan-3α,17α,21-triol-20-one
Cortol	5β-pregnan-3α,11β,17α,20α,21-pentol
Cortolone	5β-pregnan-3α,17α,20α,21-tetrol-11-one
Pregnanetriol	5β-pregnan-3α,17α,20α-triol
Pregnanediol	5β-pregnan-3α,20α-diol
Pregnanolone	5β-pregnan-3β-ol-20-one
Pregnenolone	5-pregnen-3β-ol-2-one

To denote unsaturation, the symbol Δ followed by a number indicating location (example: Δ^4 denotes a double bond in positions 4-5) has been replaced by use of a suffix; a saturated nucleus is designated by the suffix -ane (androstane), one double bond by -ene (5-androstene-3β, 17β-diol), and two double bonds by -diene (1,4-androstadien-3,11,17-trione). Stereoisomerism provides for a variety of steroids, and a slight structural change may produce a compound with very different physiologic effect. Isomerism describes molecules having the same number and kinds of atoms but arranged in different configurations. Some of the terms employed in steroid nomenclature to indicate different spatial configurations are the following: cis, trans, α, β, allo, epi, and iso. Geometric or cis-trans isomerism describes the position of a

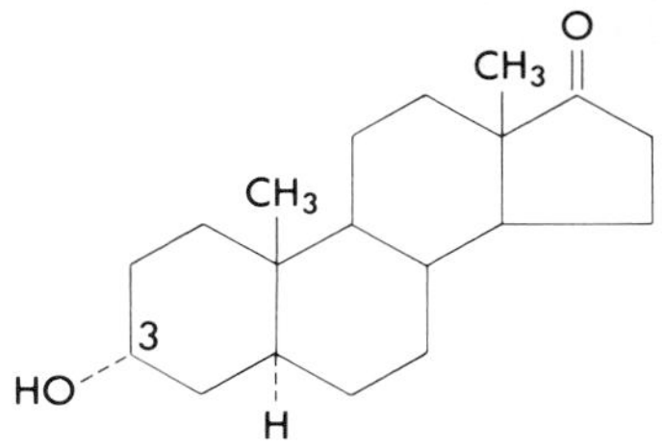

Androsterone
3α-hydroxy-5α-androstan-17-one

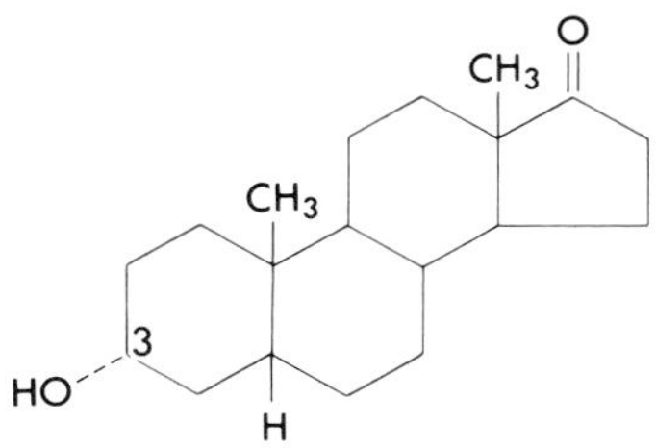

Etiocholanolone
3α-hydroxy-5β-androstan-17-one

Fig. 14-2. Steroids differing only by orientation of hydrogen atom on C-5.

substituent relative to the plane of the molecule. Two substituents of a cyclohexane ring can be either on the same side of the plane of the ring (cis) or on opposite sides (trans). Two saturated rings may be fused in two ways, cis or trans. For all naturally occurring steroids four rings are fused and both isomers are found in the attachments of rings A and B, whereas rings BC and CD are attached trans. Steroids exist in what is known as the chair conformation.

When a substituent atom is projected to the rear or underside of a steroid molecule, it is considered to be alpha-oriented to the methyl group at C-10 and is indicated by a dotted bond. In contrast, a beta-oriented substituent will be to the front or above the molecule and is indicated by a solid bond. For example, the only difference between androsterone and etiocholanolone is the orientation of the hydrogen atom on C-5 (Fig. 14-2). It is alpha-oriented in the androsterone molecule and beta-oriented in the etiocholanolone molecule. In the term 3-trans-hydroxy, the relationship is expressed that the OH group attached to C-3 is trans with respect to the methyl group attached to C-10. The prefix allo- (denotes one of two possible isomers) is applied to 5α-compounds as opposed to 5β-compounds which are considered belonging to a normal series. For example, allo-pregnanediol is 5α-pregnan-3α,20α-diol and pregnanediol is 5β-pregnan-3α,20α-diol. Epi- or iso- indicates an isomer that differs only in spatial arrangement of a group on one carbon atom, examples of which are:

Androsterone	5α-androstan-3α-ol-17-one
Epi(iso)androsterone	5α-androstan-3β-ol-17-one
Dehydroepi(iso)androsterone	5-androsten-3β-ol-17-one
Cortisol	4-pregnen-11β, 17α,21-triol-3,20-dione
Epicortisol	4-pregnen-11α,17α,21-triol-3,20-dione

Differences between the various compounds may result from replacement of a hydrogen atom by a hydroxyl or a ketone group, or the reverse, or by different spatial arrangements of substituents. A compound may be referred to as a dehydro compound if two hydrogen atoms are lost. For example, when two hydrogen atoms are lost from C-11 of corticosterone, the new compound is dehydrocorticosterone. If an oxygen atom is lost, the new compound may be termed a deoxy compound. Deoxycorticosterone is produced when an oxygen atom is lost from the hydroxy group on C-11 of corticosterone (Fig. 14-3). The presence of various groups in the compounds are frequently indicated by the following terms: for a hydroxyl group, hydroxy- and oxy- (prefix) and -ol (suffix), and for a ketone group oxo- and keto- (prefix) and -one (suffix).

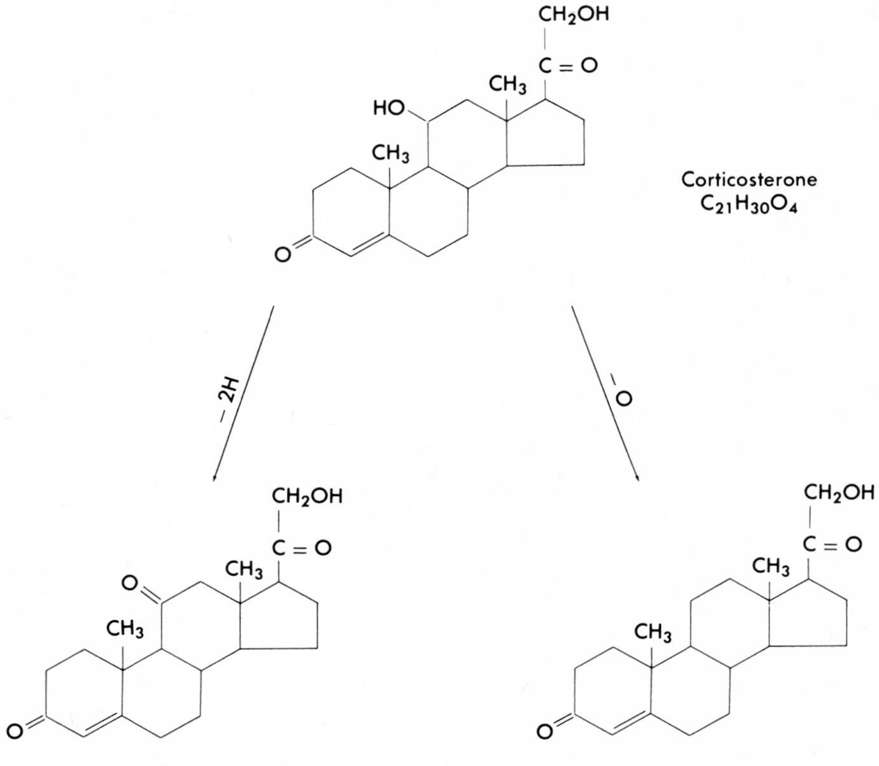

Corticosterone
$C_{21}H_{30}O_4$

Dehydrocorticosterone
$C_{21}H_{28}O_4$

Deoxycorticosterone
$C_{21}H_{30}O_3$

Fig. 14-3. Steroids formed by loss of hydrogen or oxygen from corticosterone.

The synthesis of steroids proceeds by enzymatic action from acetate by way of cholesterol (Fig. 14-4). The pathways are similar regardless of whether the hormone is produced in the adrenal cortex or in ovaries or testes. The steroids secreted from the adrenal cortex in significant amounts are cortisol and corticosterone (glucocorticoids), dehydroepiandrosterone sulfate (androgen), and aldosterone (mineralocorticoid). Small quantities of testosterone and estradiol are also secreted.

In the biosynthesis of these steroids, cholesterol is converted to 20α-hydroxycholesterol, which becomes 20α,22-dihydroxycholesterol. The latter undergoes cleavage of its side chain by mitochondrial enzymes to yield pregnenolone. Progesterone is formed from pregnenolone by the action of the enzymes 3β-hydroxydehydrogenase and Δ^5,Δ^4-isomerase. The hydroxylation of progesterone can occur in two ways and usually proceeds in the order of C-17, C-21, and C-11. If hydroxylation occurs at C-17, then hydroxylation will follow on C-21 and C-11β to yield cortisol. If it does not take place at C-17, hydroxylation at C-21 and C-11β will form corticosterone. By the action of a 17α-hydroxylase system progesterone is converted to 17α-hydroxyprogesterone which by 21-hydroxylation becomes 17α,21-dehydroxyprogesterone (substance S). The action of 11β-hydroxylase transforms substance S into cortisol. Progesterone may be hydroxylated at C-21 position to form 11-deoxycorticosterone, which is then hydroxylated at the 11 position to form corticosterone. It undergoes

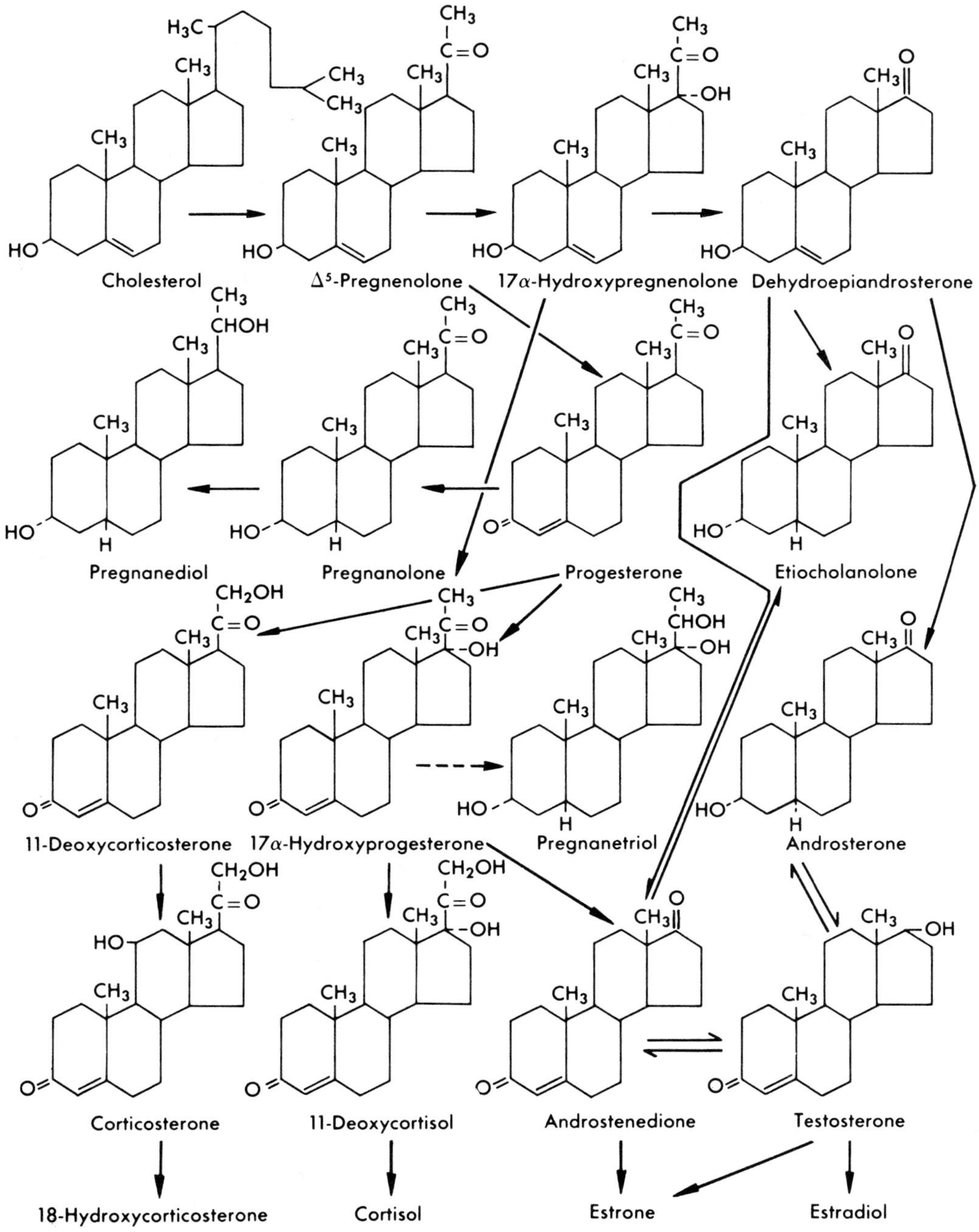

Hormones 379

Fig. 14-4. Probable major pathways of steroid biosynthesis and metabolism.

Continued.

Fig. 14-4, cont'd. Probable major pathways of steroid biosynthesis and metabolism.

18-hydroxylation and then conversion of the hydroxyl group at C-18 to an aldehyde group to form aldosterone. By cleavage of its C-20 and C-21 side chains, 17α-hydroxypregnenolone forms dehydroepiandrosterone.

The hydroxylation sequence is of clinical importance since deficiencies in any of the enzyme systems will produce an increase in the concentration of abnormal metabolites and result in a pathologic condition. For example, 21-hydroxylase deficiency is the most common cause of congenital adrenal hyperplasia.

ACTH from the anterior pituitary in response to CRF from the hypothalamus stimulates growth and increases secretions of the adrenal cortex. It acts on a membrane receptor of the adrenal cortex, activating adenyl cyclase and increasing formation of cyclic AMP, the intracellular mediator of steroid synthesis in the gland. The principal glucocorticoids secreted by the normal adrenal gland are cortisol and corticosterone. Cortisol has an important role in the basal rate of ACTH secretion (negative feedback action). When free cortisol is decreased, ACTH increases; an increase in free cortisol will decrease ACTH secretion. Other controlling factors are stress and biologic rhythm, which is related to the sleep pattern.

In the absence of stress, about 10% of the plasma cortisol is free, approximately 75% bound to transcortin, an alpha globulin (glycoprotein) also known as corticosteroid-binding globulin (CBG), and the remainder bound to albumin. The concentration of CBG is increased during estrogen administration and late pregnancy but free cortisol is elevated only slightly. Corticosterone, deoxycorticosterone and progesterone are also bound to CBG in a similar manner.

There are a number of pathways for inactivating adrenocortical steroids. Metabolism occurs mainly in the liver, where reduction of the A ring between carbon atoms 4 and 5 produces the dihydro derivatives, which are rapidly reduced to the tetrahydro compounds. Additional reduction occurs that involves the keto group on C-20, yielding cortol and cortolone. Approximately 5% of cortisol is converted to 17-ketosteroids. This is accomplished by the action of a desmolase, which can bring about this conversion on steroids having 17α-hydroxy and 20-oxo groups. The biologic half-life of plasma cortisol is 60 to 120 minutes.

The metabolites of cortisol are conjugated with glucuronic acid at position C-3 and to a slight extent with phosphoric and sulfuric acids, water-soluble forms in which they are excreted in the urine. The major metabolites found in the urine are tetrahydrocortisol, tetrahydrocortisone, allotetrahydrocortisol, cortol, and cortolone. A very small amount of the 17-hydroxycorticosteroids appear in the urine in the active, unconjugated form.

The renin-angiotensin system is the major stimulus for the secretion of aldosterone by the zona glomerulosa. Additional stimuli are sodium depletion, ACTH (production of aldosterone precursors), and elevated serum potassium. Normally, aldosterone is present in the 11-18 hemiacetal form, rather than the 18-aldehyde form, which is transformed in the liver to tetrahydroaldosterone by reduction of the A ring. This metabolite is conjugated with glucuronic acid. In the liver and kidneys an acid-labile conjugate of aldosterone is also formed, an 18-glucuronide. A small amount of free aldosterone is excreted in the urine along with the conjugates tetrahydroaldosterone 3-glucuronide and aldosterone 18-glucuronide.

The major androgenic compound secreted by the adrenal cortex is dehydroepiandrosterone (free and ester, some being conjugated with sulfuric acid before secretion) and most of it is converted to metabolic products as shown in Fig. 14-4. Small amounts of testosterone, androstenedione, and 11β-hydroxyandrostenedione are secreted. They are usually bound by globulin in the plasma. Metabolism occurs in the liver, and the metabolites are conjugated to form water-soluble glucuronides and sulfates.

Small amounts of estrogens and progesterone are normally secreted by the adrenal cortex. The estrogens are apparently derived from testosterone and androstenedione. They circulate in plasma bound to albumin and an estrogen-binding globulin. Conjugates are formed with sulfuric acid and glucuronic acid, with excretion in urine and bile. High concentrations of the free estrogens appear in bile, indicating this as one of the pathways for excretion. Estrogens can be metabolized to 2-methoxy derivatives and 16-oxygenated derivatives.

Some progesterone in circulation is derived from adrenal pregnenolone and pregnenolone sulfate. In plasma, it is primarily in the bound form associated with transcortin and albumin. Metabolism takes place chiefly in the liver. The principal metabolite is pregnanediol, which is excreted in the urine conjugated with glucuronic acid. Other metabolites are allopregnanediol, pregnanolone, and allopregnanolone.

Laboratory tests of adrenocortical function

Measurements of the different hormones secreted by the adrenal cortex and of their metabolites found in blood and urine are used as a means of evaluating adrenocortical function. The analysis may be for a group of structurally or chemically related compounds (17-hydroxycorticosteroids, 17-ketosteroids) or for a specific steroid. With the introduction of radioassay techniques (competitive protein binding or radioimmunoassay), it became possible to analyze biologic fluids for individual steroids and polypeptide hormones, limited by specificity of the antibody or binding protein.

The usual specimen for analysis is either a 24-hour urine collection or blood serum or plasma. The use of serum or plasma has been preferred because of the ease of obtaining the specimen and the presence of fewer interfering substances (as opposed to urine). It has the disadvantage of only measuring the level of the circulating hormone at the time of the blood drawing and does not reflect diurnal variations or any temporary physiologic changes that may occur. The disadvantages of a 24-hour urine collection are that it contains a number of metabolites which complicate methodology in isolating specific steroids, there is difficulty in obtaining a properly collected specimen, and minor alterations in metabolism may be obscured.

ANDROGENS AND 17-KETOSTEROIDS (17-OXOSTEROIDS)

Androgens are substances that are capable of stimulating male genitalia, axillary and pubic hair, and certain other secondary sex characteristics. For a number of years, the only chemical method for testing adrenocortical function was the measurement of adrenal steroids as 17-ketosteroids (17-KS). There was a tendency to consider 17-ketosteroids in a group as androgens, but the terms are not synonymous. The most active androgen, testosterone (4-androsten-17β-ol-3-one), is not a 17-ketosteroid.

Neutral 17-ketosteroids are a group of steroid compounds having nineteen carbon atoms and a ketone group at position 17 of the steroid nucleus. The major C-19 compound of the adrenal cortex, dehydroepiandrosterone (DHEA), is the principal precursor of the urinary 17-ketosteroids. Those present in highest concentration in the urine are androsterone, etiocholanolone, dehydroepiandrosterone, 11β-hydroxyandrosterone, 11-ketoandrosterone, 11β-hydroxyetiocholanolone, and 11-ketoetiocholanolone. Androsterone, epiandrosterone, and dehydroepiandrosterone are mildly androgenic but etiocholanolone is biologically inactive. The three major urinary 17-ketosteroids may be classified into two fractions: an alpha fraction (androsterone, etiocholanolone) and a beta fraction (dehydroepiandrosterone). Normally 85% to 95% of the 17-ketosteroids are in the alpha fraction.

A small portion of the DHEA is metabolized to testosterone. Metabolic products of testosterone from the testes are androsterone, etiocholanolone, and epiandrosterone. Hence measurement of urinary 17-ketosteroids reflect concentration of adrenal and testicular steroids. Because of the interconversion the source of a specific 17-ketosteroid may not be identified. For the normal male, about two thirds of the urinary 17-ketosteroids are produced by the adrenal cortex, the remainder being synthesized in the testes. In the female, most of the urinary 17-ketosteroids originate in the adrenal cortex. The conversion of androstenedione accounts for about 60% of the testosterone in the female.

Although in some areas 17-ketosteroid analysis is considered to have become obsolete, it is important in diagnosing the cause of virilizing disorders. For patients in whom virilization results from an excess of adrenal steroids, the increase in DHEA will be reflected in elevated urinary 17-ketosteroids. Adrenal virilism in the adult male may not be apparent since it will merely accentuate existing male characteristics. In the immature male, precocious development of secondary sex characteristics without testicular growth (precocious pseudopuberty) will occur. In prepubertal and adult females, increased adrenal androgens will cause masculinization (adrenogenital syndrome). For the female fetus, the result is pseudohermaphroditism, development of male type genitalia. In adrenal carcinoma, the 17-ketosteroids are increased, with the elevation being due to an increase in DHEA (beta fraction).

The 17-ketosteroids[20] are excreted into the urine as water-soluble conjugates, primarily as glucosiduronate and sulfate esters. Hydrolysis is required to obtain the extractable free steroids. This is accomplished routinely by treating a sample with a mineral acid at an elevated temperature. Hydrochloric and sulfuric acids have been used. No entirely satisfactory method for hydrolysis is available. The use of acid causes some destruction and alteration of steroids as well as formation of pigments and nonspecific chromogens that interfere in the final colorimetric reading. Methods have been developed for differential hydrolysis of the sulfate and glucuronides, but these are not practical as routine procedures.

Benzene, carbon tetrachloride, and diethyl ether have been the most widely used solvents for extraction of the free steroids from the hydrolyzate. Each has advantages and disadvantages. For example, ether may be preferred because of steroid solubility, ease of removal at low temperatures, and less tendency than other solvents to form emulsions. However, it must be free of peroxides and it is highly flammable.

For the final measurement, the Zimmermann reaction is generally employed. It is based upon the formation of a reddish purple color when a 17-ketosteroid reacts with m-dinitrobenzene in an alkaline solution. This action occurs only if the ketone group is adjacent to a CH_2 group. Maximum absorption is evident at 520 nm when the ketone is in position 17. For compounds with ketones in other positions, the color is less intense and the maximum does not occur at 520 nm.

CORTICOSTEROIDS

The term corticosteroids includes the active adrenal hormones related to cortisol and corticosterone. Although there is some overlapping of activity, this group of steroids may be divided functionally into mineralocorticoids and glucocorticoids.

Mineralocorticoids—aldosterone

Mineralocorticoids were so named because of their influence on electrolyte metabolism and water balance in the body. The most active steroid of the group is aldosterone. It causes sodium retention and potassium loss, which are enhanced by the glucocorticoids. Other but less potent mineralocorticoids synthesized by the adrenal cortex are 11-deoxycorticosterone, corticosterone, and their 18-hydroxy derivatives.

The factors chiefly involved in control of aldosterone secretion are the renin-angiotensin system originating in the kidney, pituitary ACTH, and changes of sodium and/or potassium status. In maintaining extracellular fluid volume, aldosterone-induced sodium retention is effected by the renin-angiotensin system during periods of volume deficiencies. Aldosterone affects renal tubular transport of sodium at the site of the distal convoluted tubule where reabsorption of sodium from the filtrate and secretion of potassium into the urine occur. Aldosterone production is increased when patients are placed on a low salt diet. Potassium ions, oral potassium loading, or infusion of potassium ions directly stimulates an increase in aldosterone. Secretion, excretion, and serum concentration of aldosterone are slightly increased with standing, markedly increased during pregnancy, and higher by day than by night (circadian rhythm).[32]

Aldosterone levels in serum and urine are measured by radioimmunoassay. Normal supine venous levels range from 3 to 10 ng/dl (upright: 5 to 30 ng/dl). For a patient on an adequate salt intake, the normal values for a 24-hour urine collection appear to be between 5 and 20 μg and on a low salt diet, between 10 and 40 μg/24 hr. Localization of the lesion in patients with primary aldosteronism may be accomplished by measuring the aldosterone level in adrenal vein blood from each side and comparing with peripheral blood. In circulation, aldosterone binding to protein is minimal. Since it is carried mostly in the free state, serum concentrations correlate well with physiologic activity. The biologic half-life is approximately 45 minutes.

Primary hyperaldosteronism or Conn's syndrome, a curable cause of hypertension, is usually (80% of cases) caused by adrenocortical adenoma. Because prognosis is favorable when the tumor is removed, distinguishing these patients from the large number of patients with essential hypertension is necessary. Primary hyperaldosteronism is characterized by an elevated serum aldosterone accompanied by an elevated serum sodium, hypokalemic alkalosis, renal potassium loss, and decreased renin activity. Among the patient's clinical findings are hypertension, polydipsia, polyuria, and muscular pains.

When interpreting aldosterone measurements, consideration must be given to the many stimuli that will increase its production. Secondary hyperaldosteronism as found in chronic renal disease, cardiac failure with sodium retention, cirrhosis of the liver with ascites, and renin-secreting renal tumors is characterized by increased levels of renin activity. Distinguishing hyperaldosteronism due to renal disease from that of adrenocortical origin can be achieved by plasma renin measurements. Primary aldosteronism is characterized by the finding of an elevated serum aldosterone and low plasma renin activity for an upright individual on a normal salt intake.

In infants, Bartter's syndrome, characterized by hyperaldosteronism, increased plasma renin, hypokalemia, and juxtaglomerular cell hyperplasia has been described.

Glucocorticoids

Steroids that affect carbohydrate metabolism, including promotion of gluconeogenesis from protein, stimulation of liver glycogen deposition, and increase of blood glucose concentrations, are classified as glucocorticoids. Cortisol, cortisone, corticosterone, and 11-dehydrocorticosterone are naturally occurring steroids with glucocorticoid activity. They also stimulate fat synthesis, depress response to injury, increase renal blood flow, maintain blood pressure, stimulate breakdown of protein, and suppress ACTH synthesis and secretion. Cortisol, the most active and abundant corticosteroid, is a weak mineralocorticoid, but because of its high concentration it accounts for approximately one fourth of the total mineralocorticoid effect. Cortisone and 11-dehydrocorticosterone have glucocorticoid activity because they are convertible in the body to cortisol and corticosterone. Measurement of cortisol or its metabolites in plasma and urine is used in assessing adrenocortical function.

17-Hydroxycorticosteroids, plasma

Cortisol (hydrocortisone, compound F). Cortisol is the major secretory product of the adrenal cortex. The steroid in second highest concentration, corticosterone, is approximately one tenth that of cortisol. Measurements of blood corticosteroids may include both of these or show mainly cortisol. Plasma or serum analysis is preferred to whole blood because of the difficulties encountered when extracting whole blood with organic solvents. Since the cortisol concentration is lower in the red cells, plasma must be removed as soon as possible after drawing the blood.

In clinical laboratories there are three general types of procedures in use for assaying plasma corticosteroids. One method[49] employs the sulfuric acid phenylhydrazine reagent described by Porter and Silber that determines the 17,21-dihydroxy-20-ketosteroids, which in plasma is principally cortisol. Other compounds may be determined, but they are present usually in such low concentration that the procedure is considered specific for plasma cortisol. Cortisone, compound S (11-deoxycortisol), and their tetrahydro derivatives have 17,21-dihydroxy-20-keto side chains and may be measured as Porter-Silber chromogens. Compound S, normally present in very small concentrations, is secreted in high concentrations in patients with congenital adrenal hyperplasia resulting from 11β-hydroxylase deficiency and in individuals given metopirone, an 11β-hydroxylase inhibitor.

A more sensitive technique is one that measures fluorescence[17, 19, 66] in a sulfuric acid medium of corticosteroids with the 11-hydroxy configuration. Both cortisol and corticosterone have this pattern. The procedure is susceptible to drug interference and nonspecific fluorescence found in some plasma samples.

A specific and sensitive method for measuring plasma cortisol is by competitive protein-binding (CPB) using naturally occurring corticosteroid-binding globulin (CBG or transcortin).[42] Other steroids including corticosterone, compound S, progesterone, and 17α-OH progesterone may be measured by this technique. A reagent containing CBG and radioactive tracer cortisol is added to cortisol extracted from plasma or to cortisol standard. An equilibrium is established between the labeled cortisol bound to the CBG and the cortisol of the sample, resulting in some of the labeled hormone being displaced by the hormone of the sample. Measurement of the amount of radioactivity displaced is proportional to the unlabeled cortisol from the standard or sample.

A radioimmunoassay procedure using a specific antibody has also been developed. The principle is the same as for competitive protein binding.

Compound S (11-deoxycortisol). Compound S is an intermediate in cortisol biosynthesis in the adrenal cortex. Although present in normal plasma in very low concentrations, it is important in diagnosing the type of adrenogenital syndrome caused by 11β-hydroxylase deficiency and in evaluation of pituitary function when metopirone is administered. Measurement is made using competitive protein-binding technique after extraction of the hormone from plasma with benzene.[38] The procedure is essentially the same as that for cortisol, except that the labeled tracer hormone is tritiated 11-deoxycortisol. Labeled hormone is displaced from a specific protein by unlabeled hormone (unknown or standards). The amount of radioactivity remaining bound to protein is thus a measure of the amount of unlabeled hormone.

17-Hydroxycorticosteroids, urinary. The major C-21 metabolites found in the urine are tetrahydrocortisol, tetrahydrocortisone, tetrahydro S, pregnanetriol, 11-ketopregnanetriol, cortols, cortolones, 6β-hydroxycortisol, and Δ^5-pregnanetriol. These are conjugated principally with glucuronic acid, although there are some sulfates. A small amount of free cortisol is also excreted. Since the amount of urinary corticosteroids excreted in the daytime is greater than during the night, it is important that the 24-hour collection be complete.

There are two basic procedures in general use for the determination of 17-hydroxycorticosteroids in urine. One method utilizes enzyme hydrolysis and the reaction of 17-hydroxycorticosteroids with a phenylhydrazine–sulfuric acid reagent to measure Porter-Silber chromogens. The second method, a determination of 17-ketogenic steroids, employs sodium metaperiodate to convert corticosteroids to 17-ketosteroids for final assay by the Zimmermann reaction.

Porter-Silber chromogens. Porter-Silber chromogens are steroids possessing a 17,-21-dihydroxy acetone side chain at C-17. These include cortisol, cortisone, tetrahydrocortisol, tetrahydrocortisone, and 11-deoxycortisol. Compound S and its tetrahydro metabolite are not metabolites of cortisol but have the dihydroxyacetone side chain and are therefore measured.

The steroid conjugates are hydrolyzed with β-glucuronidase and the free steroids extracted with an organic solvent (methylene chloride, chloroform). The organic extract is washed with dilute sodium hydroxide solution to remove chromogens and then shaken with the phenylhydrazine–sulfuric acid–ethanol reagent to produce the characteristic yellow color.

17-Ketogenic steroids. All urinary C-21 steroids containing a 17-OH group can be converted to 17-ketosteroids by reacting with sodium borohydride followed by oxidation with sodium periodate. The excreted 17-ketosteroids are reduced to noninterfering compounds by the borohydride, the side chains of C-21 hydroxycorticosteroids are oxidized by periodate to 17-ketonic groups, and these are measured as 17-ketosteroids by the Zimmermann reaction. The determination includes the 17-hydroxysteroids measured as Porter-Silber chromogens and, in addition, the cortols, cortolones, and pregnanetriol.

Free cortisol (urine). The unconjugated (free) cortisol in urine is a reliable index of the unbound or free cortisol in plasma. Since plasma cortisol varies throughout the day, the urinary cortisol is a better reflection of adrenal cortical activity. When the plasma cortisol increases, as in ACTH stimulation and Cushing's syndrome, the urinary cortisol will also increase, whereas increased cortisol secretion accompanied by increased CBG levels, as in pregnancy, will not be reflected in increased urinary cortisol. Measurement is made by competitive protein-binding radioassay.

Pregnanetriol. Pregnanetriol, a metabolite of 17α-hydroxyprogesterone, is excreted conjugated to glucuronic acid. Normally it occurs only in very small quantities, but increased production results from pathologic conditions such as congenital adrenal hyperplasia. When cortisol formation is impaired, hyperplasia of the adrenal cortex, arising from hypersecretion of ACTH through a negative feedback, occurs. The syndrome results from an enzyme deficiency in the biosynthesis of cortisol. The most common type is a deficiency of the enzyme responsible for hydroxylation at the C-21 position, which results in excessive amounts of pregnanetriol being produced and excreted. The measurement of urinary pregnanetriol is of diagnostic value in cases of adrenogenital syndrome. Cortisol and aldosterone production are decreased and adrenal androgens increased with clinical signs of virilization in females and early masculinization in males. The defect varies in clinical symptoms. If severe, a salt-losing form of the syndrome with resulting dehydration, and death if left untreated, may be seen. When present prior to birth, pseudohermaphroditism occurs in genetic females. In a mild form, the C-21 defect may be evident only at puberty and for the female clinically show masculinization of the body with absence of breast development and menses.

Adrenocortical function tests. Many factors influence the production and secretion of steroids. When functioning normally, the adrenal cortex is largely controlled by the ACTH secretion of the pituitary gland and a balance is maintained between the level of ACTH and cortisol in the blood. In establishing or confirming a diagnosis of hypofunction or hyperfunction of the adrenal cortex, compounds that stimulate or suppress the activity are used in combination with assays of blood and urine for steroids. The three compounds employed routinely for this purpose are ACTH, metyrapone (Metopirone), and dexamethasone.

ACTH stimulation test. The ACTH stimulation test provides a specific measure of the response of the adrenal gland to stimulation by adrenocorticotropin or of the functional cortisol reserve. Plasma and/or urinary steroids are measured before and after administration of ACTH. A number of techniques have been used with variation in quantity of ACTH, manner and length of time of administration, and specimens used for steroid assay.

For the 8-hour intravenous method, 40 units of ACTH in 500 ml of normal saline is infused at the rate of about 5 units/hr over an exact 8-hour interval. This may be carried out on 1 day or on 2 successive days. Blood for plasma cortisol levels is collected prior to the infusion and at 1- and 2-hour intervals after the infusion started. A 24-hour urine collection is obtained prior to the infusion for measurement of the concentration of 17-ketosteroids and 17-hydroxycorticosteroids (or 17-ketogenic steroids). Another 24-hour urine collection is made, beginning with the start of the infusion. The level of 17-hydroxycorticoids or 17-ketogenic steroids is measured.

For the normal subject, the plasma cortisol will increase 10 to 25 $\mu g/dl$ during the first hour, with an additional 4 to 7 $\mu g/dl$ during the second hour. The urinary 17-OHCS level is raised threefold to fivefold over the baseline. There is no increase of excretion in Addison's disease or primary adrenocortical failure, very little increase over the already high concentrations for patients with Cushing's syndrome or adrenocortical tumors, and little if any response in congenital adrenal hyperplasia.

Rapid ACTH screening procedures have been proposed whereby plasma is assayed for cortisol before and following intramuscular administration of 25 units of the drug. There is a marked increase in blood cortisol levels for normal persons but in those with adrenal insufficiency or Addison's disease the response is very slight. However, in interpreting the test it must be recognized that the efficiency with which absorption of ACTH takes place from the site of injection is a variable factor. A positive result must be confirmed by a more controlled procedure.

Metyrapone (Metopirone) test. Metopirone (metyrapone, SU 4885), an inhibitor, tests the response of the pituitary gland to a decrease in circulating cortisol. It selectively inhibits 11β-hydroxylase, the enzyme for conversion of 11-deoxycortisol to cortisol, and thus partially blocks the pathway to cortisol. Increased amounts of 11-deoxycortisol accumulate in the blood while cortisol decreases. The 11-deoxycortisol (compound S) cannot suppress the hypothalamic-pituitary axis, and because of the response to the normal feedback of decreased cortisol, increased quantities of ACTH are released in an effort to release larger amounts of cortisol.

Several methods of applying the test have been employed, including variation in methods of administering the Metopirone, the quantity, and the means of assessing response. A dose of 750 mg may be given orally every 4 hours for six doses. A blood sample is obtained before and after administering the drug for measuring plasma cortisol and compound S. Twenty-four-hour urinary collections before, during, and the day after the last dose are obtained for excretion levels of 17-OHCS (or 17-KGS) and urinary tetrahydro S.

The plasma compound S will increase from less than 1.5 μg/dl to a range from 13 to 32 μg/dl. A normal urinary response consists of a twofold to fourfold increase in total urinary 17-OHCS over the baseline level on either the day of treatment or on the subsequent day. No response will be observed when there is diminished adrenocortical reserve or adrenocortical tumor. In adrenal hyperplasia the output is increased.

Dexamethasone suppression test. An adrenal suppression test may be employed to differentiate Cushing's syndrome caused by an adrenal tumor from adrenal hyperplasia. Dexamethasone, a synthetic glucocorticoid, will suppress ACTH secretion. The hypothalamic-pituitary ACTH release mechanism responds to circulating glucocorticoids in the blood.

A rapid screening test for Cushing's syndrome is an overnight suppression test[46] whereby 1 mg of dexamethasone is administered orally between 11 PM and 12 midnight and the cortisol measured in an 8 AM blood specimen the following morning. In normal subjects this value should be less than 5 μg/dl.[57] Levels obtained for acutely ill patients and those taking estrogens may be misleading.

The drug may be administered orally in doses of 0.5 mg at 6-hour intervals on 2 successive days and 2 mg at 6-hour intervals on the 2 following days. Prior to the test, two 24-hour baseline control urine specimens are collected and a specimen on each second day of suppressive therapy (0.5 and 2 mg, respectively) for measurement of 17-OHCS or 17-KGS levels. Normal subjects show a decrease of about one half in urinary corticosteroid levels after administration of 0.5 mg of dexamethasone. In bilateral adrenal hyperplasia, suppression will be observed only after repetition of the procedure with the larger dose. Normal persons show suppression to less than 3 mg/day on 2-mg doses, but patients with Cushing's syndrome do not suppress. No suppression even with larger doses is obtained in cases of adrenocortical tumors.

ADRENOCORTICAL HYPOFUNCTION

Adrenal insufficiency may be primary (Addison's disease, hypoaldosteronism, or adrenogenital syndrome) or secondary (pituitary or hypothalamic insufficiency or iatrogenic pituitary insufficiency). Primary insufficiency is usually caused by an unexplained atrophy of the adrenal cortex. When secretion of aldosterone and cortisol is deficient, there is sodium loss, hyperkalemia, low blood pressure, and hypoglycemia. Clinically, a patient with Addison's disease presents with weakness, hypotension, and darkly pigmented skin. Diagnosis of primary adrenal insufficiency is confirmed by the finding of low concentration of steroids in plasma and urine which do not increase after administration of ACTH.

Secondary adrenal insufficiency may be the result of a pathologic condition of the pituitary gland and hypothalamus (usually induced by surgery, trauma, or infection) or of ACTH suppression by exogenous glucocorticoid administration. Alteration of ACTH release caused by hypothalamic or pituitary disease may be determined directly by assay for ACTH. The Metopirone test may be used to determine the integrity of the negative feedback mechanism of ACTH release.

ADRENOCORTICAL HYPERFUNCTION (CUSHING'S SYNDROME)

Cushing's syndrome is characterized by increased production of cortisol by the adrenal gland, and most cases result from bilateral adrenal hyperplasia or steroid-secreting tumors of the adrenal cortex. Absence of normal diurnal variation in plasma cortisol level and increase in urinary free cortisol should be determined to confirm the diagnosis of Cushing's syndrome. To distinguish between the causes of Cushing's syndrome, suppression and stimulation tests are used.

CONGENITAL ADRENAL HYPERPLASIA

Congenital adrenal hyperplasia includes a variety of pathologic conditions each associated with an adrenal enzyme deficiency blocking normal biosynthesis of steroids. Both infants and adults may be affected. In the infant or child, there is usually an inborn error of metabolism that causes a lack of specific adrenal enzyme and thus blocks the normal production of cortisol. This may result in deficient formation of end products such as cortisol, increased pituitary secretion of ACTH, excessive androgen secretion, and adrenal hyperplasia. Precursors at the site of the block tend to accumulate.

20,22-Desmolase deficiency, congenital lipoid hyperplasia, is a condition incompatible with life since no steroid hormones are produced from cholesterol and the adrenals are filled with lipids. Patients die in early infancy.

3β-Hydroxysteroid dehydrogenase deficiency blocks conversion of pregnenolone to progesterone, leading to production of DHEA and excess androsterone. Infants survive only a short time and usually die from sodium loss.

17α-Hydroxylase deficiency blocks formation of steroids such as 17-hydroxyprogesterone and 17-hydroxypregnenolone. Thus synthesis of cortisol and sex hormones is prevented. Characteristically, a decrease in plasma levels of cortisol, renin, aldosterone, estrogen, and testosterone and a decrease in urinary 17-OHCS and 17-KS are observed. There is an increase in plasma levels of ACTH, deoxycorticosterone, corticosterone, and pituitary gonadotropins. (The increased secretion of 11-deoxycorticosterone and corticosterone leads to salt retention and hypertension.)

21-Hydroxylase deficiency blocks both the glucocorticoid and mineralocorticoid pathways by failure to hydroxylate C-21 of progesterone. An accumulation of 17-hydroxypregnenolone, 17-hydroxyprogesterone, and progesterone leads to excess formation of androgens. Affected individuals with the partly compensated form have adequate aldosterone and cortisol so that there is masculinization without adrenal insufficiency. Other subjects show sodium loss (salt-losing syndrome) early in life due to lack of aldosterone. Early death will result unless the patient is treated with mineralocorticoids and hydrocortisone.

11β-Hydroxylase deficiency is the major cause of virilism with hypertension. There is a decrease in secretion of cortisol, corticosterone, and aldosterone and excess formation of deoxycorticosterone (DOC), 11-desoxycortisol, and ketosteroids. The hypertension is related to the DOC secretion. Salt depletion occurs in some infants.

18-Oxidase deficiency causes a decrease in aldosterone secretion.

ADRENAL MEDULLA

The adrenal medulla is clinically important because of the special tumors arising at the site. The most prominent is the pheochromocytoma, a chromaffin cell tumor characterized by the secretion into the blood of large quantities of epinephrine and norepinephrine. This type of tumor may also occur in the thoracolumbar sympathetic ganglia, in the organs of Zuckerkandl, or paraganglia situated along the aorta at the origin at the inferior mesenteric artery. Pheochromocytoma is not a common pathologic condition, but it is a curable cause of hypertension and death will result if the tumor is not removed. Both adults and children are affected.

The adrenal medulla secretes a group of hormones referred to as catecholamines, substances that contain a catechol nucleus and an amine. These are mostly dopamine

Catechol

Catecholamine

and its metabolic products norepinephrine (noradrenaline) and epinephrine (adrenaline). Noradrenaline is found mainly in the sympathetic nerves of the peripheral and central nervous systems and adrenaline is synthesized primarily in the adrenal medulla.

The catecholamines are synthesized from the amino acid tyrosine by a series of enzymatic steps (Fig. 14-5). Tyrosine is hydroxylated to form dopa (dihydroxyphenylalanine) and then decarboxylated to yield dopamine (diphydroxyphenylethylamine, hydroxytyramine). The side chain of dopamine is hydroxylated to give norepinephrine and a methyl group is introduced into the amino group to produce epinephrine. The conversion of norepinephrine to epinephrine is limited to those cells which have available phenylethanolamine-N-methyl transferase, and so this occurs almost exclusively in the adrenal medulla.

Epinephrine has been called the emergency hormone since production and secretion increase rapidly during conditions of stress. It acts on target tissue after its release from the adrenal medulla into the bloodstream. Norepinephrine acts on effector cells of vascular smooth muscle, adipose tissue, liver, heart, and brain as a neurotransmitter. Dopamine is a precursor of noradrenaline and it also functions as a neurotransmitter in that portion of the brain that coordinates motor activity.

Catecholamines are synthesized in the brain, sympathetic ganglia, peripheral tissue sympathetic nerve endings, and chromaffin cells. In the adrenal medulla and sympathetic nerve endings are granules or granulated vesicles that act as storage units for catecholamines. There is continuous secretion of small amounts of catecholamines, but the largest quantities are released in spurts in response to various nervous stimuli (acetylcholine, nicotine, histamine, 5-hydroxytryptamine, tyramine, reserpine) and physiologic stimuli (severe muscular work, asphyxia and hypoxia, hemorrhagic hypotension). In a reuptake process, much of the neurally released norepinephrine is taken up again in the neuronal vesicles. Catecholamines released by the adrenal medulla and part of the neurally released norepinephrine are inactivated by catechol-O-methyltransferase converting norepinephrine to normetanephrine and epinephrine to metanephrine. Hepatic monoamine oxidase converts the metanephrines into vanillylmandelic acid (**VMA**) and 3-methoxy-4-hydroxyphenylglycol. Catecholamines are rapidly metabolized, remaining in the blood for only a few circulations, and less than 5% are excreted in the urine in free or conjugated forms. In urine, the metabolite of highest concentration is 3-methoxy-4-hydroxymandelic acid (vanillylmandelic

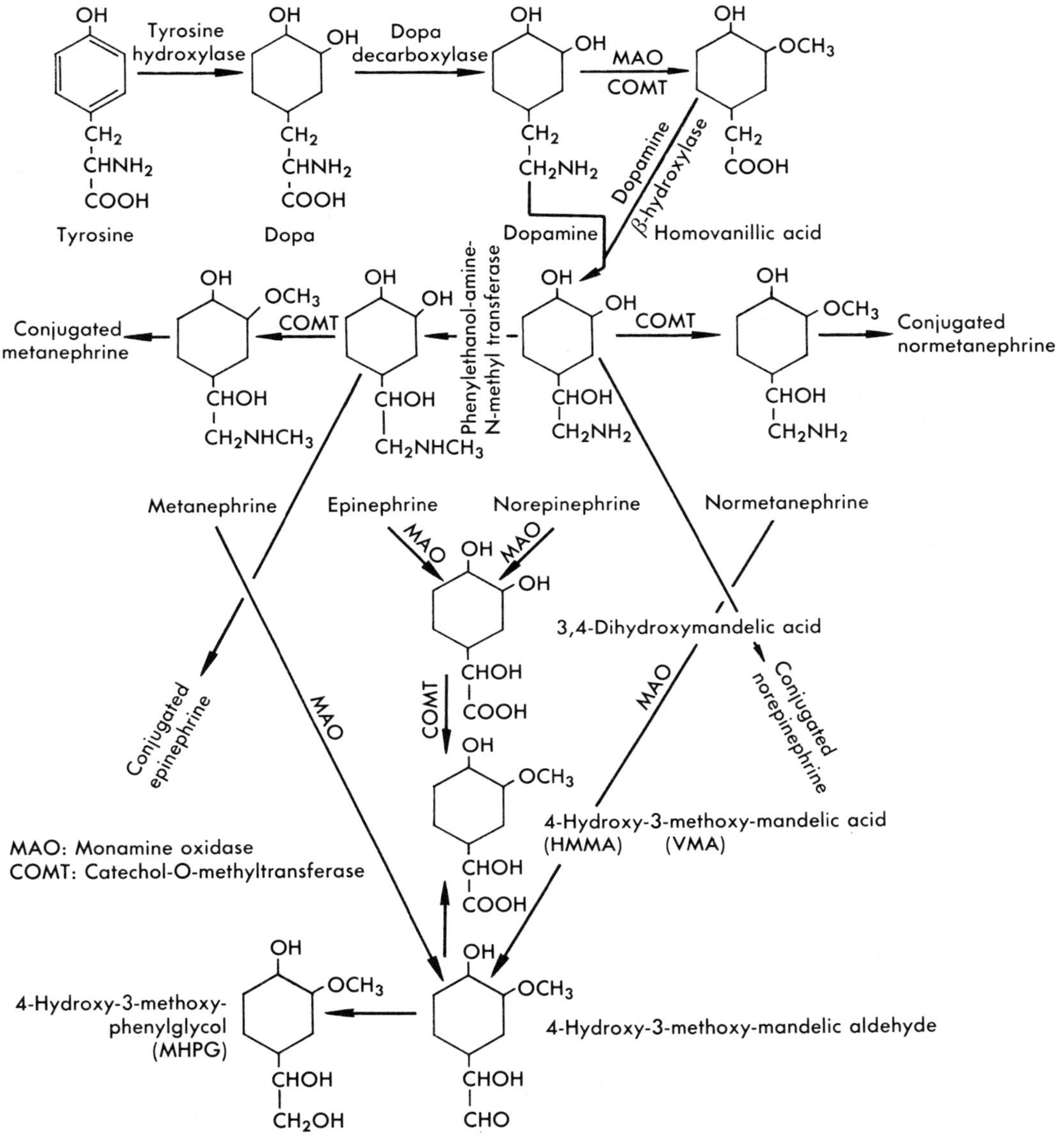

Fig. 14-5. Synthesis and metabolism of catecholamines.

acid). Metanephrine and normetanephrine in the conjugated forms are also present in measurable quantities.

Diagnosis

Even when a patient's history and clinical symptoms are strongly suggestive of pheochromocytoma, laboratory tests must be performed to support the diagnosis. Measurement of the levels of catecholamines and their metabolites in urine specimens is employed as an aid in the diagnosis of pheochromocytoma.[15] Since synthesis is controlled by the rate of secretion, measurement of the metabolites provides an index of catecholamine secretion or synthesis.

Catecholamines

The term urinary catecholamines usually refers to a measurement of the free epinephrine and norepinephrine in a 24-hour collection of urine. Although these hormones are excreted in both the free and conjugated forms, determinations that include the conjugated forms have greater interference from nonspecific substances. Analysis may also be performed on a random urine sample obtained after an episode of hypertension since an elevation of this nature could go undetected because of dilution in a 24-hour specimen.

Blood levels of catecholamines may also be measured, but the information can be misleading because of intermittent secretion of catecholamines by tumors.[8] These analyses are of value when location of a tumor is not possible by any other means, and for this purpose blood is obtained from veins draining a suspected area.[67] Potential sources of error are collection and storage of the samples; the methodology must be rigorously standardized.

After determining the presence of a pheochromocytoma, a quantitative differential determination of urinary epinephrine and norepinephrine may provide an indication of the site of the tumor.[14] Location of the lesion will usually be in the adrenal gland if both of the amines are increased. If the increase is mostly norepinephrine, the tumor may be expected to be in or near one of the adrenal glands in about two thirds of the cases and extraadrenal for the remainder. Procedures that are available for differential determination of norepinephrine and epinephrine are paper chromatography, column separation, oxidation at different pH, and application of different activation and fluorescence spectra of the lutines. The use of two different wavelength combinations eliminates two oxidation steps required when two pH levels are employed.

Fluorometric methods that are specific for catecholamines are generally employed for their measurements. Two different chemical procedures have been developed for formation of fluorescent compounds of adrenaline and noradrenaline: (1) conversion to trihydroxyindole (THI) derivatives and (2) condensation with ethylenediamine (EDA) to form quinoxaline derivatives. The THI method is more specific and is used for routine analysis. The necessary preliminary purification of catecholamines from urine is carried out using alumina or ion exchange chromatography.

The catecholamines are adsorbed onto alumina or resin column, eluted with acid, oxidized by potassium ferricyanide to adrenochrome and noradrenochrome, and rearranged to adrenolutine and noradrenolutine (trihydroxyindoles) in a strongly alkaline solution. The latter compounds give a yellowish green fluorescence in ultraviolet light.

3-Methoxy-4-hydroxymandelic acid (VMA)

The quantitative determination of the catecholamine metabolite 3-methoxy-4-hydroxymandelic acid provides a more reliable measure of catecholamine secretion than epinephrine and norepinephrine since it is present in larger quantities in urine and is less subject to false elevation by medication. Some of the urinary pigments are removed by adsorption on activated magnesium silicate; the VMA is extracted with ethyl acetate and oxidized to vanillin, using periodate. By extracting the oxidized mixture with toluene many of the interfering substances are left in the aqueous layer.

$$\text{3-Methoxy-4-hydroxymandelic acid} \xrightarrow{NaIO_4} \text{Vanillin}$$

Metanephrines

After acid hydrolysis to free the conjugated fractions metanephrine and normetanephrine may be separated on resin and converted to trihydroxyindoles for fluorometric assay or oxidized to vanillin and assayed spectrophotometrically.

In approximately 90% of the patients with pheochromocytoma, all three of the preceding tests will have elevated values. The upper limits of normal (24-hour urine) for free catecholamines is 100 μg; for VMA, 7 mg; and for metanephrines, 1 mg. In the remainder of the patients, usually one of the tests will have increased values.

Other tumors of sympathetic and adrenal medullary tissue origin

Neuroblastoma and ganglioneuroma are malignant sympathetic tumors usually found in children. They occur most frequently in the adrenal gland but may arise from other neural crest tissue. They are dopamine-secreting tumors and the urine level of homovanillic acid, a metabolite of dopamine, is frequently increased. The increased production of catecholamines appears to be the result of tyrosine hydroxylase activity of the tumor causing conversion of tyrosine to dopa. Epinephrine, which is not found in nerve tissue, is not increased. In the majority of cases, the VMA concentration in the urine is elevated and dopamine may also be increased.

Other hormones and glands
GONADAL HORMONES

The testes and ovaries elaborate gonadal hormones that, along with small amounts from the adrenal cortex, produce and maintain secondary sex characteristics, control the reproductive cycle, and promote the growth and development of secondary sex glands. They also have a generalized anabolic effect, an increase in synthesis and decrease in breakdown of protein, leading to an increase in the rate of growth.

Testosterone

The testes consist of a series of convoluted seminiferous tubules along whose walls spermatozoa are formed from primitive germ cells. Between the tubules are groups of cells containing lipid granules, the interstitial cells of Leydig, which produce and secrete testosterone into the bloodstream. Small amounts of testosterone present in blood of females are formed by conversion of precursors secreted by the ovary, such as androstenedione.

Testosterone is a C-19 steroid with an OH group in position 17. It is synthesized from cholesterol in the Leydig cells where 17α-hydroxylase is present (Fig. 14-4). Pregnenolone is hydroxylated in the 17 position and subjected to side-chain cleavage to form 17-ketosteroids which are converted to testosterone. Besides testosterone, the testes secrete two other androgens, androstenedione and dehydroepidandrosterone, in small quantities. Small amounts of estrogens are also formed in the testes.

Luteinizing hormone (LH) or interstitial cell-stimulating hormone (ICSH) from the pituitary stimulates the Leydig cells to synthesize and release testosterone. Follicle-stimulating hormone (FSH) along with testosterone stimulates the seminiferous tubules to produce sperm. The LH is controlled by the circulating testosterone through a negative feedback mechanism operating by way of the hypothalamus. The regulatory mechanism for FSH secretion has not been satisfactorily explained.

About two thirds of the circulating testosterone is transported in plasma bound to protein. In the liver and any of its target tissue, testosterone is converted to various metabolites, mostly 17-ketosteroids. The 17-ketosteroids are then conjugated with glucuronic and sulfuric acids and excreted in the urine. Peripheral conversion of testosterone to estrogen accounts for a large proportion of the estradiol production in the male. Testosterone is converted in tissues to a more potent androgen, dihydrotestosterone, which is localized in nuclei of target cells.

Radioimmunoassay methods are employed for measuring plasma testosterone levels. For a normal adult male, the plasma levels vary from 300 to 800 ng/dl, and for the female, from 25 to 100 ng/dl.

Direct measurement of plasma testosterone provides an evaluation of Leydig cell activity. Urinary testosterone may not reflect this since androstenedione secreted by the adrenal cortex is also converted to testosterone, resulting in overestimation of testosterone secretion. This interference is especially noticeable for hypogonadal males. For females, urinary testosterone is nonspecific and should not be used as an indication of blood testosterone concentration.

In testicular failure or deficiency of testosterone production before puberty the normal pubertal changes do not develop. In the adult male continuous production of the hormone is necessary for the maturation of the sperm and the activity of the secondary sex glands. Testosterone excretion may be increased in women with adrenogenital syndrome, some cases of hirsutism, and Stein-Leventhal syndrome associated with virilization and in children with congenital adrenal hyperplasia. Testosterone excretion may be decreased in hypogonadism and Klinefelter's syndrome in males.

Estrogens and progestins

Biosynthesis of hormones in the ovary proceeds by pathways comparable to those in other steroid-producing organs such as the adrenal cortex. The ovary secretes steroid hormones (estrogens, progestogens, and androgens) and a nonsteroidal hormone, relaxin. Control of ovarian function is effected by the anterior pituitary gonadotropins, follicle-stimulating hormone and luteinizing hormone. Follicle-stimulating hormone stimulates growth and distention of the ovarian follicle. Luteinizing hormone, acting with the follicle-stimulating hormone, promotes ovulation, following which the ruptured follicle is converted into a corpus luteum, increases production of estrogens, and promotes synthesis of progestogens. Secretion of ovarian estrogen and progesterone is coordinated with cyclic secretion of pituitary gonadotropins. There is a feedback control through the hypothalamus; estrogens decrease FSH secretion, and they may inhibit or increase LH secretion.

ESTROGENS

Estrogens are steroid hormones having eighteen carbon atoms and an aromatic A ring. They are secreted by the theca interna cells of the ovarian follicles, corpus luteum, and placenta and in small amounts by the adrenal cortex and testes. In the nonpregnant female, estrogen production occurs primarily in the ovary and estradiol is the major estrogen secreted. During pregnancy estrogens are derived mostly from the fetoplacental unit. For the postmenopausal female, conversion of androstenedione in the adrenal cortex provides most of the estrogen.

The three naturally occurring estrogenic compounds present in highest concentration in the blood are estrone, estradiol, and estriol. Estradiol is the most active biologically, with estrone second. Estradiol is the major estrogen secreted from the ovary and it is interconvertible with estrone. Estrone and estradiol are irreversibly converted to estriol. The major pathway of biosynthesis of estrone and estradiol is shown in Fig. 14-4. Suggested pathways for estriol are as follows:

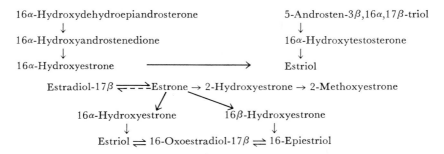

Estrogens circulate bound to an alpha$_2$ globulin. Interconversion of estradiol and estrone occurs chiefly in the liver, as does conjugation of the estrogens to sulfuric and/or glucuronic acid. These water-soluble conjugates are excreted in the urine and a small fraction in the bile.

Serum and urine estrogen levels change throughout the menstrual cycle. Estrogen production rises during the preovulatory or follicular phase to a peak at the middle of the cycle and then decreases, but it increases again during the postovulatory or luteal phase and returns to low levels during menstruation.[23] Measurement of estrogen concentrations in urine and blood provide a means of assessing estrogen secretion. In urine, total estrogen can be measured by one of the various modifications of Brown's method.[7] However, the procedure is not applicable for estrogen levels below 2 to 3 μg/24-hr specimen, a sensitivity necessary for determination of estrogens in certain pathologic states and in postmenopausal women. The final measurement using gas chromatography[68] or fluorescence[28, 70] provides more sensitivity but is subject to greater interference from nonspecific contaminants and thus requires highly purified extracts. Serum estradiol-17β can be measured directly by radioimmunoassay. Concentrations of the three estrogens in urine[23] and of estradiol in plasma are given in Table 14-4.

Estrogens influence development and maintenance of the female sex organs and tissue growth.[41] They are partly responsible for physical changes (feminization) that develop in girls at puberty as well as the enlargement of the tubes, uterus, and vagina. Estrogens affect the ovaries directly by influencing growth of ovarian follicles and indirectly by the feedback control of the pituitary gonadotropins by way of the hypothalamic-pituitary axis. Estrogens increase the levels in serum of corticosteroid-binding globulin, thyroxine-binding globulin, ceruloplasmin, and prothrombin.

In patients who present with amenorrhea, estrogen measurements may provide

Table 14-4
Estrogen levels of normal subjects

	Urine			Plasma
	Estrone (μg/24 hr)	Estradiol-17β (μg/24 hr)	Estriol (μg/24 hr)	estradiol-17β (pg/ml)
Male	3.5 (0.8-7.1)	0.9 (0.4-1.7)	6.1 (1.5-10.4)	8-35
Female, nonpregnant				
Follicular phase	3.6 (1.5-6.7)	1.4 (0.7-2.9)	7.2 (0.76-17.1)	30-75
Peak of middle cycle	19.8 (7.8-32.8)	8.4 (4.3-14.0)	25.7 (3.5-62.2)	200-340
Luteinic phase	8.6 (2.2-29.6)	3.3 (0.7-7.4)	18.9 (2.8-64.7)	60-160
Female, postmenopausal	1.2 (0.2-3.1)	0.4 (0.0-0.9)	4.1 (0.6-10.4)	9-17

an indication of whether it is the result of pituitary, ovarian, or end-organ failure. Hypoestrogenism may be caused by either ovarian or pituitary failure. If the decrease occurs after puberty there will be cessation of menstruation, sterility, and regression of some secondary sexual characteristics. Hyperestrogenism that may be the result of estrogen-producing ovarian tumors will cause early maturation in prepubertal girls. Testicular tumors in males may produce feminization and cause increase in urinary estrogens. Urinary excretion of estrogens is increased in males with breast cancer.[16]

ESTRIOL DURING PREGNANCY (FETOPLACENTAL UNIT)

Estrogen determinations are used widely to assess fetal status during pregnancy. An indication of fetal distress may be based upon decreased estriol production as demonstrated when estriol levels do not rise rapidly or begin to decrease after the second trimester of pregnancy.

Maternal cholesterol is the precursor of placental pregnenolone and progesterone. Pregnenolone is hydroxylated in the fetal adrenal glands and converted to dehydroisoandrosterone which is 16α-hydroxylated in the fetal liver to 16α-hydroxydehydroisoandrosterone. The latter is transported to the placenta where it is oxidized to 16α-hydroxyandrostenedione and converted to 16α-hydroxytestosterone, which is aromatized to estriol. The major source of estrone and estradiol formed during early pregnancy is from dehydroisoandrosterone sulfate secreted by the maternal adrenal. The dehydroisoandrosterone sulfate crosses the placenta where it is hydrolyzed to the unconjugated form, which is converted to androstenedione and then to estrone and estradiol. Steroid metabolites are conjugated with sulfuric acid in fetal tissues due to the excessive concentration of sulfokinase enzymes, and some of the steroid sulfates are hydrolyzed in the placenta.

During pregnancy, estriol excretion values in the maternal urine (Table 14-5) are relatively low during the first trimester, increase after the tenth to twelfth week, level off around the twenty-fourth to twenty-sixth week, and rise rapidly during the last 3 to 4 weeks of pregnancy. Serial urinary estriol measurements have been of value in high-risk pregnancies where there have been complications incurred by diabetes mellitus, hypertension, preeclampsia, prolonged gestation, and loss of fetus during pre-

Table 14-5
Hormone levels during pregnancy

Weeks of pregnancy	HCG in serum	Pregnanediol in urine (mg/24 hr)	Estriol in urine (mg/24 hr)	Estrogens, serum (μg/dl)	HPL/HCS in serum (μg/ml)
1	10-30 mIU/ml				
2	30-100				
3	100-1000				
6-10		5.0-15		1.3-3.9	
10-12	↓				
11-15				1.5-5.0	
13-18	6-30 IU/ml	5.0-25			
16-20	↑			1.8-7.2	
19-24		13-33			
20-25			3.7-9.9	2.0-9.2	
25-28		20-42			
26-30			7.5-14	2.3-14.7	
29-32	↓	27-47			
31-33	4-15				5.3-6.4
31-35			6.7-17	3.5-over 25	
34-36	↑				6.9-7.6
36-38			15-35	5.9-over 25	
37-38					7.7-8.3
39			Over 35	8.7-over 25	8.0-8.8
40					Over 6.6

vious pregnancies. Impairment of maternal renal function or maternal liver conjugating capacities will cause low urinary estriol levels. Estriol level reflects the function of both the fetus and its placenta. Low levels or decreasing levels indicate that the fetus is at risk. Measurement of urinary estriol for all patients during the last 4 to 6 weeks of pregnancy is becoming a routine practice in many hospitals. However, the use of urine has the limitations of prolonged period (24 hours) for specimen collection, possible incomplete collection, day-to-day variations, and influence of kidney function. In the place of urine, plasma estriol may be measured using radioimmunoassay technique. It is also affected by day-to-day variations and kidney function; in addition, it reflects the estriol level only at the moment when the blood is drawn.

Progestins

Progesterone, the major steroid possessing progestational activity, is secreted by the corpus luteum during the normal menstrual cycle and by the placenta, as its primary source, during pregnancy. For the first 12 weeks of pregnancy the corpus luteum is the major source of progesterone, but about the third month the placenta becomes the source. Smaller amounts are derived from the adrenal cortex and testes.

Progesterone is responsible for the endometrial changes that occur during the second half of the menstrual cycle, and secretion is highest between 6 and 9 days after ovulation. If fertilization occurs, the major function is to maintain pregnancy. Production of progesterone will cease about the twenty-third or twenty-fourth day of the menstrual period if the ovum is not fertilized. It is secreted throughout pregnancy, during which time it inhibits muscle contraction and subsequent ovulation. Progesterone is essential for development of the maternal portion of the placenta and of the mammary glands. The highest levels appear about the thirty-second week of pregnancy.

Plasma progesterone may be measured by radioimmunoassay. Normal values for nonpregnant females are 1 to 9 days of cycle, 0.12 to 0.79 ng/ml; 16 to 29 days of cycle, 1.4 to 17.5 ng/ml; for pregnant females, plateau of 125 ± 38 ng/ml by thirty-second week; and for males, less than 0.3 ng/ml. In circulation progesterone is bound with various proteins, has a short biologic half-life, and is rapidly metabolized by the liver.

The major metabolic products of progesterone are pregnanediol, allopregnanediol, pregnanedione, and pregnanolone. Measurement of urinary pregnanediol in nonpregnant women offers a means for determining the duration and functional activity of the corpus luteum. Pregnanediol excretion levels provide an indication of endogenous production of progesterone. It is excreted in the urine conjugated with glucuronic acid.

Chorionic gonadotropin (HCG)

Shortly after the first missed menstruation, the secretion of chorionic gonadotropin (HCG) by trophoblastic cells of the placenta begins. HCG is a glycoprotein having alpha and beta subunits, with specificity possessed by the beta subunit. It is the hormone which provides the positive reactions in the usual laboratory tests for pregnancy. HCG stimulates continued secretion of estrogen and progesterone by the corpus luteum.

About 5 weeks after the last menstrual period, a rapid increase in urinary HCG occurs, reaching a peak at the eightieth to ninetieth day followed by a decrease to a low level where it remains for the rest of the normal pregnancy.

Measurement of HCG is used for confirmation of early pregnancy since it does not normally circulate except during pregnancy (Table 14-5). The specific HCG beta subunit radioimmunoassay is the most sensitive and specific method available. During the third week after the first missed menstruation, a serum level between 100 and 1000 mIU/ml will be found.

HCG assays are important in diagnosis of hydatidiform mole and as an indicator of the response to therapy. It is excreted in large amounts in cases of chorionepithelioma and hydatidiform mole. Differential diagnosis of high HCG levels utilizes the fact that during pregnancy the normal sequence is a decreasing level following the first 3 months whereas the levels will remain high with a mole or choriocarcinoma. After expulsion of a hydatidiform mole the HCG should decrease below detectable limits. Repeated quantitative assays for serum HCG levels should be performed at frequent intervals to test for normal rate of disappearance of the hormone after evacuation of the uterus and monthly during the first year as a follow-up procedure. A variety of tumors, including those of placental origin, secrete HCG. High levels have been found in ovarian and testicular teratomas, seminomas, and embryonal carcinomas. Successful treatment of choriocarcinoma is followed by a progressive decrease of HCG to undetectable levels. In ectopic pregnancy and threatened abortion, the HCG levels will be below the normal expected values.

Placental lactogen (HPL); human chorionic somatomammotropin (HCS)[22]

Human chorionic somatomammotropin (HCS) is a polypeptide hormone secreted by chorionic tissue which possesses many of the properties of human growth hormone. It is detected in pregnancy serum from 6 weeks to term and is also found in small quantities in urine and amniotic fluid. HCS can promote growth, inhibit peripheral glucose uptake, stimulate insulin release, cause an increase in plasma fatty acids, and mobilize fat stores in pregnancy.

Measurement of plasma HCS levels is made by radioimmunoassay technique and provides an index of placental function. During pregnancy there is a continuous increase in plasma HCS with a plateau during the last 4 weeks (Table 14-5). In multiple pregnancies, the concentration is higher than in singleton pregnancy. At delivery, the concentration falls rapidly. The level will reflect fetal distress caused by placental insufficiency. There is no circadian rhythm and the biologic half-life is about 20 minutes. Complication-free diabetic pregnancies have normal HCS levels, but in unstable diabetic patients the levels are higher than normal, making it impossible to determine the status of the fetus. Low levels have been noted in molar pregnancies and in threatened abortion which ended in abortion.

SEROTONIN AND 5-HYDROXYINDOLEACETIC ACID

Serotonin (5-hydroxytryptamine) is a hormone normally produced by argentaffin cells. These cells are found in the mucous membrane of the gastrointestinal tract, appendix, bronchus, gallbladder, and ducts of the pancreas. Carcinoid tumors arising from argentaffin cells produce excessive amounts of serotonin; this is the most consistent biochemical indicator of the carcinoid syndrome. Other substances that may also be secreted are bradykinin, histamine, and ACTH. Metastatic tumors associated with the carcinoid syndrome usually originate with small primary tumors in the ileum. Carcinoid tumors may involve the liver extensively, with minimal metastatic disease elsewhere. Primary carcinoid tumors of the appendix are common, but metastasis is rare. Tumors from the large intestines may metastasize but no endocrine function is apparent.

The metabolic alteration is the increased conversion of tryptophan to serotonin. Tryptophan is hydroxylated by tryptophan hydroxylase to 5-hydroxytryptophan, which is then decarboxylated to 5-hydroxytryptamine (serotonin). Following its release from the tumor, serotonin is oxidized by monoamine oxidase to 5-hydroxyindoleacetaldehyde, which is converted to 5-hydroxyindoleacetic acid (5-HIAA) by aldehyde dehydrogenase. The 5-HIAA is rapidly excreted in the urine.

In the circulating blood, serotonin is carried by the platelets. As an index of increased production of the hormone, determination of the concentration of the metabolic product, 5-HIAA, in the urine is used. Normally, excretion of 5-HIAA ranges between 1 and 7 mg daily. Elevations in the range of 10 to 500 mg may be seen with carcinoid syndrome. Measurement of serotonin in blood or platelets may be performed but has less diagnostic value than determination of the level of 5-HIAA in urine. Analysis of the tumor tissue for serotonin may be useful in some cases.

PINEAL GLAND

The pineal gland is a small body about 8 mm in length (120 mg) that arises from the third ventricle of the brain and remains attached to the roof of the ventricle. It appears to be an important part of a neuroendocrine control system. Under the influence of light it secretes a hormone, melatonin, an indole derivative of serotonin. Serotonin is acetylated to N-acetylserotonin and then methoxylated by hydroxyindole-

O-methyl transferase (HIOMT) to melatonin. The enzyme HIOMT is concentrated within the pineal gland and has been used to differentiate true pinealomas from pineal tumors of other origin. Using bioassay technique, melatonin in plasma and urine has been measured.

The formation of melatonin follows a circadian rhythm. It appears to inhibit some aspects of gonadal and thyroid function. Calcification of the pineal gland occurs with age, but there is no alteration of the secretory activity. Pineal hyperfunction will depress gonadal maturation, and hyperfunction will cause precocious puberty in boys below the formal age of sexual maturation. The main interest in the gland at this time is the association of tumors of which there are three types: pinealomas, gliomas, and teratomas.

THYMUS

The thymus, a gland involved in development of the immune system of the body, is located between the upper part of the sternum and the pericardium. It secretes a humoral factor essential for lymphopoiesis and reaches its greatest size in children about 9 years of age. In the neonatal and growth periods of the child the major activities of the thymus are apparent. After puberty, because of the activity of the sex glands, it undergoes involution, with loss of its lymphocytes, gradual fibrosis, and shrinkage.

The thymus has been found abnormal for 70% to 80% of patients with myasthenia gravis, and about one fourth had thymomas. Hyperplasia and/or tumors of the thymus have been associated with neuromuscular, hematologic, endocrinologic, and immunologic disorders.

RENIN-ANGIOTENSIN SYSTEM

Renin, an enzyme produced by the juxtaglomerular apparatus of the nephron, acts upon a plasma globulin (renin substrate) to cleave off a decapeptide (angiotensin I) which is further hydrolyzed to an octapeptide, angiotensin II, during passage through the lung and kidney. Angiotensin II is an effective vasoconstrictor and stimulates aldosterone secretion by the adrenal cortex. Since renin has its origin in the kidney, this system is discussed more fully in the chapter on renal function (p. 261).

GASTROINTESTINAL HORMONES

Three of the various gastrointestinal hormones that have been obtained in pure form and synthesized and for which investigation has revealed their mechanism of action and release as well as their physiologic importance are gastrin, secretin, and cholecystokinin-pancreozymin. These have been included in Chapter 11.

1,25-DIHYDROXYVITAMIN D_3[34]

1,25-Dihydroxyvitamin D_3 (1,25$[OH]_2$ D_3) is a steroid hormone metabolized from vitamin D, a group of sterols produced by the action of ultraviolet light on certain provitamins. Vitamin D_3 (cholecalciferol), produced in the skin from 7-dehydrocholesterol by the action of sunlight, is modified in the liver to 25-hydroxycholecalciferol (25-$[OH]D_3$). The latter is transported to the kidney where it is converted to either the steroid hormone 1,25-dihydroxyvitamin D_3 or one of two other metabolites, 24,25-dihydroxyvitamin D_3 or 1,24,25-trihydroxyvitamin D_3.

1,25$(OH)_2D_3$ increases the retention of calcium and phosphate by the body and controls the mineralization of bone matrix. It is apparently required for the action of PTH on bone. The major target tissue for action is the intestinal mucosa. Formation is regulated by the calcium ions in the bloodstream, being increased when the calcium is decreased and decreased when the plasma calcium ions are increased. The effect of decreased calcium ions appears to be mediated by way of parathyroid hormone.

A competitive protein binding method is available for assaying $25(OH)D_3$ circulating in plasma. The normal level is approximately 20 ng/ml with increase upon exposure of an individual to excessive sunlight.

Since both the liver and kidney participate in conversion of the vitamin D to the hormone, diseases of either organ will affect the synthesis of the hormone and thereby affect bone formation and calcium absorption. Hypoparathyroid patients lack PTH, fail to synthesize $1,25(OH)_2D_3$, and thus have low concentrations of serum calcium. Investigations in progress have demonstrated that vitamin D_3 is useful for therapeutic purposes in treatment of patients with calcium deficiencies and bone diseases.

Hormone assays

The activity of endocrine glands is reflected most accurately in measurement of rates of hormone production. However, this involves administration of radioactive isotopes to the patient with collection of blood and urine samples under controlled conditions. The most practical approach to evaluate endocrine function is the measurement of hormones or their metabolites in biologic fluids, usually blood or urine. The abnormalities present are usually recognized by increased or decreased secretion of hormones.

The low concentrations of hormones in biologic fluids have limited the methodology that can be used for their assay. Some can be measured by chemical procedures (steroids, catecholamines), whereas the polypeptide hormones are best determined by radioassay (competitive protein binding, radioimmunoassay). The chemical methods usually involve extraction, purification, and an end-point reaction. The principle of all radioassays is the same whether the binding protein is an antibody or a naturally occurring protein (corticosteroid-binding globulin, thyroxine-binding globulin) with high affinity for the substance to be measured.

RADIOASSAYS

The assay is based on competition between a small fixed amount of labeled ligand and unlabeled ligand for a limited number of specific binding sites (antibody, binding protein, tissue receptor protein). The binding sites are saturated with a radioactive agent and incubated with unlabeled material to be measured. As a result of the competition between radioactive and nonradioactive ligand, the ratio of bound to free labeled ligand is decreased as the concentration of unlabeled ligand is increased. When equilibrium has been approached the free labeled ligand is separated from labeled ligand bound to the specific binding agent.

Competitive protein binding (CPB) assay (displacement or saturation analysis) refers to methods of analysis which utilize the capability of certain plasma or tissue binding proteins to combine specifically with a ligand. This technique has been used to measure plasma levels of hormones such as cortisol, deoxycortisol, corticosterone, progesterone, estradiol, testosterone, thyroxine, and triiodothyronine. The specific plasma globulins involved are corticosteroid-binding globulin, thyroxin-binding globulin, sex hormone–binding globulin, and cellular receptors.

Radioimmunoassay defines a procedure in which a specific antibody is used to combine with the ligand. The procedure finds its greatest use in measurement of peptide and protein hormones in blood. The antibody is produced usually by injecting hormones in Freund's adjuvant into either rabbits or guinea pigs. Steroid hormones and small molecules must be coupled covalently to a protein before injecting to produce an antibody.

For laboratories performing a variety of hormone assays using radioassay techniques, it is not economically feasible to prepare antibodies and labeled hormones. These are available commercially both separately and in kit form for all hormones shown to be clinically important, which includes the following:

ACTH	Human placental lactogen
Aldosterone	Insulin
Angiotensin I	LH
Chorionic gonadotropin	Progesterone
Corticosterone	Prolactin
Cortisol	PTH
Deoxycorticosterone	Testosterone
Deoxycortisol	Thyroid-stimulating hormone
Dihydrotestosterone	Triiodothyronine
Estradiol, estriol, estrone	Thyroxine
FSH	Thyroxine, free
Gastrin	Vasopressin
Growth hormone	

Specific methodology will not be presented since it varies with each supplier or producer of reagents and since frequent changes are being made to conform to the requirements of the purchaser. Brochures accompany reagents and kits giving the suppliers' recommended procedures.

CHEMICAL ANALYSIS

Hormones with a chemically reactive group that will form characteristic chromogens or fluorescent compounds can be measured utilizing this property if present in sufficient concentration. Examples of these are the Porter-Silber reaction for 17-hydroxycorticosteroids, Zimmermann reaction for 17-ketosteroids, fluorometric assay of certain steroids (cortisol, estrogens) in acid or alkaline solutions and catechol amines, and gas-liquid chromatography for some steroids (pregnanediol, pregnanetriol, estrogens, fractionated ketosteroids).

Catecholamines, urinary[2, 13, 30, 62]
Principle

Urinary free catecholamines (epinephrine and norepinephrine) are adsorbed onto alumina at pH 8.5 and subsequently eluted with acetic acid. They are then oxidized with potassium ferricyanide in the presence of zinc ion to adrenochrome and noradrenochrome. On addition of strong alkali and ascorbic acid they are further converted to the fluorescent compounds, adrenolutine and noradrenolutine or trihydroxyindoles (THI). Measurement of fluorescence is made in a fluorometer or spectrofluorometer.

Specimen collection

A 24-hour urine specimen is collected in a glass bottle containing 15 ml of 6 N hydrochloric acid to provide a final pH below 3.0. At this pH urinary catecholamines are stable for several days at room temperature and for months in the refrigerator. The urine must be kept acid during the entire collection since catecholamines deteriorate rapidly in neutral or alkaline solutions. If a single voided specimen is collected during or immediately after an episode of hypertension it should be preserved at a pH between 2 and 3 with a small amount of hydrochloric acid.

Reagents needed

All chemicals are reagent grade.

Alumina (Brockman), activity I, 80-200 mesh for chromatography
   Wash with distilled water to remove the fine dust, dry at 100° C, and heat for 2 hours at 200° C.

Disodium ethylenediamine tetraacetate (EDTA)

Sodium hydroxide, 10 N (40%)

Sodium hydroxide, 5 N (20%)

Sodium hydroxide, 0.5 N (2%)

Acetic acid, CH_3COOH, 0.2 M
   Dilute 11.5 ml of glacial acetic acid to 1 L with distilled water.

Potassium ferricyanide, $K_3Fe(CN)_6$, 0.25%
   Dissolve 250 mg in water and dilute to 100 ml. Prepare fresh each week and store in the refrigerator.

Disodium hydrogen phosphate buffer, 0.5 M, Na_2HPO_4
   Dissolve 35.5 g of the anhydrous salt in water and dilute to 500 ml. Store in the refrigerator.

Zinc sulfate, $ZnSO_4$ 0.25%
   Dissolve 250 mg in water and dilute to 100 ml.

Ascorbic acid–sodium hydroxide reagent
   Prepare immediately before using: 20 mg of ascorbic acid + 0.2 ml of distilled water + 10 ml of 10 N NaOH.

Hydrochloric acid, 0.1 N
   Dilute 8.6 ml of concentrated (11.6 N) hydrochloric acid to 1 L.

Hydrochloric acid, 6N
   Dilute concentrated hydrochloric acid 1:1.

Stock standard (100 μg norepinephrine/ml)
   Dissolve 20.1 mg of *l*-arterenol bitartrate in 0.1 N HCl and dilute to 100 ml. Stable for at least 6 months in the refrigerator.

Working standard (0.25 μg norepinephrine/ml)
   Into a 200-ml volumetric flask pipet 0.5 ml of stock standard solution and 20 ml of 0.1 N HCl. Dilute to volume with water. Prepare fresh each week.

Procedure

1. Measure the total volume of urine and filter an aliquot.
2. Into 50-ml beakers (in duplicate) pipet
   Unknown: 25 ml of filtered urine
   Control: 25 ml of filtered urine control
3. Add to each beaker 200 mg of EDTA.
4. Adjust urine to pH 8.5, using 5 N NaOH to pH 7 to 7.5 and 0.5 N NaOH for the final adjustment, while stirring with a magnetic stirrer. A pH meter should be used although narrow-range pH paper may be employed.

Adsorption and elution

5. Prepare chromatographic columns, using 8 × 300 mm columns having a 50-ml reservoir, by adding dry alumina to a height of 4 cm. Cover with a small plug of glass wool.
6. Pour the urine onto the columns. It should flow through at a rate of 1.0 to 2.0 ml/min.
7. Wash each column with 20 ml of distilled water.
8. Elute the catecholamines with 20 ml of 0.2 M acetic acid.

Oxidation and fluorescence

9. Prepare 15 × 180 mm test tubes as follows:

Reagent blank (duplicate): 2.0 ml of 0.2 M acetic acid + 1.0 ml of distilled water.
Standard (duplicate): 2.0 ml of 0.2 M acetic acid + 1.0 ml of working standard.
Tests and control (three tubes each):
Tube 1: 2.0 ml of eluate + 1.0 ml of distilled water
Tube 2 (blank): 2.0 ml of eluate + 1.0 ml of distilled water
Tube 3 (internal standard): 2.0 ml of eluate + 1.0 ml of working standard

Group tubes into sets not to exceed twelve tubes (two determinations in duplicate).

10. To each tube add 0.8 ml of phosphate buffer, pH 8.9, to give a pH of 6.5 in the tubes (check the pH, using narrow-range paper) and 0.1 ml of zinc sulfate reagent, 0.25%.
11. To tubes containing reagent blank, control blank, and test blank add 1.0 ml of ascorbic acid–NaOH reagent.
12. To tubes containing standard, control, and test (including internal standard tubes) add 0.1 ml of potassium ferricyanide reagent and allow to stand *exactly 2 minutes* (use an interval timer); then add 1.0 ml of ascorbic acid–NaOH reagent, mix, and read the fluorescence within 10 minutes.
13. To blank tubes containing ascorbic acid–NaOH reagent add 0.1 ml of potassium ferricyanide reagent and mix.

Measurement and calculation

14. Read fluorescence in a fluorometer or spectrophotofluorometer against a water blank. Use an activation wavelength of 405 nm and a fluorescent wavelength of 510 nm.

$$\frac{F_u}{F_{std}} \times \text{Standard concentration} \times \frac{\text{Volume eluate}}{\text{Aliquot eluate}} \times \frac{\text{Total volume urine}}{\text{Aliquot urine}} =$$

$$\frac{F_u}{F_{std}} \times 0.25 \times \frac{20}{2} \times \frac{TV}{25} = \mu g \text{ catecholamines in 24-hour specimen or volume received (calculated as norepinephrine)}$$

where F_u = fluorescence of unknown (sample–blank)
F_{std} = fluorescence of standard (standard–sample)

Normal values

Urine
  Catecholamines   Up to 100 μg/24 hr
                                    Up to 18 μg/dl, random sample
    Norepinephrine   10 to 80 μg/24 hr
    Epinephrine   0 to 20 μg/24 hr
Plasma
  Catecholamines   Less than 1.5 μg/L

Comments

Adsorption occurs rapidly at pH 8.0 to 8.5 and, since catecholamines are unstable at this pH, the time for passage through the columns must be as brief as possible. Usually less than 15 minutes is required.

If urine acidified below pH 1.5 is maintained at room temperature, hydrolysis of the conjugated catecholamines may occur, resulting in increased values for free catecholamines determinations.

Some of the catecholamine derivatives, if present, will also be adsorbed on the alumina. However, acetic acid will not elute dopa, and with the use of potassium ferricyanide as the oxidizing agent no interference is obtained from dopamine, normetanephrine, and metanephrine.

The addition of EDTA prevents the formation of a gelatinous calcium-magnesium-phosphate precipitate that will interfere with the reading of fluorescence.

Interference will occur from fluorescent drugs in the urine if a patient is receiving chlortetracycline (Aureomycin) or other tetracycline antibiotics, quinine, quinidine, or large amounts of the vitamin B complex.

Exogenous catecholamines prescribed as vasopressors, brochodilators, and nasal decongestants may elevate the levels in the urine. Drugs that have been reported to interfere with the determination are acetylcholine, amphetamine, chlorpromazine, cocaine, ephedrine, epinephrine, guanethidine, isoproterenol (Isuprel), mandelamine, metanephrine, methyldopa (Aldomet), reserpine, tyramine, and methenamine (Uritone, Urotropin). Biliary pigment in urine causes interference. For patients with pheochromocytoma catecholamines may be depressed during the early administration of antihypertensive drugs. Increased catecholamine excretion may be observed following vigorous exercise and in cases of progressive muscular dystrophy and myasthenia gravis.

Fractionated catecholamines: epinephrine and norepinephrine
Principle

After determining the presence of a pheochromocytoma, a quantitative differential determination of urinary epinephrine and norepinephrine may provide an indication of the site of the tumor.[9] The location of the lesion will be usually in the adrenal gland if both of the amines are increased. If the increase is mostly norepinephrine, the tumor may be expected to be in or near one of the adrenal glands in about two thirds of the cases and extraadrenal for the remainder.

The procedure is the same as for total free catecholamines with the exception that two different filter or wavelength combinations are used and an additional internal standard (epinephrine) is included. The concentration of each amine is calculated by applying simultaneous equations.

Reagents
Same as for catecholamines
Standard solution No. 2
Stock standard (100 μg epinephrine/ml)
Dissolve 18.2 mg of *l*-epinephrine *d*-bitartrate in 0.1 N HCl and dilute to 100 ml. Stable for at least 6 months in the refrigerator.
Working standard (0.25 μg epinephrine/ml)
Into a 200-ml volumetric flask pipet 0.5 ml of stock standard solution and 20 ml of 0.1 N HCl. Dilute to volume with distilled water. Prepare fresh each week.

Procedure
1 to 8. Same as for catecholamines.
9. Prepare 15 × 180 mm test tubes as follows:
Reagent blank (**duplicate**): 2.0 ml of 0.2 M acetic acid + 1.0 ml of distilled water
Standard norepinephrine (**duplicate**): 2.0 ml of 0.2 M acetic acid + 1.0 ml of NE working standard
Standard epinephrine (**duplicate**): 2.0 ml of 0.2 M acetic acid + 1.0 ml of E working standard

Tests and controls (**four tubes each**)
  Tube 1: 2.0 ml of eluate + 1.0 ml of distilled water
  Tube 2 (blank): 2.0 ml of eluate + 1.0 ml of distilled water
  Tube 3 (NE internal standard): 2.0 ml of eluate + 1.0 ml of NE working working standard
  Tube 4 (E internal standard): 2.0 ml of eluate + 1.0 ml of E working standard
10 to 13. Same as for catecholamines.

Measurement and calculation

14. Using a spectrofluorometer and two different sets of wavelengths, read the fluorescence of the solution in all tubes against a water blank.

	Set I	Set II
Exciting	405 nm	436 nm
Emitting	510	520

The following symbols may be used for setting up the equations:
  B = Test or control blank
  U = Test or control
  N = Test or control + NE internal standard
  A = Test or control + E internal standard
  Set I: $n_1 = (N_1 - U_1) \times 4$ = Fluorescence/μg NE added
  Set II: $n_{11} = (N_{11} - U_{11}) \times 4$ = Fluorescence/μg NE added
  Set I: $a_1 = (A_1 - U_1) \times 4$ = Fluorescence/μg E added
  Set II: $a_{11} = (A_{11} - U_{11}) \times 4$ = Fluorescence/μg E added
  Set I: $u_1 = U_1 - B_1$
  Set II: $u_{11} = U_{11} - B_{11}$

$$Y = \frac{u_1 a_{11}/a_1 - u_{11}}{n_1 a_{11}/a_1 - n_{11}} = \text{Norepinephrine in 2.5 ml urine*}$$

$$X = \frac{u_{11} - Y n_{11}}{a_{11}} = \text{Epinephrine in 2.5 ml urine*}$$

Y/2.5 × Total volume = μg norepinephrine/24 hr
X/2.5 × Total volume = μg epinephrine/24 hr

Clinical findings

Adrenal medullary insufficiency does not produce any recognized clinical syndromes. Pheochromocytomas, tumors involving the chromaffin cells of the adrenal medulla and of other areas in the body, frequently synthesize and secrete large quantities of catecholamines. The physical symptoms often observed are severe or intermittent hypertension associated with hypermetabolism, headache, excessive sweating, weight loss, and elevated fasting blood glucose level. Approximately 80% of all pheochromocytomas are located in the adrenal gland.

Measurement of urinary catecholamines and/or VMA should be done on all young patients with essential hypertension and on older patients where there is some suspicion that a pheochromocytoma may exist, since a diagnosis of pheochromocytoma may not be indicated by clinical examination. It is found in less than 1% of hypertensive patients. However, hypertension that results from pheochromocytoma is reversed by removal of the tumor. The ratio of urinary catecholamines

---

*Assuming that a standard containing 0.25 μg/ml and an equivalent of 2.5 ml of urine are used in the procedure. If other quantities are employed, substitute those values in the equations.

to 3-methoxy-4-hydroxymandelic acid has been shown to indicate the size of the tumor mass.

The determination of 3-methoxy-4-hydroxymandelic acid is unaffected by many of the drugs that interfere with catecholamine assays. For screening purposes it is not necessary to withdraw reserpine, thiazides, ganglionic blocking agents, or guanethidine from patients being treated with these. Misleading values may be obtained when monamine oxidase inhibitors are administered; these will cause a decrease in urinary 3-methoxy-4-hydroxymandelic acid and an increase in metanephrines.

In pheochromocytoma, catecholamine levels may vary between 300 μg and 4 mg/24 hr or up to 150 μg/dl for a random specimen; 3-methoxy-4-hydroxy mandelic acid values may range from 11 to 250 mg/24 hr.

Neuroblastomas, highly malignant tumors of infants and young children, are characterized by a high urinary excretion of 3-methoxy-4-hydroxymandelic acid and homovanillic acid. Diarrhea is the most usual symptom but hypertension may occur. Ganglioneuromas are benign sympathetic tumors which also occur most commonly in children. Affected individuals may present with chronic diarrhea, wasting, and abdominal distention that disappear with the removal of the tumor.

Summarizing, in pheochromocytoma there are increased epinephrine, norepinephrine, metanephrine, normetanephrine, and 3-methoxy-4-hydroxymandelic acid; in neuroblastoma and ganglioneuroma, increased homovanillic acid, dopa, dopamine, and 3-methoxy-4-hydroxymandelic acid; in melanoma, increased homovanillic acid; and in familial dysautonomia, increased homovanillic acid and decreased 3-methoxy-4-hydroxymandelic acid.

11-Deoxycortisol (compound S), urinary[26]
Principle

Compound S (11-deoxycortisol, cortexolone) is present only in very small quantities in normal blood and urine. Upon administration of metyrapone (Metopirone) the 11β-hydroxylation of 11-deoxycortisol is inhibited, causing increased concentration of plasma 11-deoxycortisol and deoxycorticosterone and decreased plasma cortisol concentration. The lowered plasma cortisol level stimulates production of ACTH which is not controlled by 11-deoxycortisol. This provides a means of evaluating the feedback mechanism of the hypothalamic-pituitary-adrenal axis. Subjects with hypopituitary function will not show increased levels of ACTH and 11-deoxycortisol.

Although 11-deoxycortisol is determined with the Porter-Silber chromogens and the 17-ketogenic steroids, there is no way to designate an increase in the 17-hydroxycorticosteroid concentration as being caused by compound S. The urinary corticosteroids are hydrolyzed with β-glucuronidase in the same manner as for Porter-Silber chromogens. With the exception that the extraction is made with carbon tetrachloride, the procedure is the same as for Porter-Silber chromogens.

Specimen collection
A 24-hour urine collection is obtained preferably without a preservative and kept at about 2° to 4° C. If refrigeration is not possible, 15 ml of toluene may be added as preservative.

Reagents
Except for the following, the reagents are the same as those for Porter-Silber chromogens.
Carbon tetrachloride, reagent grade

Stock standard
  Dissolve 10 mg of tetrahydro S in 5 ml of absolute ethyl alcohol and dilute to 100 ml with distilled water (0.1 mg/ml).
Working standard
  Dilute stock standard 1:10 with distilled water (10 μg/ml.)

Procedure
1. Measure total volume of 24-hour collection of urine and filter an aliquot.

Hydrolysis
2. Into 50-ml round-bottom centrifuge tubes pipet in duplicate:
   Unknown: 2.0 ml of filtered urine
   Add 1.0 ml of β-glucuronidase solution and mix. Check the pH; it should be 6.5 (indicator paper). If not, adjust by using 0.1 N NaOH. Add 0.1 ml of chloroform.
3. Incubate for 18 hours at 37° C.

Extraction and washing
4. Into 50-ml round-bottom centrifuge tubes pipet in duplicate 2.0 ml of working standard, 10 μg/ml, and 1.0 ml of distilled water.
5. To all tubes from steps 3 and 4 add 25 ml of carbon tetrachloride and shake for 15 seconds by hand. Centrifuge for 5 minutes and discard the aqueous layer by aspirating.
6. Add 2.0 ml 0.1 N NaOH, shake for 10 seconds, and centrifuge for 5 minutes. Discard the alkali wash by aspirating.

Color development
7. Into 12-ml centrifuge tubes transfer 5.0 ml of carbon tetrachloride extract and add 1.0 ml of phenylhydrazine–sulfuric acid–methyl alcohol reagent; mix for 10 seconds on a Vortex mixer.
8. Centrifuge for 5 minutes and discard the carbon tetrachloride layer by aspirating.
9. Incubate tubes in a water bath at 60° C for 1 hour.
10. Allow to cool to room temperature and transfer samples to microcuvets, using capillary pipets.
11. Determine absorbance (A) at 370, 410, and 450 nm against a blank of phenylhydrazine–sulfuric acid reagent.

Calculations

$$\text{Corrected A} = A_{410} - \left(\frac{A_{370} + A_{450}}{2}\right)$$

$$\frac{\text{Corrected A unknown}}{\text{Corrected A standard}} \times \text{Standard concentration (mg)} \times \frac{\text{Total urine volume (ml)}}{\text{Aliquot urine used}} = \text{mg compound S in 24 hr}$$

Normal values
  Less than 2 mg/24 hr

Clinical findings

Following an infusion of metyrapone (Metopirone), normal persons excreted more than 8 mg of compound S metabolites in 24 hours whereas patients with untreated panhypopituitarism and adrenal insufficiency causing 11β-hydroxylation deficiency excreted less than 2 mg.

DHEA, Allen "blue" color test, urinary[1]
Principle

Increased amounts of dehydroepiandrosterone are found in the urine in certain pathologic conditions. The Allen color test may be used as a screening procedure for adrenocortical tumors. Hydrolysis and extraction are carried out in the same manner as for total 17-ketosteroids. Color is developed by adding an ethanol–sulfuric acid reagent that reacts with ketosteroids having a 3-hydroxy-Δ^5 structure.

Specimen collection

A 24-hour urine collection is obtained preferably without a preservative and kept at about 5° C. If refrigeration is not possible, 15 ml of toluene may be added as preservative to prevent bacterial contamination.

Reagents

All chemicals are reagent grade.
Hydrochloric acid, concentrated
Ethyl ether (Squibb) for anesthesia, or redistilled and peroxide-free
Sodium hydroxide, 10% (w/v)
Ethyl alcohol, absolute, ketone-free and aldehyde-free
  Purify by the procedure described in the method for 17-ketosteroids.
Ethyl alcohol, 95% aqueous (v/v) and 90% (v/v)
Alcohol-sulfuric acid reagent
  Add 12 volumes of concentrated sulfuric acid to 3 volumes of 90% ethyl alcohol.
Sulfuric acid, concentrated
Methyl alcohol, absolute, spectro quality
Stock standard
  Dissolve 100 mg of dehydroepiandrosterone in absolute methanol and dilute to 100 ml (1000 µg/ml).
Working standard (200 µg/ml)
  Dilute 10 ml of stock standard to 50 ml with absolute methanol.

Procedure
1. Measure total volume of 24-hour collection of urine and filter an aliquot.

Hydrolysis
2. Into 50-ml round-bottom centrifuge tubes pipet in duplicate 10 ml of filtered urine and 3 ml of concentrated HCl.
3. Heat the tubes in a water bath at 80° to 85° C for 12 minutes. Cool in running tap water.

Extraction and washing
4. To each tube add 20 ml of ethyl ether and shake by hand for 30 seconds. Discard the aqueous layer by aspirating.
5. Add 10 ml of 10% NaOH, shake for 10 seconds by hand. Discard the NaOH layer by aspirating.
6. Add 10 ml of distilled water and shake for 10 seconds. Cover tubes with foil and centrifuge for 5 minutes at 1500 rpm.

Color development
7. Into 40-ml round-bottom tubes place:
   Unknown: 10 ml of ether extract
   Standard (duplicate): 0.1 ml of working standard (20 µg)
   Blank: Dry tubes
8. Evaporate to dryness in a water bath at 40° to 45° C under a stream of air. Wash down the sides of the tubes with 0.3 ml of absolute ethyl alcohol and evaporate to dryness.

9. To each tube add 2 ml of alcohol–sulfuric acid reagent and heat for 12 minutes at 55° C.
10. Cool in running tap water for 1 minute, add 3.0 ml of 95% ethyl alcohol, and mix.
11. Measure the absorbance (A) against the reagent blank at 560, 600, and 640 nm.

Calculations

$$\text{Corrected A} = A_{600} - \left(\frac{A_{560} + A_{640}}{2}\right)$$

$$\frac{\text{Corrected A unknown}}{\text{Corrected A standard}} \times \frac{20}{1000} \times \frac{\text{Total urine volume (ml)}}{\text{Aliquot urine in extract (5 ml)}} =$$

mg 3β-hydroxy-17-ketosteroids (calculated as DHA) in 24-hour specimen

Comments

The blue color is not specific for dehydroepiandrosterone but the other compounds that react similarly are present in small amount and do not interfere significantly with the test.

Clinical findings

The test has limited application in differentiating patients with adrenal tumor from those with adrenal hyperplasia. In some cases of adrenal carcinoma the level of urinary 17-ketosteroids is elevated and dehydroepiandrosterone is primarily responsible for the increase. Allen cited four out of five cases of adrenal tumor where an extract of the urine failed only once to produce a grossly positive blue color and seven cases of female pseudohermaphroditism that gave a negative reaction even though the amount of total ketosteriod was markedly increased. In cases of adult adrenal carcinoma the ratio of beta to alpha isomers ($\beta:\alpha$) may range from 0.28 to 4.0. The beta fraction consists mostly of dehydroepiandrosterone and the alpha fraction of androsterone and etiocholanolone. Normally the $\beta:\alpha$ ratio is less than 0.2. The test is considered significant when the ratio exceeds 0.4.

Estriol (pregnancy), urinary[29]
Principle

Estrogen determinations are used to assess fetal status during pregnancy. An indication of fetal distress may be based upon decreased estriol production as demonstrated when estriol levels do not rise rapidly or begin to decrease after the second trimester of pregnancy.

The high concentration of estriol in normal pregnancy urine permits use of a small volume of specimen for analysis. The less urine needed, the lower the concentration in the hydrolysate of nonspecific urinary substances that interfere with the assay. The urinary estrogen conjugates are hydrolyzed with hydrochloric acid (if patient is spilling sugar, β-glucuronidase is used). The hydrolysate is extracted with ether and the extract washed with sodium bicarbonate (pH 10.5), a reagent of sodium hydroxide–sodium bicarbonate, saturated sodium bicarbonate, and water. Solvent partition is used to separate estriol from estrone and estradiol-17β, and the Kober reagent is added for the colorimetry. Absorbance is determined by spectrophotometry.

Specimen collection

A 24-hour urine specimen is collected in a glass or polyethylene bottle without a preservative. The analysis should be performed as soon as possible after the specimen is received. It may be stored for several days at 4° C without deterioration occurring but when it is to be held for longer periods, the specimen must be frozen.

Reagents

    All chemicals are reagent grade.
    Ethyl ether (Squibb) for anesthesia, or redistilled, peroxide-free
    n-Heptane, redistilled, saturated with distilled water
    Toluene, redistilled, saturated with distilled water
    Sodium bicarbonate, saturated solution
        Prepare fresh each day.
    Sodium bicarbonate reagent, pH 10.5
        Add 15 ml of 20% NaOH solution to 100 ml of saturated sodium bicarbonate solution. Prepare fresh daily.
    Sodium hydroxide, 20%
    Sodium hydroxide, 8%
    Ethyl alcohol, 95%
    Hydrochloric acid, concentrated
    Alundum, 60 mesh
    Sodium nitrate, crystal
    p-Quinone, purified
    Hydroquinone, purified
    Sulfuric acid, concentrated
    Kober color reagent
        Cool 200 ml of distilled water in a 2-L flask in an ice bath and add 760 ml of concentrated sulfuric acid slowly while stirring. Dilute to 1 L after cooling to room temperature. Add and dissolve 10 mg of sodium nitrate and 20 mg of p-quinone in the liter of diluted acid. Warm in a water bath at 100° C until a light green color just appears. Immediately add 20 g of hydroquinone and heat in a boiling water bath for 45 minutes, shaking occasionally until solution is complete. The reagent is placed in the dark at room temperature for about a week and then filtered through a sintered glass funnel. It is stored in an amber glass bottle at room temperaure.
    Standard solution
        Dissolve 10 mg of estriol in 95% ethyl alcohol and dilute to 100 ml (100 μg/ml).

Procedure

1. Measure total volume of 24-hour collection of urine and check for glucose, using Uristix. Filter an aliquot of the urine. If glucose is present, hydrolyze, using an enzyme (β-glucuronidase); otherwise hydrolyze with acid.

Hydrolysis (acid)

    a. Unknown and control specimens: For the last trimester of pregnancy use 2% of the total volume of urine and dilute to 80 ml with distilled water in a 250-ml round-bottom distilling flask. Add approximately 2 g of alundum and heat to boiling under a reflux condenser. Add 16 ml of concentrated HCl and allow to boil for 30 minutes.
    b. Cool under running tap water and transfer to a 100-ml volumetric flask. Dilute to volume.

Hydrolysis (enzymatic)
  c. Into 50-ml round-bottom centrifuge tubes pipet in duplicate:
     Unknown: 20 ml of filtered urine
     Control: 20 ml of filtered urine
     Add to each tube 4.0 ml of phosphate buffer (pH 6.5) containing β-glucuronidase (1200 units) and mix. Check the pH, which should be 6.5. If it is not, adjust by using 0.1 N NaOH. Refer to reagents for Porter-Silber chromogens. Incubate for 18 to 24 hours at 37° C.
  d. Dilute hydrolysate to 25 ml in a volumetric flask. Calculate the amount of hydrolysate equivalent to one fifth of 2% of the original urine diluted to 100 ml. Dilute this aliquot to 20 ml and proceed as directed for acid-hydrolyzed urine.

Extraction and washing
  2. Into 125-ml separatory funnels pipet in duplicate:
     Unknown: 20 ml of hydrolysate
     Control: 20 ml of hydrolysate
     Standard: 19 ml of distilled water + 0.4 ml of estriol standard + 1.0 ml of concentrated HCl
  3. Add to each funnel 1 g of NaCl and 40 ml of ethyl ether. Shake for 1 minute by hand and discard the aqueous layer.
  4. Wash the ether extract with 8 ml of sodium bicarbonate reagent (pH 10.5) and discard the washing.
  5. Shake the extract with 2.0 ml of 8% NaOH and add 8.0 ml of saturated sodium bicarbonate. After mixing for 30 seconds discard the aqueous layer.
  6. Remove the remaining alkali from the ether layer by washing with and then discarding 2.0 ml of saturated sodium bicarbonate solution. Wash two times with 1.0 ml of distilled water or until neutral.
  7. Into 40-ml round-bottom centrifuge tubes place 30-ml aliquot of the washed ether extract.
  8. Evaporate to dryness in a water bath at 40° to 45° C under a stream of air. Wash down the sides of the tubes with 1.0 ml of 95% ethyl alcohol, allow to stand in a 50° C water bath for 15 minutes, and then evaporate to dryness.

Partition of estriol fraction
  9. Dissolve the dried ether extract residue in 0.8 ml of hot 95% ethyl alcohol and add 5.0 ml of n-heptane saturated with water and mix; add 5.0 ml of toluene saturated with water and mix; add 20 ml of distilled water and shake for 1 minute.
  10. Centrifuge for 5 minutes and discard the organic solvent layer. The estriol is in the aqueous layer.
  11. Into 50-ml round-bottom centrifuge tubes place 10-ml aliquot of the aqueous layer, 1 g of NaCl, and 20 ml of ethyl. Shake for 1 minute.
  12. Into 40-ml round-bottom centrifuge tubes transfer 15-ml aliquot of ether extract. Evapoarte to dryness in a water bath at 40° to 45° C under a stream of air.
  13. Prepare direct standards for recovery calculations by adding 0.2 ml of standard to 40-ml tubes and evaporating to dryness.

Color development
  14. Add 3.0 ml of Kober reagent to each tube from steps 12 and 13 and to two clean tubes that serve as blanks.
  15. Heat tubes in boiling water for 20 minutes. Stir frequently with a glass rod during the first 5 minutes and occasionally thereafter.

Table 14-6
Estrogen excretion in urine of normal pregnant women (range and mean)[47]

Pregnancy (weeks)	Estrone (mg/24 hr)	Estradiol-17β (mg/24 hr)	Estriol (mg/24 hr)
20-25	0.33-1.1 (0.73)	0.16-0.30 (0.23)	3.7- 9.9 (6.5)
26-30	0.40-2.0 (0.98)	0.25-0.47 (0.36)	7.5-13.3 (10.0)
31-35	0.55-1.4 (1.0)	0.29-0.61 (0.38)	6.7-17.0 (12.5)
36-40	0.58-1.8 (1.2)	0.32-0.60 (0.45)	14.9-31.2 (19.8)

16. Cool in an ice bath and add 1.0 ml of distilled water. Mix.
17. Reheat tubes in a boiling water bath for 10 minutes and then cool rapidly in an ice bath.
18. Determine absorbance (A) at 460, 500, and 540 nm, reading against the reagent blank.

Calculations

$$\text{Corrected A} = A_{500} - \left(\frac{A_{460} + A_{540}}{2}\right)$$

$$\frac{\text{Corrected A unknown}}{\text{Corrected A extracted standard}} \times \text{Standard concentration} \times \frac{\text{Total volume urine}}{\text{Aliquot urine used}} = \text{mg estriol/24 hr}$$

Normal values

For normal values during pregnancy see Table 14-6.

Comments

Acid hydrolysis causes some destruction of estrogens, but this effect is reduced by diluting the urine prior to analysis. An enzyme preparation, Glusulase, which contains both glucuronidase and sulfatase, has been used successfully for the hydrolysis. While the time required for the action is longer than with acid, the advantages are that there is no significant destruction of the estrogens and less production of interfering substances. However, urine contains variable amounts of inhibitors for both enzymes and therefore acid hydrolysis is used in the routine procedure.

Glucose has a destructive effect on estrogens during acid hydrolysis.[27] A tenfold dilution prior to hydrolysis may be used but enzyme hydrolysis is preferable.

Drugs that have been reported as interfering with some of the estrogen procedures are phenolphthalein, cascara, senna, stilbestrol, meprobamates, large doses of cortisone, and mandelamine.

Clinical findings

Serial estriol excretion levels are necessary for evaluating fetal-placental function in abnormal pregnancies as daily values fluctuate markedly. Steroids are conjugated and metabolized by the fetus. Estriol in pregnancy is derived from 16α-hydroxydehydroepiandrosterone of fetal origin. Estriol is conjugated by the fetus.

Decreased excretion of estriol may be found in fetal distress resulting from various diseases such as diabetes mellitus, toxemia, preeclampsia, or placental insufficiency. Persistently low or decreasing levels after the thirty-second week suggest the need for premature delivery as a means of fetal survival. When diabetic women have lower than normal concentrations of estriol but show a constant tendency for increase, the pregnancy can be allowed to proceed to term without anticipating complications.[25] Estriol measurements provide a method for the management of complicated pregnancies.

Estrogens (nonpregnant), total, urinary[12]

Principle

Urine is saturated with ammonium sulfate and extracted with ether-alcohol as a preliminary conjugate extraction for purification. After evaporating the organic extract to dryness, the residue is dissolved in water and treated with borohydride, a reaction which produces a more homogeneous steroid end product. Hydrolysis is effected with hydrochloric acid and heating in a boiling water bath. The hydrolysate is made alkaline and extracted with benzene-petroleum ether to remove some nonphenolic steroids and contaminants. After adjusting the pH, the estrogens are extracted by ether which is washed successively with carbonate buffer, ammonium sulfate solution, and water. The ether extract is reduced to dryness and the residue treated with hydroquinone and sulfuric acid to form the Kober-estrogen complex. The chromogen is extracted with an ethanolic chloroform reagent containing p-nitrophenol. Fluorescence of the chloroform extract is measured using an exciting wavelength of 546 nm and an emitting wavelength of 585 nm.

Specimen collection

A 24-hour urine collection is obtained without a preservative and kept at about 2° to 4° C. It may be stored for several days at 4° C without deterioration occurring, but when holding for longer periods, the specimen must be frozen.

Reagents

Ammonium sulfate, crystals AR
Ethyl ether (Squibb) for anesthesia, or redistilled, peroxide-free
Ethyl alcohol, 95%
Ether-ethanol reagent, 3:1
Potassium borohydride, 10% aqueous (w/v)
Sodium hydroxide, 10 N and 1 N
Benzene-petroleum ether reagent, 1:1
Hydrochloric acid, concentrated, AR
Sodium bicarbonate, powder, AR
Sodium carbonate, saturated solution
Buffer, pH 10.4, 2 parts saturated sodium carbonate to 1 part saturated sodium bicarbonate
Ammonium sulfate solution, 8% aqueous (w/v)
Hydroquinone, purified (Fisher), 2% in ethanol (w/v)
Sulfuric acid, concentrated, AR
p-Nitrophenol reagent, 2%, in ethanol-chloroform (1:99)
  Prepare weekly.
Standard solution

Dissolve 10 mg estriol in 95% ethyl alcohol and dilute to 100 ml (100 μg/ml) and 10 mg estradiol in 95% ethyl alcohol and dilute to 100 ml (100 μg/ml).

Working standard

Into a 100-ml flask add 5.0 ml of estriol standard and 5.0 ml estradiol standard. Dilute to volume (10 μg/ml).

Procedure

1. Measure the total volume and check for glucose with Uristix. If glucose is present, increase the concentration of borohydride for reduction.
2. Dilute the urine specimen to 1500 ml with water if it is below this volume.

Extraction

3. Into 50-ml round bottom centrifuge tubes pipet in duplicate:
   Unknown: 20 ml filtered urine
   Control: 20 ml control urine
   Blank: 20 ml estrogen-free urine
   Extracted standard: 20 ml estrogen-free urine plus 100 μl standard
4. Add to each tube 10 g ammonium sulfate and shake until the solid is in solution.
5. Extract by hand-shaking for 1 minute with 10 ml ether-ethanol reagent. Repeat extraction for two more times, shaking for 30 seconds each time. Centrifuge after each extraction and pool the solvent phases.
6. Filter the pooled extracts into a 50-ml cylinder and note the volume.
7. Into 50-ml round bottom centrifuge tubes pipet (in duplicate) one fourth of the filtered extract from each cylinder and evaporate to dryness in a water bath at 40° to 45° C under a stream of air.

Reduction and hydrolysis

8. Dissolve each residue in 8.0 ml distilled water and add 0.25 ml of a 10% aqueous solution of potassium borohydride. Mix on a Vortex mixer.
9. Allow tubes to stand overnight at room temperature.
10. To each tube add 1.6 ml concentrated HCl and place in a boiling water bath for 30 minutes. Cool in running tap water.
11. Shake for a minute by hand and remove the organic layer. Separated organic layer is extracted with 3.0 ml N NaOH, which is added to the original aqueous phase.
12. To the aqueous phase add 0.9 ml concentrated HCl and approximately 0.5 g sodium bicarbonate. Check pH; it should be between 8 and 10.

Ether extraction and wash

14. Add 10 ml ether and shake for 30 seconds. Centrifuge and transfer ether extract to a 50-ml conical centrifuge tube.
15. Repeat ether extraction twice with 5 ml ether and pool extracts.
16. Each pooled extract is washed with:
    a. 4.0 ml of buffer pH 10.4 (saturated carbonate-bicarbonate, 2:1): Shake for 15 seconds and discard the aqueous layer by aspirating.
    b. 3.0 ml of 8% ammonium sulfate solution: Shake for 10 seconds and discard the aqueous layer by aspirating.
    c. Twice with 2.0 ml distilled water: Shake 10 seconds and discard aqueous layer by aspirating. After the second water wash, centrifuge and then discard the aqueous layer.

Color development

17. Direct standard: Into each of two 50-ml tubes place 50 μl of estrogen standard.

18. To the tubes from steps 16 and 17, add 0.5 ml of 2% hydroquinone in ethanol and evaporate to dryness in a water bath at 40° to 45° C under a stream of air.
19. Dissolve each residue in 0.4 ml distilled water with gentle warming.
20. Cool the tubes in an ice water bath and add 0.73 ml concentrated sulfuric acid while mixing.
21. Place tubes in a boiling water bath for 40 minutes and then cool in an ice bath.
22. Add 1.5 ml distilled water with mixing and remove from ice bath.
23. To each tube add 5.0 ml of 2% *p*-nitrophenol in 1% ethanolic chloroform. Mix on Vortex mixer for 1 minute and centrifuge. Aspirate aqueous layer.
24. Measure fluorescence in a spectrofluorometer using an exciting wavelength of 546 nm and an emitting wavelength of 585 nm. The instrument is set using distilled water. The fluorescence of the chloroform extract is measured without delay.

Calculations

$$\frac{F_u}{F_{std} - F_{bl}} \times \text{Concentration of standard} \times \frac{\text{Total volume of urine}}{\text{Aliquot of urine used}}$$

$$= \mu g \text{ estrogen in 24-hr specimen}$$

Normal values

Men	4 to 25 µg/24 hr
Women	4 to 100
Women, postmenopausal	1 to 16

Clinical findings

In amenorrhea resulting from pituitary or ovarian failure, urinary estrogen levels are low, but they are normal in end organ failure or uterine agenesis. Estrogen production follows a chracteristic pattern during the menstrual cycle, increases to a maximum just prior to ovulation, and raises to a second peak midway in the luteal phase. In treatment of infertility due to pituitary insufficiency with clomiphene, urinary estrogen measurements are used for monitoring the induction of ovulation to prevent ovarian hyperstimulation. Testicular tumors in males may produce feminization and cause increase in urinary estrogens. They are also increased in males with breast cancer and in females with estrogen-producing ovarian tumors.

17-Hydroxycorticosteroids, plasma
FLUOROMETRIC PROCEDURE[17, 19, 45]
Principle

Plasma is washed with petroleum ether to remove nonpolar interfering substances and the steroids extracted from the plasma with methylene chloride. The extract is washed with alkali to remove chromogens and an aliquot of the extract is shaken with a sulfuric acid–ethanol reagent to develop fluorescence. Those steroids with the 11-hydroxy configuration (cortisol and corticosterone) exhibit fluorescence under these conditions. Measurement of the fluorescence is performed in a spectrofluorometer.

Specimen collection

Two blood specimens are usually obtained from a patient, morning and afternoon. By venipuncture 10 ml is drawn into heparinized tubes. The morning specimen is drawn between 7 and 9 AM from a fasting patient and the afternoon specimen between 4:30 and 5 PM, before the evening meal. Less interference from lipids will be encountered if the blood is obtained when the patient is in a fasting state or 3 or more hours after a meal. The time of drawing should be recorded because of the normal diurnal variation of cortisol secretion. The specimen is delivered to the laboratory without delay for separation of the plasma from the cells. If the plasma is to be stored before analysis it should be kept frozen.

Reagents

All chemicals are reagent grade.

Sulfuric acid, concentrated

Methylene chloride, spectro quality, redistilled

Ethyl alcohol, absolute, purified by procedure described for 17-ketosteroids

Petroleum ether

Sodium hydroxide, 0.1 N

Dissolve 4 g of sodium hydroxide in distilled water and dilute to 1 L.

Sulfuric acid-alcohol reagent

Place a flask in an ice bath with 1 volume of ethyl alcohol and add 3 volumes of concentrated sulfuric acid slowly with mixing. The reagent is stable for about a week. The blank reading should not exceed 20% of that obtained for the average normal specimen.

Stock standard

Dissolve 50 mg of hydrocortisone in 5 ml of absolute alcohol and dilute to 250 ml with distilled water (200 μg/ml).

Diluted stock standards

Dilute 10 ml of stock standard to 200 ml with distilled water (10 μg/ml).

Working standard (0.25 μg/ml)

Place 5.0 ml of the diluted stock standard in a 200-ml volumetric flask and dilute to volume with distilled water.

Procedure

1. Into 40-ml glass-stoppered centrifuge tubes pipet 3.0 ml of plasma (unknown, control), 3.0 ml of distilled water, and 20 ml of petroleum ether. Shake vigorously for 30 seconds and centrifuge for 5 minutes. Discard petroleum ether by aspirating.

Extraction

2. Into 40-ml glass-stoppered centrifuge tubes pipet in duplicate:

   Unknown: 2.0 ml of diluted-washed plasma
   Control: 2.0 ml of diluted-washed plasma
   Standard: 2.0 ml of working standard
   Blank: 2.0 ml of distilled water

3. Add to each tube 15 ml of methylene chloride. Shake vigorously for 30 seconds and centrifuge for 5 minutes. Discard the aqueous plasma by aspirating.

Washing

4. To each tube containing the methylene chloride extract add 1.0 ml of 0.1 N NaOH. Shake for 15 seconds and centrifuge for 5 minutes. Discard the NaOH layer immediately by aspirating.

Development of fluorescence

NOTE: Since development of fluorescence requires careful timing, the fluorometer, timer, mixer, centrifuge, glassware, and reagents must be set to prevent any delay in operations.

5. Into 40-ml glass-stoppered centrifuge tubes (marking duplicate tubes 1 and 2) pipet 8.0 ml of the methylene chloride extract. Divide into sets of four tubes.

6. Develop fluorescence in blank and standard tubes first. Set the clock at 5 minutes but do not start. Add 2.5 ml of sulfuric acid–ethanol reagent to blank tube 1 and start the clock. Continue with the addition of the reagent to blank tube 2, standard tube 1, and standard tube 2.
7. Stopper tubes and mix on a Vortex mixer for 10 seconds. Centrifuge for 1 minute and stop the centrifuge, using the brake. Discard the methylene chloride (top) layer by aspirating
8. Transfer the acid-alcohol layer to a fluorometer cuvet and determine fluorescence at exactly 5 minutes. Read blank tube 1, blank tube 2, standard tube 1, and standard tube 2 in that order. With a spectrofluorometer use an exciting wavelength of 436 nm and an emitting wavelength of 520 nm. The instrument is set using distilled water.
9. Follow the same procedure as in steps 6 through 8 in developing fluorescence in the control and unknown tubes.

Calculations

$$\frac{F_u - F_b}{F_s - F_b} \times \text{Standard concentration} \times \frac{100}{\text{Volume plasma used}} =$$

μg 17-hydroxycorticosteroids/dl plasma (calculated as cortisol)

Normal values

In normal individuals the range for plasma corticosteroids drawn between 8 and 9 a.m. and assayed by the fluorometric method is 17 to 26 μg/dl. The level in the 5 p.m. specimen is approximately half that at 8 a.m.

de Moor and co-workers[17] reported the following plasma corticoid levels for normal hospitalized patients, 15 to 60 years of age, the blood drawn at the times indicated, as the mean ± S.E.M.:

8 a.m.	22.3 ± 0.27	μg/dl plasma	(673 subjects)
9 a.m.	19.5 ± 0.80		(30 subjects)
10 a.m.	16.2 ± 0.61		(30 subjects)
Noon	12.3 ± 0.25		(36 subjects)
8 p.m.	11.9 ± 1.2		(36 subjects)
Midnight	9.8 ± 0.4		(81 subjects)

These values were for persons who followed the evening sleep and day activity routine. A reversal of the day-night or activity-sleep schedule resulted in a reversal of the plasma cortisol pattern.[47] The erythrocyte corticoid level of blood drawn between 8 and 9 a.m. from 139 subjects was 6.90 ± 0.22 μg (S.E.M.)/dl.[17]

Comments

All glassware used in this procedure must be washed in 50% nitric acid.

de Moor and co-workers[17] reported that under the conditions described for this method 95% of the fluorescence resulted from cortisol and corticosterone. When read at 5 minutes, both corticoids showed the same fluorescence and interference from nonspecific substances was minimal. Washing with petroleum ether and sodium hydroxide removes some of the lipids, pigments, and other interfering material.

Fluorometric methods lack specificity and investigators have suggested various modifications to correct for or partially eliminate interfering substances extracted from plasma by methylene chloride. A reagent containing water, sulfuric acid, and ethanol has been used and fluorescence determined in 13 minutes.[37] The presence of two groups of fluorogens in the plasma extracts has been reported,[63] the one being nonspecific substances where fluorescence increased linearly with time and the other cortisol and corticosterone where fluorescence remained constant between 8 and 20 minutes after extraction into the acid reagent. By reading fluorescence after 8 and 16 minutes a corrected value for plasma corticosteroids was obtained. A reagent consisting of 30 N sulfuric acid has been used and fluorescence allowed to develop for 30 to 60 minutes.[60] To increase specificity, other workers have applied methods whereby cortisol and corticosterone are separated in water and carbon tetrachloride[71] or in benzene and water.[6] However, elimination of the problem of interference by nonspecific fluorescence has not been achieved.

In procedures measuring fluorescence elevated values will occur if patients are receiving drugs that are naturally fluorescent such as quinine, quinidine, chlortetracycline (Aureomycin) or other tetracycline antibiotics, or vitamin B complex. Patients receiving contraceptive drugs may also show elevated values as the estrogens cause an increase in the corticosteroid-binding protein.

COMPETITIVE PROTEIN BINDING ASSAY[4]
Principle

In competitive protein binding assay, plasma cortisol is extracted from plasma with ethanol. The extract is dried and a reagent consisting of corticosteroid-binding globulin (CBG) in equilibrium with tritiated cortisol is added. After an equilibrium has been approached between the labeled and unlabeled ligand, the unbound labeled and unlabeled cortisol are removed by Florisil. The radioactivity of the bound cortisol is measured by liquid scintillation counting.

Specimen collection

Same as for the fluorometric procedure.

Reagents

Ethanol, absolute

Cortisol-binding globulin

Plasma from women in the third trimester of pregnancy is pooled and stored in 1-ml aliquots at −15° C.

Tritiated cortisol—1,2-^{3}H-cortisol (specific activity 46.7 Ci/mM)

Dilute 250 μCi to 100 ml with absolute alcohol.

Hydrocortisone stock standard

Dissolve 50 mg hydrocortisone in absolute ethanol and dilute to 250 ml (200 μg/ml).

Hydrocortisone working standard

By serial dilution, prepare fresh weekly a working standard containing 0.025 μg cortisol/ml.

Florisil, 60/100 mesh

Wash until fine particles removed, dry at 85° C, and then heat at 200° C for 2 hours.

CBG reagent

Place an amount of the alcoholic solution of tritiated cortisol equivalent to 4000 cpm/ml of solution into a flask and evaporate to dryness. Add proper amount of diluted plasma to give percent protein as determined from a titration curve.

CBG titration curve

Number test tubes (12 × 75 mm) 1 through 12. Into tubes 7 to 12 pipet 30 ng cortisol standard and evaporate to dryness. Pair tubes as follows: A (1,7); B (2,8); C (3,9); D (4,10); E (5,11); F (6,12). Prepare protein dilutions as indicated in table below:

	A	B	C	D	E	F
Distilled water (ml)	9.89	9.88	9.85	9.8	9.4	8.9
Plasma (μl)	10	20	50	100	500	1000
^{3}H cortisol (.03 μCi/100 μl)	100	100	100	100	100	100

Add 1.0 ml of the above CBG reagents to the appropriate tubes and continue as for cortisol determination in step 8. Preset the counter for 10,000 counts and count all samples. Plot time versus protein concentration for both 0 and 24 μg/dl. The point showing greatest difference between the two curves is the protein concentration to be used for that particular plasma pool.

Scintillation fluid

May be purchased or Bray's solution may be used and prepared as follows: Dissolve 60 g naphthalene (scintillation grade) in 100 ml methanol (spectro grade). Add 40 ml Spectrafluor PRO-POPOP, 20 ml ethylene glycol, and 880 ml p-dioxane for scintillation.

Procedure

Extraction

1. Into 12 × 75 mm disposable test tubes (set up in duplicate) add 1.5 ml absolute ethanol and 12.5 μl plasma (control, unknown).
2. Mix each tube on a Vortex for 30 seconds and allow to stand for 10 minutes. Centrifuge for 5 minutes at 2500 rpm.
3. Transfer supernatant to another test tube.
4. Add 1.5 ml ethanol to precipitate, repeat extraction, and centrifuge.
5. Combine extracts.
6. Prepare in duplicate seven standards by pipeting into 12 × 75 mm test tubes the hydrocortisone working standard equivalent to 0.25, 0.5, 1.0, 2.0, 3.0, 4.0, and 5.0 ng cortisol.
7. Dry all tubes from steps 5 and 6 in a water bath at 40° C with a stream of air.

Protein binding: equilibration and separation

8. To all tubes add 1.0 ml of CBG reagent, place in a rotary Evapo-mix with water at 40° C, and shake for 20 minutes.
9. Cool in an ice bath for 10 minutes.
10. Add to all tubes 40 mg Florisil, place in a rotary Evapo-mix with crushed ice and water at 0° C, and shake for 20 minutes.
11. Allow Florisil to settle for 10 minutes in an ice bath.
12. Transfer 0.5 ml of supernatant from each tube to a scintillation vial.
13. Add 10 ml of Bray's solution or a suitable scintillation cocktail.
14. Include with each group of tests:
    Two blanks of 0.5 ml distilled water and 10 ml Bray's solution
    Two vials with 0.5 ml CGB reagent and 10 ml Bray's solution

Isotope counting

15. The vials are placed in a liquid scintillation counter and the bound labeled cortisol counted for 10 minutes or 10,000 counts.
16. Using channel ratios, the cpm's are converted to dpm's and a standard curve constructed using dpm versus concentration of cortisol in standards.

Calculations

17. From standard curve obtain ng/μl sample used. If 12.5 μl plasma was the aliquot, then:

$$\text{ng from curve} \times 8 \times \frac{1000}{1000} = \mu\text{g/dl}$$

Normal values
 8 to 9 a.m.   8 to 24 µg/dl
 4 to 5 p.m.   Approximately half of morning specimen

Comments

With the use of ethanol for the extraction of cortisol from plasma, 97% ± 3% is recovered. In addition to cortisol, corticosteroid-binding globulin (CBG) binds corticosterone, 11-deoxycortisol, progesterone, 17-hydroxyprogesterone, and cortisone. However, equilibration at 4° C enhances the binding of cortisol over competing steroids. When using plasma from pregnant women, the CBG is present in sufficiently high concentration that the percent of protein required yields a dilution as to eliminate nonspecific binding to albumin.

In separating the unbound or free steroids from the bound, other agents such as Fuller's earth, dextran-coated charcoal, dextran-coated Florisil, and Lloyd's reagent have been used. Florisil has less affinity for cortisol than does Fuller's earth and must be kept in contact for a longer period of time. However, when shaking has been stopped, the Florisil will settle, making centrifugation unnecessary. If Fuller's earth is used, the conditions must be rigidly controlled and centrifugation is necessary.

Clinical findings

Plasma corticosteroid concentration increases as pregnancy progresses (largely because of the increased amount of corticosteroid-binding globulin), in conditions of stress such as shock and surgery, and with administration of drugs such as ACTH and vasopressin. The levels are high in Cushing's syndrome and there is no diurnal variation.

Plasma corticosteroid levels may be lowered by dexamethasone. They are usually low in patients with Addison's disease and panhypopituitarism.

17-Hydroxycorticosteroids, urinary

The major C-21 metabolites found in the urine are tetrahydrocortisol, tetrahydrocortisone, tetrahydro S, pregnanetriol, 11-ketopregnanetriol, cortols, cortolones, 6β-hydroxycortisol, and Δ^5-pregnanetriol. These are conjugated principally with glucuronic acid although there are some sulfates. A small amount of free cortisol is also excreted.

There are two basic procedures in general use for the determination of 17-hydroxycorticosteroids. One method utilizes enzyme hydrolysis and the reaction of the 17-hydroxycorticosteroids with the phenylhydrazine–sulfuric acid reagent to measure Porter-Silber chromogens. The second method, a determination of 17-ketogenic steroids, employs sodium metaperiodate to convert corticosteroids to 17-ketosteroids for final assay by the Zimmermann reaction.

Specimen collection

A 24-hour urine collection is obtained preferably without a preservative and kept at about 2° to 4° C. If refrigeration is not possible, 15 ml of toluene may be added as preservative.

PORTER-SILBER CHROMOGENS[54, 59, 61]
Principle

The Porter-Silber chromogens are cortisol metabolites possessing a 17,21-dihydroxy acetone side chain at C-17 and they represent about one third of the urinary metabolites of cortisol. Steroids such as compound S (11-deoxycortisol) and its tetrahydro metabolite that are not metabolites of cortisol but have the dihydroxyacetone side chain are also measured.

Urinary corticosteroids are hydrolyzed with β-glucuronidase and extracted with chloroform. The extract is washed with a dilute aqueous solution of alkali to remove estrogens and interfering chromogens. An aliquot of the chlorofrom extract is shaken with a phenylhydrazine–sulfuric acid–methyl alcohol reagent. The color developed is measured spectrophotometrically.

Reagents

All chemicals are reagent grade.

Chloroform, redistilled

Sodium hydroxide, 0.1 N

Dissolve 4 g of sodium hydroxide pellets in distilled water and dilute to 1 L.

Phenylhydrazine hydrochloride, recrystallized

A product especially prepared for steroid analysis is available from Sigma Chemical Co. Otherwise, purify in the following manner. Add 100 g to 200 ml of warm (60° to 70° C) distilled water and stir frequently during the 1 to 3 hours required to dissolve the phenylhydrazine. Heat 1 L of ethyl alcohol to boiling and add to the dissolved phenylhydrazine in water. While hot, filter quickly through No. 2 filter paper. Cool in the refrigerator and collect the crystals on a sintered-glass filter (medium). Repeat the procedure twice, dissolving the crystals in less water each time. The last collection of crystals should be washed with cold ethyl alcohol and dried over $CaCl_2$ in vacuo. Determine the purity by a melting point measurement 240° to 243° C.

Sulfuric acid, 62% (22 N)

Add 310 ml of concentrated acid to 190 ml of distilled water. The flask should be kept cool under running cold water from a tap during addition of the acid.

Blank reagent

Prepare the quantity required for the daily batch by mixing 7.5 parts sulfuric acid reagent (62%) with 5 parts of methyl alcohol.

Methyl alcohol, absolute, spectro quality

Phenylhydrazine–sulfuric acid–methyl alcohol reagent

Dissolve 12 mg of phenylhydrazine hydrochloride in 12.5 ml of the blank reagent. Prepare fresh daily.

β-Glucuronidase (Sigma, type II)

Prepare the required amount daily by using 250 Sigma units/ml of 0.067 M phosphate buffer, pH 6.5.

Phosphate buffer, pH 6.5

Solution I

Dissolve 9.078 g of KH_2PO_4 in distilled water and dilute to 1 L, 0.067 M primary phosphate solution.

Solution II

Dissolve 9.465 g of Na_2HPO_4 in distilled water and dilute to 1 L, 0.067 M secondary sodium phosphate solution.

Prepare 1 L of 0.067 M phosphate buffer, pH 6.5, by mixing 650 ml of solution I with 350 ml of solution II.

Stock standard

Dissolve 50 mg of hydrocortisone in 5 ml of absolute ethyl alcohol and dilute to 250 ml with distilled water (200 μg/ml).

Working standard (10 μg/ml)

Dilute 10 ml of stock standard to 200 ml with distilled water.

Procedure
1. Measure total volume of 24-hour collection of urine and filter an aliquot.

Hydrolysis
2. Into 50-ml round-bottom centrifuge tubes pipet *in duplicate:*
   Unknown: 2.0 ml of filtered urine
   Control: 2.0 ml of filtered urine
3. Add to each tube 1.0 ml of β-glucuronidase solution and mix. Check the pH; it should be 6.5. If not, adjust using 0.1 N NaOH. Add 0.1 ml of chloroform.
4. Incubate for 18 to 24 hours at 37° C.

Extraction and washing
5. Into 50-ml round-bottom centrifuge tubes pipet in duplicate 2.0 ml of working standard, 10 μg/ml, and 1.0 ml of distilled water.
6. To all tubes from steps 4 and 5 add 25 ml of chloroform and shake for 15 seconds by hand. Discard the aqueous layer by aspirating.
7. Add 2.0 ml of 0.1 N NaOH, shake for 10 seconds, and centrifuge for 5 minutes. Discard the alkali wash by aspirating.

Color development
8. Into one set of 12-ml centrifuge tubes transfer from tubes of step 7 5.0 ml of chloroform extract. Add 1.0 ml of phenylhydrazine–sulfuric acid–methyl alcohol reagent and mix for 10 seconds on a Vortex mixer. Centrifuge for 5 minutes and discard the chloroform layer by aspirating.
9. Into a second set of 12-ml centrifuge tubes transfer from tubes of step 7, for blanks, 5.0 ml of chloroform extract. Add 1.0 ml of blank reagent and mix for 10 seconds on a Vortex mixer. Centrifuge for 5 minutes and discard the chloroform layer by aspirating.
10. Incubate tubes in a water bath at 60° C for 1 hour.
11. Allow to cool to room temperature.
12. Using capillary pipets transfer the samples to microcuvets.
13. Determine absorbance (A) at 370, 410, and 450 nm against blank reagent.
14. Subtract the blank readings from the color readings and determine corrected absorbance (A).

Calculations

$$\text{Corrected A} = A_{410} - \left(\frac{A_{370} + A_{450}}{2}\right)$$

$$\frac{\text{Corrected A unknown}}{\text{Corrected A standard}} \times \text{Standard concentration (mg)} \times$$

$$\frac{\text{Total urine volume (ml)}}{\text{Aliquot urine used}} = \text{mg 17-hydroxycorticosteroids in 24 hours}$$

Normal values

de Moor and associates[18] reported that differences in body weight accounted for most of the sex differences in corticoid excretion. It was suggested that the values should be expressed in units of mg/100 kg/24 hr. Results may also be calculated per gram of creatinine with the concentration of the latter given as mg/kg/24 hr.

Values for 17-hydroxycorticosteroids in urine have been reported as Porter-Silber chromogens—for men 3 to 15 mg/24 hr and for women 2 to 13 mg/24 hr; or as 3.1 ± 1.1 mg/m³/24 hr. The nomogram (Fig. 10-2) of DuBois gives the surface area from the subject's height and weight.

Comments

Hydrolysis is necessary to split the water-soluble glucosiduronidate complexes to free the steroids in a solvent-soluble form. Mild hydrolytic procedures such as enzyme hydrolysis with β-glucuronidase are used, since corticosteroids are very labile in hot mineral acids.

Chloroform or methylene chloride may be used as the solvent in the procedure. The wash with dilute sodium hydroxide removes some of the nonspecific substances that interfere with the final color development. However, it should be carried out rapidly to prevent loss of corticosteroids.

In the phenylhydrazine–sulfuric acid reagent the 17,21-dihydroxy-20-ketosteroids rearrange to form 21-aldehydes, which then react with the phenylhydrazine to yield a yellow phenylhydrazone having a maximum absorption at 410 nm. Phenylhydrazine also reacts at the 3 position of cortisol and with other steroids without the dihydroxyacetone side chain but with the 3-keto-Δ^4 A ring. These products have absorption maxima about 340 nm.

Some of the drugs that have been reported[72] to interfere with the determination are quinine, colchicine, paraldehyde, dextroamphetamine (Dexederine), meprobamate (Equanil), chlordiazepoxide (Librium), etryptamine acetate (Monase), seconal, reserpine (Serpasil), chlorpromazine (Thorazine), ethinamate (Valmid), dexamethasone, ACTH, hydroxyzine (Atarax), and metheramine mandelate (Mandelamine).

Clinical findings

Cushing's syndrome or hyperadrenocorticism is the result of hyperfunction of the adrenal cortex. This may be caused by adrenal hyperplasia, adenoma, or carcinoma. The clinical picture of the patient may reveal typical physical characteristics such as obesity with thin extremities, moon facies, thin skin with bruising and striae, hirsutism, menstrual disturbances, testicular atrophy, hypertension, skin pigmentation, mental depression, osteoporosis, and often diabetes mellitus. Usually the adrenals are enlarged, and a tumor is found in about 30% of the patients.

In adrenal hyperplasia the primary source of the disorder may be either the pituitary or the adrenal cortex. Congenital cases, precocious genital development (male) and pseudohermaphroditism (female), result from genetic defects that cause complete or partial absence of enzymes required for steroid synthesis. Several of the defects that have been recognized are a C-20 block with lipoid hyperplasia of the adrenals, 3β-hydroxysteroid dehydrogenase deficiency, C-11 block that leads to an increase in 11-deoxycortisol and 11-deoxycorticosterone, C-18 hydroxylation defect with decrease in aldosterone secretion, and C-21 block, the most common type, with increased excretion of 17-ketosteroids and pregnanetriol. The C-21 hydroxylase defect accounts for most of the patients with the syndrome of virilization alone. It is not complete, and in mild cases the levels of cortisol are normal or nearly normal. When the deficiency is severe, little or no cortisol is formed. Adult adrenal virilizing syndrome is usually unrecognized in males. In females, a woman previously normal shows development of masculine characteristics. These patients may have adrenal tumors or a delayed congenital adrenal hyperplasia with an associated increased excretion of 17-ketosteroids. Simple or idiopathic hirsutism is of unknown cause and usually does not appear to have the above described defects in cortisol synthesis.

Adrenocortical insufficiency or Addison's disease is characterized physically by weakness, hypotension, increased pigmentation, and sensitivity to stress. Some cases are the result of infections such as tuberculosis or neoplastic processes but mostly the cause is unknown.

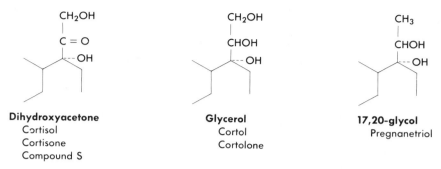

Fig. 14-6. Chemical groups at C-17 for 17-hydroxycorticosteroids and examples.

In the liver most adrenal steroids are inactivated and conjugated with glucuronic or sulfuric acid. Therefore when there is liver damage, an increase in free and a decrease in conjugated adrenal steroids occur in the blood and urine. Kidney damage causes an accumulation of the conjugates in the blood and thereby influences the rate of conjugation taking place in the liver.

17-KETOGENIC STEROIDS[21, 39, 40]
Principle

The most useful chemical measurement of hydroxycorticosteroid excretion is by the procedure for Porter-Silber chromogens or compounds with the 17,21-dihydroxy-20-keto configuration (Fig. 14-6). However, this group does not include all 17-hydroxycorticosteroids, and there are pathologic conditions where the total level is important. By chemical action all 17-hydroxycorticosteroids may be converted to and measured as 17-ketosteroids. These are referred to as 17-ketogenic steroids.

Addition of borohydride reduces endogenous 17-ketosteroids to hydroxyl compounds and thus eliminates them from the final Zimmermann reaction. The 17-hydroxycorticosteroids are oxidized to 17-ketosteroids by sodium metaperiodate. Hydrolysis of glucuronides occurs during periodate oxidation. The ether extraction and color development by the Zimmermann reaction are the same as described for 17-ketosteroids.

Reagents

All chemicals are reagent grade.
Ethyl ether (Squibb) for anesthesia or redistilled, peroxide-free
Ethyl alcohol, absolute, purified
    Refer to reagents for 17-ketosteroids.
Sodium hydroxide, 4% (1 N), 40% (10 N)
Acetic acid, 5.7%, 25%
Sodium borohydride, 5% in 0.1 N NaOH, freshly prepared
Sodium metaperiodate, 10%, freshly prepared
Sodium dithionite (hydrosulfite, $Na_2S_2O_4 \cdot 2H_2O$)
m-Dinitrobenzene (Sigma)
    Specially prepared for the Zimmermann reaction. Prepare fresh daily 2% solution in absolute ethyl alcohol.
Potassium hydroxide, 5.0 N (aqueous), standardized
Standard solution
    Dissolve 20 mg of dehydroisoandrosterone in 100 ml of absolute ethyl alcohol (200 μg/ml)

Procedure

1. Measure total volume of 24-hour collection of urine and filter an aliquot.
2. Test for the presence of glucose with Uristix. If present, double the amount of borohydride for each 2% of glucose and increase accordingly the 25% acetic acid.

Reduction

3. Into 50-ml round-bottom centrifuge tubes pipet in duplicate:
   Unknown: 5.0 ml of filtered urine (2.0 ml diluted to 5.0 ml with distilled water when the 17-OHCS is above 15 mg/24 hr)
   Control: 5.0 ml of control urine
4. Adjust pH of samples to 7.0 and add 0.5 ml of 5% sodium borohydride reagent (freshly prepared) for glucose-free urine.
5. Allow to stand overnight at room temperature.
6. Add 0.25 ml of 25% acetic acid to decompose excess sodium borohydride for glucose-free urine. Allow to stand 15 minutes.

*Recovery**
Oxidation

7. Add 2.0 ml of 10% sodium metaperiodate (freshly prepared) for glucose-free urine and 8.0 ml for glucose-positive urine (6.0 ml of 10% sodium metaperiodate solution is approximately equivalent to 5.0 ml of a 2% glucose solution).
8. Neutralize by adding 0.8 ml of 1 N NaOH for glucose-free urine or 1.0 ml for glucose-positive urine (pH must be 6.7 to 7.0); check with indicator paper.
9. Incubate at 37° C for 1 hour and then add 0.12 ml of 10 N NaOH.
10. Incubate at 37° C for 15 minutes and cool to room temperature.

Extraction and wash

11. To each tube add 20 ml of ethyl ether and shake by hand for 30 seconds. Discard the aqueous layer by aspirating.
12. Add 10 ml of 10% NaOH solution containing approximately 5% sodium dithionite (hydrosulfite) and shake by hand for 10 to 15 seconds. Discard the NaOH wash layer by aspirating.
13. Wash twice with 10 ml of distilled water for 10 to 15 seconds and discard the aqueous layer by aspirating; after the final wash cover the tubes with foil and centrifuge for 5 minutes at 1500 rpm.

Color development

14. Into 40-ml round-bottom centrifuge tubes pipet:
    Unknown: 10 ml of ether extract
    Control: 10 ml of ether extract
    Standard (duplicate): 0.1 ml of standard solution
    Blank (duplicate): Dry tubes
15. Evaporate to dryness in a water bath at 40° to 45° C under a stream of air.
16. To the dried residue in each tube add 0.5 ml of alcoholic-KOH reagent (prepared immediately before use by mixing 2 parts absolute ethyl alcohol with 3 parts 5 N KOH) and 0.2 ml of 2% *m*-dinitrobenzene reagent.
17. Place tubes in a water bath at 25° C for 60 minutes. *Keep in dark,* covering with a black cloth.

---

*Recoveries: Etiocholanolone and 11-β-hydroxyetiocholanolone (alcoholic solution) may be added to reduced urine after the destruction of the excess borohydride to test their stability and separation in the presence of other urinary constituents. C-21 steroids (alcoholic solution) may be added to the urine prior to borohydride reduction.

18. Add to each tube 4.0 ml of 80% ethyl alcohol.
19. Measure absorbance (A) against the reagent blank at 480, 520 and 560 nm.

Calculations

$$\text{Corrected A} = A_{520} - \left(\frac{A_{480} + A_{560}}{2}\right)$$

$$\frac{\text{Corrected A unknown}}{\text{Corrected A standard}} \times \text{Standard concentration (mg)} \times$$

$$\frac{\text{Total urine volume (ml)}}{\text{Aliquot urine in extract}} = \text{mg 17-ketogenic steroids/24 hr}$$

Normal values

For men, normal values range from 5 to 23 mg/24 hr and for women, from 3 to 15 mg/24 hr. Between the ages of 20 and 50 the maximum rate of excretion occurs and there is some decline with increasing age.

Comments

The two reagents that have been used to oxidize 17-hydroxycorticosteroids to 17-ketogenic steroids are sodium bismuthate and sodium metaperiodate. The poor reproducibility of results using bismuthate has been attributed to nonuniformity in oxidizing capabilities of the bismuthate from batch to batch and the hot acid hydrolysis required to complete the hydrolysis of all conjugates.

Sodium metaperiodate is a mild oxidant that eliminates the need for acid hydrolysis. It is stable and specific for oxidation of adjacent hydroxyl groups. The ketosteroids formed are mostly etiocholanolone, 11-hydroxyetiocholanolone, and a small amount of 11-hydroxyandrosterone. By the borohydride periodate method the color equivalent depends on the alkali used. With potassium hydroxide the 11-hydroxyetiocholanolone gives a color approximately the same as dehydroepiandrosterone.

When the sodium borohydride is added to the samples, the pH should not be less than 7 since the compound breaks down rapidly in an acid solution and slowly in a neutral solution. The exact pH is not critical and may vary between 7 and 10.

Clinical findings

The normal range appears to be broader than for the Porter-Silber chromogens, limiting the usefulness in diagnosis of Cushing's disease. Since this is a measure of the total 17-hydroxycorticosteroids, the determination is useful in those conditions where abnormal metabolites not measured by the Porter-Silber technique are found. For example, in some cases of carcinoma of the adrenal gland and adrenal hyperplasia elevated 17-ketogenic steroids are observed while the Porter-Silber chromogens are normal. Increased output has been noted in some cases of obesity, virilism, and precocious puberty of adrenocortical origin. A 17-ketogenic steroid level more than 5 mg/24 hr over the Porter-Silber chromogens is considered significant. Decreased levels usually accompany Addison's disease and hypopituitarism.

CORTISOL (FREE), URINARY[43]
Principle

Small quantities of unconjugated cortisol appear in the urine. It is extracted into methylene chloride, the extract dried, and the cortisol measured by competition with tritiated cortisol for the binding sites on corticosteroid-binding globulin of human plasma. Florisil is used to remove the unbound cortisol.

Specimen collection

A 24-hour urine collection is obtained without a preservative and kept at about 2° to 4° C.

Reagents

Identical to plasma cortisol by competitive protein binding
Methylene chloride, analytic reagent

Procedure
1. Into 40-ml glass stoppered centrifuge tubes pipet 3.0 ml distilled water and 100 μl of 24-hour urine (in duplicate).
2. Add 3 ml methylene chloride and mix on a Vortex mixer for 30 seconds. Centrifuge for 5 minutes at 2000 rpm.
3. Transfer methylene chloride layer to a 12 × 75 mm disposable glass test tube and evaporate to dryness.
4. Reextract the diluted urine with an additional 3 ml of methylene chloride, centrifuge, and combine the methylene chloride with the first extract. Evaporate to dryness.
5. Remainder of the procedure is the same as for plasma cortisol by competitive protein binding.

Calculations

Refer to calculations for plasma cortisol by CPB. If 0.1 ml urine used then:

$$\frac{\text{ng from standard curve}}{0.1} \times \frac{\text{Urine volume (ml)}}{1000} = \mu g \text{ cortisol/volume}$$

Normal values

16 to 100 μg/24 hr

Clinical findings

Urinary free cortisol has been considered a better guide of adrenal function than the level of cortisol metabolites in the urine. In hypercortisolism, there is proportional increase in urinary cortisol when the free cortisol in plasma is markedly increased. In Cushing's disease, values greater than 120 μg/24 hr are found. Normal persons following suppression will have 0 to 9 μg/24 hr.

5-Hydroxyindoleacetic acid[3]
Principle

Measurement of urinary 5-hydroxyindoleactic acid (5-HIAA), a metabolic product of serotonin, is the best diagnostic method for carcinoid tumors. Acidified urine is extracted with chloroform to remove indoleacetic acid, which will also yield a colored product with nitrous acid. After discarding the chloroform layer, the aqueous phase is saturated with sodium chloride and extracted with ether. The 5-HIAA is reextracted from the ether with phosphate buffer (pH 8.0). Nitrosonaphthol reagent and nitrous acid reagent are added to an aliquot of the phosphate buffer extract. After incubation, the tubes are shaken with ethyl acetate which is discarded. The absorbance of the resulting violet chromophore is determined by spectrophotometry.

Specimen collection

A 24-hour urine specimen is collected in a glass bottle containing 15 ml of 6 N hydrochloric acid.

Reagents

Chloroform, redistilled

Hydrochloric acid, 6 N
 Dilute concentrated HCl 1:1 with distilled water.

Hydrochloric acid, 1 N
 Dilute 8.6 ml of concentrated (11.6 N) HCl to 100 ml with distilled water.

Sodium hydroxide, 1 N
 Dissolve 4 g sodium hydroxide pellets in distilled water and dilute to 100 ml.

Sodium chloride, crystals, AR

Ethyl ether (Squibb or Mallinckrodt) for anesthesia or redistilled and peroxide-free

Phosphate buffer, pH 8.0
 Solution I: Dissolve 9.078 g of KH_2PO_4 in distilled water and dilute to 1 L, 0.067 M primary phosphate solution.
 Solution II: Dissolve 9.465 g of Na_2HPO_4 in distilled water and dilute to 1 L, 0.067 M secondary sodium phosphate solution.
 Prepare 0.067 M phosphate buffer, pH 8.0, by mixing 26.5 ml of solution I with 473.5 ml of solution II.

Nitrosonaphthol reagent, 0.1%
 Dissolve 100 mg in 95% ethyl alcohol and dilute to volume.

Sodium nitrite, 2.5%
 Dissolve 250 mg in distilled water and dilute to 10 ml.

Nitrous acid reagent
 Mix 0.5 ml of 2.5% sodium nitrite solution with 12.5 ml of 3 N sulfuric acid.

Sulfuric acid, 3 N
 Dilute 8.4 ml of concentrated (17.8 M) acid to 100 ml with distilled water.

Ethyl acetate, AR

Sulfuric acid, 0.1 M
 Dilute 1.4 ml of concentrated acid to 250 ml with distilled water.

Standard solution
 Dissolve 20 mg of 5-hydroxyindoleacetic acid (free acid) in 0.1 M sulfuric acid and dilute to 100 ml (50 μg/250 μl). Stable if frozen. Use only the first day after thawing.

Procedure

1. Measure the urine and filter an aliquot.
2. Into 40-ml plastic centrifuge tubes pipet, in triplicate:
 Unknown: 10 ml of filtered urine
 Control: 10 ml of filtered urine control
 Standard: 250 μl of standard + 10 ml of normal saline
3. Using a pH meter, adjust the pH to 2.0 using as little 1 N HCl or 1 N NaOH as possible.

Washing and extraction

4. To each tube add 3.5 g of sodium chloride and 25 ml of chloroform. Shake on a Vortex mixer for 1 minute and centrifuge for 5 minutes at 1500 rpm.
5. Discard the chloroform layer by aspirating, being careful not to lose any of the aqueous layer.
6. Repeat the chloroform wash by adding to each tube 25 ml of chloroform.

Shake on a Vortex mixer for 45 seconds and centrifuge for 5 minutes at 1500 rpm. Discard the chloroform layer by aspirating.
7. To each tube add 3.5 g of sodium chloride and 20 ml of ether. Shake for 1 minute on a Vortex mixer and centrifuge for 5 minutes at 1500 rpm.
8. Transfer 15 ml of ether layer to 40-ml glass-stoppered centrifuge tubes and add 1.5 ml of phosphate buffer.
9. Stopper the tubes and shake on a Vortex mixer for 1 minute. Centrifuge for 5 minutes at 1500 rpm. Discard the ether layer by aspirating.

Color development
10. Set up three direct standard tubes by pipeting into 40-ml glass-stoppered centrifuge tubes 1.5 ml of phosphate buffer and 250 µl of standard solution.
11. Divide tubes from steps 9 and 10 into two groups:
   a. Group I:
      Blanks: one tube from the set of three for each standard, extracted standard, control, and unknown.
      Reagent blank: set up by adding 1.5 ml of phosphate buffer to a 40-ml glass-stoppered centrifuge tube.
   b. Group II: Other two tubes from the set of three for each standard, extracted standard, control, and unknown.
12. To tubes of Group I (blanks) add 0.5 ml of nitrosonaphthol reagent and mix; add 0.5 ml 3 N sulfuric acid and mix.
13. To tubes of Group II (5-HIAA measurement) add 0.5 ml of nitrosonaphthol reagent and mix; add 0.5 ml of nitrous acid reagent and mix.
14. Incubate all tubes at 37° C for 10 minutes and allow to cool to room temperature.
15. Add to each tube 3.3 ml of ethyl acetate. Shake, centrifuge, and discard ethyl acetate by aspirating.
16. Repeat the ethyl acetate washing as described in step 15.
17. Transfer the aqueous phase to a microcuvet and determine absorbance (A) at 540 nm., reading against a reagent blank.

Calculations

$$\frac{A \text{ unknown} - A \text{ unknown blank}}{A \text{ extracted standard} - A \text{ extracted standard blank}} \times \text{Standard concentration } (0.05 \text{ mg}) \times \frac{\text{Total volume urine}}{\text{Aliquot urine used } (10 \text{ ml})} = \text{mg 5-HIAA in 24 hr specimen}$$

Normal values
   1 to 7 mg/24 hr

Comments

In this procedure it is considered important to have the urine sample adjusted to pH 2 and salt saturation during the first $CHCl_3$ wash.

In determining urinary 5-HIAA, the following have been reported as interfering: acetanilide, bananas, chlorpromazine, mephenesin, methocarbamol, and phenothiazine.

Clinical findings

Excessive production of serotonin is characteristic of carcinoid tumors and determination of the concentration of 5-HIAA in the urine is the most useful method of detection. However, some patients with the carcinoid syndrome do not have increased levels of 5-HIAA in the urine and some patients without the syndrome may exhibit increased concentrations.

17-Ketosteroids, urinary[50, 55]
Principle

The 17-ketosteroids are excreted into urine as water-soluble sulfate and glucuronide conjugates. They are hydrolyzed with concentrated HCl at 80° to 85° C and the free steroids extracted with ether. The extract is washed with 10% NaOH to remove phenolic compounds (estrogens) and with distilled water to remove excess base. An aliquot of the ether extract is evaporated to dryness. Color development by the Zimmermann reaction is produced with an alkaline solution of m-dinitrobenzene. The intensity of the characteristic reddish purple color is measured in a spectrophotometer. The reaction is specific for steroids having a ketone group adjacent to an unsubstituted carbon atom.

Specimen collection

The urinary 17-ketosteroids appear to be relatively stable. The collection may be made without a preservative if it is kept in a refrigerator. Otherwise, a 24-hour urine specimen is collected in a glass bottle containing 15 ml of toluene to prevent bacterial contamination. Significant deterioration will not occur for several weeks if a specimen is stored at 0° to 4° C. For longer storage periods, preserve at −10° C.

Reagents

All chemicals are reagent grade.
Hydrochloric acid, concentrated
Ethyl ether (Squibb), for anesthesia, or redistilled and peroxide-free
  Each batch should be tested for peroxides by dissolving a few crystals of potassium iodide in 5 ml of water and shaking for 1 minute with 5 ml of ether. A yellow color indicates the presence of peroxides.
Sodium hydroxide, 10% (w/v)
Dichloromethane, redistilled
Methyl alcohol, absolute, spectro quality
m-Dinitrobenzene (Sigma)
  Specially prepared for the Zimmermann reaction in steroid determinations, 1% in absolute ethanol. Prepare fresh daily.
Benzyl trimethyl ammonium methoxide, 40% in methanol
Ethyl alcohol, absolute, ketone-free and aldehyde-free
  To purify add 4 g of m-phenylenediamine hydrochloride to 1 L of absolute alcohol and allow to stand in the dark about a week with occasional shaking. Distill, using an all-glass apparatus, and discard the first and last 10%. Store the purified alcohol in the dark.
Stock standard
  Dissolve 100 mg of dehydroepiandrosterone in absolute methanol and dilute to 100 ml (1000 μg/ml).
Working standard (200 μg/ml)
  Dilute 10 ml of stock standard to 50 ml with absolute methanol.

Procedure

1. Measure total volume of 24-hour collection of urine and filter an aliquot.

Hydrolysis

2. Into 50-ml round-bottom centrifuge tubes pipet *in duplicate:*
   Unknown: **5.0 ml of filtered urine**
   Control: **5.0 ml of control urine**

3. Add to each tube 1.5 ml of concentrated hydrochloric acid.
4. Heat the tubes in a water bath at 80° to 85° C for 12 minutes. Cool in running tap water.

Extraction and washing

5. To each tube add 20 ml of ethyl ether and shake by hand for 30 seconds. Discard the aqueous layer by aspirating.
6. Add 10 ml of 10% NaOH and shake for 10 seconds by hand. Discard the NaOH layer by aspirating.
7. Add 10 ml of distilled water and shake for 10 seconds. Cover tubes with foil and centrifuge for 5 minutes at 1500 rpm.

Color development

8. Into 40-ml round-bottom tubes place:
   Unknown: 10 ml of ether extract
   Control: 10 ml of ether extract
   Standard (duplicate): 0.1 ml of working standard solution (20 µg)
   Blank (duplicate): Dry tubes
9. Evaporate to dryness in a water bath at 40° to 45° C under a stream of air. Wash down the sides of the tubes with 0.3 ml of absolute ethyl alcohol and evaporate to dryness.
10. To the dried residue in each tube add 0.1 ml of 1% *m*-dinitrobenzene in absolute ethyl alcohol and mix to dissolve the residue; add 0.2 ml of benzyl trimethyl ammonium methoxide and mix.
11. Place tubes in a water bath at 25° C for 90 minutes. Keep in the dark by covering with a black cloth.
12. Add to each tube 3.0 ml of 50% ethanol in water and mix; add 3.0 ml of dichloromethane and mix vigorously for 10 seconds.
13. Allow the tubes to stand in the dark until the two layers separate and the bottom layer becomes crystal clear.
14. Measure absorbance (A) of the lower layer against the reagent blank at 480, 520, and 560 nm.

Calculations

$$\text{Corrected A} = A_{520} - \left(\frac{A_{480} + A_{560}}{2} \right)$$

$$\frac{\text{Corrected A unknown}}{\text{Corrected A standard}} \times \text{Standard concentration (mg)} \times$$

$$\frac{\text{Total urine volume (ml)}}{\text{Aliquot urine in extract}} = \text{mg 17-ketosteroids in 24-hr specimen}$$

17-KETOSTEROIDS, URINARY—ALTERNATE METHOD[55]

Reagents

Refer to previous method.
Potassium hydroxide, 5 N
   This solution must be standardized.
Ethyl alcohol, 80% aqueous solution (v/v)
   Prepared from purified reagent.
m-Dinitrobenzene, 2%, in absolute ethyl alcohol
   Prepare fresh daily.

Procedure

Steps 1 through 9 and 14 and 15 are the same as for the previous method.
10. To the dried residue in each tube add 0.5 ml of alcoholic-KOH reagent

Table 14-7
Data on 17-ketosteroids in urine*

Age (years)	Men (mg/24 hr)	Women (mg/24 hr)
10-15	3.8- 7.8	4.3- 8.7
15-20	6.3-12.7	5.8-11.6
20-25	7.4-14.8	6.1-12.2
25-30	7.2-14.6	5.8-11.7
30-35	6.8-13.7	5.4-10.8
35-40	6.3-12.7	4.8- 9.7
40-45	5.8-11.6	4.3- 8.7
45-50	5.2-10.6	3.8- 7.7
50-60	4.2- 9.4	3.2- 6.9
60-70	3.4- 7.6	2.9- 6.2

*Values from Kaiser.

(prepared immediately before use by mixing 2 parts absolute ethyl alcohol with 3 parts 5 N KOH) and 0.2 ml of 2% m-dinitrobenzene reagent.
11. Place tubes in a water bath at 25° C for 60 minutes. *Keep in dark;* cover with a black cloth.
12. Add to each tube 4.0 ml of 80% ethyl alcohol.
13. Measurement of the absorbance and calculations are the same as described for the first method.

Normal values

Men	6 to 18 mg/24 hr
Women	5 to 15 mg/24 hr

Using a modified Zimmermann procedure, Kaiser obtained, for 519 female and 948 male healthy subjects, the values in Table 14-7.

Comments

The free steroids are relatively insoluble in an aqueous solution but soluble in organic solvents. A solvent consisting of a mixture of equal parts of benzene and petroleum ether[50] may be used in place of ether to extract the ketosteroids. (CAUTION: Benzene is highly toxic and, if used, adequate ventilation must be provided.)

The presence of nonspecific substances that contribute color to that produced by the 17-ketosteroids in the Zimmermann reaction is a major source of error. The alkaline wash removes acidic and phenolic substances including estrogenic pigments and chromogens from the crude extract. Thus the material remaining in the organic solvent has been referred to as the neutral fraction. Upon extraction of the Zimmermann color complex with dichloromethane, nonspecific brown pigments are retained in the water-alcohol layer.

The addition of formalin to the urine before hydrolysis has been used to decrease the chromogens formed that interefere with the Zimmermann reaction. However, a recent study by Goldzieher and de la Pena[24] showed that in the presence of formalin the destruction of free dehydroepiandrosterone is increased and the recovery of ketosteroid sulfates is lower.

Spectrophotometric readings are made at three wavelengths *(Allen correction)*, one on each side of the maximum for the specific constituent being assayed. Maximum absorbance is at 520 nm for the Zimmermann chromogens when the ketone group is at C-17. The corticosteroids testosterone and progesterone develop a slight color with the reagent but the maximum absorbance is near 420 nm. Corticosterone, pregnanedione, and pregnanolone yield complexes that absorb maximally at 490 nm. The Allen correction is valid as long as the absorbance of the interfering substances is linear. This has been shown to be not completely true for urine extracts free of 17-ketosteroids.

The use of benzyl trimethyl ammonium methoxide in place of KOH in the Zimmermann reaction makes possible the extraction of the color complex by dichloromethane without the turbidity that is difficult to remove. In the presence of this base, the color complex formed by the principal 17-ketosteroids has approximately the same intensity. Fading is less than with KOH.

Administration of the following drugs has been reported[73] to interfere with the 17-ketosteroid determination: cortisol, dexamethasone, meprobamate (Equanil), chlordiazepoxide (Librium), etryptamine acetate (Monase), methyprylon (Noludar), paraldehyde, phenazopyridine hydrochloride (Pyridium), quinine, seconal, reserpine (Serpasil), chlorpromazine (Thorazine), and ethinamate (Valmid).

Clinical findings

The determination of 17-ketosteroids is one of the tests used in assessing some adrenocortical, pituitary, and gonadal activity. In the normal male approximately two thirds of the urinary 17-ketosteroids are derived from the adrenal cortex and the remainder from the testes. In the normal female almost all of this group of hormones arises from the adrenal cortex with little or none from the ovaries. Hence the level of 17-ketosteroids in the urine of the female is lower than that of the male. The level of excretion also varies with age.

Extensive studies on the excretion of 17-ketosteroids in various pathologic states have been carried out because for many years this was the only method available to assess adrenal cortex function. However, it is possible to make only generalizations because of the wide variations in the values reported. Investigation has shown that the method used for routine testing whereby a group of compounds is assayed may yield a total 17-ketosteroid value consisting of up to 90% nonspecific chromogens. The test is useful in screening patients for marked changes in adrenocortical activity.

A decrease in excretion of total 17-ketosteroids is usually associated with panhypopituitarism, pituitary dwarfism, Addison's disease, myxedema, Klinefelter's syndrome, anorexia nervosa, and advanced liver disease. When Cushing's syndrome is caused by a benign tumor, the urinary 17-ketosteroids tend to be low; when the tumor is malignant, the level is usually high. An increased excretion of total 17-ketosteroids is generally found in adrenocortical hyperplasia, congenital adrenal hyperplasia, and hirsuitism. The determination has been shown to be of little diagnostic value in thyrotoxicosis, hypogonadism, Stein-Leventhal syndrome, Turner's syndrome, and mammary carcinoma.

In general, there is agreement between the urinary 17-ketosteroid and 17-hydroxycorticosteroid levels. The exceptions arise in certain pathologic conditions such as benign tumor of the adrenal gland where the urinary 17-ketosteroids are usually within normal limits and the 17-hydroxycorticosteroids are elevated.

Metanephrines, urinary[51]
Principle

Included in the group of major metabolites of catecholamines are metanephrine and normetanephrine. They are excreted mainly as conjugated derivatives and their assay provides another measurement of catecholamine secretion. Urine is acidified to pH 0.9 and the metanephrines hydrolyzed in a boiling water bath. They are adsorbed on Amberlite CG-50, eluted with ammonium hydroxide, and oxidized to vanillin by periodate. The absorbance is measured spectrophotometrically and the values are reported as total metanephrines.

Specimen collection

A 24-hour urine specimen is collected in a glass bottle containing either 15 ml of 6 N hydrochloric acid or 15 ml of toluene.

Reagents

All chemicals are reagent grade.

Sodium metaperiodate, $NaIO_4$ 2%
  Stable for 1 week at 25° C or for 1 month at 4° C.

Sodium metabisulfite, $Na_2S_2O_5$, 10%
  Prepare fresh each day.

Hydrochloric acid, 6 N
  Dilute concentrated hydrochloric acid 1:1 with distilled water.

Ammonium hydroxide, NH_4OH, 4 N
  Dilute 28 ml of concentrated (14.3 N) ammonium hydroxide to 100 ml with distilled water.

Amberlite CG-50, 100-200 mesh, pH 6.0 to 6.5
  Suspend 100 g of the resin in 300 ml of distilled water, stir for 10 minutes, allow to settle for 30 minutes, and discard the supernatant solution. The washing and settling procedure is repeated (approximately 5 times) until the supernatant is clear after a 15-minute settling period. The resin is then cycled through the sodium form. It is suspended in 300 ml of distilled water and over a period of 15 minutes 200 ml 10 N NaOH is added with constant stirring and the stirring continued for 2 hours. The resin is allowed to settle for 15 minutes, the supernatant is decanted, and the resin is washed with distilled water several times by decantation. It is converted back to the acidic form by adding 500 ml of 6 N HCl and stirring for 30 minutes. The excess acid is removed by washing several times with distilled water and decanting the supernatant. The resin is reconverted to the sodium form in the manner described above. The washed sodium form of the resin is suspended in 100 ml of distilled water, and with continuous stirring glacial acetic acid is added until the solution pH is 6.0 to 6.5. The pH must remain constant during 30 minutes of continuous stirring. The resin is added to the columns in this form. After use, the resin may be regenerated by converting to the acidic form and then the sodium form.

Stock standard (100 μg/ml)
  Dissolve 10 mg of metanephrine or normetanephrine in distilled water and dilute to 100 ml. The solution is stable at 4° C for several weeks.

Working standard (10 μg/ml)
  Dilute the stock standard 1:10 with distilled water. Store in refrigerator and prepare fresh each week.

Procedure

Hydrolysis of conjugated metanephrines

1. Into 50-ml beakers place in duplicate:

Test: 10 ml of filtered urine
Control: 10 ml of control urine
2. Adjust samples to pH 0.9 with 6 N HCl, using a pH meter.
3. Transfer to 40-ml glass-stoppered tubes.
4. Place in a boiling water bath for 20 minutes.
5. Cool solutions, adjust to pH 6.0 to 6.5 with 1 N NaOH using a pH meter, and dilute to approximately 20 ml with distilled water.

Adsorption and elution
6. Prepare chromatographic columns: Amberlite CG-50 is added to columns to make beds 1 × 5 cm. Wash with 7.0 ml of water to remove the buffer.
7. Pass through columns, at a rate of 0.5 ml/min:
   a. Hydrolyzed samples
   b. 1.0 ml of working standard in approximately 20 ml of water
8. Wash columns with 15 to 20 ml of water.
9. Elute metanephrines with 15 ml of 4 N ammonium hydroxide. Discard the first 2.0 ml and collect the next 10 ml.

Oxidation
10. Into test tubes place:
    Test (unknown, control, standard): 4.0 ml of eluate and 0.1 ml of periodate solution
    Blank (unknown, control, standard): 4.0 ml of eluate and 0.1 ml of distilled water
    Reagent blank: 4.0 ml of 4 N NH_4OH and 0.1 ml of periodate solution

Measurement and calculations
11. Determine the absorbance (A) of each solution at 360 nm against water.

$$\frac{A\ test - A\ blank - A\ reagent\ blank}{A\ standard - A\ blank - A\ reagent\ blank} \times \mu g\ standard \times \frac{24\text{-hr urine volume}}{\text{Sample volume}} = \mu g\ \text{metanephrines/24 hr}$$

Normal values
Metanephrines, total 0.3 to 0.9 mg/24 hr

Comments

The oxidation of the metanephrines by periodate in ammonium hydroxide is completed in about 30 seconds and the vanillin formed is stable for several hours.

This is a rapid screening test for pheochromocytoma but there is intereference from such drugs as chlorpromazine, imipramine, methyldopa, and pargylone.

Clinical findings

Refer to p. 393 and to catecholamines, p. 407.

Pregnanediol, urinary[33]
Principle

The excretion levels of pregnanediol, a major metabolic product of progesterone, provide an indication of endogenous production of progesterone. It is excreted in the urine conjugated with glucuronic acid. Hydrolysis is effected with β-glucuronidase. The free steroid is extracted with methylene chloride, separated by chromatographing on a column of alumina, and eluted by using 2% ethanol in benzene. It is converted to a sulfuric acid chromogen and assayed by reading the peak absorption at 425 nm.

Specimen collection

A 24-hour urine specimen is collected in a glass or polyethylene bottle preferably without a preservative. The assay should be performed as soon as possible, although during storage for several days at 4° C no appreciable deterioration appears to take place. For longer periods of storage, the specimen must be frozen.

Reagents

> Reagents are the same as those used for pregnanetriol determinations with the addition of the following:
> Ethyl alcohol, 0.05% in benzene
> 5β-Pregnane-3α,20α-diol standard
> Stock standard
>> Dissolve 40 mg in absolute ethyl alcohol and dilute to 100 ml (400 μg/ml). Store in an amber bottle at room temperature.
>
> Working standard
>> Dilute 5.0 ml of stock standard with 15 ml of absolute ethyl alcohol (100 μg/ml). Prepare monthly and store in an amber bottle at room temperature.

Procedure

Hydrolysis, extraction, and washing

Proceed as in steps 1 through 11 for pregnanetriol (p. 439) except that a standard containing 50 μg/ml (0.5 ml of working standard) pregnanediol is carried through the procedure. If the patient is pregnant, use only 5.0 ml of urine for the analysis.

Chromatography

Prepare the columns in the same manner as for pregnanetriol.
12. Allow the extract to pass through the column and rinse the column with 10 ml benzene. Discard all benzene that passed through the column.
13. Rinse the column with 40 ml of 0.05% ethanol in benzene and discard the washing.

Elution

14. Add 40 ml of 2% ethyl alcohol in benzene to the column. Collect the eluate in a 50-ml round-bottom centrifuge tube.
15. Evaporate the eluate to dryness under a stream of air in a water bath at 35° to 45° C. Wash down the sides of each tube with 2 ml of ethyl alcohol and evaporate to dryness to remove all traces of benzene.

Color development

16. Prepare two standard tubes for recovery calculations by pipeting into 50-ml round-bottom tubes 0.5 ml of working standard (50 μg) and evaporating to dryness.
17. To all tubes from steps 15 and 16 add 15 mg of sodium sulfite and 3.5 ml of concentrated sulfuric acid. *Without mixing* allow the tubes to stand at room temperature for 30 minutes.
18. After the 30-minute period mix the contents in the tubes. Allow to stand an additional 60 to 90 minutes and determine absorbance (A) at 390, 425, and 460 nm against concentrated sulfuric acid as the blank. If the concentration of pregnanediol in the unknown is too high, dilutions may be made at this step, using concentrated sulfuric acid.

Table 14-8
Normal values for urinary pregnanediol

Male	0.38-1.42 mg/24 hr
Female	
Nonpregnant	0.5 -7.0
Proliferative phase	0.5 -1.5
Luteal phase	2.0 -7.0
Postmenopausal	0.2 -1.0
Pregnant: weeks of amenorrhea	
10-12	5-15
12-18	5-25
18-24	13-33
24-28	20-42
28-32	27-47

Calculations

$$\text{Corrected A} = A_{425} - \left(\frac{A_{390} + A_{460}}{2}\right)$$

$$\frac{\text{Corrected A unknown}}{\text{Corrected A extracted standard}} \times \text{Standard concentration} \times \frac{\text{Total volume urine}}{\text{Aliquot urine used}} = \text{mg pregnanediol/24 hr}$$

Normal values

Normal values of urinary pregnanediol concentration are given in Table 14-8.

Comments

Pregnanediol and pregnanetriol may be performed on the same hydrolysate and extract. Pregnanediol is eluted from the column first by using a different concentration of alcohol in benzene.

Clinical findings

Pregnanediol excretion is increased during the secretory (luteal) phase of the menstrual cycle and reaches a maximum during the third trimester of pregnancy. Abnormalities in menstruation and pregnancy may be associated with decreased secretion of pregnanediol.

Progesterone synthesis that occurs in the placenta does not appear to involve the fetus. Thus fetal disorders may be present when the pregnanediol excretion levels are normal but can be indicated by altered estriol excretion. In pregnancy low values are often followed by abortions and in fetal death pregnanediol levels decrease rapidly.

Pregnanediol excretion is usually decreased in amenorrhea, anovular menstruation, and preeclamptic toxemia of pregnancy. Elevated values may be caused by

plasma corticosteroids ad modum de Moor and Steeno, Scand. J. Clin. Lab. Invest. **20:** 185, 1967.
46. Nugent, C. A., Nichols, T., and Tyler, F. H.: Diagnosis of Cushing's syndrome; single dose dexamethasone suppression test, Arch. Intern. Med. **116:**172, 1965.
47. Paulsen, C. A., editor: Estrogen assays in clinical medicine, Seattle, 1965, University of Washington Press.
48. Perkoff, G. T., Eik-Nes, K., Nugent, C. A., Fred, H. L., Nimer, R. A., Rush, L., Samuels, L. T., and Tyler, F. H.: Studies of diurnal variation of plasma 17-hydroxycorticosteroids in man, J. Clin. Endocrinol. Metab. **19:**432, 1959.
49. Peterson, R. E.: Measurement of plasma or serum cortisol. In Sunderman, F. W., and Sunderman, F. W., Jr., editors: Lipids and the steroid hormones in clinical medicine, Philadelphia, 1960, J. B. Lippincott Co.
50. Peterson, R. E.: Determination of urinary neutral 17-ketosteroids. In Seligson, D., editor: Standard methods of clinical chemistry, vol. 4, New York, 1963, Academic Press, Inc.
51. Pisano, J. J.: A simple analysis for normetanephrine and metanephrine in urine, Clin. Chim. Acta **5:**406, 1960.
52. Pisano, J. J., Crout, J. R., and Abraham, D.: Determination of 3-methoxy-4-hydroxymandelic acid in urine, Clin. Chim. Acta **7:**285, 1962.
53. Porter, B. A., Rosenfeld, R. L., and Lawrence, A. M.: The levodopa test of growth hormone reserve in children, Am. J. Dis. Child. **126:** 589, 1973.
54. Porter, C. C., and Silber, R. H.: A quantitative color reaction for cortisone and related 17,21-dihydroxy-20 ketosteroids, J. Biol. Chem. **185:**201, 1950.
55. Rappaport, F., Fischl, J., and Pinto, N.: A rapid method for the estimation of urinary 17-ketosteroids, Clin. Chem. **6:**16, 1960.
56. Reiss, E., and Alexander, F.: The tubular reabsorption of phosphate in the differential diagnosis of metabolic bone disease, J. Clin. Endocrinol. Metab. **19:**1212, 1959.
57. Sawin, C. T., Bray, G. A., and Idelson, B. A.: Overnight suppression test with dexamethasone in Cushing's syndrome, J. Clin. Endocrinol. Metab. **28:**422, 1968.
58. Schally, A. V., Arimura, A., and Kastin, A. J.: Hypothalamic regulatory hormones, Science **179:**341, 1973.
59. Silber, R. H.: Free and conjugated 17-hydroxycorticosteroids in urine. In Seligson, D., editor: Standard methods of clinical chemistry, vol. 4, New York, 1963, Academic Press, Inc.
60. Silber, R. H.: Fluorimetric analysis of corticoids. In Glick, D., editor: Methods of biochemical analysis, vol. 14, New York, 1966, Interscience Publishers, Inc.
61. Silber, R. H., and Porter, C. C.: Determination of 17,21dihydroxy-20-ketosteroids in urine and plasma. In Glick, D., editor: Methods of biochemical analysis, vol. 4, New York, 1957, Interscience Publishers, Inc.
62. Sobel, C., and Henry, R. J.: Determination of catecholamines (adrenalin and noradrenalin) in urine and tissue, Am. J. Clin. Pathol. **27:**240, 1957.
63. Spencer-Peet, J., Daly, J. R., and Smith V.: A simple method for improving the specificity of the fluorometric determination of adrenal corticosteroids in human plasma, J. Endocrinol. **31:**235, 1965.
64. Sutherland, E. W.: On the biological role of cyclic AMP, J.A.M.A. **214:**1281, 1970.
65. Sutherland, E. W.: Studies on the mechanism of hormone action, Science **177:**401, 1972.
66. Sweat, M. L.: Sulfuric acid induced fluorescence of corticosteroids, Anal. Chem. **26:** 773, 1954.
67. Täättelä, A.: Plasma catecholamines in the diagnosis and localization of pheochromocytoma, Ann. Clin. Res. **4:**78, 1972.
68. Thijssen, J. H. H., and Veeman, W.: A gas chromatographic method for the measurement of small amounts of estrogens in urine, Steroids **11:**369, 1968.
69. Utiger, R. D.: Serum triiodothyronine in man, Ann. Rev. Med., p. 289, 1974.
70. van Baelen, H., Heyns, W., and deMoor, P.: Measurement of urinary estrogens using adsorption on sephadex, J. Clin. Endocrinol. Metab. **27:**1056, 1967.
71. van der Vies, J.: Individual determination of cortisol and corticosterone in a single small sample of peripheral blood, Acta Endocrinol. **38:**399, 1961.
72. White, W. L., Erickson, M. M., and Stevens, S. C.: Practical automation for the clinical laboratory, St. Louis, 1972, The C. V. Mosby Co.
73. Young, D. S., Thomas, D. W., Friedman, R. B., and Pestaner, L. C.: Effects of drugs on clinical laboratory tests, Clin. Chem. **18:**1041, 1972.

General references

Bayliss, R. J. S., editor: Investigations of endocrine disorders, Clin. Endocrinol. Metab. **3**(3): 1974.
Curry, A. S., and Hewitt, J. V., editors: Biochemistry of women: methods for clinical investigation, Cleveland, 1974, CRC Press.
Dillon, R. S.: Handbook of endocrinology, diagnosis and management of endocrine and metabolic disorders, Philadelphia, 1973, Lea & Febiger.

Jaffe, B. M., and Behrman, H. R., editors: Methods of hormone radioimmunoassay, New York, 1974, Academic Press, Inc.

Lindholm, J.: Studies on some parameters of adrenocortical function, Acta Endocrinol., Suppl. 172, 1973.

Loraine, J. A., editor: Reproductive endocrinology and world population, Clin. Endocrinol. Metab. 2(3): 1973.

Odell, W. D., and Daughaday, W. H., editors: Principles of competitive protein-binding assays, Philadelphia, 1971, J. B. Lippincott Co.

Williams, R. H., editor: Textbook of endocrinology, ed. 5, Philadelphia, 1974, W. B. Saunders Co.

15

Toxicology

MARY ANN MACKELL*

The intent of this chapter is to describe a few techniques and methods of analyses of toxic materials or drugs that can be performed in the clinical laboratory employing equipment that is often already available. Since more and more laboratories are performing ultraviolet enzyme tests, a section has been included on ultraviolet spectrophotometry.

The methods described are not presented as the most recent or the most sophisticated but ones that are possible to perform with a minimum of stress on personnel and budget. However, it must be emphasized that time will be required to develop techniques and to learn to evaluate results.

NOTE: All reagents, unless otherwise specified, are AR quality. Water is distilled or deionized.

Arsenic, Reinsch test, modified[10]
Principle

Certain metals such as arsenic, antimony, and bismuth in an acid medium and with heat can be deposited on a copper wire. Arsenic can be differentiated from antimony and bismuth.

This test is sensitive to 10 μg of arsenic.

Specimen

Urine, gastric content, or tissue homogenate may be used. Tissue specimens (liver or kidney) are finely macerated with water or may be homogenized in a kitchen blender using a weight of tissue to an equal volume of water.

Reagents

Hydrochloric acid, concentrated
Copper wire, 20 gauge
   Make a 7-mm coil of the wire by wrapping it around a glass stirring rod.
Ethyl alcohol
Ether
Potassium cyanide, 10%
   Dissolve 10 g of KCN in water and dilute to 100 ml.

---

*I wish to acknowledge the technical assistance of Lynne K. Cupples, who was my assistant for 7 years.

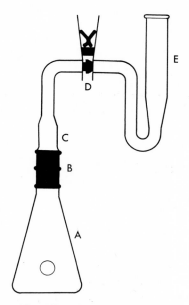

Fig. 15-1. Arsine generator.

Procedure
1. Into a 50-ml Erlenmeyer flask place 20 ml of specimen and 4.0 ml of concentrated HCl.
2. Wash the copper coil successively with alcohol and ether; add to the flask.
3. Heat the flask and contents in a boiling water bath (100° C) for 1 hour.
4. After cooling remove copper coil and examine for a dark to black deposit which may be due to arsenic, bismuth, or antimony.
5. To confirm the presence of arsenic, place the copper wire into a test tube containing 2.0 ml of 10% KCN. Arsenic will dissolve in the cyanide solution, but a bismuth or antimony deposit will remain on the coil.

Arsenic, a quantitative colorimetric method[6, 13]
Principle

Biologic fluids or tissue specimens are wet ashed. Arsine is liberated from the acid digests and reacted with silver diethyldithiocarbamate to form a red complex. The maximum absorbance of the resulting color complex is at 560 nm and is linear up to 10 μg/ml (1 mg/dl).

Special apparatus
Arsine generator (see Fig. 15-1)
Available from Fisher Scientific Co., FSCo. #1-405.
Kjeldahl flasks, 300-ml capacity, and digestion rack
(If extreme caution is observed, the acid digestion may be done in Erlenmeyer flasks on a hot plate in a fume hood.)
Safety goggles
Fume hood

Specimen
Vomitus, blood, or urine may be used. In cases of chronic exposure or if analysis is performed several days after onset of symptoms, urine is the specimen of choice.

Reagents

 Sulfuric acid, concentrated
 Nitric acid, concentrated
 Perchloric acid, concentrated (70% to 72%)
 Potassium iodide, 15%
  Dissolve 15 g of KI in water and dilute to 100 ml.
 Stannous chloride solution
  Dissolve 20 g of $SnCl_2 \cdot 2H_2O$ in 25 ml of hot water. After cooling, filter and carefully add 25 ml of concentrated HCl.
 Lead acetate, 10%
  Dissolve 10 g of lead acetate in water and dilute to 100 ml.
 Sodium hydroxide, 40% (10 N)   (D-230)
 Phenolphthalein indicator, 1%   (D-9)
 Stock arsenic standard, 100 mg/dl*
  In a 1-L flask dissolve 1.32 g of arsenic trioxide in 10 ml of 10 N sodium hydroxide. Add 500 ml of water and a few drops of 1% phenolphthalein indicator. While stirring slowly, add concentrated HCl until pink color disappears. Dilute to volume.
 Working arsenic standard, 10 µg/ml
  Dilute 1.0 ml of stock to 100 ml in volumetric flask.
 Pyridine
 Color reagent
  Dissolve 500 mg of silver diethyldithiocarbamate in 100 ml of pyridine. (If pyridine has become yellow, use a fresh bottle.)
 Glass wool
 Zinc, granular, 20 to 30 mesh

Procedure

1. *Acid digestion:* To a Kjeldahl or Erlenmeyer flask containing the specimen (5.0 ml blood or 10 ml of urine or vomitus) add 5.0 ml of concentrated sulfuric acid, 25 ml of concentrated nitric acid, and several glass beads. Another flask in which water has been substituted for the specimen is carried through the entire procedure to be used as a blank. Since the acids contain trace amounts of arsenic, it is necessary to add the same quantity of acid to the blank as to the test. After the flasks stand for 10 minutes, gentle heat is applied. As the reaction subsides the heat is increased until the contents of the flask becomes a dark brown but not charred. Remove the source of heat and cool flasks to room temperature. After cooling add 20 ml of nitric acid, allow to cool, and carefully add 2.0 ml of percholoric acid. Apply heat to flasks until white fumes appear. If the contents of the flask are not colorless or light straw color, cool and add nitric acid and perchloric acid in the same proportions and with the same care as previously stated and again heat. The addition of the acids with subsequent heating is continued until the digest is clear and the final volume is approximately 5.0 ml. When the digestion is completed all the carbonaceous material should have been destroyed, resulting in a solution of heavy metals in sulfuric acid. After cooling digestion flasks, quantitatively transfer contents to a 50-ml volumetric flask by repeated rinsings of the digestion flask with water. Dilute to volume.
2. Label three arsine generators as follows: *blank, test, standard.* Referring to Fig. 15-1, add a 10-ml aliquot of blank and test to respective flasks *(A)*; to standard add 1.0 ml of working standard. To blank and test add 25 ml of

---

*Commercially available from Harleco, #7692.

water and to standard 34 ml of water. To each flask add 5.0 ml of concentrated HCl and 2.0 ml of potassium iodide solution.
3. Allow flasks to stand 1 minute and then add stannous chloride dropwise until solution becomes colorless (approximately 0.5 ml). Swirl flasks and let stand for 15 minutes to ensure reduction of arsenic to the trivalent form.
4. Moisten plugs of glass wool with a few drops of lead acetate and place in scrubber section (C) of generators.
5. Add 3.0 ml of the color reagent to the absorber tubes (section E) and connect section C and D at the ball and socket joint with pinch clamp D.
6. Add 5 g of granular zinc to flask and quickly place assembled parts C and E into flask at section B. After several minutes bubbling will be observed in tube E. If the bubbling is too rapid (more than one bubble/second) some of the arsine may escape before being absorbed. The reaction may be slowed down by placing the flasks in an ice bath.
7. Allow reaction to continue for 30 minutes. Transfer the color reagent from tube E to cuvets and read absorbance at 525 nm. Use the blank to adjust instrument to zero before reading test. Read standard and unknowns against color reagent.

Calculations

Test = 0.55 at 525 nm
Standard 10 µg/ml = 0.65 at 525 nm
10-ml aliquot of test was used.

$$\frac{0.55}{0.65} \times \frac{50}{10} \times \frac{1}{10} \times 10 \text{ µg/ml} = 4.2 \text{ µg arsenic/ml of urine or 4.2 mg/L}$$

Normal values

Urine	170 µg/L
Blood	19 µg/dl

Ethyl alcohol (ethanol)

Contrary to popular belief, ethanol is a depressant. It is undoubtedly the drug of most common abuse. For these reasons ethyl alcohol analysis is one of the more frequently requested analyses. When alcohol is used along with other central nervous system depressants, such as barbiturates, the effect is additive. A high therapeutic level of one of the hypnotic drugs in combination with alcohol can produce toxic symptoms.

Two quantitative methods are described: an oxidation-reduction titrimetric procedure (Harger's method) and an enzymatic method. If the laboratory has a spectrophotometer capable of transmitting at 340 nm, the enzymatic method is recommended because of its speed and greater specificity.

QUANTITATIVE DETERMINATION OF ETHANOL (ENZYMATIC)[3]
Principle

Alcohol dehydrogenase (ADH) causes the conversion of ethanol to acetaldehyde with the concomitant reduction of nicotinamide adenine dinucleotide (NAD). At 340 nm NAD absorbs very little energy, while the reduced form, NADH, is highly absorptive.

The conversion of NAD to NADH in presence of ADH is proportional to the concentration of alcohol.

Specimen

1.0 ml of whole blood, serum, or plasma; keep stoppered and refrigerated.

NOTE: **Blood should be collected with disposable equipment and venipuncture sight must not be cleaned with an alcohol solution. A nonalcoholic germicide such as a 1:1000 solution of mercuric chloride may be used.**

Reagents
  NAD-ADH vials*
  Buffer, pH 9.2*
  Ethanol standard*
  Trichloroacetic acid, 6.25%
     Dissolve 6.25 g of trichloroacetic acid in water and dilute to 100 ml.

Procedure
1. **Prepare a protein-free filtrate.** Add 0.5 ml of specimen to a tube containing 2.0 ml of trichloroacetic acid solution. Mix on Vortex for 30 seconds, breaking up clumps. Centrifuge for 5 minutes. Observe that centrifugate is clear.
2. Dilute 0.5 ml of ethanol standard (80 to 150 mg/dl w/v) with 2.0 ml of trichloroacetic acid solution.
3. *Preparation of NAD-ADH vials:* **Bring vials (which are stored in freezer) to room temperature. Label vials** *standard* **and** *unknown;* **to each add 3.0 ml of buffer and invert to mix.**
4. With spectrophotometer set at 340 nm, use a water blank to set absorbance to zero. Read absorbance of each NAD-ADH vial and return to respective vial. Expected absorbance should range from 0.05 to 0.1 units. Record this as A_1.
5. Add 0.1 ml of protein-free centrifugate to test vial and 0.1 ml of diluted ethanol standard to standard vial. Replace caps on all vials and invert. Let stand at room temperature for 30 minutes.
6. After incubation period, read absorbance of each vial at 340 nm. Record this as A_2.

Calculations
  $A_2 - A_1 = A$ for unknown and standard.

  $$\frac{A \text{ unknown}}{A \text{ standard}} \times \text{Concentration of standard} = \text{mg of ethanol/dl (w/v)}$$

  Example: Unknown   **Standard, 80 mg/dl**
  $A_1 = 0.05$ $A_1 = 0.05$
  $A_2 = 0.90$ $A_2 = 0.75$
  $A\ = 0.85$ $A\ = 0.70$

  $$\frac{0.85}{0.70} \times 80 \text{ mg/dl} = 97 \text{ mg/dl (w/v)}$$

Comments

If a wide band pass instrument such as a Coleman 6/20 is used, absorption is linear to approximately 100 mg/dl concentration. However, if a standard curve ranging from 50 to 160 mg/dl is constructed the range may be expanded. Regardless of instrumentation the upper limit of concentration is 160 mg/dl. If the level is higher the test must be repeated using only 0.05 ml of centrifugate and the results multiplied by 2.

The methodology presented is general, and it is suggested that reagents be obtained commercially and that the manufacturer's procedure, which may be at some variance, be followed.

---

*These items are commercially available. Two sources are Sigma Chemical Co. and Boehringer Mannheim Corp.

Normal values

As a generalization, results up to 30 mg/dl may be considered inconclusive of ethanol ingestion; levels of 100 mg/dl and above are considered "under the influence." For more detailed information refer to American Medical Association: Alcohol and the impaired driver, 1968.

ALCOHOL, A QUANTITATIVE TITRIMETRIC METHOD (HARGER'S MICRO METHOD)[9]

Special apparatus

Glass distilling flask, 500-ml (Pyrex No. 3360) with Graham Condenser (Pyrex No. 2540)

Specimen

5.0 ml whole blood, plasma, or serum, using the same precautions during collection as for enzymatic procedure (pp. 452-453)

Reagents

Potassium dichromate solution, 0.0434 N

In a 1-L volumetric flask dissolve 2.129 g of $K_2Cr_2O_7$ in water and dilute to volume (1.0 ml will oxidize 0.5 mg of ethanol). Store in a ground glass stoppered bottle.

Ferrous sulfate solution, 20%

Add 25 g of $FeSO_4 \cdot 7H_2O$ to 75 ml of water; add 15 ml of concentrated sulfuric acid; stir until dissolved. After solution has cooled to room temperature, dilute to 125 ml with water. Store in a brown ground glass stoppered bottle.

Methyl orange, 0.1%

Dissolve 1 g of methyl orange in water, dilute to 1 L. Add 1 g of sodium hydroxide, stir until dissolved. Filter.

Sulfuric acid, 50% (v/v)

In small increments add 100 ml of concentrated sulfuric acid to 100 ml of water, stirring and allowing solution to cool between additions.

Red reducing fluid

Make fresh immediately before use. To 17.5 ml of 50% H_2SO_4 add 2.5 ml of methyl orange and 0.5 ml of ferrous sulfate solution. Mix and cool to room temperature.

Sulfuric acid, $\frac{2}{3}$ N

Slowly add 18.6 ml of H_2SO_4, concentrated, to 800 ml of water. After cooling to room temperature dilute to 1 L.

Sodium tungstate, 10%

Procedure

A standard of known concentration or preferably a control blood* is carried through the entire procedure along with test specimen.

1. Preparation of Folin-Wu filtrate (1:10 dilution): Add 3.0 ml of blood (or other specimen) to 21 ml of water in a 50-ml Erlenmeyer flask. While swirling flask add 3.0 ml of $\frac{2}{3}$ N H_2SO_4 and 3.0 ml of sodium tungstate solution. Let stand 5 minutes; filter through Whatman No. 42 filter paper. Each milliliter of filtrate is equivalent to 0.1 ml of specimen.
2. To a distilling flask add 20 ml of water, a few glass beads, and 5.0 ml of filtrate. Begin distillation.
3. In a 25-ml graduated cylinder collect approximately 15 ml of distillate. Adjust volume with water to exactly 25 ml. Invert to mix.

---

*Available from Lederle Diagnostics.

4. Into a 25 × 150 mm test tube labeled *blank*, add 5.0 ml of water, and into another tube labeled *test* add 5.0 ml of distillate. To each tube, using a volumetric pipet, add 1.0 ml of potassium dichromate solution and 5.0 ml of concentrated sulfuric acid. Mix and allow to cool to room temperature.
5. Fill the microburet with reducing fluid, making sure the tip of the buret is full and the meniscus is at 10 ml.
6. *Titration W:* While constantly swirling, titrate reducing fluid into the blank tube until the first permanent pink color is achieved. (This requires 2.0 to 2.5 ml). Record all titration figures.
7. *Titration B:* After completing step 6, add 1.0 ml of dichromate solution to blank and titrate as above. The second addition of dichromate and titration of the blank will compensate for trace amounts of reducing material that may be present in the sulfuric acid. This titration should require approximately 2.5 ml. If it exceeds ± 0.2 ml, the age and/or the preparation of the oxidizing reagent ($K_2Cr_2O_7$) and the reducing fluid should be suspect. If both are fresh and have been properly prepared, the technique of titration should be practiced.
8. *Titration U:* Titrate the test solution as in step 6. The color of the unknown will vary from yellow to pale green to colorless, the color intensity diminishing with the concentration of alcohol. If the solution is very pale green it is advisable to make a 1:2 dilution of the distillate before proceeding with step 4.

Calculations

$$\frac{W \text{ (step 6)} - U \text{ (step 8)}}{B \text{ (step 7)} \times Q \text{ (quantity of blood in aliquot)}} \times 0.5 \text{ mg (weight of ethanol oxidized by 1 ml of dichromate solution} = \text{mg alcohol/ml}$$

or

$$\frac{W-U}{B \times 0.1} \times 0.5 = \frac{W-U}{B} \times 5 = \text{mg/ml}$$

Example:
W = 2.1 ml
B = 2.5 ml
U = 1.4 ml

$$\frac{2.1 - 1.4}{2.5} \times 5 = 1.4 \text{ mg alcohol/ml} \times 100 = 140 \text{ mg/dl}$$

If a 1:2 dilution of distillate was made, then results must be multiplied by 2.

Comments

This procedure measures naturally occurring reducing substances as well as exogenous ones; levels up to 50 mg/dl may be found in abstaining subjects.

Methanol or formaldehyde will give false positive results. To rule these out see the following method for qualitative determination of these agents.

Methanol and formaldehyde screening test

Specimen

Refer to alcohol determination, Harger's method, p. 454.

Reagents

Potassium permanganate reagent
To 3 g of $KMnO_4$ add 15 ml of phosphoric acid (85%) and 85 ml of water.

Sodium bisulfite

Chromatropic acid (4,5 dihydroxy-2,7-naphthalenedisulfonic acid, disodium salt), Eastman Kodak #1613

Procedures
1. Prepare a Folin-Wu filtrate as in step 1, Harger's method, p. 454.
2. Using 10-ml capacity test tubes label No. 1 (formaldehyde) and No. 2 (methanol). To each add 1.0 ml of filtrate.
3. To tube No. 2 add 2 to 3 drops of potassium permanganate solution. Wait 2 minutes; add a small amount of solid sodium bisulfite to decolorize the permanganate (normally the amount that will adhere to the tip of an applicator stick). The permanganate oxidizes any methanol present to formaldehyde.
4. To both tubes Nos. 1 and 2, using an applicator stick as in step 3, add solid chromatropic acid. Mix both tubes.
5. To each tube add 3.0 ml of concentrated sulfuric acid. Tilt the tube and slowly pipet the acid down the side so that it forms a layer in the bottom. A purple ring at the interface of the two liquids in tube No. 1 indicates the presence of formaldehyde, in tube No. 2 the presence of methanol. At low levels it may take as long as 20 minutes for the color to develop.

Comments

Since sulfuric acid has an affinity for formaldehyde fumes, it is well advised to run a parallel reagent blank, substituting 1.0 ml of water for the blank filtrate.

Carbon monoxide

Carbon monoxide (CO) is generated when incomplete combustion occurs—whenever there is insufficient oxygen to form carbon dioxide (CO_2). Common sources of carbon monoxide are automobile exhaust, smoldering fires, faulty flues or exhaust systems in coal burning heating systems, and gas manufactured from coal. Natural gas, which is the source of most gas supply in the United States, does not contain carbon monoxide.

Toxicity results from the greater affinity of hemoglobin for carbon monoxide over oxygen, resulting in the formation of carboxyhemoglobin (HbCO). As the amount of hemoglobin that is normally combined with oxygen (oxyhemoglobin) is replaced by carboxyhemoglobin, cellular metabolism is interfered with, resulting in symptoms ranging from giddiness accompanied by nausea to anoxia to death.

The measurement of carbon monoxide is usually termed in percent saturation. If the subject has a total hemoglobin of 14 g/dl and carbon monoxide level is reported at 20% saturation, then 2.8 g/dl of the total hemoglobin is carboxyhemoglobin.

Symptoms begin to occur at 15% to 20% saturation, coma and possible death at 40% and above; however, most fatalities range between 60% and 80% saturation.

CARBON MONOXIDE, A QUICK SCREENING METHOD SENSITIVE TO 20% SATURATION[9]

Specimen

1.0 ml of blood collected in a tube containing an anticoagulant, or a heparinized capillary tube collected from a finger stick.

Reagents

Negative control blood

May be collected from a nonsmoker in the same manner as the specimen, or a quantity of fresh blood containing an anticoagulant* may be oxygenated by bubbling stream of O_2 through it for approximately 1 hour (0% CO).

Positive control blood

Bubble CO through a quantity of fresh anticoagulated blood* for 1 hour. A lecturer size tank of CO may be obtained commercially, or CO can be generated by dropping concentrated sulfuric acid in formic acid. If the CO is produced in the latter manner it should be passed through a $CaCl_2$ dryer before reaching the blood. Gassing of the blood must be performed in a fume hood with utmost precaution. Remember, carbon monoxide is odorless, colorless, and toxic.

Standard controls

With the negative and positive controls intermediate controls may be prepared:

20% control	1 part positive to 4 parts negative
50% control	Equal parts of positive and negative
80% control	4 parts positive to 1 part negative

One milliliter portions of these controls as well as the negative control may be pipeted into small test tubes, tightly stoppered, and frozen for future use.

Sodium hydroxide, 10% (2.5 N) (D-83)

Procedure

1. Five test tubes (15-ml capacity) are labeled as follows: *negative, unknown*, 20%, 50%, 80% standard control. Add 1 drop of negative, patient, and standard control blood to respective tubes.
2. Add 10 ml of water to each tube. Invert.
3. To each tube add 5 drops of sodium hydroxide solution. Invert. The negative control will turn straw colored (yellow) immediately. The positive controls will remain pink for a period of time ranging from 5 to 60 seconds. The intensity of the pink color and the length of time that intervenes before the color change occurs are related to the concentration of HbCO.

Comments

By comparing both intensity of color and time for the color change of the unknown to the controls, a good estimation of the percent of saturation of carboxyhemoglobin can be determined.

After a few practice runs to familiarize oneself with the distinctive color and the time lapse before the color change, a determination of acute carbon monoxide poisoning can be made within minutes. This test is specific for carbon monoxide.

CARBON MONOXIDE DETERMINATION, SEMIQUANTITATIVE, SENSITIVE TO 5% SATURATION (METHOD OF GETTLER AND FREIMUTH)[7]

Principle

Carbon monoxide reduces palladium ions to metallic palladium, producing a gray color which is proportional to the quantity of liberated CO.

---

*Blood used for preparing controls may be 50 ml of pooled whole blood from the hematology laboratory.

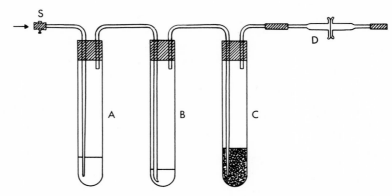

Fig. 15-2. Apparatus for carbon monoxide determination by method of Gettler and Freimuth.

Special apparatus
   Three test tubes (150 × 19 mm) fitted with rubber stoppers having two holes
   Glass tubing
   Glass flanges available from Eck and Krebs
   Water aspirator
   Screw clamps
      See Fig. 15-2.

Specimen
   3.0 ml of blood collected in an anticoagulant

Reagents
   Palladium chloride
      Dissolve 0.5 g of palladium chloride in 1.0 to 2.0 ml of concentrated HCl in a small stoppered tube. Allow to stand until dissolved (usually overnight). Add water to make a volume of 50 ml. If stored in an airtight brown bottle, reagent is stable for 6 months.
   Ferricyanide solution
      Dissolve 3.2 g of potassium ferricyanide and 0.8 g of saponin in 100 ml of water.
   Lactic acid solution
      Dilute 0.8 ml of lactic acid (specific gravity 1.2) to 100 ml with water. This is stable 1 to 2 weeks.
   Lead acetate, 10%
      Dissolve 10 g lead acetate in water and dilute to 100 ml.
   Octyl (caprylic) alcohol
   Ferricyanide–lactic acid reagent
      Combine equal volumes of ferricyanide solution and lactic acid solution (5.0 ml of each for determination). Make fresh.
   Palladium chloride paper
      From Whatman No. 1 paper cut circles slightly smaller than the ground glass surface of the flange (D). Moisten the circles with several drops of the palladium chloride solution.

Procedure (refer to Fig. 15-2)
1. *Tube A:* Add 5.0 ml of palladium chloride solution. This will cleanse the air of any carbon monoxide.

2. *Tube B:* Add 1.0 ml of thoroughly mixed whole blood, 4.0 ml of ferricyanide–lactic acid solution, and 2 drops of octyl alcohol.
3. *Tube C:* Add glass beads to a depth of 4 cm and 5.0 ml of lead acetate solution. The lead acetate will remove hydrogen sulfide that may be in the blood due to putrefaction. This is particularly important in analyzing postmortem blood.
4. *Glass flanges D:* Between the flanges fasten a disk of the palladium impregnated paper.
5. Partially open the screw clamp *(S)* and turn on the water aspirator to provide gentle suction and slow bubbling in the tubes. Continue aeration for 15 minutes.
6. Remove the disk from the flanges and rinse off any unchanged palladium chloride with water. Compare the intensity of the gray stain with previously prepared standards.

Standards

Prepare negative blood (100% oxygen-saturated and 100% carbon monoxide–saturated blood) as described in quick screening method, p. 456.

From these make dilutions that range from 0% to 80% CO saturation. Subject each of these standards to the procedure already described. Mount the stained disks on heavy paper, label the percent saturation. Cover with clear plastic and protect from light. These standard stains will remain stable for 5 months.

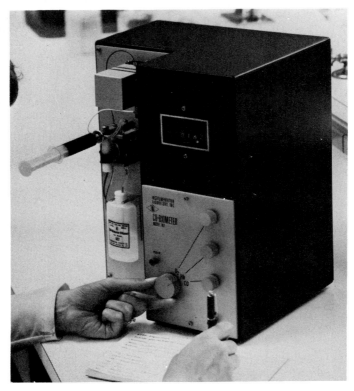

Fig. 15-3. CO-Oximeter, model 182. (Courtesy Instrumentation Laboratory, Inc., Lexington, Mass.)

CARBON MONOXIDE, A QUANTITATIVE SPECTROPHOTOMETRIC PROCEDURE
Principle

Sodium dithionite (sodium hydrosulfite) will reduce oxyhemoglobin and methemoglobin but not carboxyhemoglobin. The absorption of a hemolysate of blood after the addition of sodium dithionite is measured at 541 and 555 nm. The absorption ratio of 541/555 is determined. This ratio is compared to a standard curve.

This procedure can be precise and rapid, but it requires an accurately calibrated spectrophotometer with a band pass of less than 2 nm. See Tietz and Fiereck.[16]

THE CO-OXIMETER (IL MODEL 182) (Fig. 15-3)

The CO-Oximeter measures carbon monoxide and oxygen saturations in relation to the total hemoglobin concentration. The simplicity of the instrument has reduced to a minimum technical errors that would normally interfere with accurate measurements. If it is calibrated improperly or is not given proper maintenance, then poor results will, naturally, be the product. Measurements are independent of the total Hb concentration of the sample and are not affected by diagnostic dyes such as methylene blue or Cardio-green.

Model 182 combines an absorption spectrophotometer with a special-purpose analog computer. Three separate interference filters are used as monochromators. At selected wavelengths they monitor the absorbance of HbO_2, HbCO, and the deoxygenated Hb in the sample. The three absorbance values obtained are programmed in a matrix circuit board that instantly solves simultaneous equations for the three unknowns. These determinations may be achieved with a basic spectrophotometer but they are rather cumbersome.

Answers are presented on the digital counter in a sequence that the operator selects. The HbO_2 and HbCO values are linear over the entire 0 to 100% range.

Barbiturates

Barbiturate determinations are the most frequently requested drug analyses, which comes as little surprise when it is considered that there are over thirty individual barbiturates marketed under more than 100 trade names.

All of the barbiturates are derivatives of barbituric acid, which of itself has no hypnotic quality.

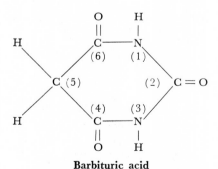

Barbituric acid

The difference in the activity or potency of the various drugs of this classification depends on the groups (R, R^1) that are substituted at the number 5 carbon. Phenobarbital, which has a therapeutic blood level of 1 mg/dl but which may range as high as 5 mg/dl, has an ethyl and phenyl group substituted at number 5 carbon.

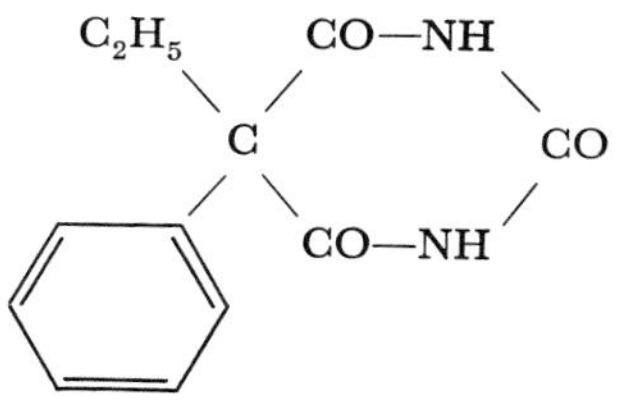

$$\underset{\text{Phenobarbital}}{\begin{array}{c}C_2H_5\\ \diagdown\\ \\ \diagup\\ \end{array}\begin{array}{c}CO-NH\\ \diagup\diagdown\\ CCO\\ \diagdown\diagup\\ CO-NH\end{array}}$$

The *toxic* blood level of secobarbital is 1 mg/dl. The structural formula is the same as for phenobarbital except that an allyl group and a 1-methylbutyl group are substituted at the number 5 position.

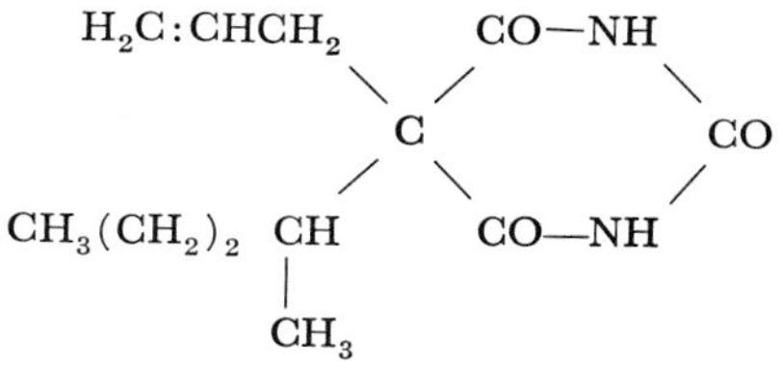

$$\underset{\text{Secobarbital}}{\begin{array}{c}H_2C{:}CHCH_2\\ \\ CH_3(CH_2)_2\;CH\\ |\\ CH_3\end{array}\begin{array}{c}CO-NH\\ CCO\\ CO-NH\end{array}}$$

Another group of barbiturates which is of moderate potency and may be used as a daytime sedative is disubstituted at the number 5 carbon as well as at one of the nitrogen atoms.

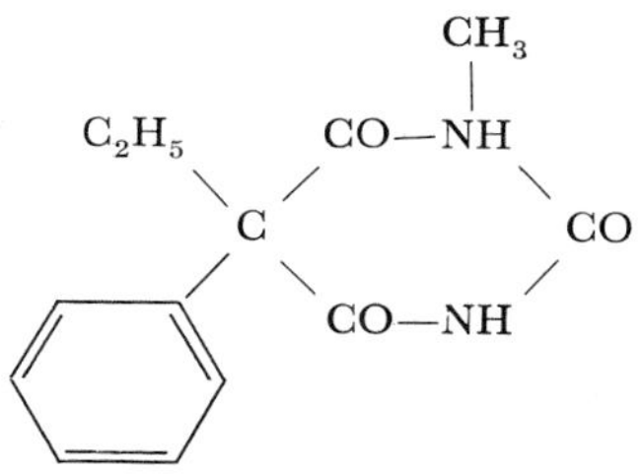

$$\underset{\text{Mephobarbital}}{\begin{array}{c}\\ C_2H_5\\ \\ \end{array}\begin{array}{c}CH_3\\ |\\ CO-NH\\ CCO\\ CO-NH\end{array}}$$

COLOROMETRIC DETERMINATION OF BARBITURATE LEVELS (GARVEY AND BOWDEN)[1, 11]
Principle

Mercuric salts will complex with barbiturates, which can then be visualized by reacting the complex with diphenylcarbazone to produce a pink color that can be quantitatively measured in a spectrophotometer at 550 nm.

Specimen

1.0 ml of serum or plasma

Reagents

All water is distilled or deionized.
Phenobarbital stock standard, 40 mg/dl
   Dissolve 44 mg of phenobarbital sodium salt in water in a volumetric flask and dilute to 100 ml. Stable for 1 month refrigerated.
Phenobarbital working standard, 0.4 mg/dl
   Add 0.1 ml of stock standard to a 10-ml volumetric flask and make to volume with water. Discard unused portion.

Sulfuric acid, 2 N (D-221)
Chloroform, AR grade
Mercuric nitrate, 2%
   To 2 g of mercuric nitrate in a 100-ml volumetric flask add 0.4 ml of concentrated HNO_3 and make to volume with water. Mix and refrigerate.
Phosphate buffer, M/15, pH 8.0
   To 8.96 g of Na_2HPO_4 (dibasic sodium phosphate) and 0.5 g KH_2PO_4 (monobasic potassium phosphate) in a 1-L flask; add water to dissolve. Make to volume with water.
Buffered mercuric nitrate
   Add 2.0 ml of 2% mercuric nitrate to a 100-ml volumetric flask. Make to volume with M/15 phosphate buffer. Mix.
Stock diphenylcarbazone reagent
   Dissolve 250 mg of diphenylcarbazone in 100 ml of chloroform. Into screw top glass tube add 1.0 ml of the diphenylcarbazone solution. Evaporate the solvent (chloroform) from the tubes in an 80° C water bath. Remove last traces of solvent under a stream of air or nitrogen. This reagent is made in advance and the tubes containing the diphenylcarbazone are stored in the dark.
Working diphenylcarbazone
   Just before using add 10 ml of chloroform to one of the stock tubes and agitate to dissolve. Discard any unused portion.

Procedure

1. Label three glass-stoppered centrifuge tubes *blank, standard,* and *unknown.* To unknown add 1.0 ml of serum, to standard 1.0 ml of working phenobarbital standard, to blank 1.0 ml of water.
2. Add 3 drops of 2 N H_2SO_4 and 10 ml of chloroform to all tubes. Stopper and shake vigorously for 2 minutes and filter through phase separating paper into correspondingly labeled glass-stoppered centrifuge tubes. The phase separating paper (Whatman 1^p_s) will filter the solvent layer while retaining the aqueous layer. In the absence of phase separating paper, allow layers to separate and with a transfer pipet remove most of the aqueous (upper) layer. Filter through Whatman No. 31 paper.
3. To all tubes add 1.0 ml of buffered mercuric nitrate and shake for 1 minute. Filter as in step 2, collecting chloroform in properly labeled test tubes.
4. Label 12 × 75 mm cuvets *blank, standard,* and *unknown* and transfer 5.0 ml of chloroform extract to each from corresponding tube. To each cuvet add 1.0 ml of diphenylcarbazone reagent. Invert and allow color to develop for 10 minutes.
5. Set spectrophotometer to 550 nm and with blank in place adjust to zero absorbance. Read absorbance of standard and unknown.

Calculations

$$\frac{\text{Absorbance unknown}}{\text{Absorbance standard}} \times \text{Concentration of standard (0.4 mg/dl)} =$$

Concentration of barbiturate of unknown, as phenobarbital

Comments

   This procedure can be performed rapidly and without special equipment. However glutethimide (Doriden) may produce false positive results.
   The different barbiturates produce varying color intensities, and for this reason,

but primarily because of wide range of toxicity of the different barbiturates, it may be important to identify the agent being measured and to rule out the presence of glutethimide by thin-layer chromatography (see p. 481).

After the particular barbiturate has been identified, the quantitation can be corrected by multiplying the concentration by a factor given in the following table:

Secobarbital	0.87
Amobarbital	0.89
Allobarbital	0.95
Butabarbital	0.96
Pentobarbital	0.96

In emergency situations, time frequently has priority over preciseness, and with a history of the drug available to the patient, this test alone can provide fast and valuable information.

Bromides, a quantitative colorimetric method[20]
Principle

A trichloroacetic acid protein-free filtrate is made using serum. Gold chloride is added to the filtrate, producing a brown color due to the formation of gold bromide. The color is obtained by the replacement of the chloride ion in the gold chloride by the bromide ion. The absorbance of the brown color is directly proportional to the bromide concentration.

Reagents
Gold chloride, 0.5% in water (D-231)
Trichloroacetic acid (TCA), 10% in water (D-95)
Bromide stock standard, 20 mg/ml (D-232)
Bromide working standard
   Pipet 1.0 ml of stock standard into a 100-ml volumetric flask, using a volumetric pipet. Dilute to mark with 10% TCA.

Procedure
1. Pipet 2.0 ml of serum into a tube containing 8.0 ml of 10% TCA. Stopper and mix by inverting.
2. Allow to stand for 10 minutes; then centrifuge for 10 minutes at 3000 rpm. Filter the supernatant.

   Reagent blank
   Pipet 5.0 ml of 10% TCA.

   Standards
   Pipet 1.0 ml of bromide working standard.
   Add 4.0 ml of 10% TCA.

   Unknown
   Pipet 5.0 ml of filtrate.
4. Add 1.0 ml of gold chloride solution to all tubes and mix.
5. Allow to stand for 5 minutes and read absorbance (A) against the reagent blank at 440 nm.

Calculation

$$\frac{A \text{ unknown}}{A \text{ standard}} \times 20 = \text{mg bromide/dl}$$

Normal range
0 to 1.5 mg/dl

Chlordiazepoxide (Librium)

A colorimetric method for the determination of chlordiazepoxide in blood or urine.[12]

Principle

Chlordiazepoxide is extracted from biologic specimens with ether. The compound is extracted from the ether with hydrochloric acid. The acid extract is hydrolyzed, converting the chlordiazepoxide to benzophenone, which after being diazotized can be reacted with N-(1-naphthyl)-ethylene-diamine dihydrochloride to form a pink color complex.

Specimen

5.0 ml of whole blood, serum, plasma, or urine

Reagents

Potassium carbonate, anhydrous
Ethyl ether, AR grade
Hydrochloric acid, 6 N (D-233)
Sodium nitrite, 0.1%
   Dissolve 100 mg of sodium nitrite in 100 ml of water. Store in refrigerator.
Ammonium sulfamate, 0.5% (D-234)
Color reagent
   Dissolve 100 mg of N-(1-naphthyl)-ethylene-diamine dihydrochloride (Matheson Coleman Bell No. NX230) in 100 ml of water. Store in refrigerator.
Stock standard, 100 mg/dl
   In a 100-ml volumetric flask add 100 mg of chlordiazepoxide; dissolve in water and dilute to volume. Stable for 6 months refrigerated.
Working standard, 1 mg/dl
   In a 10-ml volumetric flask add 100 µl of stock standard and dilute to volume with water. Make fresh.

Procedure

1. Label three (40-ml capacity) screw top tubes *blank, unknown,* and *standard.* To blank tube add 5.0 ml of water, unknown 5.0 ml of specimen, and standard 5.0 ml of working standard.
2. To each tube add 0.5 g of potassium carbonate and 20 ml of ether. Shake for 30 minutes.
3. Centrifuge for 5 minutes and remove aqueous (bottom phase) with a transfer pipet.
4. Add 2.0 ml of water to each tube and shake for 2 minutes. Centrifuge and remove water as in step 3 and repeat step 4.
5. Filter ether through Whatman No. 2 filter paper.
6. Into clean, labeled screw top tubes (40-ml capacity) place 10 ml of ether from each extraction. To each tube add 3.0 ml of 6 N HCl. Shake for 10 minutes and centrifuge.
7. With a 2.0-ml volumetric pipet and bulb withdraw 2.0 ml of HCl (lower) layer and place into a properly labeled 10-ml capacity Pyrex test tube.
8. Place tubes in a heating block or oil bath set between 110° and 125° C for 30 minutes. Remove and cool.
9. After tubes are cool add 3.0 ml of water to each tube. Invert.
10. To each tube add 0.5 ml of sodium nitrite. Stopper, invert, and let stand 3 minutes.

11. To each tube add 0.5 ml of ammonium sulfamate. Stopper, invert, and allow to stand 10 minutes.
12. To each tube add 0.5 ml of the color reagent, stopper, and invert. After 10 minutes measure absorbance at 550 nm, using the blank to set the spectrophotometer to zero absorbance.

Calculations

$$\frac{\text{Absorbance unknown}}{\text{Absorbance standard}} \times 1 \text{ mg/dl} = \text{Concentration of chlordiazepoxide}$$

Comments

Daily dosage of 30 mg will produce blood levels of 0.13 to 0.2 mg/dl.

Salicylates, a quantitative colorimetric method for the determination of the salicylate radical[17]

Principle

A reagent containing mercuric chloride and ferric nitrate in a solution of hydrochloric acid simultaneously precipitates proteins of the specimen and reacts with salicylates to form a color ranging from gold to purple. This can be quantitated by measuring absorbance or transmission at 540 nm and comparing to standards.

Specimen

Plasma, serum, whole blood, cerebrospinal fluid, urine*

Reagents

Color reagent (ferric mercuric reagent)	(D-235)
Salicylate stock standard, 1 mg/ml	(D-236)
Salicylate standard solution, 0.20 mg/ml	(D-237)

Procedure

1. Label 10-ml capacity test tubes *blank, standard,* and *test*. To blank add 1.0 ml of water, to standard add 1.0 ml of working standard solution, and to test add 1.0 ml of specimen.
2. To all tubes add 5.0 ml of color reagent.†
3. Centrifuge all tubes for 5 minutes and transfer the clear supernatant to 75 × 100 mm cuvets.
4. Read absorbance (A) at 540 nm.

Calculations

$$\frac{\text{A unknown}}{\text{A standard}} \times 0.2 \times 100 = \text{mg/dl of salicylate}$$

A standard curve may be prepared by making proper dilutions of the stock standard. The standard should range from 0.02 to 0.4 mg/ml.

Comments

For adults toxicity may appear at 30 mg/dl of serum.

If this test is to be performed on gastric content the specimen must first be hy-

---

*Urine should be boiled for a few seconds to expel any ketones that may be present.
†Urine specimens frequently exhibit a turbidity; therefore, a special urine blank should be prepared as follows. To 1.0 ml of previously boiled urine add 0.5 ml of phosphoric acid, 85%, and 4.5 ml of color reagent. Before measuring absorption of urine specimen set spectrophotometer or colorimeter to zero with this blank.

drolyzed. Add 0.5 ml of hydrochloric acid (concentrated) to 2.0 ml of specimen; place in boiling water bath for 10 minutes, centrifuge, and then continue with steps 1 to 4.

Ultraviolet spectrophotometry

When radiation falls on a transparent body or solution three things happen to the energy: some will be absorbed, some will be reflected, and some will be transmitted. At some wavelengths transmission is nearly complete; at some, absorption is nearly complete and transmission almost zero; and at other wavelenghs transmission and absorption are almost equal. Therefore, a graph of the percentage of energy absorbed (absorbance, A) or percentage transmitted (transmittance, T) plotted against wavelength gives the absorption spectrum of a material which, under proper conditions, may be considered a physical constant.

The absorption of energy by molecules in solution shows a characteristic absorption spectrum with an area of greater or lesser absorption at different wavelengths, depending on the solvent and the solute. The internal energy of a molecule is considered as the sum of the energy derived from the vibrational motions of the constituent atoms and the rotational motion of the molecule as a whole. The absorption spectra in the ultraviolet region are generally associated with displacement of outer electrons within the molecule, giving rise to a new energy state.

An ultraviolet spectrophotometer consists of a source of radiant energy covering the wavelength range from 200 to 350 nm, a monochromator for isolating a very narrow band of energy, a container for the material being examined (usually a quartz cuvet), a detector, and a device for measuring the unabsorbed energy. The method of operation is general for all instruments. Desired wavelength is set, and a cuvet holding the solvent to be used in the analysis is placed in the path of the monochromatic beam and the galvanometer adjusted to 100% transmission (or 0.00 absorbance). The cuvet containing the solvent is replaced by one containing the sample and the reading of the galvanometer is recorded. This procedure is repeated at each wavelength over the region desired.

Many newer instruments are double beam, that is, the blank (solvent) and the unknown (solvent plus solute) are in the beam path at the same time and the absorbance of solvent is automatically being subtracted from the absorbance of the solute.

Whether the absorption or transmission of radiant energy by a solution can be used as a means of quantitation depends on several laws of absorption.

1. *Bouguer-Lambert law:* The proportion of light absorbed by a transparent medium is independent of the intensity of the incident light. Each successive layer of the medium absorbs an equal fraction of light passing through it.

$$\log (I_0/I) = ab$$

where I_0 = intensity of incident light
I = intensity of transmitted light
b = thickness of layer (cm)
a = constant depending only on medium examined and the wavelength under consideration, referred to as absorptivity

2. *Beer's law:* The light absorption is proportional to the number of molecules of absorbing substance through which the light passes. If the absorbing substance is dissolved in a transparent medium, the absorption of the solution will be proportional to its molecular concentration.

3. *Combined law (Beer law):*

$$\log(I_0/I) = \varepsilon cb$$

where ε = absorptivity, molar
c = concentration in gram-moles/liter

Substituting A for the expression $\log(I_0/I)$, then: $\varepsilon = A/cb$.

If a solution containing 10 μg/ml of phenobarbital (molecular weight 232) in a 1-cm cell at pH 10 has an absorbance of 0.120 at 240 nm, molar absorptivity is then:

$$\varepsilon = \frac{0.120}{0.01 \text{ g/L} \times 1.0} \times 232 = 2784$$

The validity of these laws should be tested for each system being quantitatively investigated.

The transmittance or the absorbance may be plotted against wavelength, but the absorbance is usually used because the absorption maxima are magnified, thus facilitating calculations from one light path distance to another and affording direct measurement of concentration at a specific maximum because: $A = Ecb$.

Intensity of absorption and location of absorption maxima change with a change in pH. This property provides one of the most direct methods of qualitative determinations. Change in spectrum with change in pH may be due to any one or a combination of the following factors:

1. Formation or suppression of a negatively charged ion

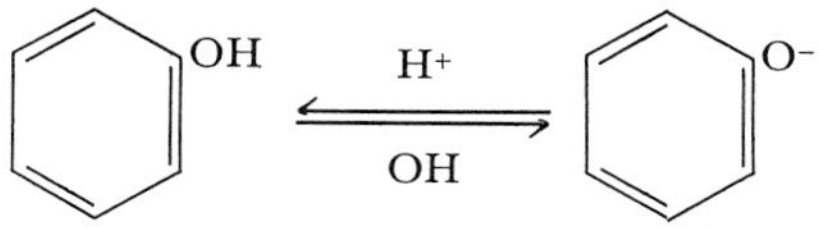

[Reaction diagram: phenol (C₆H₅OH) ⇌ phenolate (C₆H₅O⁻), with H⁺ / OH⁻]

2. Formation or suppression of a positively charged ion

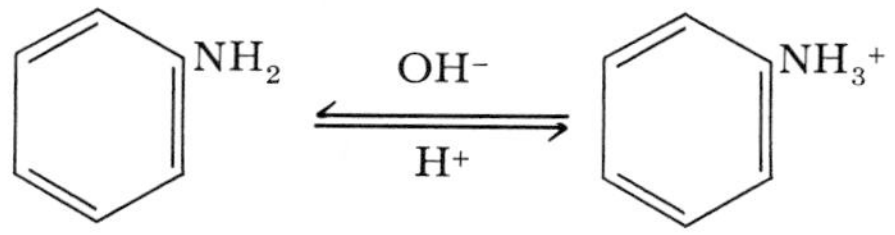

[Reaction diagram: aniline (C₆H₅NH₂) ⇌ anilinium (C₆H₅NH₃⁺), with OH⁻ / H⁺]

3. Hydrolysis

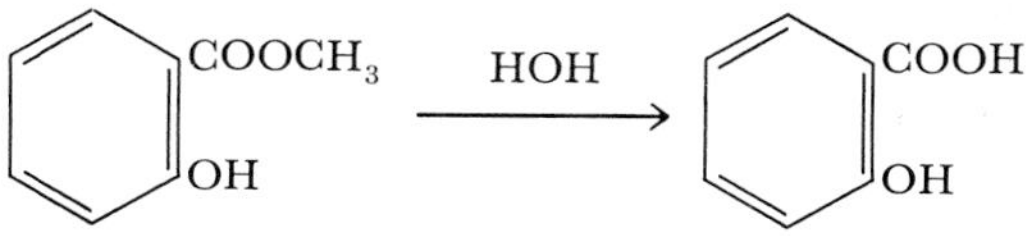

[Reaction diagram: methyl salicylate (C₆H₄(COOCH₃)(OH)) + HOH → salicylic acid (C₆H₄(COOH)(OH))]

4. Change in structure

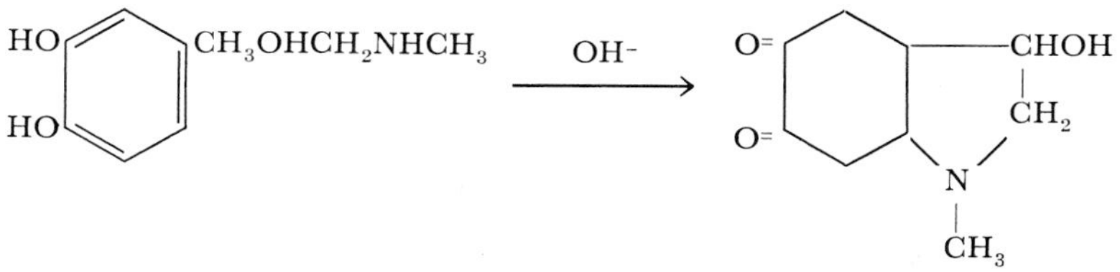

[Reaction diagram: dihydroxyphenyl-CH₃OHCH₂NHCH₃ + OH⁻ → cyclic diketone structure with CHOH, CH₂, N-CH₃]

One of the applications of this property is demonstrated in the procedure for the determination of barbiturates.

Simply stated, organic compounds which absorb ultraviolet light will, at a specific pH, always have a particular maximum absorption area, and the degree of absorbance for that compound at a specific concentration is constant.

In ultraviolet spectrophotometry the term $A_{cm}^{1\%}$ is often encountered. When it is understood it can provide very valuable information. This notation indicates that the absorbance of a 1 g/dl solution of a specific compound in a particular solvent in a 1-cm cell is constant for that compound.

Quinine in a concentration of 1 g/dl of 0.5 N sulfuric acid has an absorbance of 853 at 250 nm, but amphetamine sulfate at the same concentration in the same diluent has at its maximum (257 nm) an absorbance of only 26. This conveys that 0.6 mg/dl of quinine sulfate in 0.5 N sulfuric acid will produce an absorbance of 0.5 at its maximum absorption peak, but at the same concentration and under the same conditions the optimum absorption of amphetamine will be negligible (less than 0.02 absorbance). From the information the analyst can determine the limits of concentration for a given compound or the efficacy of undertaking an ultraviolet analysis.

While these are constants, their numerical value depends on the particular instrument and its wavelength calibration, so that concentration cannot be calculated from a table of these constants any more accurately than a standard curve made from data obtained on one instrument can be used to determine accurate concentrations on a different spectrophotometer. The problem is somewhat further compounded in that the material must first be extracted from a biologic specimen, and 100% extraction cannot be assumed. Therefore, it is necessary to process a standard and/or a control along with the unknown.

ULTRAVIOLET SPECTROPHOTOMETRIC DETERMINATION OF BARBITURATE (GOLDBAUM, MODIFIED)[8]

Principle

Barbiturates are extracted with chloroform from blood or other biologic material (urine, gastric content, tissue homogenates) at pH 6.5 to 7.0. After separation the chloroform phase is extracted with 0.5 N sodium hydroxide, and this extract is scanned from 220 to 320 nm.

If a barbiturate is present the absorption spectrum will show a characteristic maximum at 255 nm, with a minimum at approximately 235 nm.

The sodium hydroxide solution is buffered to pH 10 to 10.5 and rescanned over the same range. The change in pH converts the barbiturate to a different resonance form, and the maximum absorbance is shifted to 240 nm with no minimum. The greatest difference in the spectrum occurs at 260 nm. (See Fig. 15-4.) The difference at this wavelength (Δ absorbance at 260 nm) follows Beer's law for concentrations ranging from 0.2 to 3.0 mg/dl. When compared to standard solutions they may be used to quantitate the level of barbiturate.

Special apparatus

Ultraviolet spectrophotometer, preferably a double beam instrument attached to a recorder

Specimen

5 ml of whole blood, plasma, serum, gastric lavage, or urine.

Reagents

Chloroform, spectroanalyzed grade

Each new bottle (or new lot) should be tested for absorbance before use. Shake 75 ml of the chloroform in question with 5.0 ml of 0.5 N sodium

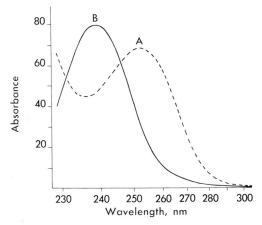

Fig. 15-4. Absorption spectrum of barbiturate.

hydroxide in a separatory funnel for 3 minutes. Allow layers to separate. Collect the sodium hydroxide layer (upper layer) in a tube and centrifuge to remove drops of the solvent. With a transfer pipet place 3.0 ml in a 1-cm cuvet. Use 3.0 ml of 0.5 N sodium hydroxide that has not been extracted with chloroform as the blank, scan from 230 to 310 nm. If any significant absorption occurs between 240 and 300 nm the lot must be rejected or all the chloroform washed before use.

Washing procedure: In a 1-L separatory funnel shake 500 ml of chloroform with 50 ml of 0.5 N NaOH for 5 minutes. Filter chloroform through Whatman No. 2 filter paper. Discard aqueous phase. Wash solvent as above with 50 ml of 0.5 N sulfuric acid (14 ml of concentrated H_2SO_4 diluted to 1 L with water). Filter solvent and wash again with 50 ml of water. Filter into a glass bottle and reserve for all ultraviolet analyses.

Sodium hydroxide, 0.5 N (D-29)
Ammonium chloride, saturated

Add 16 g of ammonium chloride to 100 ml of water.

Stock standard, 100 mg/dl (1 mg/ml)

In a 100-ml volumetric flask dissolve 100 mg of barbital [diethyl barbituric acid, $HNC:ONHC:OC(C_2H_5)_2C:O$] in 50 ml of water and make to volume. Refrigerate. Stable for 3 months.

Working standard, 1 mg/dl

Pipet 100 µl of stock standard into a 10-ml volumetric flask and dilute to volume.

Activated charcoal (Norit-A)

Procedure

1. Label three separatory funnels *blank, standard,* and *unknown.* Into blank funnel pipet 5.0 ml of water, into standard 5.0 ml of working standard, and into unknown 5.0 ml of specimen. If specimen is other than blood, plasma, or serum, adjust pH between 6.5 and 7.0.
2. To each funnel add 75 ml of chloroform. Shake each funnel for 3 minutes. Allow layers to separate and filter chloroform (lower layer) through Whatman No. 2 paper into labeled flasks. Discard aqueous phase.
3. To each flask add a small pinch of activated charcoal (the amount that will adhere to the tip of an applicator stick). The charcoal acts as a decolorizing

470 Chemistry for the clinical laboratory

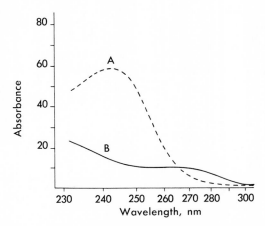

Fig. 15-5. Absorption spectrum of N-substituted barbiturate.

agent. Swirl flasks and filter contents of each through Whatman No. 2 filter paper into graduated cylinders. Collect 60 ml of each.

4. Into clean, labeled separatory funnels add the chloroform collected in step 3. To each funnel add 5.0 ml of 0.5 N sodium hydroxide and shake each funnel for 3 minutes.

5. After layers have separated remove and discard the solvent (lower layer). Drain the aqueous layer into labeled tubes and centrifuge for 5 minutes. Transfer most of the aqueous layer to clean labeled tubes, making sure not to transfer any of the chloroform.

6. Into a 1-cm cuvet add 3.0 ml of the blank solution and place into the reference holder of the spectrophotometer. Into a second cuvet add 3.0 ml of the standard extract and place in spectrophotometer. Scan from 220 to 320 nm. This is scan A.

7. Leaving blank in place, remove cuvet containing the standard and *save*. To a clean cuvet add 3.0 ml of the test extract and place in spectrophotometer. Adjust recorder for a new chart and scan the same as step 6. If the test solution shows a maximum peak at 255 or 243 nm (see Figs. 15-4 and 15-5) remove cuvet and save. If test solution exhibits an absorption maximum at 255 nm, proceed to step 8. If maximum absorption occurs at 243 nm, refer to *N*-substituted barbiturates, p. 471.

8. To blank and standard add 0.5 ml of saturated NH_4Cl and mix thoroughly with a transfer pipet and bulb. Return chart paper to the original recording of the standard and rescan as in step 6. This is scan B.

9. With blank still in place, add 0.5 ml NH_4Cl to cuvet containing test solution. Advance chart paper to original scan of test solution and continue as in step 8.

Calculations

For both standard and unknown read the absorbance of A at 260 nm; at the same wavelength read the absorbance of B. Multiply the reading of B for both standard and unknown by 3.5/3 (this compensates for the dilution with NH_4Cl). This is absorbance B, corrected.

Subtract B corrected from A (Δ absorbance).
Then

$$\frac{\Delta \text{ absorbance unknown}}{\Delta \text{ absorbance standard}} \times 1 \text{ mg/dl} = \text{mg/dl barbital in unknown}$$

Example:

Standard scan B = 0.04 absorbance at 260 nm
Standard scan A = 0.22 absorbance at 260 nm
Unknown scan B = 0.12 absorbance at 260 nm
Unknown scan A = 0.54 absorbance at 260 nm

$$0.22 - \left(0.04 \times \frac{3.5}{3}\right) = 0.173, \Delta \text{ absorbance of standard}$$

$$0.54 - \left(0.12 \times \frac{3.5}{3}\right) = 0.400, \Delta \text{ absorbance of unknown}$$

$$\frac{.400}{.173} \times 1 \text{ mg/dl} = 2.3 \text{ mg/dl barbital}$$

N-Substituted barbiturates

The sodium hydroxide extract of the unknown may be acidified and used for thin-layer chromatographic identification (see p. 481). As stated on p. 463 identification is recommended.

If a scanning spectrophotometer is not available, the absorbance of the solutions can be read against the blank at wavelengths 228, 232, 235, 240, 247, 249, 252, 260, and 270 nm. The blank and solutions are buffered and again read at the same wavelengths. The absorbances are plotted on linear graph paper versus wavelength. Calculations are performed as previously stated.

Reagents

Stock standard, 100 mg/dl
    Dissolve 100 mg of mephobarbital in ethyl alcohol and bring to volume in a 100-ml volumetric flask.
Working standard, 1 mg/dl
    Add 100 µl to a 10-ml volumetric flask and dilute to volume with water.

Procedure

Follow steps 1 to 7, substituting step 10 for steps 8 and 9.
10. Add 1 drop of concentrated H_2SO_4 to both blank and test solution, mix thoroughly, and rescan as in step 8. This is scan B.

Calculations

Subtract absorbance of B (acidified solution) from absorbance of A at 243 nm for both standard and test solutions. Because the dilution of B is so small, it is not necessary to correct.

Absorbance A − Absorbance B at 243 nm = Δ absorbance

$$\frac{\Delta \text{ absorbance unknown}}{\Delta \text{ absorbance standard}} \times 1 \text{ mg/dl} = \text{mg/dl of N-substituted barbiturate}$$

Diazepam (Valium), a quantitative ultraviolet spectrophotometric procedure[4]
Principle

Diazepam is extracted from blood at a neutral pH with ether and is reextracted from the ether phase with hydrochloric acid. Diazepam in acid media exhibits a maximum absorbance at 240 nm, with a smaller absorbance peak at 285 nm and a minimum absorbance at 265 nm. By comparing the absorption peak at 240 nm to a known standard, the concentration of diazepam can be determined (Fig. 15-6).

472 *Chemistry for the clinical laboratory*

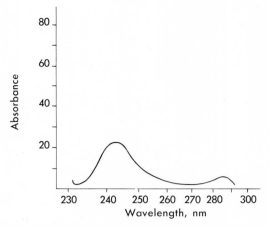

Fig. 15-6. Absorption spectrum of diazepam, 0.5 mg/100 ml.

Specimen

  5 ml of oxalated blood (very little or no unchanged diazepam is excreted in the urine)

Reagents

  Hydrochloric acid, 1 N (D-120)
  Hydrochloric acid, 2 N (D-238)
  Sodium hydroxide, 0.1 N (D-161)
  Phosphate buffer, pH 7.0 (D-239)
  Ether, AR grade
  Stock standard
    In a 100-ml volumetric flask dissolve 100 mg of diazepam in a minimal amount of 1 N HCl. After completely dissolved dilute to volume with water.
  Working standard, 0.5 ml/dl
    Add 50 µl of stock standard to a 10-ml volumetric flask and dilute with water to volume.

Procedure

1. Label 25 × 150 mm screw top tubes as follows: *blank, standard,* and *test.* To blank tube add 4.0 ml of water. Add 4.0 ml of working standard and 4.0 ml of blood to respective tubes.
2. To all tubes add 8.0 ml of phosphate buffer, 8.0 ml of water, and 20 ml of ether. Shake for 10 minutes and centrifuge for 5 minutes.
3. Transfer ether layer to another set of 25 × 150 mm screw top tubes and save.
4. To aqueous layer in each add 10 ml of ether. Shake and centrifuge as in step 2. Combine ether layers with that obtained in step 3. Filter each ether extract into a graduated cylinder through Whatman No. 2 paper. Collect 25 ml of each extract.
5. Transfer filtered ether extracts to a third set of 25 × 150 mm tubes; to each add 4.0 ml of 2 N HCl. Shake for 15 minutes and centrifuge for 5 minutes.
6. Transfer hydrochloric acid layer (bottom) to 10-ml capacity test tubes. Place test tubes in a 60° C water bath for 10 minutes to expel the dissolved ether.
7. After cooling, transfer the samples to 1-cm quartz cuvets, using the blank as the reference.

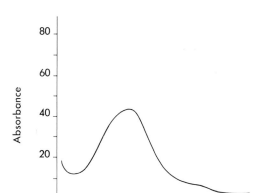

Fig. 15-7. Absorption spectrum of diphenylhydantoin product.

8. Record the absorption spectrum of the standard and of the test solution.

Calculations

$$\frac{A_{240} - A_{265} \text{ test}}{A_{240} - A_{265} \text{ standard}} \times 0.5 \text{ mg/dl} = \text{mg/dl diazepam}$$

A_{240} = absorbance at 240 nm (maximum)
A_{265} = absorbance at 265 nm (minimum)

Comments

Oral doses of 50 to 100 mg of diazepam will produce levels of 0.05 to 0.25 mg/dl.

Diphenylhydantoin (Dilantin), a quantitative ultraviolet spectrophotometric procedure[18]

Principle

Diphenylhydantoin has a very low ultraviolet absorptivity. However, after refluxing and oxidation it is converted to a highly absorptive form (benzophenone). The maximum absorption of the oxidative product is at 247 nm (Fig. 15-7).

Special apparatus

Erlenmeyer flasks, 125-ml capacity with 24/40 ground glass joints
No. 4 rubber stopper fitted with a length of glass tubing which will extend to within 5 mm of bottom of the flask and approximately 5 cm above top of stopper
    The end of the tubing, which protrudes from top of stopper, is connected by latex tubing to a source of clean air.
Reflux condenser with 24/40 joint
Magnetic stirrer/hot plate combination

Specimen

5 ml of serum or plasma

Reagents

Chloroform, spectroanalyzed grade
Sodium hydroxide, 1 N

(D-225)

Sodium hydroxide, 9 N
: Dissolve 180 g of NaOH in 300 ml of water. After cooling to room temperature dilute to 500 ml.

Phosphate buffer, 0.3 M
: In a 1-L volumetric flask dissolve 41.4 g of monobasic potassium phosphate (KH_2PO_4) in water. Dilute to volume.

Potassium permanganate, 5% (D-240)

n-Heptane, AR grade

Stock standard, 100 mg/dl
: In a 100-ml volumetric flask dissolve 109 mg of diphenylhydantoin sodium in a small quantity of water. Dilute to volume. Store in refrigerator.

Working standard, 1 mg/dl
: In a 10-ml volumetric flask dilute 100 µl of stock standard to volume with water. Make fresh.

Procedure

1. Label three 25 × 150 mm screw topped tubes as follows: *blank, standard,* and *test*. To respective tubes add 5.0 ml of water for blank, 5.0 ml of working standard (1 mg/dl), and 5.0 ml of specimen.
2. To each tube add 5.0 ml of phosphate buffer and 30 ml of chloroform. Shake for 3 minutes; centrifuge for 5 minutes.
3. Remove upper aqueous layer with a transfer pipet and discard. Filter chloroform through Whatman No. 2 filter paper and collect 25 ml from each tube in graduated cylinders.
4. Transfer the filtered chloroform to labeled tubes as in step 1. Add 6.0 ml of 1 N NaOH to each tube. Shake for 3 minutes and centrifuge for 5 minutes.
5. Transfer upper sodium hydroxide layer to labeled test tubes (10-ml capacity). Discard chloroform.
6. Label three Erlenmeyer flasks with ground glass joint *blank, standard,* and *test*. Pipet 4.0 ml of the sodium hydroxide to respective flask. Place flasks in a 60° C water bath; insert rubber stopper as described under special apparatus. Aerate each flask for 15 minutes.
7. After aeration to each flask, add 16 ml of 9 N sodium hydroxide, 6.0 ml of potassium permanganate solution, and 5.0 ml of n-heptane.
8. Place flask on magnetic stirrer/hot plate, attach reflux condenser, and with constant stirring apply enough heat to produce gentle refluxing. Reflux each flask for 30 minutes. Remove flask from heat and allow to cool to room temperature before removing condenser.
9. After contents of all flasks have been refluxed and cooled, transfer the heptane (upper layer) to 1-cm quartz cuvets.
10. Using the heptane from the blank flask as the reference, record the absorption spectrum of the standard and of the test solution from 350 to 200 nm.

    If a recording spectrophotometer is not available, read absorbance of standard and unknown at 247 nm, setting instrument to zero absorbance with blank.

Calculation

$$\frac{A \text{ test}}{A \text{ unknown}} (247 \text{ nm}) \times 1 \text{ mg/dl} = \text{mg/dl diphenylhydantoin}$$

Comments

The expected absorbance of 1 mg/dl standard at 247 nm should be approximately 0.45.

Morphine-heroin, a quantitative ultraviolet spectrophotometric procedure[9, 14]

The chemical structure of morphine and heroin are quite similar. Both compounds are opiates derived from the poppy. In the body heroin is converted to morphine.

Principle

Urine is hydrolyzed to convert morphine to a form that is extractable with organic solvents at pH 8.5. The absorptivity of morphine is low (a 10 mg/dl solution has a maximum absorption at 285 nm of approximately 0.55); therefore, a large sample must be used and consequently a number of "clean-up" steps undertaken. Because of the number of extractions there is an appreciable loss of morphine; therefore, it is imperative that a standard be carried throughout the entire extraction procedure (hydrolysis of the standard is not necessary). The standard and test are finally extracted into a weak hydrochloric acid solution, the absorption spectrum of each is recorded, and the absorption at 285 nm of the test solution is compared to the standard.

Specimen
50 ml of urine (preferably a 50-ml aliquot of a 24-hour specimen)

Reagents
Hydrochloric acid, concentrated
Hydrochloric acid, 0.1 N (D-32)
Sodium hydroxide, saturated solution
  Dissolve 100 g NaOH in 100 ml of water.
Sodium carbonate
Solvent, 3:1 chloroform/2-propanol
  Mix 600 ml of chloroform with 200 ml of 2-propanol.
Phosphate buffer, M/15, pH 8.5
  Dissolve 9.47 g of sodium dibasic phosphate (Na_2HPO_4) in water in a 1-L volumetric flask and dilute to volume.
Sodium hydroxide, 0.5 N (D-29)
Phosphoric acid, 85%
Ammoniacal silver nitrate reagent
  To 50 ml of 10% silver nitrate (dissolve 5 g of $AgNO_3$ in water and dilute to 50 ml) add ammonium hydroxide, 28% to 30%, dropwise. On the first addition a precipitate will form. Continue to add NH_4OH until the precipitate disappears. While checking the pH of the solution add phosphoric acid, 85%, dropwise until the pH is 8.5. Carefully add NH_4OH (28% to 30%) until the precipitate just dissolves. Transfer to a brown bottle and store refrigerated.
Stock standard, 100 mg/dl
  In a 100-ml volumetric flask dissolve 164 mg of morphine sulfate [$(C_{17}H_{19}NO_3)_2 \cdot H_2SO_4 \cdot 5H_2O$]. Dilute to volume.
Working standard, 1 mg/dl
  In a 100-ml volumetric flask dilute 1.0 ml of stock standard to 100 ml with M/15 phosphate buffer. Make fresh.

Procedure
1. *Hydrolysis:* Add 5.0 ml of concentrated HCl to a 50-ml urine specimen in a 150-ml beaker. Autoclave at 15 lb pressure for 30 minutes. If an autoclave is not available a pressure cooker can be used or, alternatively, the acidified urine can sit overnight at room temperature.
2. After the hydrolyzed specimen is cooled add 2.0 ml of saturated sodium

hydroxide. Very carefully add sodium carbonate, since the first addition will cause extreme foaming. Using a pH meter, check pH and continue to add sodium carbonate, stirring after each addition until pH is 8.5.

3. Transfer specimen to a 500-ml beaker and add 250 ml of solvent. Place on a magnetic stirrer for 15 minutes. In lieu of a stirrer, specimen may be shaken for 5 minutes in a separatory funnel. In another beaker place 50 ml of working standard (1 mg/dl). From this step on, carry the standard through the same steps as the specimen.
4. Transfer the mixture to a separatory funnel and allow to separate. Drain the solvent layer into a 500-ml beaker (discard the aqueous layer). Add 50 ml of M/15 phosphate buffer and place on stirrer for 5 minutes.
5. Separate phases as in step 4; discard the aqueous phase and drain solvent into a clean 500-ml beaker. Add 25 ml of 0.5 N NaOH; place on stirrer for 15 minutes.
6. Separate phases as in step 4. Discard solvent and *drain aqueous phase* into a 150-ml beaker. Add 1.0 ml of 85% phosphoric acid. Using a pH meter, adjust pH of solution to 8.5 with sodium carbonate as in step 2. When solution is correct pH add 2.0 ml of ammoniacal silver nitrate solution.
7. Transfer to a 250-ml beaker, add 125 ml of solvent, and place on stirrer for 15 minutes. Separate as in step 4, discard aqueous phase, and drain solvent into 250-ml beaker.
8. To solvent add 25 ml of M/15 phosphate buffer and place on stirrer for 5 minutes. Separate phases as in step 4; discard aqueous phase.
9. Filter solvent through Whatman No. 2 paper into a graduated cylinder. Record amount of solvent recovered.
10. Place solvent into a 250-ml separatory funnel, add 5.0 ml of 0.1 N HCl, and shake for 5 minutes.
11. After layers have separated, discard solvent, drain HCl into a conical tube, and centrifuge for 5 minutes. Transfer solvent-free acid layer to a 1-cm quartz cuvet. Record absorption spectrum from 200 to 350 nm. Save test solution for further confirmation.

Calculations

$$\frac{A_{285} \text{ unknown}}{A_{285} \text{ standard}} \times 1 \text{ mg/dl} \times \frac{\text{Solvent standard*}}{\text{Solvent unknown*}} = \text{mg/dl morphine}$$

Example:
$A_{285} = 0.65$ (unknown)
$A_{285} = 0.45$ (standard)
Quantity of solvent recovered in step 9 for unknown = 100
Quantity of solvent recovered in step 9 for standard = 110

$$\frac{0.65}{0.45} \times 1 \text{ mg/dl} \times \frac{110}{100} = 1.4 \text{ mg/dl}$$

Comments

Since many of the alkaloid drugs absorb in the range of 285 nm, the final acidic extract of the test solution should be made alkaline (pH 8.5) with sodium carbonate, extracted, and chromatographed to confirm the presence of morphine (see p. 478).

Because of the time involved in performing this analysis, it is advisable under

---

*Quantity of solvent recovered in step 9.

most circumstances to first screen the specimen for morphine. Refer to following section on chromatography.

Chromatography

The method of chromatography, whether it is column, paper, thin-layer, gas-liquid or liquid-liquid, is a technique used to separate a mixture into its component parts and subsequently to identify the separated entities.

Frequently a person will have ingested several drugs, all of which are extractable from the biologic specimen at the same pH and with the same solvent. If such an extract is subjected to ultraviolet spectrophotometric analysis, the resulting spectrum will be a pattern of absorption of all the components and consequently not specific for any one compound. In such instances it is necessary to separate the mixture into its component parts.

As discussed in the section dealing with barbiturates, the variously substituted derivatives have a wide range of toxicity, which often necessitates their identification as well as quantitation.

Thin-layer chromatography, in particular, offers a method whereby a specimen can be screened for a large number of drugs in a relatively short period of time. When using this technique as a screening method, negative results can rule out certain agents, but positive results should be confirmed by another method of analysis.

Gas-liquid chromatography and thin-layer chromatography (TLC) are both widely used in toxicologic analysis. However, since the paraphernalia for TLC is relatively inexpensive and requires a minimum amount of space, it can more conveniently be incorporated into the clinical laboratory; therefore only this technique will be discussed.

The term thin-layer is derived from the preparation of the plate or chromatogram, which consists of a coating of an adsorbent material, approximately 0.2 to 0.25 mm, on a rigid or semirigid support. This coating, usually silica gel, is of itself inert but it has incorporated in it a small quantity of water, approximately 4%, which is referred to as structural water. It is this structural water that acts as the stationary phase.

Known compounds or standards and extracts of biologic material are applied 2 cm from one edge of the plate in small discrete spots. The chromatogram is then placed into an airtight container (developing tank) which holds the developing solvent, referred to as the mobile phase. The depth of the solvent must be less than 2 cm so that it does not cover the sample spots.

As the mobile phase ascends the plate or sheet by capillary action, it moves the components of the sample.

The migration of a component depends upon its partition coefficiency, that is, its affinity for the mobile phase versus the stationary phase. The more soluble the compound is in the mobile phase, the further up the plate it will migrate. Conversely, components with a greater affinity for the stationary phase will travel a shorter distance. Components of a mixture will be separated according to the partition coefficiency as the mobile phase ascends the plate.

The separated components are usually colorless. Therefore, after development of the chromatogram it is necessary to spray them with a reagent that will react with the substances in question to form a color complex.

The chromatogram is evaluated by comparing the color reactions and the Rf values of the tests to the standards.

The Rf value is the distance traveled by the substance/distance traveled by the solvent. Both the color reaction and Rf value of the standard and unknown must coincide. In urine specimens many naturally occurring materials as well as nicotine present in specimens from smokers are extracted and give color reactions with the spray reagents. The mere presence of a colored spot on the plate does not, therefore, necessarily indicate the presence of a drug.

As the water content of the silica gel exceeds 4%, the quality of the separation deteriorates. The plates should be stored in a desiccator, and in a particularly humid climate it may be necessary to "activate" it before use. "Activation" is achieved by heating the plate at 110° C for 60 minutes.

ALKALOID DRUGS (MORPHINE, CODEINE, ETC.) AND AMPHETAMINES, A QUALITATIVE METHOD EMPLOYING THIN-LAYER CHROMATOGRAPHY[2, 5]

Principle

When ethanol is added to a urine specimen which is saturated with potassium carbonate, the alkaloid drugs and amphetamines are concentrated in the alcohol. As a clean-up step the ethanol layer is shaken with ether, and when the mixture is washed with a buffer solution the alcohol and ether separate into two layers. The drugs are preferentially soluble in the ether layer.

The ether is evaporated and the residue dissolved in a small volume of methanol. The extract is chromatographed on silica gel plates and finally subjected to a series of sprays to visualize the components.

A high percentage of morphine is excreted as a water soluble form; therefore, it is necessary to hydrolyze the specimen to increase its recovery.

Standards must be included on every chromatogram. The Rf value plus the color reaction of any components of the specimen must be carefully compared to these known drugs.

Any positive results should be considered presumptive. If there are any forensic implications the results should be substantiated by an analysis employing a different technique (ultraviolet spectrophotometry, gas-liquid chromatography, fluorometry).

Special apparatus

16 × 150 mm screw top tubes
Evaporation tubes and rack
  Available from Brinkman Instruments, Inc.: No. 3500420-2 (tubes); No. 3500450-5 (rack)
Developing tank to accommodate 20 × 20 cm chromatographic plates
  Available from Brinkman Instruments, Inc. No. 25 10250-9, or Quantum Assay Corp., No. CDC-12
Chromatographic plates, 20 × 20 cm
  Precoated and prescored plates are available from Quantum Assay Corp., LQD plates No. 5090
  Precoated plates are available from EM Laboratories, No. 5763. These plates are not prescored; a template to mark plates is necessary; it is available from Brinkman Instruments, Inc., No. 25 09200-7.
Micropipet set
  Used to apply samples to plate (Camag, Inc. No. 27-112).
Sprayer
  Either a glass sprayer or aerosol cans may be used. Chromomist spray cans are available from Gelman, No. 51901.
Ultraviolet lamp, 254 nm (Mineralite or ultraviolet cabinet)
Oven, variable temperature

Specimen

Urine, 10 ml

Reagents

Ethanol
Methanol
Ethyl ether
Potassium carbonate, anhydrous
Buffer, pH 8.5

Dissolve 53.5 g of ammonium chloride in 950 ml of water; add 18 ml of ammonium hydroxide (28% to 30% concentration). Using a pH meter adjust the pH of the solution to 8.5 by the addition of NH_4OH or 6 N HCl.

Chloroform/2-propanol

Mix 300 ml of chloroform with 100 ml of 2-propanol.

Hydrochloric acid, 6 N   (D-233)

Alkali mixture

Thoroughly mix 30 g of anhydrous potassium carbonate with 20 g of anhydrous sodium bicarbonate.

Acetic acid, 20% (v/v)   (D-162)

Solvent systems for chromatogram development:

System 1: To 85 ml of ethyl acetate add 10 ml of methanol and 5.0 ml of ammonium hydroxide (28% to 30% concentration). Prepare fresh.

System 2: Mix 60 ml of ethyl acetate, 42 ml of ethanol, 6.0 ml of 2-butanol, and 1.2 ml of ammonium hydroxide (28% to 30% concentration).

The Rf value of cocaine and methadone are very close in system 1; if the presence of either of these drugs is indicated, an aliquot of the unknown and standard should be developed in this solvent.

Visualizing reagents

Use only in fume hood.

Ninhydrin, 0.2%, in 2-propanol

Dissolve 100 mg of ninhydrin in 50 ml of 2-propanol. Stable for 2 weeks refrigerated in a tightly sealed bottle.

Iodoplatinate reagent

Dissolve 250 mg platinic chloride in 5.0 ml of water; dissolve 4.5 g of potassium iodide in 45 ml of water. Mix the two solutions and dilute with 50 ml of water. Store refrigerated in a brown bottle. Just before use mix equal parts of iodoplatinate and 2 N HCl. A total volume of 15 ml is sufficient quantity to spray one 20 × 20 cm chromatogram.

Ammoniacal silver nitrate

Ammonium hydroxide, 5 N

Dilute 230 ml of 28% to 30% NH_4OH to 500 ml with water.

Silver nitrate, 50%

Dissolve 50 g of $AgNO_3$ in water and dilute to 100 ml. Store in a brown bottle.

Just before use mix equal parts of ammonium hydroxide solution and silver nitrate solution; dropwise add more 5 N ammonium hydroxide until precipitate dissolves.

Standards, 200 mg/dl

Dissolve 10 mg of each of the following in 5.0 ml of methanol: morphine, quinine, cocaine, methadone, dihydromorphinone (Dilaudid), codeine, amphetamine, methamphetamine.

If other basic compounds, such as strychnine, meperidine (Demerol), and the like are of interest, a standard may be prepared of these.

Store each standard in a well-stoppered test tube in the refrigerator and they will remain stable indefinitely.

Control urine

Lyophilized urine specimens containing drugs of abuse are commercially available. It is recommended that a control be included on each chromatogram.

Procedure

1. To labeled screw-topped tubes add 1.0 ml of ethanol, 10 ml of urine sample, and mix. While tilting the tube add solid potassium carbonate until solution is saturated, approximately 12 g. Cap tube and invert. Mix on Vortex for 30 seconds.
2. Allow tubes to stand for 5 minutes. A dark colored layer of approximately 1.0 ml will form on the top. Centrifuge tube for 5 minutes.
3. Transfer the upper layer to clean 16 × 150 mm tubes. A thin solid layer forms at the interface of the two liquids. Wash this with a few drops of ethanol and combine with rest of extract. Be careful not to transfer any of the aqueous phase and discard.
4. To each tube add 5.0 ml of ether; place on Vortex for 30 seconds. Add 1.0 ml of buffer solution (pH 8.5); mix on Vortex for 20 seconds. Centrifuge.
5. Transfer the ether (upper) layer to evaporation tubes. Add 1 drop of 20% acetic acid solution. Place in fume hood to evaporate.
6. *Hydrolyzation of morphine:* To aqueous layer remaining in test tube add 0.5 ml of 6 N hydrochloric acid. Mix and place in a boiling water bath for 1 hour. After cooling tubes to room temperature, add alkali mixture in small increments, mixing between additions until there is no further effervescence. pH should be approximately 8.5. To each tube add 10 ml of chloroform/2-propanol solvent; shake for 2 minutes. Centrifuge. Transfer solvent layer to evaporation tubes, add 1 drop of acetic acid solution, and evaporate to dryness.
7. After extracts have completely evaporated, prepare solvent tank. Place developing solvent as described under system 1 into tank and allow it to equilibrate for 1 hour. Solvents must be prepared fresh.
8. Using the template and starting 2 cm from bottom edge of plate, lightly mark nineteen small dots with a *pencil* as indicated. Each dot is a point of application. Make small pencil marks at each side edge of plate 12 cm from bottom edge. These marks will serve as a guide to indicate when solvent has traveled 10 cm from point of sample applications. On *top* edge of plate using a pencil identify each point of application. (If prescored plates are used, it is only necessary to label the top edge of plate for identification purposes.)
9. With microliter pipet apply 10 μl of each standard to an application point.
10. Dissolve residue in evaporation tubes in 50 μl of methanol and apply a 20-μl aliquot to chromatogram. Apply in 10-μl increments, allowing the spot to dry between additions.
11. Place the plate in the developing tank with the application points being just above the level of the solvent and in such a manner that the solvent contacts the entire edge of the plate simultaneously.
12. Allow the solvent to ascend the plate to the 12-cm mark; remove from tank and allow to dry at room temperature.
13. Place plate into a 90° C oven for 10 minutes and while still warm spray with ninhydrin reagent. Place under ultraviolet lamp for 10 minutes, and

return to 90° C oven for 10 minutes. The amphetamine standard will appear as a pink spot and the methamphetamine pink with a blue tinge. Measure the Rf value of each standard and compare with unknown. Ninhydrin stains amino acids, so there will most probably be components in the urine extracts that will stain pink, but the migration will differ from the amphetamines. Antihistamines also have a strong reaction with ninhydrin.

14. Spray the plate with iodoplatinate reagent and compare the color and Rf values of the standards with any spots that appear in the unknown. After 1 hour reexamine the plate since morphine in low concentration may not appear immediately.
15. If morphine appears to be present, spray the chromatogram with ammoniacal silver nitrate; place in a 110° C oven for 5 minutes. The pink background of the plate will become white and morphine will appear as a sharp dark gray to black spot.
16. In solvent system 1, cocaine and methadone migrate close to the solvent front (the top of the plate). If either appear to be present prepare another plate and develop it in solvent 2. Spray with iodoplatinate. Cocaine will be clearly distinguishable from methadone.

BARBITURATES AND GLUTETHIMIDE, A QUALITATIVE METHOD EMPLOYING THIN-LAYER CHROMATOGRAPHY[5, 15]

Principle

Barbiturates and glutethimide are extracted into chloroform from a neutral or acid medium. The chloroform is evaporated and the residue dissolved in methanol which is spotted on silica gel plates. After development of the chromatogram, barbiturates and glutethimide may be differentiated by their Rf values and color reactions with spray reagents.

Special apparatus
See section on alkaloids, p. 478.

Specimen
3 ml of blood, serum, or plasma
10 ml of urine or extract from ultraviolet spectrophotometric analysis (p. 468)

Reagents
Chloroform
Phosphoric acid, 2 M
Dilute 13.6 ml of 85% phosphoric acid to 100 ml with water.
Methanol
Developing solvent
Mix 90 ml of chloroform with 10 ml of acetone. Prepare fresh.
Color reagents
Potassium permanganate, 100 mg/dl
Dissolve 100 mg of potassium permanganate in water and dilute to 100 ml.
Mercuric sulfate reagent
Suspend 5 g of mercuric oxide in 100 ml of water slowly, and while stirring add 20 ml of concentrated sulfuric acid. After cooling dilute to 250 ml.
Diphenylcarbazone reagent
Dissolve 10 mg of diphenylcarbazone in 100 ml of chloroform. Store refrigerated in a brown bottle.

Standards, 1 mg/ml
Prepare standards by dissolving 10 mg of each barbiturate and glutethimide in 10 ml of chloroform.

Standards of amobarbital, butabarbital, pentobarbital, phenobarbital, secobarbital, and glutethimide should be used routinely. Standards of any of the other barbiturates can be prepared and used in the same manner. If stored refrigerated in tightly stoppered test tubes the standards are stable up to 6 months.

Procedure
1. Place specimen into a 25 × 150 mm screw top tube. If specimen is other than whole blood add 1.0 ml of 2 M phosphoric acid. To all specimens add 30 ml of chloroform, shake for 5 minutes, and centrifuge for 5 minutes.
2. Remove upper aqueous layer with a transfer paper and discard; filter chloroform through Whatman No. 2 paper and collect in evaporation tube. Evaporate to dryness. Evaporation may be facilitated by placing rack in a 60° C water bath under a stream of air or nitrogen.
3. After specimens have evaporated add developing solvent to tank and allow to equilibrate for 1 hour.
4. Prepare chromatographic plate as described on p. 480. Apply 10 μl of each standard. Dissolve residue in evaporation tubes in 50 μl of chloroform and spot 20 μl of each.
5. After samples have completely dried, using the same precautions as mentioned on p. 480, place plate into tank and allow solvent to ascend 12 cm (45 to 60 minutes).
6. Remove plate from tank and allow to dry at room temperature. In a fume hood spray with mercuric sulfate reagent until standards appear as white spots on a gray background.
7. Allow plate to dry but while still slightly moist spray with diphenylcarbazone until standards appear as blue-pink to violet spots. Glutethimide will stain purple. Compare color reactions and Rf values of unknowns to standards.
8. To further distinguish barbiturates with an unsaturated carbon chain such as secobarbital or talbutal, spray the chromatogram with potassium permanganate. The unsaturated barbiturates will appear as yellow spots.

Lead

For lead by atomic absorption refer to Chapter 8.

Plasma lithium by flame photometry[19]

In recent years, lithium salts have been used in the treatment of mental patients of the manic group. Since levels above 2 mEq Li/L of blood may cause dangerous side effects, it is essential to follow such treatment with periodic plasma lithium determinations.

Methods have been described that require deproteinization of plasma. This is a time-consuming process, and there is some doubt as to whether some of the lithium ions are absorbed by the precipitated proteins. Automated methods are available that employ dialysis, but in laboratories where requests for lithium determinations are few and infrequent these are impractical.

The method described here employs a greater dilution of the plasma (1:25 instead of 1:5), which obviates the need for deproteinization. The use of a serum blank and standard minimizes background interference. The emission is linear up to 2 mEq Li/L of plasma; hence only one standard is necessary.

Since many laboratories already have a Coleman flame photometer, the only additional accessory needed is a Coleman lithium filter (21-211), which isolates the 671 nm emission line of lithium. The lower limit of detection with this filter is 0.0014 mEq Li/L.

Specimen

1.0 ml of serum or heparinized plasma

Reagents

Lithium stock standard, 10 mEq Li/L	(D-241)
Lithium working standard, 1 mEq Li/L	(D-242)
Sterox SE, 1% (Coleman reagent 1)	(D-243)
Sterox SE, 0.02% (Coleman reagent 5)	(D-244)

Procedure

1. Prepare the following in 25-ml volumetric flasks.
   Blank: 0.5 ml of reagent 1 and 1.0 ml of lithium-free serum
   Standard: 0.5 ml of reagent 1, 1.0 ml of lithium-free serum, and 1.0 ml of lithium working standard (1 mEq/L)
   Unknown: 0.5 ml of reagent 1 and 1.0 ml of unknown plasma or serum
2. Dilute each flask to volume with distilled water and mix well.
3. Place the lithium filter in position, atomize a portion of the blank solution, and set the galvanometer to zero.
4. Atomize a portion of the standard and set the galvanometer to the midpoint on the *black* scale (5 mEq on the K-Ca scale).
5. Atomize a portion of the diluted unknown plasma and record reading.

Calculation

$$\frac{\text{Reading of unknown}}{\text{Reading of standard}} \times 1 = \frac{\text{Reading of unknown}}{5} = \text{mEq Li/L plasma}$$

Comments

The blank and the standard solutions must be prepared fresh daily. Serum used in the blank and standard must be lithium-free and must have normal electrolyte and protein values. If obtaining this type of serum presents a problem, a commercial control serum such as Versatol, with values in the normal range, may be used.

When eleven replicate determinations were run on each of three specimens, 0.57, 1.05, and 1.60 mEq Li/L, coefficients of variation of 1.4%, 1.3%, and 1.3%, respectively, were obtained.

References

1. Baer, D. M.: A simple colorometric method for detection of barbiturates, suitable for the small clinical laboratory, Am. J. Clin. Pathol. 44:114, 1965.
2. Bastos, M. L., Kananen, G. E., Young, R. M., Monforte, J. R., and Sunshine, I.: Detection of basic organic drugs and their metabolites in urine, Clin. Chem. 16:931, 1970.
3. Bonnichsen, R. K., and Theorell, H.: An enzymatic method for the microdetermination of ethanol, Scand. J. Clin. Lab. Invest. 3:58, 1951.
4. de Silva, J. A. F., Roche Laboratories, Nutley, N. J., 1970.
5. Dole, V. P., Kim, W. K., and Eglitis, I.: Detection of narcotic drugs, tranquilizers, amphetamines, and barbiturates in urine, J.A.M.A. 198:349, 1966.
6. Fisher Scientific Co., Technical data T.D. No. 142.
7. Gettler, A., and Freimuth, H.: Carbon monoxide in blood: a simple and rapid estimation, Am. J. Clin. Pathol. 13:79, 1943.
8. Goldbaum, L. R.: Analytical determination of barbiturates, Anal. Chem. 24:1604, 1952.
9. Gradwohl, R. B. H.: Legal medicine, St. Louis, 1954, The C. V. Mosby Co.
10. Kaye, S.: Handbook of emergency toxicology, ed. 3, Springfield, Ill., 1970, Charles C Thomas, Publisher.

11. Lynch, M. J., Raphael, S. S., Mellor, L. D., Spare, P. D., and Inwood, M. J. H.: Medical laboratory technology and clinical pathology, ed. 2, Philadelphia, 1969, W. B. Saunders Co.
12. Roche Laboratories: Products reference manual, Nutley, N. J., 1970.
13. Schroeder, H. A., and Nason, A. P.: Trace element analysis in clinical chemistry, Clin. Chem. **17**:461, 1971.
14. Stewart, C. P., and Stoltman, A.: Toxicology mechanism and analytical methods, vol. 1, New York, 1960, Academic Press, Inc.
15. Sunshine, I., Rose, E., and LeBeau, J.: Barbiturate detection using thin layer chromatography, Clin. Chem. **9**:312, 1966.
16. Tietz, N. W., and Fiereck, E. A.: The spectrophotometric measurement of carboxyhemoglobin, Ann. Clin. Lab. Sci. **3**(1):36, 1973.
17. Trinder, P.: Rapid determination of salicylate in biological fluids, Biochem. J. **57**:301, 1954.
18. Wallace, J. E.: Microdetermination of diphenylhydantoin in biological specimens by ultraviolet spectrophotometry, Anal. Chem. **40**:978, 1968.
19. White, W., Erickson, M. M., and Stevens, S. C.: Chemistry for medical technologists, ed. 3, St. Louis, 1970, The C. V. Mosby Co.
20. Wuth, O.: Rational bromide treatment, new method for its control, J.A.M.A. **88**:2013, 1927.

16

Chromatography

JOSEPH R. SIMMLER

DEVELOPMENT OF CHROMATOGRAPHY

Chromatography is a separation process by which the components of a mixture are separated by differentially interacting between two physical phases that contact each other in a countercurrent fashion. This general definition applies to paper, thin-layer, column, and gas chromatography. The greatest application has been in the field of organic chemistry, and to an increasing extent for routine analyses in clinical and medical laboratories. The principles and application were first described by a Russian scientist, Michael Tswett, in 1906 and published in a detailed monograph in 1910. Apparently the separation technique lay dormant for many years, since there was very little chromatographic work published between 1910 and 1931. Tswett separated the complex mixtures of many botanic extracts by adsorption chromatography. Since the time of Tswett's original work several different techniques have been introduced: ion exchange, partition, emulsion, gel permeation, and gas chromatography.

Paper chromatography, which is essentially partition chromatography, was first introduced in 1944 by Consden, Gorden, and Martin. The first description of a kind of thin-layer chromatography was by Ismailov and Shraiber in 1938. The initial application of paper chromatography was much greater than that of thin-layer chromatography. While many lipophilic (nonpolar) substances were separable on adsorption columns, the many hydrophilic (polar) substances such as the nucleic acids, polysaccharides, and proteins were difficult to separate until Martin and Synge introduced partition chromatography. The development from partition chromatography on a column to the introduction of partition chromatography on paper strips took only 3 years. However, the "laying-out" of a Tswett adsorption column to an open-face column or thin-layer took many years. The development and application of thin-layer chromatography quite obviously trailed that of paper chromatography. The latter method was rapidly applied and in a short time the number of publications reporting use of this separation technique was in the hundreds. However, the application of thin-layer chromatography by Williams and Kirchner, and particularly by Stahl, soon popularized the use of thin-layer chromatograms. Today the earlier situation is reversed. This is not surprising, since thin-layer chromatography has several advantages over paper chromatography. The thin-layer method is faster and the separated components are more compact, thus giving a higher sensitivity or degree of detection. The resolution or degree of separation is better, and corrosive detecting reagents such as sulfuric acid may be used. The choice of adsorbents for thin-layer use is extensive. Also, prepared plates of uniform adsorbent thickness and of different adsorbents can now be obtained commerically at reasonable prices. It is for these reasons that thin-layer chromatography will be stressed over paper chromatography.

Paper strips that act as physical supports for silicic acid, alumina, ion exchange resins, or cellulose, having been treated chemically, have resulting properties different from those of the original paper and cannot be classified under paper chromatography. These paper-supported adsorbents have the same advantages as thin-layer chromatograms on glass plates.

Column chromatography has application where there is a need for more of the separated components than can be obtained by preparative thin-layer chromatography. About 1 g of a mixture can be chromatographed on a 20 × 20 cm thin-layer preparative plate. The automation of column chromatography and the use of fraction collectors, solvent feed, and sample zone detectors have made column chromatography very convenient but it is still a time-consuming procedure.

If the components of a mixture can be volatilized at a reasonable temperature or can be converted to volatile compounds, it may be best to separate the mixture by gas chromatography. The desirable features of gas chromatography are the component detectors and the readout. The recordings of the separated components can be quantitated by measuring the area under the curve or tracing. The resolved components on a thin-layer chromatogram can in many cases be automatically read by use of densitometers or reflectometers, but the results are not as accurate or as quantitative as in gas chromatography. Good recoveries, 95% or more, of the separated components on a thin-layer chromatogram can be obtained by removing the adsorbent and component from the plate, separating the component from the adsorbent, and measuring the component by some analytic method—usually spectrophotometry. The recovery of separated components on a column would be done in essentially the same fashion, that is, the components would be moved from the column by continued elution and the collected fractions analyzed, usually by evaporating the solvent and weighing the residue. The purity of the residues can be ascertained by thin-layer chromatography.

THEORY OF CHROMATOGRAPHY

In paper, thin-layer, column, and gas chromatography, three different principles are applied: adsorption, partition, and ion exchange. The development of the chromatographic system, or the process of separation of the components of a mixture, is primarily by elution and displacement (p. 494) and very rarely by frontal displacement. In almost every instance the components of a mixture are in contact with a solid adsorbent or stationary liquid phase on some kind of support medium, and the components are caused to migrate along the support medium by being confronted by massive (relative to the amount of solutes) gas or liquid mobile phase. The components reside in each phase for some part of the overall time required to develop the chromatogram. The amount of time spent in the respective phases is related to the partition coefficient or adsorption coefficient of the solute or component.

In adsorption chromatography the solutes interact between the solid stationary phase, such as silicic acid or alumina, and a mobile liquid phase. In partition chromatography the solutes interact between a stationary liquid phase, on an inert solid support, and a mobile liquid phase. The separation of gaseous components of a volatile mixture by interaction with a solid adsorbent in a column and a mobile gas phase flowing through the column is called gas-solid chromatography (GSC) and when the stationary phase is a liquid, it is called gas-liquid chromatography (GLC).

The solutes of a mixture will reside for a portion of the overall time of development in one phase and the remaining amount of time in the other phase. During the time the solutes are in the stationary phase, it is assumed that they are stationary and when they are in the mobile phase, they move at the same speed as the mobile phase. The distance the solutes move is equal to the time spent in stationary phase multiplied by the rate of movement of the mobile phase: $D_s = t_s \times V_m$.

The distance the solute moves divided by the distance the mobile phase moves on a thin-layer or paper chromatogram is defined as the R_f value; therefore $R_f =$ D solute/D solvent. The R_f value is a characteristic property of a solute and is related to its molecular structure. Thus solutes having different molecular composition and structure will have different R_f values. In gas chromatography the retention time, R_t, is a characteristic value of the volatile solute and is measured from the time the solute is injected into the system until the time it leaves the column and is detected. Separation of those solutes of a mixture having different R_f or R_t values should then be feasible. The R_f value, while a characteristic and intrinsic value of a solute under a given set of conditions, is difficult to reproduce and the cause of the variation is unpredictable.

Another term that is related to the R_f value is the R value. The R value is equal to the fraction of the solute in the mobile phase at any one time. The R value is also related to the velocity of migration of the solute zone and to the partition coefficient $\alpha = C_s/C_m$ in which C_s and C_m are the concentrations of the solute in the stationary phase and the mobile phase, respectively. In any volume element, the amount of solute in the mobile phase divided by the amount of the solute in the stationary phase is $A_m C_m / A_s C_s$ or $A_m / \alpha A_s$ where A_m/A_s is the relative volume of the mobile phase compared to the stationary phase (partition chromatography). The ratio $A_m / \alpha A_s$ is equal to the fraction R of solute in the mobile phase divided by 1 — R in the stationary phase or $\frac{R}{1-R}$. Equating the ratios $\frac{R}{1-R} = \frac{A_m}{\alpha A_s}$ or $\alpha = \frac{A_m(1-R)}{A_s R}$ and solving for R gives $R = \frac{A_m}{A_m + \alpha A_s} = R_f$.

This equation is commonly employed for R_f. One error in relating the R_f value to the R via the equation is that the velocity of the solvent front may be different from the velocity of the solvent at the solute zone; this is usually the case. Another error is assuming that A_m (the amount of solvent) is the same through the chromatogram or paper strip. This is actually not the case.

Fortunately the solvent movement and solvent concentration along the chromatogram have compensating effects and the overall error contribution to the validity of the equation $R_f = \frac{A_m}{A_m + \alpha A_s}$ is very small. From this equation it can be seen that the R_f value of the solute is inversely related to the partition coefficient. The R_f value of a solute in a particular two-phase liquid system can be predicted by determining the partition coefficient of the solute. For example, the amount of solute distributed between equal volumes of water and benzene gives a ratio of 10; that is, ten times as much solute is found in the water phase as compared to the amount of solute found in the benzene phase. The partition coefficient then is equal to 10. If we assume $A_m = A_s$, we find the volume of the stationary phase equal to the volume of the mobile phase in the volume element; the R_f is equal to 0.10 by calculation. A low R_f value of the solute is what would be predicted if the solute were more soluble in the water phase and the water phase were the stationary phase.

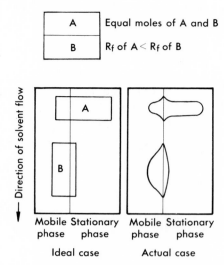

Fig. 16-1. Distribution of the solutes of an ideal and an actual condition.

The calculation of the R_f value in adsorption systems is more complicated than in partition systems. This is due to the difficulty of measuring the distribution coefficients as well as the fact that distribution coefficients between solids and liquids or gases are not constant. As the amount of solute is increased, the partition coefficient remains constant; but this is usually not true of the distribution coefficient. The variation of the distribution coefficient on a thin-layer plate is evident by the shape of the solute spot as it moves up the plate. In some cases there is a tailing of the spot or a trailing of the solute behind the major concentration of the solute. The reverse condition, in which the major concentration of the solute spot lags behind, is called bearding. These effects are primarily due to the association and dissociation of the solute in the two different media with a corresponding difference of solubility or adsorptive properties. More comprehensive discussions of the theory of chromatography are available.[7, 17, 46]

A plot of the concentration of the solute in the stationary phase versus the concentration in the mobile phase seldom results in a linear relationship because of molecular association, change in degree of ionization, and diffusion in the liquid phase or liquid phases. A picture of an ideal and an actual situation would appear as shown in Fig. 16-1. A profile of the resulting zone or spot assumes the shapes as indicated in Fig. 16-2.

The effectiveness of separation is related to the zone shape and size. The zone spreading results primarily from ordinary diffusion, eddy diffusion, and local nonequilibration.
   1. Ordinary diffusion is related directly to the length of time the chromatogram is allowed to develop. Ideally the zone is elliptic in shape, with the larger axis in the direction of the solvent flow (Fig. 16-3).
   2. Eddy diffusion occurs as the liquid moves past the particles of the adsorbent. At one moment the liquid passes readily and in another moment it is greatly restricted. This phenomenon is related to the square root of the particle size, $\sqrt{dp}$.
   3. Local nonequilibration is caused by (a) rate of flow of the mobile phase and (b) rate of attainment of equilibration between the two phases.

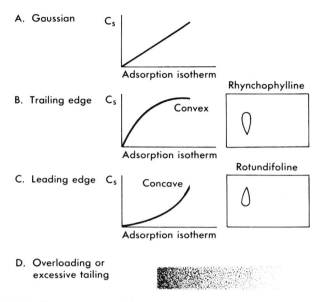

Fig. 16-2. Concentration of the solute versus its adsorption isotherm.

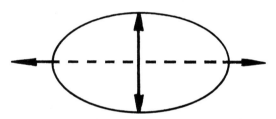

Fig. 16-3. Elliptic shape of zone.

Other parameters affecting these causes of zone or spot shape are temperature, pressure, thickness of the adsorbent layer, amount of solute, and degree of saturation of the atmosphere in the chamber with the mobile solvent. In addition to determining the zone shape, these effects also influence the R_f value. As the plates become less active with storage time, the R_f value will also change (adsorption chromatography).

Since the zone shape is the sum of the parameters listed, this compromise should be attained for maximum results: the solvent velocity should not be too fast, and the adsorbent particle size should not be small enough to excessively inhibit the solvent flow.

In partition chromatography it appears that the solute zones are even more vulnerable to diffusion effects than in adsorption chromatography because two liquid phases are involved.

Another cause for distortion of zone shape and change in R_f is related to the displacement effect of a large amount of one component on a lesser amount of another component when the two have very similar R_f values.

It is actually these same parameters, which operate to a lesser degree on thin-layer chromatograms, that account for better results than on paper strips. The superiority is not due to some intrinsic property of the cellulose. When paper is powdered and the cellulose spread as a thin-layer plate, the paper acts chromatographically very much like other thin-layer adsorbents in that it has a much shorter development time and the detection threshold is higher.

GENERAL ASPECTS OF CHROMATOGRAPHY

In the separation technique called chromatography there are two physical phases, a stationary phase and a mobile phase. These will be considered separately, although no chromatographic system can be described without considering both phases, as well as the components of the mixture to be separated.

Stationary phase

The stationary phase in adsorption chromatography is a solid, such as silicic acid or alumina, and the mobile phase is a liquid. In partition chromatography the stationary phase is a liquid, usually affixed to a supporting solid phase, such as silica gel or cellulose, and the mobile phase is a liquid. The two liquid phases will have limited solubility in each other.

While different solids—calcium sulfate, magnesium oxide, calcium carbonate, and activated charcoal, to name a few—have been used as adsorption media, silicic acid, alumina, and magnesium silicate are the adsorption solid phases most widely used. Experience has shown that the particle size of the adsorbents for thin-layer chromatography should be between 5 and 70 μ in diameter. For column chromatography the particle size should be between 50 and 200 μ in diameter. If the particles are too large, the zones on the column or the spots on the plate are diffused and poorly separated. When the particles of the adsorbent are too small, development is very slow. Silicic acid, silica gel, and silica are synonymous terms for powdered solids having the general formula $SiO_2 \cdot XH_2O$. Silica powders consist of three-dimensional siloxane structures (Si-O-Si) with surface silanol groups (Si-OH). On the silanol groups there is a mono- or multimolecular layer of adsorbed water. The water may be reversibly removed by heating at 100° C. At temperatures above 170° C an irreversible thermal decomposition of the silanol groups occurs, liberating fixed water. The adsorbent porperty or activity of the silicic acid depends on the content of silanol groups, free water, pore size, shape, and particle size. Diatomaceous earth is a completely hydrated silicic acid having practically no adsorption characteristics and it is used exclusively for partition chromatography.

The activity of an adsorbent is roughly related to its hardness; the harder the material, the greater the activity. Talc is a very weak adsorbent but silicic acid, which is much harder than talc, is an active adsorbent. The adsorptive properties of silicic acid are due to the electrostatic forces that are responsible for the crystal lattice structure of crystalline silicic acid. The electrostatic forces are, for adsorption purposes, on the surface of the crystal and result from residual ionic forces of more or less uncombined ions. The activity sites on the crystal are along the edges, corners, cracks, and defects of the crystal. The electrostatic forces of the adsorbent induced dipole moments* and increase the dipole moment of polar compounds. The binding of compounds onto the surfaces of adsorbents thus results from the ion-dipole or to dipole-dipole forces.

Alumina used as an adsorbent has several forms. The adsorption activity sites of alumina may be caused by residual ionic alumina, normal basic Al-OH, and alu-

---

*Dipole moment. A chemical dipole arises when an unequal charge distribution exists between the atoms within a molecule. This results in the molecule's having a positive and a negative charged end.

minate Al-O⁻ or Al-O-H⁺, depending on how the alumina is prepared. The latter two forms act as ion exchangers. The acidic form of the alumina is used to separate organic acids and dyestuffs while basic alumina is used to separate amino acids, alkaloids, and amines. The basic grade is also used as a dehydrating agent, such as for the removal of water from ether. The alumina also removes peroxides in the ether. Neutral alumina is used for the separation of ketosteroids, vitamins, carotenoids, ketols, and lactones. Basic alumina very likely has the greatest application. Acetone as an eluent should not be used with basic alumina, since it causes the acetone to condense to form diacetone alcohol. However, acetone can be used with neutral alumina. The latter is used to dehydrate actone.

Although the various forms of alumina and silicic acid can be prepared in the laboratory for chromatographic application, it is much more convenient and less expensive to purchase the prepared adsorbents. Sources and the names of some of the distributors or manufacturers of adsorbents are indicated in footnotes and addresses listed in Appendix E. The adsorbents from a given supplier can usually be used for both thin-layer and column chromatography. This is convenient, since the chromatographic system developed for thin-layer chromatography will generally apply to column chromatography.

The activity of the silicic acid and alumina and other adsorbents is a function of their water content. Techniques for establishing the activity of alumina have been described which involve the use of several organic dyes, such as azobenzene, p-methoxyazobenzene, sudan yellow, sudan red, and p-aminoazobenzene.

The activity grade, running from grade I (highly active) to grade V (least active), is established by the R_f values of these dyes when a mixture of the dyes is developed on alumina plates, without a binder and with carbon tetrachloride as the eluent. The R_f values of the dyes have been previously established on alumina of known designated activity. This is called the Brockman-Schodder scale of activity. Activated silicic acid has an activity comparable to alumina of grade II or III. Grade I alumina is prepared by heating the alumina to redness, and silicic acid is usually activated by heating to 110° C for 30 to 60 minutes. When a water slurry is made of the adsorbent, it is completely deactivated and must be activated by heating the prepared plates. Various ratios of adsorbents having different activities can be combined to give desired activities. To partially deactivate silicic acid, for instance, diatomaceous earth is added to the silicic acid.

Glass fiber paper, obtainable from Whatman and other sources, can be prepared by adsorption chromatography. The glass fiber paper is impregnated with silicic acid by a process of dipping the paper into a dilute solution of potassium silicate, removing the excess liquid, drying, and then dipping the paper into hydrochloric acid. After it is washed until neutral, the paper strip is dried at 100° C. By the same method, but with an aluminum chloride solution and ammonium hydroxide, a glass fiber paper impregnated with alumina can be prepared. ChromARsorb* is essentially the same as this product, in that the silica and alumina adsorbents are supported by a network of glass fibers intimately incorporated with the adsorbents to give the quality of a sheet of paper.

Various other materials can be added to the adsorbents for special applications. For the separation of very low polar compounds or unsaturated hydrocarbons, a silver nitrate solution is used in making the adsorbent slurry to prepare the thin-layer plates. The silver ions interact with the unsaturated bonding of the compounds.

---

*Mallinckrodt Chemical Works.

A binder is usually added to the adsorbents. The binders aid in affixing the adsorbents to the plates and in giving some degree of mechanical strength to the thin-layer plate. Some binders are glue, starch, and plaster of paris. Plaster of paris, a calcium sulfate hemihydrate, is by far the most commonly used binder and can readily be made by heating a reagent grade of calcium sulfate dihydrate for several hours between 140° and 180° C.

Sometimes other chemicals are added to the adsorbent to impart desired acidic or basic properties, or buffers are added to maintain a given pH value during the development of the plate. Boric acid, sodium borate, and sodium arsenite, for instance, have been incorporated into the adsorbent to aid in the resolution of glycols, lipids, and triglycerides. Ortho-hydroxyquinoline has been used to resolve some metal ions which form a complex with the o-hydroxyquinoline. The addition of silver nitrate to the adsorbent permits the separation of compounds containing double bonds.

A desirable additive to practically all the adsorbents is a fluorescing agent. Many organic compounds (all aromatic compounds) will adsorb shortwave ultraviolet light, and their presence on the plate can be readily ascertained as a dark area against the fluorescing background. In some cases, as little as 0.1 μg can be detected. There is also the added feature that compounds can be detected without being chemically altered.

The stationary phase in partition chromatography is a liquid supported on a solid medium. For the most part, the stationary liquid phase is water with the supporting medium being cellulose, kieselguhr, silica, or alumina. The activity sites of the alumina or silica when used as the support for partition chromatography are completely saturated with the highly polar stationary water phase and their adsorptive characteristics are thus neutralized. The stationary liquid phase for the usual or normal partition chromatography is a very polar liquid, such as water, while the mobile phase is an essentially nonpolar liquid, such as benzene. A few systems are given in Table 16-1. When the stationary phase is an essentially nonpolar liquid and the mobile phase a polar liquid, the system is termed reverse phase partition chromatography. Examples of this type of partition are presented in the second half of Table 16-1.

Mobile phase

In any chromatographic system there are two phases. These phases relate to each other in a countercurrent fashion, with the solute residing in one phase or the other for a given period of time. The distribution of the solutes of a mixture between the two phases is a function of their polarity or dipole moment. In adsorption chromatography, once the solid adsorption phase has been established, it is usually a simple matter to find the proper solvent, the mobile liquid phase, that will resolve the solutes. As the polarity of the solvent is increased, the greater the elution and/or displacement effect it has. A list that shows increasing polarity of solvents can be arranged and is called an elutropic series (Table 16-2). Depending on the dipole moments of the solutes, a solvent can be found that will separate the solutes. However, if there is a significant difference of the dipole moments of the solutes, no one system of solvent or solvents will satisfactorily separate the solutes by a single development. Other development techniques which can separate solutes having very different dipole moments can be resorted to and will be discussed in the practical aspects of chromatography.

The tendency of a solute to dissolve in and move with the mobile phase is called elution. For elution to take place the polarity of the solvent should be such that its

Table 16-1
Solvent systems for partition chromatography

Stationary phase	Mobile phase
Normal partition	
Water	Benzene-ethanol
Water	n-Butanol
Water and sodium hydroxide	Chloroform-ethanol
Water buffered at pH 5	Water-alcohol-butanol
Formamide	Benzene-heptane-chloroform
Water buffered at pH 6.5	Ethanol-water
Water buffered with sodium acetate	Ethyl acetate–propanol–water
Polyethylene glycol	Isopropyl ether–formic acid–water
Reverse phase partition	
Undecane	Acetic acid
Tetradecane	Acetic acid-water
Paraffin	Acetic acid-water
Silicone oil	Methyl cyanide–acetic acid–water
Petroleum	Acetone–methyl cyanide

Table 16-2
Elutropic solvent series

Solvent	Dielectric constant	
Water	80	
Methanol	33	
Ethanol	24	↑
Acetone	21	
n-Propanol	20	
n-Butanol	17	
Ethyl acetate	6.0	Increasing elution power
Chloroform	4.8	
Diethyl ether	4.3	
Trichloroethylene	3.4	
Toluene	2.4	
Benzene	2.3	
Carbon tetrachloride	2.2	
Cyclohexane	2.0	
n-Hexane	1.9	

solvating effect on the solute can satisfactorily overcome the adsorptive power of the adsorbent. Movement of the solutes along the adsorbent can also take place by displacement. In this case, the polarity of the solvent is such that it competes with the solute for the adsorption sites of the adsorbent, a case of pushing the solute along. Elution is usually thought of as occurring when the solvents are nonprotonic such as the hydrocarbons, ethers, and carbonyl compounds. Displacement usually occurs when the more protonic solvents such as the alcohols and amines are used as the developing solvent. No doubt both phenomena occur during development, and from a practical point of view the elution concept suffices for interpretation of the results.

The dielectric constant of the mobile liquid phase can be adjusted to almost any value by mixing one or more of the solvents, provided they are miscible. The dielectric values are not arithmetically additive; that is, the dielectric constant of equal volumes of acetone and benzene is not one half the sum of the dielectric constants of these two solvents. A small amount of acetone added to the benzene increases the dielectric constant to a greater extent than might be expected. The dielectric constant of a combination of solvents is more an exponential relationship than an arithmetic one. As might be expected, there are several combinations of solvents that will have the same dielectric constant; practically, however, it is best to get the desired dielectric constant by combining the proper proportions of the solvents having the nearest dielectric constant values.

The choice of the mobile liquid phase for partition chromatography is somewhat less involved than it is for adsorption chromatography. The first consideration is that the two liquid phases must be only slightly miscible in one another. For two liquids to be immiscible, their dielectric constants should be quite different, with one solvent being much more polar than the other solvent.

The movement of the solutes of a mixture by partition chromatography is a function of the solubility of the solute in the liquid phases. The ratio of the concentrations of the solute in the two liquid phases is termed the partition coefficient. If the solute is more soluble in the stationary polar liquid phase, it will spend a greater proportion of the overall development time in that phase and consequently have a low R_f value. Conversely, if the solute is more soluble in the mobile less polar liquid phase, it will have a high R_f value.

MOLECULAR STRUCTURE AND CHROMATOGRAPHIC RELATIONSHIP

The chromatographic behavior of a compound is related to its polarity. The polarity is due to the elemental and the structural composition of the compound. Thus the polarity is a function of the number and kind of functional groups that make up the compound. The very polar compounds move through a chromatographic system slowly or have low R_f values; conversely, compounds having little or no polar properties move rapidly through the chromatographic systems and thus have high R_f values. The hydrocarbons are essentially nonpolar (lipophilic) and are hardly adsorbed at all. Unsaturated hydrocarbons are more strongly adsorbed the greater the number of double bonds they contain. Hydrocarbons having conjugated double bonds are more strongly adsorbed than hydrocarbons having isolated double bonds. The hydrocarbons can be separated on highly active adsorbents with nonpolar developing solvents or by reverse phase chromatography. By incorporating silver ions in the silicic acid adsorbent it is possible to separate the hydrocarbons and, in some cases, the members of a homologous series. The adsorption affinity of the hydrocarbons increases as functional groups are added. The adsorption affinity increases with the functional group in the following sequence: —CH$_3$, —O— alkyl, C=O, —NH$_2$, —OH, —COOH. As an example, if a mixture of organic compounds were chromatographed on silicic acid with benzene used as the developing solvent, esters and ethers should be in the upper part of the plate, the aldehydes and ketones in the middle of the plate, and the alcohols and acids in the bottom of the plate.

Hydrogen bonding also plays a part in the chromatographic behavior of molecules. Molecules that have intrahydrogen bonding are retained less strongly than isomeric compounds without intrahydrogen bonding.

CHOICE OF KIND OF CHROMATOGRAPHY

Initially, when confronted with the problem of what kind of chromatography should be used to separate the components of a mixture, reliance on experience, literature reference, or possibly some shrewd and careful preliminary experimentation may be necessary. If the mixture is volatile and thermally stable, gas chromatography should be the method of choice.

Since thin-layer and column methods are closely related and involve the same kind of chromatography, the choice of one or the other would be determined by the end purpose. The preliminary investigation to find a suitable chromatographic system would be thin-layer chromatography (TLC). The chromatographic system found best by thin-layer chromatography could then be directly applied to column chromatography in the event that larger amounts of the separated components were needed for further study.

The kind of chromatography to be used depends on the type of compounds being separated. If the compounds are closely related structurally but differ only in molecular weight, as in a homologous series, it would be best to try to separate them by partition chromatography, relying on the differences of solubility that usually prevail in a homologous series. On the other hand, the adsorption characteristics of compounds having similar composition, such as isomers, but not having the same structural arrangement can be quite different, and this could be the basis of their separation. However, if a mixture is very complex, preliminary separation of major groups may have to be made and the separated groups further analyzed. Experimentally, adsorption chromatography is quicker and has a higher capacity than does partition chromatography.

Generally very polar compounds such as carbohydrates, amines, and nucleotides are best separated by partition chromatography while the less polar compounds are best separated by adsorption chromatography. To generalize, as the number of functional groups increases, the polarity of the molecule also increases but as the molecular weight and carbon content increase, the polarity of the molecule decreases.

As a first approximation to the kind of chromatography to be used some generalities apply and can be used as a guide to preliminary experimentation. These are given in Table 16-3.

If they are known, the structural differences of the compounds of a mixture should also be considered in selecting a chromatographic method. The structural differences and a rating of the chromatographic methods are given in Table 16-4. Application of thin-layer chromatography to the classes of compounds is given in Table 16-5. This table includes the kind of chromatography, compound class, adsorbent, developing solvent system, and visualization technique of the separated components. An idea of the kind of thin-layer chromatography that may be advantageous to try may be obtained from the table.

In some respects it may be just as easy to try a few systems instead of doing ex-
Text continued on p. 503.

Table 16-3
Choice of chromatographic system

Kind of compound	Adsorption	Partition
Long-chain aliphatic ketones	+	
Aliphatic and aromatic aldehydes and ketones	+	
C_2 to C_{12} aliphatic acids		+
Aromatic aldehydes and ketones	+	
C_4 to C_6 dicarboxylic acids		+
Monohydric alcohols		+
Vanillin and other substituted benzaldehydes	+	
Alkaloids	+	
Phenols and cresols		+
Dihydric and trihydric alcohols		+
Alkaloids and barbiturates	+	
Cardiac glycosides	+	
Sulfonamides	+	
Amino acids		+
Penicillins		+
Sugars and derivatives		+
Miscellaneous lipids	+	
Glycerides	+	

Table 16-4
Sensitivity of methods of separation to structural differences in the compounds to be separated*

Structural differences	Adsorption	Partition	Ion exchange
Molecular size	+ to (+)	+ to ++	+ to ++
Structural isomers			
Chains, rings	+ to (+)	+	−
Chains, branching	+ to (+)	(+)	(+)
Position of double bonds	+ to (+)	−	−
Steric isomers			
Cis-trans at a double bond	+ to (+)	+	+
Cis-trans at a ring	++ to (+)	+	+
Optic isomers	(+)	(+)	+ to (+)
Number of double bonds	++	+	−
Conjugation of double bonds	++ to +	+	+
Number of nonpolar substituents	+ to (+)	+ to (+)	+ to (+)
Number of polar substituents	++	++	++
Polarity of substituents	++	++	++

Ratings: ++, very good; +, good; (+), fair; −, not investigated.
*From Turba, F: Chromatographische Methoden in der Proteinchemie, Berlin, 1954, Springer-Verlag.

Table 16-5
Some applications of thin-layer chromatography by classes of compounds

Compounds	Adsorbent	Developing solvent system	Detection	Reference
		Alkaloids		
Fifty-four alkaloids	Silica gel G	$CHCl_3$-acetone-diethylamine (5:4:1) $CHCl_3$-diethylamine (9:1) Cyclohexane-$CHCl_3$-diethylamine (5:4:1) Cyclohexane-diethylamine (9:1) Benzene-ethylacetate-diethylamine (7:2:1)	Potassium platinate	59
	Alumina G	$CHCl_3$ Cyclohexane-$CHCl_3$ (3:7) plus 0.05% diethylamine		
	Silica gel G–NaOH	Methyl alcohol		
Indoles	Silica gel G	Methylacetate-2-propanol-NH_4OH (45:35:20) $CHCl_3$–ethyl ether (95:5)	Van Urk reagent Prochaska reagent 2,4-dinitrophenylhydrazine	47
Alkaloids and barbiturates	Silica gel G	Methanol $CHCl_3$–ethyl ether (85:15)	Dragendorff's reagent I_2 and KI in acetic acid Fluorescein $FeCl_3$ in acetic acid $KMnO_4$ in acetic acid	26
Opium alkaloids	Alumina buffered at pH 5 with acetate	Water-ethanol-butanol (1:1:9)	UV light	30
Opium alkaloids	Silica gel G	Benzene-methanol (8:2)	Dragendorff's reagent	32
		Aldehydes and ketones and their derivatives		
Long-chain aliphatic ketones	Silica gel G	Benzene-ether mixtures Toluene-ether mixtures Petroleum ether–ether mixtures	Phosphomolybdic acid	29
Vanillin and other substituted benzaldehydes	Silica gel G	Petroleum ether–ethyl acetate (2:1) Hexane-ethyl acetate (5:2) $CHCl_3$-ethyl acetate (98:2) Decalin-CH_2Cl_2-methyl alcohol (5:4:1)	Hydrazine sulfate in HCl Methanolic KOH Methanolic KOH followed by diazotized sulfonic acid	53

Continued.

Table 16-5
Some applications of thin-layer chromatography by classes of compounds—cont'd

Compounds	Adsorbent	Developing solvent system	Detection	Reference
Aldehydes and ketones and their derivatives—cont'd				
2,4-dinitrophenyl hydrazones of aldehydes	Silica gel G	Benzene-petroleum ether (3:1) Benzene-ethyl acetate for aromatic aldehydes	Colored compounds	10
2,4-dinitrophenyl hydrazones of hydroxyaldehydes and ketones	Aluminum oxide G Silica gel G	Toluene-ethyl acetate (3:1), (1:1)	Colored compounds	1
Amines and amine derivatives				
Aliphatic amines	Silica gel G	Ethanol (95%)-NH_4OH (25% NH_3) (4:1) Phenol-water (8:3) Butanol–acetic acid–water (4:1:5)	Ninhydrin	55
	Silica gel G buffered at pH 6.8	Ethanol-water (70:30)		
Diphenylamine derivatives	Silica gel G	$CHCl_3$ Toluene Benzene	$NaNO_2$ and H_2SO_4	16
Pteridines	Silica gel G	Dimethylformamide-water (190:10), (20:180) 5% citric acid NH_4OH (25% NH_3) Many others	UV light	33
Carbohydrates and glycosides and their derivatives				
Simple sugars	Silica gel G buffered with boric acid	Benzene–acetic acid–methanol (1:1:3) Methyl ethyl ketone–acetic acid–methanol (3:1:1)	Naphthoresorcinol	38
Simple sugars	Silica gel G	Propanol-concentrated NH_4OH-water (6:2:1)	Ninhydrin Aniline hydrogen Phthalate	60
Sugar acetates and inositol acetates	Silica gel G	Benzene with 2% to 10% methanol	Water Ferric hydroxamate	54
Cardiac glycosides	Silica gel G	$CHCl_3$-methanol (9:1) $CHCl_3$-acetone (65:35)	H_2SO_4 and acetic anhydride	48

Table 16-5
Some applications of thin-layer chromatography by classes of compounds—cont'd

Compounds	Adsorbent	Developing solvent system	Detection	Reference
Carboxylic acids and their derivatives				
Dicarboxylic acids	Silica gel G	Benzene-methanol-acetic acid (45:8:4) Benzene-dioxane-acetic acid (90:25:4)	Bromphenol blue acidified with citric acid	40
p-Hydroxybenzoic acid esters	Silica gel GP	Pentane-acetic acid (88:12)	UV light	13
Dyes				
Food dyes	Cellulose powder MN 300 G and 300 MN	Aqueous sodium citrate (2.5%)-NH_4OH (25% NH_3) (4:1) Propanol-ethyl acetate-acetic acid (6:1:3) Tertiary butanol-propanoic acid-water (50:12:38)	Colored compounds	64
Essential oils				
Essential oils	Silica gel G	Benzene-$CHCl_3$ (1:1)	$SbCl_2$ in $CHCl_3$	49
Oils of sage	Silica gel G	Benzene	Vanillin in H_2SO_4	6
Lipids—class separation				
Miscellaneous lipids	Silica gel G	Petroleum ether-ethyl ether-acetic acid (90:10:1)	2′, 7′-dichlorofluorescein	28
Serum lipids and phospholipids	Silica gel G	$CHCl_3$-methanol-water (80:25:3) Petroleum ether-ethyl ether-acetic acid (90:10:1), (60:40:1)	2′, 7′-dichlorofluorescein	57
Plasma lipids	Silica gel G	$CHCl_3$-methanol-water (65:24:4)	Bromthymol blue-ammonia Ninhydrin Fuchsin-H_2SO_4 Ferric hydroxamate	15
Brain tissue lipids	Silica gel G	$CHCl_3$-methanol-water (24:7:1) $CHCl_3$ CCl_4	Ammonium molybdate-$HClO_4$ Phosphomolybdic acid Ninhydrin Dragendorff's reagent Diphenylamine	22 21

Continued.

Table 16-5
Some applications of thin-layer chromatography by classes of compounds—cont'd

Compounds	Adsorbent	Developing solvent system	Detection	Reference
Lipids—class separation—cont'd				
Lipids of feces and fecoliths	Silica gel G	Petroleum ether-ethyl ether-acetic acid (80:20:1)	50% H_2SO_4	61
Lipids—fatty acids and their derivatives				
Fatty acids	Silica gel G impregnated with undecane	Acetic acid-methyl cyanide (1:1) Solution 1 with 0.5% Br_2	Copper acetate–dithiooxamide	24
Fatty acid methyl esters	Silica gel G	Hexane-ethyl ether–acetic acid (85:15:1)	2′, 7′-dichlorofluorescein	42
Lipids—glycerides				
Glycerides	Silica gel G	$CHCl_3$-benzene (70:30)	Concentrated H_2SO_4	8
Mono- di- tri-glycerides	Silica gel G	Petroleum hydrocarbon (40-60 C)-diethyl ether (70:30)	50% H_2SO_4	4
Lipids—phospholipids, sulfolipids, and glycolipids				
Phospholipids Sulfolipids Glycolipids	Silica gel Silica gel G Silica gel plus 10% ammonium sulfate	Various proportions of $CHCl_3$-methanol-water Propanol-ammonia Butanol-pyridine-water	I_2 vapor 50% H_2SO_4 Ninhydrin Dragendorff's reagent Ammonium molybdate-$HClO_4$	
Phenols and alcohols and their derivatives				
Cyclohexanol derivatives	Silica gel G	Benzene Benzene-ethyl acetate (80:20)	Water	14
Phenols	Silica gel G plus oxalic acid	Hexane-ethyl acetate (4:1), (3:2) Benzene-ethyl ether (4:1) Benzene		12
Long-chain aliphatic alcohols	Silica gel G	Petroleum ether–ethyl ether-acetic acid (90:10:1)	50% H_2SO_4	52
3,5-dinitro-benzoate of alcohols and phenols	Silica gel G	Benzene-petroleum ether (1:1) Hexane-ethyl acetate (85:15), (75:25)	Colored compounds	11

Table 16-5
Some applications of thin-layer chromatography by classes of compounds—cont'd

Compounds	Adsorbent	Developing solvent system	Detection	Reference
Amino acids and proteins and their derivatives				
Amino acids	Celluose powder MN 300	Upper phase of butanol-water-acetic acid (4:5:1) Pyridine-methyl ethyl ketone-water (15:70:15) Methanol-water-pyridine (80:20:4)	Ninhydrin	63
Polypeptides and derivatives	Silica gel G	Butanol saturated with 0.1% NH_4OH	Ninhydrin	58
Steroids				
Miscellaneous steroids	Silica gel G	Benzene Benzene-ethyl acetate (9:1), (2:1) Cyclohexane-ethyl acetate (9:1), (19:1) 1,2-dichloroethane Others	$SbCl_3$ in $CHCl_3$	56
Thirteen miscellaneous steroids	Silica gel GP	$CHCl_3$-ethyl acetate (3:2), (3:1) Benzene-ethyl acetate (1:1), (9:1)	I_2 vapors	31
	Aluminum oxide GP	Hexane-ethyl acetate (1:1)		
Estrogens	Silica gel G	Benzene-ethanol (9:1)	$SbCl_5$ in $CHCl_3$	51
Pregnanetriols in urine	Unbound aluminum oxide	Benzene-ethanol (9:1)	70% H_3PO_4	50
Cholesterol	Silica gel G	Hexane-ether-acetic acid (50:50:2)	$HClO_4$	35
Cholesterol ethers	Silica gel G	Benzene-petroleum ether (40:60)	50% H_2SO_4	27
Bile acids, free and conjugated	Silica gel G	Acetic acid-isopropyl ether-amyl acetate-CCl_4-benzene (5:30:40:20:10) for free acids Propionic acid-isoamyl acetate-water-propanol (15:20:5:10) for conjugated acids	Phosphomolybdic acid	19

Continued.

Table 16-5
Some applications of thin-layer chromatography by classes of compounds—cont'd

Compounds	Adsorbent	Developing solvent system	Detection	Reference
		Steroids—cont'd		
Corticosteroids	Unactivated silica gel G	$CHCl_3$-methanol-water (485:15:1), (188:12:1), (90:10:1)	50% H_2SO_4	3
Aldosterones from urine	Silica gel G	Cyclohexane-2-propanol (7:3) $CHCl_3$-glacial acetic acid (8:2)	Tetrazolium blue	34
		Terpenoids		
Forty-nine terpenoids	Silicic acid, starch-bound	Hexane	Fluorescein H_2SO_4-HNO_3	25
Carotenoids	Silica gel G + calcium hydroxide (1:4) Silica gel G + calcium hydroxide (1:4) impregnated with paraffin	Benzene–petroleum ether (1:1) Above solution with 1% methanol Methanol saturated with paraffin	Colored compounds	62
		Vitamins		
Carotenes and fat-soluble vitamins A, D, E, and K	Aluminum oxide, unbound	Methanol Carbon tetrachloride Xylene Eleven others	70% $HClO_4$	9
Tocopherol mixtures	Silica gel G	$CHCl_3$	Ceric sulfate Phosphomolybdic acid	43
	Aluminum oxide D5	Benzene		
		Miscellaneous		
Flavonoids and coumarins of lemon	Silica gel, starch-bound, with phosphor	$CHCl_3$-acetone mixtures	UV light	23
Inorganic ions, potassium and magnesium	Silica gel G	Ethanol-methanol (1:1) plus 1% acetic acid	Acid violet 6BU	42
Insecticide, thiophosphate esters	Aluminum oxide, starch-bound	Heptane-acetone (10:1)	$H_5IO_6^-$-$HClO_4$-V_2O_5	36

tensive literature research. This can be readily done by spotting the sample on the thin-layer plate and introducing the developing solvent from a capillary pipette into the center of the spot. The developing solvent should be one taken from the top, middle, and bottom (decreasing polarity) of an elutropic solvent series. If the sample is colorless, the separated constituents can be visualized by some of the general visualizing agents such as sulfuric acid, sulfuric acid–chromate mixture, and iodine vapor. If the compounds are aromatic, the incorporation of an ultraviolet fluorescing agent in the adsorbent will reveal these compounds when the plate is exposed to ultraviolet radiation. The major constituent(s) should move about half the radius of the circle that the developing solvent makes around the spot. The same technique can of course be followed for partition chromatography. From this preliminary investigation a combination of two or more solvents, which straddles the desired distance, may be suggested. Another rapid method to check out a chromatographic system is to use small thin-layer plates made from microscopic slides and develop the plate in a Coplin staining jar containing a few milliliters of the solvent.

There are other kinds of chromatography, such as ion exchange and gel permeation chromatography, but this chapter is intended to be primarily an introduction to the most commonly used chromatographic methods and to some of their practical applications. Briefly, however, gel filtration should be considered for trying to separate very large molecules where the differences in size of the molecules can be the basis of their separation. Amino acids, peptides, and proteins can be separated according to their molecular weights on layers of Sephadex, a gel filtration medium.[20]

Components of a mixture which have different ionic characteristics can be separated by ion exchange chromatography. The nucleic acid derivatives and proteins are such a class of compounds.[18] The stationary phase in ion exchange chromatography is for the most part organic compounds that are practically insoluble in highly polar solvents such as alcohol or water. There are two main types of ion exchange media: (1) the ion exchange resins, made from polymers so synthesized as to contain an anion exchanger group such as a quaternary ammonium or a cation exchanger group such as carboxylic or sulfonic acid, and (2) the ion exchange celluloses, made by reacting cellulose with various reagents. The latter results in functional groups being attached to the cellulose by ether or ester linkages; each functional group acts as a cation or anion exchanger site, depending on the functional group. The anion exchangers called **DEAE** (diethylaminoethyl cellulose) and **ECTEOLA** (epichlorohydrin triethanolamine cellulose) are used for the separation of a number of proteins, peptides, enzymes, hormones, and nucleic acids. The cellulose cation exchangers contain carboxymethyl and phosphate as the functional groups.

The application of a high voltage, low current, through the above media is called ionophoresis and electrophoresis and has been used to separate amino acids, peptides, and other classes of compounds. Thin-layer electrophoresis combines the separation features of thin-layer chromatography and electrophoresis.

Column chromatography

In column chromatography the sample, as a solution, is placed on a column of adsorbent in a narrow band. The adsorbent is supported for the most part in a glass tube. (For clinical application refer to Chapter 14, methods for catecholamine, pregnanediol, and pregnanetriol.) The developing solvent is then placed in a reservoir over the column of adsorbent, passing through the column by force of gravity. Sometimes pressure is applied to increase the flow rate of the solvent through the column.

The components of the sample will ideally separate into narrow bands as the components preferentially move down the column under the influence of the moving solvent. Small portions or fractions of the issuing solvent, called the elutrient, are collected. If the mixture consists of wholly colored components the degree of separation can be readily ascertained as they move down the column, and it is an easy matter to collect these colored components as they come out of the column. In most cases the components are colorless and the elutrient is collected in small fractions which are analyzed. There are devices for detecting the passing of the separated components, which are based on a change in some physical property of the elutrient such as conductivity, absorbance, or refractive index, to name a few.

From a preliminary TLC investigation of the sample some rules can be presented for adapting the thin-layer chromatography to column chromatography. The solvent system for adsorption TLC should be of such a composition that the main component of the sample mixture has an R_f of not more than 0.3. Such a solvent system is used to pack the column and to resolve the sample mixture on the column. When TLC by partition chromatography is used to separate the components, a direct application of the TLC solvent system can be made to column chromatography but again the major components should have R_f of 0.5 or less. Generally the chromatographic columns are supported in glass tubes that are tapered at the bottom. The length of the tube should be no less than 10 times the diameter of the tube and may be as long as 100 times the diameter. The size of the glass tube is determined by the amount of sample to be applied. A wad of cotton or glass wool is inserted into the tube down to the taper. The operation of the column can be improved by adding a bulb at the top of the glass tube, which serves as a reservoir for the developing solvent, and instead of a cotton plug, a sintered-glass disk or a Teflon filter disk can be used. Also a stopcock at the exit end of the glass tube aids in the control of the solvent flow. For optimum flexibility a glass tube with a ball socket at one end, which can be attached to a ball socket receptacle that accommodates a sintered-glass disk and valve control, can be constructed. The ball socket joint is secured with a pinch or screw clamp. The ball socket joint must be free of particles of adsorbent; otherwise the joint will leak, especially if the column is put under pressure. Chromatographic columns of various designs can be obtained from laboratory supply houses.

Adsorbent

Different materials have been used as column packing but the most widely used are alumina and silicic acid and, to a lesser extent, cellulose. Generally the particle size of the alumina or silicic acid is larger for column chromatography than for thin-layer chromatography. The particle size is somewhere between 80 and 200 mesh. The activity and kinds of alumina and silicic acid have already been discussed (refer to p. 490).

The column for adsorption chromatography can be prepared in either of two ways: a wet method or a dry method. By the dry method a small amount of dry adsorbent is added at a time to the column and tapped or pressed firmly into the tube. The pressing is done with a long-handle plunger with a metal or glass disk at the end. Small amounts of adsorbent are added at a time until the desired column height is obtained. In the wet packing technique the adsorbent is mixed with a portion of the developing solvent or with some member of an elutropic series which is less polar than the solvent system that will be used to develop the column. Small increments of the slurry are added and the column is tapped with a rubber stopper or the end of a pencil to facilitate the settling of the adsorbent and to release entrapped air.

For partition chromatography the stationary phase—silicic acid, alumina, or cellulose—is intimately mixed with the polar solvent in a mortar or in a powder blender. A slurry is made by using the developing solvent which has been saturated with the stationary liquid phase. The amount of stationary phase to be incorporated in the supporting medium depends on several factors. Small-size particles of silicic acid will hold more liquid than large-size particles. Kieselguhr, a diatomaceous earth, consists of very small particles of water-saturated silicic acid and is therefore particularly good for partition chromatography and reverse phase chromatography, but it will hold less of the stationary phase than small particle size silicic acid. Cellulose columns are difficult to prepare. The cellulose slurry should be added in small increments to obtain uniform packing. For most purposes the weight ratio of the stationary liquid phase to the supporting medium is 0.5 to 1 part stationary phase to support medium.

Application of the sample

The sample is introduced onto the column by dissolving it in the developing solvent. The developing solvent for partition chromatography is saturated with the same liquid that is in the stationary phase. The sample solution is carefully added to the column so as not to disturb the supporting medium. A piece of filter paper or a sintered-glass disk may be placed on the surface of the support medium. If the sample does not dissolve in a small amount of the developing solvent, a larger volume of solvent can be used and evaporated to dryness in the presence of a small amount of the supporting medium. The dry powder containing the sample is then quantitatively transferred to the column. This technique of sample introduction gives tight bands.

Solvent system

The solvent system to be used for column development is determined as given in the section on the mobile phase (p. 492). From the preliminary thin-layer investigation for suitable chromatographic system, the proper solvent system is determined and can be applied directly for development on the column. The developing solvent is usually added from some reservoir which is a part of the glass column. Usually better and sharper separations take place with a slow flow rate. However, if the flow rate is too slow, diffusion becomes excessive and the bands may overlap. On the other hand, if the flow rate is too fast, there is a tendency for the solvent to channel. Also, the optimum operational conditions are so distorted by an excessive flow rate that the ideal distribution of the component between the two phases cannot be attained. The proper flow rate is best determined by experimentation. The solvent flow rate can be changed during development but it is advisable not to stop the flow rate for any length of time. If pressure is applied to increase the flow rate, a constant pressure must be maintained; otherwise the column tends to separate on decreasing of the pressure. The polarity of the solvent can be changed during development (gradient elution) but the change must be done gradually or stepwise, going down the solvents of an elutropic series. Again in partition chromatography the developing solvent must be saturated with the solvent used in the stationary phase. A convenient way to gradually change the composition of the developing solvent is to place one solvent reservoir containing the more polar solvent over the less polar developing solvent reservoir attached to the column (Fig. 16-4). The solvents are mixed in reservoir B with a magnetic stirrer. A third solvent can be introduced in a like fashion. For complete recovery of the sample the column can be given a final elution with

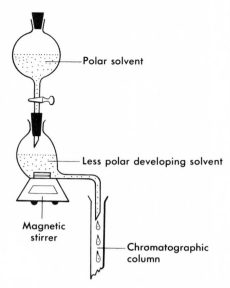

Fig. 16-4. Setup for gradual change of polarity of developing solvent.

water, methanol, or methanol containing 5% glacial acetic acid. The solvents in all cases must be of the highest purity and, in some cases, the solvent can be first purified by percolating it through a column of the same support medium used in the developing or working column.

Detection

There are several ways of detecting the components as they come out of the column. In most cases the components are colorless and therefore some chemical or physical property of the solute must be exploited to reveal its exit from the column. Portions of the effluent are collected by either volume or time factors. There are available fraction collectors that count the number of drops and, after a preset number of drops has been collected in a receiving tube, another tube should be brought up to receive the next preset number of drops. The receiving tubes are in a turntable that rotates, aligning a new tube under the column outlet after each fraction. Another device for triggering the turntable is a fixed-volume siphon that, after a certain volume of effluent has been collected, will activate the servomotor driving the turntable. A different mechanism* for detecting components in the effluent is based on measurement of the amount of light reflected from the interface of the flowing stream and a glass surface. The amount of reflected light is related to the refractive index of the liquid and the refractive index of the glass. If the liquid and glass have the same refractive index, no light is reflected and, of course, the instrument cannot detect the component in the effluent stream. If gradient elution is used to develop the columns, the baseline of the recording device must be continually adjusted to compensate for the adsorption of the changing composition of the developing solvent. Many constituents such as nucleotides, phenylalanine, steroids, and hormones can be detected as they come off the column by ultraviolet flow analyzers.† The elutrient carrying the ultraviolet-adsorbing constituent passes through a

---

*Waters Associates, Inc.
†Canal Industrial Corp.

flow cell of a spectrophotometer at the end of the column. The spectrophotometer feeds into a recorder. The recorder makes a continuous strip chart reading and indicates the test tube in the turntable collector of the constituents as they come off the column. If gradient elution is performed, the equipment will automatically compensate for the change in the ultraviolet adsorption characteristics of the solvent system.

A simple way to monitor effluent fractions is by thin-layer chromatography. Initially every fraction should be analyzed so that a qualitative profile can be obtained to indicate where the major constituents' peaks are located in relation to the fraction collected.

In some cases the components of the mixture are separated in the column and the development is stopped short of passing the first component from the column. The column is then pushed from the supporting glass tube and the position of the components identified by streaking the column with an appropriate reagent, viewing under ultraviolet radiation, or cutting the column into small portions and eluting with a very polar solvent. The subsequent extract is evaporated to dryness and the residue weighed or subjected to further analysis.

Thin-layer chromatography
PRACTICAL ASPECTS
Plate preparation

The following discussion on the preparation of thin-layer plates may be considered academic, since thin-layer plates can be purchased that are equally as good as handmade plates and in some respects are better. It is also less expensive to purchase prepared plates. Besides being readily available, prepared plates are more likely to have a uniform deposit of the adsorbent layer, which allows for a more accurate quantitative evaluation of the spot by either area measurement or densitometric or reflectrometric measurements. However, there are several reasons why it is desirable to know how to prepare thin-layer plates, such as usefulness for evaluating the feasibility of other than the usual adsorbents or a special buffered system and for special investigational work. The description of preparing thin-layer plates with silicic acid will serve as a general approach to preparing plates with practically any kind of adsorbent. For the most part, the particle size that is generally acceptable for thin-layer chromatography lies between 5 and 70 μ. Regardless of what adsorbent-solvent system is used, the weight of the solid adsorbent to the liquid should be such that a slurry results, which can be poured or pulled over a glass plate, leaving a uniform deposit. If the mixture is too fluid, lines and fissures will form; if too thick, the mixture will become firm before settling uniformly. The latter is especially true if a binding agent, such as calcium sulfate, has been added. It is a case of trial and error until the proper weight of adsorbent to liquid volume is obtained. Some suggested weight ratios of adsorbent to liquid volume are given in Table 16-6. These adsorbent to water ratios are suggested for large-plate (5 × 20 cm to 20 × 20 cm) preparation using commercially available apparatus to coat the plates, such as the Camag* or the Stahl-De Saga.† To coat microscope slides that are convenient for qualitative purposes, such as monitoring an organic reaction or checking a solvent system, the liquid used to prepare the slurry is usually other than water. Rather extensive use of microscope slides as adsorbent layer supports was made by Peifer.[39] As an exam-

---
*Gelman Instrument Co. or Arthur H. Thomas Co.
†Brinkmann Instruments, Inc.

Table 16-6
Ratio of adsorbent to water

Adsorbent	Supplier	Adsorbent/water (g/ml)
Silica gel G series	Brinkmann	30/60–65
Silica gel N series	Brinkmann	30/60–70
Silica gel H series	Brinkmann	30/80–90
Silica gel D series	Gelman	50/100
Silica gel TLC series	Mallinckrodt	30/60–70
Aluminum oxide G series	Brinkmann	30/40
Aluminum oxide, Woelm	Aluphram	35/40
Aluminum oxide D series	Gelman	20/60
Cellulose 300 and 300G series	Brinkmann	15/90
Cellulose CC41, Whatman	Reeve Angel	30/60
Kieselguhr	Brinkmann	30/60–65

ple, the silica gel adsorbents are made into a slurry by using 35 g of solid with 100 ml of chloroform-methanol (2:1, v/v). Two microscope slides, held back to back, are dipped into the slurry, separated, and allowed to air-dry. The plates may be further conditioned by exposing them to a gentle stream of steam. The steamed plates can be used as such for partition chromatography or be activated for adsorption chromatography by drying at 110° C for 30 minutes.

For most thin-layer chromatography three plate sizes are used: 5 × 20 cm, 10 × 20 cm, and 20 × 20 cm. The larger the plate, the more samples that can be analyzed. The plates must be absolutely clean. They should be washed with a detergent or cleaning solution, rinsed well with tap water and then distilled water, and dried. It is suggested that before use the plates be wiped with hexane to remove any trace of grease.

The adsorbent slurry is applied by any one of three techniques. The simplest technique for a single-plate preparation is merely to pour the slurry onto the middle of the plate and tilt the plate to evenly cover the surface. A somewhat uniform layer results. This technique requires a great deal of practice before even fair reproduction of adsorbent layer thickness can be achieved.

Another method is to place narrow strips of tape along the edges of the plate, adding more strips of tape to obtain the layer thickness desired, and to pour the slurry primarily at one end of the plate. A glass rod that extends beyond the edges of the plate is used to gently and evenly distribute the slurry over the plate. The glass rod rides on the two strips of tape. A fairly good adsorbent layer thickness can be achieved by this technique.

Generally several thin-layer plates are made at one time. The plates, up to six that are 20 × 20 cm, and combinations of narrower-size plates, but 20 cm long, are placed in an aligning tray. The aligning tray is usually made of a plastic material and is used to guide the slurry applicator over the positioned plates. The technique of pulling the slurry over the plates is called the Stahl-De Saga technique. When the

Chromatography 509

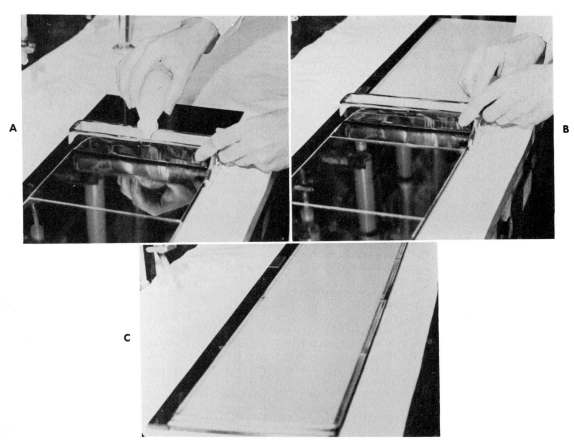

Fig. 16-5. A, Introducing silica gel slurry into applicator trough. **B,** Drawing slurry over plates, Stahl-De Saga technique. **C,** Coated plates in aligning tray.

aligning tray is used to guide the plates as they are pushed under the stationary slurry reservoir, the method is called the Camag technique of plate preparation. The Camag technique gives particularly uniform layer thickness. By the Stahl-De Saga technique the adsorbent layer is subject to variation due to warping of the aligning tray or nonuniform plate thickness. There are apparatuses, however, that accommodate for plate thickness, and a thickness variation of ± 0.01 mm can be obtained. These plate-leveling devices are obtainable commercially.* The stepwise application of the slurry by the Stahl-De Saga technique using a fixed-depth (250-μ) applicator and a regular aligning tray is shown in Fig. 16-5. For the preparation of plates having adsorbent layers of other than the usual 250-μ thickness, applicators are available that have adjustable gates permitting layer thickness up to 2 mm. Layers of this thickness would be used for preparative thin-layer chromatography.

The freshly applied slurry should be allowed to set up for about 30 minutes. If the slurry is to be used for adsorption chromatography, the plates should be activated for at least 1 hour at 110° C. After the plates have cooled, the excess adsorbent on plate edges is removed and the plates are stored in a desiccator. A conven-

---

*Quickfit, Inc., and Shandon Scientific Co., Inc.

ient desiccator can be made from a large plastic bag containing a couple of bottles of a drying agent. The carrying tray that contains the plates is placed in the plastic bag and the bag is closed with a large paper clamp.

Sample application

The sample can be applied to the plate in one of several ways, depending on the intent or purpose. Initially, in lieu of the concentrations of the unknown in the sample, it is advisable to spot 5, 20, and 100 μl. The small amount of sample will very likely give sharper separations but the large-size sample will reveal minor constituents. The sample can be conveniently spotted with disposable micropipets or capillaries.* The sample is drawn into the tube by capillary action and when touched to the adsorbent layer delivers essentially the rated volume of the capillary. Microsyringes† can be used to spot samples of practically any size. Application of large volumes of solutions from the syringe, greater than 10 μl, which have a low surface tension, such as hexane or benzene, usually have to be made in increments of 5- to 10-μl droplets. Sample spotting or streaking for preparative thin-layer chromatography is difficult to do by hand. However, there are several devices commercially available for streaking large volumes of samples in a quantitative fashion. Bobbitt and associates[5] have described a device used in their laboratory for hand streaking thin-layer preparative plates that is of interest. It consists of a 2-ml pipet. The wick tip is made by looping a fine Nichrome wire and inserting it into the end of the pipet. A piece of string is then tied to the end of the wire loop and the loop is pushed into the pipet. The string is cut to a length of 7 to 8 mm. The rate of flow can be controlled by the degree of tilt of the pipet.

Another method of adding a sample for preparative purposes is to apply a series of spots, one just touching the other, across the length of the plate. The plate is then developed a short distance with a very polar solvent, such as methanol, the spots thus being compressed into a thin straight line. The plate is dried and developed normally.

Development of the plate

For the most part the plates are developed by ascending chromatography, in which the solvent moves up the plate. There are other methods of development, such as circular, horizontal, and descending, but these methods do not have any particular advantage over the ascending development technique. Developing chambers can be any convenient-size jar or container. For rapid saturation of the atmosphere in the jar, with the developing solvent vapor, the developing chamber should be just large enough to accommodate the plate. For rapid saturation of the atmosphere, the chamber is lined with filter paper. In some cases, however, a sharper resolution will result if no attempt is made to saturate the atmosphere. The S chamber or sandwich chamber atmosphere (Fig. 16-6) is very rapidly saturated because of the very small space that exists between the cover plate and the thin-layer plate. In the S chamber the two plates are separated a distance of a few millimeters by cardboard or Teflon spacers on three sides and are held together by a couple of spring clamps. The clamped plates are then placed in a slotted trough containing the developing solvent. The order of separation of the components of a mixture is the same when develop-

---

*Available from Gelman, Helena Laboratories, and Kensington Scientific Corp.
†Hamilton Co.

Fig. 16-6. Assembled S chamber with plate in position and set in solvent trough.

ment is in a tank chamber or an S chamber, but the R_f values are usually not the same.

Several different developing techniques can be used, and it is by experimenting with these particular techniques that the best method can be determined. The developing method used must by necessity relate to the kind of sample. If the sample consists of components having a wide difference in polar properties, it may be best to resort to *multiple development*. In this case a solvent is found that will move the least polar component about half the way (R_f value of 0.5) and just about move the polar component from the origin of spotting. Subsequent drying and developing the plate again in the same solvent system will further resolve the low R_f value components without running the initial R_f value of 0.5 component into the solvent front. The multiple-development technique is equivalent to allowing the solvent to move approximately twice the distance on a larger plate. The advantage of multiple development lies in the fact that the greatest resolution of the components of a mixture takes place in the first third of the distance the solvent moves.

Continuous development, in a regular developing chamber or an S chamber, can be readily performed by merely extending the thin-layer plate outside the chamber and thus allowing the solvent to evaporate off the end of the thin-layer plate. Again a solvent system should be so chosen that the least polar component does not move up the plate too rapidly. If the components of a mixture have properties that are quite different, it may be best to resort to *two-dimensional* development. In this case a solvent system is used that will resolve the less polar components, and after drying the plate, redevelop it in a more polar solvent system to resolve the polar constituents. The plate is turned 90 degrees prior to development with the more polar solvent; thus the solvent moves at a right angle to the direction of the first solvent. The sample is spotted at one end of the plate, and only one sample can be developed per plate. Ideally the results should appear as given in Fig. 16-7.

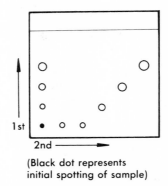

(Black dot represents initial spotting of sample)

Fig. 16-7. Idealized two-dimensional developments.

Detection of the separated components

There are two classes of detectors: the universal and the specific. The universal detectors in common usage are concentrated sulfuric acid, sulfuric acid–potassium dichromate (cleaning solution), sulfuric acid–nitric acid, perchloric acid, and iodine. The acids are sprayed onto the plate and the plate is warmed for a few minutes between 100° and 150° C. Most organic compounds will appear as black dots because of the charring by the acids. Volatile organic compounds will not be detected. Iodine is a nondestructive detector of a great many organic compounds. A developing chamber is ideal for use as an iodine chamber. A few grams of iodine in a 50-ml beaker can be placed in the developing chamber and, after the chamber is saturated with the iodine, it is ready for use. The iodine-adsorbing organic compounds appear as dark brown spots against a tan background. On removal of the plate from the iodine chamber the spots should be outlined because the iodine evaporates rapidly. Many organic compounds adsorb ultraviolet light and are readily and conveniently detected if a fluorescent agent is incorporated into the adsorbent layer. The ultraviolet-adsorbing components appear as dark spots against a green or white fluorescing background. In some instances an amount as small as 0.1 μg can be detected. With a few exceptions, the organic compounds can be detected without destruction —which is an advantage if quantitative recovery of the separated component is desired.

A few specific agents will be cited here. For a more extensive list refer to the detection column in Table 16-5. Carboxylic acids are detected by spraying with a slightly ammoniac solution of bromcresol green. The acids appear as yellow spots against a green background. Aldehydes and ketones appear as yellow to red spots when they are sprayed with an acid solution of 2,4-dinitrophenylhydrazine. A good reagent for the detection of steroids, lipids, and vitamins is a solution of antimony trichloride in chloroform. Dragendorff's reagent, which is a mixture of bismuth subnitrate and potassium iodide, is used to detect alkaloids and organic bases in general. Phenols are detected when sprayed with a solution of ferric chloride. Many inorganic cations can be detected by spraying with solutions of hydrogen sulfide, ammonium sulfide, or ammoniac 8-hydroxyquinoline.

Quantitative thin-layer chromatography

There are two approaches to quantitatively measuring the separated components. In one case the measurement is made on the plate, and in the other case the sample spot is removed from the plate and subsequently assayed.

Quantitation on the plate is achieved by measuring the spot size or by using optic densitometry. Many attempts have been made to correlate a measured parameter of the spot size and the weight of the substance. However, it was Purdy and Truter[37] who found a linear relationship that exists between the square root of the area of the spot and the logarithm of the weight (sample weight between 1 and 80 μg) of the substance:

$$\sqrt{A} = K \log_{10} M$$

where K = proportionality constant

Only two standard solutions, one made from the dilution of the other, are required to determine the equation of a straight line. Equal volumes of the two standards and the unknown are chromatographed simultaneously, and the weight of the compound in the sample is calculated from the equation:

$$\log W = \log W_s + \left[\frac{\sqrt{A} - \sqrt{A_s}}{\sqrt{A_s} - \sqrt{A_d}} \right] \log D$$

where subscripts s and d = standard and diluted standard, respectively
D = dilution factor

Great care must be exercised when performing thin-layer chromatography for quantitative evaluation. It is particularly important that the same volume of standards and sample be applied to the plate, preferably with a microsyringe of ± 1% of the rated delivery amount. The spot area must also be carefully visualized. The area of the spot is measured by placing a sheet of tracing paper over the chromatogram and tracing the area of the spots. The chromatogram can be photographed and the area of the spots measured from an enlargement of the photograph. The area can also be determined with the use of a planimeter or by taking the mean of three measurements of the diameter of the circular spot. The mean coefficient of variation by this method of Purdy and Truter has been found to be between 2% and 7%.

A very simple but not as precise a technique is to spot the sample and a series of standards. A visual comparison of the unknown spot size to the nearest standard spot size provides a quantitative estimation of the unknown to about 90% of the actual amount present.

There is a linear relationship between the extinction area of a spot on the densitometric tracing of a chromatogram and the weight of the material in the spot. This relationship allows for the direct quantitative determination of the compound on the plate by densitometry. There are instruments available* for scanning and measuring the spots. The readings are conveniently traced on recording charts. The areas under the curves are automatically integrated, which gives a rapid quantitative evaluation of the material in the spot. Careful chromatographic techniques must be adhered to if valid quantitative results are desired. Some of the scanning devices make correction for variation of the thickness and color of the plates by use of a double beam—one beam scans the sample path and the other beam scans the area adjacent to the sample path. The errors involved in densitometry vary widely and depend on several conditions, but recoveries of 95% ± 15% are feasible.

The second approach for quantitative analysis of the resolved spots involves removal of the spot material and adsorbent from the plate. The spot material is then

---

*American Instrument Co., Inc., National Instrument Laboratories, Inc., Photovolt Corp., and Schoeffel Instrument Corp.

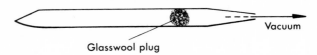

Fig. 16-8. Vacuum cleaner device.

eluted from the adsorbent and subsequently analyzed. This technique is more accurate but is time-consuming and subject to several restrictions. The plates are spotted with both the unknown and a standard sample and developed in the usual fashion. The spot positions are located, preferably by a nonchromogenic method such as fluorescence or ultraviolet adsorption, with use of a phosphor in the adsorbent. If a chromogenic method must be used, the unknown solution is spotted at one end of the plate and, after development, the plate is covered except for the end sample spot. This area is sprayed to reveal the position of the unknowns. Scoring the plate vertically every 2 cm to form channels helps to locate the unknown and to confine the path of the samples.

The unknown spot area, standard sample spot area, and an area equivalent to and adjacent to the sample spot area are quantitatively removed. One technique for the quantitative removal is to completely remove the adsorbent layer up to and around the outlined spot area. The remaining spot area, which is rectangular in shape, is carefully removed by scraping with a small flexible plastic ruler or a razor blade and a camelhair brush. The plate is slightly inclined away from the operator and the scrapings are directed down onto a piece of glassine paper under the plate. A small vacuum cleaner device (shown in Fig. 16-8) can also be used to remove the adsorbent and spot from the glass plate.

The sample is then eluted from the adsorbent with a very polar solvent such as methanol or ethanol. The elutrient is taken to volume in a small volumetric flask and read by ultraviolet spectrophotometry or some other appropriate analytic method. A blank is prepared by taking from the plate an area of adsorbent comparable to the sample area and adjacent to the sample spot. Adsorption caused by the blank is corrected for and the unknown is compared to the standard for quantitative evaluation, assuming that Beer's law prevails. Spencer and Beggs[45] further elaborate on the conditions required for quantitative spectrophotometry in working from thin-layer plates.

CLINICAL APPLICATION

Thin-layer chromatography is used to analyze many different kinds of biologic samples such as blood, urine, other body fluids, and tissue for various classes of compounds such as carbohydrates, amino acids, lipids, hormones, steroids, drugs, and drug metabolites. Most of the methods do not lend themselves to routine clinical laboratory application because the procedures are rather involved and time-consuming. In some cases it is not known if the findings by the thin-layer method have any pathologic significance. Some thin-layer procedures that have found use on a routine basis will be described.

Rapid assay for urinary porphyrins

Porphyria is a disorder of pigment metabolism characterized by an overproduction of porphyrins. The porphyrins are excreted in the urine (Chapter 10). The disease may occur as congenital porphyria or as acute porphyria. Acute porphyria may be precipitated by some toxic agent, such as barbiturate or lead poisoning. Small amounts of uroporphyrin and coproporphyrin are normally found in urine. The finding of abnormal porphyrin excretion in the urine is usually indicative of one or the other kind of porphyria and is of aid in clinical diagnosis of this disease.

The procedure is the thin-layer method described by Scott and co-workers.[44] The test has three levels of interpretation:
1. Rapid screening assay for elevated porphyrins
2. Qualitative assay for separated uroporphyrin and coproporphyrin
3. Quantitative assay for uroporphyrin and coproporphyrin

Reagents and apparatus

All the reagents used are analytic grade and further purification is not necessary.

The standard for routine fluorometric measurements is made by dissolving 5.0 ± 0.1 mg of chromatographically pure coproporphyrin in 1.5 N HCl and diluting to 1000 ml. A 5-ml quantity of this solution is diluted to 100 ml with 1.5 N HCl. A solution of 1.5 N HCl is made by diluting 125 ml of concentrated HCl to 1000 ml. The final concentration of the rountine standard is 0.05 μg/ml.

The developing solution is made by combining chloroform, methanol, ammonium hydroxide ($\approx 25\%$ NH_3), and water in the ratio 12:12:3:2, respectively.

Fluorometric measurements are made of the silica gel taken from the thin-layer plate and suspended in a mixture of 2 parts glycerol and 1 part water. The plates are made from commercially available silica gel containing calcium sulfate binder, or prepared plates can be used. Standard 20 × 20 cm plates are coated with silica gel of about 250-μ thickness, by the Stahl-De Saga method. The plates are allowed to dry overnight and can be used directly without activation. One of several brands of prepared plates can be used but the plates should be tried first to note which gives the best results.

The porphyrin spots on the thin-layer chromatogram are visualized with a high-intensity ultraviolet lamp containing a 100-watt mercury vapor spot bulb.* The fluorescence of the porphyrins is measured by using either a Farrand model A or the Turner model 111 fluorometer with a high-sensitivity door. The primary filter is a Corning No. 5113 and the secondary filter is an Eastman Kodak No. 29A. The same filter combination can be used for both instruments.

Procedure

Preparation of the urine sample
1. To 8.0 ml of urine in a 12-ml centrifuge tube, add 0.2 ml of 8 N HCl. Add 0.2 to 0.3 ml of a saturated solution of sodium acetate to the mixture, to a pH of 3 as indicated by Hydrion paper. Add 2.0 ml of n-butanol and shake vigorously. Centrifuge the mixture for 10 to 15 minutes to clearly separate the phases.
2. At this stage, as a quick screening test, the butanol layer is examined under ultraviolet light. Normal urine gives a barely perceptible fluorescence. For further quantitative analysis, smaller amounts of the sample of the elevated porphyrins should be taken and the volume made up to 8.0 ml with water.

Thin-layer chromatography
3. Spot 60 μl of the butanol layer onto the thin-layer plate, starting about 2 to 2.5 cm from the left-hand edge of the plate and about 2 cm from the bottom edge. With the aid of a stream of warm air from a hair dryer to dry the spot, apply the 60μl of butanol in increments of about 10 μl to maintain a

---
*Ultra-Violet Products, Inc.

spot size of about 7 mm or less. The spot should be completely dry before development and the spotting should be done in subdued light. As many as eight samples can be spotted on a 20 × 20 cm plate.

4. Place the plate into the developing chamber containing the solvent mixture. Allow the solvent to move up the plate 10 cm. This takes about 30 to 45 minutes. Remove the plate, or plates, since two plates can be developed at a time, and dry them in a forced-air hood for about 30 minutes. Do not dry them much longer than 30 minutes, for prolonged standing results in loss of fluorescence.

5. The separated porphyrins are visualized by illumination with the ultraviolet lamp and the spots outlined with a needle. The uroporphyrin remains at the origin of spotting and the coproporphyrin is a sharp band at an R_f of 60 to 65. At this point a more refined screening test result can be given, since the uroporphyrin and coproporphyrin can be related to a normal and to each other. The relative amount of each porphyrin is based on its intensity of fluorescence.

6. To quantitate the porphyrins, carefully scrape the spot onto a piece of glassine paper and transfer the scrapings into a 10 × 75 mm test tube used for fluorometric measurement. A similar-size area of silica gel containing no fluorescence is removed below the origin and used, as a blank. Glycerol solution (2.0 ml) is added to the test tubes, and the tubes are shaken vigorously. Allow the tubes to stand for 20 minutes and shake gently just before reading.

7. The 0.05 µg/ml of coproporphyrin in 1.5 N HCl is used as a full-scale reference to adjust the fluorometer. Since the porphyrins adsorbed on the silica gel are not stable enough to be used as standards, the fluorometric readings must be related to the coproporphyrin standard. By putting known amounts of pure uroporphyrin and coproporphyrin through the chromatographic procedure, a conversion factor for each porphyrin is determined. As an example, 0.1 µg each of pure coproporphyrin and uroporphyrin are put through the chromatographic procedure and read in the 2-ml glycerol solution with the fluorometer, which has been set at a galvanometer reading of 10.0, using 2.0 ml of the 0.05 µg/ml coproporphyrin HCl standard. If the coproporphyrin sample gave a reading of 10.0 on the galvanometer, the conversion factor would be 1.0; if the uroporphyrin sample gave a reading of 11.7, the conversion factor for uroporphyrins would be 1.17.

Calculation

To calculate the urinary porphyrin excretion use the following formula:

$$\frac{\text{Galvanometer reading}}{\text{Conversion factor}} \times 0.1 \times \frac{\text{24-hr volume (ml)}}{0.06} \times \frac{1.5}{8.0} = \mu g \text{ porphyrin}/24 \text{ hr}$$

Results

Porphyrin fluorescence is linear to a concentration of 10 µg/ml when measured in HCl. The lower limit of detection is 0.002 µg of either uroporphyrin or coproporphyrin. A reproducibility of ± 5% of the mean value can be obtained. Low recoveries, 80% to 85%, of uroporphyrin added to urine samples and high recoveries, 105% to 111%, of coproporphyrin are typical of the recovery of porphyrins that can be expected.

The amount of silica gel suspended in the glycerol solution is not important. It has little effect on the porphyrin analysis, since it contributes very little fluorescence and has no quenching effect on the porphyrin fluorescence.

The normal range for uroporphyrin is 10 to 50 $\mu g/24$ hr and for coproporphyrin it is 50 to 200 $\mu g/24$ hr. In the case of coproporphyria, the coproporphyrin content in the urine is very high and many times more than the slightly elevated uroporphyrin content. In the other cases of porphyria—such as acute intermittent, congenital erythropoietic, and acquired—the uroporphyrin content in the urine is very high and many times the elevated coproporphyrin content. In some cases significant amounts of hepta-, hexa-, and pentacarboxylic porphyrins lie between the uroporphyrin and coproporphyrin spots but they can be ignored. Some drugs and their metabolites that fluoresce and could be an interference are usually eliminated by the butanol extraction step in the procedure.

Identification of barbiturates in urine by thin-layer chromatography

On occasion it is important to rapidly identify the barbiturate in a case of suspected barbiturate poisoning. The following thin-layer chromatographic procedure* has been used to identify some of the more common barbiturates in urine. The procedure takes about 1½ hours.

Reagents and equipment

Sodium pentobarbital, sodium phenobarbital, sodium secobarbital (Seconal), barbital, and amobarbital are the *barbiturates* used in this procedure. Other barbiturates that may be of interest can be used, but their R_f values should be determined to find any relationships in regard to overlapping. Different staining characteristics that may help in differentiation may also be revealed during the initial chromatographic examination of a barbiturate different from the ones used in this procedure.

For *individual standards, 1 $\mu g/\mu l$*, weigh 0.100 ± 0.005 g of the individual barbiturate, transfer it to a 100-ml glass-stoppered volumetric flask, and add 5.0 ml of water to dissolve the sodium salts of the barbiturate. Dilute to volume with ethanol or reagent grade alcohol. Reagent alcohol is 5% by volume isopropyl alcohol and 95% ethanol by volume. The amobarbital standard is made by vigorously shaking one capsule, weight 0.172 g, with 10 ml of 1:1 water-alcohol mixture in a 100-ml flask. After complete disintegration of the capsule, dilute the mixture to volume and filter the mixture into another 100-ml flask.

For *mixed standards, 1μg each/μl*, weigh 0.100 ± 0.005 g of the individual barbiturates and transfer to a 100-ml volumetric flask. There is no point in adding the amobarbital, since it has approximately the same R_f value as the Seconal. Add 5.0 ml of water and dilute to volume with reagent alcohol. Store the standards in 4-oz polyethylene bottles.

Prepare the *developing solution* by adding 6.0 ml of acetone to 100 ml of benzene. Solvents are of analytic grade. A large volume of developing solution can be made by adding 60 ml of acetone to 1 L of benzene and storing the mixture in an amber glass-stoppered bottle.

The barbiturate spots are visualized by spraying the developed chromatograms with the following *spray reagents:*
1. Saturated solution of mercurous nitrate—place 1 g of analytic grade mercurous nitrate into a 4 oz plastic bottle, add 100 ml of water, and shake vigorously. Allow the excess mercurous nitrate to settle out and decant the supernatant liquor into another plastic bottle.
2. S-diphenylcarbazone, 0.10% (w/v)—weigh 0.100 g of S-diphenylcarba-

---
*J. R. Simmler, Kirkwood, Mo.

zone* and transfer to a 4-oz plastic bottle. Dissolve the reagent in 100 ml of reagent alcohol. Keep the solution in the refrigerator when not in use.
3. Potassium permanganate, 0.1%—dissolve 0.1 g of analytic grade potassium permanganate in 100 ml of water.

Prepared silicic acid plates designated 4GF or glass fiber impregnated sheets, ChromAR sheet 500,† can be used as the *adsorbent*.

As a *developing chamber*, the S or sandwich chamber equipment for developing plates or sheets can be obtained from Distillation Products Industries.

The detecting reagents are conveniently sprayed with an aerosol *spray apparatus* obtained from Warner-Chilcott Laboratories; however, any TLC spraying equipment can be used.

Microcapillary tubes or pipets of 5- and 10-μl volumes are obtainable from Kensington Scientific Corp. or from Hamilton Co. *Microsyringes* (10 μl) are used for spotting the standards and sample onto the adsorbent layer.

Procedure
Preparation of the sample

1. Add 1.0 ml of 0.2 N sulfuric acid (3.0 ml of concentrated sulfuric acid/L) to 25 ml of urine in a 125-ml separatory funnel. Add 25 ml of chloroform to the separatory funnel. Rock the funnel gently back and forth about 70 to 80 times, causing the chloroform and aqueous layers to intimately interact without forming an emulsion. Allow the layers to separate. Drain the chloroform layer into a filter funnel lined with Whatman No. 40 11-cm paper and filter the chloroform layer into another 125-ml separatory funnel. Repeat the extraction two more times.
2. Add 5.0 ml of 0.45 N sodium hydroxide (18 g of sodium hydroxide/L) to the combined, filtered chloroform layers in the second separatory funnel. Shake the funnel 100 times in the manner indicated before. Allow the layers to separate. Draw off the chloroform layer and discard it. To the caustic aqueous layer add 0.5 ml of 6 N hydrochloric acid (50 ml of concentrated hydrochloric acid and 50 ml of water). Extract the barbiturates with 5.0 ml of chloroform, shaking as indicated. Transfer the chloroform layer to a 10-ml conic centrifuge tube. Repeat the extraction with another 5.0 ml of chloroform. Combine the chloroform extract in the centrifuge tube.
3. Evaporate the chloroform layer to dryness. Placing the tube in a hot water bath will accelerate the evaporation. The temperature of the bath should not exceed 80° C. A stream of air directed into the tube will accelerate the evaporation of the chloroform. Under proper conditions it should not take more than 10 to 15 minutes to evaporate the chloroform. Wash down the sides of the tube with 0.5 ml of reagent alcohol and again evaporate until the volume has been reduced to about 50 μl.

Thin-layer chromatographic procedure

4. Spot the sample and standards in the following way:
5. Cut a piece of ChromAR 500 sheet 20 cm in length—the sheet adsorbent is 20 cm wide. Using either the prepared glass thin-layer plate or the sheet adsorbent, spot 5 μl of the mixed barbiturate standard with the microcapillary pipet onto the adsorbents at a point 2 cm from the left margin of the adsorbent and 2 cm from the bottom edge. See Fig. 16-9. A spotting template, obtainable from many of the scientific apparatus distributing com-

---

*Matheson Chemical Co.
†Mallinckrodt Chemical Works.

Fig. 16-9. Technique of spotting sample to thin-layer plate, using capillary pipet.

panies, is of great aid in aligning the spots. At a point 2 cm from the right of the first spot and again 2 cm from the bottom edge, spot 10 μl of the sample. Spot the sample in two 5-μl portions, allowing the first spot to evaporate before adding the second 5-μl portion. Follow the sample spot with the individual standards 2 cm apart. If more than one sample is to be analyzed, the mixed standard can be left off the plate. As many as eight spots can be placed on the plate or as many as three samples and five single standards can be placed on one plate. As a way of screening, six samples and two mixed standard spots can be placed on the plate. While spotting the solutions, direct a stream of warm air from a hair dryer onto the plate. This will help keep the spots compact. The spots should be dry before development.

6. Place the ChromAR sheet on one of the S chamber halves, aligning the lower edge of the sheet with the lower edge of the S chamber. Place the second half of the S chamber on the first half supporting the adsorbent sheet and clamp into position with the two large spring clamps provided. See Fig. 16-6 for view of the assembled chamber set in the solvent trough.

7. If the prepared plate is used, the space between the two S chambers must be increased. This is done by cutting strips of Teflon or cardboard to give a thickness of 2 to 4 mm. The thin-layer plate, spotted with the sample and standards as just indicated, is set on the inside row of glass nipples on the lower edge of one of the halves of the S chamber. The strip spacers are put in place along the outer edges of the lower half of the S chamber, and the other half of the chamber is placed on the spacers. The clamps are then put in place to hold the S chamber together.

8. The amount of solvent required to develop the ChromAR sheet is 50 ml, and the precoated prepared plate requires 100 ml. The solvent is placed in the trough and covered with the plastic plate, and the S chamber is slowly lowered into the solvent, which is allowed to move up the plate 15

cm from the point of spotting the samples. It takes about 15 to 20 minutes for the solvent to move this distance. The chamber is removed from the solvent trough and is opened. The plate or sheet is removed and is dried with the aid of a stream of hot air from the hair dryer.

9. Placing the plate or sheet in a suitable chamber and spraying it with the mercurous nitrate solution will cause the barbiturate spots to become evident. They will appear as white spots against a wet gray background. The partially dry plate may be oversprayed with the S-diphenylcarbazone solution. The purple-colored background will slowly fade away, leaving blue-violet spots.
10. Comparison of the R_f value and position of the mercurous nitrate barbiturate spot from the urine sample and the corresponding standard barbiturate will, as a first approximation, identify the barbiturate in the sample. The second spraying with the S-diphenylcarbazone, if the spot colors are the same, will further verify the relationship of the unknown barbiturate and the standard. The dry plate, after a spraying with the mercurous nitrate, will, when sprayed with the potassium permanganate solution, reveal the secobarbital as a brown-yellow spot whereas the other barbiturates are unaffected. This will help differentiate between the closely related secobarbital and amobarbital.
11. A comparison of the R_f values of barbiturates resolved on prepared plates and on sheet adsorbent and the color reactions seen after spraying are given in Table 16-7. See also Fig. 16-10.

Results

The limit of detection, using mercurous nitrate as the detecting reagent, is 0.2 μg. Therefore, starting with a 25 ml of urine sample it is possible to detect 1 μg/25 ml, or 0.04 ppm. The extraction of the barbiturates is evidently quantitative, at least to the limit of detection of the method, since urine samples to which known amounts of barbiturates are added do not reveal any barbiturates on a second extraction.

It should be possible to quantitate the barbiturates by first determining what barbiturate is present in the sample and then rerunning the analysis; compare the sample spot size against the spot size from a series of different concentrations of the barbiturate (semiquantitative) or use a recording, integrating densitometer.

Table 16-7
R_f values and identifying color of the barbiturates

Compound	R_f value		Spray reagent		
	Sheet	Plate	Mercurous nitrate	S-diphenyl-carbazone	Potassium permanganate
Barbital	0.23	0.19	White	Violet	—
Phenobarbital	0.29	0.22	White	Violet	—
Pentobarbital	0.37	0.30	White	Violet	—
Amobarbital	0.37	0.35	White	Violet	—
Secobarbital (Seconal)	0.41	0.37	White	Blue	Brown-yellow

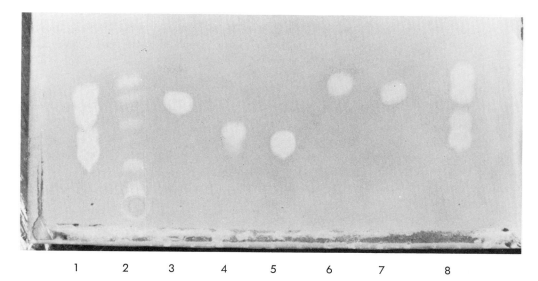

Fig. 16-10. Separated barbiturates on developed TLC plate sprayed with mercurous nitrate.

Positions 1 and 8. Barbital, phenobarbital, pentobarbital, amobarbital, Seconal mixture.

Position 2. Urine sample spiked with the above five barbiturates. First two spots nearest point of spotting are unknown.

Position 3. Pentobarbital.
Position 4. Phenobarbital.
Position 5. Barbital.
Position 6. Secobarbital (Seconal).
Position 7. Amobarbital.

Whole blood and other body fluids can also be analyzed for barbiturates by this procedure. A 2- to 10-ml quantity of whole blood or serum should be used. The extraction is the same as it is for urine samples.

Determination of 3-methoxy-4-hydroxymandelic acid in urine

3-Methoxy-4-hydroxymandelic acid, generally referred to as vanillylmandelic acid or VMA, is a product of catecholamine metabolism (Chapter 14). Patients with pheochromocytoma demonstrate large amounts of VMA in their urine. The thin-layer chromatographic procedure for VMA to be presented here is based on the procedure of Annino and co-workers.[2] It can be helpful in the diagnosis of pheochromocytoma.

Reagents

Solvent system—*n*-butanol–glacial acetic acid–water (90:20:40)
Florisil*
Silica gel G (Brinkmann)
Dye solution
   Place 0.05 g of fast red GG† in 100 ml of water. Filter through glass wool. Prepare fresh each day and protect from light.
Potassium carbonate, saturated solution

---

*Floridin Co.
†General Dyestuff.

Buffer, borate, 0.1 M
> Dissolve 38.1 g of $Na_2B_4O_7 \cdot 10\ H_2O$ in 950 ml of water and adjust pH to 10.2 ± 0.1, using concentrated (18 N) sodium hydroxide. Dilute to 1 L and store in a plastic bottle.

Eluting solution
> Mix 15 ml of borate buffer and 10 ml of methanol. Prepare just before use.

Stock standard
> Dissolve 25 mg of 3-methoxy-4-hydroxymandelic acid in 100 ml of absolute ethanol. Store in a refrigerator.

Working standard
> Dilute 4.0 ml of the stock standard to 10 ml with absolute ethanol. This standard is equivalent to a urinary VMA concentration of 11 mg/L. This value takes into account the extra dilution effected when the urine is treated with HCl.

Special apparatus

Apparatus to prepare plates by Stahl-De Saga method (Brinkmann Instruments)

Procedure

Preparation of plates
1. Using a 125-ml flask, shake vigorously, for about 1 minute, 30 g of silica gel G and 60 to 65 ml of water.
2. Pour the slurry into the trough and pull it over six 20 × 20 cm plates.
3. Set the gate of the trough applicator at 0.25 mm.
4. Dry the plates overnight and use without activation.

Preparation of urine sample
5. Into a 20 × 150 mm tube with a ground glass top, measure 10 ml of urine and 1.0 ml of concentrated HCl. Swirl to mix; then heat in a boiling water bath for 10 minutes. Cool, add 1 g of Florisil, stopper, and shake for 10 minutes. Centrifuge for 5 minutes.
6. Place a 4.0 ml portion of the urine so treated in another 20 × 150 mm tube and add 10 ml of ethyl acetate. Stopper, shake for 2 minutes, centrifuge briefly, and with a capillary pipet transfer the acetate (upper) layer to a 50-ml centrifuge tube (wide-mouth).
7. To the remaining aqueous layer add another 10 ml of ethyl acetate. Repeat the extraction as above and add to the first extract.
8. Evaporate the combined extracts to dryness in a 60° C water bath, using a stream of air to facilitate evaporation. Remove from the bath, add 0.4 ml of absolute ethanol, and mix well to redissolve the extracted VMA.

Chromatography
9. Draw a line on the plate 11.5 cm from the bottom. Apply 50 μl of extract and of standard with a syringe, pipet, or other suitable applicator, 1.5 cm from the bottom of the plate and 2 cm from the sides of the plate. Apply as streaks approximately 2 cm long, in small increments, drying with a hot-air hair dryer between applications. Six applications may be made on a 20-cm plate.
10. Place the plate in the tank with the coated side facing the wall of the chamber which has been previously lined with filter paper, cover, and allow the solvent to run until it reaches the limiting line.
11. Remove the plate from the tank and dry it with a hot-air hair dryer in a fume hood until the odor of the solvent is gone.

12. Spray with the dye solution and then dry with the hot-air hair dryer. Spray with saturated potassium carbonate solution and dry again.

Quantitation

13. Using a pointed instrument (such as a dissecting needle), outline the purple VMA spots. Outline an extra area below the standard, approximately equivalent in size to the VMA spots. This area serves as a blank and is carried through the subsequent stages. Using a flat-edged spatula, remove the silica gel around the spots; then scrape the silica gel from the outlined area into 16 × 100 mm test tubes.
14. Add 2.5 ml of eluting solution to each tube and mix well so as to break up the silicic acid gel. Add 0.5 ml of dye solution and mix again. Centrifuge for 5 minutes. Decant the supernatant into 1-cm cuvets. Read absorbance against the blank at 510 nm within 20 minutes after addition of the dye.

Calculations

$$\frac{11.0}{\text{Absorbance of standard}} \times \text{Absorbance of test} = \text{Concentration of VMA in mg/L}$$

Clinical findings

Refer to discussions in Chapter 14.

Gas chromatography

Gas chromatography is the latest chromatographic technique to be applied to the separation and analysis of biomedical and industrial specimens. It is no doubt the most sophisticated of the methods of chromatography, at least in regard to instrumentation. A great deal of mechanization has gone into what is in fact no more than the classic application of chromatography except that the mobile phase is a gas. Aside from the separation aspects of gas chromatography, the detection of the resolved components is particularly convenient. There are several different kinds of detectors, each with its degree of sensitivity and application. The detectors not only respond to a change in the composition on the mobile gas phase but also respond to the amount of change. This allows for a rapid quantitative evaluation of the resolved constituents as they come off the column.

In spite of the extensive instrumentation usually associated with gas chromatography, a quite satisfactory device can be made from rather simple equipment. Practically all gas chromatographic instruments consist of the following components:

1. Source of an inert carrier gas and a gas flow regulator
2. Injector port for introducing the sample into the system
3. Long or spiral metal or glass column for supporting the combination solid and liquid phase
4. Furnaces for controlled heating of the injector port and column
5. Detector device and associated electrometer and recorder

Of these components the most important, and the "heart" of the system, is the column. There are two kinds of columns, the packed and the capillary. The packed column is the one most in use at present. The column, made of some inert material such as stainless steel or glass, is packed with a granular solid (usually diatomaceous earth) called the solid support, which is coated with a liquid of low volatility. The supported liquid is called the stationary phase. The carrier gas is usually hydrogen, helium, argon, or nitrogen flowing through the system at a controlled rate. Between the source of the carrier gas and the column is the sample inlet. Usually this area is heated to a higher temperature than the column; thus, rapid vaporization of the injected sample is assured. It is desirable that the sample be introduced as a "plug" of vapor prior to moving into the column. The column is also heated, and the temperature of the column can be programmed to give a fixed rate of temperature increase with time. Precise temperature control of the column is essential for repetitive results, since the vapor pressure of the stationary liquid phase varies with the temperature. The vapor pressure of the liquid phase contributes to baseline effect noted on the recorder.

As the individual separated components come off the column, they pass through a detector, which then gives a signal that is related to the amount of the substance present. There are several kinds of detectors—such as the following types: (1) thermal conductivity, (2) gas density balance, (3) radioactive ionization, (4) flame ionization and thermionic, and (5) coulometric. Differential detectors are almost exclusively used. The detector measures a certain property related to the amount of the individual components in the gas stream. The resulting chromatogram is a gaussian type of curve with the peak of the curve corresponding to the greatest concentration of the substance, and the baseline is due to the pure carrier gas. The distance on the time axis of the chromatogram, from the time the sample is injected into the column to the maximum of the peak, corresponds to the retention time of the particular substance. Some detectors, the thermal conductivity and the gas density balance types, are universal detectors while the other detectors are selective. One of the characteristic properties of the thermionic detectors is its increased response to halogen- and phosphorus-bearing compounds. For further and more detailed discussion of the theory, equipment, and application refer to the various texts on gas chromatography included at the end of this chapter.

Gas chromatography of biologic materials is, almost without exception, preceded by some preliminary treatment of the sample.

Table 16-8
Some biochemical applications of gas chromatography

Class of compounds	Sample treatment
Estrogens	Hydrolysis, extraction, acetylation
Corticosteroids	Extraction, paper or thin-layer chromatography, acetylation, and oxidation
Urinary aromatic acids	Extraction, ion exchange, methylation
Amines and amino acids,	Extraction, esterification
aralkylamines and alkylamines	Extraction, injected directly
Bile acids	Extraction, esterification (usually methyl ester)
Carbohydrates	Extraction, trimethylsilyl ethers
Barbiturates	Extraction, injected directly
Organic anesthetics	Extraction, injected directly

The application of gas chromatography to biomedical studies has given great impetus to this field. Lipid metabolism has been studied by a combination of thin-layer and gas chromatography. After appropriate treatment of the sample—urine, plasma, or tissue—the lipid classes and subclasses were purified by thin-layer chromatography. The bands were removed from the thin-layer plate and the lipids methylated. The methylated fatty acids were separated and quantitated by means of gas-liquid chromatography. Other biomedical applications of gas chromatography are given in Table 16-8.

The detailed gas chromatographic conditions for the analysis of the various classes given in the table can be obtained from textbooks on gas chromatography given at the end of the chapter.

References

1. Anet, E. F.: Thin-layer chromatography of 2,-4-dinitrophenylhydrazine derivatives of hydroxycarbonyl compounds, J. Chromatogr. 9: 291, 1962.
2. Annino, J. S., Lipson, M., and Williams, L. A.: Determination of 3-methoxy-4-hydroxymandelic acid (VMA) in urine by thin-layer chromatography, Clin. Chem. 11:905, 1965.
3. Bennet, R. D., and Heftmann, E.: Thin-layer chromatography of corticosteroids, J. Chromatogr. 9:348, 1962.
4. Blank, M. L., and Laudberg, W. D.: Determination of mono, di and triglycerides by molecular distillation and thin-layer chromatography, J. Am. Oil Chem. Soc. 38:312, 1961.
5. Bobbitt, J. M., Schwarting, A. E., and Gritter, R. J.: Introduction to chromatography, New York, 1968, Reinhold Publishing Corp.
6. Brieskorn, C. H., and Wenger, E.: Constituents of *Salvia officinalis*. XI. Analysis of the ethereal sage oil by gas and thin-bedded chromatography, Arch. Pharmacol. 293(65):21, 1960.
7. Cassidy, H. G.: Fundamentals of chromatography, New York, 1957, Interscience Publishers, Inc.
8. Crump, G. B.: Rapid characterization of small amounts of oil by thin-layer silica-gel chromatography, Nature 193:674, 1962.
9. Davidek, J.: Separation of gallic acid and its esters on thin layers of polyamide powder, J. Chromatogr. 9:363, 1962.
10. Dhont, J. H., and DeRooy, C.: Chromatographic behavior of 2,4-dinitrophenylhydrazones on chromatoplates, Analyst 86:74, 1961.
11. Dhont, J. H., and De Rooy, C.: Thin-layer chromatography of 3,5-dinitrobenzoates, Analyst 86:74, 1961.
12. Furukawa, T.: Strip chromatography: X. Separation of phenols, aromatic aldehydes, ketones, and carboxylic acids by means of chromatostrips, Chem. Abstr. 54:13938, 1960.
13. Gänshirt, H., and Morianz, H.: Unterschungen zur quantitativen Auswertung der Dünnschichtchromatographie, Arch. Pharmacol. 293(65):1065, 1960.
14. Gritter, R. J., and Albers, R. J.: Non-destructive zone developing in thin-layer chromatography, J. Chromatogr. 9:392, 1962.
15. Habermann, E., Brandtlow, G., and Krusche, B.: Determination of plasma phospholipids using thin-film chromatography, Klin. Wochenschr. 39:816, 1961.
16. Hansson, J., and Alm, A.: Chromatographie sur couches minces des dérivés de la diphénylamine application à l'analyse des poudres, J. Chromatogr. 9:385, 1962.
17. Heftmann, E., editor: Chromatography, ed. 2, New York, 1967, Reinhold Publishing Corp.
18. Helfferich, F.: Ion exchange, New York, 1962, McGraw-Hill Book Co.
19. Hofmann, A. F.: Thin-layer adsorption chromatography of free and conjugated bile acids on silicic acid, J. Lipid Res. 3:127, 1962.
20. Honegger, C. G.: Dünnschicht-Ionophirese und Dünnschicht-Ionophorese-Chromatographie, Helv. Chim. Acta 44:173, 1961.
21. Honegger, C. G.: Über die Dünnschichtchromatographie von Lipidenuntersuchungen von Gehirngewebe aus der weissen Substanz Multiple-Sklerose-Kranker, Helv. Chim. Acta 45: 2020, 1962.
22. Honegger, C. G.: Über die Dünnschichtchromatographie von Lipidenuntersuchungen von Gehirngewebe Multiple-Sklerose-Kranker und Normaler, Helv. Chim. Acta 45:281, 1962.
23. Horowitz, R. M., and Gentili, B.: Flavonoids of citrus. IV. Isolation of some aglycones from the lemon *(Citrus limon)*, J. Organ. Chem. 25:2183, 1960.
24. Kaufman, H. P., and Khoe, T. H.: Thin-layer chromatography of fats. VII. Separation of fatty acids and triglycerides on plaster of paris plates, Fette Seifen 64:81, 1962.
25. Kirchner, J. G., and Miller, J. M.: Prepara-

tion of terpeneless essential oils, Ind. Eng. Chem. **44**:318, 1952.
26. Machata, G.: Dünnschichtchromatographie in der Toxikologie, Mikrochim. Acta, p. 79, 1960.
27. Mahadevan, V., and Lundberg, W. O.: Preparation of cholesterol esters of long-chain fatty acids and characterization of cholesteryl arachidonate, J. Lipid Res. **3**:106, 1962.
28. Malins, D. C., and Mangold, H. K.: Fractionation of fats, oils, and waxes on thin layers of silicic acid, J. Am. Oil Chem. Soc. **37**:576, 1960.
29. Marcuse, R.: Separation of alkanals and alkanones with carbonyl groups in different positions by thin-layer chromatography, J. Chromatogr. **7**:407, 1962.
30. Mariani, A., and Mariani-Marelli, O.: Separation of the opium alkaloids by means of aluminum plate chromatography, Rendic. Ist. Sup. Sanit. **22**:759, 1959.
31. Matthews, J. S., Pereds, V. A. L., and Aguilera, P. A.: Steroids: the quantitative analysis of steroids by thin-layer chromatography, J. Chromatogr. **9**:331, 1962.
32. Newbauer, D., and Mothes, K.: Thin-layer chromatography of poppy alkaloids, Chim. Panta Med. **9**:466, 1961.
33. Nicolaus, B. J. R.: Anwendung der Dünnschichtchromatographie auf Pteridine, J. Chromatogr. **4**:384, 1960.
34. Nishikaze, O., and Staudinger, H. J.: Dünnschichtchromatographische Isolierung von Aldosteron aus Harn, Klin. Wochenschr. **40**:1014, 1962.
35. Peifer, J. J.: A rapid and simplified method of analysis by thin-layer chromatography using microchromatoplates, Mikrochim. Acta, p. 529, 1962.
36. Petrowitz, H. J., and Pastuska, G.: Über die Kieselgelschicht-Chromatographie gesättigter aliphatischer Dicarbonsäuren, J. Chromatogr. **7**:128, 1962.
37. Purdy, S. J., and Truter, E. V.: Quantitative analysis by thin-film chromatography, Analyst **87**:802, 1962.
38. Pastuska, G.: Untersuchungen über die qualitative und quantitative Bestimmung der Zucker mit Hilfe der Kieselgelschicht-Chromatographie, Z. Anal. Chem. **179**:427, 1961.
39. Peifer, J. J.: A rapid and simplified method of analysis by thin-layer chromatography using microchromatoplates, Mikrochim. Acta, p. 529, 1962.
40. Petrowitz, H. J., and Pastuska, G.: Über die Kieselgelschicht-Chromatographie gesältigter aliphatischer Dicarbonsäuren, J. Chromatogr. **7**:128, 1962.
41. Purdy, S. J., and Truter, E. V.: Analysis of thin-film chromatography, Analyst **87**:802, 1962.
42. Ruggieri, S.: Separation of the methyl esters of fatty acids by thin-layer chromatography, Nature **193**:1282, 1962.
43. Scher, A.: Die Analyse von Tocopherol gemischen mit Hilfe der Dünnschichtchromatographie, Mikrochim. Acta, p. 308, 1961.
44. Scott, C. R., Labbe, R. F., and Nutter, J.: A rapid assay for urinary porphyrins by thin-layer chromatography, Clin. Chem. **13**:6, 493, 1967.
45. Spencer, D. R., and Beggs, B. H.: Thin-layer chromatography on silica gel, quantitative analysis by direct spectrophotometry, J. Chromatogr. **21**:52, 1966.
46. Stahl, E., editor: Thin-layer chromatography; a laboratory handbook, ed. 2 (translated by Cambridge consultants), New York, 1965, Academic Press, Inc.
47. Stahl, E., and Kaldewey, H.: Spurenanalyse physiologisch aktiver, einfacher Indolderivate, Hoppe Seyler Z. Physiol. Chem. **323**:182, 1961.
48. Stahl, E., and Kaltenbach, U.: Dünnschicht-Chromatographie: Schnelltrennung von Digitalis- und Podophyllum-Glycosidgemischen, J. Chromatogr. **5**:458, 1961.
49. Stahl, E., and Trennheuser, L.: Gas-phase chromatography of terpene and hydroxyphenyl propane substances, Arch. Pharmacol. **293**(65):21, 1960.
50. Starka, L., and Malikova, J.: A simple method for the separation of pregnanetriols by spread chromatography, J. Endocr. **22**:215, 1961.
51. Struck, H.: Trennung und Bestimmung von Oeston, 17β-Oestradiol und 16α, 17β-Oestriol mit Hilfe der Dünnschichtchromatographie, Mikrochim. Acta, p. 634, 1961.
52. Subbarao, R., Roomi, M. W., Subbaram, M. R., and Achaya, K. T.: Separation of oxygenated fatty compounds by thin-layer chromatography, J. Chromatogr. **9**:295, 1962.
53. Sundt, E., and Saccardi, A.: Thin-layer chromatography and paper chromatography of vanilla flavoring compounds, Food Technol. **16**:89, 1962.
54. Tate, M. E., and Bishop, C. T.: Thin-layer chromatography of carbohydrate acetates, Can. J. Chem. **40**:1043, 1962.
55. Teichert, K., Mutschler, E., and Rochelmeyer, H.: Plate chromatographic studies of mixtures of natural products, Z. Anal. Chem. **181**:325, 1961.
56. VanDam, M. J. O., Dekleuver, G. J., and Detteus, J. G.: Thin-layer chromatography of weakly polar steroids, J. Chromatogr. **4**:26, 1960.
57. Vogel, W. C., Doizaki, W. M., and Zieve, L.: Rapid thin-layer chromatographic separation of phospholipids, J. Lipid Res. **3**:138, 1962.
58. Vogler, K., Studer, R. O., Lergier, W., and Lanz, P.: Synthesen in der Polymyxin-Reihe

Synthesis des cyclischen Decapeptids 88, Helv. Chem. Acta **43:**1751, 1960.
59. Waldi, D., Schnackerz, K., and Munter, F.: Eine systematische Analyse von Alkaloiden aus Dünnschichtplatten, J. Chromatogr. **6:**61, 1961.
60. Weicker, H., and Brossmer, R.: The structural identification of some naturally occurring branched-chain fatty aldehydes, Klin. Wochenschr. **39:**1265, 1961.
61. Williams, J. A., Shorma, A., Morris, L. J., and Holman, R. T.: Fatty acid composition of feces and fecaliths, Proc. Soc. Exp. Biol. Med. **105:**192, 1960.
62. Winterstein, A., Studer, A., and Rüegg, R.: New results of carotenoid research, Chem. Ber. **93:**2951, 1960.
63. Wollenweber, P.: Dünnschicht-chromatographische Trennungen von Aminosauren an Cellulose-Schichten, J. Chromatogr. **9:**369, 1962.
64. Wollenweber, P. J.: Dünnschicht-chromatographische Trennungen von Farbstoffen an Cellulose-Schichten, J. Chromatogr. **7:**557, 1962.

ADDITIONAL REFERENCES

Bayer, E.: Gas chromatography, New York, 1961, American Elsevier Publishing Co., Inc.

Burchfield, H. P., and Storrs, E. E.: Biochemical applications of gas chromatography, New York, 1962, Academic Press, Inc.

Dal Nogare, S., and Juvet, R. S., Jr.: Gas-liquid chromatography, New York, 1962, Interscience Publishers, Inc.

Giddings, J. C.: Dynamics of chromatography, part I, New York, 1965, Marcel Dekker, Inc.

Hashimoto, J.: Thin-layer chromatography, Tokyo, 1962, Hirokawa Publishing Co.

James, A. T., and Morris, L. J., editors: New biochemical separations, New York, 1964, D. Van Nostrand Co., Inc.

Kromen, H. S., and Bender, S. R., editors: Theory and application of gas chromatography in industry and medicine, New York, 1968, Grune & Stratton, Inc.

Lederer, E., and Lederer, M.: Chromatography, a review of principles and applications, ed. 2, New York, 1957, American Elsevier Publishing Co., Inc.

Littlewood, A. B.: Gas chromatography, New York, 1962, Academic Press, Inc.

Mangold, H. K., Schmid, H. H. O., and Stahl, E.: Thin-layer chromatography, methods of biochemical analysis, vol. 12, New York, 1964, Interscience Publishers, Inc.

Marini-Bettolo, G. B., editor: Thin-layer chromatography, New York, 1964, American Elsevier Publishing Co., Inc.

Purnell, H.: Gas chromatography, New York, 1962, John Wiley & Sons, Inc.

Randerath, K.: Thin-layer chromatography, New York, 1966, Academic Press, Inc.

Scott, R. P. W., editor: Gas chromatography, London, 1960, Butterworth & Co., Inc.

Truter, E. V.: Thin-film chromatography, New York, 1963, Interscience Publishers, Inc.

Reference lists are available from the following manufacturers of chromatography equipment:

Camag Bibliography Service. References on TLC obtainable from Gelman Instrument Co., P. O. Box 1448, Ann Arbor, Mich. 48106; or from Arthur H. Thomas Co., P. O. Box 779, Philadelphia, Pa. 19105

Eastman Kodak Distillation Products Industries, Rochester, N. Y. 14603 (dyes for detection and adsorbents on plastic film)

Mallinckrodt Chemical Works, 2nd and Mallinckrodt Sts., St. Louis, Mo. 63160

G. K. Turner Associates, 2524 Pulgas Ave., Palo Alto, Calif. 94303

17

Electrophoresis

CARL R. JOLLIFF

Since the last edition of this book the field of protein separation and quantitation, as it pertains to the clinical laboratory, has been expanding. New methodologies, improvements in existing methods, and advances in knowledge of protein chemistry and immunochemistry have mandated continual changes to keep clinical chemistry and immunology laboratories up to date. It is the purpose of this chapter to acquaint the reader with some of the basic methodology available for the separation and quantitation of body fluid proteins and with some of the technical problems involved, and to delineate some of the more sophisticated procedures and methods that may be asked for by the clinician to better ascertain protein abnormalities exhibited in the various pathologic entities.

BASIC CONCEPTS

The original work of Reuss in 1809, with later refinements by Theorell and Tiselius, forms the background to the procedures used today in sophisticated and elaborate electrophoresis systems.

Essentially, electrophoresis is the movement of colloidal particles in an electric field. The movement of these particles depends on a number of factors: the shape of the particle, its size, and its electric charge; the strength of the electrical field; the pH of the buffer with which it comes in contact; the medium on which the separation is taking place; and the temperature.

In the classic "moving boundary" or "free" electrophoresis proposed and performed by Tiselius, no supportive medium was used. Rather, the sample was layered between buffers to which an electric potential had been applied. The difference in refractive indices between the separated protein constituents could be photographed and recorded with an elaborate optic measurement system. The cost of the equipment, the large amount of sample, and the incomplete separation necessitated establishment of a more practical method of separation.

Electrophoresis on a stabilized medium, or so-called zone electrophoresis, filled the need for a procedure that could be adapted to the clinical laboratory. This form of electrophoresis is most commonly used today. Zone electrophoresis utilizes a particular form of medium that will support the sample and allow for separation and final staining of the sample's separated components. The media most commonly used are cellulose acetate, starch gel, agar gel, agarose, polyacrylamide, and some of the ampholytes. In earlier clinical work, paper was used as a supporting medium;

however, it has largely been replaced because of its somewhat poor resolution capabilities when compared to the newer media.

Certain media such as starch and polyacrylamide utilize the phenomena of molecular sieving as well as separation by electrophoretic mobility. Molecular sieving utilizes the principle that as a solute passes through a supporting medium (stationary phase), interaction between the various solute molecular sizes and the microscopic granules in the stationary phase occurs. Molecular populations are thus separated by their affinity for these spongelike granules into a gradient similar to that seen in gel chromatography. Polyacrylamide has the advantage of serving as an inert convection medium as well as being further selective as its percentage in the stationary phase is changed.

It is possible to mix suporting media such as agarose and polyacrylamide to enhance the elucidation of certain components being separated. As many as twenty-five or more components may be observed with such media as polyacrylamide and starch. Certain polymorphisms of specific proteins may thus be delineated by these procedures.

Since the clinical laboratory is most often concerned with the separation of proteins by electrophoresis, the structure of proteins and its relationship to the electrophoretic method should be discussed.

Proteins are rather large molecules composed of amino acids. Amino acid sequence influences the properties of the protein. Since the amino acid consists of polar groups known as the amino (NH_3) and carboxyl (COOH) groups, it is capable of carrying an ionic charge. The charge carried by the amino acid depends on the pH of the medium in which it is found. In alkaline media the amino acid is negatively charged, while in acid media it becomes positively charged.

$$NH_2-\underset{R}{\overset{H}{C}}-COOH \xrightarrow{OH^-} NH_2-\underset{R}{\overset{H}{C}}-COO^- + HOH$$

$$NH_2-\underset{R}{\overset{H}{C}}-COOH \xrightarrow{H^+} NH_3^+-\underset{R}{\overset{H}{C}}-COOH$$

Amino acids behaving this way are therefore considered to be amphoteric. Since proteins are multiple amino acids, they too may be classified as amphoteric in nature. Protein molecules may have a balanced number of positive and negative charges, depending on the pH of the media in which they exist. When a balance of charges exists (isoelectric point), the molecule will not migrate in an electric field. As the pH value is raised above the isoelectric point, there is an increase in the net negative charge on the protein which allows it to migrate further toward the positive (anode) electrode. Conversely, if the pH is below the isoelectric point, the migration is toward the negative (cathodic) electrode. If a buffer of pH 8.6 is used, those protein molecules with an isoelectric point (pI) below that value will migrate toward the anode. Such is the case with albumin, which has a pI of 4.9.

Table 17-1 gives the isoelectric points of the better known proteins. Note that molecular size contributes greatly to the migration of the protein molecule.

The factors influencing electrophoresis are many. The more important ones are as follows:

The buffer   The electric field

Table 17-1
Some physical constants of the proteins separated by electrophoresis

Area	Proteins	Approximate molecular weights	pI
Albumin	Prealbumin		4.7
	Albumin	69,000	4.9
Alpha$_1$ globulins	Alpha$_1$-antitrypsin		4.0
	Alpha$_1$-acid glycoprotein	40,000	2.7
	Haptoglobin 1-1	400,000	4.1
	Alpha$_1$-lipoprotein		
Alpha$_2$ globulins	Alpha$_2$-glycoprotein		3.8-4.1
	Ceruloplasmin		4.4
	Alpha$_2$-macroglobulin		5.4
	Haptoglobin 2-2	41,000	
	Haptoglobin 1-1	20,000,000	
Beta globulins	Beta-lipoprotein	40,000	5.4
	Transferrin		5.9
	Fibrinogen	3,000,000	5.8
Gamma globulins	IgG		5.8-7.3
	IgA	156,000	
	IgM	1,000,000	
	IgD		
	IgE		

The support medium
Electroendosmosis
Wick flow

Buffer

The buffer acts as an agent to vary the charge on different proteins. Because of their varying isoelectric points, they may be separated from one another during the process of electrophoresis. The two most common buffers are the barbital (diethylbarbiturate) and TRIS (tris[hydroxymethyl]aminomethane) systems. The barbital buffers (pH range 8 to 9) have probably been used most successfully because there is minimal denaturation of the proteins during electrophoresis. The higher the pH of the buffer, the more denaturation takes place. The barbital buffers are for the most part monovalent, and thus they contribute less to the ionic strength than do the polyvalent ions. Low ionic strength buffers permit fast migration rates and low heat development. Barbital ions are very large and migrate slowly as compared with the borate ions, which will form complexes and thus increase conductivity and migration, causing a heat effect.

Ionicity of 0.025 to 0.075 for the barbital buffers and 0.12 to 0.03 for the TRIS buffers are generally acceptable. Ionic strength of a buffer solution is given by the formula:

$$\mu = \tfrac{1}{2} \Sigma \ (i \times n^2)$$
where μ = ionic strength (current carrying capacity of dissolved electrolytes)
  i = molal concentration of particular ion
  n = valence of ion

For actual determination of ionic strength of a buffer solution, electric conductance studies should be performed. Such factors as electrolyte dissociation constants, actual salt concentrations, influence of one ion on the dissociation of another, and pH must enter into the calculation of true ionic strength.

The electric field

The support medium will offer a resistance to the current passing through it and thus heat will be generated. The current being carried by the ions controls the migration of the proteins being separated. The resistance of the medium generally remains constant; therefore, to increase migration rates the voltage must be increased. The increased voltage causes heat and evaporation from the medium, which in turn will cause a concentration of ions within the medium and a change in resistance with increased migration. This variance may be controlled by a constant current or constant voltage power supply. In general, for short runs as in agarose or cellulose acetate, constant voltage will be sufficient. However, in the case of longer runs, as in polyacrylamide or starch gels, constant current is preferred since voltage will drop throughout the run and less heat will thereby be generated.

Short running times are preferred for the simple separations of serum proteins on agarose or cellulose acetate since longer times tend to diffuse the protein bands and sharp separations are not obtained.

Support media
Cellulose acetate

Cellulose acetate membranes are produced by the action of acetic anhydride on cellulose. The membrane consists of a three-dimensional structure composed of interlocking chambers whose walls are cellulose acetate polymer. The interior spaces consist of 80% void space which is capable of holding the electrolyte.

The membranes have a great many advantages, some of which are rigid manufacturing control over pore size, chemical content, thickness, and physical qualities. They may also be converted from their opaque quality to one of complete transparency through the use of organic solvents or certain oils as with a 1.47 refractive index.

The disadvantage is mainly that the membranes will vary from one manufacturer to another, which results in differences in handling, chemical stability, separation quality, and ease of clearing.

Agar

Agar is a polysaccharide obtained from the cell walls of red algae. It will form a colloidal gel in concentrations as dilute as 1%, thereby rendering it capable of drying on glass or cronar surfaces, becoming colorless and transparent. It is ideal for quantitation after staining by elution or densitometry. It is relatively inexpensive and can be preserved for long periods of time.

Agarose

Agarose is produced by separation from the undesired polysaccharide agaropectin component of agar. All of the advantages of agar as mentioned before are retained, plus agarose possesses practically no charge, does not inactivate enzymes as does agar, does not complex with antigens, forms clearer and stronger gels, can be held at 40° C before solidification, can be used with a wider variety of buffers, and can be decolorized more easily after storing than agar.

The disadvantages are its expense and, as will be discussed, it develops very little electroendosmosis.

Starch

Starch, which was introduced by Smithies in 1955, appeared to have superior resolving power compared to all other media then used. Today it is rivaled only by polyacrylamide in its superior separation qualities. About twenty proteins can be resolved in the separation of human serum. The starch as used today is partially hydrolyzed potato starch. Each batch is accompanied with specific directions as to the preparation of the gels.

Starch is difficult to work with. Inexperienced workers are often discouraged, especially if they are not meticulous laboratory workers. The patterns require special treatment for densitometric visualization and quantitation and are not easily stored or preserved.

Polyacrylamide

The gels are prepared by the polymerization of acrylamide polymer with methylene bisacrylamide. They are chemically inert, transparent, and stable. Separation of serum proteins is accomplished by electrophoresis as well as molecular sieving. They can be prepared in different pore sizes by varying the percentage concentrations. Disk type or flat bed polyacrylamide gels may be utilized.

Disadvantages are much the same as for starch. Polyacrylamide is difficult to work with, to cut, and to store, since shrinkage or swelling occurs as water is lost or gained.

Electroendosmosis

Support media, when placed in an alkaline buffer, develop a negative charge. The support media cannot migrate because of the rigidity of their structure; however, next to every negative site in the support media is a positively charged buffer ion which may migrate during the electrophoretic process. Movement of these ions is toward the cathode. Therefore, any protein fraction having a very low electrophoretic mobility can be displaced toward the cathode by this movement of positively charged buffer ions.

In some instances, the gamma and beta globulins may end up cathodic to the point of application because the electroendosmotic effect is greater than the migration capability of this group of proteins. The effect of electroendosmosis is most pronounced in agar gels; there is less effect in agarose and cellulose acetate. This effect is particularly valuable, however, in such procedures as counterimmunoelectrophoresis, as discussed later.

Wick flow

Wick flow is most pronounced in cellulose acetate electrophoresis. As heat is generated during the electrophoretic process, evaporation of solvent from the support medium occurs. To compensate for this loss, solvent flows into the medium from the cathodic and anodic buffer vessels to maintain an equilibrium of moisture (solvent) within the medium. The wick flow will greatly influence the position of the fractions

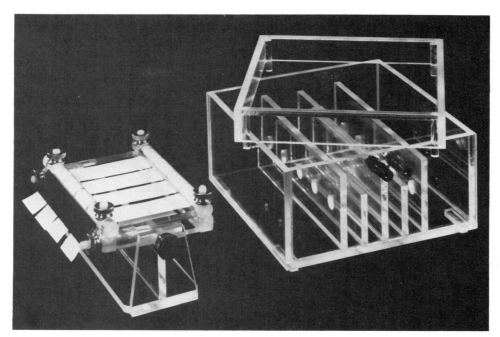

Fig. 17-1. Cell and cellulose acetate frame with stand. (Courtesy Reclin Co., Lincoln, Neb.)

and thus mandate a proper sample application site depending on the separation desired. This application site should remain the same for each run if the investigator hopes to maintain any degree of reproducibility in electrophoretic separations and quantitations.

PROTEIN SEPARATION

The following are procedures that have been or are currently being used in our laboratories for separation of serum proteins by electrophoresis. Total serum protein values are determined by refractometry, which in our hands has been quite reliable for the establishment of this value. Its advantages are the small amount of sample required, short amount of time required for measurement, and ability to retrieve the sample after measurement if needed. One must guard against insufficient volumes of serum, air bubbles, or samples with excessive amounts of solids, such as urea, glucose, lipoproteins, and bilirubin. As in any chemical procedure, strict cleanliness of the apparatus is a necessity.

Macro–cellulose acetate electrophoresis

Equipment

Cell, Reclin (No. 117A) (Fig. 17-1) or other types by Gelman Instruments or Consolidated Laboratories
Frame (Reclin No. 120F) (Fig. 17-1)
Power supply, 175 volts (Instrumentation Specialties Company No. 490) (Fig. 17-2)
Trays, plastic (Gelman No. 7211)
Trays, stainless, 2×8 inches, with lids
Glass slab, 6×12 inches

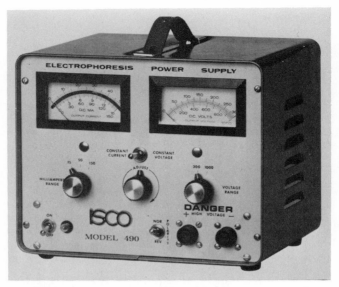

Fig. 17-2. Power supply for constant current as well as constant voltage. Polarity may be switched. (Courtesy Instrumentation Specialties Co., Lincoln, Neb.)

Photovolt densitometer with cellulose acetate attachment, 0.1-mm slit and 525-nm filter (Photovolt Corp. No. 5073-5069) (Fig. 17-3)
Blotting paper (Gelman No. 51290)
Sepraphore III, 1 × 6 inches (Gelman No. 51003)
Beveled microscope slide (see procedure)
Hamilton micro syringe, 10 μl cap in 1-μl divisions, No. 701
Teflon tubing tips (Turner No. 310-001)
Marking pen (Scientific Products No. P-1226)

Reagents

High-resolution buffer (Gelman No. 51104)
  Dissolve one packet in 1200 ml of water.
Ponceau S stain 0.1% in 5% acetic acid, aqueous (Gelman No. 51281)
Destaining solution: acetic acid, 5% v/v, in water
Dehydrating solution: absolute methanol, AR
Clearing solution: acetic acid, 15%, in absolute methanol, AR

Procedure

1. Place about 100 ml of high-resolution buffer into a plastic tray. With the aid of forceps, remove the cellulose acetate strips from the box and gently allow strip to fall into the buffer so that strips soak from bottom. Do not push the strips into the buffer; entrapped air pockets will cause an irregular pattern. As soon as strips are soaked, immerse in buffer for at least 5 minutes.
2. Remove strips and place on blotting paper. Carefully blot. Number strips with marking pen at the end of the strips. (These marks will remain visible throughout the entire procedure and will not wash off.)
3. Place strips on rack so that ends are equidistant from the frame on each side (see Fig. 17-1). Tighten the knurled nuts on frame bars to hold strips firmly in place and loosen knurled nuts on side to make strips taut.
4. Place high-resolution buffer into cell to maximum level line. Tilt cell so

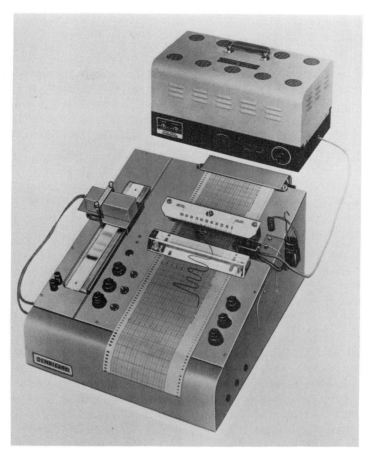

Fig. 17-3. Photovolt densitometer with integrator attachment. (Courtesy Photovolt Corp., New York, N.Y.)

that buffer level is equalized on both sides. Place rack into cell, making certain strip ends are in contact with buffer on both sides.

5. A microscope slide is beveled on both sides at one end so that the width is no more than 15 mm. This works as an excellent applicator. With the aid of a microsyringe fitted with a Teflon tubing tip carefully remove 2 μl of serum from the specimen tube. Carefully place the sample on the edge of the slide, being certain no serum spreads off the edge onto the side of the slide. With care place the slide on the strip 20 mm in from the cathodic end of the cell frame bar. After all strips are so loaded place the cover on the cell and run at a constant voltage of 175 volts for 25 minutes.

6. Remove strips from the cell and place, one at a time, into ponceau S stain as in the buffer, allowing strips to float until uniformly soaked. Then allow to stain for 5 minutes.

7. Remove from stain and remove excess stain by swishing strips in aqueous 5% acetic acid for several minutes in each of three successive trays. Place the washed strips on blotting paper and blot until barely moist.

8. The strips may be photographed or used for densitometer tracings at this point; however, we find that clearing is much preferred for densitometer evaluation.

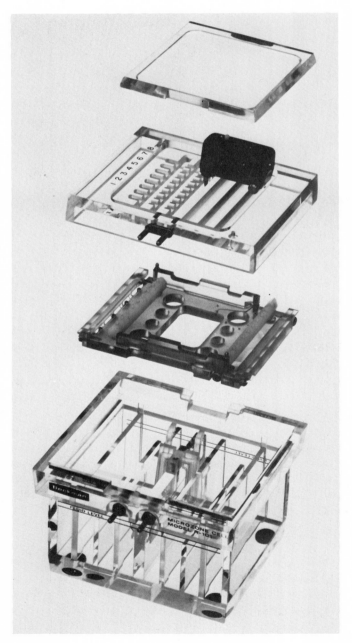

Fig. 17-4. Microzone cell, frame, cover, and application for micro-cellulose acetate electrophoresis. (Courtesy Beckman Instruments, Inc., Fullerton, Calif.)

9. The strips may be cleared by placing them (no more than two at a time) in two successive trays of absolute methanol for 5 minutes each and finally in a tray of 15% acetic acid in absolute methanol v/v for 5 minutes. After this they are carefully removed and placed on a clean glass slab by allowing the bottom of the strip to first touch the glass and letting the strip gently adhere to the plate. *Do not attempt to blot or wipe the strip until dry.*
10. The strip can be removed for densitometer evaluation by placing a strip of masking tape on each end of the strip and carefully pulling the strip from the glass plate.
11. The strips are then placed on the densitometer carriage and quantitated. (See agarose electrophoresis for quantitation method.) The cleared strips may be stored in a laboratory workbook by simply holding them in place with masking tape at each end on the page.

Micro–cellulose acetate method (Microzone system) (Beckman Instruments)

*Equipment**

Microzone electrophoresis cell (No. 324210) (Fig. 17-4)
Duostat power supply (No. 320800) (Fig. 17-5)
Sample applicator (No. 324399)

---

*Model numbers for Beckman Instruments unless otherwise noted.

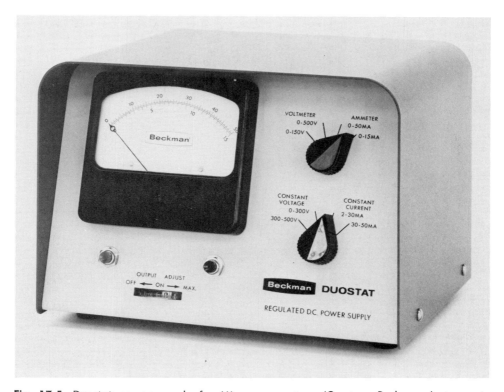

Fig. 17-5. Duostat power supply for Microzone system. (Courtesy Beckman Instruments, Inc., Fullerton, Calif.)

Fig. 17-6. Beckman Recording, Integrating Densitometer, CDS 100. (Courtesy Beckman Instruments, Inc., Fullerton, Calif.)

Fig. 17-7. Microzone applicator shown in position for sample pick-up. (Courtesy Beckman Instruments, Inc., Fullerton, Calif.)

Electrophoresis membranes (No. 324330)
Blotters (No. 324326)
Seven plastic trays with covers
Squeegee (No. 324352)
Forceps
Plastic envelopes (No. 326189)
Densitometer (No. R112 and R113) (Fig. 17-6)
Parafilm (American Can Co.; Neenah Wisc)
Glass plate, 2 × 6 inches, washed in desiccate (No. 018772)

Reagents

B_2 buffer, pH 8.6 ($\mu = 0.075$)
Diethylbarbitaric acid, 2.76 g
Sodium diethylbarbiturate, 15.4 g
  Dilute to 1000 ml with distilled water.
Fixative dye solution
  Ponceau S stain, 0.2% w/v
  Trichloroacetic acid, 3% w/v
  Sulfosalicylic acid, 3% w/v
Rinse solution: glacial acetic acid, 5% v/v
Alcohol dehydration solution, 100% ethyl alcohol
  Laboratory reagent alcohol will be satisfactory.
Clearing solution: reagent grade cyclohexanone, 30% in denatured ethanol
  Mix fresh batch for each use. Percentage needed may vary with room humidity from 29% to 32%.

Procedure

1. Fill seven plastic trays each with 100 ml of the following solutions.
   a. Reconstituted B_2 buffer
   b. Fixative-dye solution
   c. Aqueous acetic acid, 5%
   d. Aqueous acetic acid, 5%
   e. Aqueous acetic acid, 5%
   f. Alcohol dehydration solution
   g. Clearing solution
2. Fill Microzone cell with B_2 buffer solution to mark. Equalize fluid between both compartments.
3. Remove membrane from box with forceps and carefully place on top of buffer, being careful not to entrap air bubbles under or within the membrane.
4. With forceps remove saturated membrane and place between two blotters and gently blot to remove excess buffer.
5. With forceps remove membrane from blotter and place onto bridge. Line up holes in membrane with pegs on bridge. Keep reference hole on lower left. Seat membrane securely with forceps—*do not touch separation portion of membrane.*
6. Place bridge in cell, making certain that ends of membrane are in contact with the cell buffer. Replace cell cover.
7. On a Parafilm strip place 1 drop each of eight samples. One sample should be that of a control serum. Depress white button on applicator and bring applicator tip across sample so that the surface tension is barely broken. Raise applicator and depress red button (Fig. 17-7).
8. With applicator positioned in center groove on the cell, cover index at the desired sample position and depress white button, (Fig. 17-8). Wait 5 to

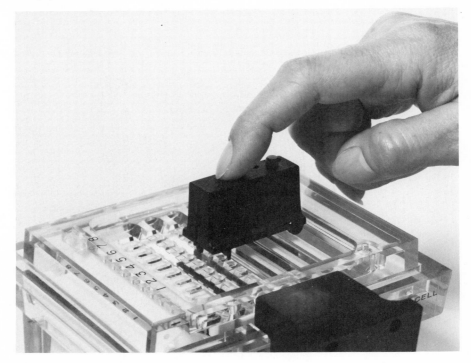

Fig. 17-8. Application of serum sample to Microzone strip. (Courtesy Beckman Instruments, Inc., Fullerton, Calif.)

      10 seconds, depress red button, and remove applicator. Rinse applicator with distilled water and repeat process until all eight samples are applied. Replace lid.
9. Plug Duostat into cell and adjust voltage to 250 volts. Starting current should be 4 to 6 ma. Perform electrophoresis for 16 to 18 minutes. Turn power switch to off and disconnect cell.
10. Remove membrane from bridge and place in fixative-dye solution for 10 minutes.
11. Place in three successive acetic acid rinses until no more dye flows from membrane.
12. (Perform under a hood) Agitate membrane in the dehydration solution for 1 minute and remove to a clean glass slide that has been submerged in the clearing solution. Agitate for 1 minute and remove membrane, which is now on glass plate.
13. Gently squeegee membrane on glass plate to remove any bubbles or excess clearing solution.
14. Place plate and membrane in 100° C preheated oven and allow to dry. Remove from oven in about 15 minutes and allow to cool to room temperature.
15. Peel membrane from glass plate (Fig. 17-9) and place into plastic envelope. Trim off ends with scissors. Label each pattern with identifying numbers.

Micro–starch gel electrophoresis

The method presented here is a modification of the Smithies technique. It is presented because of its simplicity. It may be used in specialized protein studies as well as for hemoglobin electrophoresis.

Electrophoresis   541

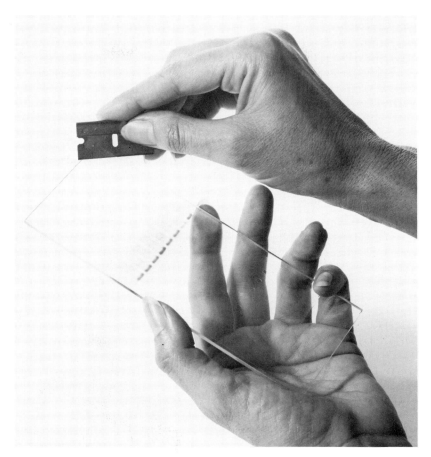

Fig. 17-9. Removal of cellulose acetate membrane from glass plate after clearing. (Courtesy Beckman Instruments, Inc., Fullerton, Calif.)

Fig. 17-10. Tray for micro–starch gel electrophoresis. (Courtesy Reclin Co., Lincoln, Neb.)

542 *Chemistry for the clinical laboratory*

Equipment
>Electrophoresis cell (Reclin No. 117 A) (same as for cellulose acetate method)
>Micro–starch gel tray (Reclin No. 39 S) (Fig. 17-10)
>Power supply (Instrumentation Specialties Co. No. 490)
>Destainer (Reclin No. 12 D) (Fig. 17-11)
>Washer (Reclin No. 130 SW) (Fig. 17-12)
>Slicing tray (Reclin No. 110 ST) (Fig. 17-13)
>Slicer (ordinary cheese cutter with fine recorder wire or piano wire)
>Filter paper wicks (Beckman No. 319329)
>Razor blade stainless steel type
>>Grind off both sides of blade so that it is approximately ½ inch wide.
>
>Filter paper strips (made from Whatman 3 mm), 2 × 20 mm.
>Small nosed forceps
>Petri dishes

Reagents
>Starch gel hydrolyzed (Connaught Laboratories, Fisher Scientific)
>Stock buffer, pH 8.8
>
>| | |
>|---|---:|
>| Tris(hydroxymethyl)aminomethane | 30.25 g |
>| Ethylenediamine tetraacetic acid | 3.00 g |
>| Boric acid | 2.30 g |
>
>Distilled water to final volume of 1000 ml

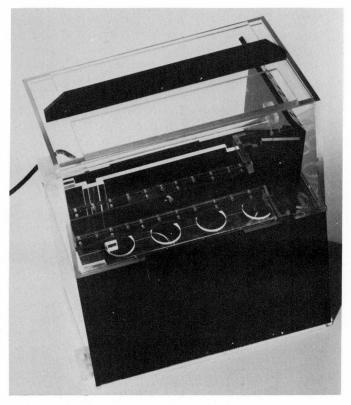

Fig. 17-11. Circulating destaining unit. May be used for starch gel procedure and immunoelectrophoresis. (Courtesy Reclin Co., Lincoln, Neb.)

Electrophoresis 543

Fig. 17-12. Washing unit. Separate chambers provided for water and saline. May be used for any washing procedure. (Courtesy Reclin Co., Lincoln, Neb.)

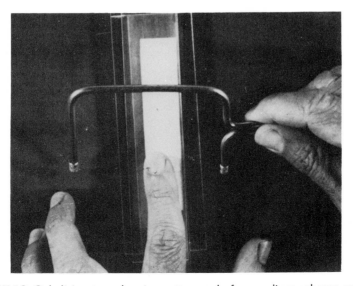

Fig. 17-13. Gel slicing tray showing cutter made from ordinary cheese cutter.

Gel buffer
Dilute 1 part stock buffer with 2 parts distilled water.
Electrode chamber buffer, pH 8.2

Boric acid	18.5 g
Sodium hydroxide	2.5 g

Distilled water to a final volume of 1000 ml
A few crystals of bromphenol blue are added to the buffer in order to visualize the borate buffer front during electrophoresis.
Starch gel, 13.3 g/dl dilute gel buffer

Procedure

1. *Preparation of the starch gel trays:* **Mix 200 ml of gel buffer and 26.6 g of hydrolyzed starch in a 500-ml Pyrex suction flask, cook, and degas as Smithies. The mixture is poured into the starch gel tray, allowing the starch mixture to overflow. With a straight edge (one time only) start at one end of the tray and scrape from one end gate to the other to remove excess starch. As soon as the gel has set (approximately 20 minutes) the gels are covered smoothly with Saran Wrap. Gels may be used when cool or they may be stored overnight at room temperature wrapped in Saran Wrap.**

2. *Application of the sample:* **Remove the Saran Wrap from the tray just far enough so that the sample may be inserted 3 cm from the cathodic end of the gel. Introduce serum into the gel by soaking a 2 × 20 mm filter paper strip in the serum sample. Place strip on the edge of a special razor blade so that the paper remains above the gel as the edge of the blade is forced into the gel at right angles to the trough edge and to the gel. The sample is placed approximately 3 cm from the cathodic end of the gel. Gently pull the blade back so that the incised gel is opened. With a pair of fine nosed forceps push the serum-impregnated filter paper into the slit formed. Carefully pull out the blade, allowing the serum sample strip to remain in the gel. Be certain there are no bubbles or gaps between the gel edges and the strip.**

3. *Electrophoresis:* **After all the samples have been introduced into the gel, recover the gel tray end with Saran Wrap. Place the tray into the cell. Remove the end gates carefully and place the wick paper vertically on the wick holders so they fit tightly against the bare starch gel. The end gates are replaced and the current is adjusted on the power supply to 50 ma (approximately 200 volts). After 10 minutes turn the power off, fold back the Saran Wrap, and remove filter paper strips from the gel. No air pockets or gaps should remain in the gel slit. Place Saran Wrap back over the gel, set current at 50 ma, and fill cell cover with ice cubes and place on the cell. As the electrophoresis progresses, the bromphenol blue–containing buffer front passes through the gel, staining the albumin component. The albumin band is allowed to migrate 4.5 to 5 cm from the point of application. The current is turned off and the tray removed from the cell.**

4. *Slicing the gels:* **The gels are removed from the tray by first running a sharp edge along the sides of the trough. (A small platinum spatula works quite well.) Remove the end gates and, with a squeeze type wash bottle, gently force water under the gel, allowing the gel to slide at a slight angle into the cutting tray. The slabs may be marked at this point using serum on the end of a pointed applicator stick. Use a series of dots to designate the gel number. Be certain to mark both ends of the gel as the gels will break at the site of serum application. Cut the gels once longitudinally with the wire cutter (Fig. 17-13). Plastic wedges of proper thickness placed**

under the gel in the slicing tray allow various thicknesses of the gels to be cut. The top slice is usually the most smeared and the lower slices are best used for staining. As many as three slices of the slab can be cut without difficulty.

5. *Staining and destaining the gels:* The gels are stained in a large Petri dish containing Amido black (10 g/1000 ml) dissolved in the following solvent: absolute methanol, 5 parts; water, 5 parts; glacial acetic acid, 1 part. The gels are allowed to stain for 4 minutes. They are then transferred to a destainer with the same solvent as above, or if a destainer is not available to a series of Petri dishes containing the solvent.

6. *Scanning the gels:* A scan of the gels is possible using the starch gel tray provided for the Photovolt densitometer (No. 5063). The slow drive used for cellulose acetate scans is also used. A 525-nm filter is used. No attempt has been made to quantitate serum patterns using this procedure for protein electrophoresis but rather to establish patterns for specific disease states (Fig. 17-14). We have not found it necessary to clear the gels in glycerol before scanning.

ELECTROPHORESIS OF BODY FLUID CONSTITUENTS
Serum proteins
AGAROSE OR PFIZER POL-E-FILM OR ACI/CORNING METHOD

Equipment

Cassette electrophoresis cell and power supply (No. 470136) (Fig. 17-15)
Microsyringe, Hamilton 10 with Chaney adaptor (No. 470152)
Disposable (Teflon) sample tips (No. 470154)
Stir-stain dishes (No. 470160)
Disposable liners for dishes (No. 470162)
Magnetic stir bars (No. 470164)
Mag-Mix
Agarose universal electrophoresis film (No. 470102) (Fig. 17-15)
Film cutter (No. 470170)
Electrophoresis Process Center (Reclin)
Densitometer (Photovolt or Beckman No. 112)
Marking pen (Scientific Products No. P-1226)

Reagents

Barbital buffer, pH 8.6, 0.05 M with 0.035% EDTA
Amido black 10B stain, 0.2% in 5% acetic acid v/v
Acetic acid, 5% v/v
Glycerol, 2% in water, v/v

Procedure for electrophoresis

1. Label back to identify specimens.
2. Elevate corner of thin plastic backing and carefully peel the agarose gel with its plastic backing from the plastic template.
3. Place film on flat surface and let any moisture on surface of film disappear before loading sample.
4. Add exactly 0.9 µl of serum to the sample well using micropipet with Teflon-coated tip.
5. With the cassette cover on its back and the power supply trip extension to the top, insert the film into the cassette so that the plastic borders of the

---

*ACI/Corning model numbers unless otherwise marked.

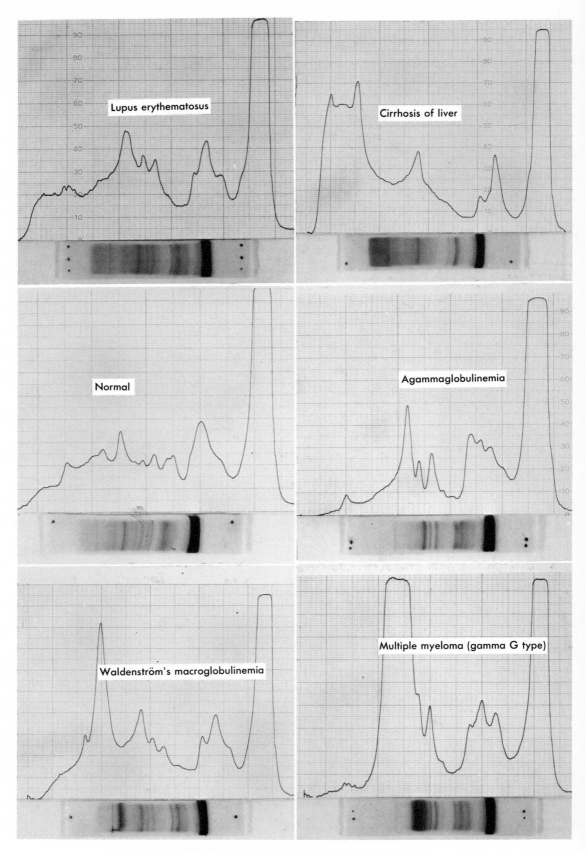

Fig. 17-14. For legend see opposite page.

anode end are on the right-hand side and the cathodic edge (nearest sample wells) is on the left-hand side. Be certain the film fits snugly into the gripper edges on both sides and is centered in the cover.

6. To the buffer chambers add 200 ml of cold barbital electrophoresis buffer. Tilt cell base upward on one end until buffer flows from one chamber to the other so that there is an equal amount of buffer in each chamber.
7. Wipe chamber divider dry of any excess buffer.
8. Plug cell base into power supply.
9. Place cell cover containing agarose film onto cell and observe red pilot light which is switched on as cover is secured to base.
10. Perform electrophoresis for approximately 45 minutes. (Time should be the same for each run to maintain quality control between runs.)
11. At end of run remove film from cover.
12. Place film in storing tray with liner containing Amido black 10B stain or in Electrophoresis Process Center staining rack and then into tray.
13. Allow film to stain 15 minutes. Drain stain from film on paper towel and place film in 5% acetic acid for 1 minute.

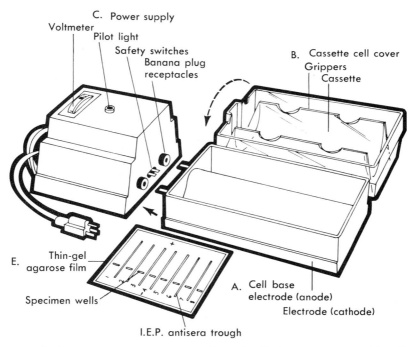

Fig. 17-15. ACI/Corning agarose power supply and cell showing agarose film. (Courtesy ACI/Corning, Palo Alto, Calif.)

Fig. 17-14. Densitometer scan of serum on starch gel. (From Marsh, C. L., and Jolliff, C. R.: Micro starch gel electrophoresis. In Smith, I., editor: Chromatographic and electrophoretic technique, vol. 2, Zone electrophoresis, ed. 2, London, 1968, William Heinemann Medical Books, Ltd.)

14. Remove from 5% acetic acid and place in Electrophoresis Process Center or under a warm air current such as from a hair dryer. Allow film to dry completely.
15. Place dried film in three changes of 5% acetic acid until background is perfectly clear.
16. Place film in 2% glycerol in H_2O for a few moments and then allow film to air dry.

Procedures for quantitation

1. The film may be cut into separate strips by cutting with the film cutter. The strips may then be scanned at 580 nm after setting zero on the unstained clear portion. Follow instructions for densitometer used.
2. If the densitometer has the capability of integration with calculation of components, automatically follow the directions provided with the instrument.
3. If the integration simply allows for counts, the pattern may be calculated as in the following example:

Total protein (8 g/dl)

Components	Total count	Percent	g/dl
Albumin	100	57.1	4.57
Alpha$_1$ globulin	5	3.0	0.24
Alpha$_2$ globulin	20	11.4	0.91
Beta globulin	20	11.4	0.91
Gamma globulin	30	17.2	1.38
TOTALS	175	100.1	8.01

Elution

1. For those who do not have a densitometer, the Amido black 10B stained strips may be quantitated by elution. This procedure is satisfactory and in some instances may be more exact than the densitometer method.
2. Cut strip at lighter region between each band, just ahead of the albumin band and just behind the gamma globulin band. Obtain a blank (background) just beyond the gamma globulin band. Be certain pre-albumin is not missed just ahead of the albumin band.
3. Place each segment plus the blank segment (including the plastic backing) into a test tube containing 2.0 ml of 2% NaOH. Boil each tube for 1 minute. Allow to cool. Pour into 1-cm light path cuvet. It may be necessary to elute the albumin band in twice the volume of 2% NaOH. The exact volume of 2% NaOH must be recorded. Remove plastic backing.
4. Calibrate spectrophotometer with solutions of a known concentration of Amido black 10B in 2% NaOH. Read standards at 580 nm against 2% NaOH as a blank. Read eluates against background eluate as the blank.
5. Convert absorbance readings to concentration by using calibration curve made with standard reading. Determine the amount of Amido black 10B eluted from each segment by multiplying the concentration by the volume of eluent. Divide the amount of Amido black 10B bound to each component by the total bound by all components to obtain the percentage of each component. Having found the percentages for each component, perform calculations as in the scanning section for determining the g/dl value.

POLYACRYLAMIDE GEL ELECTROPHORESIS

Disk electrophoresis as described by Ornstein utilizes the properties shown by that medium—stability, transparency, chemical inertness, nonionicity, and allowance for a strong gel with a large range of pore sizes. Ornstein found a sieving effect based

on molecular size of the protein relative to the gel pore size as well as more discrete, narrower bonds. He gave two reasons for these differences:
1. Frictional resistance as the particle moves through the gel, which depends on the radius of the particle and the polymer structure of the gel, thereby influencing the viscosity
2. A thin starting zone in the direction of the electric field produced by an electrophoretic step which concentrates the ions

The disk method involves a column of three layers.
1. *Sample gel* (large pores), which contain the sample ions where the electrophoretic concentration of these ions is initiated
2. *Spacer gel* (large pores), allowing for electrophoretic concentration of the sample ions
3. *Separation gel* (small pores), in which the actual electrophoretic separation takes place

The column is placed in a vertical position and the electrophoresis is accomplished by a potential applied from a buffer chamber above, which makes contact with the sample gel to a lower buffer chamber into which the separation gel end is submerged.

Voltage is applied for a specified time. The gel is then removed from the glass tube in which it has been formed, fixed, and stained for protein and decolorized by electrophoresis or washing.

Flat bed or *vertical gel slab polyacrylamide gel electrophoresis* was introduced by Raymond in 1959. He was able to obtain reproducible and uniform patterns by departing from the Ornstein and Davis method, rejecting the use of a sample and spacer gel above the supporting gel and rejecting the discontinuous buffer system. He further modified the technique by employing a prerun before introduction of the sample, thereby establishing a steady state flux of buffer ions throughout the gel. These changes, plus the fact that a series of separate serum samples could be run at the same time, allowed for inter-gel comparisons. This is of marked benefit to the clinical chemist when comparing patterns such as hemoglobin, haptoglobin, and proteins exhibiting polymorphism.

The methods of Ornstein and Davis and Raymond have been constantly modified. To present all of them here would be of little practical value for the clinical chemistry worker. References to the techniques as well as their applications to specific substances are available in the literature. Polyacrylamide gel electrophoresis has its limitations from the standpoint of time, technique, preparation, quantitation, and interpretation in the general clinical chemistry laboratory. However, like starch gel it remains one of the truly definitive procedures for certain protein separations and should remain a part of the procedural armamentarium in the clinical chemistry laboratory.

Two techniques offered here are utilized in our laboratories. Each employs the Beckman Microzone cell, which is found in most clinical laboratories. The first technique employs the use of a single percentage (7%) gel. The other technique, a modification by Marsh of the step-gradient method of Wright, Farrell, and Roberts, offers separation of more protein fractions and in many cases is superior to the single gradient method.

Microzone acrylamide gel method

Equipment and supplies*

Microzone electrophoresis cell (**No. 324210**)
Duostat power supply (**No. 320800**)

---

*Beckman model numbers.

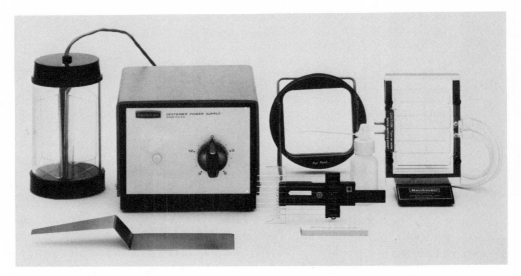

Fig. 17-16. Microzone acrylamide gel apparatus showing destaining jar, destaining power supply, drying frame, applicator, and gel mold with bridge assembly. (Courtesy Beckman Instruments, Inc., Fullerton, Calif.)

Gel mold with bridge assembly (No. 333819) (Fig. 17-16)
Gel slot former (No. 331798) (Fig. 17-16)
Multiple sample applicator (No. 333820) (Fig. 17-16)
Destainer tank (No. 334274) (Fig. 17-16)
Wicks and blotters (Nos. 332505 and 332531)
Plastic storing tray
Saran Wrap
Vacuum grease (No. 301062)
Micropipet, 30 or 40 µl
Connector for faucet

*Reagents**

Serum protein fixative dye (No. 332508)
Gel kit (No. 332509)
B-3 buffer (No. 334534)
TRIS buffer concentrate (No. 333786)
Boric acid concentrate (No. 333785)
Disodium EDTA concentrate (No. 333784)
Tracking dye (No. 320047)

Procedure

1. *Preparation of gel mold–bridge assembly*
   a. Connect water lines and check water supply. Turn off water and allow mold to come to room temperature.
   b. Lubricate rubber plugs lightly with vacuum grease and insert into slots. Wipe off any excess grease.
   c. Lightly lubricate both surfaces of V gasket and assemble mold, gasket, and stand. Tighten all four screws *lightly* and then tighten snugly using fingers and no tools. The mold may be tested with water for leakage, then inverted and drained onto a blotter.

2. *Preparation of the gel*
   a. Mix gel just prior to pouring. Wash all glassware after preparation in deionized water.
   b. Dissolve 1 vial (2.5 g) powdered acrylamide (5% methylene-bis-acrylamide cross linking agent in 95% acrylamide monomer) in 35 ml of T-G buffer. Filter through Whatman filter paper No. 1.
   c. To the entire filtrate in a 125-ml Erlenmeyer flask add 40 μl of TMED (gelling primer); mix well. Pour 11 ml of this solution into a 25-ml graduated cylinder and use later to make the top gel.
   d. Into a 50-ml beaker weigh out 20 mg of ammonium persulfate (gelling accelerator). In a 30-ml beaker weigh out 10 mg of the ammonium persulfate for mixing the top gel.
   e. Pour the remaining gelling solution (approximately 22 ml) into the 50-ml beaker containing the ammonium persulfate. Mix.
   f. Pour enough gelling solution into the mold so that the meniscus is at least halfway up the teeth of the slot former that has been previously inserted in the mold. No bubbles should form on the slot former. If they do, withdraw, wipe clean and reinsert.
   g. To keep oxygen from the surface of the gel, add several drops of distilled water on either side of the slot former. The water will form a gradient line; if it does not, carefully loosen the two top screws and allow the gradient line to form.
   h. Allow gelling to take place. This usually takes 20 minutes or more. An interface will be visible between the formed gel and the water. Now carefully remove the slot former with a slow but deliberate motion.
   i. To the 30-ml beaker containing the ammonium persulfate add the 11 ml of gelling solution. Mix well.
   j. Grab the gel mold and carefully tilt so that it is possible to flip the water out of the preformed slots. Add a small amount of gel solution to rinse out the slots and flip out. Repeat this step once more. Now fill the mold to about $\frac{1}{8}$ inch from the top of the mold with the top gel solution.
3. *Application of samples*
   a. Fill the applicator by depressing the plunger first and then insert each capillary into the designated well of the sample tray. Slowly release the plunger to allow the serum samples to fill the capillaries. Blot the capillaries with a Kem wipe or blotting material, being careful not to touch the ends.
   b. With the applicator in a perfectly vertical position, lower it until the base rests on the back lower edge of the gel mold. This will allow the capillaries to be approximately 0.040 inch from the bottoms of the slots.
   c. Slowly depress the plunger and the serum will gradually displace the liquid gel in the formed slots. Remove applicator slowly while still depressing the plunger. Allow to gel for 20 minutes.
   d. Clean the applicator with distilled water by slowly drawing and expelling several times. Keep in beaker of water. If not used afterward, dry wires and piston after removing capillaries.
   e. Inspect the gel mold to be certain no bubbles or other defects are noticeable in the gel.
   f. After top gel has set, turn on cool water so that approximately 1 L/min flows through the cell.
4. *Preparation of the cell and mold for electrophoresis*
   a. Fill Microzone cell slightly above the fill line with buffer.
   b. Remove the rubber plugs, top one first. Place mold face down with the plug cavities uppermost. Cover gel cavities with buffer and gently rock it back and forth to wet exposed surfaces completely.

c. Place two wicks together (slightly concave sides facing one another) and insert wick into plug cavities. Push in gently until the edges make contact with the surface of the gel.
   d. Place mold on Microzone cell with water lines to rear and slots to the left. Be certain buffer level is sufficient.
5. *Electrophoresis*
   a. Attach red Duostat plug into the interconnecting cable plug (for migration to the anode on the right).
   b. Turn on Duostat and adjust to 450 volts. Starting current should be between 40 and 50 ma and ending current between 32 and 36 ma. Run for 60 to 70 minutes. Migration time will depend on cold water temperature.
6. *Storing gel*
   a. Turn off the power supply and cooling water. Dismantle the bridge and carefully remove the gel with a moistened spatula that is furnished with the gel kit.
   b. Place gel in storing tray which is half filled with protein fixative-dye solution. Carefully agitate and stain for 90 minutes.
   c. With the spatula transfer the gel to a premoistened (2% acetic acid) destaining blotter. Invert and apply another moistened blotter on the other surface.
   d. Insert the sandwiched gel between the destainer electrode plates. Next to these apply the plastic pressure plates. Place into the destaining cylinder and fill to top with 2% acetic acid.
   e. Place lid on top and set timer for 10 minutes. When bell rings, remove gel from chamber and place in 2% acetic acid.
   f. Gels may now be photographed or placed in Saran Wrap for keeping. Drying of gels has not proved too satisfactory for us, and it is our belief that too much distortion takes place in this process.

Microzone step-gradient polyacrylamide gel electrophoresis

The use of a step-gradient procedure greatly improves the electrophoretic resolution of serum proteins in polyacrylamide gels. More than thirty fractions may be identified. The following procedure is a modification of the step-gradient method of Wright and associates and can be conveniently adapted to the Beckman Microzone cell and polyacrylamide accessory (Fig. 17-17).

Equipment and supplies

Same as for Microzone polyacrylamide gel procedure

Reagents

Gel buffer, pH 8.9
    TRIS   45.75 g
    HCl (1 N)   60.0 ml
    Distilled water to make   1 L

Chamber buffer, pH 8.3
    Glycine   14.4 g
    TRIS   3.0 g
    Distilled water to make   1 L

Ammonium persulfate, 7% w/v in distilled water
    Make up fresh each week.

Cap Gel, 8%
    Acrylamide   7.84 g
    "Bis"   0.16 g

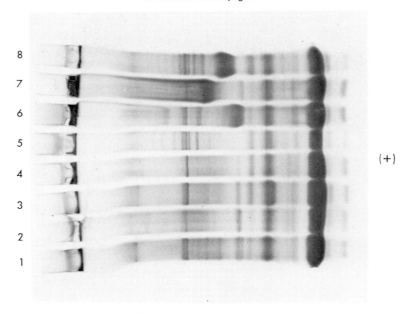

Fig. 17-17. Serum protein separation, step gel polyacrylamide electrophoresis by Microzone system.

<pre>
Gel buffer   20 ml
TMED   40 μl
Distilled water to make   100 ml
</pre>
Gel stock, 10%
<pre>
Acrylamide   9.7 g
"Bis"   0.3 g
TMED   40 μl
Gel buffer to make   100 ml
</pre>
Gel, 7%
  Dilute 7.0 ml of 10% stock gel to 10 ml with distilled water containing 50 μl TMED/dl.
Gel, 4.75%
  Dilute 4.75 ml of 10% stock gel to 10 ml with distilled water.
Gel, 4%
  Dilute 4.0 ml of 10% stock to 10 ml with distilled water.
Stain stock
  Mix 1 g coomassie brilliant blue with distilled water to make 100 ml.
Staining solution
  CBB stock, 50 ml
  Trichloroacetic acid, 20%, to 1 L

Procedure

1. *Gel layering* (Fig. 17-18): The outside of the gel mold is marked (starting at the bottom of the actual molding portion) at 5.5, 7, 8, 9.5, and 9.7 cm. This will give a 5.5-cm layer of 10% gel, a 1.5-cm layer of 7% gel, a 1-cm layer of 4.75% gel, and a final 1.5-cm layer of 4% gel. The 9.7-cm mark is to allow for shrinkage of the 4% gel during polymerization.

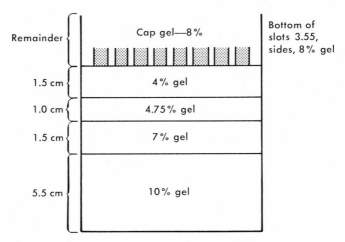

Fig. 17-18. Quantities of the various concentrations of gels used to perform the four-step Microzone acrylamide gel electrophoresis procedure. (Courtesy Dr. Connell Marsh, College of Dentistry, University of Nebraska.)

2. *Gel pouring:* **Turn on the cooling water and leave running.**
   a. Add 0.1 ml of the 7% persulfate solution to 15 ml of 10% stock gel solution in a small beaker. Mix well and pour at once into the mold up to the 5.5-cm mark.
   b. Immediately add 0.1 ml of the 7% persulfate to 10 ml of the 7% gel solution mix, and *carefully* layer the solution (side of the mold) onto the 10% gel solution to the 7-cm mark. This may be done with a syringe and a long needle or with a long-tipped Pasteur pipet.
   c. Add 0.15 ml of the 7% persulfate solution to 10 ml of the 4.75% gel solution and layer it up to just above the 8.0-cm mark.
   d. Add 0.2 ml of the 7% persulfate solution to 10 ml of the 4% gel solution and layer it up to just above the 9.5-cm mark.
   e. Immediately water layer the 4% solution to a depth of about 0.5 cm and allow at least 30 minutes for the gel layers to polymerize completely.
   f. After polymerization is complete, the water is removed from the top layer by means of rolled up absorbent tissues (Kleenex, Kimwipes, etc.) inserted at the sides of the mold. (Shaking off the water may damage the soft 4% gel layer.)
   g. *Carefully* check to see that the slot-former teeth will rest directly on the top surface of the 4% gel when inserted in the mold, and then remove it.
   h. Add 0.2 ml of 7% persulfate solution to 10 ml of Cap Gel solution. Mix and pour enough into the mold to cover the teeth of the slot former almost to the base. Now carefully insert the slot former, making sure it rests lightly on the 4% gel. Add a few drops of water carefully at the sides of the mold so the Cap Gel is water layered for smooth polymerization. Allow 20 minutes for polymerization and then *carefully and slowly* remove the slot former. Extra care is necessary at this point so that the slots are not damaged or collapsed. Now carefully dry each slot with the tip of an absorbent tissue. (Siliconizing slot former helps.)
   i. Add 0.2 ml of 7% persulfate solution to 10 ml of Cap Gel solution and pour it into the mold well *above* the rubber insert.

3. Before the gel can polymerize, deliver the samples (containing a little sucrose so they will layer under the 8% gel) with the automatic applicator. Be sure that the tips of the capillaries *do not* touch the bottom of the sample wells. This can happen if the applicator is completely lowered into position as in the standard gel procedure. Just insert the applicator far enough that the capillary tips are near the bottom of the well, and then deliver the sample and remove the applicator as usual.
4. *Notes*
   a. It is absolutely essential that the bottom of the sample slot be in the 4% gel. Otherwise the resolution will be very poor.
   b. Care *must* be exercised in layering the gels. It is a good idea to alternate sides with successive layers. It is also helpful to remove any droplets of the preceding layer from the sides of the mold before proceeding to the next layer. This is easily done by inserting a folded paper towel into the mold and picking up droplets which could fall through the next layer.
   c. It may be advantageous to slightly increase the amount of serum delivered when using the step gradient system. A delivery stroke of 15 to 18 mm is about optimum for this system, providing enough sample for good visualization of both major and minor serum components.
5. *Electrophoresis*
   a. When the Cap Gel containing the samples has completely polymerized (15 to 20 minutes), the contact wicks are *carefully* applied dry as in the usual single gel procedure. They are then saturated with the chamber buffer *before* positioning the gel mold on the Microzone cell. A few crystals of bromphenol blue may be added to the chamber buffer as a tracking dye to stain the buffer front as it moves through the gel during electrophoresis.
   b. Connect the cell to the Duostat power supply, turn on the power, and adjust to 450 volts. This should give a starting current of about 30 ma. (The current will increase somewhat during the progress of the run.)
   c. Within the first 2 to 3 minutes the chamber buffer front will be seen moving through the gel—quite rapidly through the 4% and 4.75% layers and then more slowly as it reaches the 7% and 10% layers.
6. Electrophoresis is continued until the buffer front just reaches the anode wick, or alternatively until the albumin band, which binds some of the tracking dye, has migrated about 5.0 cm (approximately 1 hour). Observation of the buffer front is more generally useful because it can be seen when separating protein solutions not containing stained albumin.

Staining and destaining

1. When the electrophoresis is completed, turn off the power and the cooling water, remove the gel mold from the cell, and then *carefully* open the mold and *carefully* transfer the gel to the desired stain with the stainless steel spatula.
2. Stain the gel as usual with the conventional Beckman stain (Buffalo black) provided in the acrylamide kit and then destain manually.
3. The gels are completely stained in the stain mixture for 30 minutes and then may be destained manually in several changes of 12.5% trichloroacetic acid. The manual procedure eliminates any possible damage to the gels during the staining and destaining.
4. The gels may be stored in 5% acetic acid and photographed for records. Gels may be kept wrapped in Saran Wrap moisturized with 5% acetic acid for some time.

Table 17-2
Normal values for serum proteins in 2000 patients*

Group	Number	Total protein	Albumin	Alpha$_1$ globulin	Alpha$_2$ globulin	Beta globulin	Gamma globulin
Total	2000	7.04 ± 1.00	4.10 ± 0.78	0.20 ± 0.14	0.62 ± 0.26	0.85 ± 0.32	1.28 ± 0.60
Newborn	2	6.50 ± 1.42	3.93 ± 0.04	0.21 ± 0.08	0.57 ± 0.30	0.62 ± 0.26	1.19 ± 0.12
Age 1	5	6.46 ± 0.64	4.31 ± 0.56	0.18 ± 0.06	0.61 ± 0.24	0.70 ± 0.28	0.66 ± 0.42
Age 2	14	6.39 ± 1.14	4.00 ± 0.86	0.20 ± 0.20	0.67 ± 0.30	0.73 ± 0.36	0.79 ± 0.42
Age 3-4	17	6.63 ± 0.94	4.06 ± 0.94	0.21 ± 0.18	0.76 ± 0.36	0.72 ± 0.26	0.89 ± 0.62
Age 5-9	25	6.80 ± 0.92	4.18 ± 0.50	0.17 ± 0.10	0.68 ± 0.36	0.77 ± 0.34	1.03 ± 0.58
Age 10-12	16	6.98 ± 0.84	4.26 ± 0.64	0.17 ± 0.10	0.61 ± 0.22	0.79 ± 0.36	1.14 ± 0.48
Age 13-59	1192	7.05 ± 0.98	4.31 ± 0.78	0.20 ± 0.14	0.61 ± 0.26	0.84 ± 0.30	1.27 ± 0.58
Age 60-69	378	7.08 ± 0.94	4.09 ± 0.70	0.19 ± 0.12	0.62 ± 0.24	0.87 ± 0.30	1.31 ± 0.60
Age 70-79	262	7.06 ± 0.96	4.03 ± 0.68	0.19 ± 0.12	0.64 ± 0.26	0.86 ± 0.34	1.33 ± 0.58
Age 80-89	72	7.01 ± 1.14	3.87 ± 0.94	0.21 ± 0.11	0.67 ± 0.26	0.89 ± 0.40	1.37 ± 0.64
Age 90+	17	6.91 ± 1.14	3.86 ± 0.98	0.20 ± 0.18	0.67 ± 0.32	0.81 ± 0.24	1.37 ± 0.90

*Mean value ± 2 SD by Microzone cellulose acetate electrophoresis.

CLINICAL SIGNIFICANCE

Many investigators believe that serum protein electrophoresis affords the clinician with an excellent screen of the patient's serum proteins. In our institution we utilize the Microzone system as part of our routine chemistry profile. Many institutions no longer utilize the densitometer scan of the completed strips but simply visually scan for evidence of dysproteinemias. We still believe the densitometer scan to be of value in that changes from established normals (Table 17-2) alert the technologist to changes in the pattern otherwise not always observed by the person charged with the visual examination of the strips. With modern technology and instrumentation, the Microzone procedure has offered no great problems or greatly added expense to the patient.

Serum protein electrophoresis has now become a screening procedure for us. The more definitive procedures such as immunoelectrophoresis and the immunoquantitation of specific proteins, by whatever methods one chooses, allows the clinician a tremendous advantage in the diagnosis of the many dysproteinemias associated with the occult as well as evident disease states (Table 17-3). Its use as a monitor of diseases reflected by changes in serum proteins has been well established in our institution.

Electrophoresis of proteins in cerebrospinal fluid

Separation of CSF proteins has been a useful procedure in the detection of neurologic diseases. It is paramount that the CSF be concentrated in most cases

Table 17-3 Electrophoretic fractionations of serum proteins in a variety of diseases

Category	Number	Protein concentrations (g/dl; mean ± standard deviation)					
		Total protein	Albumin	Alpha$_1$ globulin	Alpha$_2$ globulin	Beta globulin	Gamma globulin
Normal controls	26	7.14 ± 0.33	3.65 ± 0.31	0.42 ± 0.10	0.67 ± 0.12	0.91 ± 0.13	1.62 ± 0.18
Multiple myeloma	67	7.96 ± 1.15**	2.60 ± 0.73**	(total globulins = 5.35 ± 1.30**)			
Macroglobulinemia	13	7.84 ± 0.62**	2.59 ± 0.51**	0.32 ± 0.16	0.58 ± 0.21	0.82 ± 0.25	3.54 ± 0.52
Hodgkin's disease	14	6.38 ± 0.77**	2.45 ± 0.74**	0.59 ± 0.16**	0.85 ± 0.16**	0.81 ± 0.15	1.74 ± 0.55
Lymphatic leukemia and lymphomas	27	6.04 ± 0.82**	3.24 ± 0.71**	0.40 ± 0.13	0.73 ± 0.17	0.77 ± 0.15**	0.88 ± 0.49**
Myelogenous and monocytic leukemias	12	6.78 ± 0.70	2.64 ± 0.68**	0.44 ± 0.10	0.82 ± 0.25**	0.73 ± 0.13**	2.15 ± 0.58**
Hypogammaglobulinemia	18	5.86 ± 0.87**	3.67 ± 0.56	0.38 ± 0.19	0.74 ± 0.26	0.84 ± 0.23	0.17 ± 0.15**
Analbuminemia	1	4.8	0.1	0.4	1.0	1.5	1.8
Peptic ulcer	9	6.31 ± 0.57**	2.78 ± 0.42**	0.41 ± 0.10	0.79 ± 0.17	0.88 ± 0.17	1.43 ± 0.44
Ulcerative colitis	8	6.58 ± 0.32**	3.22 ± 0.57	0.45 ± 0.17	0.80 ± 0.16	0.73 ± 0.16**	1.35 ± 0.33
Exudative enteropathy	3	5.07 ± 1.23**	1.76 ± 0.10**	0.57 ± 0.10*	1.20 ± 0.30**	0.80 ± 0.36	0.73 ± 0.50**
Acute cholecystitis	8	6.68 ± 0.42**	2.90 ± 0.47**	0.45 ± 0.16	0.84 ± 0.33	0.95 ± 0.21	1.53 ± 0.21
Nephrosis	24	4.54 ± 2.48**	1.44 ± 0.59**	0.40 ± 0.17	1.20 ± 0.46**	0.79 ± 0.22**	0.72 ± 0.35**
Glomerulonephritis	24	5.84 ± 0.64**	2.08 ± 0.56**	0.41 ± 0.14	1.02 ± 0.29**	0.90 ± 0.26	1.48 ± 0.35
Hepatic cirrhosis	95	6.54 ± 1.71**	2.31 ± 0.67**	0.38 ± 0.14	0.62 ± 0.20	0.85 ± 0.22	2.42 ± 0.70**
Viral hepatitis	18	6.30 ± 2.32	2.93 ± 1.19*	0.29 ± 0.17**	0.52 ± 0.24	0.78 ± 0.35	1.74 ± 0.92
Sarcoidosis	32	7.54 ± 2.07	3.14 ± 0.58**	0.42 ± 0.14	0.84 ± 0.20**	1.07 ± 0.20**	2.24 ± 0.41**
Alpha$_3$-globulinemia	7	6.10 ± 1.37	2.07 ± 0.54**	0.50 ± 0.20	0.53 ± 0.14*	0.80 ± 0.26	1.41 ± 0.54
				(alpha$_3$ globulin = 0.64 ± 0.43)			
Lupus erythematosus	37	6.78 ± 0.83*	2.48 ± 0.55**	0.41 ± 0.12	0.84 ± 0.30**	0.83 ± 0.19	2.18 ± 0.80**
Polyarteritis nodosa	9	6.22 ± 0.79**	2.61 ± 0.78**	0.49 ± 0.17	0.87 ± 0.19**	0.78 ± 0.14	1.42 ± 0.42
Rheumatoid arthritis	27	6.98 ± 0.55	2.81 ± 0.47**	0.45 ± 0.13	0.95 ± 0.22**	0.89 ± 0.13	1.85 ± 0.66*
Scleroderma	13	6.84 ± 1.11	3.54 ± 0.99	0.33 ± 0.17	0.62 ± 0.17	0.83 ± 0.13	1.57 ± 0.75
Rheumatic fever	23	6.99 ± 0.42	3.16 ± 0.69**	0.45 ± 0.15	0.84 ± 0.22**	0.87 ± 0.15	1.64 ± 0.32
Essential hypertension	17	6.92 ± 0.45	3.55 ± 0.70	0.41 ± 0.15	0.73 ± 0.18	0.84 ± 0.17	1.36 ± 0.32*
Congestive heart failure	39	6.50 ± 0.71**	2.92 ± 0.56**	0.41 ± 0.11	0.71 ± 0.30	0.92 ± 0.19	1.49 ± 0.36
Carcinomatosis	37	6.69 ± 0.62**	2.81 ± 0.55**	0.50 ± 0.16*	0.86 ± 0.22	0.89 ± 0.20	1.67 ± 0.54
Bacterial pneumonia	16	6.60 ± 0.62**	2.60 ± 0.62**	0.51 ± 0.15*	0.94 ± 0.23**	0.92 ± 0.12	1.65 ± 0.39
Bacterial meningitis	7	6.23 ± 0.58**	2.74 ± 0.80**	0.44 ± 0.15	0.87 ± 0.24*	0.79 ± 0.15	1.39 ± 0.37
Bronchiectasis	10	6.97 ± 0.50	3.20 ± 0.57	0.45 ± 0.16	0.74 ± 0.20	0.87 ± 0.16	1.76 ± 0.50
Osteomyelitis	11	6.68 ± 0.86	2.73 ± 0.84**	0.48 ± 0.14	0.89 ± 0.23**	0.94 ± 0.24	1.65 ± 0.36
Myxedema	11	6.44 ± 0.68**	3.29 ± 0.78*	0.32 ± 0.12*	0.65 ± 0.16	0.81 ± 0.18	1.35 ± 0.24*
Thyrotoxicosis	14	6.67 ± 0.82*	3.07 ± 0.74**	0.43 ± 0.10	0.68 ± 0.15	0.80 ± 0.13	1.61 ± 0.27
Diabetes mellitus	36	6.40 ± 1.65**	2.96 ± 0.97**	0.34 ± 0.16	0.81 ± 0.27**	0.87 ± 0.28	1.41 ± 0.51

Reproduced with kind permission of the author and publisher from Sunderman, F. W., and Sunderman, F. W., Jr.: Serum proteins and the dysproteinemias, Philadelphia, 1964, J. B. Lippincott Co., p. 324.
*p ≦ < 0.05. **p ≦ < 0.01.

Table 17-4
Normal values for cerebrospinal fluid electrophoresis utilizing vacuum ultrafiltration, Microzone cellulose acetate, and ponceau S stain*

	Percent of total
Prealbumin	4.9 ± 1.2
Albumin	61.5 ± 5.3
Alpha$_1$ globulin	4.5 ± 1.4
Alpha$_2$ globulin	6.7 ± 1.8
Beta globulin	13.7 ± 3.6
Gamma globulin	8.8 ± 2.6

*Based on data from Kaplan, A.: Electrophoresis of cerebrospinal fluid proteins, Am. J. Med Sci. **253**:549, 1967.

where there is a low total protein, particularly if the separation is to be performed on cellulose acetate or agarose. (See section on protein concentration.)

Separation can be adequately performed as for serum proteins either on cellulose acetate or agarose. The normal values for CSF proteins should be established by the laboratory with a set procedure established in the laboratory for this purpose. Percentage values should be reported rather than mg/dl values, since concentration methods are not that quantitative. As a guide, values obtained by Kaplan[52] are presented in Table 17-4 for cellulose acetate.

Urinary proteins

Because of the selective filtration of the glomerulus, the only normal proteins found in the urine in any appreciable amounts are those with a molecular weight of less than 200,000. Two thirds of those of a molecular weight greater than 30,000 are not found in serum.

Depending on the protein content of the urine (normally 2 to 4 mg/dl), concentration methods may or may not be necessary. In normal urine it is usually necessary to concentrate at least 100 times, while urines showing an excretion of 1 g protein/day may necessitate concentration of 10 to 50 times.

Normally urine contains the following proteins over 30,000 MW: albumin, alpha$_1$ acid glycoprotein, alpha$_1$-antitrypsin, alpha$_1$-lipoprotein, alpha$_2$-HS-glycoprotein, Zn-alpha$_2$-glycoprotein, alpha$_2$ GC globulin, transferrin, hemopexin, C4, IgG, and IgA.

Proteins of the 10,000 to 30,000 MW classes are also found in all of the electrophoretic zones. The presence of these proteins may account for the poor resolution of the slow mobility proteins. Among these are the gonadotropins, corticotropin, and erythroprotein. Also found are vitamin B$_{12}$-binding component blood group substance, urinary glycoprotein, as well as the urinary enzyme proteins such as amylase, trypsin, pepsin, plasmin, and lactic dehydrogenase.

Proteinuria may occur from the following:
1. Increased glomerular permeability, may be found in toxic nephritis, subacute glomerulonephritis, amyloidosis, lipoid necrosis, and diabetic glomerulosclerosis. Here one observes an elevated albumin with possible elevation of alpha$_1$ acid glycoprotein, alpha$_1$-antitrypsin, and transferrin.

2. Prerenal origin with normal glomerular function is seen in the monoclonal gammopathies, intravascular hemolysis, crush syndrome, electrocution, arterial occlusions, muscular dystrophy myoglobinuria, inflammation cancer, and autoimmune disease. Such proteins as the monoclonal and Bence-Jones proteins, hemoglobin, myoglobin, and alpha$_1$ globulin may be present.
3. Renal tubule disorders may result from drug or metal poisoning as well as Wilson's disease, galactosemia, or other hereditary diseases such as Fanconi syndrome. Alpha$_2$ globulin and beta globulins are present as well as some of the gamma components. Albumin is usually decreased.
4. Chyluria from lymphatic obstruction results in a high lipid pattern along with increases in fibrinogen as well as all plasma proteins. The albumin peak is larger in proportion to the other plasma proteins.
5. Reduced renal flow may result from postural changes and thus albumin, alpha$_1$ globulin, and beta globulin may be present. If immunoglobulins are present it is usually IgG.
6. Exercise may cause increase in albumin and the beta globulins as well as ceruloplasmin.

PROTEIN CONCENTRATION PROCEDURES

Two methods are offered here for the concentration of cerebrospinal fluid and urine preparatory to electrophoresis or immunoelectrophoresis. It is possible to semi-quantitatively determine small amounts of protein by either of these methods providing an exact amount of material being concentrated is recorded as well as the resultant concentrate. Quantitations on the concentrate should then be divided by the times concentration is occurring. EXAMPLE: If 100 ml of urine is concentrated to 1 ml, the amount of protein in the 1 ml of concentrate must be divided by 100 to obtain the amount of protein in the original 100 ml.

Concentration of fluids of low protein content such as CSF and urine is required before electrophoresis can be performed. Molecular sieving methods such as starch and polyacrylamide gel do not require as much or any concentration of these substances as do ordinary zone electrophoresis on paper, cellulose acetate, and agar gel.

Collodion bag ultrafiltration

We have found the collodion bag ultrafiltration apparatus manufactured by Carl Schleicher and Schuell Co. to meet our needs. The collodion bag retains proteins of a molecular weight of approximately 70,000 to 100,000 but may pass small compounds of 30,000 MW and below.

The bag is removed from the 20% alcohol storage container and washed with distilled water. It is forced over the end of the plastic sleeve of the inner glass tube for about 1 cm. This is fastened tightly by the aid of the outer tube (Fig. 17-19).

The assembled parts are fitted into the suction tube which contains 0.9% NaCl. Take care that the water level inside the suction tube does not exceed the lower level of the inner and outer tubes. Attach the unit to a clamp on a ring stand.

Fill the bag with the fluid to be concentrated via pipet through the inner glass tube. Either 10 ml of centrifuged urine or 5 ml of CSF is satisfactory.

Apply vacuum to the suction flask by means of a suitable vacuum pump. The specimen is concentrated to a volume of 0.1 ml or less and if desired may be somewhat reconstituted with a small amount of buffer which is used for electrophoresis. Remove the concentrated fluid with the aid of a pipet. The bag may be reused if washed and replaced in the saline-containing suction tube.

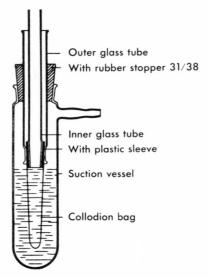

Fig. 17-19. Ultrafiltration apparatus for concentration of urine and CSF. (Courtesy Schleicher & Schuell, Inc., Keene, N.H.)

Minicon concentration method

This disposable device relies on concentration by the use of a membrane of selective permeability backed by an absorbent pad. As solute is placed in the sample well, it passes through the selective membrane into the absorbent pad leaving the concentrated specimen at the bottom of the sample well, protected by an impermeable seal (Fig. 17-20).

Graduated lines allow for a semiquantitative measure of sample concentration. Minicon concentrators come in various molecular weight cut-off configurations— B-15, 15,000 MW cut-off; S-125, 125,000 MW cut-off; A-25, 25,000 MW cut-off; and A-75, 75,000 MW cut-off. The B-15 is widely used for CSF and urine protein concentration. The S-125 has been used with success in the concentration of serum for hepatitis associated antigen testing in counterimmunoelectrophoresis. A-75 is used for the concentration of serum specimens in isoenzyme determinations having low concentration of enzyme activity such as creatine phosphokinase, alkaline phosphatase, and lactic dehydrogenase.

Serum lipoproteins
AGAROSE ELECTROPHORESIS

Equipment

    Same as for serum proteins

Reagents

    Barbital buffer, pH 8.6, 0.05 M with 0.035% EDTA (ACI/Corning No. 1-5100)
    Fat Red 7B stain and stabilizer (ACI/Corning No. 1-2500)
    Destain:
        Methanol/water, 3:1
        Glycerol, 2%, in water

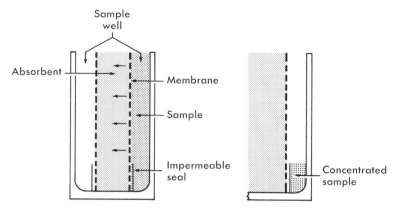

Fig. 17-20. Minicon microconcentration principle. (Courtesy Amicon Corp., Lexington, Mass.)

Procedure

1. Place 2.0 ml of serum* (obtained after 16-hour fast) in the film sample slot. Usually two applications of 1.0 ml each are required.
2. Perform electrophoresis for 40 minutes at 90 volts.
3. Remove film from cover and allow to dry in forced warm air.
4. Place dried film in stain. Stain is prepared as follows:
   a. *Fat Red 7B stock stain:* Place 0.9 g (1 vial) fat red 7B stain into a gallon glass container. Rinse vial three times with absolute methanol and add to a final volume of 4 L of absolute methanol to the stain-containing gallon glass jug. Add 8.0 ml Triton 100×. Cap and mix with a magnetic stirrer for 24 hours.
   b. *Working stain:* Just before use add 40 ml of 0.1 N NaOH to 200 ml of stock stain. Use at once. Be certain NaOH is 0.1 N. Fresh 1 N NaOH should be kept on hand and diluted to 0.1 N before use. Stain cannot be continually reused.
5. Allow film to stain completely submerged in the working stain for at least 15 minutes. Remove and allow to drain.
6. Destain in a mixture of methanol/water 3:1 by placing the entire film in the mixture for only a few seconds. Carefully observe the prebeta band; be certain that destaining does not cause this band to disappear. This process usually takes under 20 seconds.
7. Immediately place the film in a tray containing 2% glycerol in H_2O. Remove and drain.
8. Dry at room temperature. The lipoprotein patterns may be scanned in a densitometer at 520 nm if so desired.

Normal values

The normal values usually given for lipoprotein electrophoresis are as follows:

Chylomicrons	0%
Beta-lipoproteins	50% to 70%

---

*Preservation of lipoprotein specimens can be accomplished if 1 drop of the following solution is added to 1.0 to 3.0 ml plasma or serum after separation of the clot. Refrigerated sera may be run after several weeks or more. Use 0.75 g thimerosal in 30 ml of distilled water and adjust to pH 8 with 6% sodium borate. Dilute mixture to 50 ml with distilled water.

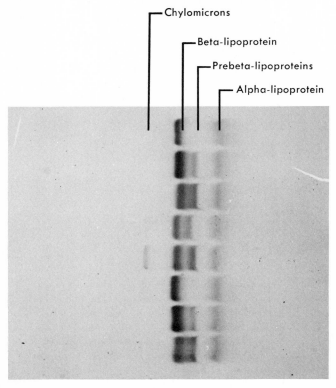

Fig. 17-21. Lipoprotein electrophoresis on agarose. (Courtesy Helena Laboratories, Beaumont, Tex.)

Prebeta-lipoproteins 11% to 29%
Alpha-lipoproteins 12% to 28%

Comments

A notation of the appearance of the serum (classified as to clear, cloudy, milky, or creamy) should be included with each report. A cholesterol, triglyceride, and glucose value should also be reported with each specimen. With this information the physician has a better chance to properly interpret the electrophoretogram for the evaluation of the patient's lipoprotein classification.

It is my belief that either agarose (Fig. 17-21) or polyacrylamide gel electrophoresis offers the best separation qualities for lipoproteins.

CLINICAL SIGNIFICANCE

The normally insoluble lipids demonstrate soluble characteristics when bound to proteins. The protein-bound lipids, lipoproteins, have become one of the most widely studied biochemical substances, during the past decade, following the classification of hyperlipidemia by Fredericksen and co-workers. Electrophoretic separation of the lipoprotein fractions has afforded a method by which a systematic approach to diseases of lipid metabolism may be delineated.

Cholesterol, cholesterol esters, triglycerides, phospholipids, sterols, and fatty acids are transported as water soluble lipoproteins in the plasma. Five main lipid particles

Table 17-5
Comparison of lipoproteins as to electrophoretic separation, ultracentrifuge, density, and chemical composition

Electrophoretic classification	Chylomicrons	Beta-lipoprotein	Prebeta-lipoprotein	Alpha-lipoprotein
SF ultracentrifuge	400	0-20	20-400	Sinks; does not float
Density	1.006	1.063-1.006 LDL	1.006 VLDL	1.210-1.063 HDL
Particle or complex size (Å)	2000	100-300	2000	70-100
Triglyceride content (percent)	90	7	64-80	7
Cholesterol content (percent)	6	35	8-13	17
Phospholipids content (percent)	4	25	6-15	27
Protein content (percent)	0.5	32	2-13	49
Apoproteins present	May exist	Major constituent—Apolp Ser	Apolp Val Apolp Gin Apolp Ala—50%	Major constituents—Apolp Thr Apolp Gln Minor constituents—Apolp Val Apolp Gln Apolp Ala

or complexes are known. They are: chylomicrons, very low density lipoproteins (VLDL), low-density lipoproteins (LDL), high-density lipoproteins (HDL), and nonesterified fatty acids (NEFA). The density classification is based on the relative proportion of lipid molecules—as lipid molecules increase, density decreases. Ultracentrifugal studies by Gofman and associates classified these densities by SF or Svedberg flotation units. The low-density lipoproteins such as chylomicrons have a high SF value, whereas the high-density lipoproteins possess a low SF value. Electrophoretic classification is based on the position of lipoprotein in the electrophoretic pattern in relation to the area of protein separation, hence the terminology alpha-, prebeta-, and beta-lipoproteins.

The lipids are carried on proteins formed in the liver known as apoproteins or apolipoproteins. They are designated by the notation Apolp, and depending on the amino acid residue at the carboxy terminal may be designated for example as Apolp The or Apolp Val.

Table 17-5 illustrates the various classifications in relation to one another.

Table 17-6
The hypolipoproteinemias

Condition	Deficient lipoproteins	Other clinical and laboratory findings
Tangier disease	Alpha (HDL)	Apolp Thi/Apolp Gln = 1:11 Normal is 3:1 Splenomegaly RES tissues have deposition of cholesteryl ester Neurologic abnormalities
Abetalipoproteinemia	Beta (LDL) Prebeta (VLDL) Chylomicrons	Retinitis pigmentosa Fat malabsorption Atrophic neuropathic disease Acanthocytosis of erythrocytes Triglycerides (low)
Hypobetalipoproteinemia	Beta (LDL) 10%-20% of normal Prebeta (VLDL) lowered slightly	HDL normal Benign condition

Hypolipoproteinemia

This clinical condition is usually detected in patients with hypocholesterolemia. Three inherited types are known at this time. See Table 17-6.

Hyperlipoproteinemia

Hyperlipoproteinemias are now classified by their phenotype as is seen in Table 17-7. Although a genetic origin is implied, many other coexisting clinical conditions may be reflected by the phenotypes. Table 17-8 lists the various clinical conditions which may accompany the various phenotypes.

Hemoglobin
CELLUOSE ACETATE (BECKMAN MICROZONE)

This method resolves hemoglobins A, F, S, and C.

Equipment
Same as for protein electrophoresis utilizing Microzone system
Centrifuge (Microzone)
Densitometer (Beckman R112)
Whatman filter paper, No. 42 or 44
Glass funnel, small

Reagents
Sodium chloride, 0.85% w/v
Toluene, AR

Table 17-7
Lipoprotein phenotypes

	Lipoprotein abnormality	Triglyceride level	Cholesterol level	Appearance of serum	Carbohydrate sensitive	Fat sensitive	Occurrence
Normal		150 mg/dl or less	250 mg/dl or less	Clear			
Type I	Chylomicrons elevated	Elevated	Normal or slightly elevated	Creamy top layer over clear plasma	No	Yes	Rare
Type II IIA	Beta-lipoproteins elevated	Normal	Elevated	Clear	No	Yes	Common
IIB	Beta- and prebeta-lipoproteins elevated	Moderate elevation	Elevated	Slightly cloudy	No	Yes	Common
Type III	Beta- and prebeta-lipoproteins elevated; smeary pattern	Elevated	Elevated	Moderately cloudy to very cloudy	Yes	Yes	Relatively rare
Type IV	Prebeta-lipoproteins elevated	Elevated	Normal or slightly elevated	Clear to very cloudy or milky	Yes	No	Common
Type V	Chylomicrons beta- and prebeta-lipoproteins elevated	Elevated	Elevated	Creamy top over milky plasma	Yes	Yes	Relatively rare

TEB buffer—prepared as follows:
Tris(hydroxymethyl)aminomethane   10.2 g
EDTA   0.6 g
Boric acid   3.2 g
Make to 1 L with distilled water and adjust to pH 8.4. Ionic strength = 0.13.

Table 17-8
Diseases that may be associated with the various lipoprotein phenotypes

Type	Nongenetic causes
Type I	Acute alcoholism Dysglobulinemia Hypothyroidism Oral contraceptives Pancreatitis Uncontrolled diabetes mellitus
Type II A and B	Biliary obstruction Dietary imprudence Dysglobulinemia Hypothyroidism Porphyria
Type III	Dysglobulinemia Myxedema
Type IV	Alcoholism Diabetes mellitus Dysglobulinemia Gaucher's disease Glycogen storage disease Hypothyroidism Nephrotic syndrome Niemann-Pick disease Pancreatitis Progestin administration Werner's syndrome
Type V	Alcoholism Chronic pancreatitis Dysglobulinemia Insulin-dependent diabetes mellitus Nephrosis

Stain

Ponceau S	0.5 g
Trichloroacetic acid, 5%	up to 100 ml

Hemoglobin standards (Helena Laboratories No. 5331)

Preparation of hemolysate

1. Centrifuge 1.0 to 5.0 ml of oxalated blood, remove plasma, and wash red blood cells 3 times with 0.85% NaCl. After last washing remove saline and record volume of packed erythrocytes.
2. For each milliliter of erythrocytes add 1.5 ml of deionized water. Mix well and place in freezer. Remove and thaw frozen specimen.
3. Add 0.4 ml toluene for each milliliter of packed cells. Shake vigorously for 5 minutes and centrifuge at 3000 rpm for 15 minutes.
4. The hemolysate is now in the aqueous layer. Filter through Whatman No. 42 or 44 filter paper into a clean tube.

5. Determine g/dl hemoglobin in the hemolysate. Adjust filtrate to 10 g/dl hemoglobin. Example: If hemoglobin concentration is 14 g/dl, add 0.4 ml distilled water to each 1.0 ml of filtrate. The hemolysate is now ready for electrophoresis and may also be used in the alkali denaturation test. After it is stored in the refrigerator for some time, the hemolysate may produce sharper patterns if it is cleaned up by adding toluene, centrifuging, and then removing clear hemolysate. The same applies to hemoglobin standards.

Procedure
1. Fill chambers with 4° C TEB buffer solution.
2. Wet membrane, blot, and place on the bridge. Allow cell and membrane to equilibrate for several minutes.
3. Apply hemolysates with applicator (0.25 µl of sample) to each application site. Allow applicator to contact membrane for at least 10 to 15 seconds.
4. Place cover on cell and place cell in refrigerator.
5. Perform electrophoresis at 450 volts for 30 minutes.
6. Stain, rinse, and dehydrate membrane by standard protein procedure.
7. Clear membrane in 30% cyclohexanone in reagent alcohol as in protein procedure.
8. Scan in the R112 densitometer at 520 nm.

CITRATE AGAROSE ELECTROPHORESIS

This method is recommended where clear separation of hemoglobins S, D, and G as well as of hemoglobins C and E is mandated.

Equipment
- Microzone cell and power supply (Beckman)
- Glass microscope slides, 2 × 3 inches, or Zip Zone-Titan IV citrate agar plate (Helena Laboratories No. 2400)
- Zip Zone applicator (Helena Laboratories No. 4080)
- Zip Zone applicator Base (Helena Laboratories No. 4082)
- Agarose applicator and multiple applicator tips (Beckman Nos. 655300 and 655313)

Reagents
Stock buffer, citrate buffer, 0.5 M
Dissolve 147 g sodium citrate, AR ($C_6H_5O_7$ Na $2H_2O$), in about 800 ml deionized water in a 1-L volumetric flask. Adjust pH to 6 with 30% citric acid. Then add water to make 1 L. Solution should be stable 4 to 6 weeks at 4° C.

Working buffer
Dilute stock buffer 1:10 with deionized water. Make fresh day of use. Titrate the pH to 6.0 to 6.2 with 30% citric acid. Maintain solution at 4° C.

If Zip Zone Titan IV citrate agar plate is not used, prepare a 1% agar as follows: Add 1 g Difco purified agar and 0.025 g Xanthum gum* to 100 ml of the working buffer. Heat over bunsen burner until all of the agar is dissolved. This usually takes 2 to 3 minutes of gentle boiling. Cool and add 1 drop of a 5% KCN solution. Use microscopic slides that have been pre-coated with 0.1% agar in distilled water. Spread 4.0 to 5.0 ml of hot agar on the precoated 2 × 3 inch microscope slide placed on a level surface. Spread evenly and smoothly on the slide.

---

*Keltrol KTL 24166, Kelco Co.

Hemolysate of 1 to 3 g/dl hemoglobin
Benzidine stain (Make just before staining.)
    Benzidine, 0.2% in 95% ethanol — 5 ml
    Acetic acid, 3% — 10 ml
    Sodium nitroferricyanide, 1% — 1 ml
    H_2O_2, 3% — 1 ml
    (o-dianisidine stain may also be used)
Hemoglobin standards
    Known hemoglobin standards may be purchased from Helena Laboratories (No. 5331) or Schering Laboratories.

Procedure

1. After slides have cooled, application of the hemolysates can be made utilizing the Zip Zone applicator (on the Zip Zone IV plate) or the agarose electrophoresis applicator (Beckman) on the prepared agarose plate. Leave tips in gel for 20 seconds to maximize application of sample.
2. Microzone cell can be used for one 2 × 3 inch glass slide. Bibulous paper wicks are required to make contact with inverted gel slide and buffer chambers.
3. Buffer chambers should be filled with 4° C buffer. Chamber should be placed in 4° C refrigerator or in crushed ice.
4. Perform electrophoresis at 60 volts, yielding 15 ma per slide for 65 minutes (Beckman cell) or 40 minutes at 50 volts or 40 ma per plate in Helena cell. If Helena cell is used with Titan IV plates, use sponges for buffer wicks as recommended by the manufacturer.
5. Remove slides from chamber and place in benzidine stain solution until blue color appears. Rinse at once with distilled water and then place in tray of distilled water to rinse thoroughly. Read results at once.
6. Slides may be air dried or dried in Reclin dryer.
7. Compare patient's sample with standards.

STARCH GEL ELECTROPHORESIS

Equipment

Same as for serum protein electrophoresis

Reagents and supplies

Cell buffer
    TRIS — 4.7 g
    Glycine — 22.6 g
    Add distilled water to make 1000 ml. Add several grains of bromphenol blue.
Buffer for preparation of gel
    TRIS — 36.3 g
    HCl — 48.0 ml
    EDTA — 1.0 g
    Add distilled water to make 1000 ml.
Amido black 10B stain
    Amido black 10B — 5 g
    Methanol — 450 ml
    Distilled water — 450 ml
    Glacial acetic acid — 90 ml
    First dissolve the stain in the methanol, then add the distilled water. Mix well and filter through glass wool into a dark bottle.

Preparation of gel

Dilute the buffer with 4 parts of distilled water.

Weigh 13.3 g of starch, hydrolyze, and add 100 ml of the diluted buffer.

Procedure

Procedure is same as for serum proteins. Hemolysate prepared for cellulose acetate method is satisfactory. It has been found that a concentration of 5 g/dl hemoglobin is most satisfactory. Higher concentrations tend to cause blurring of the samples on migration. Hemoglobins are allowed to migrate until the A_1 band is approximately 4.5 or 5.0 cm from the starting slot.

The gels are stained in amido black 10B for about 5 minutes and are placed in the destainer overnight for adequate removal of the stain.

The hemoglobin bands may be stained with the amido black stain or the benzidine stain described by Smithies.

PROCEDURE FOR THE DETERMINATION OF HEMOGLOBIN F BY ALKALI DENATURATION (MODIFIED FROM SINGER, CHERNOFF, AND SINGER)

Reagents

Stock KOH, N/4

Keep in polyethylene bottle in the refrigerator.

KOH, N/12

Dilute N/4 KOH just before use.

Ammonium sulfate, 50% saturated

Dilute 500 ml of saturated ammonium sulfate with 500 ml distilled water. Add 2.5 ml of concentrated HCl.

Procedure

1. Use the hemolysate as prepared in hemoglobin electrophoresis section. The hemolysate should be at least 10 g/dl hemoglobin.
2. Determination of denatured hemoglobin (do in triplicate)
   a. Add 1.6 ml N/12 KOH to a 100 × 13 mm test tube.
   b. Add 0.1 ml of the hemolysate, starting the stopwatch simultaneously. Rinse pipet several times.
   c. Shake 10 seconds.
   d. Exactly 1 minute after addition of hemolysate, add 3.4 ml of the ammonium sulfate solution.
   e. Invert several times to ensure mixture, centrifuge 5 minutes at 1500 rpm, and filter the supernatant through Whatman No. 3, 55-mm paper.
   f. Transfer to a suitable spectrophotometer cuvet and read the absorbance at 540 nm against reagent blank.
3. *Determination of undenatured hemoglobin* (do in triplicate)
   a. Add 5.0 ml distilled water to spectrophotometer cuvet.
   b. Add 20 µl hemolysate. Let stand 10 minutes.
   c. Read against blank of distilled water at 540 nm.

Calculation of fetal hemoglobin

$$\frac{D}{U \times 5} \times 100 = \text{Fetal hemoglobin}$$

where D = average absorbance readings of denatured hemoglobin
U = average absorbance readings of undenatured hemoglobin

Normal values

Adult	Less than 2% total hemoglobin
Birth	50% to 85% total hemoglobin

1 year   Less than 15% total hemoglobin
Up to 2 years   Less than 5% total hemoglobin

HEMOGLOBIN A_2 QUANTITATION BY USE OF DE-52 (DEAE) CELLULOSE MINI-COLUMM (AFTER EFREMOV)

The quantitation of hemoglobin A_2 by electrophoresis with subsequent quantitation by densitometry is not recommended. The following procedure allows for adequate analytic precision and quantitation of this component.

Reagents

Stock 1 M tris(hydroxymethyl)aminomethane buffer
Dissolve 121.1 g TRIS primary standard in a 1-L flask with about 500 ml distilled, deionized water, then as needed to volume.

Working buffers, 0.05 M TRIS/HCl, pH 8.3 and 8.5
Into two separate 2-L flasks add 100 ml of 1 M TRIS stock and 200 mg KCN. To each add 1800 ml distilled, deionized water. Adjust one flask to pH 8.3 with concentrated HCl; adjust the other pH to 8.5; make both to 2 L with distilled, deionized water.

Working buffer, 0.05 M TRIS/HCl, pH 7
Into a 2-L volumetric flask add 100 ml of 1 M TRIS and 200 mg KCN. Add 1800 ml distilled, deionized water and bring pH to 7 with concentrated HCl; make to 2 L with distilled, deionized water.

Anion exchanger, DE-52 (diethylaminoethyl cellulose)*
Prepare the DE-52 by washing several times with large volumes of 0.05 M TRIS/HCl, pH 8.5. Allow material to settle from each wash and decant supernatant as well as some of the fine material on top. Store DE-52 as slurry in covered container with the supernatant volume of buffer equal to about 0.7 that of the settled anion exchanger.

Columns
Five 3/4-inch disposable Pasteur pipets are set up vertically. Place a small cotton plug in the tapered part of the tube. Do not pack tightly. Moisten the column with the prepared DE-52. One filling should produce a column 6 to 7 cm in height. Cap the tube and store until used with a Critoseal cap.†

Procedure

1. Prepare a 12 g/dl hemolysate as in hemoglobin electrophoresis procedure. Dilute 1 drop of this with 6 drops of 0.05 M TRIS/HCl, pH 8.5, buffer from a Pasteur pipet. Add 1 drop of a 3% KCN solution. Mix.
2. Remove buffer from top of column. Remove Critoseal cap.
3. Slowly apply diluted hemolysate to the top of column from another Pasteur pipet.
4. When hemoglobin solution has settled into the top 5 mm of column, add pH 8.3 buffer to the top of the column and attach rubber tubing from funnel. Fill funnel with about 20 ml of pH 8.3 buffer.
5. Catch the effluent from the bottom into a 10-ml volumetric flask.
6. When all of the hemoglobin A_2 is eluted (about 1 hour), change the supernatant buffer on the top and funnel to the pH 7 buffer.
7. Catch the effluent and remaining eluted hemoglobin in a 25-ml volumetric flask.
8. Adjust volume of all flasks to mark, mix, and read the absorbance of each at 415 nm.

---

*Manufactured by Whatman, Ltd., England, marketed by Reeve Angel Co.
†Scientific Products.

Calculations

Calculate the percentage of hemoglobin A_2 from the following formula:

$$\% \text{ Hgb } A_2 = \frac{A_2}{A_2 + (2.5 \times R)} \times 100$$

where R = remaining hemoglobin

CLINICAL SIGNIFICANCE

Pauling's hypothesis that the sickling phenomena of red blood cells might be caused by a molecular disease led to the discovery that patients with sickle cell anemia did in fact have hemoglobin with a different electrophoretic mobility than normal hemoglobin. Today numerous molecular forms of hemoglobin are known.

The hemoglobin A molecule consists of four polypeptide chains. Two chains are designated as the alpha type and two the beta type $(\alpha_2\beta_2)$. The alpha chain consists of 141 amino acids and the beta chain contains 146 amino acids. Four different chains have been found in the three normal hemoglobins A, F, and A_2 and are referred to as alpha, beta, gamma, and delta chains. If one amino acid is substituted for another in either chain an abnormal hemoglobin is obtained.

Singer, Chernoff, and Singer in 1951 used an alkali denaturation method (see procedure) to establish the resistance of hemoglobin F (fetal) to denaturation. Hemoglobin F is the most common hemoglobin in the fetus and is gradually replaced by hemoglobin A in the normal developing infant. If an individual continues to produce hemoglobin F beyond the 1% level, the clinician should be alert to a possible thalassemia or acquired anemia.

Hemoglobin A_2 is found in normal individuals as a mean level of 2.6%. A study of twelve patients with thalassemia minor reveals Hb A_2 to be in a concentration of 5% to 10% with no overlap between normal values.

In the thalassemias a defect in the synthesis of the beta chains results in the doubling of the A_2 component. The difference between hemoglobin A and A_2 resides in the nonalpha chains, that is, in thirty-four amino acids or 6% of the molecules, which explains the possibility of their difference exhibited by immunologic methods.

In summary, the normal hemoglobins possess the following amino acid chain combinations:

Hb A $(\alpha_2\beta_2)$
Hb F $(\alpha_2\gamma_2)$
Hb A_2 $(\alpha_2\delta_2)$

Hunt and Ingram found in the abnormal hemoglobins S, C, and E the difference from the normal hemoglobin molecule was simply limited to a single amino acid in the beta chain, 0.34% of the entire molecule. These molecules cannot be differentiated immunologically, however, as can the A_2 from the A. If the abnormality exists in the alpha chain, such hemoglobins as I_{Norfolk}, M_{Boston}, and $G_{\text{Philadelphia}}$, to mention a few, occur.

When a child receives genes from both parents from the hemoglobin abnormality, a homozygous state exists. If the sickle cell gene is involved, the condition resulting is termed sickle cell anemia. If the thalassemia gene is involved the child suffers from thalassemia major. If, on the other hand, only one parent transmits the abnormal gene, a heterozygous state evolves, and the above states are then referred to as sickle cell trait and thalassemia minor.

Death usually results early in the child with sickle cell disease or thalassemia major, in contrast to those children with the heterozygous condition. Patients homozygous for the other hemoglobin types do not appear to be so critically affected.

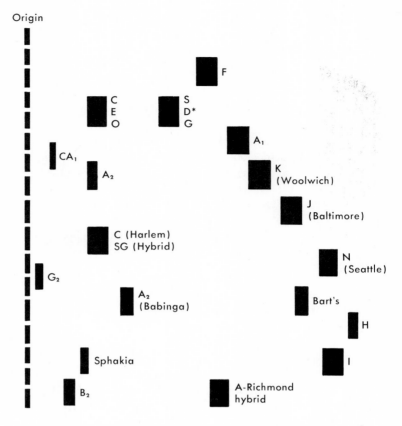

*Note: see table showing other hemoglobins that migrate like S.

Fig. 17-22. Relative mobilities of hemoglobin on cellulose acetate (TEB, pH 8.4). (After Schmidt and Brosius.)

Figs. 17-22 and 17-23 depict the relative migration of some of the various hemoglobins by cellulose acetate and agarose electrophoresis. Fig. 17-24 shows the migration pattern and differentiation by starch gel electrophoresis, which in our hands is the preferred method. Table 17-9 lists some of the known diseases caused by hemoglobin abnormalities. A flow sheet for the procedures used in the diagnosis of hemoglobinopathies is offered in Fig. 17-25.

Isoenzymes

CREATINE PHOSPHOKINASE ISOENZYMES (PFIZER POL-E-FILM OR ACI/CORNING AGAROSE METHOD)

Equipment

Same as for serum protein procedure
Auto-Scanner Flur-Vis (Helena Laboratories No. 1202)
Ultraviolet light enclosed in view box
Glass stirring rod, 8 inches long, 5 mm in diameter
37° C incubator
Dryer unit (Reclin)

Electrophoresis 573

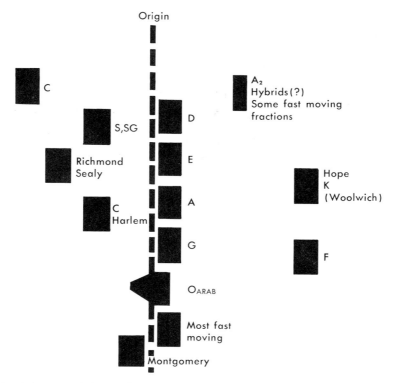

Fig. 17-23. Relative mobilities of some hemoglobins on citrate agar (pH 6.0 to 6.2). (After Schmidt and Brosius.)

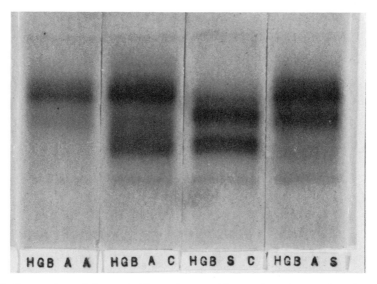

Fig. 17-24. Separation of hemoglobin variants with starch gel procedure in text. (From Marsh, C. L., and Joliff, C. R.: Micro starch gel electrophoresis. In Smith, I., editor: Chromatographic and electrophoretic technique, vol. 2, Zone electrophoresis, ed. 2, London, 1968, William Heinemann Medical Books, Ltd.)

Table 17-9
Laboratory findings in the hemoglobinopathies*

Disease	Hemoglobin type	Fetal hemoglobin (% of total)	Sickling	Severity of anemia
Normal	A	Up to 2	0	0
Thalassemia major	A	10 to 100	0	Severe
Thalassemia minor	A	Up to 10	0	Mild
Sickle cell anemia	S	Up to 20	40 to 100	Severe
Sickle cell trait	A and S	Up to 2	5 to 95	0
Sickel cell–thalassemia	S and A	Up to 40	+ to +++	Severe
Sickle cell–hereditary spherocytosis	A and S	Up to 2	+	Mild to severe
Sickle cell–Hb C	S and C	Up to 40	+	Mild to severe
Sickle cell–Hb D	S and D	Up to 12	+	Moderate to severe
Sickle cell–Hb G	S and G	Up to 2	+	0
Hb C disease	C	Up to 7	0	Mild
Hb C trait	A and C	Up to 2	0	0
Hb C thalassemia	C and A	Up to 20	0	0 to mild
Hb C elliptocytosis	A and C	Up to 2	0	0 to mild
Hb D	A and D	Up to 2	0	0
Hb E disease	E	Up to 9	0	0 to mild
Hb E trait	A and E	Up to 2	0	0
Hb E thalassemia	E and A	20 to 40	0	Moderate to severe
Hb G disease	G	Up to 2	0	0
Hb G trait	A and G	Up to 2	0	0
Hb G thalassemia	G and A	Up to 2	0	Mild
Hb H thalassemia	H and A	Up to 2	0	Moderate
Hb I (trait?)	A and I	Up to 2	0	0

*From Miale, J. B.: Laboratory medicine—hematology, ed. 4, St. Louis, 1972, The C. V. Mosby Co.

Reagents

Barbital buffer, pH 8.6 (ACI/Corning No. 470182)
Agarose special purpose electrophoresis film (ACI/Corning No. 470104)
Creatine phosphokinase isoenzyme substrate set (ACI/Corning No. 470114)

Procedure

1. Fill sample wells with 2.0 µl of serum. It will be necessary to fill well twice with 1.0 µl of serum. Allow first application to absorb before adding the next.
2. Perform electrophoresis for 20 minutes, utilizing 95 ml of buffer in each chamber.
3. During electrophoresis prepare substrate after all components of the set have come to room temperature. Add 1.0 ml of the MES buffer to one vial of the lypholized reagents. Mix well.
4. At this same time prepare a tray with filter paper in the bottom soaked in water. Place tray in incubator at 37° C for a 30-minute warmup period.
5. Remove film from cover and allow to drain onto a blotter for several minutes. Wipe any remaining buffer from the edges of the film.
6. Place the film on a flat surface with the cathodic edge nearest you. Place a

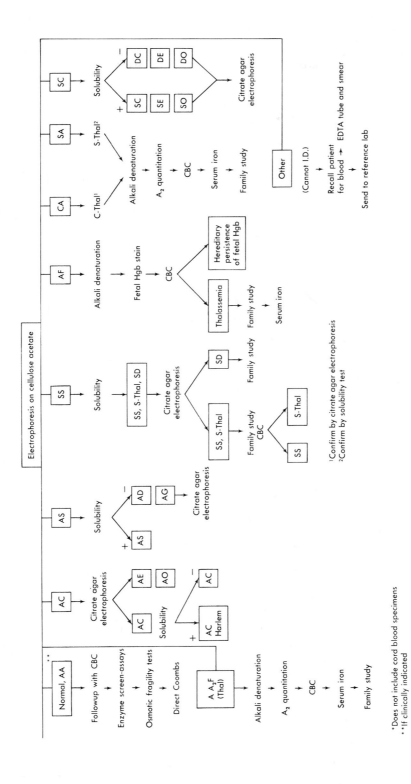

Fig. 17-25. Flow sheet for laboratory diagnosis of hemoglobinopathies.

glass stirring rod at the anodic edge. Dispense the contents of the substrate bottle along the glass rod. Carefully and with a single motion gently pull the rod across the agarose surface, keeping the remaining substrate just ahead of the rod until reaching the cathodic side of the film. The agarose should now have a uniform cast of the substrate.
7. Place film on tray in incubator for 20 minutes.
8. Remove film and allow to dry in Reclin dryer unit.
9. Patterns can be observed under ultraviolet light for visual interpretation or may be scanned in the Flur-Vis scanner with a 360-nm excitation wavelength and a 460-nm emission wavelength.

ALKALINE PHOSPHATASE ISOENZYMES (BECKMAN MICROZONE CELLULOSE ACETATE)

Equipment

Same as for cellulose Microzone acetate separation of serum proteins
Plastic trays as supplied with Dade kit.

Reagents

Buffer, 0.2 M TRIS-glycine, pH 9.3
    Prepare by adding 24.55 g TRIS and 5.25 g glycine to 1 L distilled water.
Alkaline phosphatase substrate (Dade No. B 5312-50)
Alkaline phosphatase buffer (Dade No. B 5312-50)
Special noble agar, 2% (Difco)
Alkaline phosphatase enzyme (Intestine-placenta marker) (Dade No. B 5312-52)

Procedure

1. Prepare alkaline phosphatase substrate by adding 5.0 ml alkaline phosphatase buffer and allow to dissolve. Heat vial of noble agar in boiling water until melted. Cool to 70° C and mix with substrate. Pour immediately into plastic tray and allow to solidify at room temperature.
2. Presoak strip in cell buffer, blot, and place on membrane bridge assembly.
3. If heat inactivation to reduce the bone alkaline phosphatase isoenzyme activity is desired, both a heated and an unheated specimen should be used. Place the heated specimen in a 56° C ± 0.1° C water bath for *exactly* 10 minutes and then place in an ice water bath *immediately*. Add 1.0 µl serum (four applications from the Microzone applicator) onto the strip. Heated and unheated specimens should be run with a control serum after each.
4. Place the cell in a refrigerator at 4° C. It is best to have the cell buffer at 4° C before placing the cell in the refrigerator.
5. Perform electrophoresis at 250 volts for 60 minutes.
6. Remove strip and trim off edges of membrane. Place face down on agar-substrate mixture, being careful not to entrap air bubbles. Incubate at 37° C for 1 hour.
7. Remove membrane and wash in tap water, removing all traces of agar adhering to the strip.
8. Wash strip in several changes of 5% acetic acid. Blot and dry between two blotters under a weighted object so that the strip will not wrinkle.
9. The strips may be scanned at 600 nm in a densitometer without clearing.

LDH ISOENZYMES
Cellulose acetate Microzone method

Equipment

Same as for electrophoresis of serum proteins
Special trays provided in Dade Iso-Form LDH kit

Reagents
>Dade: Iso-Form LDH Kit consisting of the following:
>1. Color reagent (nitroblue tetrazolium, phenazine methosulfate, and nicotinamide adenine dinucleotide)
>2. Buffered substrate (lactate salt in pyrophosphate buffer)
>3. Difco special noble agar, 2%
>4. Nitric acid, 3%
>
>Buffer, 0.025 M TRIS
>>Dissolve 3.025 g TRIS in 500 ml distilled water. Adjust to pH 8.6 with 5 N HCl and dilute to 1 L. Add 1 L of Beckman B2 buffer and mix well. Ionic strength of final mixture is 0.05.

Procedure
1. Heat the vial of agar in a boiling water bath until liquid. Cool to 70° C and add vial of reconstituted color reagent prepared by adding 5.5 ml buffered substrate and letting dissolve. Pour mixture into plastic tray and allow to cool on a flat surface at room temperature away from light.
2. Soak the Microzone strips in the TRIS buffer, blot, and place in the membrane bridge assembly.
3. Apply 1.0 µl serum* to each area. This is equivalent to four Microzone applicator loads.†
4. Perform electrophoresis at 250 volts for 20 minutes.
5. Remove the Microzone cellulose acetate strip from the bridge assembly and trim off excess ends. Lay strip face down on the agar, being careful not to entrap air bubbles underneath. Leave in dark at 37° C for at least 30 minutes. Check strip at this time to see whether longer time is needed to adequately develop it.
6. Wash membrane in tap water. Be certain to rub off any adhering gel with the fingers.
7. Wash in several changes of 3% nitric acid. Background should be white.
8. Heat and air dry at 50° C with pattern between blotters.
9. Patterns may be scanned at 550 to 600 nm uncleared in the R110 or with the R111 computer for semiquantitative results.

Normal values
>These should be established by each laboratory for the individual fractions; however, as a guide these values are presented here.

Heart	HHHH	1	16%-32%
	HHHM	2	26%-44%
	HHMM	3	12%-27%
	HMMM	4	3%-18%
Liver	MMMM	5	1%-14%

Agarose method (Pfizer Pol-E-Film or ACI/Corning)

Equipment
>Same as for serum protein electrophoresis utilizing Pfizer Pol-E-Film or ACI/Corning system
>Mohr pipet, 5.0 ml
>Covered tray

---

*Serum should be free of red blood cells or hemolysis. Sample should not be frozen prior to electrophoresis.

†If total LDH of sample is in excess of 250 Wacker units or 1000 Wroblewski units, this is reduced to one to three applications.

Reagents
> Dade Iso-Form LDH kit
>> See method A.
>
> Universal barbital buffer (ACI/Corning No. 470180)

Procedure
1. Prepare substrate–color reagent as in method A, except add only 3.0 ml buffered substrate and allow to dissolve. It is not necessary to utilize the agar in this method.
2. Fill sample wells with 2.0 μl of serum each.
3. Perform electrophoresis for 35 minutes.
4. Remove film from cell and place in a plastic covered dish.
5. With a 5-ml pipet add the 3.0 ml of buffered substrate to the surface of the film and allow to develop at 37° C in the dark for 30 to 60 minutes.
6. Remove and wash several times in 5% acetic acid. Dry until clear.
7. Films may be cut and scanned on the densitometer at 550 to 600 nm.

Normal values

Each laboratory should establish normal values. These values are given as a reference.

LDH 1	30%	± 5%
LDH 2	34%	± 4%
LDH 3	27%	± 5%
LDH 4	4.5%	± 2%
LDH 5	3.6%	± 2%

CLINICAL SIGNIFICANCE
Alkaline phosphatase

Some five different bands have been observed in cellulose acetate electrophoresis for alkaline phosphatase isoenzymes. The bands observed may not all be present at the same time. Liver isoenzyme appears to be the fastest migrating (anodal) and appears as a narrow sharp band. Bone isoenzyme migrates following the liver band and appears as a broad diffuse band. Placental isoenzyme migrates next, with intestinal last or closest to the cathode. The fifth band that is sometimes observed on cellulose acetate appears to be derived from liver and is located between the intestinal and placental band (Fig. 17-26).

There is considerable difficulty distinguishing between liver and bone isoenzymes. Since the liver is less heat-labile than bone, the heat treatment of serum may be a means of distinguishing between these two isoenzymes.

It appears that the isoenzyme in normal adult serum is derived from the liver whereas that in children appears to be derived from bone. Some patients show isoenzymes to be derived mainly from intestinal origin; this appears to be under genetic control. This isoenzyme appears more frequently in patients with blood groups O and B and may be influenced by fat intake.

Intestinal alkaline phosphatase isoenzyme may be inhibited by L-phenylalanine in a concentration of 0.005 M and thus may be differentiated from other sources except for placental which is also L-phenylalanine sensitive. Placental isoenzyme is, however, quite heat stable even after heating at 56° C for 30 minutes. Measurement of placental isoenzyme may be important during pregnancy as a measure of placental function since low values are encountered in placental insufficiency.

Creatine phosphokinase

The isoenzymes of CPK are classified as follows:
CPK 3 (MM) Skeletal muscle

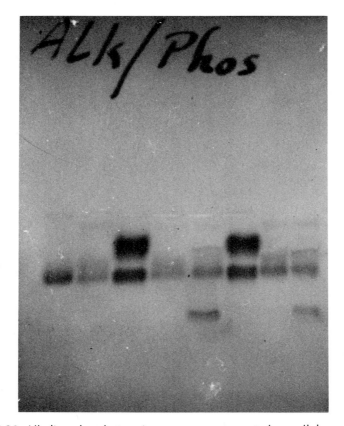

Fig. 17-26. Alkaline phosphatase isoenzymes as separated on cellulose acetate.

CPK 2 (MB) Heart
CPK 1 (BB) Nervous system and lung

CPK 1 migrates the greatest distance toward the anode into the prealbumin-albumin zone. CPK 2 is found in the alpha$_2$ region and CPK 3 in the gamma region (Fig. 17-27).

Myocardial infarction and heart surgery are responsible for the elevation of CPK 2 (MB). This isoenzyme should be studied in cases of recent surgery where chest pain may suggest the possibility of myocardial infarction. Elevated total CPK activity may alert the physician to possible myocardial infarction whereas the presence of CPK 3 (MM) might well be the cause of the elevated CPK.

CPK 3 (MM) elevations may occur in rhabdomyolysis, polymyositis, myopathy, muscular dystrophy, myasthenia gravis, heavy exercise, hypothyroidism, trauma from surgery, muscular injections, and tetanus. Neurologic disorders such as encephalitis, subarachnoid hemorrhage, brain tumors, head injury, or stroke may show elevated CPK 1 (BB) isoenzyme bands. Pulmonary disease and carbon monoxide poisoning may show patterns typical of myocardial infarction elevations. In malignant hyperthermia all bands may be increased.

Anido and associates report that an increase in the CPK (MB) band may be seen in cases of myocardial infarction, but it is not pathognomonic because of transient increases which may disappear within 48 hours after infarction. He also states that

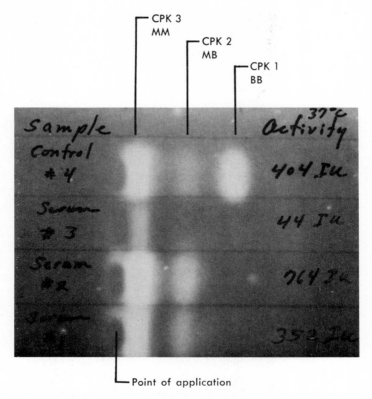

Fig. 17-27. Creatine phosphokinase isoenzymes as separated on agarose and visualized by ultraviolet (360-nm) excitation.

unless total CPK activity is twice normal, electrophoretic patterns are not of much value.

The isoenzymes are stable under refrigeration for at least 1 week. CPK 3 (MM) is stable at room temperature (25° C) for several days whereas CPK 2 (MB) may be destroyed by 37° C temperature for 2 hours.

Lactic dehydrogenase

In myocardial infarction the LDH activity may remain elevated some 10 days or more. The increase is confined mainly to LDH_1 and LDH_2. This pattern may last for over 2 weeks even though total LDH may return to normal. A normal pattern is usually observed in cardiac failure without infarction. Hypotensive liver damage and red cell damage resulting from pulmonary embolism may give rise to fractions LDH_4 and LDH_5 as well as LDH_2; however, these fractions usually exceed the LHD_1 (Fig. 17-28).

Diseases of the liver such as hepatocellular damage caused by viral hepatitis, drug toxicity, and extrahepatic biliary cirrhosis are reflected by considerable increases in the LDH_5 band. Muscular disease may exhibit varying isoenzyme activity depending on the muscle tissue affected. In general CPK activity is greater in muscular pathology and abnormal metabolism than is LDH. Diseases and conditions affecting the red fibers, such as are concerned with continued contractibility, mainly show elevations of LDH_1 and LDH_2 while those associated with the white fibers generally show

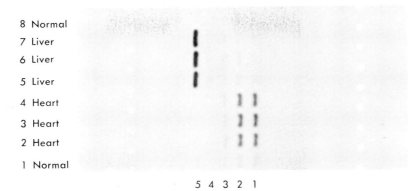

Fig. 17-28. Lactic dehydrogenase isoenzymes as separated on cellulose acetate. (Courtesy Beckman Instruments, Inc., Fullerton, Calif.)

elevations of LDH_4 and LDH_5. Muscular dystrophy, polymyositis, and muscle trauma are the principal diseases responsible for these increases.

Hemopoietic diseases exhibit dramatic increases in the total serum LDH, particularly in cases of pernicious anemia when the hemoglobin falls below 8 g/dl. Hemolytic anemia and megaloblastic anemia may also show a rise in the LDH_1 fraction of the isoenzyme pattern.

In neoplastic disease there is a tendency toward a shift to the slow cathodic isoenzymes. This may be explained by the fact that these isoenzymes are associated with anaerobic glycolysis rather than oxidative metabolism. Because of the poor vascular supply to tumors there is a dependence on anaerobic glycolysis for energy to these rapidly growing tissues.

LDH isoenzyme patterns in renal disease are of limited value. However, LDH_5 increases may be found in acute tubular necrosis, polynephritis, and chronic glomerulonephritis.

Serum haptoglobin

Two procedures are offered here for serum haptoglobin. The first method, utilizing cellulose acetate, is for the determination of total serum haptoglobin. The polyacrylamide gel method is for the determination of haptoglobin phenotypes.

CELLULOSE ACETATE METHOD

Equipment

Microzone electrophoresis system
Wash box (Beckman No. 330105)
Magnetic stirrer
Vortex mixer
Micropipets, 10, 20, and 40 µl
Micro test tubes
Dispenser, 1 µl (Drummond Scientific Co.)
Drying frame and blotters
Drying oven

Reagents

Phosphate buffer, 0.05 M, pH 7
1. Sodium phosphate, 7.1 g Na_2HPO_4 (dibasic), anhydrous, in 1 L of distilled water
2. Sodium phosphate, 3.45 g $NaH_2PO_4 \cdot H_2O$ (monobasic) in 500 ml of distilled water

Mix 1 and 2 in a 2-L flask.

Acetate buffer, pH 4.7 (for staining mixture)
Mix 1.46 g sodium acetate and 0.52 ml glacial acetic acid; make to 500 ml with distilled water.

Glacial acetic acid, 5%

Stain reaction mixture (Prepare just before staining.)
1. 20 mg o-dianisidine (J. T. Baker) 3,3' dimethoxybenzidine in 50 ml of ethanol.
2. Place the above in plastic staining dish and add 30 ml of pH 4.7 acetate buffer and 0.2 ml of the 30% hydrogen peroxide.
3. Mix and allow to equilibrate at room temperature for 5 minutes.

Normal control serum

Haptoglobin free serum

Preparation of sample

1. Obtain unhemolyzed serum sample from naturally retracted clot.
2. Prepare hemoglobin hemolysate and determine concentration in g/dl.
3. Prepare hemolysate-serum tubes as follows:

Tube	Serum (μl)	Hb g/dl (μl)	Hb/serum (mg/dl)
1	40	1	25
2	20	1	50
3	10	1	100
4	10	2	200

4. Mix well in Vortex mixer and incubate at 37° C for 30 minutes.

Electrophoresis

1. Fill cell with 4° C phosphate buffer.
2. Wet cellulose acetate strips with phosphate buffer, blot, and place on bridge.
3. Make three applications (0.75 μl) of each sample on the membrane. Place sides of sample applicator in middle groove (one step toward cathode).
4. Perform electrophoresis at 150 volts for 40 minutes. (Keeping cell at 4° C will help.)
5. Prepare fresh stain reaction mixture.
6. Remove strip from cell and stain for 5 minutes. Rinse in running tap water and place in wash box.
7. Wash in 5% acetic acid for 10 minutes utilizing magnetic stirrer.
8. Place in drying frame with blotters. Dry at 50° C or at room temperature, being careful not to use excessive heat, which will distort the membrane.
9. The electrophoretic pattern (Fig. 17-29) will show three components. In order of migration from cathode to anode they are: free hemoglobin, haptoglobin-hemoglobin complex, and methemalbumin. The first two bands are significant.
10. Permanent records can be made by scanning on the R112 Microzone densitometer at 450 nm.

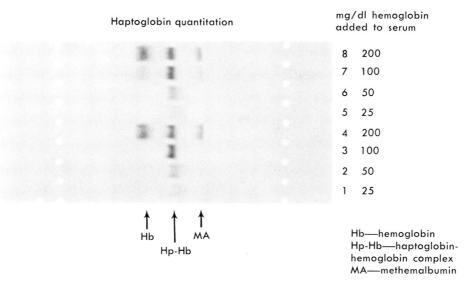

Fig. 17-29. Haptoglobin separation on cellulose acetate for quantitation. (Courtesy Beckman Instruments, Inc., Fullerton, Calif.)

Interpretation

Normal serum will bind up to 140 mg/dl hemoglobin. The 25, 50, and 100 mg/dl hemoglobin-serum mixtures may show no free hemoglobin bands and the 200 mg/dl mixture does, thus normal haptoglobin binding is present.

Hemoglobin/serum mixtures (mg/dl)	Free Hb band present	Haptoglobin
25	Yes	Absent
50	Yes	Reduced
100	Yes	Possible deficiency
200	Yes	Normal

POLYACRYLAMIDE METHOD

Equipment

Same as for serum protein electrophoresis

Reagents

Same as for polyacrylamide separation of serum proteins.
o-Dianisidine acetate stain pH 4.7 (see hemoglobin electrophoresis)
Known haptoglobin types (for reference and control)*

Procedure

1. Separate serum containing haptoglobin-bound hemoglobin hemolysate (as prepared under the cellulose acetate electrophoresis) as under separation of serum proteins either by the 7.5% gel or in the step gel system (see Fig. 17-30).
2. After electrophoresis the strips are stained in the dark for approximately ½ to 1 hour in the o-dianisidine stain. The haptoglobin types can be observed clearly and the gels can be washed in 5% acetic acid, photographed, and stored wrapped in Saran Wrap in the dark for many months.

---

*Available from Behring Diagnostics.

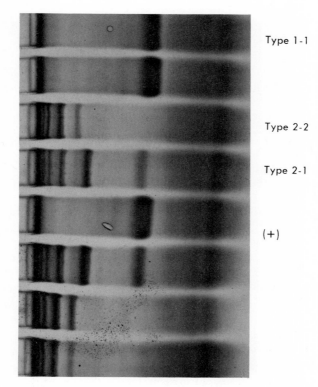

Fig. 17-30. Haptoglobin typing by polyacrylamide gel electrophoresis.

CLINICAL SIGNIFICANCE

Polonovski and Jayle first described the existence of the alpha$_2$ globulin hemoglobin-binding proteins. In their studies they found haptoglobin not to be a single molecular species but a multimolecular component exhibiting polymorphism. Smithies demonstrated in starch gel the three haptoglobin types 1-1, 2-1, and 2-2, which vary among individuals as genetic phenotypes.

The main function of haptoglobin is to combine with free hemoglobin as it is released into the circulation. The complex is catabolized by the reticuloendothelial system. It is believed that haptoglobin plays only a minor role in the normal catabolism of hemoglobin, which takes place mostly in the extravascular compartment.

Haptoglobin is decreased in hemolytic conditions and in those diseases of the hemopoietic system such as megaloblastic anemia in which marrow turnover is excessive. Blood cell destruction by transfusion reaction and by faulty prosthetic devices may also decrease haptoglobin levels, as may faulty synthesis due to hepatocellular disease. In 4% to 6% of the black population a haptoglobin type OO is found in which ahaptoglobinemia or hypohaptoglobinemia occurs.

Elevated values may occur in infections, ulcerative conditions, tissue destruction, and necrosis as well as in diabetic angiopathy, pregnancy, and estrogen therapy.

Haptoglobin typing has revealed that persons with HP 1-1 have a 3 to 4 times greater risk of leukemia than do those of the other types. This may in some way be due to decreased immunologic competency of the phenotype 1-1 individual.

Phenotyping plays an extremely important part in forensic work involving blood stains and offers yet another parameter in the identification of blood stains with regard to genetic comparisons.

Serum glycoproteins

Equipment

   Microzone electrophoresis system
   Drying frame
   Forced air oven

Reagents

   Buffer as for Microzone cellulose acetate separation of serum proteins
   Schiff's reagent—special Hotchkiss-McManus stain*
      Dissolve 6 g of basic fuchsin in 1200 ml of 90° C water; cool to 50° C and filter. Add 30 ml of 2 N HCl and 4 g potassium metabisulfite to the filtrate. Stopper and store in cool dark place overnight. Add 3 g of powdered animal charcoal to mixture, mix, and allow to stand for 1 minute. Filter through E & D 192, 385-cm filter paper. Add 40 ml of 2 N HCl. Store in refrigerator when not in use. Bring to room temperature before staining. Discard when stain turns pink.
   Trichloroacetic acid (fixative), 5%
   Periodic acid, 0.8%, in 0.3% sodium acetate (oxidizing agent)
   Nitric acid (rinse), 3%
   HCl, 0.01 N, for rinsing

Procedure

   In this procedure the separated glycoproteins are stained by the action of periodic acid on the carbohydrates, which converts them to aldehydes. Aldehydes are stained with Schiff's reagent, which renders them a purple-red color (see Fig. 17-31).

---

*Available from Hartman-Leddon Co.

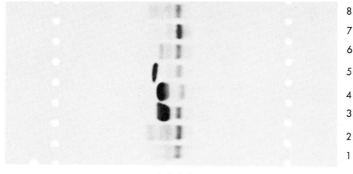

Glycoproteins

8 Liver disease
7 Amyloidosis
6 Idiopathic paraproteinemia
5 Gamma G multiple myeloma
4 Gamma M acroglobulinemia
3 Gamma A multiple myeloma
2 Hyperglobulinemic purpura
1 Control serum

Gamma — Beta — Alpha$_2$ — Alpha$_1$ — Albumin

Fig. 17-31. Glycoprotein separation on cellulose acetate stained by Schiff's reagent. (Courtesy Beckman Instruments, Inc., Fullerton, Calif.)

1. Fill Microzone cell to lower mark with 4° C B_2 buffer.
2. Wet membrane with B_2 buffer, blot, and place on bridge.
3. Apply one to three applicators full of sample to membrane in midposition.
4. Perform electrophoresis for 20 minutes at 250 volts constant voltage. This should not exceed 8 ma. It is advisable to keep cell at 4° C.
5. Remove membrane and place in 5% trichloroacetic acid for 5 minutes.
6. Oxidize in periodic acid–acetate solution for 5 minutes.
7. Wash twice in distilled water.
8. Place in Schiff's reagent (in plastic dish at room temperature) for a minimum of 10 minutes.
9. Rinse three times in 3% HNO_3, 5 minutes for each rinse.
10. Place in 0.01 N HCl for less than 1 minute.
11. Air dry in drying frame between blotters for 3 to 4 hours in 50° C oven or incubator for 40 to 60 minutes. Pattern now should remain stable.
12. Patterns may be scanned in the R112 Microzone densitometer or other suitable densitometer at 550 nm.

Normal values

Each laboratory should establish its own normal values. As a guide, values obtained by Van Neerbos and colleagues may be used.

Albumin	2.6% ± 1.3%
$Alpha_1$-glycoproteins	21.9% ± 3.0%
$Alpha_2$-glycoproteins	38.8% ± 3.1%
Beta-glycoproteins	24.2% ± 2.4%
Gamma-glycoproteins	12.6% ± 2.6%

CLINICAL SIGNIFICANCE

The following glycoproteins are found in the various fractions:

Prealbumin and albumin glycoproteins	Tryptophane-rich prealbumin
$Alpha_1$-glycoproteins	$Alpha_1$ acid glycoprotein (orosomucoid)
	Thyroxin and cortisol binding proteins
	Trypsin inhibitor
	$Alpha_1$-lipoprotein
	Inhibitor of viral agglutination
$Alpha_2$-glycoproteins	Haptoglobin
	Ceruloplasmin
	$Alpha_2$-macroglobulin
	$Alpha_2$-lipoprotein
	Zn-$alpha_2$-glycoprotein
	Thyroid-binding globulin
	Transcortin
	Viral inhibitors
	GC globulins
Beta-glycoproteins	Transferrin
	$Beta_2$-glycoprotein
	Complement C3
	$Beta_2$-macroglobulin
	IgA
Gamma-glycoproteins	IgG
	IgM
	Probably IgD and IgE

Orosomucoid appears to increase in inflammatory and neoplastic diseases. The alpha$_2$-glycoproteins are generally increased in malignant tumors, rheumatic fever, active rheumatoid arthritis, pulmonary disease, and diabetes. Beta-glycoproteins are increased in iron deficiency anemia, late pregnancy, and hepatitis while being decreased in acute infections and chronic renal disease. Increases in glycoproteins of the gamma region may be found in the monoclonal gammopathies, autoimmune disease, and other conditions responsible for the hypergammaglobulinemias.

References
BASIC CONCEPTS

1. Bier, M., editor: Electrophoresis–theory, methods and applications, vol. 2, New York, 1967, Academic Press, Inc.
2. Cawley, L. P.: Electrophoresis and immunoelectrophoresis, Boston, 1969, Little, Brown and Co.
3. Gebott, M. D.: Microzone electrophoresis manual, Beckman Instruments, Inc., 1973, Fullerton, Calif.
4. Gray, G. W.: Electrophoresis, Sci. Am. **185:** 45, 1951.
5. Nerenberg, S. T.: Electrophoresis—a laboratory manual, Philadelphia, 1966, F. A. Davis Co.
6. Reuss, F. F.: Mem. Soc. Imp. Nat. Moscow **2:**327, 1809.
7. Smith, I.: Chromatographic and electrophoretic techniques, London, 1968, William Heinemann Medical Books, Ltd.
8. Theorell, H. J.: Studies on cytochrome c. IV. The magnetic properties of ferric and ferrous cytochrome c, Am. Chem. Soc. **63:** 1820, 1941.
9. Tiselius, A.: A new apparatus for electrophoresis: analysis of colloidal mixtures, Trans. Faraday Soc. **33:**524, 1937.
10. Tiselius, A., Pedersen, K. O., and Svedberg, T.: Analytical measurements of ultracentrifugal sedimentation, Nature **140:**848, 1937.
11. Wieme, R. J.: Agar gel electrophoresis, Amsterdam, 1965, Elsevier Publishing Co.

PROTEIN SEPARATION
Cellulose acetate electrophoresis

12. Grunbaum, B. W., and Kirk, P. L.: Design and use of a refined microelectrophoresis unit, Anal. Chem. **32:**564, 1960.
13. Grunbaum, B. W., Lyons, M., Carroll, N., and Zec, J.: Microelectrophoresis cellulose acetate membranes, Anal. Chem. **32:**1361, 1960.
14. Grunbaum, B. W., Zec, J., and Durrum, E. L.: An improved microelectrophoresis and immunoelectrophoresis technique using cellulose acetate anticonvectant media, Am. Chem. Soc. Mem. Symposium, Atlantic City, Sept. 11, 1962.
15. Grunbaum, B. W., Zec, J., and Durrum, E. L.: Application of an improved microelectrophoresis technique and immunoelectrophoresis of the serum proteins on cellulose acetate, Microchem. J. **7:**41, 1963.
16. Kohn, J.: A new supporting medium for zone electrophoresis, Biochem. J. **65:**9, 1957.
17. Kohn, J.: A cellulose acetate supporting medium for electrophoresis, Clin. Chim. Acta **2:**297, 1957.
18. Kohn, J.: Membrane filter electrophoresis. In Peeters, H., editor: Protides of the biological fluids, Proceedings 5th Colloquium, Bruges, 1958.
19. Kohn, J.: Cellulose acetate: a micro-electrophoretic method, Nature **181:**839, 1958.
20. Kohn, J.: Small-scale membrane filter electrophoresis and immunoelectrophoresis, Clin. Chim. Acta **3:**450, 1958.
21. Kohn, J.: Small scale and micro-membrane filter electrophoresis and immunoelectrophoresis. In Peeters, H., editor: Protides of the biological fluids. Proceedings 6th Colloquium, Bruges, 1959.

Starch gel electrophoresis

22. Marsh, C. L., Jolliff, C. R., and Payne, L. C.: A rapid micromethod for starch-gel electrophoresis, Am. J. Clin. Pathol. **41:**217, 1964.
23. Smithies, O.: Zone electrophoresis in starch gels. Grouped variations in the serum proteins of normal human adults, Biochem. J. **61:**629, 1955.
24. Smithies, O.: Zone electrophoresis in starch gels and its application to studies of serum proteins, Adv. Protein Chem. **14:**65, 1959.

Agarose method

25. Barteling, S. J.: A simple method for the preparation of agarose, Clin. Chem. **15:**1002, 1969.
26. Elevitch, F. R., Aronson, S. B., Feichtmeier, T. V., and Enterline, M. L.: Thin gel electrophoresis in agarose, Am. J. Clin. Pathol. **46:**692, 1966.
27. Hjerten, S.: Agarose as an anticonvection agent in zone electrophoresis, Biochem. Biophys. Acta **53:**514, 1961.

28. Russel, B., Levitt, J., and Polson, A.: A method of preparing agarose, Biochim. Biophys. Acta 86:169, 1964.

Polyacrylamide gel electrophoresis

29. Davis, B. J.: Disc electrophoresis II: method and application to human serum proteins, Ann. N.Y. Acad. Sci. 121:404, 1964.
30. Marsh, C. L.: Personal communication, 1974.
31. Marsh, C. L., and Jolliff, C. R.: In Smith, I., editor: Chromatographic and electrophoretic techniques, vol. 2, London, 1968, William Heineman Medical Books, Ltd.
32. Ornstein, L.: Disc electrophoresis: background and theory, Ann. N. Y. Acad. Sci. 121:321, 1964.
33. Raymond, S.: Acrylamide gel electrophoresis, Ann. N. Y. Acad. Sci. 121:350, 1964.
34. Ritchie, R. F., Harter, J. G., and Bayles, T. B.: Refinements of acrylamide electrophoresis, J. Lab. Clin. Med. 68:842, 1966.
35. Wright, G. L., Farrell, K. B., and Roberts, D. B.: Gradient polyacrylamide gel electrophoresis of human serum proteins: Improved discontinuous gel electrophoretic technique and identification of individual serum components, Clin. Chim. Acta 32:285, 1971.

ELECTROPHORESIS OF PROTEINS IN CEREBROSPINAL FLUID

36. Brackenridge, C. J.: Cerebrospinal fluid protein fractions in health and disease, J. Clin. Pathol. 15:206, 1962.
37. Clausen, J., Matzke, J., and Gerhardt, W.: Agar-gel microelectrophoresis of proteins in the cerebrospinal fluid. Normal and pathological findings, Acta Neurol. Scand. 40 (Supp. 10):49, 1964.
38. Evans, J. H., and Quick, D. T.: Polyacrylamide gel electrophoresis of spinal-fluid proteins, Clin. Chem. 12:28, 1966.
39. Goa, J., and Tueten, L.: Electrophoresis of CSF proteins in certain neurological diseases, Scand. J. Lab. Clin. Invest. 15:152, 1963.
40. Greenhouse, A. H., and Speck, L. B.: The electrophoresis of spinal fluid proteins, Am. J. Med. Sci. 248:333, 1964.
41. Hunter, R., Jones, M., and Malleson, A.: Abnormal cerebrospinal fluid total protein and gamma-globulin levels in 256 patients admitted to a psychiatric unit, J. Neurol. Sci. 9:11, 1969.
42. Ivers, R. R., McKenzie, B. F., McGuckin, W., and Goldstein, N. P.: Spinal fluid gamma globulin in multiple sclerosis and other neurologic disease: electrophoretic patterns in 606 patients, J.A.M.A. 176:515, 1961.
43. Kaplan, A.: Electrophoresis of cerebrospinal fluid proteins, Am. J. Med. Sci. 253:549, 1967.
44. Schneck, S. A., and Claman, H. N.: CSF immunoglobulins in multiple sclerosis and other neurologic disease, Arch. Neurol 20:132, 1969.
45. Schultze, H. E., and Heremans, J. F.: Molecular biology of human proteins, New York, 1966, Elsevier Publishing Co.

ELECTROPHORESIS OF URINARY PROTEINS

46. Berggard, I.: The plasma proteins in normal urine, Nature 187:776, 1960.
47. Berggard, I.: Studies on the plasma proteins in normal human urine, Clin. Chim. Acta 6:413, 1961.
48. Gebbott, M. D.: Microzone electrophoresis manual, Beckman Instruments, Inc., 1973, Fullerton, Calif.
49. Grant, G. H.: The proteins of normal urine, J. Clin. Pathol. 10:360, 1957.
50. Rigas, D. A., and Heller, G. G.: The amount and nature of urinary proteins in normal human subjects, J. Clin. Invest. 30:853, 1951.
51. Schultz, H. E., and Heremans, J. F.: Molecular biology of human proteins, New York, 1966, Elsevier Publishing Co.

PROTEIN CONCENTRATION PROCEDURES

52. Amicon Corp.: Pre-concentration and deproteinization in the clinical laboratory, Publication No. 434, 1973.
53. Amicon Corp.: Publication No. 439, 1974.
54. Burrows, S.: Simple method for concentration of cerebrospinal fluid for protein electrophoresis, Clin. Chem. 11:1068, 1965.
55. Pollack, V. E., Gaizutis, M., and Rezaian, J.: Serum proteins in urine: Examination of new method for concentrating urine, J. Lab. Clin. Med. 71:338, 1968.

SERUM PROTEIN ELECTROPHORESIS: CLINICAL SIGNIFICANCE

56. Brackenridge, C. J., and Csillag, E. R.: A quantitative electrophoretic survey of serum protein fractions in health and disease, Acta Med. Scand. 172(Supp. 383): 1962.
57. Dimopoullos, G. T.: Plasma proteins in health and disease, Ann. N.Y. Acad. Sci. 94:1, 1961.
58. Haurowitz, F.: The chemistry and function of proteins, ed. 2, New York, 1963, Academic Press, Inc.
59. Korngold, L.: Plasma proteins: methods of study and changes in disease. In Stefanini, M., editor: Progress in clinical pathology, vol. I, New York, 1966, Grune & Stratton, Inc.
60. Papadopoulos, N. M., and Kintzios, J. A.: Differentiation of pathological conditions by

visual evaluation of serum protein electrophoretic patterns, Proc. Soc. Exp. Biol. Med. **125**:927, 1967.
61. Putnam, F. W.: The plasma proteins, vol. II, New York, 1960, Academic Press Inc.
62. Sandor, G.: Serum proteins in health and disease, Baltimore, 1966, The Williams & Wilkins Co.
63. Simmons, P., Penny, R., and Goller, I.: Plasma proteins: a review, Med. J. Aust. **2**: 494, 1969.
64. Sunderman, F. W., and Sunderman, F. W., Jr.: Serum proteins and the dysproteinemias, Philadelphia, 1964, J. B. Lippincott Co.

SERUM LIPOPROTEINS

65. Classification of hyperlipidemias and hyperlipoproteinemias, Bull. W.H.O. **43**:891-915, 1970.
66. Elphick, M. C.: Microscope slide electrophoresis of serum lipoproteins in agarose gel, J. Clin. Pathol. **24**:83, 1971.
67. Fredrickson, D. S., and Lees, K. S.: A system for phenotyping hyperlipoproteinemia, Circulation **31**:321, 1965.
68. Fredrickson, D. S., Lees, K. S., and Levy, R. I.: Genetically determined abnormalities in lipid transportation. In Paoletti, R., Kritchevsky, D., and Steinberg, D., editors: Progress in biochemical pharmacology, vol. II, Basel, 1966, S. Karger.
69. Fredrickson, D. S., Levy, R. I., and Lees, R. S.: Fat transport in lipoproteins—an integrated approach to mechanism and disorders, N. Engl. J. Med. **276**:32, 1967.
70. Hulley, S. B., Cook, S. G., Wilson, W. S., Nichaman, M. Z., Hatch, F. T., and Lindgren, F. T.: Quantitation of serum lipoproteins by electrophoresis on agarose gel: standardization in lipoprotein concentration units (mg/100 ml) by comparison with analytical ultracentrifugation, J. Lipid Res. **12**:420, 1971.
71. Lees, R. S., and Hatch, F. T.: Sharper separation of lipoprotein species by paper electrophoresis in albumin containing buffer, J. Lab. Clin. Med. **61**:318, 1963.
72. Levy, R. I., Lees, R. S., and Fredrickson, D. S.: The clinical and biochemical forms of two forms of familial hyperbetalipoproteinemia, J. Clin. Invest. **46**:1086, 1967.
73. Noble, R. P.: Electrophoretic separation of plasma lipoproteins in agarose gel, J. Lipid Res. **9**:693, 1968.
74. Papadopoulos, N. M., and Kintzios, J. A.: Varieties of human serum lipoprotein pattern: evaluation by agarose gel electrophoresis, Clin. Chem. **17**:427, 1971.

HEMOGLOBIN ELECTROPHORESIS (METHODOLOGY)

75. Bartlett, R. C.: Rapid cellulose acetate electrophoresis: qualitative and quantitative hemoglobin fractionation, Clin. Chem. **9**: 325, 1963.
76. Gebbott, M. D.: Microzone electrophoresis manual, Beckman Instruments, Inc., 1973, Fullerton, Calif., Chap. 8.
77. Gebbott, M. D.: Personal communication.
78. Robinson, A. R., Robson, M., Harrison, A. P., and Zuelzer, W. W.: A new technique for differentiation of hemoglobin, J. Lab. Clin. Med. **50**:745, 1957.
79. Schmidt, R. M., and Brosious, E. M.: Basic laboratory methods of hemoglobinopathy detection, HEW Publication No. CDC 74-8266, Atlanta, 1974.

PROCEDURE FOR DETERMINATION OF HEMOGLOBIN F BY ALKALI DENATURATION

80. Singer, K., Chernoff, A. I., and Singer, L.: Studies on abnormal hemoglobins. I. Their demonstration in sickle cell anemia and other hematologic disorders by alkali denaturation, Blood **6**:413, 1951.

HEMOGLOBIN A_2 QUANTITATION

81. Efremov, C. D., Huisman, T. H. J., Bowman, K., Wrightstone, R. N., and Shroeder, W. A.: Microchromatography of hemoglobins: a rapid method for the determination of hemoglobin A_2, J. Lab. Clin. Med. **83**:657, 1974.

HEMOGLOBINS: CLINICAL SIGNIFICANCE

82. Bhagavan, N. V., and Bloor, J. H.: Hemoglobin and porphyrin metabolism in biochemistry, a comprehensive review, Philadelphia, 1974, J. B. Lippincott Co., pp. 505-563.
83. Heller, P.: The molecular basis of the pathogenicity of abnormal hemoglobins—some recent developments, Blood **25**:110, 1965.
84. Heller, P.: Hemoglobins and heredity, Postgrad. Med. **43**:63, 91, 1968.
85. Hunt, J. A., and Ingram, V. M.: Allelomorphism and the chemical differences of the hemoglobins A, S, and C, Nature **181**: 1062, 1958.
86. Ingram, V. M.: Chemistry of the abnormal hemoglobins in hematology and blood groups, papers from the British Medical Bulletin, Chicago, 1961, University of Chicago Press, pp. 26-31.
87. Pauling, L., Itano, H. A., Singer, S. J., and Wells, I. C.: Sickle-cell anemia, a molecular disease, Science **110**:543, 1949.

88. Perutz, M. F.: The hemoglobin molecule, Sci. Am. **211**:64, 1964.
89. Schneider, R. G.: Developments in laboratory diagnosis. In Abramson, H., Bertles, J. F., and Wethers, D. L., editors: Sickle cell disease—diagnosis, management, education and research, St. Louis, 1973, The C. V. Mosby Co., p. 230.

CREATINE PHOSPHOKINASE ISOENZYMES

90. Dalal, F. R., Cilley, J., and Winsten, S.: A study of the problems of inactivation of creatine kinase in serum, Clin. Chem. **18**:330, 1972.
91. Hess, J. W., MacDonald, R. P., Fredrick, R. J., Jones, R. N., and Gross, D.: Serum creatine phosphokinase (CPK) activity in disorders of heart and skeletal muscle, Ann. Intern. Med. **61**:1015, 1964.
92. Klein, M. S., Shell, W. E., and Sobel, B. E.: Serum creatine phosphokinase (CPK) isoenzymes after intramuscular injections, surgery, and myocardial infarction, Cardiovasc. Res. **7**:412, 1973.
93. Kontinnen, A., and Somer, H.: Determination of serum creatine kinase isoenzymes in myocardial infarction, Am. J. Cardiol. **29**:817, 1972.
94. Nevins, M. A., Saran, M., Bright, M., and Lyon, L. J.: Pitfalls in interpreting serum creatine phosphokinase activity, J.A.M.A. **224**:1382, 1973.
95. Roberts, R., and Sobel, B. E.: Isoenzymes of creatine phosphokinase and diagnosis of myocardial infarction, Ann. Intern. Med. **79**:741, 1973.
96. Roe, C. R., Limbird, L. E., Wagner, G. S., and Nerenberg, S. T.: Combined isoenzyme analysis in the diagnosis of myocardial injury: application of electrophoretic methods for the detection and quantitation of the creatine phosphokinase M. B. isoenzyme, J. Lab. Clin. Med. **80**:577, 1972.
97. Somer, H., and Konttinen, A.: Demonstration of serum creatine kinase isoenzymes by fluorescence technique, Clin. Chim. Acta **40**:133, 1972.
98. Takahashi, K., Ushikubo, S., Oimami, M., and Shinko, T.: Creatine phosphokinase isozymes of human heart muscle and skeletal muscle, Clin. Chim. Acta **38**:285, 1972.
99. Wagner, C. S., Roe, C. R., Limbird, L. E., Rosatid, R. A., and Wallace, A. G:. The importance of identification of the myocardial-specific isoenzyme of creatine phosphokinase (MB form) on the diagnosis of myocardial infarction, Circulation **47**:263, 1973.
100. Wolf, P. L., Kearns, T., Neuhoff, J. and Lauredin, J.: Identification of CPK isoenzyme MB in myocardial infarction, Lab. Med. **5**:48, 1974.
101. Zsigmond, E. K., Starkweather, W. H., Duboff, G. S., and Flynn, K. A.: Abnormal creatine-phosphokinase isoenzyme pattern in families with malignant hyperpyrexia, Anesth. Analg. **51**:827, 1972.

ALKALINE PHOSPHATASE ISOENZYMES

102. Bergman, J., and Blethen, S.: Determination of alkaline phosphatase isoenzymes, Clin. Chim. Acta **36**:389, 1972.
103. Green, S., Cantor, F., Inglis, N. R., and Fishman, W. H.: Normal serum alkaline phosphatase isoenzymes examined by acrylamide and starch gel electrophoresis and by isoenzyme analysis using organ-specific inhibitors, Am. J. Clin. Pathol. **57**:52, 1972.
104. Johnson, R. B., Ellingboe, K., and Gibbs, P.: A study of various electrophoretic and inhibition techniques for separating alkaline phosphatase isoenzymes, Clin. Chem. **18**:110, 1972.
105. Kreisher, J. H., Close, V. A., and Fishman, W. H.: Identification by means of 1-phenylalanine inhibition of intestinal algaline phosphatase components separated by starch gel electrophoresis of serum, Clin. Chem. Acta **11**:122, 1965.
106. A Microzone method for alkaline phosphatase isoenzymes, adapted from the method developed by Dade Division, American Hospital Supply Corp., Miami, Fla. Nos. CH415/1271 and CH 65-DA. Reprint available from Beckman Instruments, Inc., Fullerton, Calif.
107. Moss, D. W., and King, E. J.: Properties of alkaline-phosphatase fractions separated by starch-gel electrophoresis, Biochem. J. **84**-192, 1962.
108. Muiris, X.M.F., Fénnély, J. J., and McGreeney, K.: The value of differential alkaline phosphatase thermostability in clinical diagnosis, Am. J. Clin. Pathol. **51**:194, 1969.
109. Posen, S., Neal, F. C., and Clubb, J. S.: Heat inactivation in the study of human alkaline phosphatases, Ann. Int. Med. **62**:1234, 1965.
110. Romel, W. C., LaMancusa, S. J., and DuFrene, J. K.: Detection of serum alkaline phosphatase isoenzymes with phenolphthalein monophosphate following cellulose acetate electrophoresis, Clin. Chem. **14**:47, 1968.
111. Rosalki, S. B.: Diagnostic enzymology, ed. 2, No. CH-3-1, Dade Division, American Hospital Supply Corp., Miami, Fla., 1974.
112. Warwick, R. R. G., Shearman, D. J. C.,

Percy-Robb, I. W., and Smith, A. F.: Electrophoretic separation of alkaline phoshpatase isoenzymes: a clinical evaluation, Scot. Med. J. **17:**172, 1972.
113. Winkelman, J., Nadler, S., Demetriou, J., and Pileggi, V.: The clinical usefulness of alkaline phosphatase isoenzyme determinations, Am. J. Clin. Pathol. **57:**625, 1972.

LDH ISOENZYMES

114. ACI/Corning Operations/procedures manual, Catalogue No. 470166, 1974.
115. Cawley, L. P.: Electrophoresis and immunoelectrophoresis, Boston, 1969, Little, Brown and Co.
116. Everse, J., and Kaplan, N.: Lactic dehydrogenase: structure and function, Adv. Enzymol. **37:**61, 1973.
117. Gebott, M. D.: Microzone electrophoresis manual, Beckman Instruments, Inc., 1973, Fullerton, Calif., Chap. 9.
118. Opher, A. W., Collier, C. S., and Miller, J. M.: A rapid electrophoretic method for the determination of the isoenzymes of serum lacate dehydrogenase, Clin. Chem. **12:**300, 1966.
119. Starkweather, W. H., Spencer, H. H., Schwartz, E. L., and Schoch, H. K.: The electrophoretic separation of lactate dehydrogenase isoenzymes and their evaluation in clinical medicine, J. Lab. Clin. Med. **67:**329, 1966.
120. Webb, E. C.: Nomenclature of multiple enzyme forms, Nature **203:**821, 1964.
121. Wolf, D.: Colorimetric LDH isoenzymes using prepared agarose film, Lab. Med. **4:**18, 1973.
122. Wroblewski, F. C., Ross, C., and Gregory, K.: Isoenzymes and myocardial infarction, N. Engl. J. Med. **263:**531, 1960.

SERUM HAPTOGLOBIN

123. Gebbott, M. D.: Microzone electrophoresis manual, Beckman Instruments, Inc., 1973, Fullerton, Calif., Chap. 9.
124. Giblett, E. R.: Genetic markers in human blood, Oxford, 1969, Blackwell Scientific Publications.
125. Marsh, C. L.: Serum haptoglobins in clinical medicine, No. CR-15, Fullerton, Calif., 1972, Beckman Instruments, Inc.
126. Marsh, C. L.: Personal communication, 1974.
127. Marsh, C. L., and Jolliff, C. R.: In Smith, I., editor: Chromotographic and electrophoretic techniques, vol. 2, London, 1968, William Heinemann Medical Books., Ltd.
128. Polonovski, M., and Jayle, M. F.: Existence dans le plasma sanguin d'une substance activant l'action peroxydasique de l'hémoglobine, C. R. Soc. Biol. **129:**457, 1938.
129. Polonovski, M., and Jayle, M. F.: Sur la préparation d'une nouvelle fraction des protéines plasmatiques, l'haptoglobine, C. R. Acad. Sci. **211:**517, 1940.
130. Smithies, O., and Connell, G. E.: In Wolstenholme, G. E. W., and O'Connor, C. M., editors: Biochemistry of human genetics, Boston, 1959, Little, Brown and Co., p. 178.
131. Sutton, H. E.: The haptoglobins, Progr. Med. Genet. **7:**163, 1970.
132. Wendt, G. G., Kruger, J., and Kinderman, I.: Serumgruppen und Krankheit, Humangenetik **6:**281, 1968.

SERUM GLYCOPROTEINS

133. Gebbott, M. D.: Microzone electrophoresis manual, Beckman Instruments, Inc., 1973, Fullerton, Calif., Chap. 7.
134. Gerbarg, D. S.: The clinical significance of serum glycoproteins. In Sunderman, W. F., and Sunderman, W. F., Jr., editors: Serum proteins and the dysproteinemias, Philadelphia, 1964, J. B. Lippincott Co.
135. Schultze, H. E., and Heremans, J. F.: Molecular biology of human proteins, New York, 1966, Elsevier Publishing Co.
136. Van Neerbos, B. R., and Vries-Lequin, I. E.: Electrophoresis and glycoproteins with special regard to periodic acid-Schiff staining, Clin. Chim. Acta **26:**271, 1969.

18

Immunoelectrophoresis and immunoquantitation procedures

CARL R. JOLLIFF

Immunoelectrophoresis (IEP) has become a formidable tool in the armamentarium of clinical laboratory investigators. Its history extends from the pioneer work in the field of electrophoresis by Tiselius, who in 1937 first separated human proteins in an electric field into the four major components: albumin, alpha globulin, beta globulin, and gamma globulin. From the separation of proteins in agar gel medium of Gordon and associates in 1950 and the gel immunodiffusion studies of Ouchterlony in 1953 came the pioneer work of Grabar and Williams, who combined the latter two techniques into what is known today as immunoelectrophoresis.

Many modifications have resulted in a vast number of techniques which in one way or another still rely on the basic principles as set forth in the original paper by Grabar and Williams. Electrophoretic support media have varied extensively. With the refinement in the preparation and isolation of plasma proteins, antisera to these fractions have evolved, giving the investigator extremely precise information about human plasma protein constituents.

In the classic IEP method, proteins, in this case acting as antigen (Ag), are separated by electrophoresis on a medium capable of allowing for their diffusion once separated. Antibody (Ab) from a trough parallel to the separated Ag (protein) is allowed to diffuse into the medium in all directions. As Ab and Ag meet, an area of equivalence between each is reached, at which point a zone or line of precipitate forms in an arc. Intermediate precipitation areas may allow for the varying densities and shapes of the arcs (Figs. 18-1 and 18-2).

The assignment of antibody activity to the gamma globulins by Tiselius and Kabat has led to an explosion of scientific literature on the subject. The establishment of at least five immunoglobulin classes and their relationship to the disease process has been the subject of hundreds of papers in immunology over the past 20 years. The "new immunology" which has emerged has found its way into the clinical laboratories with many recent establishments of departmnts of clinical immunology. These departments, which once performed only serologic procedures for antibody activity and blood banking, are now performing sophisticated procedures such as radioimmunoassay, tissue typing, tests for cell-mediated immunity, cancer immunology, and immunochemistry. IEP is an integral part of the laboratory for its use in its original form as well as in the newer adaptive forms. It is a requisite for most immunologic investigations.

Fig. 18-1. First stage of immunoelectrophoresis: separation of serum proteins in agar.

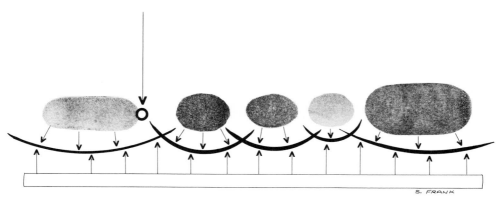

Fig. 18-2. Second stage of immunoelectrophoresis: diffusion of antibody and separated serum proteins with formation of precipitin arc at areas of antigen/antibody equivalence.

The clinical laboratory worker has a great amount of information with which to be familiar to undertake IEP analysis. It is one thing to perform the procedure properly, but to interpret the results in light of existing clinical conditions and to utilize the proper sources of materials available are indeed an entirely different challenge.

To understand the IEP method one must first become familiar with certain immunologic principles governing the procedure. A knowledge of what antigen (immunogen) is and how it complexes with antibody is essential to the comprehension of the procedure itself as well as to the immunologically related clinical conditions that are to be studied. Proteins may act as antigens or antibodies. Antibodies may act as antigens; however, not all immune responses are related to antibody (when we designate antibody as immunoglobulin). It is the purpose of this chapter to answer some of the questions which may confront the clinical laboratory worker in IEP.

THE IMMUNE SYSTEM

To appreciate the antigen-antibody complex and its relationship to IEP, one must first be familiar with what antigens and antibodies are.

Antigens

Antigens are any substances that may provoke an immune response. This response may take the form of antibody production by the so-called B cells or it may result in the development of cell-mediated (T cell) immunity. It is not necessary for circulating antibody to be evident after antigenic stimulation. It has been suggested that the term *antigen* be assigned to a substance capable of reacting with antibody while the term *immunogen* be reserved for an excitation substance of the immune system.

The surface groupings on antigens which protrude from the structure such as amino acids or sugar side chains allow for attachment to antibody and are known as *determinants* or *haptens*. The most effective determinant is the *immunodominant point*. Haptens, by themselves, are not immunogenic; however, when combined with a larger carrier molecule they may become so.

Certain requirements are necessary for a substance to be classified as an immunogen. It must be foreign to the host, be of a molecular size in most instances greater than 10,000 MW, be relatively insoluble, have a structural nature which conforms to the antibody being produced, possess a net charge, and have accessible determinant groups and a definite chemical nature. In most cases the immunogen must be capable of being phagocytosed.

Antibodies

The lymphocyte population of man provides the two main systems of immunity, humoral and cellular. Lymphocytes can be morphologically and immunologically divided into either B (bursal) dependent cells or T (thymus) dependent cells. The B cells are designated as such since they were first described in the bursa of Fabricius of the chicken. The T cells are thymus dependent and are found in the peripheral blood and lymph nodes.

The T cells form in the deep cortical areas of the lymph nodes and in certain areas of other outlying lymphoid tissues. These cells are long lived. Certain populations migrate to inflammatory areas while others serve as memory cells. T cells constitute approximately 65% of the lymphocyte population and are responsible for cellular immunity. There are excellent tests for cell-mediated immunity and they have become a very necessary addition to the immunology laboratory.

The B cells are probably differentiated by the bone marrow in the human. B cells are found mainly in the peripheral blood and lymph nodes. They constitute approximately 35% of the lymphocytes in the peripheral circulation. The lymph node germinal centers are associated with the B cells.

In the presence of a specific antigen, a B cell population responds with the production of plasma cells which produce antibody (immunoglobulin) directed toward that specific antigen. The immunoglobulins produced may fall into one of the following classes: gamma G, gamma A, gamma M, gamma D, or gamma E, or a portion of the specific immunoglobulin molecule such as its heavy or light chains. Over- or underproduction of immunoglobulins by these cells, as well as the production of an immunoglobulin from a single B cell clone, may result in dysgammaglobulinemia, which can be mirrored in the IEP pattern.

IMMUNOGLOBULINS
Structure

The immunoglobulin molecules are glycoproteins, which are heterogeneous in nature, consisting of two identical heavy polypeptide chains and two identical light polypeptide chains held together by interchain disulfide bonds and noncovalent forces

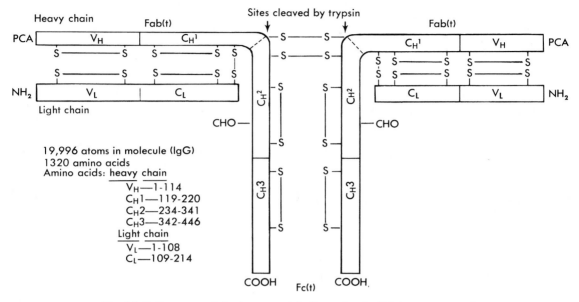

Fig. 18-3. Structure of the immunoglobulin molecule (Edelman proposal).

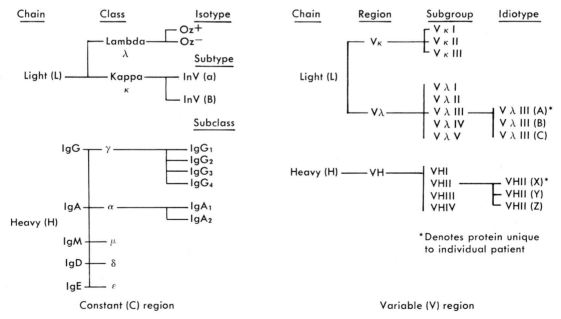

Fig. 18-4. Classification of the constant and variable portions of heavy and light chains of the immunoglobulin molecule.

(Fig. 18-3). The immunoglobulins may be classified according to three parameters (Fig. 18-4):
1. Heavy chain constant (C) regions constitute the basis for different classes and subclasses.
2. Light chain constant (C) regions specify type or subtype.
3. Variable (V) regions of heavy and light chains may be divided into different groups or subgroups.

Heavy (H) chain

To date there are five known classes of heavy chains in the human—gamma (γ), alpha (α), mu (μ), delta (δ), and epsilon (ϵ). These in turn mandate the immunoglobulin class: gamma G, gamma A, gamma M, gamma D, or gamma E. Each heavy chain consists of some 446 amino acids. Amino acids are numbered from the end that is first produced (the amino terminus). Amino acids 1 to 114 are variable between heterogeneous immunoglobulin molecules of the same class, whereas amino acids 119 to 446 are constant for all immunoglobulins of a specific class. This amino acid sequencing is the basis for the terminology of the variable and constant portions of the heavy chain. The constant portions of the heavy chains have three homologous regions denoted C_{H1}, C_{H2}, and C_{H3}. It is the constant portion of the heavy chain which is the site of many biologic activities such as complement fixation, cell membrane receptor interaction, passive cutaneous anaphylaxis, and transplacental transfer. The variable portion of the heavy chain has a PCA or pyrollidone-carboxylic terminal and the constant portion of COOH (carboxyl) terminal. The variable portion of the heavy chain is involved in attachment to antigen.

To date four subclasses of IgG have been described. They are IgG1, IgG2, IgG3, and IgG4. The subclasses of IgA heavy chain are designated IgA1 and IgA2.

Light (L) chains

The light chains are of two types designated either kappa (κ) or lambda (λ). Each immunoglobulin molecule has either two kappa or two lambda light chains, never one of each. The light chain consists of approximately 214 amino acids divided into a variable (V) and constant (C) portion; the variable portion has an NH_2 terminus and the constant portion a COOH terminus. Amino acids 1 to 108 appear to be variable between each individual kappa or lambda immunoglobulin molecule light chain. Amino acids 110 to 214 remain constant between each kappa or lambda light chain attached to the molecule.

Subtypes of the kappa light chains are observed on the constant portion at amino acid residue position 191. This allotypic marker is known as the InV genetic locus. If the amino acid at this locus is leucine, the subtype is designated as InV (a^+). If the residue is valine the subtype is InV (b^+).

The Oz marker, originally thought to be an allotypic marker, is now considered to be an isotypic marker on the lambda chain, since all normal sera studied appear to have both forms of the lambda chain. If the amino acid at position 190 on the lambda chain is lysine, the designation is Oz+; if it is arginine the designation is Oz−.

It is the variable portion of the heavy and light chains that allows for the many possibilities of amino acid sequences and thereby antigenic binding specificity.

The immunoglobulin molecule is subject to degradation and may be separated into fragments such as F(ab)'2, Fc, and Fd. This may be accomplished by hydrolysis and reduction methods (Fig. 18-5). The F(ab)2 fragments obtained from an individual immunoglobulin molecule are so designated because of their two antigen com-

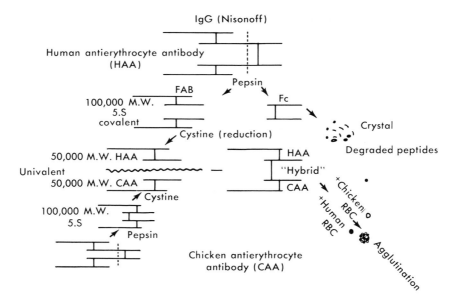

Fig. 18-5. Schematic representation of the production of F(ab') fragment hybrids. (Based on data from Nisonoff and associates.)

bining sites. Each immunoglobulin molecule in the monomer form has two such sites. The Fc fragment (crystalline) is the portion of the immunoglobulin molecule involved in complement binding. The Fd fragment, part of the heavy chain, is primarily involved in determining antibody specificity and carries certain well-defind genetic factors in man and rabbits.

Nisonoff and his associates were able to produce antibody hybrids (Fig. 18-5) by the recombination of Fab fragments from two different antibody molecules. Such a procedure allows for many possibilities directed toward the elucidation of antigen-antibody complexes as well as complexes of antibody and other biochemical constituents.

Immunoglobulin classes
IgG

Some 80% of all antibodies produced as immunoglobulins by the B cells fall into the IgG class. This class has a mean serum concentration of 1240 mg/dl and a molecular weight of 140,000. Most bacterial, viral, and toxin type antibodies occur in this class. IgG antibodies are capable of passing the placental barrier and thereby serve as the neonate's early protection.

The four subclasses of IgG and their individual biologic activities are noted in Table 18-1. There is considerable allotypic heterogeneity in the molecule with some twenty-three Gm heavy chain variants. (See Table 18-2.)

IgA

IgA comprises 10% to 15% of the total B cell–produced serum antibodies. However, this molecule, which occurs as a diamer form attached to a secretory piece and J chain, becomes an extremely important immunoglobulin found in the secretions of the respiratory, genitourinary, and gastrointestinal tracts.

An allotypic heavy chain marker Am allows for the two subclasses of IgA (IgA1

Table 18-1
IgG subclasses and biologic activity

Activity	IgG1	IgG2	IgG3	IgG4
Complement fixation	Yes	Slight	Yes	No
PCA activity	Yes	No	Yes	Yes
Placental transfer	Yes	Yes	Yes	Yes
Macrophage receptor	Yes	No	Yes	No
Reaction to staphylococcal protein A	Yes	Yes	No	Yes
Antibody activity	Anti-Rh	Anti-dextran and levan	Anti-Rh	Anti-factor III

Table 18-2
Immunoglobulin heavy-chain allotypes and relation to immunoglobulin classes and subclasses*

Nomenclature		Subclass of heavy chain
New	Original	
Gm markers		
1	a	IgG1
2	x	IgG1
3	b^w or b^2	IgG1
4	f	IgG1
5	b and b^1	IgG3
6	c	IgG3
7	r	IgG1
8	e	α
9	p	α
10	b^α	IgG3
11	b^β	IgG3
12	b^γ	IgG3
13	b^3	IgG3
14	b^4	IgG3
15	s	IgG3
16	t	IgG3
17	z	IgG1
18	Rouen 2	IgG1
19	Rouen 3	?
20	San Francisco 2	IgG1
21	g	IgG3
22	y	a
23	n	IgG2
	b^0	IgG3
	b^5	IgG3
	c^3	IgG3
	c^5	IgG3
Am markers		
1	1 or †	IgA2

*From Natvig, J. B., and Kunkel, H. G.: Human immunoglobulins, Adv. Immunol. **16**: 1973.
†May partly measure non-a.

and IgA2). The latter shows a genetic variant, Am⁺. This variant lacks a disulfide linkage between the heavy and light chain.

IgA does not fix complement, but it does play an important part in antimicrobial activity. It is believed that the secretory piece is synthesized in the epithelial cell, and as the dimeric IgA molecule passes through the cell the piece is added and the combined molecule is secreted onto the mucous surface as secretory IgA weighing approximately 370,000. Some 85% of the cells of the lamina propria lining the gastrointestinal tract are involved in IgA synthesis. The main immunoglobulin of saliva, tears, and gastrointestinal secretions is secretory IgA.

Both serum and secretory IgA have the ability to develop protein complexes. This has been seen in the case of the combination of IgA with antihemolytic factor which may result in a deficiency state severe enough to cause abnormal coagulation.

IgM

Approximately 7% of the humoral immunoglobulin is IgM, which occurs most often as a pentamer structure. However, 7S monomers have been found. Called macroglobulins, these pentamer units can be reduced by thiol reagents into subunits, and with alkylation of the sulfhydryl groups they can be prevented from reaggregation back into their 19S complex. The molecular weight of IgM has been reported from 870,000 to 970,000. The subunits are arranged in a circle (Fig. 18-6) and are joined by a single disulfide bond at the C terminus of the mu chain. Each IgM antibody possesses ten antigen combining sites, although at times only five are expressed. IgM antibodies are found in the circulation quite rapidly after primary antigenic stimula-

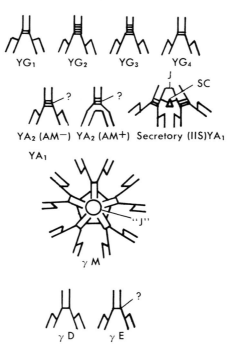

Fig. 18-6. Possible configurations of the immunoglobulin molecules. (From Hong, R.: In Bach, F. H., and Good, R. A., editors: Clinical immunobiology, vol. 1, New York, 1972, Academic Press, Inc.)

tion. It is known that IgM can bind complement and that a single molecule of antibody on the surface of an antigenic cell is able to bind C1, whereas two molecules of certain subclasses of IgG are required. Another characteristic feature of the IgM molecule is the carbohydrate moiety, which comprises about 10% of the structure by weight and is found for the most part on the mu chain of the Fc fragment. The J chain is found on this molecule as it is in the polymeric form of IgA. Most IgM appears to be located in the intravascular system, and in cases of selective IgA deficiency IgM appears to be secreted by mucous surfaces, allowing the large molecular weight molecule to bind antigen at these sites more effectively.

IgD

IgD is found in a concentration of less than 1% of the total serum immunoglobulins, with a mean serum concentration of 0.03 mg/dl. Seventy-three percent of IgD is found in the intravascular compartment, which is a considerably higher percentage than for IgG or IgA. IgD does not bind complement, but it does possess antibody activity. The 7S molecule occurs as a monomer with a molecular weight of approximately 150,000. It is found in normal human serum and in patients with IgD monoclonal gammopathy, of which some fifty cases have been reported.

IgE

This 8.2S immunoglobulin occurs as a monomer possessing, as other immunoglobulin molecules, four polypeptide chains: two heavy epsilon chains and two light chains. The size of the epsilon chains with their covalently linked carbohydrate prosthetic groups contributes to the high molecular weight, 190,000. The polypeptide portion of the epsilon chain possesses about 100 more amino acid residues than the gamma or alpha chains. It is possible that four constant portions of the epsilon chain exist. The Fcϵ fragment consists of the Cϵ2, Cϵ3, and Cϵ4 regions and carries the skin-fixing properties peculiar to this class which inhibits the Prausnitz-Kustner (PK) test in man and the passive cutaneous anaphylaxis (PCA) test in primates. Serum values exceeding 500 ng/dl are considered elevated.

COMPLEMENT SYSTEM

The complement system is composed of a group of specific enzymes found in normal blood which in combination with immunoglobulins (antibodies) or other specific substances mediate immunologic and allergic reactions. As the antibody molecule combines with the antigen molecule on the cell surface, the complement system is activated, starting a series of reactions leading to the destruction of the antigenic substance and cell cytolysis (Fig. 18-7). During these reactions chemical substances such as histamine and kinins are released, which in turn results in the anaphylactoid reaction and cell destruction as seen in the immune complex diseases. IEP plays an important role in the study of the complement components as increases or decreases in disease states.

Evaluation of complement system

The complement system may be activated by three pathways: (1) via $C1_q$ (Fig. 18-7), (2) by properdin on the alternate pathway, and (3) by proteolytic enzymes.

In the so-called classic pathway, as depicted in Fig. 18-7, the action is initiated by the interaction of one molecule of IgM or two molecules of IgG, providing the IgG molecule belongs to a specific class (see immunoglobulin structure). Through

Immunoelectrophoresis and immunoquantitation procedures 601

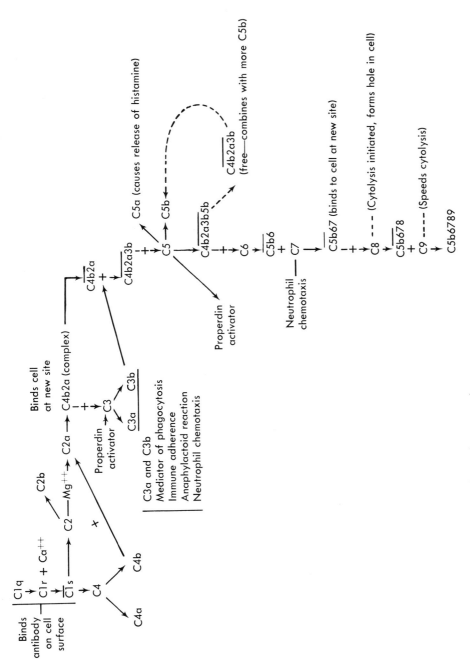

Fig. 18-7. Schematic drawing of complement pathway.

a cascading series of enzymatic reactions ultimate destruction of the offending cell occurs by lytic destruction.

An alternate method or the properdin pathway of activation occurs. C3 is activated by the C3 activator substance or properdin factor B. C3 activator appears to be released by such substances as polysaccharide endotoxins of bacteria, aggregates of immunoglobulins such as IgA and IgE (not normally capable of combining with complement at the $C1_q$ point), and other substances such as zymosan and inulin. Patients with sickle cell disease (s-s) appear to be deficient in this system, which may somewhat explain the reason for their increased incidence of infection.

The third pathway by which the system may be activated may again occur at the C3 level and result from the direct cleavage of C3 by such proteolytic enzymes as plasmin, thrombin, trypsin, and certain other tissue proteases. Such conditions as myocardial infarction and disseminated intravascular coagulation (DIC) may be responsible for the activation by this series of events.

A combination of immunoelectrophoresis and immunoquantitation of the complement factors will allow for the elucidation of any errors or deficiencies in these pathways, which in turn will mirror certain disease processes. A necessary prerequisite to any complement determination is knowledge of the specimen. All blood specimens for complement analysis should be collected in EDTA tubes and plasma removed from the cells immediately. If the analysis is not performed within 4 hours, the plasma should be frozen. In vitro activation of the complement may occur, making interpretation of the results invalid.

Antisera to known C3-C4 and C3 activators may be obtained. Radial immunodiffusion plates are commerically available for these components. Immunoelectrophoretic studies of plasmas utilizing anti-C3 (B1C/1A) and C3 activator (C3A) can be used to determine the possibility of in vivo activation of the system which may occur before lowered levels are ascertained by radial immunodiffusion (RID) in vitro.

In our laboratory immunoelectrophoretic plates are set up as follows:

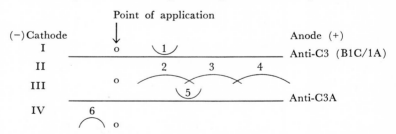

I shows no in vivo or in vitro activation of arc 1 C3 either by the $C1_q$ or alternate pathway. Arc occurs in beta globulin area.

II indicates activation of C3 into B1C (arc 2) with arc 3 B1A and arc 4 α_2D (may be detectable as an additional breakdown product). The B1C/1A arcs are found in the beta globulin area while the α_2D is found in the alpha globulin area.

In III arc 5 is the normal position for C3PA (C3 preactivator) which is found in the beta region.

In IV arc 6, found in the gamma region, denotes alternate pathway activation of the complement system.

An adequate assessment of complement activity can be made using the RID assays of C3, C4, and C3A. This assessment will not necessitate hemolytic assays of complement activity unless more complement deficiencies are to be studied. Table 18-3 will indicate the general type of complement response occurring.

Table 18-3
Complement values by radial immunodiffusion and immunoelectrophoresis as they relate to clinical manifestations

	Radial immunodiffusion quantitation				Region of IEP migration	
	C3	C4	C3A		C3(B1C/1A)	C3A
Normal level (mg/dl)	80-150	20-60	15-30	Normal mobility	B2	B2
In vitro aged specimen					B1	
Classic response via Cl_q immunologic activation	Dec	Dec	N		B1	B2
Nonimmunologic activation —alternate properdin pathway	Dec	N	Dec		B2	
C3 cleavage; tissue injury; proteolytic enzymes	Dec	N	N		B1	B2
Classic and properdin pathway response	Dec	Dec	Dec		B1	
C1 esterase deficiency	N	Dec	N		B2	B2

N, Normal; Dec, decreased.

IMMUNOELECTROPHORESIS

Various manufacturers have introduced precast as well as rehydratable IEP sheets. These procedures offer a tremendous time-saving device as well as an excellent method of separation in the busy laboratory that is called on to perform these procedures nonroutinely. The materials were previously mentioned in the section on serum protein separation by electrophoresis. It is only necessary to place the serum to be separated on the precast or rehydratable gel as a drop (approximately 0.9 μl) and to migrate following instructions for protein electrophoresis. Antibody troughs are premolded on some prepared gels and need to be cut on others according to the technique recommended by the manufacturer.

For those who wish to prepare their own plates, we have successfully used for a number of years the agar plate method offered here. Given proper training and experience, the technologist can perform the method quite well in a manner of days.

Routine method for immunoelectrophoresis

Materials
Lantern slide, 3¼ x 4 inches
Immunoelectrophoresis pattern cutter (14 P.D.) (Reclin Deluxe pattern cutter) (Fig. 18-8)
Immunoelectrophoresis cell (Reclin No. 117A)
Power supply (Beckman No. R-120) (Fig. 18-9)
Electrophoresis process center (Reclin) (Fig. 18-10)

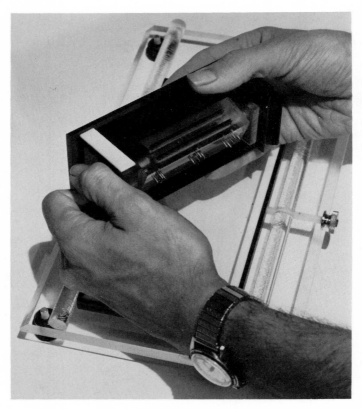

Fig. 18-8. Immunoelectrophoresis and immunodiffusion pattern cutter. (Courtesy Reclin Co., Lincoln, Neb.)

Fig. 18-9. Constant voltage power source Beckman Model R-120. (Courtesy Beckman Instruments, Inc., Fullerton, Calif.)

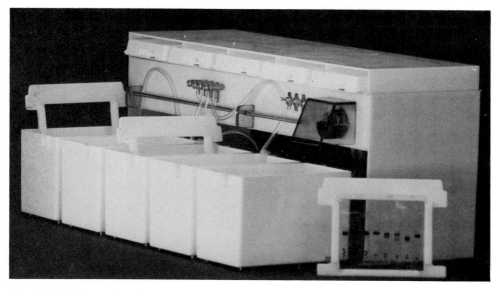

Fig. 18-10. Electrophoresis process center with facilities for washing, destaining, and drying agarose, cellulose acetate, or Ionagar electrophoresis and immunoelectrophoretic patterns. (Courtesy Reclin Co., Lincoln, Neb.)

Humidity cabinet (Reclin No. 113H) (Fig. 18-11)
Pattern viewer, darkfield IEP box (Reclin)
Polaroid MP_3 camera assembly
Petri dishes, large, for staining
Meat keeper, plastic (Tupperware)
Diamond marking pencil
Bibulous paper booklet

Reagents

Buffer for immunoelectrophoresis chamber

Veronal	1.38 g
Sodium vernol	8.70 g
Calcium lactate	0.38 g
Distilled water to	1000 ml

Solution for preparation of Ionagar

Veronal	1.66 g
Sodium veronal	10.50 g
Calcium lactate	1.54 g
Distilled water to	1000 ml

Add aqueous thimerosal to a concentrate of 1:10,000 to prevent bacterial and mold growth.

Preparation of Ionagar slides

Two parts of solution for preparation of Ionagar are mixed with 1 part distilled water. We usually use 66 ml and add distilled water to the 100-ml mark on a cylinder graduate. This is poured into a 250-ml flask to which has been added 2g Ionagar 2.* The agar is partially dissolved and then heated to boiling

---

*Consolidated Laboratories, Inc.

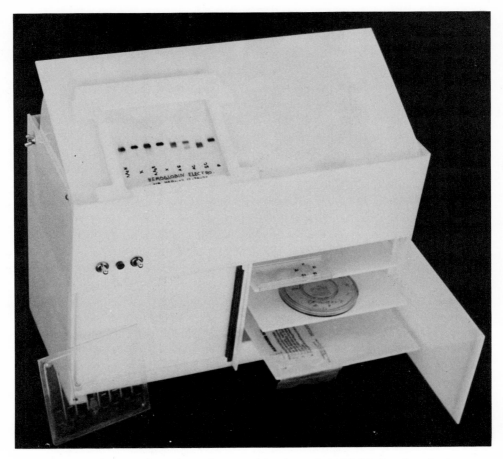

Fig. 18-11. Combination humidity cabinet and bright and darkfield view box for immunodiffusion and immunoelectrophoretic procedures. (Courtesy Reclin Co., Lincoln, Neb.)

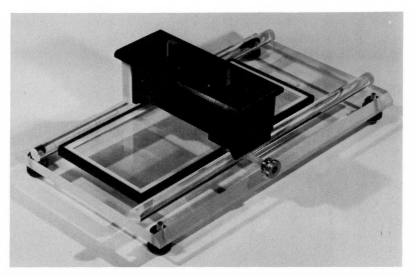

Fig. 18-12. Tray and jig for holding IEP plates in place during cutting. (Courtesy Reclin Co., Lincoln, Neb.)

with constant swirling. Heat until the mixture foams almost to the top of the flask 3 times to ensure proper dissolution of the Ionagar. With a 10-ml pipet, transfer 10 ml of the hot mixture to a 3¼ × 4 inch lantern slide (previously cleansed so that it is free of all grease) lying on a level table.

Carefully spread the agar on the slide so that the entire surface is covered. The slide will easily hold up to 12 ml of agar, we usually added this much instead of 10 ml.

After the slides have solidified, transfer them to a plastic (air-tight) container which contains several paper towels soaked in water. Seal the container and keep slides under refrigeration until ready to use.

Antisera

Antisera for use in immunoelectrophoresis may be prepared by the individual. However, for clinical purposes and for those in laboratories where such equipment and animals are not available, commercially available antisera have been found satisfactory. A list of sources of antisera can be found in Appendix E.

It is important that these antisera be kept at 4° C when not in use or between use. If storage before use is intended, store in deep freeze and thaw before use. Do not repeat freezing and thawing, but rather thaw and keep at 4° C when using.

Procedure

1. The previously prepared 3¼ × 4 inch Ionagar plates are removed from the humidity box in the refrigerator and allowed to come to room temperature. At this time, etch the number of the test on the non–agar-containing side of the plate.

2. Place the plate in the cutting jig (Fig. 18-12) and carefully press the cutting tool into the agar at the desired location on the plate. (Fig. 18-13 shows the position of the patterns of the plate.) The end pattern should be well enough in from the edge so that distortion of the developing arcs does not occur from the smaller volume of agar at the edges of the slide. If the jig is followed correctly, this will not occur.

3. After the pattern is cut in the agar, the plate is removed from the jig. A cut-off 15 gauge needle (Fig. 18-14) is used to suck out the agar plug in the antigen (serum) well. The antibody trough agar is not removed at this time.

4. After the eight antigen wells are freed of agar, the measured antigen is introduced into the wells with the aid of a microliter syringe or capillary tube (Fig. 18-15).

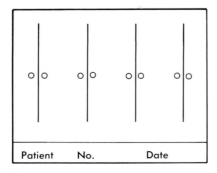

Fig. 18-13. Position of IEP patterns on 3¼ × 4 inch slide.

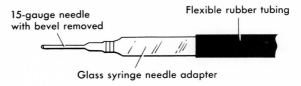

Fig. 18-14. Drawing of needle with adapter for use in withdrawing antigen well agar after cutting.

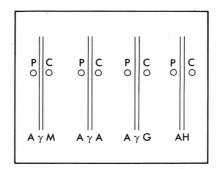

Fig. 18-15. Schematic for positions of antigen and antibody on an IEP run. *P*, Patient sera; *C*, control sera; *AH*, antihuman serum; *AγG*, anti–gamma G serum; *AγA*, anti–gamma A serum; *AγM*, anti–gamma M serum.

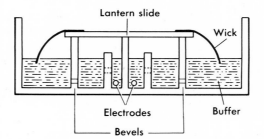

Fig. 18-16. Drawing of position of IEP slide, wicks, and buffer level in cell.

5. A pooled control serum approximating normal level for the age of the patient being investigated is used. Bromphenol blue is added to this serum until the serum is colored sufficiently to allow for notation of migration of the albumin portion of the patterns.

6. Place the slide in the electrophoresis chamber (Fig. 18-16).

7. Place filter paper wicks (two thicknesses of bibulous paper cut in half) approximately ⅜ to ½ inch long on the agar on each side and dip into the chamber buffer. Place the filter paper wicks straight on the slide (Fig. 18-17).

8. Be certain the wells of antigen are nearest the cathode since migration of albumin is toward the anode.

9. Close the cover of the electrophoresis chamber and set voltage at 175 volts, which is about 40 ma. We have used constant voltage and have had excellent results.

10. When the albumin marker in the control moves about 35 to 40 mm, shut off current and remove the plate from the chamber. With a scapel cut the agar at the ends of each trough. Holding the plate agar upside down, carefully re-

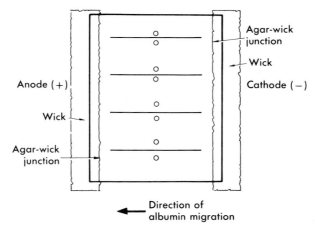

Fig. 18-17. Wick position preparatory to IEP run.

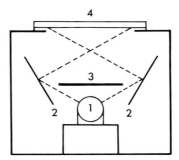

Fig. 18-18. Schematic of view box for use in photographing IEP and immunodiffusion patterns: *1*, fluorescent light; *2*, mirrors (adjustable to proper angle); *3*, metal plate, dark background; *4*, immunoelectrophoresis pattern.

move the agar in each of the four troughs. By holding it upside down the agar will fall away from the slide and prevent damage to the agar surface if the trough gel should have to be removed from the surface of the agar.

11. Fill the troughs with the appropriate antisera (Fig. 18-18), being careful not to damage the troughs or wells in doing so.

12. Carefully place the slide in the humidity cabinet for 12 to 24 hours at room temperature, being certain that sufficient water is in the cabinet.

Photography of plates

1. After adequate time has elapsed for the arcs to develop (18 to 24 hours), the plate is photographed for a permanent record of the test. This is accomplished by placing the plate on a view box so constructed as to allow a darkfield background with 45-degree angle lighting to the side (Fig. 18-19).

We have found the Polaroid MP_3 camera assembly to be the most adaptable for this purpose. A Polaroid photograph is taken of the entire plate. If special attention is necessary regarding one or more of the arcs, an enlargement of the arc can be made by simply using the proper lens combination.

Generally two photographs of each plate are taken: one for the laboratory record and the second for the physician's record. All photographs are numbered

IEP ARC No.	Protein	Electrophoretic mobility	Molecular weight	Concentrated normal plasma mg/dl	Function	Polymorphisms	Changes in disease states
1	Prealbumin	α0	61,000	10 to 40	Thyroxine binding	Yes	
2	Albumin	α0	69,000	3500 to 5500	Protein transport Lipid metabolism Regulation of osmotic pressure Storage form of amino acids and protein	Bisalbuminemia may occur	Decreased: pregnancy, analbuminemia, malnutrition, toxic hepatitis, carcinoma of liver, hemorrhage, burns, intestinal obstruction, gastroenteropathies, renal disease
3 (C.I.E.P. or RIA best test)	α1 fetoprotein	α1	65,000	5 to 10 mg/dl	(Synthesized by fetal liver cells)		Increased: hepatoma
4	α1 antitrypsin	α1	45,000	200 to 400	Proteinase inhibitor Acute phase reactant	Yes P$_i$ types	Decreased: chronic pulmonary disease Increased: inflammatory disease
5	G$_{1c}$ globulin	α1-α2	50,800	30 to 55	Unknown	Yes	Used for forensic and genetic studies
6 (can use specific stain, Prussian blue)	Ceruloplasmin	α2	160,000	20 to 45	Copper binding	Yes mainly due to copper loss	Decreased: Wilson's disease, malnutrition, nephrosis, gastroenteropathies Increased: malignancy, oral contraceptives, hypoplastic anemia
7	α2 macroglobulin	α2	820,000	105 to 350	Hormone binding (insulin) Proteinase inhibitor Acute phase reactant	Yes Xm types	Decreased: sometimes in rheumatoid arthritis, preeclampsia Increased: nephrotic syndrome, severe liver disease, pregnancy, diabetes mellitus
8	Haptoglobin	α2	100,000	100 to 300 depends on Hp type	Hemoglobin binding Acute phase reactant	Yes Hp types	Decreased: liver disease, hemolytic anemia, genetic ahaptoglobinemia Increased: chronic and acute inflammatory and neoplastic disease
9	C3/C3$_c$ (β1C β1A globulin)	β1		80 to 140 on standing C3 converts to inactive form C3c	Complement C3 factor	Not known	Decreased: autoimmune disease Increased: inflammatory disease (acute), atopic dermatitis
10	C4(β1E globulin)	β1		20 to 40	Complement C4 factor	Not known	Decreased: autoimmune disease
11	Transferrin (Note: difficult to interpret due to dissolution on standing.)	β1	90,000	200 to 400	Fe binding Fe transport	Yes	Decreased: malignant disease, portal cirrhosis, hemochromatosis, chronic hepatitis, genetic atransferrinemia Increased: late pregnancy, acute hepatitis, iron deficiency anemia, nephrotic syndrome
12	Hemopexin	β1	80,000	70 to 130	Heme transport and conservation	Not known	Decreased: liver disease and hemolytic conditions Increased: newborns of diabetic mothers
13	C1s activator	β	4s		Inhibits first component of complement	Not known	Absent in patients with hereditary angioneurotic edema
14	Fibrinogen	β1	341,000	200 to 450	Blood clotting Factor I	Not known	Decreased: fibrinogen deficiencies (acquired and congenital)

Fig. 18-19. A, Plasma electrophoresis.

IEP ARC No.	Protein	Electrophoretic mobility	Molecular weight	Concentrated normal plasma, mg/dl	Function	Polymorphisms	Changes in disease states
15	IgE	γ	188,000 to 209,000	0.03	See text	Not known	See text
16	IgD	γ	160,000	3.0	See text	Not known	See text
17	IgM	γ	900,000	120	See text	Polymers	See text
18	IgA	γ	150,000 to 600,000	280	See text	Allotypes Polymers	See text
19	IgG	γ	149,000 to 153,000	1240	See text	Allotypes and subclasses	See text

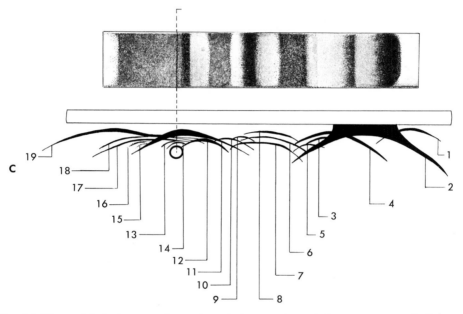

Fig. 18-19, cont'd. B, Immunoelectrophoresis (agarose); and **C,** components of diagnostic significance.

with the patient or experiment number and antisera troughs as well as antigen wells are identified. A marking pencil* works very well for this purpose.

Staining of patterns

After the photographs of the IEPs have been taken, the slide is carefully washed with the cold water so that the agar is not washed off. This is accomplished in about 10 seconds. The slide is then placed agar side up on a paper towel. Two pieces of bibulous paper are then wet with tap water. The two pieces are laid on the slide and the slide allowed to stand until all of the moisture is removed from the agar and the pattern remains as a thin film on the glass slide. This usually takes about 8 hours at room temperature but may be quickened by allowing warm dry air as from a fan or hair dryer to flow over the wet bibulous paper. Do not allow the slide to dry too fast and do not place

*Scientific Products, Inc., No. P1226.

it in an oven, since cracking of the agar and distortion of the pattern will occur.

Once the slide has dried, it is washed in 0.85% NaCl with several changes for a period of about 18 hours. A washing apparatus* is quite suitable for this purpose. The slide is then washed for several hours with several changes of distilled water. The slide washer may be used, or the slide may be washed in a large Petri dish if a slide washer is not available. Frequent changes of saline and water are imperative. After the water wash, the slide is again laid on a paper towel and two pieces of bibulous paper are placed on top of the agar surface and the paper allowed to dry.

After drying the slide the slide may be stained by any number of methods or stains depending on the components the investigator is most interested in. For routine immunoelectrophoresis we use ponceau S and amido black 10B. The formula for these solutions is as follows:

Amido black 10B		*Ponceau S*	
Amido black 10B	1 g	Ponceau S	1 g
Glacial acetic acid	100 ml	Glacial acetic acid	100 ml
Methanol	700 ml	Methanol	700 ml
Distilled water, filter through glass wool before using.	200 ml	Distilled water	200 ml

Coomassie blue	
Coomassie blue	1 g
Glacial acetic acid	100 ml
Methanol	700 ml
Distilled water	200 ml

The slides are now placed in a slide staining tray or slide stainer* for 5 minutes. The slides are allowed to drain dry of the amido black 10B, the ponceau S stain, or the coomassie blue stain and are then placed in a destaining solution for as long a period as it takes for the precipitation bands to be clear and the background stained agar become devoid of stain.

Destaining solution	
Glacial acetic acid	100 ml
Methanol	700 ml
Distilled water	200 ml

The slides are allowed to dry at room temperature. When completely dry, the pattern may be covered with a protective coating by using the coater from a Polaroid black and white film pack. The slides are now permanent and may be labeled with an ink pen, filed, photographed, or used in a regular lantern slide projector for better visualization or presentation to groups.

Proteins delineated by IEP and their importance in disease states

IEP offers to the clinical laboratory investigator a means of semiquantitatively assessing protein abnormalities. The visual inspection of the IEP pattern may be interpreted in light of several factors: how large the arc of precipitation in question is, how the distance of the arc from the antibody trough compares with the control serum, and the appearance of the arc. The larger the arc and the closer it is to the trough, the more antigen present. The appearance of the arc may give a clue to

---

*Reclin Co.

the amount of antigen as well as its character. Increases or decreases that may be exhibited in the IEP pattern will be discussed here.

The plasma proteins which can be visualized by IEP and some of the clinical conditions that may be delineated by this method are given in Fig. 18-19. More quantitative techniques such as radial immunodiffusion, electroimmunodiffusion, and radioimmunoassay are now available to the clinical laboratory and should be used for the final assessment of protein values. The investigator aware of the IEP pattern and the various arcs involved may assess the various dysproteinemias with a certain degree of accuracy.

IEP probably offers more in the assessment and diagnosis of dysgammaglobulinemias than any of the other disease states reflecting protein abnormalities. The dysgammaglobulinemias may be divided into the hypogammaglobulinemias, polyclonal gammopathies, and monoclonal gammopathies.

Hypogammaglobulinemia

Not all immunodeficiency diseases are reflected by lowered immunoglobulin levels; only those conditions that involve the B cell–related functions are implicated. Methods other than IEP or at least modifications of the method are better suited for the quantitative estimation of the immunoglobulins. However, a qualitative estimation may be made on routine IEP patterns by the observation of the particular immunoglobulin arc structure and location.

In 1952 Bruton first described an X-linked condition which he termed agammaglobulinemia. At that time sophisticated immunologic methodology for the detection and elaboration of immunoglobulin classes was not available. We now know that patients with this condition can regularly be shown to possess small amounts of immunoglobulins in each class. A delay in the production of immunoglobulins occurs in infancy, and adult normal values are not realized until about 12 years of age. Interpretation of hypogammaglobulinemia in the IEP pattern must be made with a knowledge of the age of the patient, and controls used from patients which fall within the same age group as the patient being studied must be included with each run (Fig. 18-20).

The primary immune deficiencies have been classified by a group convened under the auspices of the World Health Organization; they are listed in Table 18-4. These conditions do not include other pathologic entities such as hypercatabolic states, myelomatosis, leukemia, and protein losing diseases which are also associated with various severities of hypogammaglobulinemia.

Hypergammaglobulinemia (polyclonal gammopathy)

The conditions eliciting increased production of one or more immunoglobulin by numerous B cell clones are called *polyclonal gammopathies* (Fig. 18-21). Polyclonal gammopathy may be seen in various disease states. Table 18-5 lists those conditions, with notation of immunoglobulin classes reported to have been increased. The investigator must bear in mind the interpretation of these polyclonal gammopathy patterns in light of other possible existing conditions which might in themselves increase or decrease the immunoglobulin level. Such examples might be an immunodeficient patient with existing bacterial pneumonia or the patient with lupus erythematosus and nephrosis. The patient's age must again be correlated with immunoglobulin level, particularly in the younger and older age groups where immunoglobulin values may vary from the so-called normal values that have been furnished by the literature.

614 *Chemistry for the clinical laboratory*

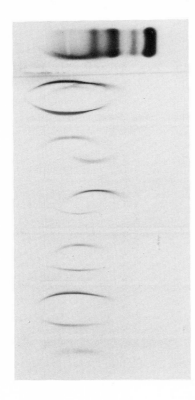

Patient
A-Ig
Control
A-IgG
Patient
A-IgA
Control
A-IgM
Patient
A-kappa
Control
A-lambda
Patient

Fig. 18-20. Hypogammaglobulinemia of *Pneumocystis carinii* pneumonia. Patient shows marked decrease in IgG and IgA. IgM was normal.

Fig. 18-21. Hypergammaglobulinemia (polyclonal gammopathy)—cirrhosis of liver.

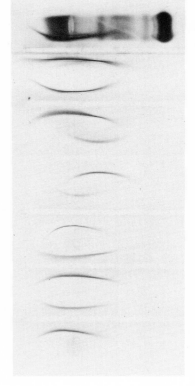

Patient
A-Ig
Control
A-IgG
Patient
A-IgA
Control
A-IgM
Patient
A-kappa
Control
A-lambda
Patient

Table 18-4
Classification of primary immunodeficiency disorders (as established by the World Health Organization)

Type	Suggested cellular defect			Immunoglobulin deficiency
	B cells	T cells	Stem cells	
Infantile X-linked agammaglobulinemia	+			All classes deficient
Selective immunoglobulin deficiency (IgA)	+*			IgA deficient
Transient hypogammaglobulinemia of infancy	+			IgG depressed
X-linked immunodeficiency with increased IgM	+*			Variable
Thymic hypoplasia (pharyngeal pouch syndrome, Di George syndrome)		+		‡
Episodic lymphopenia with lymphocytotoxin		+		‡
Immunodeficiency with normal or hyperimmunoglobulinemia	+	+†		Class variable
Immunodeficiency with ataxia-telangiectasia	+	+		IgA depressed
Immunodeficiency with thrombocytopenia and eczema (Wiskott-Aldrich syndrome)	+	+		Variable
Immunodeficiency with thymoma	+	+		All classes deficient
Immunodeficiency with short limbed dwarfism	+	+		All classes deficient
Immunodeficiency with generalized hematopoietic hypoplasia	+	+	+	Variable
Severe combined immunodeficiency				
Autosomal recessive	+	+	+	Variable
X-linked	+	+	+	Variable
Sporadic	+	+	+	Variable
Variable immunodeficiency (largely unclassified)	+	+†		Variable

*Involve some but not all B cells.
†Encountered in some but not all patients.
‡Usually normal.

Table 18-5
Reported clinical conditions exhibiting polyclonal gammopathies

	γG	γA	γM	γD	γE
Infections					
Subacute bacterial endocarditis	×		×		
Bronchitis	×	×			
Acute mycoplasma pneumonia			×		
Acute *Diplococcus* pneumonia	×				
Tuberculosis	×	×			
Pneumocystis carinii pneumonia	×		×		
Pulmonary aspergillosis					×
Actinomycosis	×	×	×		
Coccidiomycosis	×	×	×		
Toxoplasmosis (newborn)			×		
Hydatid disease	×		×		
Visceral larva migrans					×
Malaria	×	×	×		
Kala azar (leishmaniasis)	×				
Trichinosis					×
Trypanasomasis			×		
Cytomegalic inclusion disease			×		
Infectious mononucleosis	×	×	×		
Acute hepatitis			×		
Viral hepatitis (carriers)	×	×	×		
Viral hepatitis	×		×		
Mumps	×				
Rubella			×		
Cholera			×		
Shigellosis		×			
Leprosy	×	×	×		
Urinary tract infections, general	×		×		
Chronic inflammatory disease of central nervous system	×				
Septicemia	×				
Peritonitis	×	×			
Autoimmune disease					
Lupus erythematosus	×	×	×		
Glomerulonephritis	×	×			
Lupoid hepatitis	×				
Rheumatoid arthritis	×	×	×		
Hemolytic anemia	×				
Dermatitis herpetiformis		×			
Pemphigus vulgaris		×			
Herpes gestationis		×			
Sjogren's syndrome	×		×		
Dermatomyositis	×				
Scleroderma	×		×		
Rheumatic heart disease	×	×	×		
Other conditions					
Nephrotic syndrome			×		
Asthma					×
Atopic eczema					×

Table 18-5
Reported clinical conditions exhibiting polyclonal gammopathies—cont'd

	γG	γA	γM	γD	γE
Seasonal rhinitis					×
Food sensitivity					×
Multiple sclerosis					
Demyelinating disease of central nervous system	×	×			
Cystic fibrosis	×	×			
Interstitial plasma cell pneumonia	×		×		
Biologic false positive test for syphilis	×		×		
Burns					×
Celiac disease		×			×
Gastrectomy		×	×		
Splenectomy		×			
Kwashiorkor		×		×	
Postmyocardial infarction	×				
Sarcoidosis	×	×	×		
Primary pulmonary emphysema		×			
Berylliosis	×				
Gout		×			
Chronic liver disease	×	×	×		
Alcoholic cirrhosis	×	×	×		×
Biliary cirrhosis			×		
Hepatoma	×				
Thalassemia	×	×	×		
Sickle cell disease	×	×	×		
Carcinoma of cervix	×	×			
Carcinoma of breast		×			
Carcinoma of bronchus	×	×			
Hodgkin's disease	×				
Acute myeloid leukemia	×				
Chronic myeloid leukemia	×	×	×		

Hypergammaglobulinemia (monoclonal gammopathy)

IEP plays an extremely important role in the classification of the monoclonal gammopathies. In these conditions a single clone of B cells proliferates in a malignant fashion and may produce a single immunoglobulin of a specific heavy chain class and/or light chain type. The most frequently observed M or myeloma and macroglobulinemia protein, as designated by Riva, is IgG; however, any of the five classes may be implicated in the production of the now accepted term, monoclonal gammopathy (Fig. 18-22). Instances may occur where the malignant clone may produce only the heavy chain of the molecule or only the light chains; these diseases are termed heavy and light chain disease, respectively. IgM monoclonal gammopathy is classified as Waldenstrom's macroglobulinemia.

The immunoglobulin molecule produced in monoclonal gammopathy is not an abnormal one but rather of a single clone origin. It has been shown that such molecules possess normal immunoglobulin characteristics such as the capability to bind

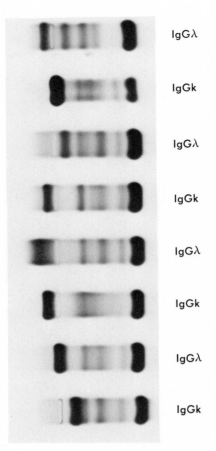

Fig. 18-22. Eight different IgG monoclonal gammopathies (four chain types) each of kappa and lambda light.

complement as well as many other functions of the normally produced immunoglobulins. Some of the heavy chain disease monoclonal molecules may be considered abnormal in that part of the amino acid sequence is missing in the otherwise complete chain sequence. Light chain disease may show the same digression from normal antibody synthesis. Osserman has suggested the following criteria to establish the presence of a monoclonal gammopathy:
  1. Production of immunologically competent cells in the absence of known antigenic stimulus
  2. The presence of an M type globulin which is electrophoretically and antigenically homogeneous or a subunit of the protein
  3. The usual presence of an associated deficiency in the synthesis of other immunoglobulin classes

Examples of the various monoclonal gammopathies are shown in Figs. 18-23 to 18-27.

The associated immunodeficiency is evident in the IEP pattern and may well be an excellent criterion in the assessment of the malignant or benign status of the disease. The quantitation of immunoglobulins by conventional methods such as radial immunodiffusion (RID), electroimmunodiffusion (EID), radioimmunoassay (RIA), and nephelometry of immune complexes alone in the diagnosis of monoclonal

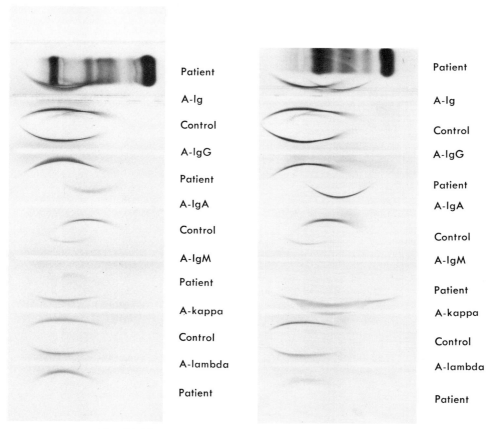

Fig. 18-23. IgGλ monoclonal gammopathy.
Fig. 18-24. IgAκ monoclonal gammopathy.

gammopathy may lead to error since quantitative estimation results in mg/dl values only and does not delineate the monoclonal peak, which must be observed visually. Many times the peak may be observed in the serum protein electrophoresis overlying the dark beta globulin band as in the case of some IgA monoclonal gammopathies.

Light chain disease may not be entirely evident on the electrophoretic pattern, nor would immunoglobulin quantitation or IEP of the serum be of value if the small molecular weight protein (Bence Jones) passed through the glomerular filtrate and into the urine. Only by concentrating the urine and performing IEP can the monoclonal disease be diagnosed (Fig. 18-28).

Routine electrophoresis and IEP were performed in a study of 2000 patients of all age groups presenting themselves to our clinic for routine physical examinations. Thirteen patients showed monoclonal peaks, two of which were of the malignant type, and one light chain disease was discovered. The incidence certainly varies with age; in older patients the percentage becomes much higher. *Civantos* and his group[28] found 121 monoclonal gammopathies in 7100 patients. We have found Bence Jones proteinuria to be present in 51% of all our monoclonal gammopathies considered to be of the malignant type, whereas Civantos' group showed 57%.

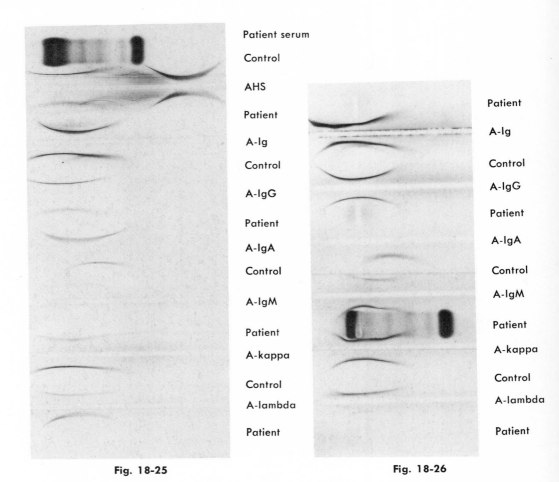

Fig. 18-25. IgA monoclonal gammopathy appearing as a polyclonal gammopathy due to intermolecular aggregation.

Fig. 18-26. IgM$_\kappa$ monoclonal gammopathy—Waldenström's macroglobulinemia.

Not all of the monoclonal gammopathies can be considered to be of the malignant type. A number of discussions on the benign monoclonal gammopathies have been presented. Natvig and Kunkel have established the following criteria in the assessment of the benign monoclonal gammopathies:

1. Band on the EP is low in concentration and rarely exceeds that of albumin.
2. There is a constant level of monoclonal protein over long periods of time.
3. Other immunoglobulins are not usually depressed.
4. Bence Jones proteins are rare.

The question arises as to the possibility of benign monoclonal gammopathy developing into the malignant form and whether or when therapy should be instituted in such cases. Although many cases have been studied where no incidence of malignant transformation has ensued, it is believed that these patients should be monitored for changes in the amount of monoclonal protein throughout life.

Biclonal (Fig. 18-29) and triclonal gammopathies have been reported as well as

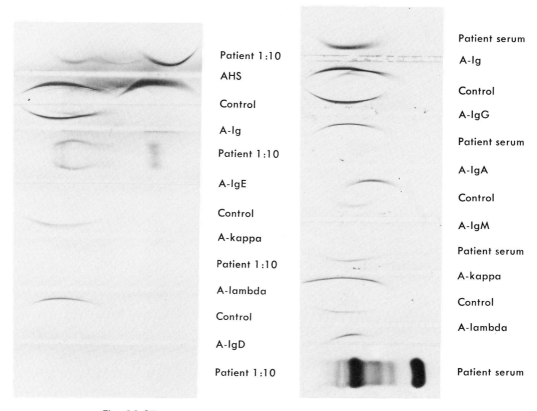

Fig. 18-27. IgE monoclonal gammopathy.
Fig. 18-28. Lambda light chain disease.

heavy chain disease of the IgG, IgA, and IgM classes. Low molecular weight IgM and half molecule monoclonal gammopathies have been described as well as multiple myeloma without monoclonal gammopathy where production of the immunoglobulins is either nil or so slight as to not be evident by conventional methods of detection.

Pseudo M proteins are usually associated with peaks in the $alpha_2$ region and most often seen in cases of nephrosis and in some instances of thermal burns. Excessive lipoprotein, bacterial, fibrinogen, and hemoglobin contamination of the specimen may result in apparent M proteins.

IMMUNOQUANTITATION PROCEDURES

Immunoelectrophoresis offers a qualitative method by which the various serum proteins can be assessed. A quantitative procedure was necessary for the establishment of the values for protein components in the normal as well as diseased individual. Numerous methods have been developed over the past few years. Most popular in the past has been the radial immunodiffusion (RID) method. Recently electroimmunodiffusion (EID, sometimes referred to as one-dimensional single electroimmunodiffusion), two-dimensional single electroimmunodiffusion, and automated specific protein analysis by nephelometry (AIP) have become available for the

622 *Chemistry for the clinical laboratory*

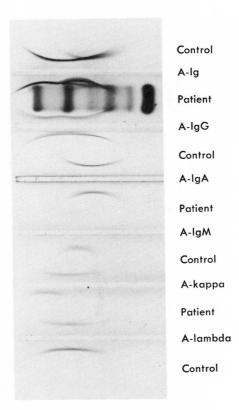

Fig. 18-29. IgG$_\kappa$ diclonal gammopathy.

quantitative assay of specific proteins. Counterimmunoelectrophoresis has been used to distinguish antigenicity in serum from patients suffering from viral, bacterial, and mycotic disease.

Radial immunodiffusion

When antigen of an unknown concentration is allowed to diffuse for a sufficient time in a radial direction from a central well into a uniform layer of agar containing the specific antibody to that antigen, the resultant precipitate ring diameter is directly proportional to the concentration of antigen employed and inversely proportional to the concentration of the specific antibody. The end point or fully developed precipitation disk is reached at final equivalence of antigen and antibody. At this point diffusion will no longer occur (Fig. 18-30).

Immunodiffusion depends on antigen concentration, antibody concentration, temperature, time, and the medium in which separation takes place.

A great deal of controversy has arisen in the past as to the best method for radial immunodiffusion of serum proteins. It is pretty well agreed now that the method of choice is that of Mancini, Carbonara, and Heremans. Numerous commercial immunodiffusion plates are now available which utilize this technique. It has been our experience that the M-Partigen and Tri-Partigen plates* have given us the most

---
*Behring Diagnostics.

Fig. 18-30. Hyland immunoplate immunoglobulin quantitation. Numbers, *1, 2,* and *3* are standards. γG quantitation.

reproducible and accurate results when compared with other kits. For that reason the method is not given here, but rather the clinical chemist is referred to the instructions provided with each group of plates for the specific protein to be analyzed. According to the latest information available, Behring Diagnostics has plates for radial immunodiffusion for the following specific proteins:

Prealbumin	Fibrinogen
Alpha$_1$ acid glycoprotein	Antithrombin III
Gc globulin	Ceruloplasmin
Alpha$_2$-HS-glycoprotein	Plasminogen
Haptoglobin	Beta-lipoprotein
Alpha$_2$-macroglobulin	Albumin
Hemopexin	C3 activator (beta$_2$-glycoprotein II)
C3c (B$_1$A globulin)	Retinol binding protein
C4 (B$_1$E globulin)	IgG
Beta$_2$-glycoprotein I	IgA
Alpha$_1$-antitrypsin	IgM
Transferrin	

Each year newer plates are being added. Other commerical sources offer plates for IgD and IgE. IgE, however, is better measured by radioimmunoassay in a kit provided by Phadebas. Those who wish to prepare their own immunodiffusion plates are referred to a manual entitled *Radial Immunodiffusion Test and the Immunoelectrophoresis Test for Qualitation and Quantitation of Immunoglobulins* by Dan F. Palmer and Roy Woods.*

Plates for radial immunodiffusion of serum proteins may be purchased from many suppliers. Under no circumstances should an investigator attempt immunoglobulin immunodiffusion quantitations without checking the results with World Health

---

*DHEW Publication No. (H5M) 72-8102. It may be obtained by writing the U. S. Government Printing Office.

Table 18-6. Serum immunoglobulin levels of adults and of children from birth to 13 years*

Age	No.	Level of IgG† mg/dl (± 1 SD)	Significance of difference from adult level‡	Percent of adult level	Level of IgA† mg/dl (± 1 SD)	Significance of difference from adult level‡	Percent of adult level	Level of IgM† mg/dl (± 1 SD)	Significance of difference from adult level‡	Percent of adult level
Newborn	20	1145 ± 252 (750-1500)	N.S.	100%	N.D.	N.D.	1%	10.8 ± 8 (0-23)	P < 0.001	10%
1-3 mo.	7	371 ± 132 (270-780)	P < 0.001	31%	28 ± 21 (6-68)	P < 0.001	14%	66 ± 34 (12-87)	P < 0.01	65%
4-6 mo.	7	556 ± 111 (190-860)	P < 0.001	33%	56 ± 14 (10-96)	P < 0.001	28%	70 ± 23 (25-120)	P < 0.01	69%
7-12 mo.	7	721 ± 208 (350-1180)	P < 0.001	63%	72 ± 22 (35-165)	P < 0.001	40%	76 ± 16 (52-104)	P < 0.01	75%
2 yr.	26	705 ± 106 (520-1080)	P < 0.001	62%	69 ± 33 (36-165)	P < 0.001	35%	96 ± 28 (72-160)	P < 0.01	83%
3 yr.	21	866 ± 213 (500-1360)	P < 0.001	76%	81 ± 21 (45-135)	P < 0.001	40%	93 ± 37 (46-190)	P < 0.1	91%
4 yr.	21	1026 ± 208 (540-1440)	P < 0.01	92%	108 ± 42 (52-210)	P < 0.001	54%	121 ± 63 (52-200)	N.S.	118%
5 yr.	38	924 ± 234 (640-1420)	P < 0.001	81%	111 ± 41 (52-220)	P < 0.001	56%	112 ± 36 (40-180)	N.S.	110%
6 yr.	26	1031 ± 213 (650-1410)	P < 0.01	90%	126 ± 39 (83-217)	P < 0.001	63%	108 ± 33 (55-210)	N.S.	106%
7 yr.	19	1000 ± 203 (570-1320)	P < 0.05	87%	117 ± 49 (65-240)	P < 0.001	58%	121 ± 29 (60-175)	N.S.	118%
8 yr.	18	1040 ± 244 (730-1410)	N.S.	91%	136 ± 56 (74-260)	P < 0.001	68%	118 ± 33 (68-175)	N.S.	116%
9 yr.	20	975 ± 131 (760-1330)	P < 0.001	85%	144 ± 34 (108-200)	P < 0.001	72%	101 ± 38 (55-160)	N.S.	99%
10 yr.	22	1073 ± 181 (730-1350)	N.S.	94%	139 ± 34 (70-222)	P < 0.001	69%	106 ± 30 (80-200)	N.S.	106%
11 yr.	19	1082 ± 147 (850-1300)	N.S.	95%	141 ± 51 (91-255)	P < 0.001	70%	110 ± 23 (66-155)	N.S.	107%
12-13 yr.	26	1143 ± 228 (770-1510)	N.S.	100%	188 ± 46 (108-325)	P < 0.01	94%	114 ± 32 (70-150)	N.S.	112%
Adult	72	1142 ± 195 (770-1510)			201 ± 44 (134-297)			102 ± 30 (67-208)		

*From Uffelman, J. A., Engelhard, W. E., and Jolliff, C. R.: Quantitation of immunoglobulins in normal children, Clin. Chim. Acta 28:185, 1970.
†Geometric mean. ‡Student 2 sample t test. N.D.—Only 2 samples with detectable IgA (22 and 60 mg/dl).

Table 18-7
Values for specific proteins as determined by the radial immunodiffusion method*

Protein	Mean value (mg/dl)
Prealbumin	25
Albumin	4400
Alpha$_1$ acid glycoprotein	90
Alpha$_1$-lipoprotein	360
Alpha$_1$-antitrypsin	290
Gc globulin	35
Inter-α-trypsin inhibitor	50
Haptoglobin	160
Ceruloplasmin	30
Cholinesterase	1
Beta$_2$-macroglobulin	240
Beta$_2$-HS-glycoprotein	60
Beta-lipoprotein	350
B$_1$A globulin	100
Hemopexin	117
B$_1$E globulin	30
Transferrin	320
Plasminogen	30
Beta$_2$-glycoprotein I	24

*From Störiko, N.: Normal values for 23 different human plasma proteins determined by single radial immunodiffusion, Blut **16:**200, 1968.

Organization Immunoglobulin Standard 67/95.* Cotrols and standards may be purchased from a number of sources for other specific proteins.

Several words of warning should be given to those who undertake radial immunodiffusion for the quantitation of serum proteins. The amount of serum placed in the well should remain constant between patients and plates. The overflowing of wells with serum should not occur. Certain proteins are very heat labile, and precautions as to their storage should be observed (example: serum complement fractions). Various proteins may show polymorphisms, and the measurement by RID must take into account the polymorphic form being measured (example: haptoglobin). The sample well should not be gouged or in any way made larger or smaller than the wells of the other serum wells in a given plate. Always include a known standard or series of at least three standards with each determination as well as a control serum to maintain quality control in this area of the immunology laboratory.

Normal values for the serum immunoglobulins as determined by RID in our laboratories are given in Table 18-6. Normal values for specific proteins as obtained by Störiko are given in Table 18-7.

---

*This material can be obtained by writing the Director, National Cancer Institute, Immunoglobulin Reference Center, 6715 Electronics Drive, Springfield, Va. 22151.

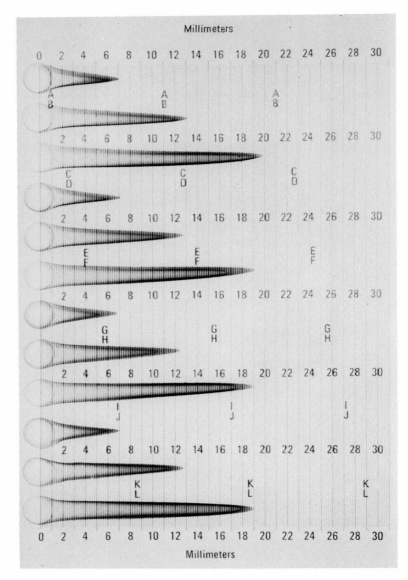

Fig. 18-31. Example of electroimmunodiffusion. (Reprinted with permission from Millipore Corp., Bedford, Mass.)

Electroimmunodiffusion (EID)

In electroimmunodiffusion agarose gel–containing antibodies to the specific protein being quantitated becomes a matrix into which a serum is electrophoresed. As the serum antigen migrates through the antibody-containing matrix, the antibody remains stationary because of the adjustment of the pH in the gel to correspond to the isoelectric point of the antibody. With only the movement of antigen, antigen-antibody equivalence complexes form in narrow, stationary precipitates. These precipitates form first as sidelines enclosing the antigen path, and as consumption and finally complete consumption of antigen occur a convergence of these precipitin lines

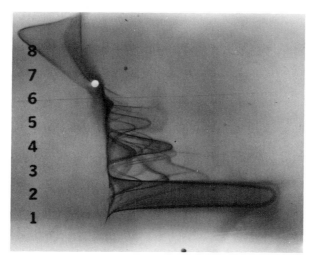

Fig. 18-32. Two-dimensional single electroimmunodiffusion performed on Beckman rehydratable agarose film.

occurs to form a peak. The height of the peak from the center of the antigen well to its tip is proportional to the concentration of antigen applied (Fig. 18-34).

An excellent system is commercially available for this procedure.* At the present time about eight specific proteins are available. If you wish to pour your own plates for EID, the directions may be found in Crowle or C. B. Laurell.

Two-dimensional single electroimmunodiffusion (Laurell crossed electrophoresis)

This method was first described by Ressler and later by Laurell. Clarke and Freeman offered certain modifications, with Weeke giving specific protein labels to the separated peaks so obtained. Serum proteins are first separated by electrophoresis in a 1% agarose. The agarose-separated proteins are then allowed to diffuse into an antibody (antihuman serum) containing agarose and electrophoresis is performed again in a 90-degree angle from the first electrophoresis pathway. A series of peaks (Fig. 18-32) is observed. The area under each peak is proportionate to the antigen concentration. Some forty alpha and beta globulins can be observed and semiquantitated by this method.

Automated specific protein analysis by nephelometry—automated immunoprecipitins (AIP)

With the advent of automation in the clinical laboratory has come the concept of specific protein analysis by automated nephelometric methodology. The diseases reflected by specific protein analysis certainly make screening of those proteins as much a necessity as screening for other blood constituents such as cholesterol, BUN, and calcium.

These tests are based on the original work of Libby, Schultze and Schwick, Boyden and co-workers, Goodman and colleagues, and Ritchie, who measured immunoglobulins as well as other specific proteins in body fluids at concentrations as

---

*Electro-Agaro Slide, Millipore Biomedica.

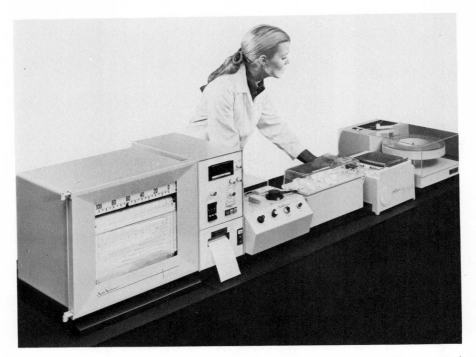

Fig. 18-33. Automated immunoprecipitin (AIP-PEG) AutoAnalyzer system. (Courtesy Technicon Instruments Corp., Tarrytown, N. Y.)

low as 0.5 μg/dl. The method was later automated by Kahan and Sundblad, Eckman and associates, Larson, and Gitlin and Edelhock.

Undiluted serum, cerebrospinal fluid, or other protein-containing solutions are aspirated into the automated immunoprecipitin system. Dilution according to the specific protein is made and the protein is then reacted with specific antiserum to that protein for 15 minutes. The antigen (protein) antibody complex thus formed is measured quantitatively by the nephelometric technique (Figs. 18-33 and 18-34). A recorder will record sequential standard, control, and unknown peaks delineating concentrations of the proteins in question.

Further modification of this method can be expected to be available in the near future. Enhancement of the antigen-antibody complex with the use of polymers is now being used, as reported by Lizana and Hellsing. Centrifugal analyzers are now utilized in a kinetic nephelometric methodology.

Counterimmunoelectrophoresis

This procedure utilizes an agar slide into which two wells have been placed close to one another. Antigen is placed in the well closer to the cathode, and antibody is placed in the well closer to the anode. Since antibodies have no significant net charge, they move toward the cathode by electroendosmosis (see electrophoresis). The antigen being negatively charged will migrate toward the anode.

As the two meet, an antigen-antibody complex is formed which takes the form of a precipitate line between the two wells.

Since other factors such as pH, ionic strength, gel composition, and antigen-antibody complexing characteristics have effects on electrophoresis, the net mobility

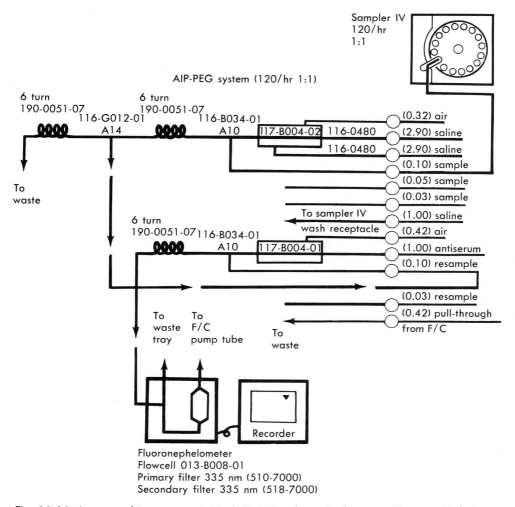

Fig. 18-34. Automated immunoprecipitin (AIP-PEG) schematic diagram. (Courtesy Technicon Instruments Corp., Tarrytown, N. Y.)

characteristics of the system must be known. This may be termed as the migration ratio $(-M_r)$, and it may be calculated with a substance such as albumin tagged with bromphenol blue and an antibody possessing no net charge which will migrate by electroendosmosis to the cathode.

Utilizing definite voltage and time in a specified system, the investigator measures the distance traveled by the antibody possessing no net electric charge. This distance is $-D_1$. The distance from the center of this separated substance to the center of the albumin is D_{total}.

$$\frac{-D_1}{D_{total}} = -M_r$$

This ratio is the expression of the degree of electro-endosmosis of the system. One can predict the endosmotic mobility of the antibody and thereby match it with the electrophoretic mobility of the antigen.

Selection of the proper buffer system will allow for precipitation arcs to be formed equidistant between the antigen and antibody wells.

Millipore Biomedica in its CIEP system has prepared three types of plates, thereby allowing proper buffer systems to be used with various antigen sources such as antinuclear antibodies, certain mycotic antigens, bacterial antigens, fibrinogen, and fibrin products as well as for forensic examination of blood stains.

Various suppliers have made available test kits for the examination of hepatitis associated antigen by this methodology. More recently, however, an even more sensitive test, radioimmunoassay, has been employed.

QUALITY CONTROL IN ELECTROPHORETIC PROCEDURES

Electrophoretic techniques have been considered to be within the realm of black magic, alchemy, and art. In actuality, the movement of charged particles via an imposed potential follows predictable scientific principles, which, if understood fully and by imposing simple standardized practices of quality control, can greatly improve the repeatability and performance of virtually all electrophoretic procedures, irrespective of the type of medium employed.

The following discussion of quality control in electrophoresis procedures has been developed for those individuals currently employing cellulose membrane methodologies. However, the basic principles apply to other electrophoretic media as well. It is incumbent upon each technologist practicing the science of electrophoresis to acquaint the laboratory with the practices that will guarantee better performance in electrophoretic separations.

Systems considerations

Any discussion of electrophoresis quality control necessarily begins with an overall view of the entire electrophoresis system and the techniques required to ensure continuance of operation from one time period to another. The following considerations should be a part of the education of every individual involved in practicing electrophoretic technique.

Electrophoresis cell

The electrophoresis cell should be checked periodically for cracks between the cathodic and anodic chambers. The electrodes, irrespective of composition, should be checked at least once a month for signs of decomposition, pitting, cracks (as in the case of carbon rods), or actual breaks. The most likely place for any discontinuity to occur is at the juncture between the electrode and the pins or sockets employed to connect the cell to the power supply. At the same time, careful inspection should reveal whether or not excessive quantities of precipitated buffer salts have built up on either the electrodes or the divider between cathodic and anodic chambers. Cracks between the buffer chambers, broken or defective electrodes, and buffer buildup on electrodes or center dividers can all result in irregular, irreproducible migration of the moving proteins.

Most conventional electrophoretic cells should never be cleaned with organic solvents such as alcohols. Alcohols tend to remove plasticizers from the plastic, resulting in a cracking or crazing phenomenon which permanently affects the cell. The most effective way to maintain cleanliness within the electrophoresis cell is to:

1. Never leave buffer in the cell for long periods of time when the cell is not in use.
2. At the end of each day's electrophoresis, dump all the cell buffer and rinse the

cell several times with tap water followed by at least two rinses with distilled water.

3. Never attempt to remove salt buildup on electrodes with wiping or with spatulas, etc.

The most effective way to remove electrode salts is to fill the cell with distilled or deionized water and leave it alone for 12 to 24 hours. The salts will readily dissolve from the electrode, leaving a clean surface.

The most overlooked portion of the electrophoresis cell is the cell cover. Make certain that it has not been warped by too much heat or physical abuse. This is a very common problem with the flat or "vegetable tray" type electrophoresis cells. Make certain that the cell and cover mate sufficiently tightly so that if a blast of air were directed over the top of the cell by a fan or a heat register, discontinuities in electrophoretic mobility would not occur. The best way to eliminate irregular protein mobilities resulting from drafts, temperature differences, and the like is to make certain that the electrophoresis cell is not operated close to fans, heat registers, or windows which may be subjected to seasonal variations in temperature.

The last major consideration relative to the electrophoresis cell is to maintain the same buffer levels within the cell as closely as possible. The level of buffer on one side of the cell should be equal to that on the other side of the cell. The former is accomplished by actually measuring the amount of buffer placed into the electrophoresis cell in order to bring the levels up to their normal operating quantity. Levels below this will result in fewer volts per centimeter and shorter migration lengths. Levels greater than this will result in more volts per centimeter and longer migration lengths. The amount of buffer placed into the cell should be controlled within $\pm 1\%$. Equalization of buffer levels is easily accomplished by tilting the cell until buffer runs over the center divider, then returning the cell to its normal horizontal position. The center divider should always be wiped dry when this method of leveling is employed.

Power supply

Since the power supply is the source of the potential to drive the proteins, it is important that the output voltage be checked on an external voltage meter every month. The power supply meter should then be adjusted to coincide with the actual voltage output under a given load. If the output voltage and meter reading do not coincide, lesser or greater migration distances may be encountered during the electrophoresis.

It is also an excellent idea to monitor the initial current levels produced in the system when the voltage is set to its proper operating level. This enables one to determine whether the ionic strength or type of buffer being employed is the same as employed previously for the given technique. If the initial current level is excessively higher than previously obtained for a given technique, there is sufficient evidence to presume that some substantial change has occurred. Typically, this results from dissolving the buffer in water which has been improperly or insufficiently deionized. This can be evaluated by dissolving new buffer salts in commercially produced distilled water. If the current level is still too high, then the electrophoresis cell buffer or power supply is suspect. The center divider of the cell may be allowing electron flow either over the top of the divider or through the center divider via a small crack. If the electron flow is over the top of the divider, it can be caused by droplets of buffer adhering to the divider or it can be a result of decomposition of a portion of the divider resulting from carbonizing of the plastic. This carbonizing is virtually impossible to repair since it results in a small track of carbon from one side of the divider to the other. Electron flow will probably occur through the carbon track rather than through the support medium. Likewise, if the center divider is cracked, electrons will choose to move through the path of least resistance, which would be the crack. In either of these two cases, the cell is no longer useable. If the divider is wet with buffer solution, wiping with a dry towel should readily eliminate the buffer solution from the divider. If the buffer is faulty, dilution of the suspect buffer or preparation of a fresh buffer should depict the problem.

Periodically, the operator should check the contacts between the electrophoresis cell and the power supply. Generally, a combination of pins and sockets are employed such as with the so-called banana plug connectors. The very nature of these types of connectors results in the pins loosening quite readily and the sockets breaking away from their respective cables, generally through excessive bending of the cords leading to the power supply. In either case, the imposed potential may be interrupted during the course of an electrophoretic run. The hazards of bare wires from a connecting cable should be obvious to the operator.

It is important that the power supply have a relatively stable voltage output once set at a given level. For short electrophoretic runs, operation at constant voltage over the course of the run is totally adequate. When current levels are exceptionally high, such as with some gel methodologies, or electrophoresis running times are going to be in excess of an hour or two, operation of the system under constant current conditions is desirable. The goal in either case is to maintain as low a heating effect as is possible in order to reduce destruction of such heat-labile substances as enzymes. An acceptable alternative to constant voltage or constant current is to operate the electrophoresis cell and procedure under refrigerated conditions, about 4° C. This is accomplished by placing the buffer-containing cell into the freezer until the cell and buffer are extremely cold. The medium is prepared with chilled buffer and the electrophoresis is carried out within the refrigerator or with the cell immersed in ice. Electrophoresis running times, irrespective of the voltage conditions employed, should be maintained within ± 1 minute or should be carried out until a marker dye, such as bromthymol blue coupled to albumin, has migrated a given distance from the point of application.

Electrophoresis buffer

The pH, ionic strength, and level of buffer within the electrophoresis cell should all be controlled very closely. Every freshly prepared buffer solution should have a pH check and appropriate adjustments made to ensure sameness in pH from one run to another. If a commercial buffer is employed, then lot number and pH should be recorded in a laboratory manual in case problems develop. When possible, each new lot of buffer solution should have a conductivity reading made to ensure the same relative ionicity of the buffer from one day or lot to another. Generally, conductivities varying as much as ± 5% do not dramatically affect electrophoretic separations. Discrepancies in the water purity employed for dissolving of the buffer salts are readily determined by conductivity readings. These readings should be logged in the laboratory manual along with other pertinent information relative to a given buffer preparation. Needless to say, proper labeling and dating of all buffer reagents are imperative to good, standardized laboratory technique. (It also prevents someone else from grabbing the wrong solution by mistake.) If buffer solutions are made from the raw materials rather than from a commercial source, the operator must be certain of the grade, impurities, and characteristics of each raw material employed in the formula. A rule of thumb is to always employ reagent grade or better raw chemicals where the operator is not in a position to ascertain the composition of the substances within the facility.

Ideally, an electrophoretic buffer should be employed for only a single electrophoretic separation. The rationale for this is quite simple. After buffer has been placed into the cell and appropriate leveling done, the composition of buffer ions on the cathode side and on the anode side is identical. During the first electrophoretic run, positive ions will tend to move toward the cathode side of the cell and negative ions will move toward the anode side of the cell, that is, the cell buffer becomes polarized. A second electrophoretic separation carried out in the same buffer, even with polarity reversal, is not equivalent to the first run since the second run is starting with unequal distributions of buffer ions in the two chambers. Simultaneously, at each electrode, some electrolytic decomposition of buffer ions occurs. These decomposition products have the potential of interacting with the proteins in the electrophoretic medium resulting in complexing, denaturing, or inactivating. A second run in the same buffer increases the probability of interaction between separating proteins and buffer decomposition products. For all intents and purposes, polarity reversal as a means of extending the life of the electrodes is not required in systems employing platinum electrodes, since their electrolytic decomposition requires a considerable length of time. However, systems with electrodes composed of other than platinum may require frequent changing, even if polarity reversal is employed.

Other reagents employed in the procedure

Since it is nearly impossible to list all of the chemicals employed in the multitude of electrophoretic procedures currently fashionable in clinical laboratories throughout the world, some basic guidelines are recommended:

1. Wherever practical, employ only reagent grade or better chemicals for all final detection reagents or enzyme substrates. If possible, obtain a certificate of composition (purity) from the supplier. Employ practical grades of reagents only where financial restrictions are imposed or when complete knowledge relative to impurities is available. These impurities must not contribute error to the system.
2. All dyestuffs employed directly in the detection of proteins or indirectly in the coupling to proteins or the products of their reactions should be ordered by both part number and color index number. In addition, each new lot of dye material minimally should have a visible absorption spectrum in a given solvent performed and recorded as part of the history of that dye. Where possible, a thin-layer chromatogram of each new lot of dye substance should be run and compared to previous lots. Also, where refrigeration is not required, all dyestuffs should be stored in a cool, dry, dark place in order to reduce or totally eliminate photometric decomposition—a phenomenon quite common to all dye materials.
3. All liquid chemicals employed as stain solvents, rinsing solutions, etc. should be the best grade possible. In addition, they should be employed only at the recommended concentration levels. The date of their receipt should be recorded both on the container and in an incoming raw materials log book.
4. If possible, all incoming chemicals to be employed in electrophoretic procedures should be labeled with their date of receipt and logged in an incoming chemicals logbook. It is strongly recommended that they be given an incoming reference number such that a history of its use and performance can be maintained during its lifetime within the laboratory. Should instability or perform-

ance problems be experienced at a future date, the history of the chemical is readily available to both the laboratory director and any industrial representatives of the supplied chemical(s).

5. All control sera employed as either markers, quantitative calibrators, etc. should conform to the latest regulations for such materials. The Food and Drug Administration and the National Committee for Clinical Laboratory Standards should serve as two excellent reference sources for the latest regulations relative to control sera employed in laboratory testing.

Electrophoresis support media

Since the types and sources of various electrophoretic support media are so extensive, several guidelines for assessment of performance are recommended. The guidelines should serve generally for all types of support; however, all may not be applicable to a particular type of support material. Greater detail relative to cellulose acetate membranes can be found throughout the context of this presentation.

1. Make certain that a given lot of electrophoresis support medium has been evaluated and compared to what had been employed previously. This evaluation can be performed with control sera, homemade protein solutions, or dye markers; however, it is imperative to assess within lot and lot-to-lot variability in performance.

2. Never interchange one type of support medium with another commercial source of the same support medium without having first demonstrated their equivalence in performance. This is especially true with cellulose acetate membranes where seldom are two different sources of membrane identical in composition, pore size, distribution, properties, etc.

3. With handcast gel media such as agar, agarose, starch, and polyacrylamide, make certain that standard formulas, mixing times, casting conditions, etc, are maintained by all electrophoresis technologists.

4. With commercially prepared media, at the first hint of irregularities in performance of the product, contact the manufacturing representative for assistance. Usually, the manufacturer knows more about the product than can be known by the technologist. Therefore, there is little reason for the technologist to randomly attempt to troubleshoot a medium performance problem.

5. Record all lot numbers of media employed and the approximate date of that lot number's beginning use in the laboratory. This can serve as a useful aid to both the laboratory supervisor and the manufacturer in pinpointing when and how a given problem began.

6. Store all surplus support media not requiring refrigeration in a cool, dry place. Most electrophoretic support media can be adversely affected by extremely humid environments. Humidity can cause wrinkled or distorted media and very diffuse electrophoretic separations.

7. Always perform electrophoresis on some type of control specimen with each electrophoretic run. For routine serum proteins, a serum pool or a control serum will suffice. For various isoenzyme procedures, a tissue homogenate or extract is virtually mandatory. If animal sera or tissue homogenates and extracts are employed as markers, make certain that their mobility characteristics closely parallel those of the test biological.

Technique considerations

After the electrophoresis technologist has ascertained that the components of the electrophoresis system are operating optimally, careful evaluation and control of the technique portions of a given analytic procedure should be taken into consideration.

Electrophoretic conditions

As was stated previously, electrophoresis running times should be controlled within ± 1 minute of the designated time as called for in the procedure. Alternatively, a marker dye such as bromthymol blue complexed to albumin may be migrated a fixed distance from the point of application. Either approach ensures the highest probability of sameness in pattern migration length. Initial and final current levels should be monitored during each electrophoretic separation. The value of the former has been described previously. Monitoring final current levels can provide a clue to a problem having occurred during the course of the electrophoretic separation. If there is no final current flow, then something has occurred which resulted in a discontinuance of the electron flow. Typically, this is seen where loose connections are at play. If final current levels are too high, then bridging of the center divider is to be suspected.

Electrophoresis under constant voltage or constant current conditions should be determined prior to experimentation with a given technique. A rule of thumb is that short running times (an hour or less) with relatively low current levels (less than 10 ma) (primarily for cellulose acetate membranes) require only constant voltage conditions. Longer running times and higher current levels would require constant current conditions. When various types of gels are employed, such as starch, PVC, agar, agarose, polyacrylamide, and so on, then constant current or refrigerated conditions may yield more repetitive results. The major consideration here is to operate at the lowest current levels possible without sacrificing resolution of protein fractions. Lower current levels result in less heat being generated within the system and less denaturing or inactivation of the protein(s) of interest. Additionally, once a technique is employed, do not alternate between different conditions and expect to obtain the same electrophoretic results.

When one is employing an electrophoretic procedure for the first time, it is incumbent upon the technologist to establish the final electrophoretic conditions most suitable to the needs of that particular laboratory. If a procedure calls for voltage levels of 250 volts, then the operator should be aware of what happens to protein migration when the system is run at 150, 200, 300, and 350 volts as well. If the running time is supposed to be 15 minutes, then the operator should evaluate 5-, 10-, 20-, and 25-minute running times. Often it is helpful to plot the relative migration in millimeters of some protein fraction such as albumin against both imposed voltage and running time. The optimal conditions can then be determined from the resulting graphs. If a technique calls for the use of a buffer of an ionic strength of 0.05, then the operator should determine what happens in the system if the ionic strength is 0.01, 0.025, 0.075, etc. Likewise, the investigator should plot current levels against ionic strength of the buffer to determine at which ionic strength the best pattern resolution occurs while maintaining the least difference between beginning and final current levels. Finally, if the procedure calls for use of a barbital type buffer, the operator should determine the performance of the technique in other buffers such as TRIS/barbital or TRIS/glycine. Instructions supplied with commercial equipment should serve only as guidelines, not as optimal running conditions.

Staining

If direct staining of separated proteins is to be employed, the composition of the staining solution should be controlled as closely as possible. Staining times should be controlled to within approximately ± 1 minute from that called for in the procedure. A minimum quantity of dye solution should be used to adequately stain the separation, and it should be used once, then discarded. Each time an electrophoretic support medium is placed into a solution of stain, residual buffer leaches from the medium and affects the pH or denaturing capability of the denaturants generally composing direct staining solutions. Continuous use of the same staining solution cannot help but interfere with the staining efficiency of subsequent electrophoretic separations.

The operator should be aware of the dye-binding capacity of the particular dye being employed in the technique. In most cases, proteins such as albumin do not bind dyes in the same ratio as they bind globulins.

The operator should assess linearity of the dye-protein complex. One method of doing such an evaluation is to make up solutions containing at least two proteins, for example, albumin and transferrin. The relative concentrations of these two should span from 3 times their normal concentration in serum down to 10% of their normal serum concentration. These control samples are then applied to the electrophoretic medium as though they were normal sera. Electrophoresis and subsequent detection are carried out as usual. First, quantitation should be carried out by elution of the relative fractions with buffer from the medium and read in a spectrophotometer. Optic density versus concentration should be graphed. Concentrations of protein-bound dye which are outside the linear range should be diluted prior to evaluation. Second, densitometric assessment of these control protein solutions should be carried out. Integration counts per given protein should be plotted against their respective concentrations. Ranges of linearity should be determined. No biologic sample should be evaluated if any of its protein components exceeds the linearity capability of the densitometer. This is usually between 1.5 and 2.0 optic density units.

As a rule, if a protein fraction concentration is greater than its upper limit of linearity, the quantitating instrument will underestimate its relative concentration when evaluated in a medium containing other proteins of lower concentration. Simultaneously, the detecting device will overestimate the lower concentration proteins in the pattern (often seen as an inverse A/G ratio).

Destaining

Most direct staining procedures impose excessive amounts of stain on the electrophoretic medium. This excess must be removed with a destaining solution, usually composed of an organic solvent, such as acetic acid and water. Destaining times should be controlled to ± 1 minute of the times called for in a given procedure. Excessive rinsing times or excessive concentration of the strongest solvent (in the above case, the acetic acid) increase the possibility that protein-dye complex may be removed from the medium. This is particularly true in gel media when proteins or lipids are detected directly by staining methods. The dye-protein or dye-lipid complex can be totally leached from the medium, resulting in light staining or nonvisible fractions.

When setting up an electrophoretic procedure for the first time, evaluate different staining times and determine the minimal and the maximal staining times. Typically, it is found that excessive staining produces fewer problems in repeatability than do staining times that are too short. Also, destaining times should be optimized such that the greatest repeatability in technique results. Rinsing solutions should be employed as few times as possible in order to reduce the possibility of contamination from one electrophoretic pattern to another.

Clearing

Prior to evaluation by conventional densitometers, cellulose acetate membranes must be rendered transparent. Except where the chemistry prohibits use of clearing solutions, evaluation of uncleared acetate membranes is a poor quantitative procedure since the background optic density of the membrane, coupled with the optic density of the protein-dye complex, often results in pushing the most intense peak above its limits of linearity. The net result, quantitatively, is to underestimate the intense peak(s) and overestimate the lesser peaks. This is readily demonstrated by taking a routine serum protein electrophoretogram, scanning it uncleared, then carrying out the conventional clearing procedure and rescanning the membrane. Comparison of values will be quite enlightening.

Clearing solutions are generally not interchangeable from one manufacturing source of cellulose acetate membranes to another. The most generally accepted, universal clearing solution for acetate membranes is clear mineral oil, which has a refractive index of about 1.4. Once the pore structure of the membrane is filled with mineral oil, the membrane is rendered transparent, and, if the messiness of the oil can be overcome, densitometric evaluation can be attempted. Most acetate membranes are rendered transparent by placing them into a mixture of organic solvents, laying them onto a solid support such as a glass plate, and heating to a temperature of about 100° C. The organic solvents, coupled with the elevated temperatures, result in partial dissolving of the cellulose acetate polymer. However, before total solution can occur, the elevated temperatures drive off excessive solvent, resulting in a transparent film.

These organic clearing solutions are generally composed of at least one or more fast and slow solvents. A fast solvent is defined as one which, if used at 100% concentration, would result in total dissolving of the cellulose acetate polymer. The slow solvent is defined as one which, if used at 100% concentration, would not result in dissolving of the cellulose acetate polymer. When the fast and slow solvent(s) are combined in the proper ratio, a uniformly transparent acetate film results. A membrane that does not clear fully manifests this property by being either cloudy or having portions of the membrane cloudy to opaque. Increasing the amount of the fast solvent by 1% or 2% generally will result in better clearing. Conversely, a membrane that clears in an irregular, runny, gummy mess generally has excessive fast solvent. Reduction in the amount of fast solvent or increasing the amount of slow solvent will generally result in a properly transparent acetate film. With cellulose acetate membranes, it is to be expected that the optimal clearing solution:solvent ratios will vary somewhat from lot to lot. The optimal ratio of fast and slow solvents should be reestablished on each new lot of membranes received. This is best accomplished by cutting a small portion from the first processed membrane of a given lot. Place the membrane into the previously employed clearing solution and, if proper transparency results, clear the remainder of the membrane containing the separation with that formula of clearing solution. If a cloudy membrane results, then a slight increase in fast solvent is indicated. In this way, there is virtually no reason for wasting a processed membrane by experimenting with clearing solutions. Time in the clearing solution should be controlled to about ± 30 seconds of the time called for in the given procedure. Membranes should be dried in an oven at a temperature of within ± 3 degrees of that called for in the procedure and left until the smell of organic solvents no longer exudes from the surface of the film.

The ease with which a cleared cellulose acetate film is removed from its glass supporting plate depends on the cleanliness of the glass plate. It is imperative that the glass plate be cleaned periodically with chromic–sulfuric acid solution to remove oils and other contaminants from fingers or soapy rinsing solutions. An alternative solution for films sticking to glass plates is to first clean the plate with chromic–sulfuric acid, then coat the plate with a thin layer of silicone stopcock grease. Wiping away all excess grease leaves a very thin layer of this release agent on the surface of the glass. The cleared membrane cannot adhere strongly to a silicone-treated glass plate.

Densitometry

Most laboratories employ some type of commercially available transmission densitometer to evaluate a stained electrophoretic pattern. Basically, the densitometer consists of a tungsten light source which is usually focused onto the surface of the electrophoretic medium. A filter is placed between the light source and the position of the electrophoretic separation. Monochromatic light, at the absorption maximum of the dyestuff, is impinged upon the electrophoretic separation. The electrophoretic pattern is placed onto a movable carriage which allows the pattern to move through the beam of monochromatic light. On the side of the pattern opposite the source of the light beam, there is placed a photodetector which monitors the amount of light passing through the electrophoretic separation. As each stained protein fraction passes between the light beam and the detector, some amount of light is absorbed by the dye-protein complex. The greater the intensity of the dye-protein complex, the greater the amount of light absorbed. The detector is coupled to an amplifier circuit and to an analog recorder such that the greater the amount of light absorbed by the dye-protein complex, the greater is the analog pen deflection on the recorder chart paper. Hence, a densitometric evaluation of an electrophoretic pattern consists of a series of large and small connected peaks. The area under their curves is proportional to the amount of dye-protein complex absorbing the monochromatic light, which in turn is proportional to the amount of protein in the given fraction.

The operator must be aware of several operating conditions prior to intelligent employment of the densitometer. First, the operator must have some concept of the ability of the instrument to repeat itself within a given day and from day to day on a given protein separation. This is most efficiently accomplished as follows.

Take any appropriate serum protein electrophoresis pattern and scan it twenty consecutive times on a given day. Repeat the twenty scans on the same pattern, but remove the membrane from the carriage between each scan. Perform mean, standard deviation, and coefficients of variation for each of the two sets of scans (Tables 18-8 and 18-9). Compare the coefficients of variation. Between additional runs, store the electrophoretic pattern in a dark place in order to reduce decomposition from the light. Then each day for 5 successive days, scan the same pattern twenty times and compute means, standard deviations, and coefficients of variation for each of the 5 days, then for all 100 patterns over the total of 5 days (Table 18-10). Record values in a densitometer log book. At least once per month, remove the electrophoretic pattern from its protective dark place and rescan as described. Repeatability should be quite similar for the various scan times (where the pattern is scanned without removal from the carriage). Note, however, that removing the pattern from the carriage between scans results in higher coefficients of variation. One can readily conclude from this that care in placement of the membrane on the carriage is critical to good repeatability with a densitometer. This operation of assessing instrument repeatability should be performed by every individual asked to operate the densitometer (Table 18-11).

The second type of test that should be performed is one which determines the ability of the electrophoretic technique, the operator, and the densitometer to reproduce given values for an electrophoretic pattern. This is conducted as follows: On three different cellulose acetate membranes from the same lot number, apply several single applications of a control serum sample or a serum

Table 18-8
Typical within-day repeatability of a transmission densitometer employed in the repetitive scanning of a single routine serum protein electrophoresis pattern*

	Protein fraction				
	Albumin	Alpha$_1$	Alpha$_2$	Beta	Gamma
Mean (percent)	63.8	3.1	8.6	11.9	12.6
Standard deviation	0.285	0.099	0.141	0.141	0.170
Coefficient of variation (percent)	0.45	3.23	1.63	1.18	1.35

*Based upon twenty repetitive scans of the same protein pattern, conducted at three different times intervals on a given day (total of 60 replicate scans). Scans performed on a Beckman CDS-100 Computing Densitometer System. Courtesy Beckman Instruments, Inc., Fullerton, Calif.

Table 18-9
Typical within-day repeatability of a transmission densitometer employed in the repetitive scanning of a single serum protein electrophoresis pattern twenty times with removal and repositioning of the membrane between successive runs*

	Protein fraction				
	Albumin	Alpha$_1$	Alpha$_2$	Beta	Gamma
Mean (percent)	62.9	3.2	8.8	11.4	13.8
Standard deviation	1.475	0.167	0.294	0.320	0.95
Coefficient of variation (percent)	2.35	5.30	3.33	2.81	6.87

*Scans not conducted on same day as those in Table 18-8; however, the same serum pattern was employed for data in both Table 18-8 and Table 18-9. Scans performed on a Beckman CDS-100 Computing Densitometer System. Courtesy Beckman Instruments, Inc., Fullerton, Calif.

sample from one patient. Twenty or more applications of the same serum sample or control should be sufficient. Perform electrophoresis and process under the controlled conditions as described in previous sections. Quantitate by densitometry and compute mean, standard deviation, and coefficient of variation for the twenty or more replicates (Table 18-12). Then each day for 5 successive days, apply the same serum sample five or more times to a single membrane from a single lot of acetate membrane. Process, quantitate, and compute means, standard deviations, and coefficients of variation for each of the 5 days and for all 5 days together. Repeat this last process for several different lot numbers of cellulose acetate. The standard deviations and coefficients of variation represent a given operator's ability to repeat the electrophoretic technique. Each electrophoresis technician from a given laboratory should perform this

Table 18-10
Typical day-to-day repeatability of a transmission densitometer employed in the repetitive scanning of a single serum protein electrophoresis pattern twenty times on each of 5 successive work days*

Day	Protein fraction				
	Albumin	$Alpha_1$	$Alpha_2$	Beta	Gamma
Day 1					
Mean (percent)	64.34	3.03	8.46	11.89	12.28
Standard deviation	0.131	0.057	0.150	0.190	0.140
Coefficient of variation (percent)	0.20	1.89	1.78	1.60	1.14
Day 2					
Mean	63.85	3.00	8.71	11.88	12.57
Standard deviation	0.061	0.89	0.147	0.116	0.104
Coefficient of variation	0.10	2.96	1.69	0.98	.83
Day 3					
Mean	64.10	2.95	8.66	11.83	12.51
Standard deviation	0.069	0.076	0.147	0.113	0.123
Coefficient of variation	0.11	2.58	1.69	0.95	0.99
Day 4					
Mean	63.93	3.01	8.65	11.90	12.54
Standard deviation	0.064	0.072	0.176	0.173	0.110
Coefficient of variation	0.10	2.39	2.04	1.46	0.87
Day 5					
Mean	65.07	2.78	8.28	11.36	12.51
Standard deviation	0.049	0.077	0.180	0.193	0.100
Coefficient of variation	0.08	2.76	2.18	1.70	0.80
Summary: all 5 days					
Mean	64.25	2.95	8.55	11.77	12.48
Standard deviation	0.448	0.117	0.226	0.262	0.153
Coefficient of variation	0.70	3.96	2.65	2.22	1.23

*Single day values based upon twenty repetitive scans of the same protein pattern. Five-day totals based upon 100 repetitive scans of the same protein pattern. Scans performed on a Beckman CDS-100 Computing Densitometer System. Courtesy Beckman Instruments, Inc., Fullerton, Calif.

type of analysis. The results will provide an index of membrane within lot repeatability of the electrophoretic technique, lot-to-lot repeatability of the electrophoretic technique, and the ability of the electrophoretic technique to repeat a given analysis. If a commercial control serum is employed, then comparing the values obtained by the technician with those stated by the manufacturer of the commercial control will provide an index of the ability of the technique to recover a given value.

The next evaluation that should be undertaken is the determination of normal values and, subsequent to this, the production of a pooled serum control. Collect a minimum of 100 serum samples from a random hospital population. Perform total protein and albumin determinations by conventional wet chemistry methods. Likewise, perform routine serum protein electrophoresis on all 100 specimens. Compute means and standard deviations on all wet chemistry values and separated protein fractions. These first values can serve as a guide-

Table 18-11
Typical operator-to-operator repeatability of a transmission densitometer in the scanning of a single serum protein pattern*

	Protein fraction				
Operator	Albumin	Alpha$_1$	Alpha$_2$	Beta	Gamma
Operator 1					
Mean (percent)	64.1	3.0	8.5	11.9	12.5
Standard deviation	0.302	0.114	0.214	0.255	0.248
Coefficient of variation (percent)	0.47	3.83	2.53	2.14	1.98
Operator 2					
Mean	64.0	3.10	8.7	12.1	12.2
Standard deviation	0.051	0.138	0.389	0.433	0.416
Coefficient of variation	0.08	4.44	4.48	3.57	3.43
Operator 3					
Mean	63.9	3.0	8.6	11.7	12.9
Standard deviation	0.067	0.095	0.253	0.328	0.273
Coefficient of variation	0.11	3.14	2.94	2.82	2.12
Operator 4					
Mean	63.8	3.0	8.7	12.0	12.5
Standard deviation	0.051	0.102	0.175	0.224	0.188
Coefficient of variation	0.08	3.37	2.01	1.87	1.50
Operator 5					
Mean	63.8	3.1	8.4	12.5	12.2
Standard deviation	0.049	0.134	0.322	0.439	0.416
Coefficient of variation	0.08	4.37	3.84	3.50	3.41
Summary: all operators					
Mean	63.9	3.1	8.6	12.0	12.5
Standard deviation	0.182	0.125	0.303	0.446	0.411
Coefficient of variation	0.29	4.11	3.70	3.54	3.30

*Values based upon a twenty replicate scan of a single serum protein sample by each different operator. Scans performed on a Beckman CDS-100 Computing Densitometer System. Courtesy Beckman Instruments, Inc., Fullerton, Calif.

line for a normal range of values. Continue to collect random serum samples and obtain values until at least a 1000 random samples have been collected and processed. Continue to update the means and standard deviations to coincide with all new data collected. Once means and standard deviations have been collected on the 1000 random patients, divide the population by sex into various age groups.

From the first 100 random patient sera, accurately pool 1 or 2 ml from each specimen and collect in a common vessel such as a sterile volumetric flask. Slowly mix the various aliquots of serum until thorough mixing has been obtained. Then dispense 0.25- to 0.5-ml aliquots of the pooled serum into a conveniently sized, sealable test tube or vial. Label as pooled serum and place all tubes or vials into a freezer at −20 to −80° C. Each day that electrophoresis is carried out, remove one vial of pooled serum and use as a control sample. Compare values obtained from the pooled control with the statistically derived values

Table 18-12
Typical repeatability of serum protein electrophoresis on cellulose acetate membranes while separating a single serum protein sample*

	Protein fraction				
	Albumin	Alpha$_1$	Alpha$_2$	Beta	Gamma
Summary of a single sample electrophoresed twenty-four times on three membranes from a single lot number on a single day					
Mean 24 (percent)	63.3	2.7	8.2	9.7	13.0
Standard deviation	1.66	0.33	0.56	0.54	0.81
Coefficient of variation (percent)	2.50	12.2	6.8	5.6	6.2
Summary of a single serum sample electrophoresed forty times on five membranes from a single lot number over 5 successive days					
Mean 40	63.6	3.2	8.7	10.5	14.0
Standard deviation	2.52	0.50	0.60	0.93	1.39
Coefficient of variation	4.0	15.6	6.9	8.8	9.9
Summary of single serum sample electrophoresed eight times on a single membrane from each of ten different lots of cellulose acetate membrane					
Mean 80	62.2	3.1	9.0	10.4	13.2
Standard deviation	1.43	0.32	0.67	1.05	0.88
Coefficient of variation	2.2	10.2	7.4	10.1	6.7

*All values obtained with a Beckman Microzone Electrophoresis System and Beckman Cellulose Acetate Membranes. Scans performed on a Beckman CDS-100 Computing Densitometer System. Courtesy Beckman Instruments, Inc., Fullerton, Calif.

for the 100 initial patients. Electrophoresis values from the pool and statistically determined values from the 100 patients should agree quite closely.

In concluding the discussion on densitometers, it is noteworthy to state that the densitometer is generally the least variable portion of the entire electrophoretic technique. Since the sample handling, sample application, electrophoresis, and postelectrophoresis processing contribute the greatest amount of error in the electrophoretic technique, it stands to reason that control of all timed steps and reagents employed in these processes should be maintained within very close tolerances. Only then can the resultant values begin to have substantial clinical relevance.

QUALITY CONTROL OF IMMUNOLOGIC PROCEDURES

It is apparent from the foregoing descriptions of the methodology involved in the determination of proteins by immunoelectrophoresis and immunoquantitative procedures that certain quality control measures are necessary. The controls used for electrophoresis are well outlined in the preceding section and apply to IEP, RID, EID, CIEP, two-dimensional single electroimmunodiffusion, and any other procedure utilizing a specific medium in which protein separation takes place.

Media selection and preparation, buffer composition, pH, ionic strength, temperature, cell and power supply maintenance, and systems utilized for the delivery of specimens onto the media must all be subjected to the quality control procedures previously explained.

The parameter of antibody must now be considered since this source of possible error is introduced into the immunologic system. It is imperative that standards for each specific protein being examined or quantitated, when available, be obtained. The specificity of each antigen (protein) can be checked by immunoelectrophoresis against specific antiserum. Likewise the specificity of antiserum and its ability to produce sharp, discernible precipitin lines can be determined utilizing known isolated specific proteins. Such standards at this time are difficult to obtain; however, with increasingly definitive methods of protein isolation becoming available, it is hoped these standards will soon become more plentiful. Known examples of monoclonal gammopathy serum and urine controls are available for purchase.* It behooves the technologist to prepare a series of known specimens for control studies from the abnormal sample population in the individual laboratory. Such specimens may be maintained for a number of years if kept frozen in aliquots.

Quality control charts should be maintained for all RID, EID, and AIP methods. Normal sera can be obtained from the chemistry laboratory and pooled. Four fifths of the pool can be aliquoted (0.25 ml) and frozen to be used as the normal internal control serum. The remaining one fifth is diluted 1:5 with a 1% bovine albumin to prepare a low internal control serum, aliquoted (0.25 ml), and frozen.

To prepare a quality control chart at least thirty determinations should be made of each normal and low serum for the specific protein to be examined. The mean and standard deviation should be determined and a quality control chart established with limits of ± 2 SD for each specific protein. As mentioned previously, a World Health Organization Immunoglobulin Standard 67/95 may be obtained from the National Cancer Institute.

Immunoelectrophoresis, as well as other electrophoretic procedures coupled with immunodiffusion resulting in antigen-antibody equivalence precipitates, is difficult to control if the technologist is preparing individual plates. Agarose or Ionagar delivered to each plate must remain constant; however, depth of agar is difficult to control. Placement of the template or gel cutter must be exact, with distance between antibody trough and antigen wells remaining constant. Premolded or prepoured plates as well as rehydratable systems offer many advantages in controlling the IEP or related procedure. As in all methods, known controls should be run with each procedure as well as standards if at all available.

The time of immunodiffusion should be constant as well as the temperature and humidity in which it takes place. A closed chamber which will not allow outside air currents to dry the slides or allow changes in temperature should be used for this purpose.

Antibody should be utilized from the same source, if at all possible. The source should be recorded in a log book giving manufacturer, type of antibody, type of animal used, lot number, and available protein nitrogen units whenever possible. It is advisable to utilize antibody prepared in rabbits or goats rather than horses since horse antiserum precipitates occur over a much narrower region between antigen or antibody excess. A discussion of the precipitin reaction with the variables of the

---

*Meloy Laboratories.

reaction may be found in Crowle and should be read by the technologist charged with the responsibility of performing these tests.

Washing of the patterns after immunodiffusion in physiologic saline (0.85%) must be done with constant agitation or longer periods and changes of saline must be utilized. Adequate washing is necessary; otherwise background staining of non-reacted protein will diminish the appearance of light or very faint precipitin arcs that might otherwise be visible on perfectly cleared patterns. Several changes of saline, even with agitation, are recommended.

Staining of the patterns should be rigidly controlled as to time, type, and preparation of stain. As in the staining of protein electrophoretic patterns, stain should not be reused after a pattern has been stained.

References

IMMUNOELECTROPHORESIS (GENERAL AND METHODOLOGY)

1. Deodhar, S.: Lab. Management 11:20, 1973.
2. Gordon, A. H., Keil, B., and Sebesta, K.: Electrophoresis of proteins in agar jelly, Nature 164:498, 1949.
3. Gordon, A. H., Keil, B., Sebesta, K., Knessl, O., and Sorm, F.: Electrophoresis of proteins in agar jelly, Coll. Czech. Chem. Commun. 15:1, 1950.
4. Grabar, P., and Williams, C. A., Jr.: Methode perméttant l'étude conjuguée des propriétés electrophorétiques et immunochimiques d'un melonge des proteines: applications au serum sanguin, Biochim. Biophys. Acta 10:193, 1953.
5. Ouchterlony, O.: Antigen-antibody reactions in gels, Acta Pathol. Microbiol. Scand. 26:507, 1949.
6. Ouchterlony, O.: Antigen-antibody reaction in gels and the practical application of this phenomenom in the laboratory diagnosis of diphtheria, Stockholm, 1949.
7. Ouchterlony, O.: Antigen-antibody reactions in gels. II. Factors determining the site of the precipitate, Arkh. Kemi 1:43, 1949.
8. Ouchterlony, O.: Antigen-antibody reactions in gels. III. The time factor, Arkh. Kemi 1:55, 1949.
9. Ouchterlony, O.: Antigen-antibody reactions in gels. IV. Types of reactions in coordinated systems of diffusion, Acta Pathol. Microbiol. Scand. 32:231, 1953.
10. Ouchterlony, O.: Diffusion in gel methods for immunological analysis, Progr. Allergy 5:1, 1958.
11. Ouchterlony, O.: Handbook of immunodiffusion and immunoelectrophoresis with appendices by Hirshfield, J., Clousen, J., and Sewell, M. M. H., Ann Arbor, Mich., 1968, Ann Arbor Science Publishers.
12. Tiselieus, A.: Electrophoresis of serum globulins: electrophoretic analysis of normal and immune sera, Biochem. J. 31:1464, 1937.
13. Tiselieus, A.: A new apparatus for electrophoresis: analysis of colloidal mixtures, Trans. Faraday Soc. 33:524, 1937.
14. Tiselieus, A., and Kabat, E. A.: Electrophoresis of immune serum, Science 87:416, 1938.
15. Williams, C. A., Jr., and Grabar, P.: Immunoelectrophoretic studies on serum proteins, J. Immunol. 74:158, 397, 404, 1955.

THE IMMUNE SYSTEM

16. Amos, B., editor: Progress in immunology, 1st International Congress of Immunology, New York, 1971, Academic Press, Inc.
17. Bellanti, J. A.: Immunology, Philadelphia, 1971, W. B. Saunders Co.
18. Cooper, M. D., and Lawton, A. R.: The development of the immune system, Sci. Am. 231:58, 1974.
19. Good, R. A.: Structure-function relations in the lymphoid system. In Bach, F. H., and Good, R. A., editors: Clinical immunobiology, New York, 1972, Academic Press, Inc.
20. Good, R. A., and Fisher, D. S., editors: Immunobiology, Stamford, Conn., 1971, Sinauer Associates, Inc.
21. Gordon, B. L., II: Essentials of immunology, Philadelphia, 1974, F. A. Davis Co.
22. Holborow, E. J.: The ABC of modern immunology, Boston, 1968, Little, Brown & Co.
23. Kabat, E. A.: Structural concepts in immunology and immunochemistry, New York, 1968, Holt, Rinehart and Winston, Inc.
24. Kabat, E. A., and Mayer, M. M.: Experimental immunochemistry, Springfield, Ill., 1961, Charles C Thomas, Publisher.
25. Katz, D. H., and Benacerraf, B.: The regulatory influence of activated T cells on B cell responses to antigen, Adv. Immunol. 15:2, 1972.

26. Maddison, S. E.: Delayed hypersensitivity and cell-mediated immunity. A survey of current insights into these responses of the body to antigens, Clin. Pediatr. 12:529, 1973.
27. Nossal, G. J. V., and Ada, G. L.: Antigens, lymphoid cells and the immune response, New York, 1971, Academic Press, Inc.
28. Lin, Sun Peck, Cooper, A. G., and Wortis, H. H.: Scanning electron microscopy of human T-cell and B-cell rosettes, N. Engl. J. Med. 289:548, 1973.
29. Sela, M.: Antigenicity: some molecular aspects, Science 166:1365, 1969.
30. Sell, S.: Immunopathology and immunity, New York, 1972, Harper & Row, Publishers.
31. Weiss, L.: The cells and tissue of the immune system—structure, function, interactions. In Osler, A. G., and Weiss, L., editors: Foundations of immunobiology, Englewood Cliffs, N. J., 1972, Prentice-Hall, Inc.
32. Williams, C. A., Jr., and Chase, M. W.: Methods in immunology and immunochemistry, vol. I: Preparation of antigens and antibodies, New York, 1967, Academic Press, Inc.

IMMUNOGLOBULINS

33. Appela, E., and Ein, D.: Two types of lambda polypeptide chains in human immunoglobulins based on an amino acid substitution at position 190, Proc. Nat. Acad. Sci. 57:1449, 1967.
34. Bernnich, H., and Johansson, S. G. O.: Immunoglobulin E and immediate hypersensitivity, Vox Sang. 19:1, 1970.
35. Brandtzaeg, P., Fjellanger, I., and Gjenuldsen, S. T.: Human secretory immunoglobulins: salivary secretions from individuals with normal or low levels of serum immunoglobulins, Scand. J. Hematol., Supp. 12, 1970.
36. Dayton, D. H., Small, P. A., Chanock, R. M., Kaufman, H. E., and Tomasi, T. B., Jr., editors: The secretory immunologic system, proceedings of a conference on the Secretory Immunologic System, U.S. Department of Health, Education and Welfare Bethesda, Md., 1969, Public Health Service–National Institutes of Health.
37. Deutsch, H. F., and Morton, J. I.: Human serum macroglobulins and dissociation units: physicochemical properties, J. Biol. Chem. 231:1107, 1958.
38. Edelman, G. M.: The structure and functions of antibodies, Sci. Am. 223:34, 1970.
39. Edelman, G. M.: Antibody structure and molecular immunology, Ann. N. Y. Acad. Sci. 190:5, 1971.
40. Edelman, G. M.: Antibody structure and molecular immunology, Science 180:830, 1973.
41. Ein, D., and Fahey, J. L.: Two types of lambda polypeptide chains in human immunoglobulins, Science 156:947, 1967.
42. Fleishman, J. B., Porter, R. R., and Press, E. M.: The arrangement of the peptide chains in γ-globulin, Biochem. J. 88:220, 1963.
43. Frangione, B., Prelli, F., and Franklin, E. C.: The structure of Fd fragments of G myeloma proteins, Immunochemistry 4:95, 1967.
44. Frank, M. M., and Humphrey, J. H.: The subunits in rabbit anti-Forssman IgM antibody, J. Exp. Med. 127:967, 1968.
45. Freeman, M. J., and Stavitsky, A. B.: Radioimmuno-electrophoretic study of rabbit anti-protein antibodies during the primary response, J. Immunol. 95:981, 1965.
46. Fudenberg, H. H., Wang, A. C., Pink, J. R. L., and Levin, A. S.: Studies of an unusual biclonal gammopathy: implications with regard to genetic control of normal immunoglobulin synthesis, Ann. N. Y. Acad. Sci. 190:501, 1971.
47. Gleich, G. J., Bieger, R. C., and Stankievic, R.: Antigen combining activity associated with immunoglobulin D, Science 165:606, 1969.
48. Grant, J. A., and Hood, L.: N-terminal analysis of normal immunoglobulin light chains —a study of thirteen individual humans, Immunochemistry 8:63, 1971.
49. Grey, H. M., Abel, C. A., and Zimmerman, B.: Structure of IgA proteins, Ann. N. Y. Acad. Sci. 190:37, 1971.
50. Hong, R.: The immunoglobulins. In Bach, F. H., and Good, R. A., editors: Clinical immunobiology, vol. 1, New York, 1972, Academic Press, Inc.
51. Hood, L., and Ein, D.: Immunoglobulin lambda chain structure: two genes, one polypeptide chain, Nature 220:764, 1968.
52. Hood, L., and Talmage, D. W.: Mechanism of antibody diversity: germ line basis for variability, Science 168:325, 1970.
53. Ishizaka, K.: The identification and significance of gamma E, Hosp. Pract. 4:70, 1969.
54. Ishizaka, K., and Ishizaka, T.: Identification of γE-antibodies as a carrier of reaginic activity, J. Immunol. 99:1187, 1967.
55. Ishizaka, K., Ishizaka, T., and Lee, E.: Biologic function of the Fc fragments of E myeloma protein, Immunochemistry 7:687, 1970.
56. Johansson, S. G. O., and Bennich, H.: Immunological studies of an atypical (myeloma) immunoglobulin, Immunology 13:381, 1967.
57. Johansson, S. G. O., Bennich, H., and Wide, L.: A new class of immunoglobulins in human serum, Immunology 14:265, 1968.
58. Kochwa, S., and Kunkel, H. G.: Immuno-

globulins, Ann. N. Y. Acad. Sci. **190:**1971.
59. Kochwa, S., Terry, W. D., Capra, J. D., and Young, N. L.: Structural studies of immunoglobulin E: physiochemical studies of IgE molecule, Ann. N. Y. Acad. Sci **190:**94, 1971.
60. Merler, E.: Immunoglobulins—biologic aspects and clinical uses, Washington, D.C., 1970, National Academy of Sciences.
61. Merler, E., Karlin, L., and Matsumato, S.: The valency of human αM immunoglobulin antibody, J. Biol. Chem. **243:**386, 1968.
62. Mihaesco, C., and Seligmann, N. M.: Papain digestion fragments of human IgM globulins, J. Exp. Med. **127:**431, 1968.
63. Muller-Eberhard, H. J., Kunkel, H. G., and Franklin, E. C.: Two types of α-globulin differing in carbohydrate content, Proc. Exp. Biol. Med. **93:**146, 1956.
64. Natvig, J. B., and Kunkel, H. G.: Genetic markers of human immunoglobulins. The Gm and Inv systems, Ser. Haematol. **1:**66, 1968.
65. Natvig, J. B., and Kunkell, H. G.: Human immunoglobulins, classes, subclasses, genetic variants, and idiotypes, Adv. Immunol. **16:** 1973.
66. Natvig, J. B., Turner, M. W., and Michaelsen, T. E.: Genetic control of the constant homology regions of immunoglobulin G heavy chain subclasses, Ann. N. Y. Acad. Sci. **190:**150, 1971.
67. Nisonoff, A.: Molecules of immunity, Hosp. Pract., June, 1967, p. 19.
68. Nisonoff, A., Wessler, F. C., and Lipman, L. N.: Properties of the major component of a peptic digest of rabbit antibody, Science **132:** 1770, 1960.
69. Porter, R. R.: Structural studies of immunoglobulins, Science **180:**713, 1973.
70. Putnam, F. W.: The plasma proteins, New York, 1960, Academic Press, Inc.
71. Putnam, F. W.: Immunoglobulin structure: variability and homology, Science **163:**633, 1969.
72. Putnam, F. W.: Fractions #1, Beckman Instruments Monograph, 1973.
73. Putnam, F. W., Shimizu, A., Paul, C., Shinoda, T., and Kohler, H.: The amino acid sequence of human macroglobulins, Ann. N. Y. Acad. Sci. **190:**83, 1971.
74. Putnam, F. W., Titani, K., Wikler, M., and Shinoda, T.: Sympos. Quant. Biol. **32:**9, 1967.
75. Ritzman, S. E., Daniels, J. C., and Levin, W. C.: Leukemia-lymphoma, 14th Clinical Conference on Cancer, University of Texas, M. D. Anderson Hospital and Tumor Institute, Houston, 1969, Chicago, 1969, Year Book Medical Publishers.

76. Rogentine, G. N., Rowe, D. S., Bradley, J., Waldman, T. A., and Fahey, J. L.: Metabolism of human immunoglobulin D (IgD), J. Clin. Invest. **45:**1467, 1966.
77. Rowe, D. S., and Fahey, J. L.: A new class of human immunoglobulins, J. Exp. Med. **121:**171, 185, 1965.
78. Schubert, D., and Cohn, M.: Immunoglobulin biosynthesis. V. Light chain assembly, J. Molec. Biol. **53:**305, 1970.
79. Schultze, H. E., and Heremans, J. F.: Molecular biology of human proteins, vol. I, New York, 1966, Elsevier Publishing Co.
80. Schur, P. H.: Human gamma G: subclasses, Prog. Clin. Immunol. **1:**71, 1972.
81. Smith, G. P., Hood, L., and Fitch, W. M.: Antibody diversity, Ann. Rev. Biochem. **40:** 969, 1971.
82. Smith, R. T.: Human immunoglobulins—a guide to nomenclature and clinical application, Pediatrics **37:**822, 1966.
83. Steinberg, A. G.: Globulin polymorphisms in man, Ann. Rev. Genet. **3:**25, 1969.
84. Sullivan, A. L., Grimley, P. M., and Metzger, H.: Electron microscopic localization of immunoglobulin E on the surface membrane of human basophils, J. Exp. Med. **134:**1403, 1971.
85. Warner, N.: The expanding family of human immunoglobulins, Med. J. Aust. **2:**506, 1969.
86. Wu, T. T., and Kabat, E. A.: An analysis of the sequences of the variable regions of Bence Jones proteins and myeloma light chains and their implications for antibody complementarity, J. Exp. Med. **132:**211, 1970.

COMPLEMENT SYSTEM

87. Gewurz, H.: The immunologic role of complement, Hosp. Pract., Sept., 1967, pp. 45-56.
88. Gotze, O., and Muller-Eberhard, H. J.: J. Exp. Med. **135:**68, 1972.
89. Mayer, M. M.: Mechanism of cytolysis by complement, Proc. Nat. Acad. Sci. **69:**2954, 1972.
90. Mayer, M. M.: The complement system, Sci. Am. **229:**54, 1973.
91. Muller-Eberhard, H. J.: Complement, Adv. Immunol. **8:**1-80, 1968.
92. Ruddy, S., Gigli, I., and Austin, K. E.: The complement system of man, N. Engl. J. Med. **287:**489, 545, 592, 642, 1972.
93. Wolstenholme, G. E. W., editor: Ciba foundation symposium on complement, Boston, 1965, Little, Brown and Co.

PROTEINS IN DISEASE STATES

94. Adinolfi, M., and Gardner, B.: Synthesis of beta-1E and beta-1C components of com-

plement in human foetuses, Acta Pediatr. Scand. 56:450, 1967.
95. Alper, C. A., and Johnson, A. M.: Commentary: alpha$_1$-antitrypsin deficiency and disease, Pediatrics 46:837, 1970.
96. Alpert, E., Hershberg, R., Shur, P. H., and Isselbacker, K. J.: α-Fetoprotein in human hepatoma: improved detection in serum, and quantitative studies using a new sensitive technique, Gastroenterology 61:137, 1971.
97. Andrieu, J., Breart, G., Rodier, B., and Robert, P. E.: A propos de l'existence de l'alpha 1-foeto-protéine en dehors de l'hépatome, Presse Med. 79:1595, 1971.
98. Austen, K. F.: Inborn and acquired abnormalities of the complement system of man, Johns Hopkins Med. J. 128:57, 1971.
99. Azen, E. A., and Smithies, O.: Genetic polymorphism of C'3 (β_1C-globulin) in human serum, Science 162:905, 1968.
100. Becker, W., Schwick, H. G., and Storiko, K.: Immunologic determination of proteins found in low concentrations in human serum, Clin. Chem. 15:649, 1969.
101. Bruninga, G. L.: Complement—a review of the chemistry and reaction mechanisms, Am. J. Clin. Pathol. 55:273, 1971.
102. Carrico, R. J., Deutsch, H. F., Beinert, H., and Orme-Johnson, W. H.: Some properties of an apoceruloplasmin-like protein in human serum, J. Biol. Chem. 244:4141, 1969.
103. Carpenter, C. B., Gill, T. J., Merrill, J. P., and Dammin, G.: Alterations in human serum beta-1C-globulin (C'3) in renal transplantation, Am. J. Med. 43:854, 1967.
104. Dubois, R. S., and Hambridge, K. M.: Wilson's disease: identification of an abnormal copper-binding protein, Science 181:1175, 1973.
105. Efremov, G., and Braend, M.: Serum albumin: polymorphism in man, Science 146:1679, 1964.
106. Eriksson, S.: Studies in α_1-antitrypsin deficiency, Acta Med. Scand. 177 (Suppl. 432): 1, 1965.
107. Fagerhol, M. K., and Braend, M.: Serum prealbumin: polymorphism in man, Science 149:986, 1965.
108. Fagerhol, M. K., and Laurell, C. B.: The polymorphism of "prealbumins" and α_1-antitrypsin in human sera, Clin. Chim. Acta 16:199, 1967.
109. Fagerhol, M. K., and Laurell, C. B.: The Pi system-inherited variants of serum alpha 1-antitrypsin, Progr. Med. Genet. 7:96, 1970.
110. Giblett, E. R.: Genetic markers in human blood, Oxford, 1969, Blackwell Scientific Publications, Ltd.
111. Gitlin, D., and Biasucci, A.: Development of γG, γA, γM, β_1c/β_1a, C'1 esterase inhibitor, ceruloplasmin, transferrin, hemopexin, haptoglobin, fibrinogen, plasminogen, $>$ α_1-antitrypsin, orosomucoid, β-lipoprotein, α_2-macroglobulin, and prealbumin in the human conceptus, J. Clin. Invest. 48:1433, 1969.
112. Gitlin, D., and Boesman, M.: Serum α-fetoprotein, albumin and globulin in the human conceptus, J. Clin. Invest. 45:1826, 1966.
113. Gotoff, S. P., Isaacs, E. W., Muehrcke, R. C., and Smith, R. D.: Serum beta$_{1/c}$ globulin in glomerulonephritis and systemic lupus erythematosus, Am. Int. Med. 71:327, 1969.
114. Gottlieb, A., Wisch, N., and Ross, J.: Familial hypo-haptoglobinemia. A genetically determined trait segretating from glucose-6-phosphate dehydrogenase deficiency, Blood 21:129, 1963.
115. Grieco, M. H., Capra, J. D., and Paderon, H.: Reduced serum beta lc/la globulin levels in extrarenal disease, Am. J. Med. 51:340, 1971.
116. Hepper, N. G., Block, L. F., Gleich, G. J., and Kueppers, F.: The prevalence of alpha$_1$-antitrypsin deficiency in selected groups of patients with chronic obstructive lung disease, Mayo Clin. Proc. 44:697, 1969.
117. Hever, O.: Relations entre des phenotypes d'haptoglobine dans diverses maladies, Presse Med. 77:1081, 1969.
118. Hornung, M., and Arquembourg, R. C.: Beta 1c-globulin: "an acute phase" serum reactant of human serum, J. Immunol. 94:307, 1965.
119. Housley, J.: Alpha$_2$-macroglobulin levels in disease, in man, J. Clin. Pathol. 21:27, 1968.
120. Hughes, N. R.: Serum group-specific (Gc) protein concentrations in patients with carcinoma, melanoma, sarcoma, and cancers of hematopoietic tissues as determined by radial immunodiffusion, J. Nat. Cancer Inst. 46:665, 1971.
121. Hughes, N. R.: Serum transferrin and ceruloplasmin concentrations in patients with carcinoma, melanoma, sarcoma and cancers of hematopoietic tissues, Aust. J. Exp. Biol. Med. Sci. 50:97, 1972.
122. James, K.: The papain digestion of human serum α_2-macroglobulin, Biochim. Biophys. Acta 100:509, 1965.
123. James, K., Johnson, G., and Fudenberg, H. H.: The quantitative estimation of α_2-macroglobulin in normal, pathological and cord sera, Clin. Chim. Acta 14:207, 1966.
124. Johnson, A. M., and Alper, C. A.: Deficiency of α_1-antitrypsin in childhood liver disease, Pediatrics 46:921, 1970.
125. Kauder, E., and Mauer, A. M.: The phys-

iology and clinical significance of haptoglobin, J. Pediatr. **59**:286, 1961.
126. Kaufman, H. S., Frick, O. L., and Fink, D.: Serum complement (beta 1C) in young children with atopic dermatitis, J. Allergy **42**:1, 1968.
127. Kluthe, R., Hagemann, U., and Kleine, N.: The turnover of α_2-macroglobulins in the nephrotic syndrome, Vox Sang. **12**:308, 1967.
128. Kohler, P. F., and Muller-Eberhard, H. J.: Immunochemical quantitation of the third, fourth and fifth components of human complement: Concentrations in the serum of healthy adults, J. Immunol. **99**:1211, 1967.
129. Krauss, S., Schrott, M., and Sarcione, E. J.: Haptoglobin metabolism in Hodgkin's disease, Am. J. Med. Sci. **252**:184, 1966.
130. Kueppers, F.: Alpha-1-antitrypsin: physiology, genetics and pathology, Humangenetik **11**:177, 1971.
131. Kueppers, F., Briscoe, W. A., and Bearn, A. G.: Hereditary deficiency of serum α_1-antitrypsin, Science **146**:1678, 1964.
132. Lahrent, B.: Non-specific haemoglobin binding of human serum globulins in immunoelectrophoresis, Nature **202**:1121, 1964.
133. Laurell, C. B.: Electroimmunoassay, Scand. J. Clin. Lab. Invest. **29** (Suppl. 124):21, 1972.
134. Laurell, C. B., and Eriksson, S.: The electrophoretic α_1-globulin pattern of serum in α_1-antitrypsin deficiency, Scand. J. Clin. Lab. Invest. **15**:132, 1963.
135. Laurell, C. B., Lindegren, J., Malmros, I., and Mortensson, H.: Enzymatic and immunochemical estimation of C'1 esterase inhibitors in sera from patients with hereditary angioneurotic edema, Scand. J. Clin. Lab. Invest. **24**:221, 1969.
136. Laurell, C. B., and Nilehn, J. E.: A new type of inherited serum albumin anomaly, J. Clin. Invest. **45**:1935, 1966.
137. Leikola, J., Fudenberg, H. H., Kasukawa, R., and Milerom, F.: A new genetic polymorphism of human serum; α_2 macroglobulin (AL-M), Am. J. Hum. Genet. **24**:134, 1972.
138. Lytle, R. I., Rosenbaum, M. J., Nuller, L. F., and Rosenthal, S.: Detection of a prealbumin in sera of children following extensive burns, J. Lab. Clin. Med. **64**:117, 1964.
139. Marsh, C. L.: Serum haptoglobins in clinical medicine CR-15, Beckman Instruments Publication, 1973.
140. Mawas, C., Buffe, D., Schweisguth, O., and Burtin, P.: I feotoprotein and children's cancer, Rev. Europ. Etudes Clin. Biol. **16**:430, 1971.
141. Miller, F., and Kuschner, M.: Alpha$_1$-antitrypsin deficiency, emphysema, necrotizing angiitis and glomerulonephritis, Am. J. Med. **46**:615, 1969.
142. Miyasato, F., Pollak, V. E., and Barcelo, R.: Serum β_1A (β_1C) globulin levels in systemic lupus erythematosus: their relationship to clinical and renal histologic findings, Arthritis Rheum. **9**:308, 1966.
143. Morell, A. G., and Scheinberg, I. H.: Heterogeneity of human ceruloplasmin, Science **131**:930, 1960.
144. Northway, J. D., McAdams, A. J., Forristal, J., and West, C. D.: A "silent" phase of hypocomplementemic persistent nephritis detectable by reduced serum B1C-globulin levels, J. Pediatr. **74**:28, 1969.
145. Oudin, J., and Stoltz, F.: Les concentrations des antigènes du sérum human dans des états physiologiques ou pathologiques différents. I. Etude de 19 antigènes comparativement chez des noveau-nés et chez leurs mères, Ann. Inst. Past. **121**:42, 1971.
146. Owen, J. A., Smith, H., Padanyi, R., and Martin, J.: Serum haptoglobulin in disease, Clin. Sci. **26**:1, 1964.
147. Pensky, J., and Schwick, H. G.: Human serum inhibitor of C'1 esterase: identity with α_2-neuraminoglycoprotein, Science **163**:698, 1969.
148. Priolisi, A., and Guuffré, L.: Immunoelectrophoretic and autoradiographic pattern of Hb-binding serum proteins, Acta Haematol. **33**:210, 1965.
149. Propp, R. P., and Alper, C. A.: C'3 synthesis in the human fetus and lack of transplacental passage, Science **162**:672, 1968.
150. Rabinovitz, B. A., and Schen, R. J.: A simple immunoelectrophoretic method for demonstrating deficiency of the B1a-c component of complement in serum, J. Pediatr. **70**:617, 1967.
151. Reimann, H. A., Coppola, E. D., and Villegas, G. R.: Serum complement defects in periodic diseases, Ann. Intern. Med. **73**:737, 1970.
152. Robinson, J. C., Blumberg, B. S., Pierce, J. E., Cooper, A. J., and Harnes, C. G.: Studies on inherited variants of blood proteins: familial segregation of transferrin, B_{1-2} B_2, J. Lab. Clin. Med. **62**:762, 1963.
153. Rosen, F. S.: The complement system and increased susceptibility to infection, Sem. Hematol. **8**:221, 1971.
154. Rosen, F. S., Charache, P., Pensky, J., and Donaldson, V.: Hereditary angioneurotic edema: two genetic variants, Science **148**:957, 1965.
155. Ruddy, S., Gigli, I., and Austin, K. E.: The complement system of man, N. Engl. J. Med. **287**:489, 545, 592, 642, 1972.
156. Sandor, G., and Orley, C.: Alpha-2-M-glob-

ulin pathology, Res. Comm. Chem. Pathol. Pharmacol. 3:655, 1972.
157. Sauger, F., Duval, C., Fondimare, A., Matray, F., and Buffe, D.: Systematic analysis α-feto protein in 5,000 sera, Rev. Europ. Etudes Clin. Biol. 16:471, 1971.
158. Saunders, R., Dyce, B., Vannier, W., and Haverbach, B.: The separation of alpha-2 macroglobulin into five components with differing electrophoretic and enzyme-binding properties, J. Clin. Invest. 50:2376, 1971.
159. Schen, R. J., and Rabinovitz, M.: Differential staining of ceruloplasmin as an aid to interpretation in immunoelectrophoresis, Clin. Chim. Acta 13:537, 1966.
160. Schienberg, I. H., and Gitlin, D.: Deficiency of ceruloplasmin in patients with hepatolenticular degeneration (Wilson's disease), Science 116:484, 1952.
161. Schreffler, D. C., Brewer, G. J., Gall, J. C., and Honeyman, M. S.: Electrophoretic variation in human serum ceruloplasmin: A new genetic polymorphism, Biochem. Genet. 1:101, 1967.
162. Scolari, L., Picard, J. J., and Heremans, J. F.: Quantitative determination of human alpha$_2$-macroglobulin by means of immunoelectrophoresis in antibody-containing gel, Clin. Chim. Acta 19:25, 1968.
163. Sharp, H. L.: Alpha-1-antitrypsin deficiency, Hosp. Pract., May, 1971, p. 83.
164. Sharp, H. L., Bridges, R. A., Krivit, W., and Freier, E.: Cirrhosis associated with alpha-1-antitrypsin deficiency: a previously unrecognized inherited disorder, J. Lab. Clin. Med. 73:934, 1969.
165. Shim, B., and Bearn, A.: Immunological and biochemical studies on serum haptoglobin, J. Exp. Med. 120:611, 1964.
166. Shinton, N. K., Richardson, R. W., and Williams, J. D. F.: Diagnostic value of serum haptoglobin, J. Clin. Pathol. 18:114, 1965.
167. Simmons, P., Penny, R., and Goller, I.: Plasma proteins: a review, Med. J. Aust. 2:494, 1969.
168. Sizaret, P. P., McIntire, K. R., and Princler, G. L.: Quantitation of human α-fetoprotein by electroimmunodiffusion, Cancer Res. 31:1899, 1971.
169. Smith, J. B.: Alpha-fetoprotein: Occurrence in certain malignant diseases and review of clinical applications, Med. Clin. N. Am. 54:797, 1970.
170. Smithies, O.: Variations in human serum β-globulins, Nature 180:1482, 1957.
171. Stanbury, J. B., Wyngaarden, J. B., and Fredrickson, D. S.: The metabolic basis of inherited disease, New York, 1960, McGraw-Hill Book Co.
172. Sternlike, I., and Schrenberg, I. H.: Ceruloplasmin in health and disease, Ann. N. Y. Acad. Sci. 94:71, 1961.
173. Störiko, N.: Normal values for 23 different human plasma proteins determined by single radial immunodiffusion, Blut 16:200, 1968.
174. Sutton, E. H.: The haptoglobins, Progr. Med. Genet. 7:163, 1970.
175. Tina, L. U., D'Albora, J. B., Antonovych, T. T., Bellanti, J. A., and Calcagno, P. L.: Acute glomerulonephritis associated with normal serum B$_1$C-globulin, Am. J. Dis. Child. 115:29, 1968.
176. Torisu, M., Sonozaki, H., Inai, S., and Arata, M.: Deficiency of the fourth component of complement in man, J. Immunol. 104:728, 1970.
177. Vallota, E. H., Forristal, J., Davis, N. C., and West, C. D.: The C3 nephritic factor and membranoproliferative nephritis: correlation of serum levels of the nephritic factor with C3 levels, with therapy, and with progression of the disease, J. Pediatr. 80:947, 1972.
178. Veneziale, C. M., McGuckin, W. F., Hermans, P. E., and Mankin, H. T.: Hypohaptoglobinemia and valvular heart disease: association with hemolysis after insertion of valvular prostheses and in cases in which operation had not been performed, Mayo Clin. Proc. 41:657, 1966.
179. West, C. D., Davis, N. C., Forrestal, J., Herbst, J., and Spitzer, R.: Antigenic determinants of human β1C-globulins, J. Immunol. 96:650, 1966.
180. West, C. D., Northway, J. D., and Davis, N. C.: Serum levels of beta$_{1c}$ globulin, a complement component, in the nephritides, lipoid nephrosis, and other conditions, J. Clin. Invest. 43:1507, 1964.
181. Wieme, R. J., and Demeulenaere, L.: Genetically determined electrophoretic variant of the human complement component C3, Nature 214:1042, 1967.
182. Wilding, P., Adham, N. F., Mehl, J. W., and Haverbach, B. J.: Alpha-2-macroglobulin concentrations in human serum, Nature 214:1226, 1967.

GAMMOPATHIES

183. Arquembourg, P. C., Salvaggio, J. E., and Bickers, J. N.: Primer of immunoelectrophoresis, with interpretation of pathologic human serum patterns, Basel, 1970, S. Karger.
184. Axelesson, U., Bachmann, R., and Hällén, J.: Frequency of pathological proteins (M-components) in 6,995 sera from an adult population, Acta Med. Scand. 179:235, 1966.
185. Axelsson, U., and Hällén, J.: Review of fifty-

four subjects with monoclonal gammopathy, Br. J. Haematol. **15**:417, 1968.
186. Axelsson, U., and Hállén, J.: A population study on monoclonal gammopathy, Acta Med. Scand. **191**:111, 1972.
187. Bachmann, R.: The diagnostic significance of the serum concentration of pathological proteins (M-components), Acta Med. Scand. **178**:801, 1965.
188. Buckley, C. E.: Immunologic evaluation of older patients, Postgrad. Med., March, 1972, p. 235.
189. Carrell, R. W., Colls, B. M., and Murray, J. T.: The significance of monoclonal gammopathy in a normal population, Aust. N. Z. J. Med. **1**:398, 1971.
190. Cawley, L. P.: Electrophoresis and immunoelectrophoresis, Boston, 1969, Little, Brown & Co.
191. Civantos, F., Dominquez, C. J., Rywlin, A. M., and Di Bella, J.: Protein immunoelectrophoresis, Lab. Med. **4**:37, 1973.
192. Clauvel, J. P., Danon, F., and Seligmann, M.: Immunoglobulines monoclonales décelées en l'absence de myélome ou de macroglobulinémie de Waldenström, Nouv. Rev. Fr. Hematol. **11**:677, 1971.
193. Cochrane, C. G., and Koffler, D.: Immune complex disease in experimental animals and man, Adv. Immunol. **16**:185, 1973.
194. Cooke, K. B.: Essential paraproteinaemia, Proc. R. Soc. Med. **62**:777, 1969.
195. Dammacco, F., and Waldenström, J.: Bence Jones proteinuria in benign monoclonal gammopathies. Incidence and characteristics, Acta Med. Scand. **184**:403, 1968.
196. Ellman, L. L., Pachas, W., Pinals, R. S., and Block, K. J.: M-components in patients with chronic liver disease, Gastroenterology **57**:138, 1969.
197. Engle, R. L., Jr., and Wallis, L. A.: Immunoglobulinopathies, immune globulins, immune deficiency syndromes, Springfield, Ill., 1969, Charles C Thomas, Publisher.
198. Franklin, E. C., Fragione, B., and Cooper, S.: Heavy chain diseases, Ann. N. Y. Acad. Sci. **190**:457, 1971.
199. Fudenberg, H. H., Wang, A. C., Pink, J. R. L., and Levin, A. S.: Studies of an unusual biclonal gammopathy: implications with regard to genetic control of normal immunoglobulin synthesis, Ann. N. Y. Acad. Sci. **190**:501, 1971.
200. Good, R. A.: Immunoglobulin deficiency syndromes in man. In Merler, E., editor: Immunoglobulins, Washington, D.C., 1970, National Academy of Sciences.
201. Grabar, P., and Burtin, P.: Immunoelectrophoretic analysis, Amsterdam, 1964, Elsevier Publishing Co.
202. Hállén, J.: Frequency of "abnormal" serum globulins (M-components) in the aged, Acta Med. Scand. **173**:737, 1963.
203. Hállén, J.: Discrete gamma globulin (M-) components in serum: clinical study of 150 subjects without myelomatosis, Acta Med. Scand., Suppl. 462, 1966.
204. Heremans, J. F.: Biochemical features and biologic significance of immunoglobulin A. In Merler, E., editor: Immunoglobulins, Washington, D.C., 1970, National Academy of Sciences.
205. Hobbs, J. R.: Immunoglobulins in clinical chemistry, Adv. Clin. Chem. **14**:219, 1971.
206. Isobe, T., and Osserman, E. F.: Pathologic conditions associated with plasma cell dyscrasias: a study of 806 cases, Ann. N. Y. Acad. Sci. **190**:507, 1971.
207. Janeway, C. A., Rosen, F. S., Merler, E., Alper, C. A.: The gamma globulins, Boston, 1967, Little, Brown & Co.
208. Johansson, S. G. O., Bennich, H. H., and Berg, T.: The clinical significance of IgE, Progr. Clin. Immunol. **1**:157, 1972.
209. Kim, I., Harley, J. B., and Weksler, B.: Multiple myeloma without initial paraproteins, Am. J. Med. Sci. **264**:267, 1972.
210. Korngold, L.: Plasma proteins: methods of study and changes in disease, Progr. Clin. Pathol. **1**:340, 1966.
211. Kyle, R. A., and Bayrd, E. D.: Benign monoclonal gammopathy: a potentially malignant condition? Am. J. Med. **40**:426, 1966.
212. Kyle, R. A., Bieger, R. C., and Gleich, G. J.: Diagnosis of syndromes associated with hyperglobulinemia, Med. Clin. N. Am. **54**:917, 1970.
213. Laurell, C. B.: The diagnostic significance of immunoglobulin M-components in clinical work, Scand. J. Clin. Lab. Invest. **22**:83, 1968.
214. Lee, B. J., Pinsky, C., and Miller, D. G.: The management of plasma cell neoplasms, Med. Clin. N. Am. **55**:703, 1971.
215. Levin, W. C., and Ritzmann, S. E.: The immunoglobulins. In Mengel, C. E., Frei, E., and Nachman, R., editors: Textbook of hematology, principles and practice, Chicago, 1972, Year Book Medical Publishers.
216. Lindström, F. D., Williams, R. C., Jr., Swaim, W. R., and Freier, E. F.: Urinary L-chain excretion in myeloma and other disorders—an evaluation of the Bence-Jones test, J. Lab. Clin. Med. **71**:812, 1968.
217. Meischner, P. A., and Muller-Eberhard, H. J.: Textbook of immunopathology, New York, 1968, Grune and Stratton, Inc.
218. Michaux, J. L., and Heremans, J. F.: Thirty cases of monoclonal immunoglobulin dis-

orders other than myeloma or macroglobulinemia, Am. J. Med. **46**:562, 1969.
219. O'Loughlin, J. M.: Immunologic deficiency states, Med. Clin. N. Am. **56**:747, 1972.
220. Osserman, E. F., Takatsuki, K., and Talal, N.: The pathogenesis of "amyloidosis": studies on the role of abnormal gamma globulins and gamma globulin fragments of the Bence-Jones (L-polypeptide) types in the pathogenesis of "primary" and "secondary amyloidosis" and the "amyloidosis" associated with plasma cell myeloma. In Miescher, P. A., editor: Multiple myeloma, New York, 1964, Grune & Stratton, Inc.
221. Palmer, D. F., and Woods, R.: A procedural guide to the performance of the radial immunodiffusion test and the immunoelectrophoresis test for qualitation and quantitation of immunoglobulins, DHEW Publication No. 72:8102, Washington, D.C., 1972.
222. Petite, J., and Cruchaud, A.: Qualitative and quantitative abnormalities of immunoglobulins in relatives of patients with idiopathic paraproteinemia: a study of 12 families, Helv. Med. Acta **35**:248, 1966.
223. Pruzanski, W., and Ogryzlo, M. A.: The changing pattern of diseases associated with M components, Med. Clin. N. Am. **56**:371, 1972.
224. Ritzmann, S. E., Daniels, J. C., and Lawrence, M. C.: Monoclonal gammopathies present status, Tex. Med. **68**:91, 1972.
225. Ritzmann, S. E., and Levin, W. C.: Polyclonal and monoclonal gammopathies. In Dettelbach, H. R., and Ritzmann, S. E., editors: Lab synopsis, 2nd revised edition, New York, 1969, American Hoeschst Corp.
226. Rosen, F. S.: Immunological deficiency disease. Clin. Immunobiol. **1**:271, 1972.
227. Samter, M., editor: Immunological diseases, ed. 2, Boston, Little, Brown & Co.
228. Schneider, Von W.: Zur diagnostik seltener paraproteine, Blut **20**:4, 1970.
229. Seligmann, M., Mihaesco, E., and Frangione, G.: Studies on alpha chain disease, Ann. N. Y. Acad. Sci. **190**:487, 1971.
230. Sleeper, C. A., and Cawley, L. P.: Detection and diagnosis of monoclonal gammopathy, Am. J. Clin. Pathol. **51**:395, 1969.
231. Snapper, I., and Kahn, A.: Myelomatosis, fundamental and clinical features, Baltimore, 1971, University Park Press.
232. Terry, W. D.: Antibody activity of myeloma proteins. In Merler, E., editor: Immunoglobulins, Washington, D.C., 1970, National Academy of Science.
233. Waldenstrom, J. G.: Monoclonal and polyclonal hypergammaglobulinemia: clinical and biological significance, Nashville, 1968, Vanderbilt University Press.
234. Williams, R. C., Jr., Bailey, R. C., and Howe, R. B.: Studies of "benign" serum M-components, Am. J. Med. Sci. **257**:275, 1969.
235. Williams, R. C., Jr., Brunning, R. D., and Wollheim, F. A.: Light-chain disease. An abortive variant of multiple myeloma, Ann. Intern. Med. **65**:471, 1966.
236. Wunderly, C.: Immunoelectrophoresis: methods, interpretation, results, Adv. Clin. Chem. **4**:207, 1961.
237. Zawadzki, Z. A., and Edwards, G. A.: Nonmyelomatous monoclonal immunoglobulinemia, Progr. Clin. Immunol. **1**:105, 1972.

IMMUNOQUANTITATION PROCEDURES (GENERAL)

238. Crowle, A.: Immunodiffusion, ed. 2, New York, 1973, Academic Press, Inc.

RADIAL IMMUNODIFFUSION

239. Berne, B. H.: Differing methodology and equations used in quantitating immunoglobulins by radial immunodiffusion—A comparative evaluation of reported and commercial techniques, Clin. Chem. **20**:61, 1974.
240. Fahey, J. L., and McKelvey, E. M.: Quantitative determination of serum immunoglobulins in antibody-agar plates, J. Immunol. **94**:84, 1965.
241. Grant, G. H.: Present day practice—principles of protein estimation by radial immunodiffusion, J. Clin. Pathol. **24**:89, 1971.
242. Harboe, M.: Quantitation of human serum immunoglobulins, Scand. J. Clin. Lab. Invest. **28**:241, 1971.
243. Hosty, T. A., Hollenbeck, M., and Shane, S.: Intercomparison of results obtained with five commercial plates supplied for quantitation of immunoglobulins, Clin. Chem. **19**:524, 1973.
244. Mancini, G., Carbonara, A. O., and Heremans, J. F.: Immunochemical quantitation of antigens by single radial immunodiffusion, Immunochemistry **2**:235, 1965.
245. Palmer, D. F., and Woods, R.: A procedural guide to the performance of radial immunodiffusion tests and the immunoelectrophoresis test for qualitation and quantitation of immunoglobulins, Department of Health, Education and Welfare Publication (HSM) 72-8102, Atlanta, 1972.
246. Störiko, N.: Normal values for 23 different human plasma proteins determined by single radial immunodiffusion, Blut **16**:200, 1968.

ELECTROIMMUNODIFFUSION

247. Grubb, A.: Quantitation of immunoglobulin G by electrophoresis in agarose gel contain-

ing antibodies, Scand. J. Clin. Lab. Invest. 26:249, 1970.
248. Kroll, J.: Changes in the B_{1c}-B_{1a}-globulin during the coagulation process demonstrated by means of a quantitative immunoelectrophoresis method, Protides Biol. Fluids 17: 529, 1970.
249. Laurell, C. B.: Quantitative estimation of proteins by electrophoresis in agarose gel containing antibodies, Anal. Biochem. 15: 45, 1966.
250. Laurell, C. B.: Electroimmunoassay, Scand. J. Clin. Lab. Invest. 29 (Suppl. 124:21, 1972.
251. Miesch, F., Bieth, J., and Metais, P.: Application de la methode d'electro-immunodiffusion de Laurell dosage de quelques proteins seriques, Ann. Biol. Clin. 26:939, 1968.
252. Procedures manual for rocket electroimmuno assay system, PM 712, Bedford, Mass. 1974, Millipore Biomedica.
253. Salvaggio, J. E., Arquembourg, P. C., and Sylvester, G. A.: A comparison of the sensitivity of electroimmunodiffusion and single radial diffusion in quantitation of immunoglobulins in dilute solutions, J. Allergy 46: 326, 1970.

TWO-DIMENSIONAL SINGLE ELECTROIMMUNODIFFUSION

254. Clarke, H. G. M., and Freeman, T.: A quantitative immuno-electrophoresis method (Laurell electrophoresis), Protides Biol. Fluids 14:503, 1966.
255. Clarke, H. G. M., and Freeman, T.: Quantitative immunoelectrophoresis of human serum proteins, Clin. Sci. 35:403, 1968.
256. Laurell, C. B.: Antigen-antibody crossed electrophoresis, Anal. Biochem. 10:358, 1965.
257. Ressler, N.: Two dimensional electrophoresis of antigens with an antibody containing buffer, Clin. Chim. Acta 5:795, 1960.
258. Weeke, B.: The serum proteins identified by means of the Laurell cross electrophoresis, Scand. J. Clin. Lab. Invest. 25:269, 1970.

AUTOMATED SPECIFIC PROTEIN ANALYSIS BY NEPHELOMETRY

259. Boyden, A., Bolton, E., and Gemeroy, D.: Precipitin testing with special reference to the photoelectric measurement of turbidity, J. Immunol. 57:211, 1947.
260. Boyden, A., and DeFalco, R. J.: Report on the use of the photonreflector in serological comparisons, Physiol. Zool. 16:229, 1943.
261. Buffone, G. J., Savory, J., and Cross, R. E.: Use of a Laser-equipped centrifugal analyzer for kinetic measurement of serum IgG, Clin. Chem. 20:1320, 1974.
262. Eckman, I., Robbins, J. B., van den Hamer, C. J. A., Lentz, J., and Scheinberg, I. H.: Automation of quantitative immunochemical microanalysis of human serum transferrin: a model system, Clin. Chem. 16:558, 1970.
263. Gitlin, D., and Edelhock, H.: A study of the reaction between human serum albumin and its homologous equine antibody through the medium of light scattering, J. Immunol. 66: 67, 1951.
264. Goodman, M., Ramsey, D. S., Simpson, W. L., Remp, D. G., Basinsky, D. H., and Brennan, M. J.: The use of chicken antiserum for the rapid determination of plasma protein components, J. Lab. Clin. Med. 49:151, 1957.
265. Kahan, J., and Sundblad, L.: Technicon symposium, Paris, 1966, Automation in analytical chemistry, 1967, pp. 361-364.
266. Killingsworth, L. M., and Savory, J.: Nephelometric studies of the precipitin reaction: A model system for specific protein measurements, Clin. Chem. 19:403, 1973.
267. Larson, C., Orenslein, P., and Ritchie, R. F.: An automated method for quantitation of proteins in body fluids. In Barton, C. E., and others, editors: Advances in automated analysis, Technicon International Congress, 1970, Miami, 1971, Thurman Association, p. 9.
268. Libby, R. L.: A new and rapid quantitative technic for the determination of the potency of Types I and II antipneumococcal serum, J. Immunol. 34:269, 1938.
269. Lizana, J., and Hellsing, K.: Manual immunonephelometric assay of proteins, with use of polymer enhancement, Clin. Chem. 20: 1181, 1974.
270. Ritchie, R. F.: A simple, direct, and sensitive technique for measurement of specific protein in dilute solution, J. Lab. Clin. Med. 70:512, 1967.
271. Ritchie, R. F., Alper, C. A., Graves, J., Pearson, N., and Larson, C.: Automated quantitation of proteins in serum and other biologic fluids, Am. J. Clin. Pathol. 59:151, 1973.
272. Schultze, H. E., and Schurck, G.: Quantitative immunologische Beslimmung von Plasmaproteinen, Clin. Chim. Acta 4:15, 1959.

COUNTERIMMUNOELECTROPHORESIS

273. Counterimmunoelectrophoresis system, No. PM713, Acton, Mass., 1974, Millipore Biomedica.
274. Crowle, A.: Immunodiffusion, ed. 2, New York, 1973, Academic Press, Inc.
275. Culliford, B. J.: The examination and typing of blood stains in the crime laboratory;

National Institute of Law Enforcement and Criminal Justice, 1971, p. 62.

QUALITY CONTROL IN ELECTROPHORESIS PROCEDURES

276. Crowle, A.: Immunodiffusion, ed. 2, New York, 1973, Academic Press, Inc.
277. Gebott, M. D.: Microzone methods manual, Fullerton, Calif., 1973, Beckman Instruments, Inc.
278. Gebott, M. D., Beckman Instruments, Inc.: Personal communication, March, 1975.
279. Jordan, W. C., and White, W. W.: Controlling the variables of immunoelectrophoresis of serum proteins, Am. J. Med. Tech. **31:** 169, 1965.
280. Nerenberg, S. T.: Electrophoresis: practical laboratory manual, Philadelphia, 1966, F. A. Davis Co., pp. 9-11.
281. Palmer, D. F., and Woods, R.: A procedural guide to the performance of the radial immunodiffusion test and the immunoelectrophoresis test for qualitation and quantitation of immunoglobulins, Department of Health, Education, and Welfare Publication (HSM) 72-8102, Atlanta, 1972.
282. Shore, S. L., Phillips, D. J., and Reimer, C. B.: Umbrella effect: pitfall in analysis of specificity of antisera to immunoglobulins, Immunochemistry **8:**562, 1971.

Appendixes

A Tables
B Fire precautions, chemical hazards, and antidotes of poisons
C Basic information for the student
D Reagents
E Company directory

A

Tables

Table A-1. 1968 table of relative atomic weights ($^{12}C = 12$)*

Element	Symbol	Atomic No.	Atomic Weight	Element	Symbol	Atomic No.	Atomic Weight
Actinium	Ac	89	[227]†	Fermium	Fm	100	[257]†
Aluminum	Al	13	26.9815	Fluorine	F	9	18.9984
Americium	Am	95	[243]†	Francium	Fr	87	[223]†
Antimony	Sb	51	121.75	Gadolinium	Gd	64	157.25
Argon	Ar	18	39.948	Gallium	Ga	31	69.72
Arsenic	As	33	74.9216	Germanium	Ge	32	72.59
Astatine	At	85	[210]†	Gold	Au	79	196.967
Barium	Ba	56	137.34	Hafnium	Hf	72	178.49
Berkelium	Bk	97	[247]†	Helium	He	2	4.0026
Beryllium	Be	4	9.0122	Holmium	Ho	67	164.930
Bismuth	Bi	83	208.980	Hydrogen	H	1	1.00797ª
Boron	B	5	10.811ª	Indium	In	49	114.82
Bromine	Br	35	79.904ᵇ	Iodine	I	53	126.9044
Cadmium	Cd	48	112.40	Iridium	Ir	77	192.2
Calcium	Ca	20	40.08	Iron	Fe	26	55.847ᵇ
Californium	Cf	98	[252]†	Krypton	Kr	36	83.80
Carbon	C	6	12.01115ª	Lanthanum	La	57	138.91
Cerium	Ce	58	140.12	Laurentium	Lr	103	[256]†
Cesium	Cs	55	132.905	Lead	Pb	82	207.19
Chlorine	Cl	17	35.453ᵇ	Lithium	Li	3	6.939
Chromium	Cr	24	51.996	Lutetium	Lu	71	174.97
Cobalt	Co	27	58.9332	Magnesium	Mg	12	24.305
Copper	Cu	29	63.546ᵇ	Manganese	Mn	25	54.9380
Curium	Cm	96	[247]†	Mendelevium	Md	101	[257]†
Dysprosium	Dy	66	162.50	Mercury	Hg	80	200.59
Einsteinium	Es	99	[254]†	Molybdenum	Mo	42	95.94
Erbium	Er	68	167.26	Neodymium	Nd	60	144.24
Europium	Eu	63	151.96	Neon	Ne	10	20.179ᵇ

*By permission of the International Union of Pure and Applied Chemistry and Butterworths Scientific Publications.
†Value in brackets denotes the mass number of the isotope of longest known half-life (or a better known one for Cf, Po, Pm, and Tc).
ªAtomic weight varies because of natural variation in isotopic composition: B, ±0.003; C, ±0.00005; H, ±0.00001; O, ±0.00001; Si, ±0.001; S, ±0.003.
ᵇAtomic weight is believed to have following experimental uncertainty: Br, ±0.001; Cl, ±0.01; Cu, ±0.001; Fe, ±0.003; Ne, ±0.003; Ag, ±0.001. (For other elements, the last digit given for the atomic weight is believed reliable to ±0.5.)

Continued.

Table A-1. 1968 table of relative atomic weights ($^{12}C = 12$)—cont'd

Element	Symbol	Atomic No.	Atomic Weight	Element	Symbol	Atomic No.	Atomic Weight
Neptunium	Np	93	[237]†	Selenium	Se	34	78.96
Nickel	Ni	28	58.71	Silicon	Si	14	28.086ª
Niobium	Nb	41	92.906	Silver	Ag	47	107.868ᵇ
Nitrogen	N	7	14.0067	Sodium	Na	11	22.9898
Nobelium	No	102	[255]†	Strontium	Sr	38	87.62
Osmium	Os	76	190.2	Sulfur	S	16	32.064ª
Oxygen	O	8	15.9994ª	Tantalum	Ta	73	180.948
Palladium	Pd	46	106.4	Technetium	Tc	43	[99]†
Phosphorus	P	15	30.9738	Tellurium	Te	52	127.60
Platinum	Pt	78	195.09	Terbium	Tb	65	158.924
Plutonium	Pu	94	[244]†	Thallium	Tl	81	204.37
Polonium	Po	84	[210]†	Thorium	Th	90	232.038
Potassium	K	19	39.102	Thulium	Tm	69	168.934
Praseodymium	Pr	59	140.907	Tin	Sn	50	118.69
Promethium	Pm	61	[147]†	Titanium	Ti	22	47.90
Protactinium	Pa	91	[231]†	Tungsten	W	74	183.85
Radium	Ra	88	[226]†	Uranium	U	92	238.03
Radon	Rn	86	[222]†	Vanadium	V	23	50.942
Rhenium	Re	75	186.2	Xenon	Xe	54	131.30
Rhodium	Rh	45	102.905	Ytterbium	Yb	70	173.04
Rubidium	Rb	37	85.47	Yttrium	Y	39	88.905
Ruthenium	Ru	44	101.07	Zinc	Zn	30	65.37
Samarium	Sm	62	150.35	Zirconium	Zr	40	91.22
Scandium	Sc	21	44.956				

Table A-2. Temperature conversion—centigrade to Fahrenheit

The values in the body of the table give, in degrees Fahrenheit, the temperatures indicated in degrees centigrade at the top and side. $1°C = 1.8°F$; $-273.16°C = -459.72°F =$ Absolute zero.

Temp. °C	0	1	2	3	4	5	6	7	8	9
0	+32.0	30.2	28.4	26.6	24.8	23.0	21.2	19.4	17.6	15.8
−10	+14.0	12.2	10.4	8.6	6.8	5.0	3.2	+1.4	−0.4	−2.2
−20	−4.0	5.8	7.6	9.4	11.2	13.0	14.8	16.6	18.4	20.2
−30	−22.0	23.8	25.6	27.4	29.2	31.0	32.8	34.6	36.4	38.2
−40	−40.0	41.8	43.6	45.4	47.2	49.0	50.8	52.6	54.4	56.2
−50	−58.0	59.8	61.6	63.4	65.2	67.0	68.8	70.6	72.4	74.2
−60	−76.0	77.8	79.6	81.4	83.2	85.0	86.8	88.6	90.4	92.2
−70	−94.0	95.8	97.6	99.4	101.2	103.0	104.8	106.6	108.4	110.2
−80	−112.0	113.8	115.6	117.4	119.2	121.0	122.8	124.6	126.4	128.2

For temperatures below 0° C

Table A-2. Temperature conversion—centigrade to Fahrenheit—cont'd

The values in the body of the table give, in degrees Fahrenheit, the temperatures indicated in degrees centigrade at the top and side. 1° C = 1.8° F; −273.16° C = −459.72° F = Absolute zero.

For temperatures below 0° C

Temp. °C	0	1	2	3	4	5	6	7	8	9
−90	−130.0	131.8	133.6	135.4	137.2	139.0	140.8	142.6	144.4	146.2
−100	−148.0	149.8	151.6	153.4	155.2	157.0	158.8	160.6	162.4	164.2
−110	−166.0	167.8	169.6	171.4	173.2	175.0	176.8	178.6	180.4	182.2
−120	−184.0	185.8	187.6	189.4	191.2	193.0	194.8	196.6	198.4	200.2
−130	−202.0	203.8	205.6	207.4	209.2	211.0	212.8	214.6	216.4	218.2
−140	−220.0	221.8	223.6	225.4	227.2	229.0	230.8	232.6	234.4	236.2
−150	−238.0	239.8	241.6	243.4	245.2	247.0	248.8	250.6	252.4	254.2
−160	−256.0	257.8	259.6	261.4	263.2	265.0	266.8	268.6	270.4	272.2
−170	−274.0	275.8	277.6	279.4	281.2	283.0	284.8	286.6	288.4	290.2
−180	−292.0	293.8	295.6	297.4	299.2	301.0	302.8	304.6	306.4	308.2
−190	−310.0	311.8	313.6	315.4	317.2	319.0	320.8	322.6	324.4	326.2
−200	−328.0	329.8	331.6	333.4	335.2	337.0	338.8	340.6	342.4	344.2
−210	−346.0	347.8	349.6	351.4	353.2	355.0	356.8	358.6	360.4	362.2
−220	−364.0	365.8	367.6	369.4	371.2	373.0	374.8	376.6	378.4	380.2
−230	−382.0	383.8	385.6	387.4	389.2	391.0	392.8	394.6	396.4	398.2
−240	−400.0	401.8	403.6	405.4	407.2	409.0	410.8	412.6	414.4	416.2
−250	−418.0	419.8	421.6	423.4	425.2	427.0	428.8	430.6	432.4	434.2
−260	−436.0	437.8	439.6	441.4	443.2	445.0	446.8	448.6	450.4	452.2
−270	−454.0	455.8	457.6	459.4	..	..	..	..	..	..

For temperatures above 0° C

Temp. °C	0	1	2	3	4	5	6	7	8	9
0	32.0	33.8	35.6	37.4	39.2	41.0	42.8	44.6	46.4	48.2
10	50.0	51.8	53.6	55.4	57.2	59.0	60.8	62.6	64.4	66.2
20	68.0	69.8	71.6	73.4	75.2	77.0	78.8	80.6	82.4	84.2
30	86.0	87.8	89.6	91.4	93.2	95.0	96.8	98.6	100.4	102.2
40	104.0	105.8	107.6	109.4	111.2	113.0	114.8	116.6	118.4	120.2
50	122.0	123.8	125.6	127.4	129.2	131.0	132.8	134.6	136.4	138.2
60	140.0	141.8	143.6	145.4	147.2	149.0	150.8	152.6	154.4	156.2
70	158.0	159.8	161.6	163.4	165.2	167.0	168.8	170.6	172.4	174.2
80	176.0	177.8	179.6	181.4	183.2	185.0	186.8	188.6	190.4	192.2
90	194.0	195.8	197.6	199.4	201.2	203.0	204.8	206.6	208.4	210.2
100	212.0	213.8	215.6	217.4	219.2	221.0	222.8	224.6	226.4	228.2
110	230.0	231.8	233.6	235.4	237.2	239.0	240.8	242.6	244.4	246.2
120	248.0	249.8	251.6	253.4	255.2	257.0	258.8	260.6	262.4	264.2
130	266.0	267.8	269.6	271.4	273.2	275.0	276.8	278.6	280.4	282.2
140	284.0	285.8	287.6	289.4	291.2	293.0	294.8	296.6	298.4	300.2
150	302.0	303.8	305.6	307.4	309.2	311.0	312.8	314.6	316.4	318.2
160	320.0	321.8	323.6	325.4	327.2	329.0	330.8	332.6	334.4	336.2
170	338.0	339.8	341.6	343.4	345.2	347.0	348.8	350.6	352.4	354.2
180	356.0	357.8	359.6	361.4	363.2	365.0	366.8	368.6	370.4	372.2
190	374.0	375.8	377.6	379.4	381.2	383.0	384.8	386.6	388.4	390.2

Table A-3. Metric units

Quantity	Unit	Symbol	Relationship of units
Length	millimeter	mm	1 mm = 0.001 m
	centimeter	cm	1 cm = 10 mm
	decimeter	dm	1 dm = 10 cm
	meter	m	1 m = 100 cm
	kilometer	km	1 km = 1000 m
Area	square centimeter	cm^2	1 cm^2 = 100 mm^2
	square decimeter	dm^2	1 dm^2 = 100 cm^2
	square meter	m^2	1 m^2 = 100 dm^2
	are	a	1 a = 100 m^2
	hectare	ha	1 ha = 100 a
	square kilometer	km^2	1 km^2 = 100 ha
Volume	cubic centimeter	cm^3	1 cm^3 ⎱ = 0.001 L
	milliliter	ml	1 ml ⎰
	cubic decimeter	dm^3	1 dm^3 ⎱ = 1000 ml
	liter	L	1 L ⎰
	cubic meter	m^3	1 m^3 = 1000 L
Mass	milligram	mg	1 mg = 0.001 g
	gram	g	1 g = 1000 mg
	kilogram	kg	1 kg = 1000 g
	metric ton	t	1 t = 1000 kg

Table A-4. Metric decimal prefixes

Multiplication factors	Prefix	Symbol
1 000 000 000 000 = 10^{12}	tera	T
1 000 000 000 = 10^9	giga	G
1 000 000 = 10^6	mega	M
1 000 = 10^3	kilo	k
100 = 10^2	hecto	h
10 = 10^1	deka	da
1		
0.1 = 10^{-1}	deci	d
0.01 = 10^{-2}	centi	c
0.001 = 10^{-3}	milli	m
0.000 001 = 10^{-6}	micro	μ
0.000 000 001 = 10^{-9}	nano	n
0.000 000 000 001 = 10^{-12}	pico	p
0.000 000 000 000 001 = 10^{-15}	femto	f
0.000 000 000 000 000 001 = 10^{-18}	atto	a

Table A-5. Conversion factors

U. S. and metric units

Each unit in bold face type is followed by its equivalent in other units of the same quantity.

Acre—0.0015625 square mile (statute); 4.3560×10^4 square feet; 0.40468564 hectare

Bushel—(U. S.)—1.244456 cubic feet; 2150.42 cubic inches; 0.035239 cubic meter; 35.23808 liters

Centimeter—0.0328084 foot; 0.393701 inch

Circular mil—7.853982×10^{-7} square inches; 5.067075×10^{-6} square centimeters

Cubic centimeter—0.061024 cubic inch; 0.270512 dram (U. S. fluid); 16.230664 minims (U. S.); 0.999972 milliliter

Cubic foot—0.803564 bushel (U. S.); 7.480520 gallons (U. S. liquid); 0.028317 cubic meter; 28.31605 liters

Cubic inch—16.387064 cubic centimeters

Cubic meter—35.314667 cubic feet; 264.17205 gallons (U. S. liquid)

Foot—0.3048 meter

Gallon (U. S. liquid)—0.1336816 cubic foot; 0.832675 gallon (British); 231 cubic inches; 0.0037854 cubic meter; 3.785306 liters

Grain—0.06479891 gram

Gram—0.00220462 pound (avoirdupois); 0.035274 ounce (avoirdupois); 15.432358 grains

Hectare—2.471054 acres; 1.07639×10^5 square feet

Inch—2.54 centimeters

Kilogram—2.204623 pounds (avoirdupois)

Kilometer—0.621371 mile (statute)

Liter—0.264179 gallon (U. S. liquid); 0.0353157 cubic foot; 1.056718 quarts (U. S. liquid)

Meter—1.093613 yards; 3.280840 feet; 39.37008 inches

Mile (statute)—1.609344 kilometers

Ounce (U. S. fluid)—1.804688 cubic inches; 29.573730 cubic centimeters

Ounce (avoirdupois)—28.349523 grams

Ounce (apothecary or troy)—31.103486 grams

Pint (U. S. liquid)—0.473163 liter; 473.17647 cubic centimeters

Pound (avoirdupois)—0.453592 kilogram; 453.59237 grams

Pound (apothecary or troy)—0.3732417 kilogram, 373.24172 grams

Quart (U. S. dry)—1.10119 liters

Quart (liquid)—0.946326 liter

Radian—57.295779 degrees

Rod—5.0292 meters

Square centimeter—0.155000 square inch

Square foot—0.09290304 square meter

Square inch—645.16 square millimeters

Square meter—10.763910 square feet

Square yard—0.836127 square meter

Ton (short)—907.18474 kilograms

Yard—0.9144 meter

Table A-6. Square roots of numbers from 1 to 100 in steps of 0.2

n	$\sqrt{n}$	n	$\sqrt{n}$	n	$\sqrt{n}$	n	$\sqrt{n}$	n	$\sqrt{n}$	n	$\sqrt{n}$	n	$\sqrt{n}$	n	$\sqrt{n}$
1.0	1.00	6.0	2.45	11.0	3.32	16.0	4.00	21.0	4.58	26.0	5.10	31.0	5.57	36.0	6.00	41.0	6.40	46.0	6.78
1.2	1.09	6.2	2.49	11.2	3.35	16.2	4.02	21.2	4.60	26.2	5.12	31.2	5.59	36.2	6.02	41.2	6.42	46.2	6.80
1.4	1.18	6.4	2.53	11.4	3.38	16.4	4.05	21.4	4.63	26.4	5.14	31.4	5.60	36.4	6.03	41.4	6.43	46.4	6.81
1.6	1.26	6.6	2.57	11.6	3.41	16.6	4.07	21.6	4.65	26.6	5.16	31.6	5.62	36.6	6.05	41.6	6.45	46.6	6.83
1.8	1.34	6.8	2.61	11.8	3.44	16.8	4.10	21.8	4.67	26.8	5.18	31.8	5.64	36.8	6.07	41.8	6.47	46.8	6.84
2.0	1.41	7.0	2.65	12.0	3.46	17.0	4.12	22.0	4.69	27.0	5.20	32.0	5.66	37.0	6.08	42.0	6.48	47.0	6.86
2.2	1.48	7.2	2.68	12.2	3.49	17.2	4.15	22.2	4.71	27.2	5.22	32.2	5.67	37.2	6.10	42.2	6.50	47.2	6.87
2.4	1.55	7.4	2.72	12.4	3.52	17.4	4.17	22.4	4.73	27.4	5.23	32.4	5.69	37.4	6.12	42.4	6.51	47.4	6.88
2.6	1.61	7.6	2.76	12.6	3.55	17.6	4.20	22.6	4.75	27.6	5.25	32.6	5.71	37.6	6.13	42.6	6.53	47.6	6.90
2.8	1.67	7.8	2.79	12.8	3.58	17.8	4.22	22.8	4.77	27.8	5.27	32.8	5.73	37.8	6.15	42.8	6.54	47.8	6.91
3.0	1.73	8.0	2.83	13.0	3.61	18.0	4.24	23.0	4.80	28.0	5.29	33.0	5.74	38.0	6.16	43.0	6.56	48.0	6.93
3.2	1.79	8.2	2.86	13.2	3.63	18.2	4.27	23.2	4.82	28.2	5.31	33.2	5.76	38.2	6.18	43.2	6.57	48.2	6.94
3.4	1.84	8.4	2.90	13.4	3.66	18.4	4.29	23.4	4.84	28.4	5.33	33.4	5.78	38.4	6.20	43.4	6.59	48.4	6.96
3.6	1.90	8.6	2.93	13.6	3.69	18.6	4.31	23.6	4.86	28.6	5.35	33.6	5.80	38.6	6.21	43.6	6.60	48.6	6.97
3.8	1.95	8.8	2.97	13.8	3.71	18.8	4.34	23.8	4.88	28.8	5.37	33.8	5.81	38.8	6.23	43.8	6.62	48.8	6.99
4.0	2.0	9.0	3.0	14.0	3.74	19.0	4.36	24.0	4.90	29.0	5.39	34.0	5.83	39.0	6.24	44.0	6.63	49.0	7.00
4.2	2.05	9.2	3.03	14.2	3.77	19.2	4.38	24.2	4.92	29.2	5.40	34.2	5.85	39.2	6.26	44.2	6.65	49.2	7.01
4.4	2.10	9.4	3.07	14.4	3.79	19.4	4.40	24.4	4.94	29.4	5.42	34.4	5.87	39.4	6.28	44.4	6.66	49.4	7.03
4.6	2.14	9.6	3.10	14.6	3.82	19.6	4.43	24.6	4.96	29.6	5.44	34.6	5.88	39.6	6.29	44.6	6.68	49.6	7.04
4.8	2.19	9.8	3.13	14.8	3.85	19.8	4.45	24.8	4.98	29.8	5.46	34.8	5.90	39.8	6.31	44.8	6.69	49.8	7.06
5.0	2.24	10.0	3.16	15.0	3.87	20.0	4.47	25.0	5.00	30.0	5.48	35.0	5.92	40.0	6.32	45.0	6.71	50.0	7.07
5.2	2.28	10.2	3.19	15.2	3.90	20.2	4.49	25.2	5.02	30.2	5.50	35.2	5.93	40.2	6.34	45.2	6.72	50.2	7.09
5.4	2.32	10.4	3.22	15.4	3.92	20.4	4.52	25.4	5.04	30.4	5.51	35.4	5.95	40.4	6.36	45.4	6.74	50.4	7.10
5.6	2.37	10.6	3.26	15.6	3.95	20.6	4.54	25.6	5.06	30.6	5.53	35.6	5.97	40.6	6.37	45.6	6.75	50.6	7.11
5.8	2.41	10.8	3.29	15.8	3.97	20.8	4.56	25.8	5.08	30.8	5.55	35.8	5.98	40.8	6.39	45.8	6.77	50.8	7.13

n	√n	n	√n	n	√n	n	√n	n	√n	n	√n	n	√n
51.0	7.14	56.0	7.48	61.0	7.81	66.0	8.12	71.0	8.43	76.0	8.72	81.0	9.00
51.2	7.16	56.2	7.50	61.2	7.82	66.2	8.14	71.2	8.44	76.2	8.73	81.2	9.01
51.4	7.17	56.4	7.51	61.4	7.84	66.4	8.15	71.4	8.45	76.4	8.74	81.4	9.02
51.6	7.18	56.6	7.52	61.6	7.85	66.6	8.16	71.6	8.46	76.6	8.75	81.6	9.03
51.8	7.20	56.8	7.54	61.8	7.86	66.8	8.17	71.8	8.47	76.8	8.76	81.8	9.04
52.0	7.21	57.0	7.55	62.0	7.87	67.0	8.19	72.0	8.49	77.0	8.77	82.0	9.06
52.2	7.22	57.2	7.56	62.2	7.89	67.2	8.20	72.2	8.50	77.2	8.79	82.2	9.07
52.4	7.24	57.4	7.58	62.4	7.90	67.4	8.21	72.4	8.51	77.4	8.80	82.4	9.08
52.6	7.25	57.6	7.59	62.6	7.91	67.6	8.22	72.6	8.52	77.6	8.81	82.6	9.09
52.8	7.27	57.8	7.60	62.8	7.92	67.8	8.23	72.8	8.53	77.8	8.82	82.8	9.10
53.0	7.28	58.0	7.62	63.0	7.94	68.0	8.25	73.0	8.54	78.0	8.83	83.0	9.11
53.2	7.29	58.2	7.63	63.2	7.95	68.2	8.26	73.2	8.56	78.2	8.84	83.2	9.12
53.4	7.31	58.4	7.64	63.4	7.96	68.4	8.27	73.4	8.57	78.4	8.85	83.4	9.13
53.6	7.32	58.6	7.66	63.6	7.97	68.6	8.28	73.6	8.58	78.6	8.87	83.6	9.14
53.8	7.33	58.8	7.67	63.8	7.99	68.8	8.29	73.8	8.59	78.8	8.88	83.8	9.15
54.0	7.35	59.0	7.68	64.0	8.00	69.0	8.31	74.0	8.60	79.0	8.89	84.0	9.17
54.2	7.36	59.2	7.69	64.2	8.01	69.2	8.32	74.2	8.61	79.2	8.90	84.2	9.18
54.4	7.38	59.4	7.71	64.4	8.02	69.4	8.33	74.4	8.63	79.4	8.91	84.4	9.19
54.6	7.39	59.6	7.72	64.6	8.04	69.6	8.34	74.6	8.64	79.6	8.92	84.6	9.20
54.8	7.40	59.8	7.73	64.8	8.05	69.8	8.35	74.8	8.65	79.8	8.93	84.8	9.21
55.0	7.42	60.0	7.75	65.0	8.06	70.0	8.37	75.0	8.66	80.0	8.94	85.0	9.22
55.2	7.43	60.2	7.76	65.2	8.07	70.2	8.38	75.2	8.67	80.2	8.96	85.2	9.23
55.4	7.44	60.4	7.77	65.4	8.09	70.4	8.39	75.4	8.68	80.4	8.97	85.4	9.24
55.6	7.46	60.6	7.78	65.6	8.10	70.6	8.40	75.6	8.69	80.6	8.98	85.6	9.25
55.8	7.47	60.8	7.80	65.8	8.11	70.8	8.41	75.8	8.71	80.8	8.99	85.8	9.26

n	√n	n	√n	n	√n
86.0	9.27	91.0	9.54	96.0	9.80
86.2	9.28	91.2	9.55	96.2	9.81
86.4	9.30	91.4	9.56	96.4	9.82
86.6	9.31	91.6	9.57	96.6	9.83
86.8	9.32	91.8	9.58	96.8	9.84
87.0	9.33	92.0	9.59	97.0	9.85
87.2	9.34	92.2	9.60	97.2	9.86
87.4	9.35	92.4	9.61	97.4	9.87
87.6	9.36	92.6	9.62	97.6	9.88
87.8	9.37	92.8	9.63	97.8	9.89
88.0	9.38	93.0	9.64	98.0	9.90
88.2	9.39	93.2	9.65	98.2	9.91
88.4	9.40	93.4	9.66	98.4	9.92
88.6	9.41	93.6	9.67	98.6	9.93
88.8	9.42	93.8	9.68	98.8	9.94
89.0	9.43	94.0	9.70	99.0	9.95
89.2	9.44	94.2	9.71	99.2	9.96
89.4	9.46	94.4	9.72	99.4	9.97
89.6	9.47	94.6	9.73	99.6	9.98
89.8	9.48	94.8	9.74	99.8	9.99
90.0	9.49	95.0	9.75	100.0	10.00
90.2	9.50	95.2	9.76		
90.4	9.51	95.4	9.77		
90.6	9.52	95.6	9.78		
90.8	9.53	95.8	9.79		

Table A-7. Interpretation of chemical resistance*

The chart should be taken as a general guide only. It states that, for most compounds of the class indicated, the specific material is rated either Excellent, Good, Fair, or Not recommended.

Because so many factors can affect the chemical resistance of a given product, we recommend you test under your own conditions if any doubt exists. The combination of compounds of two or more classes may cause an undesirable chemical effect. Other factors affecting chemical resistance include temperature, pressure and other stresses, length of exposure, and concentration of the chemical. As the maximum useful temperature of the plastic is approached, resistance to attack decreases.

Chemical resistance chart

Class of substances	CPE	LPE	PA	PP	TPX†	FEP	PC	NO	PS	SA	PVC
Acids, inorganic	E	E	E	E	E	E	E	G	N	E	G
Acids, organic	E	E	E	E	E	E	G	E	G	E	G
Alcohols	E	E	E	E	E	E	G	E	G	G	G
Aldehydes	G	G	G	G	G	E	F	G	N	G	G
Amines	G	G	G	G	G	E	N	F	G	F	F
Bases	E	E	E	E	E	E	N	E	G	G	E
Dimethyl sulfoxide (DMSO)	E	E	E	E	E	E	N	E	N	E	E
Esters	E	E	E	E	E	E	N	F	N	N	N
Ethers	G	G	G	G	G	E	F	N	F	N	F
Foods	E	E	E	E	E	E	E	G	E	N	F
Glycols	E	E	E	E	E	E	G	E	G	G	G
Hydrocarbons, aliphatic	G	G	G	G	G	E	F	G	N	G	F
Hydrocarbons, aromatic	G	G	G	G	F	E	N	N	N	E	F
Hydrocarbons, halogenated	G	G	G	G	F	E	N	N	N	N	N
Ketones	G	G	G	G	G	E	N	N	N	N	N
Mineral oil	E	E	E	E	E	E	E	E	G	G	E
Oils, essential	G	G	G	G	G	E	G	F	N	F	N
Oils, lubricating	G	E	E	E	E	E	G	E	G	G	E

	CPE	LPE	PA	PP	TPX†	FEP	PC	NO	PS	SA	PVC
Oils, vegetable	E	E	E	E	E	E	E	G	E	E	E
Proteins, unhydrolyzed	E	E	E	E	E	E	E	G	G	E	G
Salts	E	E	E	E	E	E	E	E	E	E	E
Silicones	G	E	E	E	E	E	E	G	G	G	G
Water	E	E	E	E	E	E	E	E	E	E	E

E = Excellent. Long exposures (up to one year) at room temperature have no effect.
G = Good. Short exposures (less than 24 hours) at room temperature cause no damage.
F = Fair. Short exposures at room temperature cause little or no damage under unstressed conditions.
N = Not recommended. Short exposures may cause permanent damage.

Resins code
CPE—Conventional polyethylene
LPE—Linear polyethylene
PA—Polyallomer
PP—Polypropylene
TPX—Polymethylpentene
FEP—Teflon FEP
PC—Polycarbonate
NO—Noryl
PS—General purpose polystyrene
SA—Styrene-acrylonitrile
PVC—Polyvinyl chloride

Physical properties

	CPE	LPE	PA	PP	TPX†	FEP	PC	NO	PS	SA	PVC
Temperature limit, °C	80	120	130	135	175	205	135	135	70	95	70
Specific gravity	0.92*	0.95	0.90	0.90	0.83	2.15	1.20	1.06	1.07	1.07	1.34
Tensile strength, psi	2000	4000	2900	5000	4000	3000	8000	9600	6000	11,000	6500
Brittleness temperature, °C	−100	−100	−40	0		−270	−135		‡	−25	−30
Water absorption, %	<0.01	<0.01	<0.02	<0.02	<0.01	<0.01	0.35	0.07	0.05	0.23	0.06
Flexibility	Excellent	Rigid	Slight	Rigid	Rigid	Excellent	Rigid	Rigid	Rigid	Rigid	Rigid
Transparency	Translucent	Opaque	Translucent	Translucent	Clear	Translucent	Clear	Opaque	Clear	Clear	Clear
Relative O₂ permeability	0.40	0.08	0.20	0.11	2.0	0.59	0.15	0.11	0.11	0.03	0.01
Autoclavable	No	With caution	Yes	Yes	Yes	Yes	Yes	Yes	No	No	No

*Courtesy Nalge Co.
†Trademark of Imperial Chemical Industries, Ltd., for their brand of methylpentene polymer.
‡Normally somewhat brittle at room temperatures.

Table A-8. Values of t for given degrees of freedom (n) and at specified levels of significance (P)*

	Level of significance (P)															
n	.90	.80	.70	.60	.50	.40	.30	.25	.20	.10	.05	.025	.02	.01	.005	.001	n
1	.158	.325	.510	.727	1.000	1.376	1.963	2.414	3.078	6.314	12.706	25.452	31.821	63.657	127.32	636.619	1
2	.142	.289	.445	.617	.816	1.061	1.386	1.604	1.886	2.920	4.303	6.205	6.965	9.925	14.089	31.598	2
3	.137	.277	.424	.584	.765	.978	1.250	1.423	1.638	2.353	3.182	4.176	4.541	5.841	7.453	12.941	3
4	.134	.271	.414	.569	.741	.941	1.190	1.344	1.533	2.132	2.776	3.495	3.747	4.604	5.598	8.610	4
5	.132	.267	.408	.559	.727	.920	1.156	1.301	1.476	2.015	2.571	3.163	3.365	4.032	4.773	6.859	5
6	.131	.265	.404	.553	.718	.906	1.134	1.273	1.440	1.943	2.447	2.969	3.143	3.707	4.317	5.959	6
7	.130	.263	.402	.549	.711	.896	1.119	1.254	1.415	1.895	2.365	2.841	2.998	3.499	4.029	5.405	7
8	.130	.262	.399	.546	.706	.889	1.108	1.240	1.397	1.860	2.306	2.752	2.896	3.355	3.832	5.041	8
9	.129	.261	.398	.543	.703	.883	1.100	1.230	1.383	1.833	2.262	2.685	2.821	3.250	3.690	4.781	9
10	.129	.260	.397	.542	.700	.879	1.093	1.221	1.372	1.812	2.228	2.634	2.764	3.169	3.581	4.587	10
11	.129	.260	.396	.540	.697	.876	1.088	1.214	1.363	1.796	2.201	2.593	2.718	3.106	3.497	4.437	11
12	.128	.259	.395	.539	.695	.873	1.083	1.209	1.356	1.786	2.179	2.560	2.681	3.055	3.428	4.318	12
13	.128	.259	.394	.538	.694	.870	1.079	1.204	1.350	1.771	2.160	2.533	2.650	3.012	3.372	4.221	13
14	.128	.258	.393	.537	.692	.868	1.076	1.200	1.345	1.761	2.145	2.510	2.624	2.977	3.326	4.140	14
15	.128	.258	.393	.536	.691	.866	1.074	1.197	1.341	1.753	2.131	2.490	2.602	2.947	3.286	4.073	15
16	.128	.258	.392	.535	.690	.865	1.071	1.194	1.337	1.746	2.120	2.473	2.583	2.921	3.252	4.015	16
17	.128	.257	.392	.534	.689	.863	1.069	1.191	1.333	1.740	2.110	2.458	2.567	2.898	3.222	3.965	17
18	.127	.257	.392	.534	.688	.862	1.067	1.189	1.330	1.734	2.101	2.445	2.552	2.878	3.197	3.922	18
19	.127	.257	.391	.533	.688	.861	1.066	1.187	1.328	1.729	2.093	2.433	2.539	2.861	3.174	3.883	19
20	.127	.257	.391	.533	.687	.860	1.064	1.185	1.325	1.725	2.086	2.423	2.528	2.845	3.153	3.850	20
21	.127	.257	.391	.532	.686	.859	1.063	1.183	1.323	1.721	2.080	2.414	2.518	2.831	3.135	3.819	21
22	.127	.256	.390	.532	.686	.858	1.061	1.182	1.321	1.717	2.074	2.406	2.508	2.819	3.119	3.792	22
23	.127	.256	.390	.532	.685	.858	1.060	1.180	1.319	1.714	2.069	2.398	2.500	2.807	3.104	3.767	23
24	.127	.256	.390	.531	.685	.857	1.059	1.179	1.318	1.711	2.064	2.391	2.492	2.797	3.090	3.745	24
25	.127	.256	.390	.531	.684	.856	1.058	1.178	1.316	1.708	2.060	2.385	2.485	2.787	3.078	3.725	25
26	.127	.256	.390	.531	.684	.856	1.058	1.177	1.315	1.706	2.056	2.379	2.479	2.779	3.067	3.707	26
27	.127	.256	.389	.531	.684	.855	1.057	1.176	1.314	1.703	2.052	2.373	2.473	2.771	3.056	3.690	27
28	.127	.256	.389	.530	.683	.855	1.056	1.175	1.313	1.701	2.048	2.368	2.467	2.763	3.047	3.674	28
29	.127	.256	.389	.530	.683	.854	1.055	1.174	1.311	1.699	2.045	2.364	2.462	2.756	3.038	3.659	29
30	.127	.256	.389	.530	.683	.854	1.055	1.173	1.310	1.697	2.042	2.360	2.457	2.750	3.030	3.646	30
40	.126	.255	.388	.529	.681	.851	1.050	1.167	1.303	1.684	2.021	2.329	2.423	2.704	2.971	3.551	40
60	.126	.254	.387	.527	.679	.848	1.046	1.162	1.296	1.671	2.000	2.299	2.390	2.660	2.915	3.460	60
120	.126	.254	.386	.526	.677	.845	1.041	1.156	1.289	1.658	1.980	2.270	2.358	2.617	2.860	3.373	120
∞	.126	.253	.385	.524	.674	.842	1.036	1.150	1.282	1.645	1.960	2.241	2.326	2.576	2.807	3.291	∞

*From Fisher, R. A., and Yates, F.: Statistical tables for biological, agricultural, and medical research, ed. 6, Edinburgh, 1963, Oliver & Boyd; and from Merrington, M.: Table of percentage points of the *t*-distribution, Biometrika 32:300, 1942.

Table A-9. Greek alphabet

Greek letter	Greek name	Greek letter	Greek name
A α	Alpha	N ν	Nu
B β	Beta	Ξ ξ	Xi
Γ γ	Gamma	O o	Omicron
Δ δ	Delta	Π π	Pi
E ϵ	Epsilon	P ρ	Rho
Z ζ	Zeta	Σ σ s	Sigma
H η	Eta	T τ	Tau
Θ θ	Theta	Υ υ	Upsilon
I ι	Iota	Φ ϕ	Phi
K κ	Kappa	X χ	Chi
Λ λ	Lambda	Ψ ψ	Psi
M μ	Mu	Ω ω	Omega

Table A-10. Sörensen's $\frac{1}{15}$ M phosphate buffer for pH 5.8 to 8.2 ($\frac{1}{15}$ M primary phosphate, 9.08 g KH_2PO_4/L; $\frac{1}{15}$ M secondary phosphate, 9.47 g Na_2HPO_4 [anhydrous]/L)

pH 20° C*	1/15 M Na_2HPO_4 (ml)	1/15 M KH_2PO_4 (ml)
5.8	8.0	92.0
5.9	9.0	90.1
6.0	12.2	87.8
6.1	15.3	84.7
6.2	18.6	81.4
6.3	22.4	77.6
6.4	26.7	73.3
6.5	31.8	68.2
6.6	37.5	62.5
6.7	43.5	56.5
6.8	49.6	50.4
6.85	52.5	47.5
6.90	55.4	44.6
6.95	58.2	41.8
7.00	61.1	38.9
7.05	63.9	36.1
7.10	66.6	33.4
7.15	69.2	30.8
7.20	72.0	28.0
7.25	74.4	25.6
7.30	76.8	23.2
7.35	78.9	21.1
7.40	80.8	19.2
7.45	82.5	17.5
7.50	84.1	15.9
7.55	85.7	14.3
7.60	87.0	13.0
7.65	88.2	11.8
7.70	89.4	10.6
7.75	90.5	9.5
7.80	91.5	8.5
7.90	93.2	6.8
8.0	94.7	5.3
8.1	95.8	4.2
8.2	97.0	3.0

*Subtract 0.03 from pH at 38.°

From Sörensen, S. P. L.: Über die Musung und Bedeutung der Wasserstaffanenkanzentration bei biologischen Prozessen, Ergeb. Physiol 12:393, 1912.

Table A-11. Chromium 51 (half-life 27.8000 days)

Days	Precalibration decay factor	Postcalibration decay factor	Days	Precalibration decay factor	Postcalibration decay factor
1	1.025	.9753	19	1.606	.6226
2	1.051	.9513	20	1.646	.6073
3	1.077	.9279	21		.5923
4	1.104	.9050	22		.5777
5	1.132	.8827	23		.5635
6	1.161	.8610	24		.5496
7	1.190	.8398	25		.5361
8	1.220	.8191	26		.5229
9	1.251	.7989	27		.5100
10	1.283	.7793	28		.4975
11	1.315	.7601	29		.4852
12	1.348	.7414	30		.4733
13	1.382	.7231	31		.4616
14	1.417	.7053	32		.4502
15	1.453	.6879	33		.4391
16	1.490	.6710	34		.4283
17	1.527	.6545	35		.4178
18	1.566	.6383			

Radiations	Beta	Gamma	Other	$\overline{E\beta}$	Γ
	none	0.32(~8%)	EC	0.005	0.18

Table A-12. Cobalt 57 (half-life: 270.0000 days)

Days	Precalibration decay factor	Postcalibration decay factor	Days	Precalibration decay factor	Postcalibration decay factor
1	1.002	.9974	141		.6962
11	1.028	.9721	151		.6786
21	1.055	.9475	161		.6614
31		.9235	171		.6446
41		.9000	181		.6283
51		.8772	191		.6124
61		.8550	201		.5968
71		.8333	211		.5817
81		.8122	221		.5670
91		.7916	231		.5526
101		.7716	241		.5386
111		.7520	251		.5249
121		.7329	261		.5116
131		.7144	271		.4987

Radiations	Beta	Gamma	Other	$E\beta$	Γ
	none	0.14 (0.6%)	EC	0.007	—
		.121 (92%)			
		.01 (9%)			

Table A-13. Cobalt 60 (half-life: 1925.0000 days [5.27 years])

Days	Precalibration decay factor	Postcalibration decay factor	Days	Precalibration decay factor	Postcalibration decay factor
1	1.0004	.9996	541	1.2151	.8230
31	1.0112	.9889	571	1.2283	.8142
61	1.0222	.9783	601	1.2416	.8054
91	1.0333	.9678	631	1.2551	.7968
121	1.0445	.9574	661	1.2687	.7882
151	1.0559	.9471	691	1.2825	.7797
181	1.0673	.9369	721	1.2964	.7713
211	1.0789	.9268	751	1.3105	.7631
241	1.0907	.9169	781	1.3247	.7549
271	1.1025	.9070	811	1.3391	.7468
301	1.1145	.8973	841	1.3537	.7387
331	1.1266	.8876	871	1.3684	.7308
361	1.1388	.8781	901	1.3832	.7229
391	1.1512	.8687	931	1.3983	.7152
421	1.1637	.8593	961	1.4134	.7075
451	1.1763	.8501	991	1.4288	.6999
481	1.1891	.8410	1021	1.4443	.6924
511	1.2020	.8319			

Radiations	Beta	Gamma	$\overline{E}\beta$	Γ
	0.306	1.17 1.33	0.093	12.9

Table A-14. Iodine 125 (half-life: 60 days)

Days	Precalibration decay factor	Postcalibration decay factor	Days	Precalibration decay factor	Postcalibration decay factor
1	1.0116	.9885	37		.6522
3	1.0353	.9659	39		.6373
5	1.0595	.9439	41		.6227
7	1.0842	.9223	43		.6085
9	1.1096	.9013	45		.5946
11	1.1355	.8807	47		.5810
13	1.1620	.8606	49		.5678
15	1.1892	.8409	51		.5548
17		.8217	53		.5421
19		.8029	55		.5297
21		.7846	57		.5176
23		.7667	59		.5058
25		.7492	61		.4943
27		.7320	63		.4830
29		.7153	65		.4719
31		.6990	67		.4612
33		.6830	69		.4506
35		.6674			

Radiations	Beta	Gamma	X-ray	$\overline{E}\beta$	Γ
	none	0.0354	0.0274	0.027	0.6

Table A-15. Iodine 131 (half-life: 8.0800 days)

Days	Precalibration decay factor	Postcalibration decay factor	Days	Precalibration decay factor	Postcalibration decay factor
1	1.089	.9177	11	2.569	.3892
2	1.187	.8423	12	2.799	.3572
3	1.293	.7730	13	3.050	.3278
4	1.409	.7095	14	3.324	.3008
5	1.535	.6512	15	3.621	.2761
6	1.673	.5976	16	3.946	.2534
7	1.823	.5485	17	4.299	.2326
8	1.986	.5034	18	4.686	.2134
9	2.164	.4620	19	5.104	.1959
10	2.358	.4240	20	5.561	.1798

Relations	Beta	Gamma	$\overline{E}\beta$	Γ
	0.250 (2.8%)	0.080 (2.2%)	0.188	2.2
	0.335 (9.3%)	0.163 (0.7%)		
	0.608 (87.2%)	0.284 (5.3%)		
	0.815 (0.7%)	0.364 (80%)		
		0.637 (9%)		
		0.722 (3%)		

Table A-16. Iron 59 (half-life: 44.3000 days)

Days	Precalibration decay factor	Postcalibration decay factor	Days	Precalibration decay factor	Postcalibration decay factor
1	1.015	.9844	31	1.624	.6156
3	1.048	.9541	33	1.675	.5967
5	1.081	.9247	35	1.729	.5783
7	1.115	.8962	37	1.784	.5604
9	1.151	.8686	39	1.840	.5432
11	1.187	.8418	41	1.899	.5264
13	1.225	.8159	43	1.960	.5102
15	1.264	.7908	45	2.022	.4945
17	1.304	.7664	47	2.086	.4793
19	1.346	.7428	49	2.152	.4645
21	1.389	.7199	51	2.221	.4502
23	1.433	.6977	53	2.292	.4363
25	1.478	.6762	55	2.364	.4229
27	1.525	.6554	57	2.440	.4098
29	1.574	.6352	59	2.517	.3972

Radiations	Beta	Gamma	$\overline{E}\beta$	Γ
	0.271 (46%)	0.191	0.118	6.8
	0.462 (54%)	1.098		
	1.560 (0.3%)	1.289		

Table A-17. Molybdenum 99 (half-life: 67.0000 hours)

Hours	Precalibration decay factor	Postcalibration decay factor	Hours	Precalibration decay factor	Postcalibration decay factor
6	1.063	.940	78	2.242	.446
12	1.132	.883	82	2.336	.428
18	1.204	.830	88	2.487	.402
24	1.282	.780	94	2.645	.378
30	1.364	.733	100	2.816	.355
36	1.451	.689	106	2.994	.334
42	1.543	.648	112	3.184	.314
48	1.642	.609	118	3.389	.295
54	1.748	.572	124	3.610	.277
60	1.858	.538	130	3.831	.261
66	1.980	.505	136	4.081	.245
72	2.105	.475			

Radiations	Beta	Gamma	$\bar{E}\beta$	Γ
	1.23	0.181 (7%) 0.740 (12%) 0.780 (4%)		

Table A-18. Phosphorus 32 (half-life: 14.3000 days)

Days	Precalibration decay factor	Postcalibration decay factor	Days	Precalibration decay factor	Postcalibration decay factor
1	1.049	.9526	11	1.704	.5867
2	1.101	.9076	12	1.789	.5589
3	1.156	.8646	13	1.877	.5325
4	1.214	.8237	14	1.971	.5073
5	1.274	.7847	15	2.069	.4833
6	1.337	.7476	16	2.172	.4604
7	1.404	.7122	17	2.279	.4386
8	1.473	.6785	18	2.392	.4179
9	1.547	.6464	19	2.511	.3981
10	1.623	.6158	20	2.637	.3792

Radiations	Beta	Gamma	$\bar{E}\beta$	Γ
	1.701	none	0.7	—

Table A-19. Selenium 75 (half-life: 120.0000 days)

Days	Precalibration decay factor	Postcalibration decay factor	Days	Precalibration decay factor	Postcalibration decay factor
1	1.005	.9942	66		.6830
6	1.035	.9659	71		.6635
11	1.065	.9384	76		.6446
16		.9117	81		.6263
21		.8857	86		.6085
26		.8605	91		.5911
31		.8360	96		.5743
36		.8122	101		.5579
41		.7891	106		.5421
46		.7666	111		.5266
51		.7448	116		.5116
56		.7236	121		.4971
61		.7030	126		.4829

Radiations	Beta	Gamma	Other	$\bar{E}\beta$	Γ
	none		EC	0.011	1.84

	Mev	Relative intensity
	0.265	100
	0.136	95
	0.280	43
	0.121	28
	0.400	22

Table A-20. Strontium 85 (half-life: 64.0000 days)

Days	Decay factor	Days	Decay factor
1	.9892	37	.6698
3	.9680	39	.6554
5	.9472	41	.6414
7	.9269	43	.6276
9	.9071	45	.6142
11	.8876	47	.6010
13	.8686	49	.5881
15	.8500	51	.5755
17	.8318	53	.5632
19	.8140	55	.5511
21	.7965	57	.5393
23	.7795	59	.5278
25	.7627	61	.5165
27	.7464	63	.5054
29	.7304	65	.4946
31	.7148	67	.4840
33	.6994	69	.4736
35	.6845		

Radiations	Beta	Gamma	$\bar{E}\beta$	Γ
	none	0.51 Mev	0.014	3.2

Table A-21. Sodium 22 (half-life: 949.0000 days [2.6 years])

Days	Precalibration decay factor	Postcalibration decay factor	Days	Precalibration decay factor	Postcalibration decay factor
1	1.000	.9992	541	1.484	.6735
31	1.022	.9776	571	1.517	.6589
61	1.045	.9564	601	1.551	.6447
91	1.068	.9356	631	1.585	.6307
121	1.092	.9154	661	1.620	.6170
151	1.116	.8955	691	1.656	.6036
181	1.141	.8761	721	1.693	.5905
211	1.166	.8571	751	1.731	.5777
241	1.192	.8385	781	1.769	.5652
271	1.218	.8204	811	1.808	.5530
301	1.245	.8026	841	1.848	.5410
331	1.273	.7852	871	1.889	.5293
361	1.301	.7682	901	1.931	.5178
391	1.330	.7515	931	1.973	.5066
421	1.360	.7352	961	2.017	.4956
451	1.390	.7193	991	2.062	.4848
481	1.421	.7037	1021	2.108	.4743
511	1.452	.6885			

Radiations	Beta	Gamma	Other	$\overline{E}\beta$	Γ
	B+ (89%) .54	.511 1.28	EC (11%)	0.193	1.32

Table A-22. Technetium 99m (half-life: 6.0000 hours)

Hours	Precalibration decay factor	Postcalibration decay factor	Hours	Precalibration decay factor	Postcalibration decay factor
1	1.122	.891	13		.223
2	1.259	.794	14		.198
3	1.414	.707	15		.177
4	1.587	.630	16		.157
5	1.782	.561	17		.140
6	2.000	.500	18		.125
7	2.247	.445	19		.111
8	2.518	.397	20		.099
9	2.824	.354	21		.088
10	3.174	.315	22		.079
11	3.558	.281	23		.070
12	4.000	.250			

Radiations	Beta	Gamma	Γ
	none	0.140 (98.4%) 0.142 (1.6%)	0.72

Table A-23. Weight distribution in the average human adult (total body weight: 70 kg)*

Organ	Weight (g)	Organ	Weight (g)
Muscles	30,000	Heart	300
Skeleton		Lymphoid tissue	700
Bones (without marrow)	7000	Brain	1500
Red marrow	1500	Spinal cord	30
Yellow marrow	1500	Bladder (urinary)	150
Blood	5400	Salivary glands	50
Gastrointestinal tract†	2000	Eyes	30
Fat	10,000	Teeth	20
Lungs	1000	Prostate gland	20
Liver	1700	Adrenals or suprarenals	20
Kidneys	300	Thymus	10
Spleen	150	Skin and subcutaneous tissues	6100
Pancreas	70	Other tissues and organs (including ovaries)	2034
Thyroid	20		
Testes	40		

*Adapted from International Commission on Radiological Protection, Report of Committee 2 (1959).
†Excluding contents of gastrointestinal tract.

B

Fire precautions, chemical hazards, and antidotes of poisons

FIRE PRECAUTIONS AND CHEMICAL HAZARDS

acetone. Dilute with a spray of water to avoid spread of burning liquid. Use suitable gas mask.

alcohol. See **acetone.**

ammonia. Use water and dilute acid. Use a suitable gas mask.

benzol or benzene. Use water to cool containers that are endangered; extinguish flame with sand, earth, fire-foam or carbon tetrachloride fire extinguishers. Use a suitable gas mask.

calcium carbide. Do not use water since this generates acetylene, an inflammable and explosive gas; cut off electric current to avoid ignition of gas. Remove containers to a dry place. Use a gas mask.

carbon disulfide. Use water to cool containers that are endangered; extinguish blaze with sand, earth, fire-foam or carbon tetrachloride fire extinguishers. Use a suitable gas mask.

carbon tetrachloride. When a carbon tetrachloride type extinguisher is used on a fire in a confined space, the fire should be attacked from outside the enclosure, if possible, or the area should be vacated immediately after the fire is out. No one should return to the enclosure until the air is cleared of smoke and fumes. These precautions should be observed regardless of the means by which the fire is extinguished, however, since fire in a confined space rapidly produces a toxic atmosphere. Do not put carbon tetrachloride on a sodium fire; violent explosions may be caused.

celluloid. Use large volumes of water and sand. The smoke contains oxides of nitrogen which are injurious. Use a suitable gas mask.

chlorine. Spray with water. The pungent nature of the gas makes the use of a gas mask imperative.

collodion. See **carbon disulfide.**

ether. See **carbon disulfide.**

gasoline. See **carbon disulfide.**

hydrochloric acid. Use large volumes of water; also chalk or soda. Use a gas mask.

hydrocyanic or prussic acid. A suitable gas mask is essential because of the extremely poisonous nature of the vapors. Provide ventilation.

lacquer solvents. See **carbon disulfide.**

magnesium. Do not use water. Use sand or earth to extinguish flames. Remove containers to a dry place.

nitric acid and oxides of nitrogen. Use large volumes of water. Do not use sand or earth. Use a gas mask.

potassium. Do not use water. Remove containers to dry place. Extinguish flames with sand or earth. For storage, potassium is kept immersed in petroleum.

potassium hydroxide. Use large volumes of water or dilute acids.

phosphorus. Use water and wet sand. Use gas mask. For storage, white phosphorus must be kept immersed in water. Red phosphorus is less dangerous.

sodium. See **potassium.**
sodium hydroxide. See **potassium hydroxide.**
sulfur. Extinguish with water or sand. Use a gas mask.
sulfuric acid. See **hydrochloric acid.**
turpentine. See **acetone.**

ANTIDOTES OF POISONS

acetic acid. Emetics, magnesia, chalk, soap, oil.
acetylene. Same as for carbon monoxide.
arsenic, rat poison, paris green. Milk, raw egg, sweet oil, lime water, flour and water.
carbolic acid. Any soluble nontoxic sulfate, after provoking vomiting with zinc sulfate; uncooked white of egg in abundance, milk of lime, olive or castor oil with magnesia in suspension, ice, washing the stomach with equal parts water and vinegar; give alcohol or whiskey or about 4 fluidounces of camphorated oil at one dose.
carbon monoxide. Remove to fresh air immediately and call for pulmotor; apply artificial respiration for at least 1 hour or until the pulmotor arrives. Administration of oxygen containing 5% of carbon dioxide is beneficial; inhalation of ammonia or amyl nitrite is often of value.
chloroform, chloral, ether. Dash cold water on head and chest, artificial respiration.
ethylene. Same as for carbon monoxide.
gas (illuminating). Same as for carbon monoxide.
hydrochloric acid. Magnesia, alkali carbonates, albumin, ice.
hydrocyanic or prussic acid. Hydrogen peroxide internally, and artificial respiration, breathing ammonia or chlorine from chlorinated lime, ferrous sulfate followed by potassium carbonate, emetics, warmth.
iodine. Emetics, stomach siphon, starchy foods in abundance, sodium thiosulfate.
lead acetate. Emetics, stomach siphon, sodium, potassium, or magnesium sulfates, milk, albumin.
mercuric chloride or corrosive sublimate. Zinc sulfate, emetics, stomach siphon, raw white of egg, milk, chalk, castor oil, table salt, reduced iron.
nitrate of silver. Salt and water.
nitric acid. Same as for hydrochloric acid.
opium, morphine, laudanum, paregoric, etc. Strong coffee, hot bath. Keep awake and moving at any cost.
phosphoric acid. Same as for hydrochloric acid.
sodium hydroxide or potassium hydroxide. Vinegar, lemon juice, orange juice, oil, milk.
sulfuric acid. Same as for hydrochloric acid with the addition of soap or oil.
sulfurous acid or sulfur dioxide. Mustard plaster on chest, narcotics, expectorants.
wood alcohol (methyl alcohol or methanol). Emetic or wash out stomach (stomach tube) with a solution of 10 gr of sodium citrate per ounce of water. Give milk, raw white of egg, or flour in water; purgative of magnesium sulfate (15 g); stimulate, and combat collapse. In case of cardiac or pulmonary failure use artificial respiration. Physicians may administer atropine, digitalis, or strychnine as stimulants; to cause perspiration and elimination of the poison use 0.1 gr of pilocarpine hydrochloride.

C

Basic information for the student

TYPES OF SOLUTIONS
Saturated solutions

When a solid is dissolved in a liquid, the solid is known as the solute and the liquid is known as the solvent. For example, if sugar is added to coffee, the sugar is the solute and the coffee is the solvent.

A saturated solution is one that contains as much solute as it can hold in solution. To illustrate, if 25 teaspoons of salt are added to a glass of water, some salt would remain undissolved. The supernatant would then be a saturated solution.

Percent solutions

A percent solution contains a measured amount of the solute in a specified volume of solution. The solute may be either a solid or a liquid. For example, sodium chloride is a solute that is a solid, whereas acetic acid is a solute that is a liquid.

In most percent solutions, distilled water is used as the solvent. However, in some instances, alcohol or another liquid may be used.

Molar solutions

A molar (M) solution contains 1 gram-molecular weight per liter of solution.

Molal solutions

Whereas the *molar* solution contains 1 mole (molecular weight in grams) of solute in 1000 ml of solution, the *molal* solution contains 1 mole of solute in 1000 g (1 kg) of solvent. A more technical definition of molal solution states that it contains 1 gram-molecular weight of solute dissolved in 1 kg of solvent.

An easy way to remember the difference is to note that the molar solution is always on a solute per solution basis and the molal solution is always on a solute per solvent basis.

Normal solutions

Normal solutions are used extensively in the laboratory. Therefore, the student should thoroughly master the material of this section. The discussion covers the following: composition of normal solutions, equivalent weights, and standardization.

COMPOSITION OF NORMAL SOLUTIONS

The word "normal" is abbreviated with the capital letter N.

The normality of a solution is expressed with numbers. For example, 0.1 N NaOH is one tenth normal sodium hydroxide.

Unfortunately for the student, some chemists write the numbers as fractions. Thus they may write 0.1 as N/10. Or they may write 0.01 N as N/100. In order to avoid confusion, the student should note the type of labeling used in the particular laboratory.

A replaceable hydrogen atom is one that is capable of ionizing. Hydrochloric acid (HCl) has 1 replaceable hydrogen atom. Sulfuric acid (H_2SO_4) has 2 replaceable hydrogen atoms. Acetic acid ($HC_2H_3O_2$) has 4 hydrogen atoms but only 1 is capable of ionizing. Therefore, acetic acid has only 1 replaceable hydrogen atom. The number of replaceable hydrogen atoms in an acid is an inherent property, which the chemist has found through experimentation and a study of molecular structure.

An equivalent weight of an *acid* is the molecular weight divided by the number of replaceable hydrogen atoms. For example, hydrochloric acid (HCl) has 1 replaceable hydrogen atom. Its equivalent weight is, therefore, the molecular weight divided by 1:

$$\text{Equivalent weight HCl} = \frac{\text{Molecular weight}}{\text{Number of replaceable hydrogen atoms}}$$

$$= \frac{36.461 \text{ g}}{1} = 36.461 \text{ g}$$

Sulfuric acid (H_2SO_4) has 2 replaceable hydrogen atoms, and, consequently, its equivalent weight is the molecular weight divided by 2:

$$\text{Equivalent weight } H_2SO_4 = \frac{\text{Molecular weight}}{\text{Number of replaceable hydrogen atoms}}$$

$$= \frac{98.078 \text{ g}}{2} = 49.039 \text{ g}$$

The equivalent weight of a *base* is the molecular weight divided by the number of replaceable hydroxyl ions. For example, sodium hydroxide (NaOH) has 1 replaceable hydroxyl ion. Therefore, its equivalent weight is the molecular weight divided by 1:

$$\text{Equivalent weight NaOH} = \frac{\text{Molecular weight}}{\text{Number of replaceable hydroxyl ions}}$$

$$= \frac{39.997 \text{ g}}{1} = 39.997 \text{ g}$$

Calcium hydroxide (Ca[OH]$_2$) has 2 replaceable hydroxyl ions. Therefore, its equivalent weight is the molecular weight divided by 2:

$$\text{Equivalent weight Ca(OH)}_2 = \frac{\text{Molecular weight}}{\text{Number of replaceable hydroxyl ions}}$$

$$= \frac{74.0947 \text{ g}}{2} = 37.047 \text{ g}$$

The equivalent weights of *salts*, such as silver nitrate and potassium permanganate, are determined by the number of electrons which the salts give or take during a reaction. For example, silver nitrate ($AgNO_3$) gives 1 electron. Therefore, the

Table C-1. Equivalent weights of compounds

Compound	Formula	Molecular weight	Equivalent weight
Hydrochloric acid	HCl	36.461	36.461
Sulfuric acid	H_2SO_4	98.078	49.039
Oxalic acid	$H_2C_2O_4 \cdot 2H_2O$	126.067	63.033
Sodium hydroxide	NaOH	39.997	39.997
Potassium hydroxide	KOH	56.109	56.109
Silver nitrate	$AgNO_3$	169.874	169.874
Potassium permanganate	$KMnO_4$	158.038	31.607

equivalent weight is the molecular weight divided by 1. Potassium permanganate ($KMnO_4$) takes 5 electrons. Therefore the equivalent weight is the molecular weight divided by 5.

The equivalent weights of the compounds that the student is likely to meet are given in Table C-1.

An equivalent weight is contained in 1 L of a 1 normal (1 N) solution. For example, 49.039 g of sulfuric acid is present in a liter of 1 N H_2SO_4, and 39.997 g of sodium hydroxide is present in a liter of 1 N NaOH.

It follows that 2 equivalent weights will be present in 1 L of a 2 N solution, and 0.1 equivalent weight in 1 L of an 0.1 N solution.

Some prefer to discuss equivalent weights in terms of milliequivalent weights. The prefix "milli" means 1/1000. Consequently, a milliequivalent weight is 1/1000 of an equivalent weight. The word "millequivalent" is often abbreviated mEq (or meq). Milliliter is abbreviated ml. There is 1 mEq in 1 ml of every 1 N solution. This may be expressed by the following equation:

$$\text{Volume (in ml)} \times \text{Normality} = \text{Milliequivalent}$$

or

$$V \times N = mEq$$

EQUIVALENT WEIGHTS

The system of using equivalent weights is a useful tool in chemistry calculations because the differences in the weights (or masses) of various atoms need not be considered. For example, 3 gram atoms of sodium have a combined weight of 69 g, whereas 2 gram atoms of chlorine have a weight of 71 g. Even though their weights are nearly the same, it is known that only 2 atoms of sodium can combine with the 2 atoms of chlorine. This, then, leaves an excess of sodium even though there are fewer grams of sodium than there are of chlorine.

On the other hand, if it were stated that there are 3 equivalents of sodium and only 2 equivalents of chlorine, it would be readily apparent that there is an excess of sodium.

Another example of the usefulness of the equivalent system is in the neutralization of acids and bases. Keep in mind, however, that the same scheme may be used in any situation where chemical reaction occurs.

By definition 1 ml of 1 N solution contains 1 mEq of material. Therefore, 2 ml contains 2 mEq, and 10 ml contains 10 mEq. Generally then, volume in milliliters times normality equals milliequivalents. Stated mathematically, $V \times N = mEq$. If one has, for example, 10 ml of a 1 N acid, it will contain 10 mEq of acid. This amount of acid will exactly neutralize 10 mEq of any alkali. If one wants to calculate the normality of an alkali and it is found that 5 ml of the alkali is required to neutralize the 10 ml of 1 N acid, it is known that there must be 10 mEq of alkali in the 5-ml volume. The normality of the alkali must then be 2, since 5 ml of a 2 N alkali would contain 10 mEq. Examples that follow will illustrate this principle of $V \times N = mEq$.

Since the equivalent weight of acid is the same as that of alkali: Volume of acid $\times$ Normality of acid $=$ Volume of base $\times$ Normality of base.

This same system applies to many other types of calculation. For example, for a mixture of 10 ml of 1 N HCl with 5 ml of 2 N NaOH the following calculations may be made:

$$10 \times 1 = 10 \text{ mEq of HCl}$$
and
$$5 \times 2 = 10 \text{ mEq of NaOH}$$

Therefore, the alkali exactly neutralizes the acid.

In addition, the amount of NaCl that could be produced in such a combination is easily determined. Since 1 mEq of HCl and 1 mEq of NaOH produce 1 mEq of NaCl, the mixture just described would yield 10 mEq of NaCl. Since 1 mEq requires 58.5 mg, 10 mEq would represent 585 mg of NaCl.

At this point it might be brought out that all quantities need not necessarily be discussed in terms of specific volumes. For example, if one had a flask that contained exactly 40 mg of NaOH, the amount of water in which it was dissolved would not alter the absolute quantity. When titrating this amount of NaOH with a standard acid it would not matter whether the NaOH was dissolved in 5 ml or 500 ml of water. Consider the following examples:

1. Assume that 40 mg of NaOH is dissolved in 5 ml of H_2O. Therefore there is 1 mEq in a volume of 5 ml. Since $V \times N = mEq$, you can calculate that the value of N will be 0.2. It would require 1 mEq of acid to neutralize the NaOH. Suppose that 0.1 N HCl is used. Then:

$$V_A \times N_A = V_B \times N_B$$
$$V \times 0.1 = 5 \times 0.2$$
$$V = 10 \text{ ml}$$

Therefore 10 ml of 0.1 N HCl would be required to neutralize 40 mg of NaOH dissolved in 5 ml of H_2O.

2. If 40 mg of NaOH is dissolved in 10 ml of H_2O, $10 \times N = 1$ or $N = 0.1$. Suppose the same HCl solution is used. Then:

$$V_A \times 0.1 = 10 \times 0.1$$
$$V = 10 \text{ ml}$$

Note that in both cases 10 ml of the 0.1 N HCl is required. This should be evident, since 10 ml of 0.1 N HCl constitutes 1 mEq. In other words, the 1 mEq of HCl neutralizes the 1 mEq of NaOH regardless of the amount of H_2O used to dissolve the NaOH.

The following problem demonstrates the usefulness of the equivalent system. Suppose 20 ml of 0.2 N HCl is added to 10 ml of 0.3 N NaOH: (1) Is the resulting solution acid or alkaline? (2) What is the final normality? (3) How much NaCl is produced?

The amounts of the acid and alkali are calculated with the equation $V \times N = mEq$ as follows:

$$20 \times 0.2 = 4 \text{ mEq of HCl}$$
$$10 \times 0.3 = 3 \text{ mEq of NaOH}$$

Since there is more HCl than required to neutralize the NaOH, the resulting solution will be acid. In this situation, 3 mEq of HCl will be utilized to neutralize the 3 mEq of NaOH. Therefore, there will be 1 mEq of HCl in the final solution, which has a volume of 30 ml. There is, therefore, 1 mEq of HCl in a volume of 30 ml, or $30 \times N = 1$. The final N of the HCl is then $\frac{1}{30}$. Further, 3 mEq of NaOH with 3 mEq of HCl will produce 3 mEq of NaCl. Since 1 mEq of NaCl weighs 58.5 mg there would be 3×58.5 or 175.5 mg of NaCl produced.

Many other uses of the mEq system, too numerous to itemize, make this an essential tool for use in quantitative chemistry.

By way of summary—an equivalent weight of an acid is the molecular weight divided by the number of replaceable hydrogen atoms, whereas the equivalent weight of a base is the molecular weight divided by the number of replaceable hydroxyl ions. A 1 normal solution of an acid or base contains 1 equivalent weight per liter of solution.

STANDARDIZATION

Suppose a hydrochloric acid solution is of unknown normality, and the problem is to determine the exact normality of this solution. The following two steps are necessary: titration or standardization, and calculation.

Titration. Titration is a volumetric operation measuring the concentration of a solution or a constituent within a solution by measuring the reaction of this unknown against a standard of known concentration. The proper indicator is used to aid in determining the end point. In this discussion hydrochloric acid is the solution to be standardized. First, 10 ml of the HCl is pipeted into a flask or beaker. A few drops of the indicator phenolphthalein are then added. Next, 0.1 N NaOH is slowly added from a buret until the indicator changes from colorless to a faint pink, thus signifying the completion of neutralization. The number of milliliters required to reach this end point is recorded and used in the calculation given below. Assume, for the sake of illustration, that 20 ml of the 0.1 N NaOH was used. Consequently, from the above titration, the data are as follows:

Volume of acid (HCl) used:	10 ml
Normality of base (NaOH) used:	0.1 N
Volume of base (NaOH) used:	20 ml

Calculation. The normality of the hydrochloric acid is found by writing down the following equation, filling in the blanks with the available data, and then solving the equation.

$$\text{Volume of acid} \times \text{Normality of acid} = \text{Volume of base} \times \text{Normality of base}$$

For example, we first write down the equation as indicated above and then proceed to fill in the blanks in the following manner. Since we are looking for the normality of the acid, a question mark is put under the Normality of acid label. Because the volume of acid used was 10 ml, a 10 is put below Volume of acid. Since the normality of the base was 0.1 N, and 0.1 is put below the Normality of base. Then, as we are assuming the volume of the base was 20 ml, a 20 is placed below the Volume of base. Thus, we have:

Volume of acid × Normality of acid = Volume of base × Normality of base
10 × ? = 20 × 0.1

To solve the equation, we must first eliminate the number 10 on the side of the unknown. Since there is a rule in algebra which states that both sides of an equation can be divided by the same number, we divide both sides by 10:

$$\frac{10 \times \text{Normality of acid}}{10} = \frac{20 \times 0.1}{10}$$

Canceling the 10's, we have:

$$\text{Normality of acid} = 2 \times 0.1$$

Multiplying, we have:

$$\text{Normality of acid} = 0.2$$

Consequently, the exact normality of the HCl is 0.2 N.
Another way to calculate the normality is shown below:

Volume of acid × Normality of acid = Volume of base × Normality of base
10 × x = 20 × 0.1
10 × x = 20 × 0.1
10 x = 2.0
x = 0.2

IONIZATION CONSTANTS

According to the law of mass action, the rate of a chemical reaction is proportional to the molecular concentration of the reactants. As the concentration of any weak electrolyte becomes less, the percent of ionization becomes greater. These changes react proportionally to each other, or in other words, offset each other, and the ionization constant remains the same. These weak electrolyte concentrations are expressed in moles per liter; when applied to an ionization, equilibrium is usually defined as the ionization constant of K_i. The law of mass action does not hold for strong electrolytes. An example would be HCl and H_2SO_4 which have no usable ionization constant.

In solving problems concerning these constants it must be remembered that the constant involved is not always known to a precision greater than two or three significant figures. Therefore, in various books you may see examples with the word "approximate" written to the side.

The concentration value of an aqueous solution needed for the calculation of the ionization constant may be obtained by measuring the concentration of ions in the solution by means of electric current; the more ions there are in a solution, the more current flows through the solution.

The following examples are given to help you understand various reactions that may take place with the weak acids and bases. Acetic acid will be the main example used for calculating various problems. Remember that opposing arrows in the equations indicate reversible reactions—meaning that the reaction will go in both directions, depending on the conditions of the reaction, which will then reach a point of equilibrium where no further change is apparent. The ions are still in constant motion, separating as rapidly as they are bound together, only there is no change in the balance or equilibrium. The brackets [ ] enclosing the constant factor equations are used as a symbol of molecular concentration.

In relation to their ionization, weak acids may be classified as monobasic (monoprotic), dibasic (diprotic), or polybasic (triprotic, etc.). These are classifications according to the number of positive hydrogen ions (protons) that can be replaced per molecule.

Monobasic acid

A monobasic acid is an acid having 1 replaceable hydrogen atom per molecule. Example—acetic acid:

$$HC_2H_3O_2 \rightleftarrows H^+ + C_2H_3O_2^-$$

$$\frac{[H^+] \times [C_2H_3O_2^-]}{[HC_2H_3O_2]} = K_1$$

Dibasic acid

A dibasic acid is an acid having 2 replaceable hydrogen atoms per molecule. The ionization takes place in two steps and each reaction has its own equilibrium constant (K_i), designated K_1 or K_2. Example—carbonic acid:

1. $H_2CO_3 \rightleftarrows H^+ + HCO_3^{-1}$

$$\frac{[H^+] \times [HCO_3^-]}{[H_2CO_3]} = K_1$$

2. $HCO_3^{-1} \rightleftarrows H^+ + CO_3^{-2}$

$$\frac{[H^+] \times [CO_3^{-2}]}{[HCO_3^-]} = K_2$$

Polybasic acid

An acid having more than 2 replaceable hydrogen atoms per molecule, thus requiring more than two steps in order to obtain an ionization equilibrium, is termed a polybasic acid. Each step has its K value. Example—phosphoric acid:

1. $H_3PO_4 \rightleftarrows H^+ + H_2PO_4^{-1}$

$$\frac{[H^+] \times [H_2PO_4^-]}{[H_3PO_4]} = K_1$$

2. $H_2PO_4^{-1} \rightleftarrows H^+ + HPO_4^{-2}$

$$\frac{[H^+] \times [HPO_4^{-2}]}{[H_2PO_4^-]} = K_2$$

3. $HPO_4^{-2} \rightleftarrows H^+ + PO_4^{-3}$

$$\frac{[H^+] \times [PO_4^{-3}]}{[HPO_4^{-2}]} = K_3$$

Examples

A 0.1 molar (M) solution of acetic acid is 1.34% ionized. This has been established by the measurement, mentioned earlier, of electric conductance. Since 1.34% of the hydrogen acetate is ionized, the remaining covalent molecules would be 98.66%. As has been discussed, the ion and molecule concentrations remain constant although there is a continual reaction taking place. Water is not included in the equilibrium constant equation because water concentration is practically unchanged in a dilute aqueous solution. However, because of the importance of water and its ions, the numerical value of that K_i (constant) should be remembered (Table C-2).

Table C-2. Ionization constants at 25° C (other temperatures noted in parentheses)

Name	Ionization reaction	K_i
Acetic acid	$HC_2H_3O_2 \rightleftarrows H^+ + C_2H_3O_2^-$	1.8×10^{-5}
Ammonium hydroxide	$NH_3 + H_2O \rightleftarrows NH_4^+ + OH^-$	1.8×10^{-5}
		$(0°)\ 1.4 \times 10^{-5}$
Arsenic	$H_3AsO_4 \rightleftarrows H^+ + H_2AsO_4^{-1}$	$K_1\ 5 \times 10^{-3}$
	$H_2AsO_4^{-1} \rightleftarrows H^+ + HAsO_4^{-2}$	$K_2\ 8.3 \times 10^{-8}$
	$HAsO_4^{-2} \rightleftarrows H^+ + AsO_4^{-3}$	$K_3\ 6 \times 10^{-10}$
Arsenious acid	$H_3AsO_3 \rightleftarrows H^+ + H_2AsO_3^-$	6×10^{-10}
Benzoic acid	$HC_7H_5O_2 \rightleftarrows H^+ + C_7H_5O_2^-$	6.3×10^{-5}
Boric acid	$H_3BO_3 \rightleftarrows H^+ + H_2BO_3^-$	5.8×10^{-10}
n-Caprylic alcohol	$HC_8H_{15}O_2 \rightleftarrows H^+ + C_8H_{15}O_2^-$	1.27×10^{-5}
Carbonic acid	$H_2CO_3 \rightleftarrows H^+ + HCO_3^{-1}$	$K_1\ 4.5 \times 10^{-7}$
	$HCO_3^{-1} \rightleftarrows H^+ + CO_3^{-2}$	$K_2\ 5.6 \times 10^{-11}$
Creatine (anhydrous)	$C_4H_9O_2N_3 \rightleftarrows H^+ + C_4H_8O_2N_3^-$	$(40°)\ 1.9 \times 10^{-11}$
Creatinine	$C_4H_7ON_3 \rightleftarrows H^+ + C_4H_6ON_3^-$	$(40°)\ 3.7 \times 10^{-11}$
Formic acid	$HCHO_2 \rightleftarrows H^+ + CHO_2^-$	1.8×10^{-4}
Hydrochloric acid	$HCl \rightleftarrows H^+ + Cl^-$	No constant
Lactic acid	$HC_3H_5O_3 \rightleftarrows H^+ + C_3H_5O_3^-$	1.387×10^{-4}
Phosphoric acid	$H_3PO_4 \rightleftarrows H^+ + H_2PO_4^{-1}$	$K_1\ 7.5 \times 10^{-3}$
	$H_2PO_4^{-1} \rightleftarrows H^+ + HPO_4^{-2}$	$K_2\ 6.2 \times 10^{-8}$
Phosphorous acid	$H_3PO_3 \rightleftarrows H^+ + H_2PO_3^{-1}$	$K_1\ 1.6 \times 10^{-2}$
	$H_2PO_3^{-1} \rightleftarrows H^+ + HPO_3^{-2}$	$K_2\ 7 \times 10^{-7}$
Picric acid	$HC_6H_2O_7N_3 \rightleftarrows H^+ + C_6H_2O_7N_3^-$	4.2×10^{-1}
Sulfuric acid	$H_2SO_4 \rightleftarrows H^+ + HSO_4^{-1}$	No constant
	$HSO_4^{-1} \rightleftarrows H^+ + SO_4^{-2}$	$K_2\ 1.2 \times 10^{-2}$
Sulfurous acid	$H_2SO_3 \rightleftarrows H^+ + HSO_3^{-1}$	$K_1\ 1.7 \times 10^{-2}$
	$HSO_3^{-1} \rightleftarrows H^+ + SO_3^{-2}$	$K_2\ 5 \times 10^{-6}$
Trichloroacetic acid	$HC_2O_2Cl_3 \rightleftarrows H^+ + C_2O_2Cl_3^-$	1.3×10^{-1}
Urea	$CH_4ON_2 \rightleftarrows H^+ + CH_3ON_2^-$	1.5×10^{-14}
Uric acid	$C_5H_4O_3N_4 \rightleftarrows H^+ + C_5H_3O_3N_4^-$	1.3×10^{-4}
Water	$H_2O \rightleftarrows H^+ + OH^-$	1.0×10^{-14}

686 Appendix C

1. Calculate the ionization constant (K_i) for 0.1 M acetic acid at 25° C:

$$\begin{array}{ccc} HC_2H_3O_2 & \rightleftarrows \quad H^+ \quad + & C_2H_3O_2^- \\ 98.66\% \text{ of } 0.1 \text{ M} & 1.34 \text{ of } 0.1 \text{ M} & 1.34\% \text{ of } 0.1 \text{ M} \\ \text{or} & \text{or} & \text{or} \\ 0.09866 \text{ M} & 0.00134 \text{ M} & 0.00134 \text{ M} \end{array}$$

$$\frac{[H^+] \times [C_2H_3O_2^-]}{[HC_2H_3O_2]} = K_i$$

$$\frac{0.00134 \times 0.00134}{0.1^* - 0.00134} = \frac{0.00000179}{0.09866}$$

$$\frac{0.00000179}{0.09866} = 0.0000182 \text{ or } 1.8 \times 10^{-5} = K_i$$

2. Calculate the percent ionization in a 0.2 M solution of acetic acid at 25° C:

$$K_i \text{ constant} = 1.8 \times 10^{-5}$$

Let x = hydrogen concentration in a 0.2 M solution

$$\frac{[H^+] \times [C_2H_3O_2^-]}{[HC_2H_3O_2]} = 1.8 \times 10^{-5}$$

$$\frac{x \cdot x}{0.2 - x} = 1.8 \times 10^{-5}$$

$$x^2 = 0.000018 \times 0.2$$
or
$$(1.8 \times 10^{-5}) \times 0.2, \text{ since } 0.2 - x = \text{approximately } 0.2$$
$$x^2 = 0.0000036 \text{ or } 3.6 \times 10^{-6}$$
$$x = 0.0019 \text{ or } 1.9 \times 10^{-3}$$

$$\frac{0.0019}{0.2} \times 100 = 0.95\% \text{ ionization of a 0.2 M solution of acetic acid}$$

3. A normal solution of acetic acid is 0.42% ionized. What is the ionization constant? Referring to problem 1, solve as follows:

$$\frac{0.0042 \times 0.0042}{1 - 0.0042} = 0.0000177 \text{ or } 1.8 \times 10^{-5} = K_i$$

4. Sufficient sodium acetate was dissolved in a 0.1 M solution of acetic acid to bring the molarity to a 1.5 solution. What is the hydrogen ion concentration of this 1.5 M solution?

$$\frac{[H^+] \times 1.5}{0.1} = 1.8 \times 10^{-5} = K_i$$

$$15 [H^+] = 0.000018 \text{ or } 1.8 \times 10^{-5}$$
$$[H^+] = 0.0000012 \text{ or } 1.2 \times 10^{-6} \text{ M}$$

Problems

1. Calculate the ionization constant of formic acid, $HCHO_2$, which ionizes 4.2% in a 0.1 M solution.

   Answer: 1.8×10^{-4} or 0.00018.

2. Refer to Table C-2 for the ionization constant of ammonium hydroxide at

---

*0.1 M acetic acid.

25° C. (a) Determine the degree of ionization. (b) Find the OH⁻ concentration of a 0.08 M solution of NH_4OH.

Answers: (a) 1.5%. (b) 1.2×10^{-3} mole/L.

3. A solution of acetic acid is 1.0% ionized. (a) What is the molarity of the solution? (b) What is the hydrogen concentration of the solution?

Answers: (a) 0.18 M. (b) 1.8×10^{-3} or 0.0018 mole/L.

CHEMICAL INDICATORS

Indicators show color changes according to the concentration of the hydrogen (hydroxyl) ion. These are used to determine the strength of solution and the end points of reactions. A few types of indicators are:

1. Achromatic—a mixture of indicators chosen for their complementary colors so that at the end of the reaction the end point is colorless
2. Absorption—an indicator that shows the end point of a titration by changing the indicator's colloidal condition, so that the precipitate, which has been colored by the indicator, will either lose or gain the color
3. External (outside)—an indicator that is used outside the solution being titrated, with the indicator placed in a small beaker while the solution being titrated is added dropwise into the beaker to obtain the end point
4. Fluorescent—an indicator that shows the end point of a reaction by either an increase or a decrease in fluorescence
5. Internal—an indicator that is added to the solution being analyzed
6. Neutralization—the indicator that undergoes a distinct color change at the equivalence point where the acid and base solutions are neutralized in titrations
7. Oxidation-reduction—an indicator that undergoes a distinct color change at a definite state of oxidation of a solution
8. Turbidity—indicators of a type that shows thet end point by an increase in turbidity or precipitation; usually colloidal solutions
9. Universal—a selected mixture of indicators giving a wide range of pH values

Five common indicators

Litmus paper. Litmus paper is a strip of chemically treated paper. It turns red in an acid solution and blue in a basic solution.

Nitrazine paper. Nitrazine paper is also a strip of chemically treated paper. This indicator turns from green to yellow in an acid solution and blue in a basic solution. It is used to test urine for acid or basic properties.

Töpfer's reagent. Töpfer's reagent is a dye dissolved in alcohol. It is red up to a pH of 2.8. As the pH value increases, it turns light orange and finally yellow at pH 4.0. Töpfer's reagent is used to test gastric juices for hydrochloric acid and to determine the strength of weak acid and basic solutions.

Phenolphthalein. Phenolphthalein is an organic salt dissolved in alcohol. It is colorless up to pH 8.3. As the pH value increases beyond 8.3, it turns pink and finally bright red at pH 10.0. Phenolphthalein is used to test the strength of weak acid and basic solutions.

pHydrion. pHydrion is a chemically treated paper strip that gives specific pH values. The color of the paper is compared to the color chart in the container.

Table C-3. Indicators

Indicator	Color change		pH range	Preparation	
	Acid	Alkaline			
Thymol blue					
Acid	Red	Yellow	1.2-2.8	Thymol blue (thymolsulfonphthalein)	40.0 mg
Alkaline	Yellow	Blue	8.2-9.6	Absolute ethyl alcohol	100.0 ml
Bromphenol blue	Violet	Blue	3.0-3.6	Tetrabromophenolsulfonphthalein	0.1 g
				Ethyl alcohol (20%)	100.0 ml
Töpfer's reagent	Red	Yellow	3.0-4.0	p-Dimethylaminoazobenzene	0.5 g
				Ethyl alcohol (95%)	100.0 ml
Congo red	Blue	Red	3.0-5.0	Congo red	0.5 g
				Distilled water	90.0 ml
				Ethyl alcohol (95%)	10.0 ml
Methyl orange	Orange-red	Yellow	3.1-4.4	Methyl orange	100.0 mg
				Distilled water	100.0 ml
Bromcresol green	Yellow	Blue	3.8-5.4	Tetrabromometacresolsulfonphthalein	40.0 mg
				Absolute ethyl alcohol	100.0 ml
Methyl red	Red	Yellow	4.2-6.3	Methyl red	1.0 g
				Ethyl alcohol (95%)	300.0 ml
				Dilute to 500 ml with distilled water	
Chlorphenol red	Yellow	Red	4.8-6.4	Chlorphenol red	40.0 mg
				Absolute ethyl alcohol	100.0 ml
Alizarin red	Yellow	Red	5.5-6.8	Sodium alizarin monosulfonate	1.0 g
				Distilled water	100.0 ml
Bromthymol blue	Yellow	Blue	6.0-7.6	Dibromthymolsulfonphthalein	40.0 mg
				Absolute ethyl alcohol	100.0 ml
Neutral red	Red	Yellow	6.8-8.0	Neutral red	0.5 g
				Ethyl alcohol (95%)	300.0 ml
				Dilute to 500 ml with distilled water	
Phenol red	Yellow	Red	6.8-8.4	Phenol red (phenolsulfonphthalein)	20.0 mg
				Ethyl alcohol (95%)	100.0 ml
Phenolphthalein	Colorless	Red	8.3-10.0	Phenolphthalein	1.0 g
				Ethyl alcohol (95%)	100.0 ml

Reason for the color changes of indicators

Indicators are usually organic dyes whose color in a solution changes according to the degree of ionization of the indicator. In a basic indicator there will be more negative ions than in a neutral solution. The weak acid indicator has even fewer ions. In the case of the indicator methyl red, since the negative ion differs from the molecule, the acid color of red changes to the base color of yellow according to the ionization taking place in water solution these indicator ions are in balance (equilibrium) with the molecule, symbolized as follows:

$$H\ Ind \rightleftarrows H^+ + Ind^-$$
$$\text{Color A} \quad\quad \text{Color B}$$

If acid with more H^+ is added, the equilibrium will be pushed to the left; color A is dominant. If base is added, uniting with the H^+ ions, the equilibrium will move to the right; color B is dominant.

In such an equilibrium there is always a constant ratio between the product of the ions and the molecule:

$$\frac{[H^+] \times [Ind^-]}{[H\ Ind]} = K\ \text{(constant)}$$

or

$$[H^+] \times [Ind^-] = K\ [H\ Ind]$$
$$\text{Color A} \quad\quad\quad \text{Color B}$$

At the end point of a reaction, the two color forms will be in equal concentrations, and the amount of $[H^+]$ will equal the ionization K of the indicator.

• • •

Since each indicator has its own K value, the indicator that is to be used must be selected to fit the ionization value of the mixture being measured. A list of indicators and their pH range is given in Table C-3.

D

Reagents

PREPARATION OF REAGENTS

In preparing the reagents in this section only the purest chemicals available should be used unless another grade is specifically designated. In some cases a specific supplier of a chemical is recommended. These recommendations are generally those of the person or persons who developed the individual method.

An analytic balance should always be used for weighing chemicals unless the amount to be weighed exceeds the load limit of the balance; then a torsion balance may be used. Material for standards and buffers should be weighed accurately to the fourth decimal place (on an analytic balance).

When measuring liquid materials, use class A volumetric pipets unless otherwise specified in the directions. Do not pipet strong acids or alkalies by mouth. Attach a rubber bulb or vacuum line to the mouthpiece when drawing up these materials into a pipet. Hot and cold liquids should be allowed to come to room temperature before measurement.

Only distilled or deionized water is to be used in preparing any of the aqueous reagents listed. The source of the water supply should be checked periodically to be sure that water of the highest purity is being delivered.

When using a magnetic stirring bar be sure to remove the bar from a volumetric flask before diluting a solution to volume. After dilution to volume, a solution or standard should be mixed thoroughly. One of the most common errors in preparing standards, particularly, is not mixing thoroughly after dilution.

Some reagents require adjusting to a certain pH. This must be done by using a pH meter. The pH adjustment can be simplified somewhat by placing the solution in a beaker on a magnetic stirrer. With the stirring bar spinning and a combination electrode suspended in the solution, acid or alkali may be added dropwise and the change in pH noted almost immediately.

Some reagents are light-sensitive, and it is recommended that these be stored in amber or brown bottles. Polyethylene bottles are recommended for storing most other reagents, mainly for reasons of safety. All bottles should be clearly labeled as to content and the date of preparation. If the reagent has an expiration date, this should be included on the label.

The following reagents are listed in alphabetic order with their indentification numbers. Directions for preparing the reagents are then given in numerical order. For example, the first reagent mentioned in the text is listed as "(D-1) Lithium nitrate stock, 1000 mEq Li/L in 2 N H_2SO_4."

Alphabetic list of reagents

Acetic acid, 1%	(D-218)
Acetic acid, 5%	(D-194)
Acetic acid, 10% (v/v)	(D-210)
Acetic acid, 20% (v/v)	(D-162)
Acetic acid, 30%	(D-276)
Acetic acid, 50%	(D-186)
Acid antifoam diluent	(D-7)
Acid buffer, 0.09 M citric acid, pH 4.8	(D-159)
Acid buffered substrate, pH 4.8 to 4.9	(D-160)
Acid molybdate	(D-46)
Acid reagent, 2 M NaH_2PO_4	(D-143)
Albumin stock standard, 10 g/dl	(D-77)
Albumin standard, stock solution, 10 g/dl	(D-127)
Albumin standard, working solutions, 2, 3, 4, 5, 6, and 8 g/dl	(D-128)
Alcoholic KOH hydroxide wash solution, 2%	(D-53)
Alkali-hypochlorite reagent	(D-93)
Alkaline buffer, pH 10.6 to 10.7	(D-142)
Alkaline buffer, 0.1 M glycine, pH 10.5	(D-155)
Alkaline buffered substrate, pH 10.3 to 10.4	(D-157)
Alkaline iodide	(D-70)
Alkaline potassium ferricyanide	(D-18)
Ammonium chloride stock standard, 1 M	(D-16)
Ammonium hydroxide, 5%	(D-121)
Ammonium hydroxide, 10% (w/v)	(D-209)
Ammonium hydroxide, dilute (0.5%)	(D-112)
Ammonium oxalate, 0.1 M	(D-212)
Ammonium oxalate, 4%	(D-111)
Ammonium oxalate, 10%	(D-119)
Ammonium sulfamate, 0.5%	(D-234)
Ammonium sulfate, saturated	(D-228)
AMP stock buffer, 50% (w/v)	(D-54)
AMP working buffer, 0.625 M, pH 10.25	(D-55)
Ascorbic acid, 4%	(D-68)
Azoene fast red dye diluent	(D-51)
Azoene fast red dye solution	(D-52)
Barium chloride, 5%	(D-213)
Barium chloride, 10%	(D-220)
Benedict's qualitative reagent	(D-195)
Benedict's quantitative reagent	(D-196)
Benzidine, saturated, in glacial acetic acid	(D-219)
Benzoic acid, saturated	(D-25)
Biuret reagent	(D-126)
Biuret stock solution	(D-71)
Biuret working solution	(D-72)
Bromcresol green stock solution, 0.6 mM	(D-130)
Bromcresol green working solution	(D-131)
Bromide stock standard, 20 mg/ml	(D-232)
Bromine water	(D-205)
BSP standard, 5 mg/dl	(D-144)
Buffer, stock, 0.8 M TRIS	(D-151)
Buffer, working, 0.2 M, pH 8.0 at 27° C	(D-152)
BUN acid, working	(D-24)
BUN color reagent, working	(D-21)
Caffeine diluent	(D-63)
Calcium base, 3.75% diethylamine	(D-41)
Calcium chloride, 2.5%	(D-135)
Calcium chloride, 10%	(D-202)
Calcium standard (1 ml = 100 µg Ca)	(D-118)
Calcium stock standard, 50 mg Ca/dl	(D-42)
Cal-red indicator	(D-116)
Carbonate-bicarbonate buffer	(D-12)
Cephalin-cholesterol stock ether solution	(D-133)
Cephalin-cholesterol emulsion	(D-134)
Chloride color reagent	(D-6)
Citrate buffer	(D-50)
CO_2 color reagent	(D-13)
Color reagent (ferric-mercuric reagent)	(D-235)
CPK blank substrate	(D-181)
Creatine substrate	(D-182)
Creatinine stock standard, 1 mg/ml	(D-31)
Cresolphthalein complexone, 0.01%, with 8-hydroxyquinoline	(D-40)
Cyanide-urea working solution	(D-37)
Diacetyl monoxime stock, 2.5%	(D-19)
Diazo blank solution	(D-138)
Diazo reagent, combined	(D-66)
2,4-Dinitrophenylhydrazine, 1mM	(D-168)
Diphenylcarbazone indicator	(D-104)
EDTA solution	(D-115)
Ehrlich's diazo reagent	(D-201)
Ehrlich's reagent	(D-199)
Esbach's reagent	(D-190)
Ethyl alcohol, 95%, containing 0.4% capryl alcohol	(D-222)
Ferric chloride, 5%	(D-214)
Ferric chloride, 10%	(D-197)
Ferric chloride–phosphoric acid stock	(D-22)
Ferric nitrate, 0.5 M	(D-4)
Glucose standard, 100 mg/dl	(D-80)
Glucose stock standard, 10 mg/ml	(D-26)
Glycerin-silicate reagent	(D-87)
Glycine, 7.5%	(D-100)
Gold chloride, 0.5% in water	(D-231)
GOT substrate, α-ketoglutarate, 2 mM/L; dl-asparate, 200 mM/L	(D-166)
GPT substrate, α-ketoglutarate, 2 mM/L; dl-alanine, 200 mM/L	(D-167)
Gram's iodine solution	(D-217)

Appendix D

HABA dye stock solution, 6×10^{-3} M	(D-74)	Phosphate buffer, pH 7.4	(D-179)
HABA dye working solution, 6×10^{-4} M	(D-76)	Phosphate buffer, 0.067 M, pH 7.4	(D-59)
Hydrazine sulfate, 0.2%, in 1 N H_2SO_4	(D-44)	Phosphate buffer, 0.1 M, pH 7.4	(D-165)
Hydrochloric acid, 0.02 N	(D-33)	Phosphorus stock standard, 0.1 mg P/ml	(D-125)
Hydrochloric acid, 0.05 N	(D-69)	Phosphorus stock standard, 1 mg P/ml	(D-47)
Hydrochloric acid, 0.1 N	(D-32)	Phosphotungstic acid	(D-88)
Hydrochloric acid, 0.25 N	(D-39)	Phosphotungstic acid (AutoAnalyzer)	(D-34)
Hydrochloric acid, 1 N	(D-120)	Picric acid, 0.04 M	(D-84)
Hydrochloric acid, 2 N	(D-238)	Picric acid, saturated	(D-30)
Hydrochloric acid, 6 N	(D-233)	Polyanethol sodium sulfonate (Liquoid)	(D-86)
Hydrochloric acid, 5% (v/v)	(D-189)	Potassium chloride stock standard, 0.1 M	(D-15)
Hydrochloric acid, 25%	(D-224)	Potassium hydroxide, 1.25 N	(D-114)
		Potassium hydroxide, 33%	(D-223)
INT dye	(D-61)	Potassium permanganate, 0.1 N	(D-110)
		Potassium permanganate, 5%	(D-240)
$l(+)$ Lactate, 0.02 M in 2 M urea	(D-176)	Potassium phosphate, monobasic, 0.1 M	(D-178)
$l(+)$ Lactate, 2 M	(D-175)	Potassium standard, 4 mEq/L	(D-103)
Lactic acid, 0.1 N	(D-107)	Protein blank solution	(D-73)
Lanthanum stock solution, 5% in 25% (v/v) HCl	(D-145)	Protein standard, 3 mg/ml	(D-137)
Lanthanum working solution, 0.5% in 2.5% (v/v) HCl	(D-146)	Protein standards, working, 25, 50, and 100 mg/dl	(D-193)
LDH buffer reagent	(D-172)	PSP stock standard, 120 mg/500 ml	(D-184)
LDH color reagent	(D-171)	Pyruvate standard, 2 mM/L	(D-170)
LDH control reagent	(D-174)	Pyruvic acid standard, 0.5 μM/ml	(D-180)
LDH substrate	(D-58)		
LDH substrate, 0.1 M $l(+)$ lactate	(D-173)	Reducing reagent (aminonaphthol-sulfonic acid)	(D-124)
Liebermann-Burchard color reagent	(D-62)	Roberts' reagent	(D-188)
Lithium nitrate stock, 1000 mEq Li/L in 2 N H_2SO_4	(D-1)	Salicylate standard solution, 0.2 mg/ml	(D-237)
Lithium nitrate working solution, 125 mEq Li/L in 0.25 N H_2SO_4	(D-2)	Salicylate stock standard, 1 mg/ml	(D-236)
		Saline, 0.9%	(D-17)
Lithium stock standard, 10 mEq Li/L	(D-241)	SGOT substrate	(D-48)
Lithium working standard, 1 mEq Li/L	(D-242)	Sodium acetate, 20%	(D-208)
Lugol's solution	(D-203)	Sodium acetate, half-saturated	(D-98)
		Sodium acetate, saturated	(D-200)
Magnesium chloride, 1 M	(D-56)	Sodium bicarbonate, 1 M	(D-11)
Magnesium stock standard, 100 mEq/L	(D-147)	Sodium carbonate, 1 M	(D-10)
Magnesium stock standard, 100 mg Mg/dl	(D-43)	Sodium carbonate, 1%	(D-215)
Mercaptoethanol, 0.25 M	(D-183)	Sodium carbonate, 14%	(D-90)
Mercuric chloride, 10%	(D-221)	Sodium carbonate, 25%	(D-101)
Mercuric nitrate, 0.2 M in 0.2 N HNO_3	(D-5)	Sodium chloride, saturated	(D-185)
Mercuric thiocyanate, saturated	(D-3)	Sodium chloride stock standard, 1 M	(D-14)
Molybdate reagent	(D-122)	Sodium citrate, 0.05 M	(D-117)
		Sodium cobaltinitrite reagent	(D-97)
NAD-diaphorase	(D-60)	Sodium cyanide stock solution, 10%	(D-35)
		Sodium diethyldithiocarbamate, 1%	(D-148)
p-Nitrophenol, 10 mM/L	(D-164)	Sodium ethylenediamine tetraacetate (EDTA), 1%	(D-149)
p-Nitrophenyl phosphate stock substrate, 0.4%	(D-156)	Sodium hydroxide, 0.02 N	(D-158)
p-Nitrophenyl phosphate substrate	(D-57)	Sodium hydroxide, 0.05 N	(D-153)
		Sodium hydroxide, 0.1 N	(D-161)
Olive oil emulsion	(D-150)	Sodium hydroxide, 0.2 N	(D-136)
		Sodium hydroxide, 0.4 N	(D-169)
Pandy's reagent	(D-227)	Sodium hydroxide, 0.5 N	(D-29)
Phenol color reagent	(D-92)	Sodium hydroxide, 0.75 N	(D-85)
Phenol reagent of Folin-Ciocalteu, stock	(D-102)	Sodium hydroxide, N	(D-225)
Phenolphthalein indicator, 1%	(D-9)	Sodium hydroxide, 5 N (carbonate-free)	(D-108)
Phenyl mercuric acetate, 0.04%	(D-49)	Sodium hydroxide, 10% (2.5 N)	(D-83)
Phosphate buffer, pH 6.05	(D-75)	Sodium hydroxide, 40% (10 N)	(D-230)
Phosphate buffer, pH 7.0	(D-239)		

Sodium hydroxide wash solution	(D-8)	Sulfuric acid stock, 20%	(D-23)
Sodium nitrite, 0.5% (diazo II)	(D-65)	Tartrate acid buffer, pH 4.8	(D-163)
Sodium nitrite stock solution B, 5%	(D-140)	Tartrate buffer	(D-67)
Sodium nitrite working solution B, 0.5%	(D-141)	Thiosemicarbazide stock, 0.5%	(D-20)
Sodium nitroprusside, saturated	(D-198)	Thymol-barbital buffer, pH 7.55	(D-132)
Sodium oxalate, 0.1 N	(D-109)	Thymol blue, 0.2% in 50% ethyl alcohol	(D-226)
Sodium phosphate, dibasic, 0.1 M	(D-177)	Thymolphthalein indicator, 1%	(D-154)
Sodium standard, 140 mEq/L	(D-96)	Trichloroacetic acid, 3%	(D-229)
Sodium tungstate, 10%	(D-82)	Trichloroacetic acid, 10%	(D-95)
Standard diluent, 0.01 N H_2SO_4 containing 40 mg/L PMA	(D-27)	Trichloroacetic acid, 20%	(D-207)
		Tsuchiya's reagent	(D-191)
Standard mercuric nitrate solution	(D-106)	o-Tolidine reagent	(D-206)
Standard sodium chloride, 10 mEq/L	(D-105)	o-Toluidine reagent (stabilized)	(D-79)
Stannous chloride–hydrazine sulfate working solution	(D-45)	Tungstic acid reagent (stabilized)	(D-78)
Starch, 1%	(D-216)	Uranyl zinc acetate reagent	(D-94)
Sterox SE, 0.02%	(D-244)	Urea nitrogen stock standard, 10 mg urea N/ml	(D-28)
Sterox SE, 1%	(D-243)		
Succinate buffer, 0.1 M, pH 4.0	(D-129)	Urea stock, 20%	(D-36)
Sulfanilic acid (diazo I)	(D-64)	Urease solution, buffered	(D-91)
Sulfanilic acid, solution A, 0.1%	(D-139)	Uric acid special reagent	(D-89)
Sulfosalicylic acid, 3%	(D-192)	Uric acid stock standard, 1 mg uric acid/ml	(D-38)
Sulfosalicylic acid, 20%	(D-187)		
Sulfuric acid, 2/3 N	(D-81)	Wash solution, saturated with potassium sodium cobaltinitrite	(D-99)
Sulfuric acid, 1 N	(D-113)		
Sulfuric acid, 2 N	(D-211)		
Sulfuric acid, 10 N	(D-123)	Zinc acetate, saturated alcoholic solution	(D-204)

Reagents—directions

(D-1) Lithium nitrate stock, 1000 mEq Li/L in 2 N H_2SO_4

Weigh 69 g of lithium nitrate, $LiNO_3$, and transfer to a 1-L volumetric flask containing about 700 ml of water. Mix until dissolved. Slowly add 98.08 g (53.5 ml) of concentrated sulfuric acid with mixing. Allow to cool to room temperature. Dilute to volume and mix. Store in a polyethylene bottle.

(D-2) Lithium nitrate working solution, 125 mEq Li/L in 0.25 N H_2SO_4

Dilute 125 ml of lithium nitrate stock solution to 1000 ml with water. Add 0.5 ml of Brij 35 and mix.

(D-3) Mercuric thiocyanate, saturated

Place a 2-L Erlenmeyer flask on a magnetic stirrer. Add 1 L of water and about 2 g of mercuric thiocyanate, $Hg(SCN)_2$. Stir for several hours and then allow to sit overnight. Filter through a good grade of filter paper and store in an amber reagent bottle.

(D-4) Ferric nitrate, 0.5 M

Weigh 202 g of ferric nitrate, $Fe(NO_3)_3 \cdot 9H_2O$, and transfer to a 1-L volumetric flask. Add about 700 ml of water and mix until all the ferric nitrate is dissolved. Slowly add 31.5 g (21 ml) of concentrated nitric acid and mix. Dilute to volume and mix thoroughly. Filter through a good grade of filter paper and store in an amber reagent bottle.

(D-5) Mercuric nitrate, 0.2 M in 0.2 N HNO_3

Place about 700 ml of water in a 1-L volumetric flask. Slowly add 12.6 g (8.4 ml) of concentrated nitric acid. Swirl to mix thoroughly. Weigh 68.5 g of mercuric nitrate, $Hg(NO_3)_2 \cdot H_2O$, and transfer to the flask. Mix until completely dissolved. Dilute to volume and mix again. Filter through a good grade of filter paper and store in a polyethylene bottle.

(D-6) Chloride color reagent

Combine 700 ml of saturated mercuric thiocyanate, 300 ml of ferric nitrate solution, and 3.5 ml (see step 11 under operating sequence, electrolytes) of mercuric nitrate solution. Mix thoroughly, filter, and store in an amber reagent bottle.

(D-7) Acid antifoam diluent

To about 700 ml of water in a 1-L volumetric flask, slowly add 2.8 ml of concentrated sulfuric acid and mix. Add 1.0 ml of Dow Corning Antifoam B, dilute to volume with water, and mix thoroughly. Store in a glass or polyethylene bottle. Shake well each day before use. If bubbling in the CO_2 trap still occurs, add another 1 ml/L of Antifoam B.

(D-8) Sodium hydroxide wash solution

Weigh 40 g of sodium hydroxide and transfer to a 1-L volumetric flask containing about 700 ml of water. Mix until all sodium hydroxide is dissolved. Dilute to volume and mix. Store in a polyethylene bottle.

(D-9) Phenolphthalein indicator, 1%

Weigh 1 g of phenolphthalein powder and transfer to a 100-ml volumetric flask. Dissolve and dilute to volume with methyl alcohol. Store in a 4-ounce amber bottle.

(D-10) Sodium carbonate, 1 M

Very carefully weigh 106 g of anhydrous sodium carbonate and transfer to a 1-L volumetric flask. Add about 600 ml of water and mix until the sodium carbonate is completely dissolved. Dilute to volume and mix thoroughly. Check the molarity of the sodium carbonate by the following procedure: Measure 20 ml of 1 N HCl with a volumetric pipet and transfer to a 125-ml Erlenmeyer flask. Add 3 drops of 1% methyl orange indicator. Titrate the HCl with the sodium carbonate until the indicator changes from red to yellow. Note the volume of sodium carbonate required to reach this end point. It should require 10 ml if the sodium carbonate is 2 N (1 M). Adjust if necessary. This solution may be divided into two portions—one to be used for the carbonate-bicarbonate buffer and the other for the sodium carbonate stock standard. Store in tightly capped 500-ml polyethylene bottles. Do not interchange bottles and do not use reagent from one bottle for both purposes. There is the danger of contaminating the stock standard.

(D-11) Sodium bicarbonate, 1 M

Weigh 84 g of sodium bicarbonate and transfer to a 1-L volumetric flask. Add about 600 ml of water and mix until all the bicarbonate is dissolved. Dilute to volume with water and mix thoroughly. Check the molarity by the same procedure as that given for the sodium carbonate. This time, however, it should require 20 ml to reach an end point if the sodium bicarbonate solution is 1 N (1 M). Adjust if necessary. Store in a 1-L polyethylene bottle.

(D-12) Carbonate-bicarbonate buffer

To 35 ml of 1 M sodium carbonate solution, add 70 ml of 1 M sodium bicarbonate solution. Mix well and store in a 4-ounce polyethylene bottle.

(D-13) CO_2 color reagent

Prepare this reagent fresh on the day of use. Pipet 1.5 ml of carbonate-bicarbonate buffer* and 0.9 ml of phenolphthalein indicator* and transfer to a 1-L graduated cylinder. Dilute to 1 L with water, add 0.5 ml of Brij 35, and transfer to a 1-L bottle. Stopper and mix well. Avoid prolonged exposure to air.

(D-14) Sodium chloride stock standard, 1 M

Weigh 58.448 g of sodium chloride and transfer to a 1-L volumetric flask. Dissolve and dilute to volume with water. Store in a tightly capped polyethylene bottle. This standard contains 1000 mEq Na/L and 1000 mEq Cl/L.

(D-15) Potassium chloride stock standard, 0.1 M

Weigh 7.4557 g of potassium chloride and transfer to a 1-L volumetric flask. Dissolve and dilute to volume with water. Store in a tightly capped polyethylene bottle. This standard contains 100 mEq K/L and 100 mEq Cl/L.

(D-16) Ammonium chloride stock standard, 1 M

Weigh 53.5 g of ammonium chloride, NH_4Cl, and transfer to a 1-L volumetric flask. Dissolve and dilute to volume with water. Store in a tightly capped polyethylene bottle. This standard contains 1000 mEq Cl/L.

(D-17) Saline, 0.9%

Weigh 9 g of sodium chloride and transfer to a 1-L volumetric flask. Dissolve and dilute to volume with water. Store in a Pyrex or polyethylene bottle.

(D-18) Alkaline potassium ferricyanide

Weigh 9 g of sodium chloride and transfer to a 1-L volumetric flask containing about 600 ml of water. Mix until dissolved. Add 0.25 g of potassium ferricyanide, $K_3Fe(CN)_6$, and mix until dissolved. Slowly and while mixing (with magnetic stirrer) add 20 g of sodium carbonate, Na_2CO_3, and mix until dissolved. Dilute to volume, mix thoroughly, and filter. Add 2.0 ml of Brij 35 and mix. Store in a brown glass or polyethylene bottle. If the solution has been stored for a period of time, filter just prior to use.

(D-19) Diacetyl monoxime stock, 2.5%

Weigh 25 g of diacetyl monoxime (2,3-butanedione-2-oxime) and transfer to a 1-L volumetric

---

*These amounts may need to be adjusted as directed in operating sequence (electrolytes), steps 7 and 10.

flask. Add about 500 ml of water and mix until dissolved. Dilute to volume with water and mix thoroughly. Filter and store in a brown bottle.

(D-20) Thiosemicarbazide stock, 0.5%

Weigh 5 g of thiosemicarbazide and transfer to a 1-L volumetric flask containing about 500 ml of water. Mix until dissolved. Dilute to volume with water and mix thoroughly. Store in a brown bottle.

(D-21) BUN color reagent, working

Measure 67 ml of stock diacetyl monoxime and transfer to a 1-L volumetric flask containing about 500 ml of water. Add 67 ml of stock thiosemicarbazide and dilute to volume with water. Add 0.5 ml of Brij 35 and mix thoroughly. Store in a polyethylene or brown glass bottle.

(D-22) Ferric chloride–phosphoric acid stock

Dissolve 15 g of ferric chloride, $FeCl_3 \cdot 6H_2O$, in about 30 ml of water and transfer to a 500-ml glass-stoppered graduated mixing cylinder. Slowly add 300 ml of 85% phosphoric acid. Dilute to the 450-ml mark with water and mix. Store in a brown bottle.

(D-23) Sulfuric acid stock, 20%

Place a 1-L Erlenmeyer flask containing about 600 ml of water into a pan of ice water for about 10 minutes. Slowly add 200 ml of concentrated sulfuric acid to the flask and mix by swirling. Allow the solution to come to room temperature and then transfer to a 1-L volumetric flask. Rinse the Erlenmeyer flask 3 times with 30 ml of water and transfer washings to the volumetric flask. Dilute to volume with water and mix.

(D-24) BUN acid, working

Pipet 1.0 ml of stock ferric chloride–phosphoric acid into a 1-L volumetric flask and dilute to volume with 20% sulfuric acid. Mix thoroughly and store in a polyethylene bottle.

(D-25) Benzoic acid, saturated

Place a beaker or flask containing 2 L of water on a magnetic stirrer. Add about 8 g of benzoic acid and allow to stir overnight. Filter and store in a polyethylene bottle.

(D-26) Glucose stock standard, 10 mg/ml

Weigh exactly 10 g of dextrose on an analytic balance and transfer to a 1-L volumetric flask. Dissolve and dilute to volume with saturated benzoic acid. Mix thoroughly and store in a tightly closed polyethylene or Pyrex bottle.

(D-27) Standard diluent, 0.01 N H_2SO_4 containing 40 mg/L PMA

Weigh 0.2 g of phenylmercuric acetate (PMA) (Eastman P-4267) and transfer to a 250-ml beaker containing 100 ml of water. Heat on a steam bath until completely dissolved. Cool to room temperature and transfer to a 5-L volumetric flask (Corning No. 94450). Rinse the beaker several times with water and transfer the washings to the flask. Add 1.4 ml of concentrated sulfuric acid and mix. Dilute to volume with water and mix.

(D-28) Urea nitrogen stock standard, 10 mg urea N/ml

Weigh exactly 21.4383 g of urea and transfer to a 1-L volumetric flask. Dissolve and dilute to volume with Standard diluent (D-27). Mix thoroughly and store in a tightly closed polyethylene or Pyrex bottle.

(D-29) Sodium hydroxide, 0.5 N

Prepare by using a concentrated solution of NaOH that has been standardized.

$$ml \times N = ml \times N$$

(D-30) Picric acid, saturated

Heat 1 L of distilled water to boiling in a 2-L flask or beaker. Remove from heat and add 11.75 g of picric acid. (Keep away from an open flame.) Allow to cool to room temperature and filter. Store in a brown, glass-stoppered bottle.

NOTE: Reagent grade picric acid will have 10% to 12% water added.

(D-31) Creatinine stock standard, 1 mg/ml

Weigh 1 g of creatinine and transfer to a 1-L volumetric flask. Dissolve and dilute to volume with 0.1 N HCl. Mix and store in a polyethylene bottle.

(D-32) Hydrochloric acid, 0.1 N

Using a 10-ml serologic pipet attached to a rubber bulb or vacuum line, transfer 8.4 ml of concentrated, reagent grade hydrochloric acid to a 1-L volumetric flask containing about 700 ml of water. Dilute to volume and mix. Standardize. Be sure that neither the concentrated nor the dilute acid is exposed to ammonia fumes. *Do not pipet strong acids by mouth.*

(D-33) Hydrochloric acid, 0.02 N

Measure 1.67 ml of concentrated hydrochloric acid into a 1-L volumetric flask containing about 700 ml of water. Dilute to volume with water and mix. Store in a glass or polyethylene bottle.

(D-34) Phosphotungstic acid (AutoAnalyzer)

Measure 300 ml of water and transfer to a 1-L boiling flask with a ground glass joint top. Add 40 g of sodium tungstate and swirl until dissolved. Add several glass beads to the flask. Add 32 ml of 85% o-phosphoric acid and mix. Attach a reflux condenser to the flask and boil gently for 2 hours. Cool to room temperature. While the reflux condenser is still attached to the boiling flask, rinse the inside of the condenser with about 50 ml of water and allow the washings to run down into the flask. Transfer the contents of the flask to a 1-L volumetric flask. Rinse the boiling flask 3 times with about 30 ml of water and transfer washings to the volumetric flask. Dilute to volume with water. Add 32 g of lithium sulfate, Li_2SO_4, and mix until completely dissolved. Filter and store in an amber bottle. This reagent is stable if kept refrigerated when not in use.

(D-35) Sodium cyanide stock, 10%

Weigh 100 g of sodium cyanide, NaCN, and transfer to a 1-L volumetric flask. Add about 800 ml of water and mix until all of the cyanide is dissolved. Add 2.0 ml of concentrated ammonium hydroxide and dilute to volume with water. Filter and store in a polyethylene bottle in a refrigerator.

CAUTION: This is extremely poisonous. *Do not pipet by mouth.*

(D-36) Urea stock, 20%

Weigh 200 g of urea, NH_2CONH_2, and transfer to a 1-L volumetric flask. Dissolve and dilute to volume with water. Filter and store in a polyethylene bottle.

(D-37) Cyanide-urea working solution

This reagent should be freshly prepared on the day of use. Allow about 150 ml for 1 hour's running time. Mix 1 part sodium cyanide stock with 1 part urea stock. *Do not pipet by mouth.*

(D-38) Uric acid stock standard, 1 mg uric acid/ml

Weigh 1 g of uric acid and transfer to a 1-L volumetric flask. Weigh 0.6 g of lithium carbonate and transfer to a 250-ml beaker containing 150 ml of water. Mix until completely dissolved. Filter off any insoluble material and heat the solution to 60° C. Add this solution to the volumetric flask and swirl the flask under running hot water until the uric acid is dissolved (about 5 minutes). Then cool the flask under cold running water. Add 20 ml of 37% formalin and about 400 ml of water. With mixing add 25 ml of 1 N sulfuric acid. Dilute to volume with water and mix thoroughly. Store in an amber bottle.

(D-39) Hydrochloric acid, 0.25 N

Place about 500 ml of water in a 1-L volumetric flask. Carefully measure 21 ml of concentrated hydrochloric acid in a 25-ml graduated cylinder and add to the flask. Rinse the cylinder with water and add the washings to the flask. Swirl to mix. Dilute to volume with water and mix. Add 0.5 ml of Brij 35 and mix thoroughly. Store in a polyethylene bottle.

(D-40) Cresolphthalein complexone, 0.01%, with 8-hydroxyquinoline

Weigh 0.1 g of cresolphthalein complexone and transfer to a 1-L volumetric flask containing about 700 ml of 0.25 N hydrochloric acid. Place the flask on a magnetic stirrer and allow to stir until all the dye is dissolved. Then add 2.5 g of 8-hydroxyquinoline and mix until dissolved. Dilute to volume with 0.25 N hydrochloric acid and mix. Filter and store in a polyethylene bottle. *Do not* add any wetting agent to this reagent.

(D-41) Calcium base, 3.75% diethylamine

It is recommended that this reagent be prepared in a hood. Place a 1-L volumetric flask containing about 500 ml of water on a magnetic stirrer. Add 37.5 ml of diethylamine. Dilute to volume and mix. Store in a polyethylene bottle.

(D-42) Calcium stock standard, 50 mg Ca/dl

Weigh 1.25 g of calcium carbonate and transfer to a 1-L volumetric flask. Add about 100 ml of water. *Very slowly* add 7.0 ml of concentrated hydrochloric acid. Swirl the flask gently until all of the calcium carbonate is dissolved. Dilute to volume with water and mix thoroughly. Store in a polyethylene bottle.

(D-43) Magnesium stock standard, 100 mg Mg/dl

Weigh 8.36 g of magnesium chloride, $MgCl_2 \cdot 6H_2O$, and transfer to a 1-L volumetric flask containing about 500 ml of water. Dissolve and dilute to volume with water. Mix thoroughly and store in a polyethylene bottle.

(D-44) Hydrazine sulfate, 0.2% in 1 N H_2SO_4

Place a 2-L volumetric flask containing about 1500 ml of water on a magnetic stirrer. Slowly add 56 ml of concentrated sulfuric acid and allow to stir until thoroughly mixed. Add 4 g of hydrazine sulfate and stir until dissolved. Dilute to volume with water and mix. Store in a polyethylene or amber glass bottle.

(D-45) Stannous chloride–hydrazine sulfate working solution

Weigh 0.2 g of stannous chloride, $SnCl_2 \cdot 2H_2O$, and transfer to a 1-L volumetric flask containing about 750 ml of 0.2% hydrazine sulfate. Mix until dissolved. Dilute to volume with 0.2% hydrazine sulfate and mix thoroughly. Store in a polyethylene or amber glass bottle. This reagent is stable for at least 2 weeks at room temperature.

(D-46) Acid molybdate

Place a 1-L volumetric flask containing about 800 ml of water into a pan of ice water for about 10 minutes. Slowly add 35 ml of concentrated sulfuric acid and swirl to mix. Add 10 g of ammonium molybdate, $(NH_4)_6Mo_7O_{24} \cdot 4H_2O$, and mix until dissolved. Allow the solution to come to room temperature and dilute to volume with water. Mix thoroughly and store in a polyethylene bottle.

(D-47) Phosphorus stock standard, 1 mg P/ml

Weigh exactly 4.3937 g of monobasic potassium phosphate, KH_2PO_4, on an analytic balance and transfer to a 1-L volumetric flask. Dissolve and dilute to volume with water. Mix thoroughly and store in a polyethylene bottle.

(D-48) SGOT substrate

Place a 1-L beaker containing about 800 ml of water on a magnetic stirrer. Slowly add 33.5 g of dibasic potassium phosphate, K_2HPO_4, and mix until completely dissolved. Add 1 g of monobasic potassium phosphate, KH_2PO_4, and mix until dissolved. Add 2.66 g of l-aspartic acid and mix until dissolved. Add 1.462 g α-ketoglutaric acid and mix until dissolved. Add 1 g of tetrasodium EDTA (tetrasodium ethylenediamine tetraacetate) and mix until dissolved. Check the pH with a pH meter. The substrate should be pH 7.4. If necessary, adjust the solution to pH 7.4 with either 0.1 N HCl or 0.1 N NaOH. Transfer the solution to a 1-L volumetric flask. Rinse the beaker three times with about 20 ml of water and add washings to the flask. Dilute to volume with water and mix thoroughly. Filter and transfer to a polyethylene bottle. Add 0.5 ml of chloroform and 1.0 ml of Triton X-405. Mix and store in a refrigerator.

(D-49) Phenyl mercuric acetate, 0.04%

Weigh 0.4 g of phenyl mercuric acetate and transfer to a 1-L volumetric flask containing about 850 ml of hot water. Mix until dissolved. Allow to cool to room temperature, dilute to volume, and mix. Filter and store in a polyethylene bottle.

(D-50) Citrate buffer

Weigh 32.4 g of sodium citrate, $Na_3C_6H_5O_7 \cdot 2H_2O$, and transfer to a 1-L volumetric flask containing about 500 ml of water and 50 ml of 0.04% phenyl mercuric acetate. Mix until dissolved. Add 18.94 g of citric acid, $H_3C_6H_5O_7 \cdot H_2O$, and mix until dissolved. Dilute to volume with water and mix. Add 1.0 ml of Triton X-405 and mix thoroughly. Check the pH with a pH meter and adjust to pH 4.5, if necessary, using either 0.1 N NaOH or 0.1 N HCl. Store in a polyethylene bottle.

(D-51) Azoene fast red dye diluent

Measure 150 ml of 95% ethyl alcohol and 5.0 ml of 1 N hydrochloric acid and transfer to a 1-L volumetric flask containing about 600 ml of water. Dilute to volume and mix. Store in a polyethylene bottle.

(D-52) Azoene fast red dye solution (prepare fresh daily)

Weigh 0.4 g of azoene fast red, PDC (fast Ponceau L), and transfer to a 200-ml volumetric flask. Dissolve and dilute to volume with azoene fast red dye diluent. Mix, with as little exposure to air as possible, and filter in a refrigerator. Keep this reagent in an amber bottle in ice during use. This amount will be sufficient for $2\frac{1}{2}$ hours of running time.

(D-53) Alcoholic KOH wash solution, 2%

Dissolve 2 g of potassium hydroxide pellets in 100 ml of 95% ethyl alcohol.

(D-54) AMP stock buffer, 50% (w/v)

Weigh 500 g of 2-amino-2-methyl-1-propanol* in a beaker or Erlenmeyer flask. Transfer to a 1-L volumetric flask. Rinse the beaker or flask three times with 100 ml quantities of water. Add the washings to the volumetric flask. Dilute to volume with water and mix thoroughly. Store in a Pyrex or polyethylene bottle.

(D-55) AMP working buffer

Place a beaker containing about 700 ml of water on a magnetic stirrer. Add 118 ml of AMP stock buffer. Using a pH meter, adjust the pH to 10.25 ± 0.05 with 5.0 N hydrochloric acid. About 25 to 27 ml is required. Transfer to a 1-L volumetric flask and rinse the beaker two times with 50-ml quantities of water. Transfer washings to the volumetric flask. Dilute to volume with water. Add 1.0 ml of Triton X-405 and mix thoroughly. Store in a polyethylene bottle.

---

*Eastman P-4780.

(D-56) Magnesium chloride, 1 M

Weigh 20.331 g of magnesium chloride, $MgCl_2 \cdot 6H_2O$, and transfer to a 100-ml volumetric flask containing about 80 ml of water. Swirl until dissolved. Add 2 drops of concentrated hydrochloric acid. Dilute to volume with water and mix. Store in a Pyrex or polyethylene bottle.

(D-57) p-Nitrophenyl phosphate substrate

Measure 800 ml of AMP working buffer and transfer to a 1-L volumetric flask. Add 2 g of p-nitrophenyl phosphate and swirl to dissolve. Add 2.0 ml of 1 M magnesium chloride. Dilute to volume with water and mix thoroughly. This substrate is stable for 2 days when stored in the refrigerator. During use, the bottle should be kept in an ice bath.

(D-58) LDH substrate

Place a 1-L volumetric flask on a magnetic mixer and add about 50 ml of water. Begin mixing gently and slowly add 6 g of sodium hydroxide. When it has dissolved, slowly add 15 ml of 85% dl-lactic acid to the warm solution and continue mixing for at least 30 minutes. Add about 500 ml of water to the flask and then add 12.1 g of TRIS (tris[hydroxymethyl]aminomethane) to the mixing solution. When the TRIS is dissolved, add water to bring the volume to about 950 ml. Using a pH meter, adjust the pH of the substrate to 8.8 ± 0.05 with 1 N HCl. This will require about 7.5 ml. Dilute to volume with water, add 1.5 ml of Triton X-100, and mix well. Filter and store in a polyethylene bottle. Although this reagent is stable for at least 4 weeks at room temperature, it should be stored in a refrigerator for maximum stability.

(D-59) Phosphate buffer, 0.067 M, pH 7.4

Place a 1-L beaker containing about 800 ml of water on a magnetic mixer. Add 9.1 g of monobasic potassium phosphate, KH_2PO_4, to the mixing water. When the phosphate is dissolved, add 2.2 g of sodium hydroxide and continue mixing until it is dissolved. Allow to cool to room temperature. Using a pH meter, check the pH of the buffer. Adjust to pH 7.4, if necessary, with 0.01 N NaOH or 0.01 N HCl. Transfer the solution to a 1-L volumetric flask. Rinse the beaker with water and add washings to the flask. Dilute to volume with water and mix well. Store in a polyethylene bottle.

(D-60) NAD-diaphorase

Weigh 0.30 g of bovine albumin fraction V powder and transfer to a 250-ml beaker containing 175 ml of 0.067 M phosphate buffer, pH 7.4. Place the beaker on a magnetic mixer and mix until dissolved. Using a pH meter, check pH and adjust to 7.4 if necessary. Add 605 Worthington units of diaphorase and mix. Add 1.2 g of nicotinamide adenine dinucleotide (NAD) and mix until dissolved. Transfer to a 200-ml volumetric flask. Rinse beaker with three 5-ml portions of phosphate buffer and add washings to the flask. Dilute to volume with phosphate buffer and mix thoroughly. This solution should be divided into quantities sufficient for daily use and stored in a freezer. About 15 ml/hr is used. Thaw the reagent at room temperature and place in an ice bath during use. It is stable for 2 days when kept refrigerated or iced. This material is available from Technicon Corp. in a lyophilized form. It can be diluted with distilled water in quantities sufficient for daily use.

(D-61) INT dye

Weigh 0.625 g of INT, 3-p-nitrophenyl-1-2-p-iodophenyl-5-phenyl tetrazolium chloride, and transfer to a 1-L volumetric flask. Dissolve and dilute to volume with water. Store in an amber glass bottle. The dye is stable for at least 3 months at room temperature if it is protected from light. Mix to dissolve. Add 1.0 ml of 2 M $l(+)$ lactate and dilute to volume with water. Mix well. Adjust to pH 5.5. Transfer to a polyethylene bottle and store in a refrigerator.

(D-62) Liebermann-Burchard color reagent

Measure 350 ml of glacial acetic acid and transfer to a 2-L Erlenmeyer flask. Slowly add 550 ml of acetic anhydride. Seal to prevent absorption of atmospheric moisture. Cool to 5° C by overnight refrigeration or by placing in an ice bath. Cool 100 ml of concentrated sulfuric acid to 5° C. When both reagents are chilled, place the Erlenmeyer flask on a magnetic mixer. *Slowly* add the 100 ml of sulfuric acid to the stirring mixture. Add 10 g of anhydrous sodium sulfate and continue stirring until dissolved.

CAUTION: Reagents must be precooled to prevent charring when the sulfuric acid is added to the acetic acid–anhydride mixture. Divide into convenient aliquots and store in glass bottles in the refrigerator.

(D-63) Caffeine diluent

Place about 500 ml of moderately hot (50° to 60° C) distilled water in a 1-L volumetric flask. Add 37.5 g of caffeine, purified alkaloid, $C_{18}H_{10}N_4O_2$, 57.0 g of sodium benzoate, USP, C_6H_5COONa, and 94.5 g of sodium acetate, $CH_3COONa \cdot 3H_2O$. Mix until dissolved. Allow to cool to room temperature and dilute to volume with water. Mix thoroughly. Store in a polyethylene bottle. This reagent is stable for at least 6 months at room temperature. The formula has been increased by 50% over the formulation in previously published AutoAnalyzer methods. This was done to solve the problem of an occasional false low bilirubin due to insufficient caffeine concentration (Technicon Laboratory method file N-51a).

(D-64) Sulfanilic acid (diazo I)

Weigh 10 g of sulfanilic acid and transfer to a 1-L volumetric flask containing about 600 ml of water. Mix until dissolved and add 15 ml of concentrated hydrochloric acid. Dilute to volume with water and mix thoroughly. Store in a polyethylene bottle.

(D-65) Sodium nitrite, 0.5% (diazo II)

Weigh 0.5 g of sodium nitrite, $NaNO_2$, and transfer to a 100-ml volumetric flask. Dissolve and dilute to volume with water. Mix thoroughly and store in an amber glass bottle in a refrigerator. This reagent is stable for about 3 weeks.

(D-66) Diazo reagent, combined

Prepare fresh daily. Mix 2.5 ml of diazo II reagent with every 100 ml of diazo I.

(D-67) Tartrate buffer

Weigh 50 g of sodium hydroxide and 175 g of potassium sodium tartrate and transfer to a 1-L volumetric flask containing about 500 ml of water. Dissolve and dilute to volume with water. Mix thoroughly and store in a polyethylene bottle. This buffer is stable for at least 6 months at room temperature.

(D-68) Ascorbic acid, 4%

Prepare fresh daily. Weigh 1 g of ascorbic acid and transfer to a 25-ml volumetric flask. Dissolve and dilute to volume with water. Mix thoroughly.

(D-69) Hydrochloric acid, 0.05 N

Place about 900 ml of water in a 1-L volumetric flask and add 4.2 ml of concentrated hydrochloric acid. Dilute to volume and mix well. Store in a polyethylene or amber glass bottle.

(D-70) Alkaline iodide

Weigh 8 g of sodium hydroxide and transfer to a 1-L volumetric flask containing about 800 ml of water. Mix until dissolved. Add 5 g of potassium iodide and mix until dissolved. Dilute to volume with water and mix thoroughly. Store in a polyethylene bottle.

(D-71) Biuret stock solution

Weigh 45 g of potassium sodium tartrate, $KNaC_4H_4O_6 \cdot 4H_2O$, and transfer to a 1-L volumetric flask. Add 400 ml of 0.2 N sodium hydroxide and mix until the tartrate is dissolved. Add 15 g of copper sulfate, $CuSO_4 \cdot 5H_2O$, and mix until dissolved. Dilute to volume with 0.2 N sodium hydroxide and mix thoroughly. Filter and store in a Pyrex or polyethylene bottle.

(D-72) Biuret working solution

Thoroughly mix 200 ml of stock biuret with 800 ml of alkaline iodide solution. Store in a polyethylene bottle.

(D-73) Protein blank solution

Weigh 9 g of potassium sodium tartrate and transfer to a 1-L volumetric flask. Dissolve and dilute to volume with alkaline iodide solution. Store in a polyethylene bottle.

(D-74) HABA dye stock solution, 6×10^{-3} M

Dissolve 1.452 g of 2-(4'-hydroxyazobenzene)benzoic acid (HABA)* in 10 ml of 1 N sodium hydroxide in a 1-L volumetric flask. Dilute to volume with water and mix thoroughly. Store in an amber bottle.

(D-75) Phosphate buffer, pH 6.05

Weigh 4.662 g of dibasic potassium phosphate, K_2HPO_4, and transfer to a 2-L volumetric flask containing about 800 ml of water. Mix until dissolved. Add 16.595 g of monobasic potassium phosphate and mix until dissolved. Check pH with a pH meter. Adjust to pH 6.05 ± 0.05 if necessary. Dilute to volume and mix.

(D-76) HABA dye working solution, 6×10^{-4} M

Measure 100 ml of stock HABA dye and transfer to a 1-L volumetric flask. Dilute to volume with phosphate buffer and mix thoroughly. Check pH with a pH meter and adjust to pH 6.2 to 6.4 if necessary. Store in a brown bottle. This reagent is stable for 3 days if it is kept refrigerated between uses. Bring to room temperature before using.

(D-77) Albumin stock standard, 10 g/dl

To prepare a standard containing 10 g of protein/dl it is necessary to know the Kjeldahl nitrogen factor for the lot of bovine albumin used. Crystalline bovine albumin fraction V may be obtained from Armour Pharmaceutical Co. To calculate the amount of bovine albumin needed use the following formula:

$$10 \text{ g} \times \frac{160}{\text{N factor}} = \text{grams of albumin needed}$$

---

*Dajac Laboratories.

Weigh this amount and very carefully transfer it to a 250-ml beaker. The albumin is very light and fluffy and, unless care is taken, some of the material may be lost in the transfer. Add about 80 ml of saline and carefully stir until all the albumin is in solution. Transfer the solution to a 100-ml volumetric flask. Rinse the beaker with small quantities of saline and add washings to the flask. Dilute to volume with saline, add 0.05 g of sodium azide, and mix thoroughly. Duplicate Kjeldahl nitrogen determinations should be performed on this solution to determine the actual protein concentration.

(D-78) Tungstic acid reagent (stabilized)

Dissolve 1 g polyvinyl alcohol* in about 100 ml of water with gentle warming.

CAUTION: Do not boil. Cool and transfer to a 1-L volumetric flask containing 11.1 g sodium tungstate, $Na_2WO_4 \cdot 2H_2O$, previously dissolved in about 100 ml of water. Mix. In a separate container, add 2.1 ml concentrated H_2SO_4 to about 300 ml of water, mix, and then add to the tungstate solution in the volumetric flask. Mix and dilute to volume with water. This reagent is stable for up to 1 year at room temperature.

(D-79) o-Toluidine reagent (stabilized)

To 5 g of thiourea add 9.0 ml o-toluidine and dilute to 1 L with glacial acetic acid. This reagent is stable for up to 2 years when refrigerated stored in an amber glass bottle. As the reagent ages, it yields less color for a given concentration of glucose, but proportionality between standards and unknowns is maintained.

(D-80) Glucose standard, 100 mg/dl

Dissolve 1.000 g of anhydrous glucose (dextrose) in 1 L of saturated benzoic acid solution. This standard is stable up to 2 years in the refrigerator.

(D-81) Sulfuric acid, ⅔ N

Transfer exactly 19 ml of concentrated sulfuric acid, H_2SO_4, to a 1-L volumetric flask containing about 700 ml of water. Dilute to volume and mix. Standardize.

NOTE: With phenolphthalein as an indicator, 20 ml of the above solution should require 13.33 ml of 1 N NaOH for neutralization. If the solution is too acid, adjust the normality to 0.67 N by dilution.

(D-82) Sodium tungstate, 10%

Weigh 100 g of sodium tungstate, Na_2WO_4, and transfer to a 1-L volumetric flask. Dissolve and dilute to volume with water. Stopper and mix thoroughly. Store in a Pyrex or polyethylene bottle.

(D-83) Sodium hydroxide, 10% (2.5 N)

Dilute solutions of sodium hydroxide should be prepared by using a nearly saturated solution that has been standardized. If the normality of the concentrated solution is known, then the volume needed to make a dilute solution may be determined with the following formula:

$$ml \times N = ml \times N$$

Or: Weigh 100 g of sodium hydroxide and transfer to a 1-L volumetric flask containing 800 ml of water. Mix to dissolve, cool to room temperature, dilute to volume, and mix.

(D-84) Picric acid, 0.04 M

Weigh 9.16 g of anhydrous or 10.17 g of reagent grade picric acid (containing 10% to 12% added water) and transfer to a 1-L volumetric flask. Dissolve and dilute to volume with water. Mix well and store in an amber bottle.

(D-85) Sodium hydroxide, 0.75 N

Weigh 30 g of sodium hydroxide and transfer to a 1-L volumetric flask containing about 800 ml of water. Mix to dissolve. Dilute to volume with water and mix thoroughly. Store in a polyethylene bottle.

(D-86) Polyanethol sodium sulfonate (Liquoid)

Weigh 2 g of polyanethol sodium sulfonate (Liquoid)* and carefully transfer to a 100-ml volumetric flask. Use a funnel. This material is very light and fluffy, so care must be taken to avoid losing any. Wash any material remaining in the funnel down into the flask with about 80 ml of distilled water. When all the Liquoid is dissolved, dilute to the 100-ml mark and mix thoroughly. Transfer to a brown bottle and store in a refrigerator.

(D-87) Glycerin-silicate reagent

Weigh 50 g of crystalline sodium silicate "soluble" and transfer to a 1-L beaker containing 500 ml of hot water. Stir until dissolved. Add 100 ml of glycerin. Stir until thoroughly mixed. Allow to come to room temperature. If the solution is cloudy, filter through hardened paper. Store in a Pyrex or polyethylene bottle.

---

*Elvanol 70-05, DuPont.

*May be obtained from Hoffman-La Roche, Inc.

(D-88) Phosphotungstic acid

Weigh 50 g of sodium tungstate, $Na_2WO_4 \cdot H_2O$, and transfer to a 1-L round-bottom flask. Add 400 ml of water and 40 ml of 85% o-phosphoric acid. Boil gently under a reflux condenser for 4 hours. Allow to cool to room temperature. Transfer to a 500-ml volumetric flask. Rinse the boiling flask twice with 10 ml of water and transfer rinsings to the volumetric flask. Dilute to the 500-ml mark with water. Mix thoroughly and store in a brown bottle.

(D-89) Uric acid special reagent

To 180 ml of water, add 80 ml of phosphotungstic acid (D-88) and 100 ml of Liquoid solution (D-86). Mix well and store in a brown bottle.

(D-90) Sodium carbonate, 14%

Weigh 140 g of anhydrous sodium carbonate and transfer to a 1-L volumetric flask. Add about 800 ml of water and mix until all the carbonate is dissolved. Dilute to volume and mix again. Store in a Pyrex or polyethylene bottle.

(D-91) Urease solution, buffered

Weigh 150 mg of urease* and transfer to a 100-ml volumetric flask. Add 1 g of ethylenediamine tetraacetic acid (EDTA). Dissolve and dilute to volume with water. This solution is stable for 1 month when stored in a refrigerator.

(D-92) Phenol color reagent

Weigh 50 g of reagent grade phenol and 0.25 g of reagent grade sodium nitroprusside and transfer to a 1-L volumetric flask. Dissolve and dilute to volume with water. This solution is stable for at least 2 months if stored in a brown bottle in a cool place.

(D-93) Alkali-hypochlorite reagent

Weigh 25 g of reagent grade NaOH and transfer to a 1-L volumetric flask. Add 40 ml of Clorox (commercial bleach). Dilute to volume with water and mix. Store in an amber bottle protected from light and heat. This solution is stable for at least 2 months.

(D-94) Uranyl zinc acetate reagent

Weigh 10 g of uranyl acetate and dissolve in a beaker containing 50 ml of boiling water and 2.0 ml of glacial acetic acid. In another beaker, dissolve 30 g of zinc acetate in 50 ml of boiling water and 1.0 ml of glacial acetic acid. Mix the two solutions and heat again to the boiling point. Let the beaker stand overnight at room temperature. Filter and then mix with an equal volume of 95% ethyl alcohol. Place in a refrigerator for 2 days and then filter. This reagent is stable at room temperature.

(D-95) Trichloroacetic acid, 10%

Trichloroacetic acid (TCA) is extremely hygroscopic (capable of absorbing water from the atmosphere) and is therefore difficult to weigh accurately. It is recommended that reagent grade TCA be purchased in ¼-pound (113.4 g) bottles and that an entire bottle be dissolved and diluted to the desired concentration. To prepare 10% TCA, dissolve and dilute a ¼-pound or 113.4 g bottle of TCA (check weight on bottle) to 1134 ml. Store in a Pyrex bottle.

(D-96) Sodium standard, 140 mEq/L

Weigh 8.1820 g of sodium chloride and transfer to a 1-L volumetric flask. Dissolve and dilute to volume with water.

(D-97) Sodium cobaltinitrite reagent

a. Weigh 25 g of cobalt nitrate, $Co(NO_3)_2$, and transfer to a 400-ml beaker containing 50 ml of water. Stir to dissolve and then add 12.5 ml of glacial acetic acid.
b. Weigh 120 g of sodium nitrite, $NaNO_2$, and dissolve in 180 ml of water.

Place the beaker containing *solution a* in a fume hood and to it, add 210 ml of *solution b*. Bubble air through this solution until all fumes of nitrous oxide are driven off. Filter. This reagent is stable for about 1 month when stored in a refrigerator.

(D-98) Sodium acetate, half-saturated

Dilute saturated sodium acetate (D-200) with an equal volume of water.

(D-99) Wash solution, saturated with potassium sodium cobaltinitrite

Weigh 0.07 g of potassium sulfate, K_2SO_4, and dissolve in 100 ml of water. Slowly add 25 ml of half-saturated sodium acetate and 12.5 ml of sodium cobaltinitrite reagent. Allow to stand for about 30 minutes and then filter through a sintered glass filter. Wash the precipitate twice with 70% ethyl alcohol and once with 95% ethyl alcohol. Transfer the precipitate to an Erlenmeyer flask and prepare a saturated solution with 10% ethyl alcohol. Allow any excess of the precipitate to remain in the bottom of the flask. Remove and filter a portion of the supernatant on the day of use.

---

*Urease, Type II, about 800 to 1000 Sumner units/g, Sigma Chemical Co.

(D-100) Glycine, 7.5%

Weigh 7.5 g of glycine and transfer to a 100-ml volumetric flask. Dissolve and dilute to volume with water. Add 5 drops of chloroform as a preservative.

(D-101) Sodium carbonate, 25%

Weigh 25 g of anhydrous sodium carbonate, Na_2CO_3, and transfer to a 100-ml volumetric flask containing about 50 ml of water. Dissolve and dilute to volume with water and mix. It may be necessary to warm the solution slightly to aid in dissolving the carbonate. Store in a warm place in a polyethylene bottle.

(D-102) Phenol reagent of Folin-Ciocalteu stock

Weigh 100 g of sodium tungstate, $Na_2WO_4 \cdot 2H_2O$, and 25.0 g of sodium molybdate, $Na_2MoO_4 \cdot 2H_2O$, and transfer to a 2-L boiling flask containing 700 ml of water. Swirl to dissolve and add 50 ml of 85% o-phosphoric acid and 100 ml of concentrated hydrochloric acid. Attach a reflux condenser to the flask (apparatus must have ground glass joints) and reflux gently for 10 hours. Add 150 g of lithium sulfate, Li_2SO_4, 50 ml of water, and 2 or 3 drops of bromine. Place the flask in a fume hood without the condenser and boil the mixture for 15 minutes to remove excess bromine. Cool to room temperature and dilute to 1 L with water. This reagent should be yellow without any trace of a green tint. If it turns green during storage it can be restored by adding 2 or 3 drops more of bromine and boiling again. Store in a refrigerator.

(D-103) Potassium standard, 4 mEq/L

Weigh 0.2982 g of potassium chloride, KCl, and transfer to a 1-L volumetric flask. Dissolve and dilute to volume with water. Mix thoroughly.

(D-104) Diphenylcarbazone indicator

Weigh 0.1 g of diphenylcarbazone and transfer to a 100-ml volumetric flask. Dissolve and dilute to volume with 95% ethyl alcohol. Mix thoroughly. Store in a refrigerator in a brown bottle.

(D-105) Standard sodium chloride, 10 mEq/L

Weigh 0.585 g of dry sodium chloride and transfer to a 1-L volumetric flask. Dissolve and dilute to volume. Mix thoroughly.

(D-106) Standard mercuric nitrate solution

Weigh 3 g of mercuric nitrate and transfer to a 1-L volumetric flask containing about 500 ml of water and 20 ml of 2 N nitric acid. Swirl to dissolve and dilute to volume with water. Mix thoroughly. Standardize this solution as follows: pipet 2.0 ml of standard sodium chloride solution into a small flask, add 4 drops of diphenylcarbazone indicator, and titrate with the mercuric nitrate solution. The number of milliliters of the mercuric nitrate solution used equals the value E used in the calculations of the chloride procedure. The standard mercuric nitrate solution is stable indefinitely.

(D-107) Lactic acid, 0.1 N

Pipet 4.4 ml of concentrated lactic acid (85%) into a 500-ml volumetric flask. Dilute to volume with water. This solution is stable for about 2 weeks.

(D-108) Sodium hydroxide, 5 N (carbonate-free)

Prepare by using a nearly saturated solution of sodium hydroxide that has been standardized:

$$ml \times N = ml \times N$$

(D-109) Sodium oxalate, 0.1 N

Dry about 10 g of anhydrous sodium oxalate, $Na_2C_2O_4$, in a 110° C oven for 12 hours or overnight. Cool in a desiccator; then weigh 6.7 g and transfer to a 1-L volumetric flask. Add about 200 ml of water and 5.0 ml of concentrated sulfuric acid. Swirl to dissolve and dilute to volume with water. Mix thoroughly.

(D-110) Potassium permanganate, 0.1 N

Weigh 40 g of ammonium oxalate, $(NH_4)_2C_2O_4$, and transfer to a 1-L volumetric flask. Dissolve and dilute to volume with water. Mix and allow to stand in a dark place at room temperature for about a week. A sediment of MnO_2 will form. Being careful not to disturb the sediment, slowly pour about 700 ml of the solution into a brown, glass-stoppered bottle. Discard the remainder. This solution will be diluted and standardized before it is used in the determination of calcium. Store in a refrigerator.

(D-111) Ammonium oxalate, 4%

Weigh 40 g of ammonium oxalate, $(NH_4)_2C_2O_4$, and transfer to a 1-L volumetric flask. Dissolve and dilute to volume with water. Mix.

(D-112) Ammonium hydroxide, dilute (0.5%)

Transfer 2.0 ml of concentrated (28%) ammonium hydroxide to a flask and add 98 ml of water. This solution is stable for about a month.

(D-113) Sulfuric acid, 1 N

Transfer 28 ml of concentrated sulfuric acid to a 1-L Pyrex volumetric flask containing about 600 ml of water. Swirl to mix and allow to cool to room temperature. Dilute to volume with water and mix.

(D-114) Potassium hydroxide, 1.25 N

Weigh 70 g of potassium hydroxide pellets and transfer to a 1-L volumetric flask containing about 600 ml of water. Swirl to dissolve. Cool to room temperature and dilute to volume with water. Stopper and mix well. Store in a polyethylene bottle.

(D-115) EDTA solution

Weigh 0.395 g of disodium ethylenediamine tetraacetate dihydrate and transfer to a 1-L volumetric flask. Dissolve and dilute to volume with water. Mix well. Store in a Pyrex or polyethylene bottle.

(D-116) Cal-red indicator

Weigh 1 g of "Cal-red dilute"* and transfer to a mortar of about 50-ml capacity. Add 10 ml of distilled water and grind with the pestle until a homogeneous suspension is obtained. (Not all of the material will dissolve.) This suspension should be stored in a refrigerator and is usually stable for about 2 weeks. If, however, it does not give sharp end points, then it must be discarded. With some batches of indicator it may be necessary to prepare a new suspension daily.

(D-117) Sodium citrate, 0.05 M

Weigh 14.7 g of sodium citrate, $Na_3C_6H_5O_7 \cdot 2H_2O$, and transfer to a 1-L volumetric flask. Dissolve and dilute to volume with water. Store in a Pyrex or polyethylene bottle.

(D-118) Calcium standard (1 ml = 100 μg Ca)

Using an analytic balance, weigh 0.2472 g of calcium carbonate, $CaCO_3$, and transfer to a 1-L volumetric flask. Slowly add 7.0 ml of dilute hydrochloric acid (1 part concentrated HCl + 9 parts distilled H_2O). Allow to stand until all $CaCO_3$ is dissolved and then add about 900 ml of distilled water. Adjust the pH to 6.0 with 50% ammonium acetate. Dilute to volume with water and mix thoroughly. Store in a Pyrex or polyethylene bottle.

(D-119) Ammonium oxalate, 10%

Weigh 10 g of ammonium oxalate, $(NH_4)_2C_2O_4$, and transfer to a 100-ml volumetric flask. Dissolve and dilute to volume with water. Mix well and store in a polyethylene bottle.

(D-120) Hydrochloric acid, 1 N

Using a 10-ml serologic pipet attached to a rubber ball or vacuum line, transfer 8.4 ml of concentrated hydrochloric acid to a 100-ml volumetric flask containing about 70 ml of water. Dilute to volume and mix. *Do not pipet strong acids by mouth.*

(D-121) Ammonium hydroxide, 5%

Mix 20 ml of concentrated ammonium hydroxide, NH_4OH (28%), with 80 ml of water. Store in a glass-stoppered bottle.

(D-122) Molybdate reagent

Weigh 25 g of ammonium molybdate, $(NH_4)_2MoO_4$, and transfer to a 1-L volumetric flask containing about 300 ml of water. Swirl to dissolve. Add 300 ml of 10 N sulfuric acid (D-123). Dilute to volume with water and mix. Store in a glass-stoppered bottle.

(D-123) Sulfuric acid, 10 N

Measure 282 ml of concentrated sulfuric acid and add slowly to a 1-L Pyrex volumetric flask containing about 600 ml of water. Carefully swirl to mix. Cool to room temperature and dilute to volume with water. Mix.

(D-124) Reducing reagent (aminonaphtholsulfonic acid)

Weigh 14.64 g of sodium bisulfite, $NaHSO_3$, and 0.25 g of 1-amino-2-naphthol-4-sulfonic acid and transfer to a 100-ml volumetric flask containing about 75 ml of water. Swirl to dissolve. Add 0.5 g of sodium sulfite, Na_2SO_3. Dissolve and dilute to volume. Transfer to a brown bottle and store in a refrigerator. If a precipitate forms, filter in the refrigerator so that the reagent does not become warm.

(D-125) Phosphorus stock standard, 0.1 mg P/ml

Weigh 0.4394 g of monobasic potassium phosphate, KH_2PO_4, and transfer to a 1-L volumetric flask. Add 10 ml of 10 N sulfuric acid and dilute to volume with water. Mix thoroughly.

(D-126) Biuret reagent

Weigh 45 g of sodium potassium tartrate and transfer to a 1-L volumetric flask containing about 600 ml of 0.2 N NaOH (prepare from a concentrated solution of NaOH that has been standardized). Mix until dissolved. Add 15 g of copper

---
*Scientific Service Laboratories, Inc.

sulfate, $CuSO_4 \cdot 5H_2O$, and dissolve completely. Add 5 g of potassium iodide and mix. Dilute to volume with 0.2 N NaOH and mix thoroughly. Store in a Pyrex or polyethylene bottle.

(D-127) Albumin standard, stock solution, 10 g/dl

Dissolve 10 g of human serum albumin, Cohn fraction V,* corrected for moisture content, and 50 mg of sodium azide in water in a 100-ml volumetric flask. Dilute to volume with water and store at 4° C.

(D-128) Albumin standard, working solutions, 2, 3, 4, 5, 6, and 8 g/dl

Dilute 2.0, 3.0, 4.0, 5.0, 6.0, and 8.0 ml of stock albumin solution to 10 ml in volumetric flasks using aqueous 5 mg/dl sodium azide. Store at 4° C. These standards are stable for at least 6 months at 4° C.

(D-129) Succinate buffer, 0.1 M, pH 4.0

Dissolve 11.9 g of succinic acid in about 800 ml of water, adjust pH to 4.0 with NaOH, and dilute to 1 L with water. Store at 4° C.

(D-130) Bromcresol green stock solution, 0.6 mM

Dissolve 419 mg of bromcresol green (BCG) in 10 ml of 0.1 N NaOH in a 1-L volumetric flask. Dilute to volume with water and store at 4° C.

(D-131) Bromcresol green working solution

Dilute 125 ml of stock BCG solution to 500 ml with 0.10 M succinate buffer. Add 2.0 ml of 30% Brij 35 solution and carefully adjust the pH to 4.20 ± 0.05. Store at 4° C.

(D-132) Thymol-barbital buffer, pH 7.55

Boil 1100 ml of distilled water for about 5 minutes to drive off CO_2. Allow to cool to about 95° C. Weigh 6 g of thymol crystals and transfer to a 2-L Erlenmeyer flask. Add 300 ml of the hot water. Add 3.09 g of barbital, 1.69 g of sodium barbital, and 720 ml of hot water. Stopper and shake flask for 5 minutes. (Vigorous stirring with a magnetic stirrer works very well.) Release pressure from time to time by removing the stopper. Allow to cool to 25° C and add 20 ml of water to compensate for evaporation. Seed the solution with about 1 g of thymol crystals. Shake the flask vigorously until the supernatant solution becomes clear. Allow to stand at room temperature (25° C) until the next day. Then filter through Whatman No. 1 paper. Check pH. It must be between 7.50 and 7.60 at 25° C. If the pH is outside these limits, adjustments must be made by altering the ratio of barbital to sodium barbital in a new preparation. Store in a tightly stoppered container to protect it against changes in pH from uptake of CO_2.

(D-133) Cephalin-cholesterol stock ether solution (Difco Bacto-Cephalin Cholesterol Antigen)

Reconstitute one Difco vial volumetrically with 5.0 ml anesthetic ether. If turbidity persists, add 1 drop of distilled water. Stable for months under refrigeration if protected against evaporation.

(D-134) Cephalin-cholesterol emulsion

Warm 35 ml of distilled water to 65° to 70° C in a 50-ml Erlenmeyer flask calibrated at 30 ml with a marking pencil; add 1.0 ml stock ether solution slowly with stirring. Raise temperature slowly to boiling and allow to simmer until final volume is 30 ml. Cool to room temperature before using. The emulsion should be milky and translucent and have no trace of ether. It is stable for 2 weeks when refrigerated.

(D-135) Calcium chloride, 2.5%

Weigh 2.5 g of calcium chloride, $CaCl_2$, and transfer to a 100-ml volumetric flask. Dissolve and dilute to volume with water. Mix.

(D-136) Sodium hydroxide, 0.2 N

Dilute 20 ml of 1 N sodium hydroxide (D-225) with 80 ml of water. Mix well. Store in a polyethylene bottle.

(D-137) Protein standard, 3 mg/ml

Weigh 0.3 g of pure albumin* and transfer to a 100-ml volumetric flask containing about 50 ml of saline (D-17). Swirl flask to dissolve and dilute to volume with saline. Stopper and mix. Pour about 3-ml aliquots of the standard into Pyrex test tubes and freeze for future use. When thawing protein standards, place the tubes in a beaker of water at room temperature. *Never use hot water.* Once a standard has been thawed, do not refreeze. Discard the unused portion.

(D-138) Diazo blank solution

Transfer 1.5 ml of concentrated hydrochloric acid to a 100-ml volumetric flask. Dilute to volume with water and mix.

---

*Available from Nutritional Biochemicals Corporation.

*Sigma Chemical Co.

(D-139) Sulfanilic acid, solution A, 0.1%

Weigh 100 mg of sulfanilic acid and transfer to a 100-ml volumetric flask. Add 1.5 ml of concentrated hydrochloric acid. Dilute to volume with water.

(D-140) Sodium nitrite stock solution B, 5%

Weigh 5 g of sodium nitrite and transfer to a 100-ml volumetric flask. Dissolve and dilute to volume with water. Store in a refrigerator.

(D-141) Sodium nitrite working solution B, 0.5%

Prepare this solution just before use. Mix 1.0 ml of stock solution B (D-140) with 9.0 ml of water.

(D-142) Alkaline buffer, pH 10.6 to 10.7

Weigh 24.4 g of dibasic sodium phosphate, $Na_2HPO_4 \cdot 7H_2O$, (or 12.92 g of Na_2HPO_4) and 3.54 g of $Na_3PO_4 \cdot 12H_2O$ and 6.4 g of sodium p-toluensulfonate and transfer to a 1-L volumetric flask. Dissolve and dilute to volume with water. Check pH and adjust if necessary.

(D-143) Acid reagent, 2 M NaH_2PO_4

Weigh 27.6 g of $NaH_2PO_4 \cdot H_2O$ and transfer to a 100-ml volumetric flask. Dissolve and dilute to volume with water.

(D-144) BSP standard, 5 mg/dl

Pipet 1.0 ml of the intravenous test solution (50 mg/ml)* into a 1-L volumetric flask. Dilute to volume with water. This diluted standard is stable for 1 week.

(D-145) Lanthanum stock solution, 5% in 25% (v/v) HCl

Weigh 58.65 g of lanthanum oxide, La_2O_3, and transfer to a 1-L volumetric flask. Add about 50 ml of water. *Very slowly* add 250 ml of concentrated hydrochloric acid and mix until dissolved. Dilute to volume with water and mix well. Store in a polyethylene bottle.

(D-146) Lanthanum working solution, 0.5% in 2.5% (v/v) HCl

Transfer 100 ml of lanthanum stock solution to a 1-L volumetric flask. Dilute to volume with water, add 1.0 ml of Brij 35, and mix. Store in a polyethylene bottle.

---

*Hynson, Westcott & Dunning.

(D-147) Magnesium stock standard, 100 mEq/L

Weigh 1.216 g of pure magnesium metal and transfer to a 1-L volumetric flask. Dissolve in 9 to 10 ml of 12 N HCl. Dilute to volume with water and mix thoroughly. Store in a polyethylene bottle.

(D-148) Sodium diethyldithiocarbamate, 1%

Weigh 1 g of sodium diethyldithiocarbamate and transfer to a 100-ml volumetric flask. Dissolve and dilute to volume with water. Mix well and store in a polyethylene bottle.

(D-149) Sodium ethylenediamine tetraacetate (EDTA), 1%

Weigh 1 g of sodium ethylenediamine tetraacetate and transfer to a 100-ml volumetric flask. Dissolve and dilute to volume with water. Mix well and store in a polyethylene bottle.

(D-150) Olive oil emulsion

Add 0.2 g of sodium benzoate and 7 g of gum acacia to 100 ml of water in a high-speed blender. Mix gently until dissolved. Add 100 ml of pure (USP or better grade) olive oil and blend (emulsify) at high speed for 10 minutes. Store the emulsion in a refrigerator but do not allow it to freeze. Shake vigorously before use. If complete separation of the oil and water phases occurs, discard.

(D-151) Buffer, stock, 0.8 M TRIS

Dissolve and dilute 48.554 g of tris(hydroxymethyl)aminomethane (TRIS) to 500 ml with water. Store in a refrigerator.

(D-152) Buffer, working, 0.2 M, pH 8.0 at 27° C (pH 7.85 at 37° C)

To 50 ml of stock buffer, add 21 ml of 1 N HCl and about 90 ml of water. Cool to 25° C. Using a pH meter, adjust to pH 8.0 by careful addition of more HCl. The temperature of the solution should be at 27° C when final pH measurement is made. Dilute to 200 ml and store in a refrigerator.

(D-153) Sodium hydroxide, 0.05 N

Prepare from a concentrated solution of sodium hydroxide that has been standardized.

$$ml \times N = ml \times N$$

Standardize this dilute solution against a standard acid solution and adjust if necessary to exactly 0.05 N.

(D-154) Thymolphthalein indicator, 1%

Dissolve and dilute 1 g of thymolphthalein to 100 ml with 95% ethanol.

(D-155) Alkaline buffer, 0.1 M glycine, pH 10.5

Weigh 7.5 g of glycine and 0.095 g of magnesium chloride and transfer to a 1-L volumetric flask containing about 750 ml of water. Add 85 ml of 1 N sodium hydroxide (D-225) and dilute to volume with water. Mix thoroughly. Store in a refrigerator.

(D-156) p-Nitrophenyl phosphate stock substrate, 0.4%

Dissolve 0.1 g of p-nitrophenyl phosphate* in 25 ml of water.

(D-157) Alkaline buffered substrate, pH 10.3 to 10.4

Mix equal volumes of alkaline buffer (D-155) and stock substrate (D-156).

(D-158) Sodium hydroxide, 0.02 N

Pipet 20 ml of 1 N NaOH (D-225) into a 1-L volumetric flask and dilute to volume with water. Mix well.

(D-159) Acid buffer, 0.09 M citric acid, pH 4.8

Weigh 18.907 g of citric acid and transfer to a 1-L volumetric flask containing about 600 ml of water. Add 180 ml of 1 N sodium hydroxide (D-225) and 100 ml of 0.1 N hydrochloric acid (D-32). Dilute to volume with water and mix. Store in a refrigerator.

(D-160) Acid buffered substrate, pH 4.8 to 4.9

Mix equal volumes of acid buffer (D-159) and stock substrate (D-156).

(D-161) Sodium hydroxide, 0.1 N

Transfer 100 ml of 1 N sodium hydroxide to a 1-L volumetric flask and dilute to volume with water. Mix.

(D-162) Acetic acid, 20% (v/v)

Transfer 20 ml of glacial acetic acid to a 100 ml volumetric flask and dilute to volume with water. Mix.

---
*Available from Sigma Chemical Co.

(D-163) Tartrate acid buffer, pH 4.8

Weigh 18.9 g of citric acid, $C_6H_8O_7 \cdot H_2O$, and 6.0 g of $l(+)$ tartaric acid and transfer to a beaker containing about 600 ml of water. Stir to dissolve. Add 100 ml of 0.1 N hydrochloric acid (D-32) and mix. Add enough 1.0 N sodium hydroxide (D-225) to bring the pH to 4.8. Use a pH meter. Transfer quantitatively to a 1-L volumetric flask and dilute to volume with water. Store in a refrigerator.

(D-164) p-Nitrophenol, 10 mM/L

Weigh 0.1391 g of p-nitrophenol and transfer to a 100-ml volumetric flask. Dissolve and dilute to volume with water.

(D-165) Phosphate buffer, 0.1 M, pH 7.4

a. Disodium phosphate, 0.1 M. Weigh 14.2 g of disodium phosphate, Na_2HPO_4, and transfer to a 1-L volumetric flask. Dissolve and dilute to volume with water. Mix.
b. Potassium dihydrogen phosphate, 0.1 M. Weigh 13.609 g of potassium dihydrogen phosphate, KH_2PO_4, and transfer to a 1-L volumetric flask. Dissolve and dilute to volume with water. Mix.

Mix 420 ml of 0.1 M disodium phosphate with 80 ml of 0.1 M potassium dihydrogen phosphate.

(D-166) GOT substrate, α-ketoglutarate, 2 mM/L; dl-asparate, 200 mM/L

Weigh 0.0292 g of α-ketoglutaric acid and 2.66 g of dl-aspartic acid and transfer to a small beaker. Add 20 ml of 1 N sodium hydroxide and stir until the acids are dissolved. Using a pH meter, adjust the solution to pH 7.4 by adding 1 N sodium hydroxide drop wise. Transfer quantitatively to a 100-ml volumetric flask, using buffer D-165 to rinse the beaker. Dilute to volume with buffer and mix. Add 1 ml of chloroform as a preservative and store in a refrigerator.

(D-167) GPT substrate, α-ketoglutarate, 2 mM/L; dl-alanine, 200 mM/L

Weigh 0.0292 g of a-ketoglutaric acid and 1.78 g of dl-alanine and transfer to a 100 ml beaker. Add 20 ml of water and stir to dissolve. Using a pH meter, adjust the solution to pH 7.4 by adding 1 N sodium hydroxide dropwise. This will require about 10 drops. Transfer quantitatively to a 100 ml volumetric flask, using buffer D-165 to rinse the beaker. Dilute to volume with buffer and mix. Add 1.0 ml of chloroform as a preservative and store in a refrigerator.

(D-168) 2,4-Dinitrophenylhydrazine, 1 mM

Weigh 0.1981 g of 2,4-dinitrophenylhydrazine and transfer to a 1-L volumetric flask. Dissolve and dilute to volume with 1 N HCl. Store in a polyethylene bottle in a refrigerator.

(D-169) Sodium hydroxide, 0.4 N

a. This may be prepared from a concentrated solution of sodium hydroxide that has been standardized.
b. Or weigh 16 g of sodium hydroxide pellets and transfer to a 1-L volumetric flask. Dissolve and dilute to volume with water. Mix well and store in a polyethylene bottle.

(D-170) Pyruvate standard, 2 mM/L

Weigh 0.02 g of pure sodium pyruvate and dissolve in 100 ml of phosphate buffer.

(D-171) LDH color reagent

Weigh 0.2 g of 2-p-iodophenyl-3-p-nitrophenyl-5-phenyl tetrazolium chloride (INT)* and transfer to a 100-ml beaker containing about 75 ml of water. Place on a magnetic stirrer and mix until dissolved. Add 0.5 g of nicotinamide adenine dinucleotide (NAD) and mix until dissolved. Add 0.05 g of phenazine methosulfate (PMS) and mix until dissolved. Transfer immediately to a low-actinic 100-ml volumetric flask. Rinse the beaker with water and transfer washings to the flask. Dilute to volume with water and mix. This reagent should be protected from light at all times since it is very light-sensitive. Store in a low-actinic glass reagent bottle in a refrigerator. It is stable for several weeks.

(D-172) LDH buffer reagent

Dissolve 1 g of ethoxylated oleyl alcohol (Lipal 10-OA)† in 10 ml of water by heating to about 95° C. Add about 50 ml of water and add 12.1 g of tris(hydroxymethyl)aminomethane. Mix to dissolve. Adjust to pH 8.2 with 3 N HCl. Transfer to a 100-ml volumetric flask. Rinse the beaker with water and add washings to the flask. Dilute to volume with water and mix. Store in a refrigerator and discard if any visible growth occurs. Lipal 10-OA may be omitted if only serum is to be assayed.

(D-173) LDH substrate, 0.1 M $l(+)$ lactate

Add 5.0 ml of $l(+)$ lactic acid, 20% solution, to about 50 ml of water. Adjust to pH 5.5 with 1 N NaOH. Dilute to 120 ml with water and mix. Add a few drops of chloroform and store in a refrigerator.

---
*Dajac Laboratories.
†Drew Chemical Co.

(D-174) LDH control reagent

Weigh 0.2 g. of potassium oxalate and 0.2 g of ethylenediamine tetracetic acid, disodium dihydrate, and transfer to a 100-ml volumetric flask. Dissolve and dilute to volume with water. Store in a polyethylene bottle.

(D-175) $l(+)$ Lactate, 2 M

Dilute 60 ml of 30% $l(+)$ lactic acid (Sigma L1875) to 100 ml with water. Adjust to pH 5.5. Store in a refrigerator.

(D-176) $l(+)$ Lactate, 0.02 M in 2 M urea

Weigh 12.012 g of urea and transfer to a 100-ml volumetric flask containing about 80 ml of water. Add 1 ml of 2 M $l(+)$ lactate, dilute to volume with water, and mix. Adjust to pH 5.5. Store in a refrigerator.

(D-177) Sodium phosphate, dibasic, 0.1 M

Weigh 14.1959 g of dibasic sodium phosphate, Na_2HPO_4, and transfer to a 1-L volumetric flask. Dissolve and dilute to volume with water.

(D-178) Potassium phosphate, monobasic, 0.1 M

Weigh 13.609 g of monobasic potassium phosphate, KH_2PO_4, and transfer to a 1-L volumetric flask. Dissolve and dilute to volume with water.

(D-179) Phosphate buffer, pH 7.4

Mix 840 ml of 0.1 M Na_2HPO_4 with 160 ml of 0.1 M KH_2PO_4. Adjust to pH 7.4 if necessary. Store in a polyethylene bottle.

(D-180) Pyruvic acid standard, 0.5 µM/ml

Weigh 0.055 g of pyruvic acid, sodium salt, and transfer to a 1-L volumetric flask. Dissolve and dilute to volume with phosphate buffer. Store in an amber bottle in a refrigerator.

(D-181) CPK blank substrate

Weigh 0.73 g of adenosine-5′-triphosphate (ATP), disodium 3H_2O, crystalline (Sigma grade), 0.1864 g of phospho(enol)pyruvic acid (PEP), tricyclohexylamine salt, crystalline, 0.2964 g of magnesium sulfate, $MgSO_4 \cdot 7H_2O$, and 3.64 g of tris(hydroxymethyl)aminomethane (Sigma Trizma base) and transfer to a 200-ml volumetric flask. Add about 160 ml of water and mix to dissolve. Add 0.6 ml of pyruvate kinase, type II, crystalline (from rabbit skeletal muscle); ammonium sulfate suspension, 10 mg/ml. Mix and dilute to volume with water. Adjust to pH 9.0 if necessary, using 10% HCl or 10% NaOH.

(D-182) Creatine substrate

Weigh 0.9956 g of creatine·H$_2$O and transfer to a 100-ml volumetric flask. Dissolve and dilute to volume with CPK blank substrate. These two substrates are stable for up to 1 month when stored in a refrigerator.

(D-183) Mercaptoethanol, 0.25 M

Pipet 17.1 ml of mercaptoethanol into a 1-L volumetric flask. Dilute to volume with water and mix. Store in an amber bottle in the refrigerator. This solution is stable for about 1 month.

(D-184) PSP stock standard, 120 mg/500 ml

Weigh 0.12 g of phenolsulfonphthalein (phenol red) and transfer to a 500-ml volumetric flask. Add 1.0 ml of 10% NaOH. Dilute to volume with water. Mix thoroughly.

(D-185) Sodium chloride, saturated

Weigh 40 g of sodium chloride, NaCl, and transfer to a flask or beaker containing 100 ml of hot distilled water. Stir to dissolve as much as possible. Allow any undissolved salt to remain on the bottom of the bottle in which the solution is stored.

(D-186) Acetic acid, 50%

Add 50 ml of glacial acetic acid to 50 ml of water. Mix.

(D-187) Sulfosalicylic acid, 20%

Weigh 20 g of sulfosalicylic acid and transfer to a 100-ml volumetric flask. Dissolve and dilute to volume with water. Mix.

(D-188) Roberts' reagent

To 1 L of distilled water, add magnesium sulfate, with stirring, until no more dissolves. Add 200 ml of concentrated nitric acid. Mix.

(D-189) Hydrochloric acid, 5% (v/v)

Dilute 5.0 ml of concentrated hydrochloric acid with 95 ml of water. Mix.

(D-190) Esbach's reagent

Weigh 1 g of picric acid and 2 g of citric acid and dissolve in 100 ml of water.

(D-191) Tsuchiya's reagent

Weight 15 g of phosphotungstic acid and transfer to 2-L Erlenmeyer flask containing 1 L of 95% ethyl alcohol. Add 50 ml of concentrated hydrochloric acid and mix thoroughly.

(D-192) Sulfosalicylic acid, 3%

Weigh 3 g of sulfosalicylic acid and transfer to a 100-ml volumetric flask. Dissolve and dilute to volume with water. Mix.

(D-193) Protein standards, working, 25, 50, and 100 mg/dl

Dilute 0.25, 0.5, and 1.0 ml of stock albumin solution (D-127) to 100 ml in volumetric flasks using 5 mg/dl sodium azide. Store at 4° C. These standards are stable for at least 3 months at 4° C.

(D-194) Acetic acid, 5%

Pipet 5.0 ml of glacial acetic acid into a 100-ml volumetric flask. Dilute to volume with water and mix.

(D-195) Benedict's qualitative reagent

Weigh 100 g of anhydrous sodium carbonate, Na$_2$CO$_3$, or 200 g of hydrated sodium carbonate, Na$_2$CO$_3$·10H$_2$O. Place in a Pyrex container. Add about 800 ml of distilled water. Add 173 g of sodium citrate. Heat to dissolve. Cool to room temperature. Weigh 17.3 g of copper sulfate, CuSO$_4$·5H$_2$O. With the aid of heat, dissolve in 100 ml of distilled water. Cool to room temperature. With constant stirring, pour the copper sulfate solution slowly into the solution of carbonate and citrate salts. Make up to 1 L with distilled water. Mix. This preparation keeps indefinitely.

(D-196) Benedict's quantitative reagent

Weigh out the first four reagents listed below; place in separate labeled containers; and then prepare 5.0 ml of the 5% potassium ferrocyanide solution.

Copper sulfate (crystallized) CuSO$_4$·5H$_2$O	18.0 g
Sodium carbonate (anhydrous) Na$_2$CO$_3$ or 200 g of crystalline Na$_2$CO$_3$·10H$_2$O)	100.0 g
Sodium citrate (or potassium citrate)	200.0 g
Potassium thiocyanate (sulfocyanate)	125.0 g
Potassium ferrocyanide, 5%	5.0 ml

Pour about 600 ml of distilled water into a 1-L Pyrex beaker. Heat to about 70° C. With stirring, transfer the above weights of sodium carbonate, sodium citrate, and potassium thiocyanate to the beaker. Stir until dissolved. Filter. Dissolve the copper sulfate in about 100 ml of distilled water. With stirring, slowly add this copper solution to the 1-L beaker. Now add the 5.0 ml of the 5% potassium ferrocyanide solution. Mix. Allow the solution to cool. Transfer all of the solution to a 1-L volumetric flask. Add distilled water to the 1-L mark. Mix. Five milliliters of this reagent should be reduced by 1.0 ml of a 1% glucose solution.

(D-197) Ferric chloride, 10%

Weigh 10 g of ferric chloride, $FeCl_3$, and transfer to a 100-ml volumetric flask. Dissolve and dilute to volume with water. Mix.

(D-198) Sodium nitroprusside, saturated

Add 40 g of sodium nitroprusside (sodium nitroferricyanide) to 100 ml of water in a brown bottle. Shake to dissolve as much as possible. Allow any undissolved salt to remain in the bottom of the bottle.

(D-199) Ehrlich's reagent

Pour 100 ml of distilled water into a brown bottle. Add 0.7 g of *p*-dimethylaminobenzaldehyde. Carefully add 150 ml of concentrated hydrochloric acid. Mix.

(D-200) Sodium acetate, saturated

Heat 100 ml water to 45° C and add 130 g of sodium acetate, $NaC_2H_3O_2 \cdot 3H_2O$. Mix well and place in a 37° C water bath overnight. Cool to room temperature and allow excess sodium acetate crystals to settle out.

(D-201) Ehrlich's diazo reagent

a. Add about 500 ml of distilled water to a 1-L volumetric flask. Add 50 ml of concentrated hydrochloric acid. Weigh 5 g of sulfanilic acid and transfer to the volumetric flask. Add distilled water to the 1-L mark. Mix.

b. Weigh 0.5 g of sodium nitrite. Dissolve in 100 ml of distilled water. Pour into a brown bottle. Mix. Store in a refrigerator.

Place 25 ml of solution A in a flask. Add 0.5 ml in solution B. Mix. This reagent must be prepared just before use, since it does not keep.

(D-202) Calcium chloride, 10%

Weigh 10 g of anhydrous calcium chloride, $CaCl_2$, and transfer to a 100-ml volumetric flask. Dissolve and dilute to volume with water. Mix.

(D-203) Lugol's solution

Weigh 5 g of iodine and 10 g of potassium iodide and transfer to a brown bottle. Add 100 ml of water and mix.

(D-204) Zinc acetate, saturated alcoholic solution

Place 100 ml of ethyl alcohol in a beaker. Add zinc acetate, with stirring, until no more dissolves.

(D-205) Bromine water

Cautiously add a few drops of liquid bromine to 100 ml of distilled water. Mix. Store in a brown bottle. When the solution loses its color, prepare a fresh solution.

(D-206) *o*-Tolidine reagent

Dissolve 4 g of *o*-tolidine in 100 ml of glacial acetic acid. Keep in a brown bottle and store in a refrigerator. Prepare fresh every month.

(D-207) Trichloroacetic acid, 20%

Dissolve and dilute a ¼-pound bottle (see D-95) of trichloroacetic acid to 567 ml. Mallinckrodt trichloroacetic acid, AR, is recommended because of its extremely low iron content.

(D-208) Sodium acetate, 20%

Weigh 33 g of sodium acetate, $CH_3COONa \cdot 3H_2O$, and transfer to a 100-ml volumetric flask. Dissolve and dilute to volume with water. Mix.

(D-209) Ammonium hydroxide, 10% (w/v)

Measure 36 ml of concentrated ammonium hydroxide (28%) and transfer to a 100-ml volumetric flask. Dilute to volume with water and mix.

(D-210) Acetic acid, 10% (v/v)

Measure 100 ml of glacial acetic acid and transfer to a 1-L volumetric flask containing about 600 ml of water. Mix, dilute to volume with water, and mix again.

(D-211) Sulfuric acid, 2 N

Measure 56 ml of concentrated sulfuric acid in a 100-ml graduated cylinder and transfer to a 1-L volumetric flask containing about 600 ml of water. Swirl to mix. Cool to room temperature. Dilute to volume with water and mix.

(D-212) Ammonium oxalate, 0.1 M

Weigh 1.42 g of hydrated ammonium oxalate, $(NH_4)_2C_2O_4 \cdot H_2O$, and transfer to a 100-ml volumetric flask. Dissolve and dilute to volume with water. Mix. Store in a refrigerator.

(D-213) Barium chloride, 5%

Weigh 5 g of barium chloride and transfer to a 100-ml volumetric flask. Dissolve and dilute to the mark with distilled water. Store in a Pyrex or polyethylene bottle.

(D-214) Ferric chloride, 5%

Weigh 5 g of $FeCl_3$ or 8.3 g of $FeCl_3 \cdot 6H_2O$ and transfer to a 100-ml volumetric flask. Dissolve and dilute to volume with water. Mix.

(D-215) Sodium carbonate, 1%

Weigh 1 g of sodium carbonate, Na_2CO_3, and transfer to a 100-ml volumetric flask. Dissolve and dilute to volume with water. Mix.

(D-216) Starch, 1%

Heat 100 ml of distilled water to boiling. Add a mixture of 1 g of soluble starch in 5.0 ml of distilled water, stirring constantly until the fluid again reaches the boiling point. Let cool. If the solution is to be kept more than a few days, add a few drops of toluol or chloroform and store in a refrigerator.

(D-217) Gram's iodine solution

Weigh 0.4 g of iodine and 0.8 g of potassium iodide and dissolve in 120 ml of water. Store in a brown bottle.

(D-218) Acetic acid, 1%

Pipet 1.0 ml of glacial acetic acid into a 100-ml volumetric flask. Dilute to volume with water and mix.

(D-219) Benzidine, saturated, in glacial acetic acid

The benzidine powder must be labeled "for blood tests." Weigh 4 g of benzidine powder and transfer to a brown bottle. Add 50 ml of glacial acetic acid. Shake to dissolve as much as possible. This solution is stable for about 1 month.

(D-220) Barium chloride, 10%

Weigh 10 g of barium chloride and transfer to a 100-ml volumetric flask. Dissolve and dilute to volume with water. Mix.

(D-221) Mercuric chloride, 10%

Weigh 10 g of mercuric chloride, $HgCl_2$, and transfer to a 100-ml volumetric flask. Dissolve and dilute to volume with water. Mix.

(D-222) Ethyl alcohol, 95%, containing 0.4% capryl alcohol

Pipet 4.0 ml capryl alcohol into a 1-L volumetric flask and dilute to volume with 95% ethyl alcohol. Mix.

(D-223) Potassium hydroxide, 33%

Weigh 330 g of potassium hydroxide pellets and transfer to a 1-L Pyrex volumetric flask containing about 600 ml of water. Swirl to dissolve and allow to cool to room temperature. Dilute to volume and mix. Store in a polyethylene bottle.

(D-224) Hydrochloric acid, 25%

Measure 676 ml of concentrated hydrochloric acid (37%) and transfer to a 1-L volumetric flask containing 300 ml of water. Dilute to volume with water and mix.

(D-225) Sodium hydroxide, 1 N

Prepare from a concentrated solution of sodium hydroxide that has been standardized. Standardize the diluted solution against 1 N acid. Adjust to 1.000 N if necessary.

(D-226) Thymol blue, 0.2% in 50% ethyl alcohol

Weigh 0.2 g of thymol blue and transfer to a 100-ml volumetric flask. Add 53 ml of 95% ethyl alcohol and dilute to volume with water and mix.

(D-227) Pandy's reagent

Weigh 10 g of phenol, C_6H_5OH, and dissolve in 100 ml of water. Place in an incubator at 37° C for 2 or 3 days. Use only the clear supernatant solution.

(D-228) Ammonium sulfate, saturated

Weigh 80 g of ammonium sulfate, $(NH_4)_2SO_4$, and dissolve in 100 ml of water.

(D-229) Trichloroacetic acid, 3%

Trichloroacetic acid crystals are difficult to weigh accurately because they absorb moisture from the air very rapidly. It is recommended that dilute solutions of TCA be made from more concentrated solutions. Measure 300 ml of 10% trichloroacetic acid (D-95) and transfer to a 1-L volumetric flask. Dilute to volume with water and mix.

(D-230) Sodium hydroxide, 40% (10 N)

Weigh 40 g of sodium hydroxide pellets and transfer to a Pyrex 100-ml volumetric flask containing about 50 ml of water. Swirl to dissolve. Cool to room temperature, dilute to volume, and mix.

(D-231) Gold chloride, 0.5% in water

Weigh 0.5 g of gold chloride, $AuCl_3 \cdot HCl \cdot 3H_2O$, and transfer to a 100-ml volumetric flask. Dissolve and dilute to volume with water. Store in a refrigerator.

(D-232) Bromide stock standard, 20 mg/ml

Weigh 2.575 g of sodium bromide, NaBr, and transfer to a 100-ml volumetric flask. Dissolve and dilute to volume with water. Mix.

(D-233) Hydrochloric acid, 6 N

Slowly and with mixing add 500 ml of concentrated hydrochloric acid to 500 ml of water. Mix well and store in a Pyrex bottle.

(D-234) Ammonium sulfamate, 0.5%

Weigh 0.5 g of ammonium sulfamate, $NH_2SO_3NH_4$, and transfer to a 100-ml volumetric flask. Dissolve and dilute to volume with water. Store in a refrigerator.

(D-235) Color reagent (ferric-mercuric reagent)

Weigh 40 g of mercuric chloride, $HgCl_2$, and transfer to a 1-L volumetric flask containing about 800 ml of water. Swirl to dissolve. Add 120 ml of 1 N hydrochloric acid (D-120) and 40 g of ferric nitrate, $Fe(NO_3)_3 \cdot 9H_2O$. When all the ferric nitrate is dissolved, dilute to volume with water and mix.

(D-236) Salicylate stock standard, 1 mg/ml

Weigh 1.16 g of sodium salicylate, $C_7H_5O_3 \cdot Na$, and transfer to a 1-L volumetric flask. Dissolve and dilute to volume with water. Add a few drops of chloroform as a preservative. Store in a refrigerator. This solution is stable for about 6 months.

(D-237) Salicylate standard solution, 0.2 mg/ml

Pipet 20 ml of stock salicylate solution (D-236) into a 100-ml volumetric flask and dilute to volume with water. Add a few drops of chloroform as a preservative. Store in a refrigerator. This solution is stable for about 6 months.

(D-238) Hydrochloric acid, 2 N

Measure 167 ml of concentrated hydrochloric acid, HCl, and transfer to a 1-L volumetric flask containing about 600 ml of water. Dilute to volume and mix.

(D-239) Phosphate buffer, pH 7.0

Weigh 5.785 g of dibasic sodium phosphate, Na_2HPO_4, and 3.532 g of monobasic potassium phosphate, KH_2PO_4, and transfer to a 1-L volumetric flask. Dissolve and dilute to volume with water. Mix thoroughly.

(D-240) Potassium permanganate, 5%

Weigh 5 g of potassium permanganate and transfer to a 100-ml volumetric flask. Dissolve and dilute to volume with water. Mix and store in an amber glass bottle.

(D-241) Lithium stock standard, 10 mEq Li/L

Weigh 0.4240 g of lithium chloride which has been dried at 125 C for ½ hour and cooled in a desiccator. Dissolve and dilute to 1000 ml with distilled water and mix well. Store in a Pyrex or polyethylene bottle.

(D-242) Lithium working standard, 1 mEq Li/L

Dilute 10 ml of lithium stock standard to 100 ml with distilled water. Store in a Pyrex or polyethylene bottle.

(D-243) Sterox SE, 1%

Pipet 9.5 ml (10 g) of Sterox SE into a 1-L volumetric flask. Allow pipet to drain completely. Slowly run water down the side of the flask and dilute to volume. If excessive foaming occurs, add 1 drop of caprylic alcohol. Mix well and store in a polyethylene bottle.

(D-244) Sterox SE, 0.02%

Pipet 20 ml of 1% Sterox SE into a 1-L volumetric flask. Dilute to volume with water and mix. Store in a polyethylene bottle.

E

Company directory

Abbott Laboratories, Diagnostics Division
Abbott Park
North Chicago, Ill. 60064

Advanced Instruments, Inc.
1000 Highland Ave.
Needham Heights, Mass. 02194

Aldrich Chemical Co., Inc.
940 N. St. Paul Ave.
Milwaukee, Wis. 53233

Allied Chemical Corp.
Industrial Chemicals Division
P. O. Box 70
Morristown, N. J. 07960

Aluminum Company of America
1501 Alcoa Bldg.
Pittsburgh, Pa. 15219

Alupharm Chemicals
610-612 Commercial Pl.
P. O. Box 30628
New Orleans, La. 70130

American Cyanamid Co.
Process Chemicals Department
Wayne, N. J. 07470

American Instrument Co., Inc.
Division of Travenol Laboratories, Inc.
8030 Georgia Ave.
Silver Spring, Md. 20910

Amersham-Searle Corp.
2636 S. Clearbrook Dr.
Arlington Heights, Ill. 60005

Ames Co., Division of Miles Laboratories, Inc.
1127 Myrtle St.
Elkhart, Ind. 46514

Analabs, Inc., Subsidiary of New England Nuclear
80 Republic Dr.
North Haven, Conn. 06473

Analtech Inc.
75 Blue Hen Dr.
Newark, Del. 19711

Arco Chemical Co.
Division of AtlanticRichfieldCo.
260 S. Broad St.
Philadelphia, Pa. 19101

Armour Pharmaceutical Co.
P. O. Box 511
Kankakee, Ill.

Atlas Chemical Industries, Inc.
Wilmington, Del. 10899

Bausch & Lomb, Inc.
635 St. Paul St.
Rochester, N. Y. 14602

Beckman Instruments, Inc.
2500 Harbor Blvd.
Fullerton, Calif. 92634

Beckman Instruments, Inc.
Spinco Division
1117 California Ave.
Palo Alto, Calif. 94304

Behring Diagnostics, Inc., Subsidiary of American Hoechst Corp.
Behringwerke A. G.
Route 202-206
North Sommerville, N. J. 08876

Biological Corporation of America
(formerly Schering Diagnostics)
40 Markley St.
Port Reading, N. J. 07064

Bio-Rad Laboratories
32nd and Griffin Ave.
Richmond, Calif. 94804

BIO-RIA
10850 Rue Hamon
Montreal, Quebec, Canada H3M 3A2

Blaisdell Publishing Co.
275 Wyman St.
Waltham, Mass. 02154

Brinkmann Instruments, Inc.
Cantiaque Rd.
Westbury, N. Y. 11590

Buchler Instruments, Inc.
Division of Searle Analytic, Inc.
1327 16th St.
Fort Lee, N. J. 07024

Calbiochem
P. O. Box 12087
San Diego, Calif. 92112

Canalco Inc.
Canal Industrial Corp.
5635 Fisher Lane
Rockville, Md. 20852

Clay Adams, Division of
Becton, Dickinson & Co.
299 Webro Rd.
Parsippany, N. J. 07054

Clinical Assays, Inc.
237 Binney St.
Cambridge, Mass. 02142

Cobe Laboratories, Inc.
1201 Oak St.
Denver, Colo. 80215

Colab Laboratories, Inc.
1526 Halsted St.
Chicago Heights, Ill. 60411

Coleman Instruments,
Division of The Perkin-Elmer Corp.
2000 York Rd.
Oak Brook, Ill. 60521

Corning Scientific Instruments
Medfield, Mass. 02052

Curtis Laboratories, Inc.
1948 E. 46th Street
Los Angeles, Calif. 90058

Dade, Division of America Hospital Supply Corp.
P. O. Box 520672
Miami, Fla. 33152

Dajac Laboratories,
Division of The Borden Chemical Co.
5000 Langdon St.
Philadelphia, Pa. 19124

F. A. Davis Co.
1915 Arch St.
Philadelphia, Pa. 19103

Denver Chemical Manufacturing Co., Inc.
35 Commerce Rd.
Stamford, Conn.

Diagnostic Products
12306 Exposition Blvd.
Los Angeles, Calif. 90064

Diagnostics Biochem Canada, Inc.
249 Wortley Rd.
London, Ontario, Canada N6C 3P9

Distillation Products Industries,
Division of Eastman Kodak Co.
755 Ridge Road W.
Rochester, N. Y. 14603

The Dow Corning Corp.
P. O. Box 1592
2030 Abbott Rd. Center
Midland, Mich. 48640

Drew Chemical Co.
Boonton, N. J.

E-C Apparatus Corp.
3831 Tyrone Blvd.
North St. Petersburg, Fla. 33709

Eastman Organic Chemicals,
Division of Eastman Kodak Co.
343 State St.
Rochester, N. Y. 14650

Evergreen Scientific
2300 East 49th St.
P. O. Box 58248
Los Angeles, Calif. 90058

Fisher Scientific Co.
711 Forbes Ave.
Pittsburgh, Pa. 15219

Floridin Co.
Three Penn Center
Pittsburgh, Pa. 15235

Fluka, A. G.
Buchs, S. G.
Switzerland

Gelman Instrument Co.
600 S. Wagner Rd.
Ann Arbor, Mich. 48106

General Diagnostics,
Division of Warner-Lambert Pharmaceutical Co.
210 Tabor Rd.
Morris Plains, N. J. 07950

Gilford Instruments Laboratories, Inc.
132 Artino St.
Oberlin, Ohio 44074

W. R. Grace & Co.,
Davison Chemical Division
Charles and Baltimore Sts.
Baltimore, Md. 21203

Hamilton Co.
P. O. Box 10030
Reno, Nev. 89510

Harleco, Division of
American Hospital Supply Corp.
480 Democrat Rd.
Gibbstown, N. J. 08027

Helena Laboratories
P. O. Box 752
Beaumont, Tex. 77704

Hoffman-LaRoche Inc.
Kingsland Rd. and Bloomfield Ave.
Nutley, N. J. 07110

Hyland, Division of Travenol Laboratories, Inc.
P. O. Box 2214
Costa Mesa, Calif. 02626

Hynson, Westcott & Dunning, Inc.
Charles and Chase Sts.
Baltimore, Md. 21201

ICN Pharmaceuticals, Inc.
P. O. Box 3932
Portland, Ore. 97208

Instrumentation Laboratory, Inc.
113 Hartwell Ave.
Lexington, Mass. 02173

Instrumentation Specialties Co., Inc., (ISCO)
P. O. Box 5347
Lincoln, Neb. 68505

Johns-Manville Products Corp.
Greenwood Plaza
Denver, Colo. 80217

Kallestad Laboratories, Inc.
1000 Lake Hazeltine Dr.
Chaska, Minn. 55318

K & K Laboratories, Inc.
121 Express St.
Plainview, N. Y. 11803

Kensington Scientific Corp.
1819 67th St.
Oakland, Calif. 94608

Lederle Diagnostics
American Cyanamid Co.
N. Middletown Rd.
Pearl River, N. Y. 10965

E. Leitz, Inc.
Link Dr.
Rockleigh, N. J. 07647

Eli Lilly & Co.
740 S. Alabama St.
Indianapolis, Ind. 46206

The London Co.
811 Sharon Dr.
Cleveland, Ohio 44145

3M Company
3M Center
St. Paul, Minn. 55101

Macalaster Bicknell Co. of Conn. Inc.
181 Henry St.
New Haven, Conn. 06507

Mallinckrodt Inc.
P. O. Box 5439
St. Louis, Mo. 63147

Matheson Gas Products
P. O. Box 85
East Rutherford, N. J. 07073

Matheson, Coleman & Bell
2909 Highland Ave.
Norwood, Ohio 45212

Melpar Division, American Standard Co.
7700 Arlington Blvd.
Falls Church, Va. 22046

Merck, Sharp & Dohme
Division of Merck & Co., Inc.
West Point, Pa. 19486

Miles Chemical Co.
Clifton, N. J.

Millipore Corp.
Ashby Rd.
Bedford, Mass. 01730

The C. V. Mosby Co.
11830 Westline Industrial Drive
St. Louis, Mo. 63141

Nalgene Labware Division,
Nalge Co.
75 Panorama Creek Dr.
Rochester, N. Y. 14625

National Aniline Division,
Allied Chemical Corp.
P. O. Box 70
Morristown, N. J. 07960

National Instrument Laboratories, Inc.
4119 Fordleigh Rd.
Baltimore, Md. 21215

New England Nuclear
575 Albany St.
Boston, Mass. 02118

Nutritional Biochemicals Corp.
26201 Miles Rd.
Cleveland, Ohio 44128

Pentex, Inc.
P. O. Box 272
195 W. Birch St.
Kankakee, Ill. 60901

Perkin-Elmer Corp., Instrument Division
Main Ave.
Norwalk, Conn. 06856

Pierce Chemical Co.
P. O. Box 117
Rockford, Ill. 61105

Phipps & Bird, Inc., Manufacturing Division
P. O. Box 27324
Richmond, Va. 23261

Photovolt Corp.
1115 Broadway
New York, N. Y. 10010

Quickfit, Division of Instrutec Corp.
7 Just Rd.
Fairfield, N. J. 07006

Reclin Co.
Box 80371
Lincoln, Neb. 68502

Regis Chemical Co.
8210 N. Austin Ave.
Morton Grove, Ill. 60053

Rohm & Haas Co.
Independence Mall West
Philadelphia, Pa. 19105

Schleicher & Schuell Co.
543 Washington St.
Keene, N. H. 03431

Schoeffel Instrument Corp.
24 Booker St.
Westwood, N. J. 07675

Scientific Service Laboratories, Inc.
P. O. Box 175
Dallas, Texas

Scientific Products
Division of American Hospital Supply Corp.
1430 Waukegan Rd.
McGraw Park, Ill. 60085

**Schwarz/Mann, Division
of Becton, Dickinson & Co.**
Mountain View Ave.
Orangeburg, N. Y. 10962

Searle Analytic Inc.
2000 Nuclear Dr.
Des Plaines, Ill. 60018

Serono Laboratories, Inc.
607 Boylston St.
Boston, Mass. 02116

Shandon Scientific Co.
515 Broad St.
Sewickley, Pa. 15143

Sigma Chemical Co.
P. O. Box 14508
St. Louis, Mo. 63178

Simmler & Son Inc.
3755 Forest Park Ave.
St. Louis, Mo. 63108

G. Frederick Smith Chemical Co.
867 McKinley Ave.
P. O. Box 23344
Columbus, Ohio 43223

E. R. Squibb & Sons
P. O. Box 4000
Princeton, N. J. 08540

Technicon Instruments Corp.
511 Benedict Ave.
Tarrytown, N. Y. 10591

**Texas Instruments Inc.
Digital Systems Division**
P. O. Box 1444
Houston, Tex. 77001

Arthur H. Thomas Co.
P. O. Box 779
Vine St. at Third
Philadelphia, Pa. 19105

**Travenol Laboratories, Inc.
Artificial Organs Division**
Sanders Rd.
Deerfield, Ill. 60015

G. K. Turner Associates, Inc.
2524 Pulgas Ave.
Palo Alto, Calif. 94303

Ultra-Violet Products, Inc.
5114 Walnut Grove Ave.
San Gabriel, Calif. 91778

UniChem Corp.
3897 Stephens Court
Tucker, Ga. 30084

Union Carbide Corp.
270 Park Ave.
New York, N. Y. 10017

Warner-Lambert Co., General Diagnostics Division
201 Tabor Rd.
Morris Plains, N. J. 07950

Waters Associates, Inc.
165 Maple St.
Milford, Mass. 01757

Wheaton Scientific, Division of Wheaton Industries
1000 N. Tenth St.
Millville, N. J. 08332

Wien Laboratories, Inc.
P. O. Box 227
Succasunna, N. J. 07876

Worthington Biochemical Corp.
Hall Mills Rd.
Freehold, N. J. 07728

SUPPLIERS OF MATERIALS FOR IMMUNOLOGIC PROCEDURES

ACI/Corning
Medfield Industrial Pk.
Medfield, Mass. 02052

Antibodies, Inc.
P. O. Box 442
Davis, Calif. 95616

Behring Diagnostics, Inc.
Rt. 202-206
North Sommerville, N. J. 08876

Biological Corporation of America
(formerly Schering Diagnostics)
40 Markley St.
Port Reading, N. J. 07064

Bio-Tec, Inc.
P. O. Box 35071
Minneapolis, Minn. 55435

Bio-Ware, Inc.
Box 8152
Wichita, Kan. 67208

Boehringer Mannheim Corp.
219 E. 44th St.
New York, N. Y. 10017

Cappel Laboratories, Inc.
P. O. Box 156
Downington, Pa. 19335

Colorado Serum Co.
4950 York St.
Denver, Colo. 80216

Dade, Division of American Hospital Supply Corp.
P. O. Box 520672
Miami, Fla. 33152

Fisher Scientific Co.
711 Forbes Ave.
Pittsburgh, Pa. 15219

Gibco Diagnostics
2801 Industrial Dr.
P. O. Box 4385
Madison, Wisc. 53713

Grafar Corp.
7340 Fenkel
Detroit, Mich. 48238

Helena Laboratories
P. O. Box 752
Beaumont, Tex. 77704

Hyland, Division of Travenol Laboratories, Inc.
P. O. Box 2214
Costa Mesa, Calif. 92626

ICN Medical Diagnostics Products
P. O. Box 3932
Portland, Ore. 97208

Kallestad Laboratories, Inc.
1000 Hazeltine Dr.
Chaska, Minn. 55318

Lederle Diagnostics
N. Middletown Rd.
Pearl River, N. Y. 10965

Lee Laboratories, Inc.
Route 1, Box 37
Grayson, Ga. 30221

Meloy Laboratories, Inc.
6715 Electronic Dr.
Springfield, Va. 22151

Microbiological Associates
4733 Bethesda Ave.
Bethesda, Md. 20014

Miles Research
Elkhart, Ind. 46514

**Millipore Biomedica,
Division of Millipore Corp.**
15 Craig Rd.
Acton, Mass. 01720

Oxford Laboratories, Inc.
1149 Chess Dr.
Foster City, Calif. 94404

Pfizer Diagnostics Division
235 E. 42nd St.
New York, N. Y. 10017

Pharmacia Laboratories, Inc.
800 Centennial Ave.
Piscataway, N. J. 08854

Rowley Biochemical
Route 1
Rowley, Mass. 01969

Technicon Instruments Corp.
511 Benedict Ave.
Tarrytown, N. Y. 10591

**Virgo Reagents,
Electro Nucleonics Laboratories, Inc.**
4905 Delray Ave.
Bethesda, Md. 20014

**Wellcome Reagents
Division of Burroughs Wellcome Co.**
3030 Cornwallis Rd.
Research Triangle
Triangle Park, N. C. 27709

Company	Electrophoresis equipment				Chemical reagent kits	Antigen/antibody reagents
	Cell	Power supply	Densi-tometer	Medium		
ACI/Corning Medfield Industrial Park Medfield, Mass. 02052	Yes	Yes	No	Agarose	Yes	No
Beckman Instruments 2500 Harbor Blvd. Fullerton, Calif. 92634	Yes	Yes	Yes	Acetate Agarose	Yes	Yes
Behring Diagnostics, Inc. Route 202-206 North Sommerville, N. J. 08876	Yes	?	No	Agar Agarose	Yes	Yes
Bio-RAD Laboratories 32nd & Griffin Ave. Richmond, Calif. 94804	Yes	Yes	No	Agarose Polyacrylamide	Yes	No
Bioware, Inc. P. O. Box 8152 Wichita, Kan. 67208	Yes	Yes	No	Agarose	Yes	Yes
Brinkmann Instruments, Inc. Cantiague Road Westbury, N. Y. 11590	Yes	Yes	No	Polyacrylamide	No	No
Buchler Instruments 1327 16th St. Ft. Lee, N. J. 07024	Yes	Yes	Yes	Acetate Starch Polyacrylamide	No	No
Calbiochem Box 12087 San Diego, Calif. 92112	No	No	No	Agar Agarose Polyacrylamide	Yes	No
CANALCO, Inc. 5635 Fisher Lane Rockville, Md. 20852	Yes	Yes	Yes	Polyacrylamide	Yes	Yes
Clifford Instruments, Inc. 17 Erie Dr. Natick, Mass. 01760	No	No	Yes	No	No	No
Colab Laboratories, Inc. Chicago Heights, Ill.	Yes	Yes	No	Acetate	No	No
E-C Apparatus Corp. 3831 Tyrone Blvd. N St. Petersburg, Fla. 33709	Yes	Yes	No	Polyacrylamide Starch	Yes	No
Electrothermal, Ltd. 111 West Industry Court Deer Park, N. Y. 11729	Yes	Yes	No	Polyacrylamide	No	No
Elvi Electronic Canada, Ltd. 615 West Dorchester Room 1010 Montreal, Quebec, Canada	Yes	Yes	Yes	Acetate	Yes	No
Gelman Instruments Co. 600 S. Wagner Road Ann Arbor, Mich.	Yes	Yes	Yes	Acetate	Yes	No

	Electrophoresis equipment				Chemical reagent kits	Antigen/antibody reagents
Company	Cell	Power supply	Densitometer	Medium		
Grafar Corp. 7340 Fenkel Detroit, Mich. 48238	Yes	Yes	No	Agar Agarose	No	No
Heath/Schlumberger Scientific Instruments Benton Harbor, Mich. 49022	No	Yes	No	No	No	No
Helena Laboratories P. O. Box 752 Beaumont, Tex. 77704	Yes	Yes	Yes	Acetate Agar	Yes	Yes
Hoefer Scientific Instruments 520 Bryant St. San Francisco, Calif. 94107	Yes	Yes	No	Polyacrylamide	No	No
Hyland Division, Travenol Laboratories, Inc. P. O. Box 2214 Costa Mesa, Calif. 92626	Yes	Yes	No	Agar	Yes	Yes
Instrumentation Laboratories, Inc. Lexington, Mass. 02173	Yes	Yes	Yes	Acetate	No	No
Instrumentation Specialties Co. 4700 Superior Lincoln, Neb. 68504	Yes	Yes	Yes	Polyacrylamide (Acetate)	No	No
Isolab, Inc. Drawer 4350 Akron, Ohio 44321	Yes	Yes	No	Polyacrylamide	No	No
Kallestad Laboratories, Inc. 1000 Hazeltine Dr. Chaska, Minn. 55318	Yes	Yes	No	No	No	Yes
LKB Instruments, Inc. 12221 Parklawn Drive Rockville, Md. 20852	Yes	Yes	No	Polyacrylamide Agar	Yes	Yes
Joyce Loebl & Co., Ltd. Northwest Industrial Park Burlington, Mass. 01803	No	No	Yes	No	No	No
Meloy Laboratories, Inc. 6715 Electronic Dr. Springfield, Va. 22151	Yes	Yes	No	No	No	Yes
Miles Laboratories, Inc. P. O. Box 272 Kankakee, Ill. 60901	No	No	No	Agarose	Yes	Yes
Millipore Biomedica, Division of Millipore Corp. 15 Craig Road Acton, Mass. 01720	Yes	Yes	Yes	Acetate Agarose	Yes	Yes

Company	Electrophoresis equipment				Chemical reagent kits	Antigen/antibody reagents
	Cell	Power supply	Densitometer	Medium		
National Instrument Laboratories, Inc. 12300 Parklawn Dr. Rockville, Md. 20852	Yes	Yes	No	No	No	No
ORTEC 1000 Midland Road Oakridge, Tenn. 37830	Yes	Yes	No	Polyacrylamide	No	No
Ortho Diagnostics Raritan, N. J. 08869	Yes	Yes	No	No	No	Yes
Pfizer Diagnostics Division 235 E. 42nd St. New York, N. Y. 10017	Yes	Yes	No	Agarose	Yes	No
Pharmacia Fine Chemicals AB Box 175, S-75104 Uppsala 1, Sweden	Yes	Yes	No	Polyacrylamide	No	No
Photovolt Corp. 1115 Broadway New York, N. Y. 10010	Yes	Yes	Yes	Acetate	No	No
Reclin Co. Box 80371 Lincoln, Neb. 68502	Yes	Yes	No	All	No	No
Savant Instruments, Inc. 221 Park Ave. Hicksville, N. Y. 11801	Yes	Yes	No	Paper Polyacrylamide	No	No
Schoeffel Instrument Corp. 24 Booker St. Westwood, N. J. 07675	No	No	Yes	No	No	No
Shandon Scientific Co., Inc. 515 Broad St. Sewickley, Pa. 15143	Yes	Yes	No	Silica gel Paper Acetate Polyacrylamide	No	No
Spectra Biologicals Oxnard, Calif. 93030	Yes	Yes	No	Agarose	No	Yes
Transidyne General Corp. Ann Arbor, Mich. 48103	No	No	Yes	No	No	No

Index

A

ABB; see Air-blood barrier
Abel, Rowntree, and Turner hemodialysis apparatus, 263
Abell, Levy, Brodie, and Kendall, total serum cholesterol determination with method of, 193-194
Abetalipoproteinemia, 564
Abnormal fat catabolism, 288-289
Absorbance
  conversion of percent absorption to, 207-208
  energy, pH and, 467
Absorption
  background, in atomic absorption spectroscopy, 203
  digestion and, 308-310
  energy, determination of, 466
  light, laws for
    Beer's, 466-467
    Bouguer-Lambert, 466
  percent, conversion of, to absorbance, 207-208
  photoelectric, 333
Absorption chemical indicator, 687
Absorption spectrum
  of barbiturate, 469
  of diazepam, 472
  of diphenylhydantoin, 473
  of N-substitute barbiturate, 470
Accumulated dose of radiation, 350
Accuracy in radioassay, 339
Acetest procedure for protein in urine, 289
Acetone
  dialysis to remove, 274
  in urine, 288-289
    Acetest procedure for, 289
Achromatic chemical indicators, 687
ACI/Corning agarose power supply and cell for electrophoresis of serum proteins, 547
ACI/Corning method
  for electrophoresis or serum proteins, 545-548
  for lactic dehydrogenase isoenzyme electrophoresis, 577-578

Acid; see also specific acid
  dibasic, 684
  monobasic, 684
  polybasic, 684
Acid phosphatase
  calibration curve for, 231-232
  determination of, 227
  increase of, in women, 225
  and metastatic prostatic carcinoma, 225
  prostatic, determination of, 231
  total, determination of, 229-230
Acid urine, causes of, 277-278
Acidemia, lactic, causes of, 142, 144
Acidity, gastric, titratable, determination of, 313
Acidosis, 128
  diabetic, causing electrolyte imbalance, 116
  lactic, 144
  metabolic, 145-147, 173
    definition of, 147
    hypoxic, 147
    in renal disease, 260
  respiratory, compensation slopes for, 174
  ventilatory, 172
Acromegaly, insulin in, 374
ACTH; see Adrenocorticotropic hormone
Action lines of Shewhart chart, 19
Activation, enzymatic, 214
  factors affecting, 215-218
Activation energy, 214
Activity
  amylase, determination of, with Klein-Foreman-Searcy method, 221-222
  Brockman-Schodder scale of, 491
  enzymatic; see Enzymes, activity of
  lactate dehydrogenase, elevation of, 236
  plasma renin, 261-262
  between solute and solvent, 137
Actual bicarbonate, 128
Actual carbon dioxide tension, 127
Actual pH, 127
Acute meningitic infection, cerebrospinal fluid in, 323

Acute poisoning, dialysis for, 273-274
Acute porphyria, 514-515
Acute renal failure, 262
Acute tubular necrosis and renal failure, 262
Addison's disease, 117
　clinical findings in, 424
Additions, method of
　in atomic absorption spectroscopy, 204-205
　to minimize interferences in atomic absorption spectroscopy, 203
Adenohypophysis; see Anterior pituitary gland
ADH; see Antidiuretic hormone
Adrenal cortex
　ACTH and, 380-381
　and blood glucose level, 93
　functions of
　　laboratory tests of, 382-389
　　tests of, 387-388
　hormones of, 374-389
　hyperfunction of, 388-389
　hypofunction of, 388
　　hypoglycemia from, 93-94
　increased activity of, causing hyperglycemia, 94
Adrenal gland, 374
　hormones of, 374-393
Adrenal hyperplasia
　clinical findings in, 424
　congenital, 389
Adrenal insufficiency, primary and secondary, 388
Adrenal medulla, 389-393
Adrenocortical insufficiency, clinical findings in, 424
Adrenocortical steroid hormones; see Steroid hormones
Adrenocorticotropic hormone, 363
　and adrenal cortex, 380-381
　feedback control of, 360-361
Adrenocorticotropic hormone stimulation test of adrenocortical function, 387
Adrenogenital syndrome, 383
Adsorbents
　in chromatography, 490-492
　for column chromatography, 504-505
Adsorption affinity of hydrocarbons in chromatography, 494-495
Adsorption isotherm, concentration of solute versus, 489
Aerosol 22 as surfactant with AutoAnalyzer, 40
AES; see Automatic external standardization
Agar
　citrate, hemoglobin mobility on, 573
　as support medium in electrophoresis, 531
Agarose
　creatinine phosphokinase isoenzymes separated on, 580
　as support medium in electrophoresis, 531-532
Agarose electrophoresis of serum lipoproteins, 560-561
Agarose immunoelectrophoresis, 611
Agarose method
　of creatine phosphokinase isoenzyme electrophoresis, 572-576
　for electrophoresis of serum proteins, 545-548
　for lactic dehydrogenase isoenzyme electrophoresis, 577-578
AIP; see Automated immunoprecipitins

Air-blood barrier, 148-150
Airborne radioactivity area, sign in, 347
Alanine aminotransferase reaction in transaminase determination, 233
Albumin
　bromcresol green method of determination of, 183-184
　determination of, with AutoAnalyzer II, 73-74
　serum, changes in, in disease, 178
　total protein and, determination of, 183-184
　and total protein determination with AutoAnalyzer I, simultaneous, 70-73
Alcohol
　ethyl; see Ethyl alcohol
　thin-layer chromatography of, 500
Aldehydes, thin-layer chromatography for, 497-498
Aldosterone, 383-384
　radioimmunoassay for determination of, 384
　secretion of, stimulus for, 381
Alkali denaturation, determination of hemoglobin F by, procedure for, 569-570
Alkaline phosphatase, 224-225
　abnormal values of, conditions accompanied by, 226
　and bone disease, 224-225
　calibration curve for, 231-232
　determination of, 225-227
　　with AutoAnalyzer I, 61-63
　　with AutoAnalyzer II, 74-75
　　Bessey-Lowry-Brock method of, 227, 228-229
　　Bodansky method of, 226
　　King-Armstrong method of, 226-227
　liver and, 225
　tests with, for hepatobiliary disease, 179
Alkaline phosphatase isoenzyme electrophoresis, 576
　clinical significance of, 578
Alkaline phosphatase isoenzymes separated on cellulose acetate, 579
Alkaline urine, causes of, 278
Alkaloid drugs, thin-layer chromatographic method for determination of, 478-481, 497
Alkalosis, 128
　metabolic, 173-175
　　kidneys and, 175
　ventilatory, 155, 172
Allen "blue" color test for urinary dehydroepiandrosterone, 409-410
Allen correction, 445
　in determination of 17-ketosteroids, 433
Allowable limits of error, 20
Alpha decay, 330-331
Alpha emitters, shielding for, 348
Alpha melanocyte-stimulating hormone, 363
Alpha thyroid-stimulating hormone, 362
Alumina as adsorbent in chromatography, 490-491
Alveolar gas equation, 154-155
Alveolar hyperventilation, 155
Alveolar hypoventilation, 155
Alveolar hypoxia, pulmonary arteriolospasm from, 159
Alveolar ventilation, 150, 154-155
　minute, 154
　regulation of, 162
Alveoli, lung, hemodynamics of, 157-158
Alwall dialyzer, 264

Amanita phalloides, dialysis to remove, 274
Amines, thin-layer chromatography for, 498
Amino acids
    amphoteric, 529
    thin-layer chromatography of, 501
Ammonia, blood
    and liver, 181
    and liver function, 188-190
Ammonium oxalate as anticoagulant, 7, 8
Amphetamines
    dialysis to remove, 274
    thin-layer chromatographic method for determination of, 478-481
Amphoteric amino acids, 529
Amphoteric proteins, 529
Amylase
    abnormal values of, conditions accompanied by, 220
    activity of, determination of, Klein-Foreman-Searcy method of, 221-222
    decreased values of, in pancreatic diseases, 315
    determination of
        amyloclastic method of, 221
        chromogenic methods of, 221
        saccharogenic method of, 220-221
    in duodenal contents, 315
    in urine, 303
Amyloclastic method of amylase determination, 221
Amylopsin, determination of, in duodenal contents, 315
Analyses
    blood gas and pH, 164-165
    chemical, 402
        potassium determination by, 119-120
        sodium determination by, 117-118
    competitive protein binding, 340-341
    displacement; *see* Radioassay
    duodenal, 314-317
    fecal; *see* Feces, analysis of
    gastric, 312-314
    hormone, general comments on, 444-445
    saturation, 401
    sequential saturation, 340
Analysis report, urine, sample, 275
Anatomic dead space, 154
Anatomy of kidneys, 243-244
Androgens, testing of, 382-383
Anemia
    megaloblastic, 312
    pernicious, 312
        maximum gastric acid output in, 313
    sickle cell, 571
Anesthesia, general, causing hyperglycemia, 94
Angiotensin I, 261
Angiotensin II, 261
Anion gap, 173
Anoxia and epinephrine secretion causing hyperglycemia, 94
Anterior pituitary gland
    and blood glucose level, 93
    blood supply of, 360
    feedback control of, 360-361
    hormones of, 361-365
    hypofunction of, hypoglycemia from, 93-94
    hypothalamic control of, 360

Anterior pituitary gland—cont'd
    releasing/inhibiting factors of, 360
Antibodies, 594
    formation of, induced by antigens, 336
    heterogenous, 336
    position of, on immunoelectrophoresis run, 608
    in radioimmunoassay, 336-337
Anticoagulants, 7-8
Antidiuretic hormone, 365
Antidotes of poisons, 677
Antigen-antibody reaction, factors affecting, 336
Antigens, 594
    antibody formation induced by, 336
    bound, separation from free antigen, in radioimmunoassay, 337-339
    position of, on immunoelectrophoresis run, 608
    in radioimmunoassay, 336
Antilymphocyte globulin to suppress homograft reaction, 273
Antiseptic for venipuncture, 5
Antiserum, formation of, 337
Aphagia, 348
Apparatus
    for Gettler-Freimuth method for carbon monoxide determination, 458
    Microzone acrylamide gel, 550
    ultrafiltration, for electrophoresis of urinary proteins, 560
Application of serum sample in micro–cellulose acetate electrophoresis, 540
Approximate rule of 1200, 153
Arachnoid mater, 322
Archibald method for uric acid determination, 108-109
Areas
    airborne radioactivity, sign in, 347
    controlled, 346-347
    radiation
        high, sign in, 347
        sign in, 347
Arginine infusion test of human growth hormone, 362
Arithmetic mean, 29
Army Ordnance and Army Service Forces, sampling inspection tables for, 10
Arsenic
    quantitative colorimetric method for determining, 450-452
    Reinsch modified test for, 449-450
Arsine generator for arsenic determination, 450
Arterial catheter for blood gas and pH measurements, 164
Arteries
    hypophyseal, blood supply to anterior pituitary by, 360
    punctures of, for blood gas and pH measurements, 164
Arteriolar nephrosclerosis from primary hypertension, 263
Arteriolospasm, pulmonary, from alveolar hypoxia, 159
Artificial kidney; *see* Dialyzer
ARW-7 as surfactant with AutoAnalyzer, 40
Aspartate aminotransferase determination
    with AutoAnalyzer I, 59-61
    with AutoAnalyzer II, 82-84

Aspartate aminotransferase reaction in transaminase determination, 232-233
Assay
  competitive protein binding, 401
    for plasma 17-hydroxycorticosteroids, 419-421
  definition of, 401
  hormone, 401-445
  immunoradiometric, 340
  radioreceptor, 340
  rapid, for urinary porphyrins using thin-layer chromatography, 514-517
Astrup studies of blood pH, P_{CO_2}, and P_{O_2}, 127-128
Atelectasis, 162
Atomic Absorption Spectrophotometer, Perkin-Elmer, 204
Atomic absorption spectroscopy, 202-213
  acetylene cylinders causing interference in, 203
  calcium determination using, 205-206
  copper determination in serum and urine using, 209-211
  detection limits in, 203
  hollow cathode lamps in, 202-203
  interferences in, 203
  lead determination in blood and urine in, 211-213
  magnesium determination in serum and urine using, 209
  method of additions in, 204-205
  sensitivity in, 203
  standards for, 203-204
  zinc determination in plasma and urine using, 211
Atomic number of atom, 329
Atomic theory, Dalton's 328
Atoms
  atomic number of, 329
  electrons of, 328, 329
  isotope, 328
  mass number of, 329
  neutron number of, 329
  neutron/proton ratio of, 330
  neutrons of, 329
  particles of, 329
  protons of, 328, 329
  structure of, 328, 329
Auger electron, 332
AutoAnalyzer
  adapting manual methods to, 36-37
  chloride determination with, 41-46
  general operating procedure for, 37-39
  methods for, 36-89
  wetting agents for use with, 39-41
AutoAnalyzer I
  alkaline phosphatase determination with, 61-63
  aspartate aminotransferase determination with, 59-61
  calcium determination with, 55-57
  carbon dioxide determination with, 41-46
  creatinine determination with, 50-52
  direct cholesterol determination with, 65-67
  electrolyte determination with, 41-46
  glucose determination with, 46-47
  inorganic phosphate determination with, 57-59
  lactic dehydrogenase determination with, 63-65

AutoAnalyzer I—cont'd
  methodology using, 41-73
  methods for, 36-73
  potassium determination with, 41-46
  serum glutamic-oxaloacetic transaminase determination with, 59-61
  simultaneous direct and total bilirubin determination with, 67-70
  simultaneous glucose and urea nitrogen determination with, 46-50
  simultaneous total protein and albumin determination with, 70-73
  sodium determination with, 41-46
  uric acid determination with, 52-54
AutoAnalyzer II
  albumin determination with, 73-74
  alkaline phosphatase determination with, 74-75
  aspartate aminotransferase determination with, 82-84
  calcium determination with, 75-76
  creatinine determination with, 78-79
  direct cholesterol determination with, 77-78
  glucose determination with, 79-82
  glucose oxidase determination with, 79-81
  glutamic-oxaloacetic transaminase determination with, 82-84
  inorganic phosphorus determination with, 84-85
  methods using, 73-89
  neocuproine determination with, 81-82
  total protein determination with, 85-86
  urea nitrogen determination with, 86-87
  uric acid determination with, 88-89
Automated calcium and magnesium determination, 209, 210
Automated immunoprecipitins, 627-628, 629
Automated specific protein analysis by nephelometry, 627-628
Automatic external standardization, 344-345
Azathioprine to suppress homograft reaction, 273
Azobilirubin, 196

B

B cells, 594
Babson-Phillips method of lactate dehydrogenase determination, 235-236
Bachra-Dauer-Sobel method of calcium determination, 131-132
Bacilli, Boas-Oppler, microscopic examination of, 313, 314
Background absorption in atomic absorption spectroscopy, 203
Badges for radiation monitoring, 349
Barbital buffers in electrophoresis, 530
Barbiturates
  absorption spectrum of, 469
  determination of, 460-463
    colorimetric methods for, 461-463
    Garvey-Bowden method for, 461-463
    Goldbaum modified method for, 468-471
    using thin-layer chromatography, 481-482
    ultraviolet spectrophotometric method of, 468-471
  dialysis to remove, 274
  identifying color of, in thin-layer chromatography, 520

Barbiturates—cont'd
  N-substituted
    absorption spectrum of, 470
    determination of, 471
  R_f values of, 520
  separated, on thin-layer plate, 521
  in urine, identification of, by thin-layer chromatography, 517-521
Basal gastric secretion, determination of, 313
Base, buffer, 128
  concept of, 171
Base deficit, total, estimation of, in blood gas quality control, 166
Base excess, 127-128
Basosis, 128
BCG; see Bromcresol green
Beckman Recording Integrating Densitometer for micro–cellulose acetate electrophoresis, 538
Beer's law for light absorption, 466-467
BEI; see Butanol-extractable iodine
Bellini, papillary ducts of, 244
Bell-shaped curve, 9
  on Shewhart chart, 20
Bence Jones protein in urine, determination of, 282-283
Benedict method of glucose determination, 95
  in urine, 284-285
Benign monoclonal gammopathies, criteria for, 620
Benzidine test for fecal analysis for occult blood, 317
Berthelot reaction for urea nitrogen determination, 112
Bessey-Lowry-Brock method
  of acid phosphatase determination, modified, 227
  of alkaline phosphatase determination, 227, 228-229
Beta decay, 331
Beta emitter
  energy frequency distribution of, 341
  shielding for, 348
Beta melanocyte-stimulating hormone, 363
Beta thyroid-stimulating hormone, 362
Bicarbonate
  actual, 128
  as blood buffer, 145
  blood concentration of, in kidney disease, 260
  standard, 128
Bile
  in digestion, 309-310
  in duodenal contents, 314
    determination of, 316
  in stomach contents, 314
Bile pigment
  fecal analysis for, 318
  metabolism of, 195-200
Bile salts in digestion, 310
Biliary tract, inflammation of, microscopic examination of, 317
Bilirubin
  determination of, 195-196
    recommended standardization of, 197-198
  liver function tests for, 180
  serum, 178-179
  total and direct, determination of, 196-197
    Malloy-Evelyn modification of, 196-197

Bilirubin—cont'd
  total and direct, determination of—cont'd
    simultaneous, with AutoAnalyzer I, 67-70
  in urine, Ictotest for, 295
  van den Bergh test of, 195
Bimodal distribution, 25
Binder in adsorbent in chromatography, 492
Biologic half-life, 334-335
Biosynthesis of steroid hormones, 378-380
Biuret reaction, Kingsley, for proteins, 181-182
Blood
  bicarbonate concentration in, in kidney disease, 260
  calcium concentration in, in kidney disease, 260-261
  capillary, for blood gas and pH measurements, 164-165
  chemistry of, in renal disease, 259-262
  collection of, 1-7
    capillary puncture of finger or ear for, 1-3
    and sample preparation, 1-8
  creatinine concentration in, in kidney disease, 260
  lead in, determination of, using atomic absorption spectroscopy, 211-213
  magnesium concentration in, in kidney disease, 261
  normal adult, approximate partition of carbon dioxide in, 150
  occult
    fecal analysis for
      benzidine test for, 317
      guaiac test for, 317-318
      Hematest for, 318
      o-tolidine test for, 318
    in stool, 180
    in urine, 300
  phosphate concentration of, in kidney disease, 260-261
  potassium concentration in, in kidney disease, 260
  protein concentration in, in kidney disease, 261
  sample preparation of; see Sample preparation
  sodium concentration in, in kidney disease, 260
  urea concentration in, in renal disease, 259-260
  withdrawing
    in capillary puncture, 3
    in venipuncture, 5-7
Blood ammonia
  and liver, 181
  and liver function, 188-190
Blood buffers, 143
  bicarbonate as, 145
  changes in components in, classification, 171
  hemoglobin as, mechanism of, 145
  pathology of, changes in, 147
Blood creatinine calibration chart, 103
Blood flow, renal, 256-259
  decreased, 257
  effective, 256
  total, 256
Blood gas
  analyses of, 164-165
  measurements of, 163-169
    electrodes in, 168-169
    quality control in, 165-168

Blood gas—cont'd
  measurements of—cont'd
    temperature and, 167
Blood gas machines, 165-166
Blood glucose, control of level of, 93
Blood pH
  Astrup studies of, 127-128
  determination of, 125-128
"Blue bloater," 163
Boas-Oppler bacilli, microscopic examination of, 313, 314
Bodansky method
  of acid phosphatase determination, 227
  of alkaline phosphatase determination, 226
Bodansky unit, 226
Body fluids
  composition of, 1
  constituents of, electrophoresis of, 545-587
Body surface area, chart for estimating, 248
Bohr effect, 149
Bonding
  hydrogen, and chromatography, 495
  to solid phase to separate bound from free antigen, 338
Bone disease and alkaline phosphatase, 224-225
Boric acid, dialysis to remove, 274
Bouguer-Lambert law for light absorption, 466
Bound antigen, separation from free antigen, in radioimmunoassay, 337-339
Bowman's capsule, 243
Brain
  meninges of, 322
  ventricles of, 322
Brij 35 as surfactant with AutoAnalyzer, 39
Brockman-Schodder scale of activity, 491
Bromcresol green method for albumin determination, 183-184
  with AutoAnalyzer II, 73-74
Bromide, quantitative colorimetric method for determination of, 463
Bromine test for melanin in urine, 299
Bromsulphalein test of excretory capacity of liver, 198-200
  dye dosages for, 199
  procedure for, 199-200
  Seligson-Marino-Dodson method for, 200
Bronchiolar-bronchial disease and ventilation:perfusion ratio, 161-162
Brown method for uric acid determination, 108
BSP; see Sulfobromophthalein
Buchler-Cotlove Chloridometer, 120-121
Buffer
  blood; see Blood buffers
  in electrophoresis, 530-531, 632-633
Buffer base, 128
  concept of, 171
Buffer diseases, metabolic, 171
Burner heads for Perkin-Elmer Atomic Absorption Spectrophotometer, 204
Butanol-extractable iodine
  measurement of thyroxin by, 367
  normal values for, 370

C

Calcitonin, 371
  and parathyroid gland, 372

Calcium
  abnormal values of, conditions accompanied by, 129
  blood concentration in, in kidney disease, 260-261
  determination of, 129-135
    using atomic absorption spectroscopy, 205-206
    with AutoAnalyzer I, 55-57
    with AutoAnalyzer II, 75-76
    automated, 209, 210
    Bachra-Dauer-Sobel method of, 131-132
    Clark-Collip modification of Kramer-Tisdall method of, 129-131
  diffusible, determination of, 133-135
  ionized, determination of, 133-135
  and phosphorus, determination of, 128-129
  serum, diffusible, nomogram for estimating, 134, 135
  in urine, 300-302
    determination of, 132-133
    Wang method for determination of, 301-302
Calculations
  dpm, 344
  in radioimmunoassay, 339
  of square root, 18
  of squares, 18
  of standard deviation, 17-18
Calculi in urine
  analysis of, 304
  tests for, 303
Calibration chart
  blood creatinine, 103
  urine creatinine, 106
Calibration curve
  for alkaline and acid phosphatase, 231-232
  for glutamic oxaloacetic transaminase and glutamic pyruvic transaminase, 234
Capillaries, lung, hemodynamics of, 157-158
Capillary blood for blood gas and pH measurements, 164-165
Capillary puncture of finger or ear for blood collection, 1-3
Capsule, Bowman's, 243
Carbohydrates
  thin-layer chromatography for, 498
  in urine, 286-288
Carbon dioxide
  check of, in blood gas quality control, 166
  compartments of, as volume, 152
  content, determination of, 122-125
  determination of
    with AutoAnalyzer I, 41-46
    temperature factors in, 125
  in normal adult blood, approximate partition of, 150
  oxyhemoglobin dissociation curve of, 151
Carbon dioxide tension
  actual, 127
  Astrup studies of, 127-128
  clinical conditions of, 160
  determination of, 125-128
  nomogram for calculating, 126
  total, 128
Carbon monoxide
  determination of, 456-460
    Gettler-Freimuth method of, 457-459

Carbon monoxide—cont'd
  determination of—cont'd
    semiquantitative method for, sensitive to 5% saturation, 457-459
    quantitative spectrophotometric procedure for, 460
    quick screening method sensitive to 20% saturation of, 456-457
Carbon tetrachloride, dialysis to remove, 274
Carbonate, urinary calculi of, 304
Carbonic acid, formation of, 145-146
Carboxylic acids, thin-layer chromatography for, 499
Carcinoid tumors, 399
Carcinoma
  gastric
    basal gastric secretion in, 313
    maximum gastric acid output in, 313
  prostatic, metastatic, and acid phosphatase, 225
Catabolism, abnormal fat, 288-289
Catalyst, enzymes as, 214
Catecholamines
  determination of, 392
  fluorescent measurement of, 392
  formation of, 390-391
  fractionated, determination of, 405-407
  synthesis and metabolism of, 391
  urinary, determination of, 402-405
Catheter, arterial, for blood gas and pH measurements, 164
Cell count of cerebrospinal fluid, 324-325
Cells
  B, 594
  and cellulose acetate frame with stand for electrophoresis, 533
  electrophoresis, 630-631
  Microzone, for micro–cellulose acetate electrophoresis, 536
  power supply and, ACI/Corning agarose, for electrophoresis of serum proteins, 547
  of stomach wall, 309
  T, 594
Cellulose acetate
  alkaline phosphatase isoenzymes separated on, 579
  glycoprotein separation on, 585
  hemoglobin mobility on, 572
  lactic dehydrogenase isoenzymes separated on, 581
Cellulose acetate membranes as support media for electrophoresis, 531
Cellulose acetate method
  of hemoglobin electrophoresis, 564-567
  of serum haptoglobin electrophoresis, 581-583
Cellulose mini-column, DE-52, hemoglobin A_2 quantitation with, 570-571
Central tendency, 25, 27
  measures of, 29
Centrolobular emphysema, 163
Cephalin-cholesterol flocculation test of liver function, 186-187
Cerebrospinal fluid, 322-327
  acute meningitis infection in, 323
  cell count of, 324-325
  chemical examination of, 325-327
  chloride in, determination of, 327

Cerebrospinal fluid—cont'd
  collection of, 323
  commercially prepared, as control, 13
  composition of, 322
  enzymes in determination of, 327
  functions of, 323
  globulin in, determination of, 326
  glucose in, determination of, 327
  microscopic examination of, 324-325
  movements of, 323
  physical appearance of, with various diseases, 323-324
  pooled, controls of, 13
  protein in
    chemical examination of, 325-327
    electrophoresis of, 556, 558
      normal values for, 558
  spinal puncture to collect, 323
  in suppurative meningitis, 323
  total protein in, determination of, 326-327
  in tuberculous meningitis, 323
  in xanthochromia, 323
Chair conformation of steroid hormones, 377
Chambers, electroscope ionization, for radiation monitoring, 349
Channel ratio method of measuring quench, 343-344
Charcoal, coated, to separate bound from free antigen, 338
Chart
  calibration
    blood creatinine, 103
    urine creatinine, 106
  control, 19
    interpretation of, 22-24
    plotting of, 19
    statistical methods using, 19-24
  cusum; see Cusum chart
  for estimating body surface area, 248
  Levy-Jennings, 20
  Shewhart; see Shewhart chart
Chemical analysis, 402
  potassium determination by, 119-120
  sodium determination by, 117-118
Chemical examination of cerebrospinal fluid, 325-327
Chemical fractionation of lactate dehydrogenase, 236-237
Chemical hazards, fire precautions and, 676-677
Chemical indicators, 687-689
Chemical inhibition of lactate dehydrogenase, 236-237
Chemical precipitation to separate bound from free antigen, 338
Chemical tests of renal function, 246-262
Chemicals, pure, solutions of, as controls, 13
Chemistry, blood, in renal disease, 259-262
Cherry-Crandall method of lipase determination, 222-223
Chief cells of stomach, 309
Chi-square test of goodness of fit in quality control in radio-assay, 352
Chloral hydrate, dialysis to remove, 274
Chlordiazepoxide, determination of, 464-465

Chloride
  abnormal values of, conditions accompanied by, 121
  in cerebrospinal fluid, determination of, 327
  determination of, 120-122
    with AutoAnalyzer I, 41-46
    Schales-Schales titration method for, 122
  in urine, 303
Chloridometer, Buchler-Cotlove, 120-121
Cholecystokinin-pancreozymin, 312
Cholesterol
  abnormal values of, conditions accompanied by, 193
  direct, determination
    with AutoAnalyzer I, 65-67
    with AutoAnalyzer II, 77-78
  serum, total, determination of, with method of Abell, Levy, Brodie, and Kendall, 193-194
  total, determination of, 192-194
Chorionic gonadotropin, determination of, 398
Chorionic somatomammotropin, human, determination of, 399
Chromatoelectrophoresis to separate bound from free antigen, 338
Chromatography, 477-482, 485-525
  adsorbents in, 490-492
  choice of kind of, 495, 496, 503-525
  column; see Column chromatography
  development of, 485-486
  dielectric constant in, 494
  elution and displacement in, 492, 494
  elutropic solvent series for, 493
  gas; see Gas chromatography
  gas-liquid, 477
  general aspects of, 490-494
  hydrocarbons and, 494-495
  mobile phase of, 492-494
  molecular structure and, relationship between, 494-495
  paper, development of, 485, 486
  partition
    development of, 485
    solvent for, 493
  polarity and, 494
  sensitivity of types of, to structural differences in compounds, 496
  stationary phase in, 490-492
  theory of, 486-490
  thin-layer; see Thin-layer chromatography
  zone in, shape of, 489
  zone spreading in, 488
Chromogenic methods of amylase determination, 221
Chromogens, Porter-Silber
  measurement of, 386
  urinary 17-hydroxycorticosteroid determination using, 421-425
Chronic renal insufficiency, 262-263
Chylomicrons of liver, 190-191
Circulating destaining unit for micro–starch gel electrophoresis, 542
Cirrhosis of liver, 614
Citrate agar, hemoglobin mobility on, 573
Citrate agarose electrophoresis for hemoglobin electrophoresis, 567-568
Clark polarographic electrodes for blood gas measurements, 164
Clark-Collip modification of Kramer-Tisdall method of calcium determination, 129-131
Class separation of lipids, thin-layer chromatography of, 499-500
Classes of immunoglobulins, 597-600
  relation to heavy-chain allotypes of immunoglobulins, 598
Classification
  of enzymes, 218
  of lactated dehydrogenase isoenzymes, 237
  of primary immunodeficiency disorders, 615
Cleaning of mercury, 125
Clearance function of liver, 178-180
Clearance tests
  creatinine, 252-253
  inulin, 247, 249
  PAH, of renal blood flow, 257-258
  urea, 249-252
    Moller-McIntosh-Van Slyke method of, 250-252
Clearing in electrophoretic procedures, 636-638
Clinical development of statistical quality control, 10-11
Clinical pathology of hydrogen ion homeostasis, 141-176
Clinitest method of testing for glucose in urine, 285
Closing volume, determination of, 162
Clothing worn when handling radioisotopes, 347
Coated charcoal to separate bound from free antigen, 338
Cocktail, scintillation, 342
Coefficient of variation, 18
Cold laboratory, 346
Collection
  blood; see Blood, collection of
  specimen, and quality control, 12
  of urine, 274-275
    factors affecting, 444
Collodion bag ultrafiltration for electrophoresis of urinary proteins, 559
Color
  of barbiturates, identifying, in thin-layer chromatography, 520
  of urine, 275
    causes of, 276
Color reaction
  direct, in simultaneous glucose and urea nitrogen determination, 47
  inverse
    in electrolyte determination, 41
    in simultaneous glucose and urea nitrogen determination, 47
    in uric acid determination, 52
Colorimetric determination of barbiturate levels, 461-463
Colorimetric measurement of blood pH, P_{CO_2}, and P_{O_2}, 126
Colorimetric method
  for arsenic determination, quantitative, 450-452
  for bromide determination, quantitative, 463
  for determination of salicylate radical, quantitative, 465-466

Column chromatography, 503-507
  adsorbent for, 504-505
  application of sample in, 505
  detection of components in, 506-507
  development of, 486
  solvent for, 505-506
    setup for change in polarity of, 506
Comag method of plate preparation in thin-layer chromatography, 509
Commercially prepared sera and cerebrospinal fluid as controls, 13
Commission on enzymes of International Union of Biochemists, 218
Compensation, 145-147
Compensation slopes for respiratory acidosis, 174
Competitive equilibrium radioimmunoassay, 336
Competitive inhibition of enzymes, 217
Competitive protein binding
  for compound S determination, 385
  for cortisol determination, 385
  measurement of thyroxine by, 368
Competitive protein binding analysis, 340-341
Competitive protein binding assay, 401
  for plasma 17-hydroxycorticosteroids, 419-421
Complement system, 600-602
  classic pathway of, 601
  evaluation of, 600, 602
  pathways in, 600, 602
Complement values by radial immunodiffusion and immunoelectrophoresis, 603
Compound F, determination of, 385
Compound S, determination of, 385
  procedure for, 407-408
Compton scattering, 333
Computation
  decimal point in, 16
  of percentages, 16
  of ratios, 15-16
Concentration
  hydrogen ion, 144
  substrate, and enzyme activity, 215-216
Concentration tests of tubular function, 254-255
Congenital adrenal hyperplasia, 389
Congenital lipoid hyperplasia, 389
Conjugation in liver, 198
Conn's syndrome, 384
Constant voltage power source for immunoelectrophoresis, 604
Constants
  ionization, 683-687
  Michaelis, and enzyme activity, 215-216
Contamination, monitoring of, 345
Continuous data in frequency distribution, 29, 31
Continuous development of plates in thin-layer chromatography, 511
Control charts, 19; see also Charts
  interpretation of, 22-24
  plotting of, 19
  statistical methods using, 19-24
Controlled areas, 346-347
Controls
  commercially prepared sera and cerebrospinal fluid as, 13
  difference of, from standard, 14-15
  patient, and quality control, 13
  of pooled sera, cerebrospinal fluid, or urine, 13

Controls—cont'd
  quality control and, 12-13; see also Quality control
  shelf-life of, 13
  solutions of pure chemicals in, 13
Conversion of units, 138-139
Conversion factors, electrolyte, 139
CO-Oximeter, 459, 460
Copper in serum and urine, determination of, using atomic absorption spectroscopy, 209-211
Correction, Allen, 445
  in determination of 17-ketosteroids, 433
Cortex
  adrenal; see Adrenal cortex
  of kidney 243
Cortexolone; see Compound S
Corticosteroids, 383-388
Corticotropin, 363
Cortisol
  determination of, 385
  free, in urine, measurement of, 386
  metabolites of, 381
  Porter-Silber method of determination of, 385
  radioimmunoassay for determination of, 385
  urinary, free, determination of, 427-428
Count, cell, of cerebrospinal fluid, 324-325
Counterimmunoelectrophoresis, 628-630
Counting, liquid scintillation, 341-345
Counting window for frequency distribution, 342
CPBA; see Competitive protein binding analysis
CPK; see Creatine kinase
Creatine
  serum creatinine and, determination of, 103-105
  in urine, 306
    determination of, 106-107
Creatine kinase, determination of, Nuttall-Wedin method of, 238-239
Creatine phosphokinase; see Creatine kinase
Creatine phosphokinase isoenzymes
  electrophoresis of, 572-576
    clinical significance of, 578-580
  separated on agarose, 580
Creatinine, 101-107
  blood, calibration chart for, 103
  concentration of, in blood, in kidney disease, 260
  determination of, 101-103
    with AutoAnalyzer I, 50-52
    with AutoAnalyzer II, 78-79
  serum, and creatine, determination of, 103-105
  in urine, 306
    calibration chart for, 106
    determination of, 105-106
Creatinine clearance tests, 252-253
Cretinism, 370
Cross-linked dextrans, gel filtration on, to separate bound from free antigen, 338
Crystal, sodium iodide, in gamma ray spectrometry, 345
CSC; see Cusum chart
Culture of cerebrospinal fluid, microscopic examination of, 324
Cumulative distribution, 24, 26
Cumulative sum chart; see Cusum chart
Curves
  bell-shaped, 9
  in Shewhart chart, 20

Curves—cont'd
  calibration; see Calibration curve
  gaussian, 9
    on Shewhart chart, 20
  leptokurtic, 28
  oxyhemoglobin dissociation, of oxygen and carbon dioxide in vivo, 151
  platykurtic, 28
Cushing's syndrome, 361, 388-389
  dexamethasone suppression test for, 388
  symptoms of, 424
Cusum chart
  normal distribution on, 22
  systematic shift on, 23
  systematic trends on, 23
  tabulation of data, on, 21-22
Cyanocobalamin deficiency, 312
Cystine, urinary calculi of, 304

D

Dalton's atomic theory, 328
Data
  collecting, for quality control, 15
  continuous, in frequency distribution, 29, 31
  on cusum chart, tabulation of, 21-22
  discrete, 24
  in quality control, 15-19
    tabulating and studying, 15
  tabulation of, in frequency distribution, 31
Data reduction in radioassay, 352-353
Dead space, 150, 154
  anatomic, 154
  minute, 154
Decay
  alpha, 330-331
  beta, 331
  radioactive, 330-332
    rate of, 333-335
Decimal point in computation, 16
Dehydroepiandrosterone, 382-383
  urinary, Allen "blue" color test of, 409-410
Delivery of specimen and quality control, 12
Delivery volumes, pump tube sizes and, 37
Delta check in blood gas quality control, 166
Denaturation, alkali, determination of hemoglobin F by, procedures for, 569-570
Densitometer
  Beckman Recording Integrating, for micro-cellulose acetate electrophoresis, 538
  for electrophoretic procedures, 638-642
    repeatability of, 639, 640, 641, 642
  Photovolt, with integrator attachment, 535
Densitometer scan of serum on starch gel, 546
Densitometry in electrophoretic procedures, 638-642
  repeatability in, 639, 640, 641, 642
11-Deoxycortisol
  determination of, 385
  urinary, determination of, 407-408
20,22-Desmolase deficiency, 389
Destaining in electrophoretic procedures, 636
Detection of components
  in column chromatography, 506-507
  in thin-layer chromatography, 512
Detection limits in atomic absorption spectroscopy, 203

Detectors
  for gas chromatography, 524
  for thin-layer chromatography, 512
Determinants, 594
Detoxification and synthesis, liver function tests for, 181
Detoxifying mechanism of liver, 198
Dexamethasone suppression test of adrenocortical function, 388
Dextrans, cross-linked, gel filtration on, to separate bound from free antigen, 338
Dextrostix method of glucose determination, 97
Diabetes, insulin in, 374
Diabetes mellitus
  causing hyperglycemia, 94
  insulin for, 374
Diabetic acidosis causing electrolyte imbalance, 116
Diabetic glomerulosclerosis, chronic renal insufficiency from, 263
Diacetic acid in urine, 289-290
  Gearhardt test for, 290
Dialysate solutions
  for hemodialysis, 264-266
    formula for, 265
  for peritoneal dialysis, 271
Dialysis, 263-272
  for acute poisoning or drug intoxication, 273-274
  peritoneal
    clinical application of, 272
    dialysis fluid for, 270
    patient dialysis with, 271-272
    patient preparation for, 271
  renal insufficiency and, 262-264
  serum glutamic-oxaloacetic transaminase by, 60
Dialysis bath; see Dialysate solutions
Dialysis fluid-mixing system, online, 265-266
Dialyzer
  of Abel, Rowntree, and Turner, 263, 264
  Alwall, 264
  Kiil, 264, 268-270
  Kolff, 263-264
  Skeggs-Leonards, 264
  twin-coil
    assembly of, 266-267
    connecting patient to, 268
    hemodialysis with, 266
    schematic of, 267
    testing of, 267-268
Diazepam
  absorption spectrum of, 472
  quantitative ultraviolet spectrophotometric procedure for determination of, 471-473
Diazo substances in urine, determination of, 296
Dibasic acid, 684
Diclonal gammopathy, IgGκ, 622
Dielectric constant in chromatography, 494
Diffusible calcium, determination of, 133-135
Diffusible serum calcium, nomogram for estimating, 134, 135
Diffusion in chromatography, 488
Digestion
  and absorption, 308-310
  bile in, 309-310
  bile salts in, 310
  duodenum and, 309

Digestion—cont'd
  hydrochloric acid, 310
  intrinsic factor in, 310
  in large intestine, 310
  lipase in, 309
  liver and, 309
  mucus in, 310
  pepsinogens in, 310
  ptyalin and, 308
  rennin and, 309
  saliva and, 308
  in small intestine, 310
  in stomach, 309
1,25-Dihydroxyvitamin D, 400-401
Dilantin; see Diphenylhydantoin
Dilution tests of kidney, 256
Diphenylhydantoin
  absorption spectrum of, 473
  quantitative ultraviolet spectrophotometric procedure for determination of, 473-474
Dipole moment, 490
Direct cholesterol determination
  with AutoAnalyzer I, 65-67
  with AutoAnalyzer II, 77-78
Direct color reaction in simultaneous glucose and urea nitrogen determination, 47
Direct and total bilirubin determination, 196-197
  Malloy-Evelyn modification of, 196-197
  simultaneous, with AutoAnalyzer I, 67-70
Discrete data, 24
Disease
  Addison's, 117
    clinical findings in, 424
  bone, and alkaline phosphatase, 224-225
  bronchiolar-bronchial, and ventilation:perfusion ratio, 161-162
  buffer, metabolic, 171
  changes in prothrombin times in, 178
  Graves', 371
  hepatobiliary; see Hepatobiliary diseases
  lambda light chain, 621
  liver, evaluation of, 177-190
  kidney
    bicarbonate concentration in, 260
    blood chemistry in, 259-262
    calcium concentration in, 260-261
    creatinine concentration in blood in, 260
    increased amounts of protein with, 279
    magnesium concentration in, 261
    phosphate concentration in, 260-261
    potassium concentration in blood in, 260
    protein concentration in blood in, 261
    renin-angiotensin system and, 261-262
    sodium concentration in blood in, 260
    urea concentration in blood in, 259-260
  pancreatic
    decreased amylase values in, 315
    decreased trypsin values in, 316
  plasma pH adjustments in, 170
  polycystic, chronic renal insufficiency from, 263
  serum albumin changes in, 178
  sickle cell, 571
  Tangier, 564
  Von Gierke's, hypoglycemia from, 93
Disintegration, radioactive, 330
Disintegrations per minute, calculation of, 344

Disk electrophoresis on polyacrylamide gel, 548-549
Displacement, 401
  in chromatography, 494
Displacement analysis; see Radioassay
Distribution
  bimodal, 25
  cumulative, 24, 26
  frequency; see Frequency distribution
  normal; see Normal distribution
  unimodal, 25
Divisor, 18
Dodge-Romig sampling inspection tables, 10
L-Dopa administration test for human growth hormone, 362
Doriden; see Glutethimide
Doses of radiation, 350
  maximum permissible, 350
Dosimeters, pocket, for radiation monitoring, 349
Double antibody technique to separate bound from free antigen, 338-339
Double labeling, 344
dpm calculations, 344
Drug intoxication, dialysis for, 273-274
Drugs, alkaloid, thin-layer chromatographic method for determination of, 478-481
Ducts of Bellini, papillary, 244
Dumping syndrome causing hyperglycemia, 94
Duodenal ulcer, maximum gastric output in, 313
Duodenum
  contents of
    amylase in, 315
    analysis of, 314-317
    bile in, 314
      determination of, 316
    microscopic examination of, 317
    trypsin in, determination of, 316
  digestion and, 309
  inflammation of, microscopic examination of duodenal contents for, 317
Duostat power supply for micro-cellulose acetate electrophoresis, 537
Dura mater, 322
Dwarfism, 361
Dye excretion test of liver function, 180
Dyes, thin-layer chromatography for, 499

E

Earlobe
  preparation of, for capillary puncture, 3
  puncture of, for blood collection, 1-3
Eddy diffusion in chromatography, 488
Edelman proposal for structure of immunoglobulins, 595
EDTA; see Ethylenediamine tetra-acetate
Effective renal blood flow, 256
Efficiency measurements in radioassay, 344
Electric field in electrophoresis, 531
Electrodes for blood gas and pH measurements, 164, 168-169
Electroendosmosis in electrophoresis, 532
Electroimmunodiffusion, 626-627
  single, two-dimensional, 627
Electrolyte conversion factors, 139
Electrolytes, 115-139
  determination of, with AutoAnalyzer I, 41-46
  in human plasma, 115

Electrolytes—cont'd
  imbalance in
    causes of, 115-116
      diabetic acidosis causing, 116
      kidney damage causing, 115
      poisoning causing, 116
      uremia causing, 115
  in urine, 116
Electrometric measurement of blood pH, P_{CO_2}, and P_{O_2}, 126-127
Electron capture, 331-332
Electrons
  of atom, 328, 329
  Auger, 332
Electrophoresis, 528-587; see also Electrophoretic procedures
  of alkaline phosphatase isoenzymes, 576
    clinical significance of, 578
  basic concepts in, 528-533
  of body fluid constituents, 545-587
  buffer in, 530-531
  citrate agarose, for hemoglobin electrophoresis, 567-568
  of creatine phosphokinase isoenzymes, 572-576
    clinical significance of, 578-580
  definition of, 528
  disk, on polyacrylamide gel, 548-549
  electric field in, 531
  electroendosmosis in, 532
  factors affecting, 529-530
  "free," 528
  hemoglobin; see Hemoglobin, electrophoresis of
  isoenzyme, 572-581
    clinical significance of, 578-581
  of lactic dehydrogenase isoenzymes
    agarose method for, 577-578
    cellulose acetate Microzone method for, 576-577
    clinical significance of, 580-581
  Laurell crosses, 627
  macro-cellulose acetate for protein separation, 533-537
    power supply for, 534
  micro-cellulose acetate; see Micro-cellulose acetate electrophoresis
  micro-starch gel; see Micro-starch gel electrophoresis
  "moving boundary," 528
  plasma, 610
  polyacrylamide gel; see Polyacrylamide gel electrophoresis
  protein separation with, 533-545
    physical constants of, 530
  of proteins in cerebrospinal fluid, 556, 558
    normal values for, 558
  to separate bound from free antigen, 338
  of serum glycoproteins, 585-587
    clinical significance of, 586-587
  of serum haptoglobin
    cellulose acetate method of, 581-583
    clinical significance of, 584-585
    polyacrylamide method of, 583-584
  of serum lipoproteins
    agarose, 560-561, 562
    clinical significance of, 561-564

Electrophoresis—cont'd
  of serum proteins, 545-556
    clinical significance of, 556
  starch gel, for hemoglobin electrophoresis, 568-569
  support media for, 531-532, 634
  of urinary proteins, 558-560
    collodion bag ultrafiltration in, 559
    Minicon concentration method of, 560
    protein concentration procedures in, 559-560
  wick flow in, 532-533
  zone, 528-529
Electrophoresis buffer, 632-633
Electrophoresis cell, 630-631
Electrophoresis process center, 605
Electrophoretic fractionations of serum proteins, 557
Electrophoretic procedures; see also Electrophoresis
  clearing in, 636-638
  conditions affecting, 635
  densitometry in, 638-642
  destaining in, 636
  power supply for, 631-632
  quality control in, 630-642
  reagents in, guidelines for, 633-634
  staining in, 635-636
  systems considerations in, 630-633
  technique considerations in, 634
Electroscope ionization chambers for radiation monitoring, 349
Elements, resonance lines of, measurement of, 202
Elution in chromatography, 492, 494
Elutropic solvent series for chromatography, 493
Emergency dose of radiation, 350
Emission, positron, 331
Emitters
  alpha, shielding for, 348
  beta
    energy frequency distribution for, 341
    shielding for, 348
  gamma, shielding for, 348-349
Emphysema
  centrolobular, 163
  lung changes in, 162-163
Endocrine glands, 358; see also Hormones
  role of, in metabolism, 93
Energy
  absorbance of, pH and, 467
  absorption of, determination of, 466
  activation, 214
  frequency distribution of, for beta emitter, 341
  transfer of, by radiation, 333
Enzymatic method for ethyl alcohol determination, 452-454
Enzyme level, liver function tests for, 181
Enzymes, 214-240
  activation of, 214
    factors affecting, 215-218
  activity of
    effects of inhibitors on, 216-218
    effects of pH on, 218
    effects of temperature on, 218
    methods for determining, 219, 220-240
    Michaelis constant and, 215-216
    substrate concentration and, 215-216
  as catalysts, 214

Enzymes—cont'd
  in cerebrospinal fluid, determination of, 327
  classification of, 218
  definition of, 214
  inhibition of
    competitive, 217
    lock and key theory of, 215
    noncompetitive, 217
    substrate, 217-218
  and liver, 179-180
  nomenclature of, 218-219
  numbering system for, 218
  recommended name for, 219
  systematic name for, 218-219
  trivial name for, 219
  units of, 219
Enzyme-substrate complex, 214-215
Epinephrine
  action of, 390
  and blood glucose level, 93
  determination of, 405-407
  secretion of, and anoxia causing hyperglycemia, 94
Equation, alveolar gas, 154-155
Equipment
  hemodialysis, 266-270
  for venipuncture, 4
Equivalent weights, 680-682
Errors
  allowable limits of, 20
  law of, 9
Esbach's method for determining protein in urine, 281
Essential oils, thin-layer chromatography for, 499
Estradiol, 395
Estriol, 395
  during pregnancy, 396-397
  urinary determination of, in pregnancy, 397
    procedure for, 410-414
Estrogens, 395-396
  determination of, 395
  effects of, 395-396
  excretion of, in urine of normal pregnant women, 413
  levels of, 396
  total urinary (nonpregnant), determination of, 414-416
Estrone, 395
Ethanol; see Ethyl alcohol
Ethyl alcohol
  determination of, 452-455
    quantitative determination of, 452-454
    enzymatic method for determination of, 452-454
    Harger's micro method for determination of, 454-455
    quantitative titrimetric method for determination of, 454-455
Ethylene glycol, dialysis to remove, 274
Ethylenediamine tetra-acetate as anticoagulant, 8
Examination
  chemical, of cerebrospinal fluid, 325-327
  laboratory, for proteins, 181-190
  microscopic
    of Boas-Oppler bacilli, 313, 314
    of cerebrospinal fluid, 324-325
    of duodenal contents, 310

Examination—cont'd
  microscopic—cont'd
    of sarcinae, 314
    of stomach contents, 313-314
  of urine
    microscopic, 291-292
    routine, 274-306
Excited state of atom, 202
Excretion
  dye, liver function tests for, 180
  estrogen, in urine of normal pregnant women, 413
Excretion tests, phenolsulfonphthalein, of renal blood flow, 258-259
Excretory capacity of liver, 198-200
Excretory function, tubular, 256
External chemical indicator, 687
External standardization, automatic, 344-345
Exton-Rose test for glucose tolerance determination, 99
  procedure for, 100-101

F

Fat
  abnormal catabolism of, 288-289
  in feces, determination of, 319-321
Fatty acids
  in liver, 190
  thin-layer chromatography for, 500
FC-134 as surfactant with AutoAnalyzer, 40
Feces, 310
  analysis of, 317-321
    for bile pigment, 318
    for occult blood
      benzidine test for, 317
      guaiac test for, 317-318
      Hematest for, 318
      o-tolidine test for, 318
    fat in, determination of, 319-321
    lipids in, fecal analysis for, 319-321
Feedback control of anterior pituitary gland, 360-361
Feminization, 395
Fermentation test for glucose in urine, 287
  reading results of, 288
Ferric chloride test
  for melanin in urine, 299
  for phenylketonuria, 305
Fetus, determination of status of, by estriol determination, 396
Fibrinogen
  and liver function, 187-188
  quantitative determination of, 187-188
Fibroplasia, retrolental, prevention of, 164-165
Film badges for radiation monitoring, 349
Filtrate, glomerular, 245
Filtration
  gel, on cross-linked dextrans to separate bound from free antigen, 338
  glomerular, tests of, 246-253
  of kidneys, 244-245
Finger, preparation of, for capillary puncture, 3
Finger puncture for blood collection, 1-3
Fire precautions and chemical hazards, 676-677
Fishberg test of tubular function, 255

Fiske-SubbaRow method of determination of inorganic phosphate, 136
Fission, spontaneous, 332
Flame emission spectroscopy, 202
Flame photometry, 202
  plasma lithium determination by, 482-483
Flat bed polyacrylamide gel electrophoresis, 549
Flocculation test, cephalin-cholesterol, of liver function, 186-187
Floraquin; see Diiodohydroxyquin
Flow, vascular, lung, 156
Flow sheet for laboratory diagnosis of hemoglobinopathies, 575
Fluids
  body
    composition of, 1
    constituents of, electrophoresis of, 545-587
  cerebrospinal; see Cerebrospinal fluid
Fluorescence, measurement of
  of catecholamines, 392
  in cortisol determination, 385
Fluorescent chemical indicator, 687
Fluorescing agent as adsorbent in chromatography, 492
Fluorometric procedure for plasma 17-hydroxycorticosteroid determination, 416-419
Fluors in scintillation cocktail, 342
Folate deficiency, 312
Folic acid deficiency, 312
Folin method
  for determination of sulfates in urine, 302-303
  for uric acid determination, 108
Folin-Svedberg method for urea nitrogen determination, 112
Folin-Wu method for glucose determination, 95
Follicle-stimulating hormone, 363-364
  normal range of, in urine and serum, 364
Formaldehyde and methanol screening test, 455-456
Formalin
  antibody, induced by antigens, 336
  of antiserum, 337
Forms, graphic, of frequency distribution, 24, 26
  purpose of, 25, 27-28
Formula
  for measurement of kurtosis, 28
  for measurement of skewness, 28
  Tonks', for allowable limits of error, 20
Fractionated catecholamines, determination of, 405-407
Free cortisol in urine, measurement of, 386
"Free" electrophoresis, 528
Free thyroxine, measurement of, 369-370
  normal values of, 370
Free urinary cortisol, determination of, 427-428
Freezing point, measurement of, 137
Frequency distribution, 24-33
  continuous data in, 29, 31
  counting window for, 342
  discrete data in, 24
  graphic forms of, 24, 26
    purpose of, 25, 27-28
  kurtosis in, 28
  measures of central tendency in, 29
  symmetry of, 27
  t test in, 32-33

Frequency distribution—cont'd
  tabulation of data in, 31
  tally of, 30
  variability in, 27
FSH; see Follicle-stimulating hormone
Functional residual capacity, determination of, 162

G

Galactose in urine, mucic test for, 286-287
Galatest method of testing for glucose in urine, 285
Gamma emitters, shielding for, 348-349
Gamma ray spectrometry, 345-346
  sodium iodide crystal in, 345
Gammopathy
  diclonal, IgGκ, 622
  monoclonal; see Monoclonal gammopathy
  polyclonal, 613, 614
    clinical conditions exhibiting, 616-617
Ganglioneuroma, 393
  symptoms of, 407
Garvey-Bowden method for barbiturate determination, 461-463
Gas, blood; see Blood gas
Gas chromatography, 486, 523-525
  biochemical applications of, 524
  columns for, 523-524
  detectors for, 524
Gas equation, alveolar, 154-155
Gas-liquid chromatography, 477
Gastrectomy causing hyperglycemia, 94
Gastric acid
  maximum output of, determination of, 313
  measurement of, 312
Gastric acidity, titratable, determination of, 313
Gastric analysis, 312-314
Gastric secretion, 310-311
  basal, determination of, 313
  regulation of, 311
Gastrin, 311
  in ulcer patients, 311
Gastrointestinal hormones, 311-312, 400
Gastrointestinal system, 308-321
Gaussian curve, 9
  on Shewhart chart, 20
Gel, starch, densitometer scan of serum on, 546
Gel filtration on cross-linked dextrans to separate bound from free antigen, 338
Gel slicing tray for micro-starch gel electrophoresis, 543
General anesthesia causing hyperglycemia, 94
Generator, arsine, for arsenic determination, 450
Gentzkow method for urea nitrogen determination, 112
Gerhardt test for diacetic acid in urine, 290
Gettler-Freimuth method for carbon monoxide determination, 457-459
  apparatus for, 458
GGT; see γ-Glutamyl transpeptidase
Giardia lamblia, microscopic examination of duodenal fluid for, 317
Gigantism, 361
Glands
  adrenal, 374
    hormones of, 374-393
  endocrine, 358; see also Hormones
    in metabolism, 93

Glands—cont'd
  parathyroid, 371-372
    hormones of, 372
  pineal, 399-400
    hyperfunction of, 400
  pituitary; see Pituitary gland
  thymus, 400
Glass electrode for blood gas and pH measurement, 164
Globulin
  antilymphocyte, to suppress homograft reaction, 273
  in cerebrospinal fluid, determination of, 326
Glomerular filtrate, 245
Glomerular filtration tests, 246-253
Glomerulonephritis, chronic renal insufficiency from, 262-263
Glomerulosclerosis, diabetic, chronic renal insufficiency from, 263
Glomerulus of kidney, 243
Glossary for radioassay, 353-356
Glucagon, 374
  and blood glucose level, 93
α-1,4-Glucan,4-glucanohydrolase; see Amylase
Glucocorticoids, 384-388
Glucose, 92-101
  abnormal values of, conditions accompanied by, 95
  blood, control of level of, 93
  in cerebrospinal fluid, determination of, 327
  decreased concentration of, 93-94
  determination of
    with AutoAnalyzer I, 46-47
    with AutoAnalyzer II, 79-82
    Benedict method of, 95
    Dextrostix method of, 97
    Folin-Wu method of, 95
    glucose oxidase method of, modified, 95
    methods of, 95-96
    Somogyi-Nelson method of, 95
    o-toluidine method of, 96-97
  increased concentration of, 94
  and urea nitrogen determination, simultaneous, with AutoAnalyzer I, 46-50
  in urine, 283-285
    Benedict method of testing, 284-285
    Clinitest method of testing, 285
    fermentation test for, 287
      reading results of, 288
    Galatest method of testing, 285
    Kowarsky test for, 288
    Tes-Tape method of testing, 285
Glucose oxidase determination with AutoAnalyzer II, 79-81
Glucose oxidase method of glucose determination, modified, 95
Glucose tolerance, determination of, 98-101
  intravenous test for, 98-99
    procedure for, 100
  Janney-Isaacson test for, 98
    procedure for, 99-100
Glutamic oxaloacetic transaminase
  calibration curve for, 234
  determination of
    with AutoAnalyzer II, 82-84
    Reitman-Frankel method of, 233-234

Glutamic oxaloacetic transaminase—cont'd
  reaction of, in transaminase determination, 232-233
Glutamic pyruvic transaminase
  calibration curve for, 234
  determination of, Reitman-Frankel method of, 233-234
  reaction of, in transaminase determination, 233
γ-Glutamyl transpeptidase, determination of, 239-240
γ-Glutamyltransferase; see γ-Glutamyl transpeptidase
Glutethimide
  determination of, using thin-layer chromatography, 481-482
  dialysis to remove, 274
Glycerides, thin-layer chromatography of, 500
Glycerol released from triglycerides, determination of, 191-192
Glycerol ester hydrolase; see Lipase
Glycogen, 92; see also Glucose
Glycolipids, thin-layer chromatography of, 500
Glycoproteins, serum
  electrophoresis of, 585-587
    clinical significance of, 586-587
  separation of, on cellular acetate, 585
Glycosides, thin-layer chromatography for, 498
Gmelin test for fecal analysis for bile pigment, 318
Goblet cells of stomach, 309
Goiter, 370
Goldbaum modified method of barbiturate determination, 468-471
Gonadal hormones, 393-399
Gonadotropic hormones, 363-364
Gonadotropin, chorionic, determination, 398
GOT; see Glutamic oxaloacetic transaminase
Gout, nephropathy from, 263
GPT; see Glutamic pyruvic transaminase
Graphic forms of frequency distribution, 24, 26
  purpose of, 25, 27-28
Graves' disease, 371
Gross procedure for trypsin determination in duodenal contents, 316
Ground state of atom, 202
Guaiac test for fecal analysis for occult blood, 317-318

H

Haldane effect, 148-149
Half-life, 334
  biologic, 334-335
  of hormone in plasma, 359
Handling
  radiation, 346-349
  of radioisotopes, procedures for, summary of, 350-352
Haptens, 594
Haptoglobin, serum, electrophoresis of
  cellulose acetate method of, 581-583
  clinical significance of, 584-585
  polyacrylamide method of, 583-584
Haptoglobin typing by polyacrylamide gel electrophoresis, 584
Harger's micro method for ethyl alcohol determination, 454-455

Liebermann-Burchard reaction, 65, 66, 67
Ligands in radioassay, 340
Light absorption, laws for
  Beer's, 466-467
  Bouguer-Lambert, 466
Light chains of immunoglobulins, 596-597
Limits of error, allowable, 20
Lines
  action, of Shewhart chart, 19
  resonance, of elements, measurement of, 202
  warning, of Shewhart chart, 19-20
Lipase
  determination of, 222-224
    Cherry-Crandall method of, 222-223
  in digestion, 309
  serum, determination of, with Tietz-Fiereck method, 223-224
Lipids
  fecal, fecal analysis for, 319-320
  metabolism of
    in liver, 190-200
    liver function tests for, 181
  thin-layer chromatography for class separation of, 499-500
Lipoid hyperplasia, congenital, 389
Lipoproteins
  of liver, 190-191
  serum
    comparison of, 563
    electrophoresis of
      agarose, 560-561, 562
      clinical significance of, 561-564
    phenotypes of, 565
      diseases associated with, 566
Liquid scintillation, kinds of samples of, 342-343
Liquid scintillation counting, 341-345
Liquid scintillation counting spectrometer, 341-342
Liquids, temperature of, and peristalsis, 308-309
Lithium, plasma, determination of, by flame photometry, 482-483
Lithium oxalate as anticoagulant, 8
Litmus paper, 687
Liver
  and alkaline phosphatase, 225
  blood ammonia and, 181
  chronic dysfunction of, causing hyperglycemia, 94
  chylomicrons in, 190-191
  cirrhosis of, 614
  clearance function of, 178-180
  conjugation in, 198
  destruction of, by poisons, hypoglycemia from, 93
  detoxifying mechanism of, 198
  and digestion, 309
  disease of, evaluation of, 177-190
  enzymes and, 179-180
  excretory capacity of, 198-200
  fatty acids in, 190
  inflammation and necrosis of, 177
  lipid metabolism in, 190-200
  lipoproteins of, 190-191
  phospholipids in, 190
  role of, in metabolism, 92-93
  sterols in, 190

Liver—cont'd
  synthetic function of, 178
  triglycerides in, 190
    determination of, 191-192
Liver function tests, 180-181
  for bilirubin, 180
  for blood ammonia, 188-190
  cephalin-cholesterol flocculation, 186-187
  for detoxification and synthesis, 181
  for dye excretion, 180
  for enzyme level, 181
  for fibrinogen, 187-188
  for lipid metabolism, 181
  for protein, 189
Lobules of lung, regulation of, 159
Local nonequilibrium in chromatography, 488
Lock and key theory of enzyme inhibition, 215
Long-acting thyroid stimulating substance, 362
Long-acting thyroid stimulator, 371
Loop, Henle's, 244
Low level laboratory, 346-347
Lung
  air-blood barrier of, 148-150
  alveoli of, hemodynamics of, 157-158
  arteriolospasm in, from alveolar hypoxia, 159
  capillaries of, hemodynamics of, 157-158
  functions of, 148-163
  hemodynamics of, 156-159
  injured, regenerating, 142
  lobules of, regulation of, 159
  oxygen tension regulation by, 141
  reserve vascular capacity of, 156
  vascular flow in, 156
  vascular and ventilatory factors of, 156
  volume of, 150
    plotting of, 153
Luteinizing hormone, 363-364

M

Machines, blood gas, 165-166
Maclagen method for determination of thymol turbidity, Shank-Hoagland, 184-186
Macro–cellulose acetate electrophoresis for protein separation, 533-537
  power supply for, 534
Macroglobulinemia, Waldenström's, 617, 620
Magnesium
  blood concentration in, in kidney disease, 261
  determination of, automated, 209, 210
  in serum and urine, determination of, using atomic absorption spectroscopy, 209
Malabsorption syndrome, 319
Malloy-Evelyn modification for determination of direct and total bilirubin, 196-197
Manometric apparatus, Van Slyke, carbon dioxide determination using, 122-125
Manual methods of analysis, adaptation of, to AutoAnalyzer, 36-37
Mask, V, 21
Mass number of atom, 329
Mass spectrometers, 163
Material, radioactive, sign for, 347
Matter, photon interaction with, 333
Maximum gastric acid output, determination of, 313
Maximum permissible doses of radiation, 350

Mean
  arithmetic, 29
  percent of, 18-19
Mean-life of radioisotopes, 334
Measurements
  efficiency, in radioassay, 344
  of free thyroxine, 369-370
  of quench, 343-344
  of thyroxine; see Thyroxine, measurement of
  of triiodothyronine, 370
Median, 29
Medical dose of radiation, 350
Medical laboratories, statistical quality control in, 11
Medulla
  adrenal, 389-393
  of kidney, 243
Megaloblastic anemia, 312
Melanin in urine, 299
Melanocyte-stimulating hormone, 363
Melanotropin, 363
Melatonin, 399-400
Meninges of brain, 322
Meningitic infection, acute, cerebrospinal fluid in, 323
Meningitis, 322
  suppurative, cerebrospinal fluid in, 323
  tuberculous, cerebrospinal fluid in, 323
Meprobamate, dialysis to remove, 274
Mercury, cleaning of, 125
Metabolic acidosis, 145-147, 173
  definition of, 147
  hypoxic, 147
  in renal disease, 260
Metabolic alkalosis, 173-175
  kidneys and, 175
Metabolic buffer diseases, 171
Metabolic homeostasis, 172-175
Metabolism, 92-94
  bile pigment, 195-200
  of catecholamines, 391
  endocrine glands in, 93
  of hormones, 359
  lipid
    in liver, 190-200
    liver function tests for, 181
  liver in, 92-93
  peripheral tissues in, 93
  of steroid hormones, 381
Metabolites
  of cortisol, 381
  steroid hormones and, 376
Metals, heavy, dialysis to remove, 274
Metanephrine
  determination of, 393
  urinary, determination of, 435-436
Metastatic prostatic carcinoma, acid phosphatase and, 225
Methanol and formaldehyde screening test, 455-456
Method of additions
  in atomic absorption spectroscopy, 204-205
  to minimize interferences in atomic absorption spectroscopy, 203
3-Methoxy-4-hydroxymandelic acid
  determination of, 393, 407

3-Methoxy-4-hydroxymandelic acid—cont'd
  urinary, determination of
    procedure for, 442-444
    with thin-layer chromatography, 521-523
Methyl alcohol, dialysis to remove, 274
Metopirone; see Metyrapone
Metyrapone for determination of urinary compound S, 407-408
Metyrapone test of adrenocortical function, 387-388
Meuleman's method for determination of total protein in cerebrospinal fluid, 326-327
Michaelis constant and enzyme activity, 215-216
Micro–cellulose acetate electrophoresis
  application of serum sample in, 540
  Beckman Recording Intebrating Densitometer for, 538
  Duostat power supply for, 537
  Microzone applicator for, 538
  Microzone cell for, 536
Microgasometer, Natelson, in blood gas quality control, 166
Micro-Kjeldahl method for protein determination, 182
Microscopic examination
  of Boas-Oppler bacilli, 313, 314
  of cerebrospinal fluid, 324-325
  of duodenal contents, 317
  of sarcinae, 314
  of stomach contents, 313-314
  of urine, 291-292
Micro-starch gel electrophoresis, 540-545
  circulating destaining unit for, 542
  gel slicing tray for, 543
  removal of membrane after cleaning, 541
  tray for, 541
  washing unit for, 543
Microzone acrylamide gel apparatus, 550
Microzone acrylamide gel method of polyacrylamide gel electrophoresis, 549-552
Microzone applicator for micro–cellulose acetate electrophoresis, 538
Microzone cell for micro–cellulose acetate electrophoresis, 536
Microzone cellulose acetate method
  for alkaline phosphatase isoenzyme electrophoresis, 576
  for lactic dehydrogenase isoenzyme electrophoresis, 576-577
Microzone step-gradient polyacrylamide gel electrophoresis, 552-555
Microzone system of electrophoresis; see Micro–cellulose acetate electrophoresis
Mineralocorticoids, 383-384
Minicon concentration method for urinary protein electrophoresis, 560
Minicon concentration principle, 562
Minute alveolar ventilation, 154
Minute dead space, 154
Mitochondrial hypoxia, 173
Mobile phase of chromatography, 492-494
Mobility of hemoglobin
  on cellulose acetate, 572
  on citrate agar, 573
Mode, 29
Molal solutions, 678

Molar solutions, 678
Mole, hydatidiform, chorionic gonadotropin in, 398
Molecules, structure of, and chromatographic relationship, 494-495
Moller-McIntosh-Van Slyke method for urea clearance tests, 250-252
Moment dipole, 490
Monitoring
  contamination, 345
  radiation, 349-352
Monobasic acid, 684
Monoclonal gammopathy, 617-621
  benign, criteria for, 620
  criteria for, 618
    IgA, 620
    IgAκ, 619
    IgE, 621
    IgG, 618
    IgGλ, 619
    IgMκ, 617, 620
Morphine-heroin, quantitative ultraviolet spectrophotometric procedure for determination of, 475-477
Mosenthal test of tubular function, 254-255
"Moving boundary" electrophoresis, 528
MSH; see Melanocyte-stimulating hormone
Mucic test for galactose in urine, 286-287
Mucus in digestion, 310
Mucus-secreting cells of stomach, 309
Multiple development of plates in thin-layer chromatography, 511
Murphy-Pattee method of measuring thyroxine by competitive protein binding, 368
Myers-Fine procedure for determination of amylase in duodenal contents, 315
Myxedema, 370-371

N

N-substituted barbiturate
  absorption spectrum of, 470
  determination of, 471
Natelson Microgasometer in blood gas quality control, 166
National Bureau of Standards, 13
Necrosis
  liver, 177
  tubular, acute, and renal failure, 262
Needle
  for immunoelectrophoresis, 608
  for venipuncture, 4
    inserting, 5
Neocuproine determination with AutoAnalyzer II, 81-82
Nephelometry, automated specific protein analysis by, 627-628
Nephron, 244
Nephrosclerosis, arteriolar, from primary hypertension, 263
Neuroblastoma, 393
  symptoms of, 407
Neurohypophysis, 365-366
Neutralization chemical indicator, 687
Neutron number of atom, 329
Neutron/proton ratio of atom, 330
Neutrons of atoms, 329

Nitrazine paper, 687
Nitrogen
  nonprotein, 101-114
  urea; see Urea nitrogen
Nomenclature, enzyme, 218-219
Nomogram
  for calculating carbon dioxide tension, 126
  for estimating diffusible serum calcium, 134, 135
Noncompetitive inhibition of enzymes, 217
Nonequilibrium, local, in chromatography, 488
Nonprotein nitrogen compounds, 101-114
Norepinephrine
  action of, 390
  determination of, 405-407
Normal distribution
  in cusum chart, 22
  in Shewhart chart, 20, 22
Normal solutions, 687
  composition of, 679-680
NPN; see Nonprotein nitrogen compounds
Nuclear isomers, 332
5'-Nucleotidase test for hepatobiliary disease, 179
Nucleus, basic, of steroid group, 375
Number
  atomic, of atom, 329
  mass, of atom, 329
  neutron, of atom, 329
  small, square roots of, 251
Numbering system for enzymes, 218
Nuttal-Wedin method of creatine kinase determination, 238-239

O

Occult blood
  fetal analysis for
    benzidine test for, 317
    guaiac test for, 317-318
    Hematest for, 318
    o-tolidine test for, 318
  in stool, 180
  in urine, 300
Occultest for blood in urine, 300
Offline application in radioassay, 353
Oils, essential, thin-layer chromatography for, 499
Online application in radioassay, 353
Online dialysis fluid-mixing system, 265-266
Operating procedure for AutoAnalyzer, general, 37-39
Orabilex; see Bunamiodyl
Ordinary diffusion in chromatography, 488
Organic solvents with Perkin-Elmer Atomic Absorption Spectrophotometer, 204
Orgive, 24, 26
Orthostatic proteinuria, 279
Osmolarity, determination of, 137-138
Osteodystrophy, renal, 261
Outliers, 16, 24, 25
Oxalates, urinary calculi of, 304
18-Oxidase deficiency, 389
Oxidation-reduction chemical indicator, 687
Oximetry in blood gas quality control, 166
17-Oxosteroids, tests of, 382-383
β-Oxybutyric acid in urine, Hart method for detecting, 290-291
Oxygen, oxyhemoglobin dissociation curve of, 151

Oxygen tension
  clinical conditions of, 160
  determination of, 125-128
  partial, useful physiologic approximations of, 143
  regulation of, by lung, 141
Oxyhemoglobin dissociation curves of oxygen and carbon dioxide in vivo, 151
Oxyntic cells of stomach, 309
Oxytocin, 365-366

P

P_{CO_2}; see Carbon dioxide tension
P_{50} concept, 152
Packed column in gas chromatography, 523-524
PAH clearance tests of renal blood flow, 257-258
Pair production, 333
Pancreas
  diseases of
    decreased amylase values in, 315
    decreased trypsin values in, 316
    islets of Langerhans of, 372-374
Pandy test for globulin in cerebrospinal fluid, 326
Paper chromatography, development of, 485, 486
Papillary ducts of Bellini, 244
Parasites, examination of duodenal contents for, 317
Parathyroid glands, 371-372
  hormones of, 372
Parietal cells of stomach, 309
Particles of atoms, 329
Partition chromatography
  development of, 485
  solvents for, 493
Pathology
  blood buffer, changes in, 147
  clinical, of hydrogen ion homeostasis, 141-176
Patient controls and quality control, 13
Pattern cutter, immunoelectrophoresis and immunodiffusion, 604
Patterns, immunoelectrophoresis, position of, on slide, 607
PBI; see Protein-bound iodine
PBI test, 32
Pentose in urine, Tauber test for, 287
Pepsinogens in digestion, 310
Peptic cells of stomach, 309
Percent of mean, 18-19
Percent absorption, conversion of, to absorbance. 207-208
Percent solutions, 678
Percentages, 15
  computing, 16
Peripheral tissues, role of, in metabolism, 93
Peristalsis, 308
  temperature of liquids and, 308-309
Peritoneal dialysis, 270-272
  clinical application of, 272
  dialysis fluid in, 271
  patient dialysis for, 271-272
  patient preparation with, 271
Perkin-Elmer Atomic Absorption Spectrophotometer, 204
Pernicious anemia, 312
  maximum gastric acid output in, 313

Pfizer Pol-E-Film method
  of creatine phosphokinase isoenzyme electrophoresis, 572-576
  for electrophoresis of serum proteins, 545-548
  for lactic dehydrogenase isoenzyme electrophoresis, 577-578
pH
  actual, 127
  analyses of, 164-165
  blood
    Astrup studies of, 127-128
    determination of, 125-128
  colorimetric measurement of, 126
  effects of, on enzymatic activity, 218
  electrometric measurement of, 126-127
  and energy absorbance, 467
  measurement of, 163-169
    electrodes in, 168-169
    quality control in, 165-168
    temperature and, 167
  plasma, adjustment of, in disease, 170
  of urine, disorders and, 277-278
Phase, solid
  bonding to, to separate bound from free antigen, 338
  to separate bound from free antigen, 339
Phenistix test for phenylketonuria, 305-306
Phenolphthalein, 687
Phenols, thin-layer chromatography of, 500
Phenolsulfonphthalein excretion tests of renal blood flow, 258-259
Phenylhydrazine test for glucose in urine, 288
Phenylketonuria, tests for, 305-306
Pheochromocytoma, 389
  catecholamine levels in, 407
  diagnosis of, 391-393
  causing hyperglycemia, 94
  symptoms of, 406
Phlebotomy for blood gas and pH measurements, 164
Phosphate
  blood concentration of, in kidney disease, 260-261
  inorganic
    Fiske-SubbaRow method of determination of, 136
    urine, determination of, 136-137
  tubular reabsorption of, 441-442
  urinary calculi of, 304
Phospholipids
  in liver, 190
  thin-layer chromatography of, 500
Phosphorus
  abnormal values of, conditions accompanied by, 135
  and calcium, determination of, 128-129
  determination of, 135-137
  inorganic, determination of, with AutoAnalyzer II, 84-85
Photoelectric absorption, 333
Photometry, flame, 202
  plasma lithium determination by, 482-483
Photon interaction with matter, 333
Photons, 333
Photovolt densitometer with integrator attachment, 535

pHydrion, 687
Physiology of kidneys, 244-246
Pia mater, 322
Pigment, bile
  fecal analysis for, 318
  metabolism of, 195-200
Pineal gland, 399-400
  hyperfunction of, 400
"Pink puffers," 163
Pituitary gland, 359-366
  anterior, 360-365
    and blood glucose level, 93
    hypofunction of, hypoglycemia from, 93-94
  posterior, 365-366
Placental lactogen, determination of, 399
Plasma
  color of, indicating disease states, 8
  electrolytes in, 115
  lithium in, determination of, by flame photometry, 482-483
  pH of, adjustments of, in disease, 170
  in sample preparation, 7-8
  zinc in, determination of, using atomic absorption spectroscopy, 211
Plasma electrophoresis, 610
Plasma 17-hydroxycorticosteroids, determination of, 385
  competitive protein binding assay for, 419-421
  fluorometric procedure for, 416-419
Plasma renin activty, 261-262
Plates
  development of, in thin-layer chromatography, 510-511
  immunoelectrophoretic
    setup for, 602
    tray and jig for holding in place during cutting, 606
  preparation of, in thin-layer chromatography, 507-510
  for radial immunodiffusion, 623
Platykurtic curve, 28
Pneumocystis carinii pneumonia, hypogammaglobulinemia of, 614
Pneumonia, *Pneumocystis carinii,* hypogammaglobulinemia of, 614
Pocket dosimeters for radiation monitoring, 349
Point, immunodominant, 594
Poisoning
  acute, dialysis for, 273-274
  causing electrolyte imbalance, 116
  liver destruction caused by, hypoglycemia from, 93
Poisons, antidotes of, 677
Poisson distribution in quality control in radioassay, 352
Polarity and chromatographic behavior, 494
Polyacrylamide gel electrophoresis, 548-556
  flat bed or vertical gel slab, 549
  haptoglobin typing by, 584
  Microzone acrylamide gel method of, 549-552
  Microzone step-gradient, 552-555
Polyacrylamide gel electrophoresis disk, 548-549
Polyacrylamide method of serum haptoglobin electrophoresis, 583-584
Polyacrylamides as support media in electrophoresis, 532

Polybasic acid, 684
Polyclonal gammopathy, 613, 614
  clinical conditions exhibiting, 616-617
Polycystic disease, chromic renal insufficiency from, 263
Porphobilinogen in urine, 297-298
Porphyria, 514-515
Porphyrins in urine, 297
  rapid assay for, using thin-layer chromatography, 514-517
Portal veins, blood supply to anterior pituitary by, 360
Porter-Silber chromogens
  measurement of, 386
  urinary 17-hydroxycorticosteroid determination using, 421-425
Porter-Silber method of cortisol determination, 385
Positron emission, 331
Posterior pituitary gland, 365-366
Potassium
  abnormal values of, conditions accompanied by, 118
  blood concentration of, in kidney disease, 260
  determination of, 118-120
    with AutoAnalyzer I, 41-46
    by chemical analysis, 119-120
Potassium oxalate as anticoagulant, 7, 8
Power supply
  and cell, ACI/Corning agarose, for electrophoresis of serum proteins, 547
  constant voltage, for immunoelectrophoresis, 604
  for electrophoretic procedures, 631-632
Precipitation, chemical, to separate bound from free antigen, 338
Precision in radioassay, 339-340
Precocious pseudopuberty, 383
Prednisone to suppress homograft reaction, 273
Pregnancy
  estriol during, 396-397
  hormone levels in, 397
  urinary estriol determination in, 397
  procedure for, 410-414
Pregnanediol, urinary
  determination of, 436-439
  normal values for, 438
Pregnanetriol
  measurement of, 386-387
  urinary, determination of, 439-441
Primary adrenal insufficiency, 388
Primary hyperaldosteronism, 384
Primary hyperparathyroidism, 372
Primary hypertension, arteriolar nephrosclerosis from, 263
Primary immunodeficiency disorders, classification of, 615
Primary standard, 14
Progesterone, 397-398
  determination of, 398
Progestins, 397-398
Proinsulin, 373
Prolactin, 365
Propylene glycol, dialysis to remove, 274
Prostatic acid phosphatase, determination of, 231
Prostatic carcinoma, metastatic, and acid phosphatase, 225

Protein analysis by nephelometry, automated specific, 627-628
Protein binding, competitive
　for compound S determination, 385
　for cortisol determination, 385
　measurement of thyroxine by, 368
Protein binding analysis, competitive, 340-341
Protein binding assay, competitive, 401
　for plasma 17-hydroxycorticosteroids, 419-421
Protein-bound iodine, 367
　measurement of thyroxine by, normal values for, 370
Protein concentration procedures in electrophoresis of urinary proteins, 559-560
Protein-free filtrate in sample preparation, 7
Protein separtion with electrophoresis, 533-545
　physical constants of, 530
Proteins
　abnormal values of, conditions accompanied by, 182
　amphoteric, 529
　blood concentration of, in kidney disease, 261
　in cerebrospinal fluid
　　chemical examination of, 325-327
　　electrophoresis of, 556, 558
　　normal values for, 558
　　total, determination of, 326-327
　delineated by immunoelectrophoresis, 612-621
　determination of
　　Kingsley biuret reaction for, 181-182
　　micro-Kjeldahl method for, 182
　laboratory examinations for, 181-190
　liver function tests for, 180
　serum; *see* Serum proteins
　structure of, 529
　thin-layer chromatography of, 501
　total
　　and albumin
　　　determination of, 183-184
　　　simultaneous determination of, with AutoAnalyzer I, 70-73
　　determination of
　　　with AutoAnalyzer II, 85-86
　　　with method of Weichselbaum, 183
　in urine, 278-283
　　Bence Jones, determination of, 282-283
　　electrophoresis of, 558-560
　　　collodion bag ultrafiltration in, 559
　　　Minicon concentration method of, 560
　　　protein concentration procedures in, 559-560
　　tests for
　　　qualitative, 279-280
　　　quantitative, 281-282
　value for, determined by radial immunodiffusion, 625
Proteinuria
　causes of, 558-559
　orthostatic, 279
Proteoses in urine, determination of, 282
Prothrombin times, changes in, in disease, 178
Protons of atom, 328, 329
Pseudohermaphroditism, 383
Pseudopuberty, precocious, 383
Psychoneurosis from hypoglycemia, 93
Ptyalin and digestion, 308

Pump tube sizes and delivery volumes, 37
Punctures
　arterial, for blood gas and pH measurement, 164
　capillary, of finger or ear, for blood collection, 1-3
　spinal, to collect cerebrospinal fluid, 323
　venous, for blood gas and pH measurements, 164
Purdy test for protein in urine, 279-280
Pyelonephritis, chronic renal insufficiency from, 263

Q

Qualitative method
　of alkaloid and amphetamine determination using thin-layer chromatography, 478-481
　for barbiturate and glutethimide determination using thin-layer chromatography, 481-482
Qualitative tests
　Benedict's, for glucose in urine, 284
　for phenylketonuria, 305-306
　for protein in urine
　　Heller, 280
　　Purdy, 279-280
　　Roberts', 280
　　sulfosalicylic acid, 280
　Schmidt, for urobilin in feces, 319
Quality control, 9-35
　in blood gas and pH measurement, 165-168
　clinical development of, 10-11
　collecting data for, 15
　controls in, 12-13
　data in, 15-19
　in electrophoretic procedures, 630-642
　glossary of terms in, 33-35
　history and function of, 9
　in immunology, 642-644
　industrial development of, 10
　laboratory applications of, 12-19; *see also* Laboratory applications of quality control
　laboratory errors in, 12
　in medical laboratories, 11
　purpose of, 11
　in radioassay, 352-353
　specimen collection in, 12
　specimen delivery and storage in, 12
　standards in, 13-15
　starting a program in, 11
　statistical; *see* Statistical quality control
　statistics in, 9-11
　tabulating and studying data for, 15
Quanta, 333
Quantitative colorimetric method
　for arsenic determination, 450-452
　for bromide determination, 463
　for determination of salicylate radical, 465-466
Quantitative determination
　of ethyl alcohol, 452-454
　of fibrinogen, 187-188
Quantitative spectrophotometric procedure for carbon monoxide determination, 460
Quantitative test
　for calcium in urine, 300-302
　for glucose in urine, Benedict's, 284-285

Quantitative test—cont'd
  for protein in urine
    Esbach's method for, 281
    Shevky-Stafford rapid approximate sedimentation method for, 281
    sulfosalicylic acid turbidity method for, 282
Quantitative thin-layer chromatography, 512-514
Quantitative titrimetric method for determination of ethyl alcohol, 454-455
Quantitative ultraviolet spectrophotometric procedure
  for determination of diazepam, 471-473
  for diphenylhydantoin determination, 473-474
  for morphine-heroin determination, 475-477
Quenching, 342-344
  measuring, 343-344
Quick's method for determination of hippuric acid synthesis in urine, 298-299

R

R value in chromatography, 487
R_f values
  of barbiturates, 520
  in chromatography, 487
    calculation of, 488
Radial immunodiffusion, 622-625
  complement values by, 603
  plates for, 623
  protein values determined by, 625
Radiation
  accumulated dose of, 350
  emergency dose of, 350
  energy transfer by, 333
  handling and hazards in, 346-349
  maximum permissible doses of, 350
  medical dose of, 350
  monitoring of, 349-352
  permissible levels of, 350
  weekly dose of, 350
Radiation area
  high, sign in, 347
  sign in, 347
Radioactive decay, 330-332
  rate of, 333-335
Radioactive disintegration, 330
Radioactive iodine uptake by thyroid, 368
  normal values for, 370
Radioactive material, sign for, 347
Radioactive tracers in radioimmunoassay, 336
Radioactivity, 330
Radioactivity area, airborne, sign in, 347
Radioassay, 328-356
  accuracy in, 339
  basic theory of, 328-335
  clinical applications of, 335-346
  data reduction in, 352-353
  glossary for, 353-356
  hormone, 401-402
  ligands in, 340
  offline application of, 353
  online application in, 353
  precision in, 339-340
  quality control in, 352-353
  receptors in, 340
  reliability of, 339-340
  reproducibility in, 339-340

Radioassay—cont'd
  sensitivity in, 339
  specificity in, 339
  technique of, 340-341
Radioimmunoassay, 335-339; see also Radioassay
  for aldosterone determination, 384
  antibody in, 336-337
  antigen in, 336
  for calcitonin measurement, 371
  calculation in, 339
  competitive equilibrium, 336
  for cortisol determination, 385
  definition of, 401-402
  for glucagon measurement, 374
  immunoglobulins in, 336-337
  incubation period in, 337
  insulin measurement by, 373
  to measure human growth hormone, 361
  for progesterone determination, 398
  radioactive tracers in, 336
  reactions in, 336
  separation of bound from free antigen in, 337-339
  specimen preparation in, 336
  for testosterone determination, 394
  for triiodothyronine measurement, 370
Radioisotopes
  alpha decay in, 330-331
  atoms of, 328
  biologic half-life of, 334-335
  clothing required to handle, 347
  half-life of, 334
  hazards with, 347-349
  mean-life of, 334
  procedures for handling, summary of, 350-352
  shielding materials for, 348
Radioreceptor assay, 340
Rapid assay for urinary porphyrins, thin-layer chromatography for, 514-517
Rapid tests of urine, 276-277
Rate, 15
  of radioactive decay, 333-335
Ratio, 15
  computing, 15-16
  ventilation:perfusion, 159-163
    bronchiolar-bronchial diseases and, 161-162
Reabsorption, tubular, 253
  of phosphate, 441-442
Reaction
  antigen-antibody, factors affecting, 336
  in radioimmunoassay, 336
  of urine, 277
Reagents
  for agarose electrophoresis of serum lipoproteins, 560
  for agarose method for lactic dehydrogenase isoenzyme electrophoresis, 578
  for agarose or Pfizer Pol-E-Film of ACI/Corning method for electrophoresis of serum proteins, 545-548
  for alkaline phosphatase determination, 62
  for alkaline phosphatase isoenzyme electrophoresis, 576
  for alkaloid and amphetamine determination using thin-layer chromatography, 479-480

Reagents—cont'd
- for Allen "blue" color test of urinary dehydroepiandrosterone, 409
- for amylase determination in duodenal contents, 315
- for Archibald method of uric acid determination, 108-109
- for Babson-Phillips method of lactate dehydrogenase determination, 235
- for Bachra-Dauer-Sobel method of calcium determination, 131-132
- for barbiturate and glutethimide determination using thin-layer chromatography, 481-482
- for Benedict's method of testing for glucose in urine, 285
- for Berthelot reaction for urea nitrogen determination, 112
- for Bessey-Lowry-Brock method of alkaline phosphatase determination, 228
- for blood ammonia tests in liver function, 189
- for bromcresol green method for albumin determination, 183
- for bromine test for melanin in urine, 299
- for calcium determination, 55
  - using atomic absorption spectroscopy, 205-206
- for calibration curve
  - for alkaline and acid phosphatase, 232
  - for glutamic oxaloacetic transaminase and glutamic pyruvic transaminase determination, 234
- for carbon dioxide determination using Van Slyke manometric apparatus, 122-123
- for carbon monoxide screening test, 457
- for cellulose acetate method
  - of hemoglobin electrophoresis, 564-566
  - for serum haptoglobin electrophoresis, 582
- for cephalin-cholesterol flocculation test, 186
- for chemical fractionation of lactate dehydrogenase, 237
- for chlordiazepoxide determination, 464
- for citrate agarose electrophoresis in hemoglobin electrophoresis, 567-568
- for Clark-Collip modification of Kramer-Tisdall method of calcium determination, 130
- for Clinitest method of testing for glucose in urine, 285
- for competitive protein binding assay for plasma 17-hydroxycorticosteroids, 419-420
- for copper determination in serum and urine using atomic absorption spectroscopy, 209-210
- for creatine phosphokinase isoenzyme electrophoresis, 574
- for creatinine determination, 51, 102
- for 11-deoxycortisol determination, 407-408
- for determination of amylase in duodenal contents, 315
- for determination of Bence Jones protein in urine, 283
- for determination of diazo substances in urine, 296
- for determination of 3-methoxy-4-hydroxymandelic acid in urine with thin-layer chromatography, 521-522
- for determination of porphobilinogen in urine, 297

Reagents—cont'd
- for determination of porphyrins in urine, 297
- for determination of proteoses in urine, 282
- for determination of urobilin in urine, 296
- for direct cholesterol determination, 65
- for electrolyte determination with AutoAnalyzer I, 41-42
- in electrophoretic procedures, guidelines for, 633-634
- for Esbach's method for determining protein in urine, 281
- for fermentation test for glucose in urine, 287
- for ferric chloride test for melanin in urine, 299
- for Fiske-SubbaRow method of determination of inorganic phosphate, 136
- for fluorometric procedure for determination of plasma 17-hydroxycorticosteroids, 417
- for Folin method for determination of sulfates in urine, 302
- for fractionated catecholamine determination, 405
- for free urinary cortisol determination, 428
- for Galatest method of testing for glucose in urine, 285
- for Garvey-Bowden method for barbiturate determination, 461-462
- for Gerhardt test for diacetic acid in urine, 290
- for Gettler-Freimuth method for carbon monoxide determination, 457-459
- for gmelin test for fecal analysis for bile pigment, 318
- for guaiac test for fecal analysis for occult blood, 317
- for Harger's micro method for ethyl alcohol determination, 454
- for Hart method for detecting β-oxybutyric acid in urine, 290
- for Heller test for protein in urine, 280
- for hemoglobin A_2 quantitation with DE-52 cellulose mini-column, 570-571
- for hemoglobin F determination by alkali denaturation, 569
- for Henry method for uric acid determination, 110
- for 5-hydroxyindoleacetic acid determination, 429
- for Ictotest for bilirubin in urine, 295
- for identification of barbiturates in urine by thin-layer chromatography, 517-518
- for inorganic phosphate determination, 57
- for inulin clearance tests, 247
- for Klein-Foreman-Searcy method of determination of amylase activity, 221
- for Kowarsky test for glucose in urine, 288
- for lactic dehydrogenase determination, 64
- for lead determination in blood and urine using atomic absorption spectroscopy, 211
- list of, 690-711
- for macro–cellulose acetate electrophoresis, 534
- for magnesium determination in serum and urine using atomic absorption spectroscopy, 209
- for Malloy-Evelyn modification for determination of direct and total bilirubin, 196
- for methanol and formaldehyde screening tests, 456

Reagents—cont'd
- for Meuleman's method for determination of total protein in cerebrospinal fluid, 326
- for micro–cellulose acetate electrophoresis, 539
- for micro–starch gel electrophoresis, 542, 544
- for Microzone acrylamide gel method of polyacrylamide gel electrophoresis, 550
- for Microzone cellulose acetate method for lactic dehydrogenase isoenzyme electrophoresis, 577
- for Microzone step-gradient polyacrylamide gel electrophoresis, 552-553
- for mucic test for galactose in urine, 286-287
- for N-substituted determination of barbiturates, 471
- for Nuttal-Wedin method for creatine kinase determination, 238
- for occult blood test in fecal analysis, 317
- for Occultest for blood in urine, 300
- for Pandy's test for globulin in cerebrospinal fluid, 326
- for Phenistix test for phenylketonuria, 305
- for phenolsulfonphthalein excretion tests, 258
- for plasma lithium determination by flame photometry, 483
- for polyacrylamide method of serum haptoglobin electrophoresis, 583
- for potassium determination by chemical analysis, 119
- preparation of, 690-691
- for Purdy test for protein in urine, 279
- for quantitative colorimetric method
  - for arsenic determination, 451
  - for bromide determination, 463
  - for determination of salicylate radical, 465
- for quantitative determination of ethyl alcohol, 453
- for quantitative fibrinogen determination, 187
- for quantitative ultraviolet spectrophotometric procedure
  - for determination of diazepam, 472
  - for diphenylhydantoin determination, 473-474
  - for morphine-heroin determinatioin, 475
- for Quick's method for determination of hippuric acid synthesis in urine, 298
- for rapid assay of urinary porphyrins using thin-layer chromatography, 515
- for Reinsch modified test for arsenic, 449
- for Reitman-Frankel method of transaminase determination, 233
- for Roberts' test for protein in urine, 280
- for routine method for immunoelectrophoresis, 605
- for Rubner test for lactose in urine, 286
- for Schales-Schales titration method for chloride determination, 122
- for Schwartz-Watson test for urobilinogen in urine, 295
- for Seligson-Marino-Dodson method for sulfobromophthalein determination, 200
- for Seliwanoff test for levulose in urine, 286
- for serum creatinine and creatine determination, 103-104
- for serum glutamic oxaloacetic transaminase, 59
- for serum glycoprotein electrophoresis, 585

Reagents—cont'd
- for Shank-Hoagland modification of Maclagen method for thymol turbidity determination, 185
- for Shevky-Stafford rapid approximate sedimentation method for determining protein in urine, 281
- for simultaneous direct and total bilirubin determination, 67, 69
- for simultaneous glucose and urea nitrogen determination, 47-48
- for simultaneous total protein and albumin determination, 71
- for sodium determination by chemical analysis, 117
- for starch gel hemoglobin electrophoresis, 568
- for sulfosalicylic acid test for protein in urine, 280
- for sulfosalicylic acid turbidity method for determining protein in urine, 282
- for Tauber test for pentose in urine, 287
- for Tes-Tape method of testing for glucose in urine, 285
- for tests for calculi in urine, 303
- for Tietz-Fiereck method of serum lipase determination, 223-224
- for o-tolidine test for occult blood in urine, 300
- for o-toluidine method for glucose determination, 96
- Töpfer's, 687
- for total acid phosphatase determination, 230
- for total protein method of Weichselbaum, 183
- for total serum cholesterol determination with method of Abell, Levy, Brodie, and Kendall, 193-194
- for total urinary estrogen determination, 414-415
- for trypsin determination of duodenal contents, 316
- for utraviolet spectrophotometric determination of barbiturates, 468-469
- for uric acid determination, 53
- for urinary catecholamine determination, 403
- for urinary estriol determination in pregnancy, 411
- for urinary 17-hydroxycorticosteroid determination
  - using 17-ketogenic steroids, 425
  - using Porter-Silber chromogens, 422
- for urinary 17-ketosteroid determination, 431, 432
- for urinary metanephrine determination, 435
- for urinary pregnanediol determination, 437
- for urinary pregnanetriol determination, 439
- for urinary vanillylmandelic acid determination, 442-443
- for van de Kamer method of fecal fat determination, 320
- for Wang method for determination of calcium in urine, 301
- for zinc determination in plasma and urine using atomic absorption spectroscopy, 211

Receptors in radioassay, 340
Recirculating single-pass system for dialysate solutions, 264
Recirculating system for dialysate solutions, 264
Recommended name for enzymes, 219

Red cell count in capillary puncture, 1
Reducing substances in urine, 283-285
Regenerating injured lung, 142
Regional ventilation, normal, extremes of, 159
Reinsch modified test for arsenic, 449
Reitman-Frankel method of glutamic oxaloacetic transaminase and glutamic pyruvic transaminase determination, 233-234
Rejection of transplanted kidney, 272-273
Releasing/inhibiting factors of anterior pituitary, 360
Renal osteodystrophy, 261
Renin, plasma, activity of, 261-262
Renin-angiotensin system, 400
and kidney disease, 261-262
Rennin and digestion, 309
Report, sample urine analysis, 275
Reproducibilty in radioassay, 339-340
Reserve vascular capacity of lung, 156
Residual capacity, functional, determination of, 162
Resonance lines of elements, measurement of, 202
Respiratory acidosis, 172
compensation slopes for, 174
Respiratory alkalosis, 172
Retrolental fibroplasia, prevention of, 164-165
RIA; see Radioimmunoassay
Ring badges for radiation monitoring, 349
Roberts' test for protein in urine, 280
Routine urine examination, 274-306
Rubner test for lactose in urine, 286
Rule of 1200, 153

S

S chamber for thin-layer chromatography, 510-511
Saccharogenic method of amylase determination, 220-221
Salicylates
dialysis to remove, 274
quantitative colorimetric method for determination of, 465-466
Saliva
and digestion, 308
functions of, 308
Salt-losing syndrome, 389
Sample preparation, 7-8
blood collection and, 1-8
gross examination of, 8
plasma in, 7-8
protein-free filtrate in, 7
serum in, 7
Sample work sheet for tabulating test results, 16
Sampling inspection tables
for Army Ordnance and Army Service Forces, 10
Dodge-Romig, 10
Sarcinae, microscopic examination of, 314
Saturated solutions, 678
Saturation in steroid hormones, 376
Saturation analysis, 401
sequential, 340
Scale of activity, Brockman-Schodder, 491
Scattering, Compton, 333
Schales-Schales titration method for chloride determination, 122
Schmidt qualitative test for urobilin in feces, 319

Schwartz-Watson test for urobilinogen in urine, 295-296
Scintillation, liquid, kinds of samples of, 342-343
Scintillation cocktail, 342
Scintillation counting, liquid, 341-345
Screening method for carbon monoxide sensitive to 20% saturation, 456-457
Screening test, methanol and formaldehyde, 455-456
Secondary adrenal insufficiency, 388
Secondary hyperaldosteronism, 384
Secondary hyperparathyroidism, 372
Secondary standard, 14
Secretin, 311-312
Secretion
of aldosterone, stimulus of, 381
gastric, 310-311
basal, determination of, 313
regulation of, 311
of hormones, control of, 359
Sediments in urine
causes of, 291
microscopic examination of, 291-294
Self-absorption in quenching, 343
Seligson-Marino-Dodson method for sulfobromophthalein determination, 200
Seliwanoff test for levulose in urine, 286
Semiquantitative method for carbon monoxide determination sensitive to 5% saturation, 457-459
Sensitivity
in atomic absorption spectroscopy, 203
of chromatography to structural differences in compounds, 496
in radioassay, 339
Separation
of bound from free antigen in radioimmunoassay, 337-339
class, of lipids, thin-layer chromatography for, 499-500
of hemoglobin variants with starch gel procedure, 573
of immunoglobulins, 596-597
protein, with electrophoresis, 533-545
of serum glycoproteins on cellulose acetate, 585
Sequential saturation analysis, 340
Serotonin, determination of, 399
Serum
commercially prepared, as control, 13
copper determination in, using atomic absorption spectroscopy, 209-211
densitometer scan of, on starch gel, 546
follicle-stimulating hormone in, 364
icteric and hemolyzed, Bachra-Dauer-Sobel method of calcium determination using, 132
luteinizing hormone in, 364
magnesium in, determination of, using atomic absorption spectroscopy, 209
pooled, controls of, 93
in sample preparation, 7
Serum albumin, changes in, in disease, 178
Serum bilirubin, 178-179
Serum calcium, diffusible, nomogram for estimating, 134, 135

Serum cholesterol, total, determination of, with method of Abell, Levy, Brodie, and Kendall, 193-194
Serum creatinine and creatine determination, 103-105
Serum glutamic oxaloacetic transaminase
   determination of, with AutoAnalyzer I, 59-61
   and dialysis, 60
   test with, for hepatobiliary disease, 179
Serum glycoproteins
   electrophoresis of, 585-587
      clinical significance of, 586-587
   separation of, on cellulose acetate, 585
Serum haptoglobin electrophoresis
   cellulose acetate method of, 581-583
   clinical significance of, 584-585
   polyacrylamide method of, 583-584
Serum immunoglobulins, levels of, 624
Serum lipase, determination of, with Tietz-Fiereck method, 223-224
Serum lipoproteins
   comparison of, 563
   electrophoresis of, 560-564
      agarose, 560-561, 562
      clinical significance of, 561-564
   phenotypes of, 565
      diseases associated with, 566
Serum proteins
   electrophoresis of, 545-556
      clinical significance of, 556
   electrophoretic fractionations of, 557
   normal values for, 556
Serum sample, application of, in micro-cellulose acetate electrophoresis, 540
SGOT; *see* Serum glutamic oxaloacetic transaminase
SGPT test for hepatobiliary disease, 179
Shank-Hoagland modification of Maclagen method for determination of thymol turbidity, 184-186
Shevky-Stafford rapid approximate sedimentation method for determining protein in urine, 281
Shewhart chart, 10, 19-20
   action lines of, 19
   bell-shaped curve on, 20
   gaussian curve on, 20
   normal distribution on, 20, 22
   systematic shift on, 23
   systematic trends on, 23
   warning lines of, 19-20
Shielding materials for radioisotopes, 348
Shift, systematic, 24
   on cusum chart, 23
   on Shewhart chart, 23
Sickle cell disease, 571
Signs, radiation, 347
Silicic acid as absorbent in chromatography, 490-491
Simultaneous direct and total bilirubin determination with AutoAnalyzer I, 67-70
Simultaneous glucose and urea nitrogen determination with AutoAnalyzer I, 46-50
Simultaneous total protein and albumin determination with AutoAnalyzer I, 70-73

Single electroimmunodiffusion, two-dimensional, 627
Single-pass system for dialysate solutions, 264
Skeggs-Leonards dialyzer, 264
Skewness
   formula for measurement of, 28
   on graphs of frequency distribution, 27
Slide rule to calculate squares and square roots, 18
Small intestine, digestion in, 310
Smear of cerebrospinal fluid, microscopic examination and, 324
Sodium
   abnormal values of, conditions accompanied by, 117
   blood concentration of, in kidney disease, 260
   determination of, 117-118
      with AutoAnalyzer I, 41-46
      by chemical analysis, 117-118
Sodium chlorate, dialysis to remove, 274
Sodium fluoride as anticoagulant, 7, 8
Sodium iodide crystal in gamma ray spectrometry, 345
Solid phase
   bonding to, to separate bound from free antigen, 338
   to separate bound from free antigen, 339
Solutes
   concentration of, versus adsorption isotherm, 489
   distribution of, 488
   movement of, in chromatography, 487
Solutions
   clearing, in electrophoretic procedures, 636-638
   dialysate
      for hemodialysis, 264-266
         formula for, 265
      for peritoneal dialysis, 271
   molal, 678
   molar, 678
   normal, 678
      composition of, 679-680
   percent, 678
   of pure chemicals as controls, 13
   saturated, 678
   Tuerk, 325
   types of, 678-683
Solvent series, elutropic, for chromatography, 493
Solvents
   for column chromatography, 505-506
      setup for change of polarity of, 506
   organic, with Perkin-Elmer Atomic Absorption Spectrophotometer, 204
   for partition chromatography, 493
Somatomammotropin, human chorionic, determination of, 399
Somatotropin; *see* Human growth hormone
Somogyi-Nelson method of glucose determination, 95
Space, dead, 150, 154
Species specificity, 336
Specific detectors for thin-layer chromatography, 512
Specific gravity of urine, 278
Specificity
   in radioassay, 339
   species, 336

Specimen
  collection of
    quality control and, 12
    rules for, 12
  delivery and storage of, quality control and, 12
  preparation of, in radioimmunoassay, 336
Spectrometer
  liquid scintillation counting, 341-342
  mass, 163
Spectrometry, gamma ray, 345-346
  sodium iodide crystal in, 345
Spectrophotometer
  Atomic Absorption, Perkin-Elmer, 204
  ultraviolet, 466
Spectrophotometric procedure for carbon monoxide determination, quantitative, 460
Spectrophotometry, ultraviolet, 466-471
Spectroscopy
  atomic absorption; see Atomic absorption spectroscopy
  flame emission, 202
Spectrum, absorption; see Absorption spectrum
Spinal puncture to collect cerebrospinal fluid, 323
Spontaneous fission, 332
Spotting sample on thin-layer plate, 519
Square roots, 17, 18
  calculating, 18
  of small numbers, 251
Squares, 17
  calculating, 18
Stahl-De Saga method of plate preparation in thin-layer chromatography, 508-509
Staining in electrophoretic procedures, 635-636
Standard deviation, 17-18
Standardization, 682-683
  automatic external, 344-345
  of bilirubin determination, recommended, 197-198
  internal, to measure quench, 343
Standards
  for atomic absorption spectroscopy, 203-204
  bicarbonate, 128
  difference of, from control, 14-15
  internal, 15
  primary, 14
  in quality control, 13-15
  secondary, 14
  stock, in electrolyte determination, 42, 44
  working; see Working standards
Starch as support medium in electrophoresis, 532
Starch gel, densitometer scan of serum on, 546
Starch gel electrophoresis for hemoglobin electrophoresis, 568-569
Stationary phase in chromatography, 490-492
Statistical evaluation of daily results as control, 13
Statistical methods
  in quality control, 19-33
    control chart techniques in, 19-24
    frequency distribution in, 24-33
  in radioassay, 352-353
Statistical quality control, 10
  clinical development of, 10-11
  conditions monitored by, 11
  industrial development of, 10
  in medical laboratories, 11

Statistics in quality control, 9-11
  statistical quality control in, 10
Stereoisomerism in steroid hormones, 376
Steroid hormones, 359, 375-382
  basic nucleus of, 375
  biosynthesis of, 378-380
  chair configuration of, 377
  classification of, 375-376
  hydroxylation of, 378, 380
  isomerism in, 376
  metabolism of, 381
  and metabolites, 376
  saturation of, 376
  stereoisomerism in, 376
  structure of, 375
  terminology of, 376-377
  unsaturation of, 376
Steroids
  17-ketogenic
    measurement of, 386
    urinary 17-hydroxycorticosteroid determination using, 425-427
  thin-layer chromatography of, 501-502
Sterols, 192
  in liver, 190
Sterox SE as surfactant with AutoAnalyzer, 39
Stimulation test, ACTH, of adrenocortical function, 387
Stimulator, thyroid, long-acting, 371
Stock standards in electrolyte determination, 42, 44
Stomach
  carcinoma of
    basal gastric secretion in, 313
    maximum gastric acid output in, 313
  cells of wall of, 309
  contents of
    analysis of, 312-314
    bile in, 314
    microscopic examination of, 313-314
  digestion in, 309
  secretion in, 310-311
Stool, occult blood in, 180
Storage of specimen and quality control, 12
Strongyloides stercoralis, microscopic examination of duodenal fluid for, 317
Studying data for quality control, 15
Substrate concentration and enzyme activity, 215-216
Substrate inhibition of enzymes, 217-218
Sugar in urine, 283-285
  Benedict method of testing, 284-285
  Clinitest method of testing, 285
  Galatest method of testing, 285
  Tes-Tape method of testing, 285
Sulfate in urine, Folin method of determining, 302-303
Sulfobromophthalein test of excretory capacity of liver, 198-200
  dye dosages for, 199
  procedure for, 199-200
  Seligson-Marino-Dodson method for, 200
Sulfolipids, thin-layer chromatography of, 500
Sulfonamides, dialysis to remove, 274
Sulfosalicyclic acid test for protein in urine, 280
Sulfosalicyclic acid turbidity method for determining protein in urine, 282

Support media for electrophoresis, 531-532, 634
Suppression test, dexamethasone, of adrenocortical function, 388
Suppurative meningitis, cerebrospinal fluid in, 323
Surfactants for use with AutoAnalyzer, 39-41
Svedberg units of flotation, 190
Swipe method of contamination monitoring, 345
Syndrome
    adrenogenital, 383
    Conn's, 384
    Cushing's, 361, 388-389
        dexamethasone suppression test for, 388
        symptoms of, 424
    dumping, causing hyperglycemia, 94
    malabsorption, 319
    salt-losing, 389
    Zollinger-Ellison
        basal gastric secretion in, 313
        gastrin in, 311
        maximum gastric output in, 313
Synthesis
    of catecholamines, 391
    hippuric acid, and urine function, 298-299
    of steroid hormones, 378-380
Synthetic function of liver, 178
Syringe for venipuncture, 4
Systematic name for enzymes, 218-219
Systematic shift, 24
    on cusum chart, 23
    on Shewhart chart, 23
Systematic trends, 22, 24
    on cusum chart, 23
    on Shewhart chart, 23
Systems considerations in electrophoretic procedures, 630-633

T

T cells, 594
t test, 32-33
t value, 32, 33
T_3; see Triiodothyronine
T_3 uptake tests, 369
    normal values for, 370
T_4; see Thyroxine
Tabulating data for quality control, 15
Tangier disease, 564
Tauber test for pentose in urine, 287
Temperature
    and blood gas and pH measurements, 167
    effects of, on enzymatic activity, 218
Tendency, central, 25, 27
    measures of, 29
Tension
    carbon dioxide; see Carbon dioxide tension
    oxygen; see Oxygen tension
Terminology of steroid hormones, 376-377
Terpenoids, thin-layer chromatography of, 502
Tes-Tape method of testing for glucose in urine, 285
Testosterone, 393-394
    deficiency of, 394
    determination of, radioimmunoassay for, 394
Tests
    ACTH stimulation, of adrenocortical function, 387

Tests—cont'd
    of adrenocortical function, 387-388
        laboratory, 382-389
    alkaline phosphatase, for hepatobiliary disease, 179
    Allen "blue" color, of urinary dehydroepiandrosterone, 409-410
    of androgens, 382-383
    arginine infusion, of human growth hormone, 362
    bromine, for melanin in urine, 299
    Bromsulphalein, of excretory capacity of liver, 198-200
        dye dosages for, 199
        procedure for, 199-200
        Seligson-Marino-Dodson method for, 200
    for calcium in urine, quantitative, 300-302
    for calculi in urine, 303
    cephalin-cholesterol flocculation, of liver function, 186-187
    chi-square, in quality control in radioassay, 352
    clearance
        creatinine, 252-253
        inulin, 247, 249
        PAH, of renal blood flow, 257-258
        urea, 249-252
            Moller-McIntosh-Van Slyke method of, 250-252
    concentration, of tubular function, 254-255
    dexamethasone suppression, of adrenocortical function, 388
    for diacetic acid in urine, Gerhardt, 290
    dilution, of kidneys, 256
    excretion, phenolsulfonphthalein, of renal blood flow, 258-259
    for fecal analysis for occult blood
        benzidine, 317
    guaiac, 317-318
        Hematest, 318
        o-tolidine, 318
    ferric chloride, for melanin in urine, 299
    for galactose in urine, mucic, 286-287
    glomerular filtration, 246-253
    for glucose in urine
        fermentation, 287
            reading results of, 288
        Kowarsky, 288
        phenylhydrazine, 288
    for glucose tolerance determination
        Exton-Rose, 99
            procedure for, 100-101
        intravenous, 98-99
            procedure for, 100
        Janney-Isaacson, 98
            procedure for, 99-100
    gmelin, for fecal analysis for bile pigment, 318
    Heller, for protein in urine, 280
    for hemoglobinuria, 300
    isocitric dehydrogenase, for hepatobiliary disease, 180
    of 17-ketosteroids, 382-383
    lactic dehydrogenase, for hepatobiliary disease, 180
    for lactose in urine, Rubner, 286
    leucine aminopeptidase, for hepatobiliary disease, 180

Tests—cont'd
　for levulose in urine, Seliwanoff, 286
　liver function, 180-181
　for melanin in urine, 299
　metyrapone, of adrenocortical function, 387-388
　Mosenthal, of tubular function, 254-255
　5'-nucleotidase, for hepatobiliary disease, 179
　for occult blood in urine, 300
　of 17-oxosteroids, 382-383
　Pandy, for globulin in cerebrospinal fluid, 326
　PBI, 32
　for pentose in urine, Tauber, 287
　Phenistix, for phenylketonuria, 305-306
　for phenylketonuria, 305-306
　for protein in urine
　　qualitative, 279-280
　　quantitative, 281-282
　Purdy, for protein in urine, 279-280
　qualitative, for phenylketonuria, 305-306
　quantitative, for calcium in urine, 300-302
　Reinsch modified, for arsenic, 449
　renal function, 242-262
　　chemical, 246-262
　results of sample work sheet for tabulating, 16
　Roberts', for protein in urine, 280
　Schmidt qualitative, for urobilin in feces, 319
　screening, methanol and formaldehyde, 455-456
　sulfosalicylic acid, for protein in urine, 280
　t, 32-33
　T_3 uptake, 369
　　normal values for, 370
　o-tolidine, for occult blood in urine, 300
　transaminase, for hepatobiliary disease, 179
　of urine, rapid, 276-277
　for urobilinogen in urine, Schwartz-Watson, 295-296
　van den Bergh, of bilirubin, 195
　ventilatory:perfusion inequality, 161
Thalassemia, 571
Theory
　atomic, Dalton's, 328
　of chromatography, 486-490
　lock and key, of enzyme inhibition, 215, 217
　of radioassay, basic, 328-335
Thin-layer chromatography, 477, 507-523
　for aldehydes and ketones, 497-498
　for alkaloids, 497
　　and amphetamine determination, qualitative method employing, 478-481
　for amines, 498
　of amino acids and proteins, 501
　for barbiturates and glutethimide determination, qualitative method for, 481-482
　for carbohydrates and glycosides, 498
　for carboxylic acids, 499
　for class separation of lipids, 499-500
　clinical application of, 514-523
　detection of components in, 512
　detectors for, 512
　development of, 485-486
　for dyes, 499
　for essential oils, 499
　for fatty acids, 500
　for glycerides, 500
　for glycolipids, 500

Thin-layer chromatography—cont'd
　identification of barbiturates in urine by, 517-521
　identifying color of barbiturates in, 520
　3-methoxy-4-hydroxymandelic acid determination with, in urine, 521-523
　of phenols and alcohols, 500
　of phospholipids, 500
　plate development in, 510-511
　plate preparation in, 507-510
　practical aspects of, 507-514
　quantitative, 512-514
　for rapid assay for urinary porphyrins, 514-517
　ratio of adsorbent to water in, 508
　S chamber for, 510-511
　sample preparation in, 510
　of steroids, 501-502
　of sulfolipids, 500
　of terpenoids, 502
　for vitamins, 502
Thin-layer plate
　separated barbiturates on, 521
　spotting sample on, 519
Theocyanate, dialysis to remove, 274
Threshold of substances in urine, 253-254
Thymol turbidity
　calibration of, chart for, 186
　determination of, 184-186
Thymus gland, 400
Thyrocalcitonin, 371
Thyroid, 366-371
　and blood glucose level, 93
　hormones of, 367-371
　　causing hyperglycemia, 94
　hypofunction of, hypoglycemia from, 93-94
　iodide and, 366
　iodine and, 366
　uptake of ^{131}I by, 368
　　normal values for, 370
Thyroid-stimulating hormone, 362
　normal values for, 370
Thyroid stimulating substance, long-acting, 362
Thyroid stimulator, long-acting, 371
Thyrotropin, 362
Thyroxine, 366-367
　formula for, 366
　free, measurement of, 369-370
　　normal values for, 370
　measurement of
　　by column, 367-368
　　　normal values for, 370
　　by competitive protein-binding, 368
　　　normal values for, 370
　　by protein-bound iodine and butanol-extractable iodine, 367
Tidal volume, 150
Tietz-Fiereck method of serum lipase determination, 223-224
Tissue
　matching of, in kidney transplantation, 273
　peripheral, role of, in metabolism, 93
Titratable gastric acidity, determination of, 313
Titration, 682
Titration method for chloride determination, Schales-Schales, 122

Titrimetric method for ethyl alcohol determination, quantitative, 454-455
Tolerance, glucose; see Glucose tolerance
o-Tolidine test
 for fecal analysis for occult blood, 318
 for occult blood in urine, 300
o-Toluidine method of glucose determination, 96
Tonks' formula for allowable limits of error, 20
Töpfer's reagent, 687
Total acid phosphatase, determination of, 229-230
Total base deficit, estimation of, in blood gas quality control, 166
Total carbon dioxide tension, 128
Total cell count of cerebrospinal fluid, 324-325
Total cholesterol, determination of, 192-194
Total and direct bilirubin determination, 196-197
 Malloy-Evelyn modification of, 196-197
 simultaneous, with AutoAnalyzer I, 67-70
Total protein
 in cerebrospinal fluid, determination of, 326-327
 determination of
  and albumin, 183-184
   simultaneous, with AutoAnalyzer I, 70-73
   with AutoAnalyzer II, 85-86
  with method of Weichselbaum, 183
Total renal blood flow, 256
Total serum cholesterol, determination of, with method of Abell, Levy, Brodie, and Kendall, 193-194
Total urinary (nonpregnant) estrogen determination, procedure for, 414-416
Tourniquet for venipuncture, 5
Toxicology, 449-483
Tracers, radioactive, in radioimmunoassay, 336
Transaminase
 determination of, 232-234
  alanine aminotransferase reaction in, 233
  aspartate aminotransferase reaction in, 232-233
 tests with, for hepatobiliary disease, 179
Transfer of energy by radiation, 333
Transition, isomeric, 332
Transparency of urine, 277
Transplantation, kidney, 272-273
Transport of hormones, 359
Tray
 and jig for holding immunoelectrophoresis plates in place during cutting, 606
 for micro–starch gel electrophoresis, 541
Trends, systematic, 22, 24
 on cusum chart, 23
 on Shewhart chart, 23
Triglycerides
 changes in levels of, causes of, 191
 determination of, 191
 distribution of, 192
 in liver, 190
  determination of, 191-192
Triiodothyronine, 366-367
 formula for, 366
 measurements of, 370
TRIS buffers in electrophoresis, 530-531
Triton X-100 as surfactant with AutoAnalyzer, 40
Triton X-405 as surfactant with AutoAnalyzer, 40
Trivial name for enzymes, 219
Tropic hormones, actions of, 360

Trypsin in duodenal contents, determination of, 316
TSH; see Thyroid-stimulating hormone
Tube, pump, sizes of, and delivery volumes, 37
Tuberculous meningitis, cerebrospinal fluid in, 323
Tubular necrosis, acute, and renal failure, 262
Tubular reabsorption of phosphate, 441-442
Tubules of kidney, 243-244
 concentration tests of
  Fishberg, 255
  Mosenthal, 254-255
 excretory function of, 256
 functions of, 253-256
 reabsorption of, 253
Tuerk solution, 325
Tumors, carcinoid, 399
Turbidity, thymol
 calibration of, chart for, 186
 determination of, 184-186
Turbidity chemical indicator, 687
Twin-cell dialyzer
 assembly of, 266-267
 connecting patient to, 268
 hemodialysis with, 266
 schematic of, 267
 testing of, 267-268
Two-dimensional development of plates in thin-layer chromatography, 511, 512
Two-dimensional single electroimmunodiffusion, 627

U

Ulcers
 basal gastric secretion in, 313
 duodenal, maximum gastric acid output in, 313
 gastrin in patients with, 311
Ultrafiltration, collodion bag, for electrophoresis of urinary proteins, 559
Ultrafiltration apparatus for electrophoresis of urinary proteins, 560
Ultraviolet spectrophotometer, 466
Ultraviolet spectrophotometric method of barbiturate determination, 468-471
Ultraviolet spectrophotometric procedure, quantitative
 for diazepam determination, 471-473
 for diphenylhydantoin determination, 473-474
 for morphine-heroin determination, 475-477
Ultraviolet spectrophotometry, 466-471
Ultrawet 60L as surfactant with AutoAnalyzer, 40-41
UN; see Urea nitrogen
Unimodal distribution, 25
Units
 Bodansky, 226
 conversion of, 138-139
 enzyme, 219
 King-Armstrong, 227
Universal chemical indicator, 687
Universal detectors for thin-layer chromatography, 512
Unsaturation in steroid hormones, 376
Uptake, thyroid, of iodine 131, 368
 normal values for, 370
Uptake tests, T_3, 369
 normal values for, 370

Urates, urinary calculi of, 304
Urea, blood concentration of, in renal disease, 259-260
Urea clearance tests, 249-252
    Moller-McIntosh-Van Slyke method of, 250-252
Urea nitrogen, 111-113
    abnormal values for, conditions accompanied by, 111
    determination of, 112-113
        with AutoAnalyzer I, 47
        with AutoAnalyzer II, 86-87
        Berthelot reaction for, 112
            procedure for, 112-113
        Folin-Svedberg method of, 112
        Gentzkow method for, 112
        Karr method for, 111
        Van Slyke-Cullen method for, 111-112
    and glucose, determination of, simultaneous, with AutoAnalyzer I, 46-50
    in urine, 306
Uremia causing electrolyte imbalance, 115
Uric acid
    abnormal values of, conditions accompanied by, 108
    determination of
        Archibald method for, 108-109
        with AutoAnalyzer I, 52-54
        with AutoAnalyzer II, 88-89
        Brown method for, 108
        Folin method for, 108
        Henry method for, 109-111
        procedures for, 108-111
    urinary calculi of, 304
    in urine, 306
Urinary catecholamines, 392
    determination of, 402-405
Urinary cortisol, free, determination of, 427-428
Urinary dehydroepiandrosterone, Allen "blue" color test of, 409-410
Urinary 11-deoxycortisol, determination of, 407-408
Urinary estriol determination in pregnancy, 397
    procedure for, 410-414
Urinary estrogens (nonpregnant), total, determination of, procedure for, 414-416
Urinary 17-hydroxycorticosteroids, determination of, 385-387
    using 17-ketogenic steroids, 425-427
    using Porter-Silber chromogens, 421-425
Urinary 5-hydroxyindoleacetic acid, determination of, 428-431
Urinary 17-ketosteroids, determination of, 431-434
Urinary metanephrines, determination of, 435-436
Urinary 3-methoxy-4-hydroxymandelic acid, determination of, 442-444
Urinary porphyrins, rapid assay for, using thin-layer chromatography, 514-517
Urinary pregnanediol
    determination of, 436-439
    normal values for, 438
Urinary pregnanetriol, determination of, 439-441
Urinary vanillylmandelic acid, determination of, 442-444
Urine
    acetone in, 288-289
        Acetest procedure for, 289

Urine—cont'd
    acid, causes of, 277-278
    alkaline, causes of, 278
    amylase in, 303
    barbiturates in, identification of, by thin-layer chromatography, 517-521
    Bence Jones protein in, determination of, 282-283
    bilirubin in, Ictotest for, 295
    calcium in, 300-302
        determination of, 132-133
        Wang method for determination of, 301-302
    calculi in
        analysis of, 304
        tests for, 303
    carbohydrates in, 286-288
    chloride in, 303
    collection of, 274-275
        factors affecting, 444
    color of, 275
        causes of, 276
    copper determination in, using atomic absorption spectroscopy, 209-211
    creatine in, 306
        determination of, 106-107
    creatinine in, 306
        determination of, 105-106
    diacetic acid in, 289-290
        Gerhardt test for, 290
    diazo substances in, determination of, 296
    electrolytes in, 116
    estrogen excretion in, in normal pregnant women, 413
    excreted, biochemical constituents of, 245
    follicle-stimulating hormone in, 364
    free cortisol in, measurement of, 386
    galactose in, mucic test for, 286-287
    glucose in
        fermentation test for, 287
            reading results of, 288
        Kowarsky test for, 288
        or sugar, 283-285
            Benedict method of testing, 284-285
            Clinitest method of testing, 285
            Galatest method of testing, 285
            Tes-Tape method of testing, 285
    hippuric acid synthesis and, 298-299
    inorganic phosphate in, determination of, 136-137
    17-ketosteroids in, 433
    lactose in, Rubner test for, 286
    lead in, determination of, using atomic absorption spectroscopy, 211-213
    levulose in, Seliwanoff test for, 286
    luteinizing hormone in, 364
    magnesium in, determination of, using atomic absorption spectroscopy, 209
    melanin in, 299
    3-methoxy-4-hydroxymandelic acid in, determination of, with thin-layer chromatography, 521-523
    microscopic examination of, 291-292
    occult blood in, 300
    β-oxybutyric acid in, Hart method for detecting, 290-291
    pentose in, Tauber test for, 287

Urine—cont'd
　pH of, disorders and, 277-278
　pooled, controls of, 13
　porphobilinogen in, 297-298
　porphyrins, 297
　proteins in, 278-283
　　determination of, 282
　　electrophoresis of, 558-560
　　　collodion bag ultrafiltration in, 559
　　　Minicon concentration method of, 560
　　　protein concentration procedures in, 559-560
　　tests for
　　　qualitative, 279-280
　　　quantitative, 281-282
　rapid tests of, 276-277
　reaction of, 277
　reducing substances in, 283-285
　routine examination of, 274-306
　sediments in
　　causes of, 291
　　microscopic examination of, 291-294
　specific gravity of, 278
　sulfates in, Folin method for determination of, 302-303
　thresholds of substances in, 253-254
　transparency of, 277
　urea nitrogen in, 306
　uric acid in, 306
　urobilin in, 296-297
　urobilinogen in, Schwartz-Watson test for, 295-296
　zinc in, determination of, using atomic absorption spectroscopy, 211
Urine analysis report, sample, 275
Urine creatinine calibration chart, 106
Urobilin
　fecal analysis for, 319
　in urine, 296-297
Urobilinogen in urine, Schwartz-Watson test for, 295-296

V

V mask, 21
Vacutainer tubes for venipuncture, 4
Vacutainers of anticoagulants for analysis compatibility, 8
Vacuum cleaner device used in quantitative thin-layer chromatography, 514
Value, t, 32, 33
Van de Kamer modified method for fecal fat determination, 319-321
Van den Bergh test of bilirubin, 195
Van Slyke-Cullin method for urea nitrogen determination, 111-112
Van Slyke manometric apparatus, 123
　carbon dioxide determination using, 122-125
　care of, 124-125
Vanillylmandelic acid, urinary determination of, 442-444
Variability on graphs of frequency distribution, 27
Variation, coefficient of, 18
Vascular capacity, reserve, of lung, 156
Vascular flow, lung, 156
Vascular and ventilatory factors of lung, 156
Vasopressin, 365

Veins
　portal, blood supply to anterior pituitary by, 360
　punctures of, for blood gas and pH measurements, 164
Venipuncture, 3-7
　applying antiseptic in, 5
　equipment for, 4
　failure to obtain free-flowing blood in, 7
　hematoma from, 5, 7
　inserting needle in, 5
　needle and syringe for, 4
　preventing further bleeding in, 5, 7
　steps in, 6
　technical problems in, 7
　technique of, 5-7
　tourniquet for, 5
　Vacutainer tubes for, 4
Ventilation
　alveolar, 150, 154-155
　　minute, 154
　　regulation of, 162
　regional, normal, extremes of, 159
Ventilation: perfusion inequality test, 161
Ventilation: perfusion ratio, 159-163
　bronchiolar-bronchial disease and, 161-162
Ventilatory acidosis, 172
Ventilatory alkalosis, 172
Ventilatory and vascular factors of lung, 156
Ventricles of brain, 322
Vertical gel slab polyacrylamide gel electrophoresis, 549
View box for immunoelectrophoresis, 606, 609
Virilism, adrenal, 383
Vitamins
　B_{12}, deficiency of, 312
　D and parathyroid gland, 372
　D_3, 400
　thin-layer chromatography for, 502
VMA; see 3-Methoxy-4-hydroxymandelic acid
Volume
　carbon dioxide compartments as, 152
　closing, determination of, 162
　delivery, pump tube sizes and, 37
　lung, 150
　　plotting of, 153
　tidal, 150
Von Gierke's disease, hypoglycemia from, 93
VQI; see Ventilation: perfusion ratio

W

Waldenström's macroglobulinemia, 617, 620
Wang method for determination of calcium in urine, 301-302
Warning lines on Shewhart chart, 19-20
Washing unit for micro–starch gel electrophoresis, 543
Weekly dose of radiation, 350
Weights, equivalent, 680-682
Wetting agent A for use with AutoAnalyzer, 40-41
Wetting agents for use with AutoAnalyzer, 39-41
White cell count in capillary puncture, 1, 3
Wick for immunoelectrophoresis, 608, 609
Wick flow for electrophoresis, 532-533
Window, counting, for frequency distribution, 342

Withdrawing blood
    in capillary puncture, 3
    in venipuncture, 5-7
Work sheet for tabulating test results, sample, 16
Working standards
    for alkaline phosphatase determination, 62-63
    for Babson-Phillips method of lactate dehydrogenase determination, 235
    for calcium determination, 56
        using atomic absorption spectroscopy, 205-206
    for copper determination in serum and urine using atomic absorption spectroscopy, 210
    for creatinine determination, 51
    for direct cholesterol determination, 66
    in electrolyte determinations, 44
    in inorganic phosphate determination, 57
    for lactic dehydrogenase determination, 64
    for lead determination using atomic absorption spectroscopy
        in urine, 212
        in whole blood, 211
    for magnesium determination in serum and urine, 209
    for serum glutamic oxaloacetic transaminase, 59-60
    for simultaneous direct and total bilirubin determination, 69

Working standards—cont'd
    for simultaneous glucose and urea nitrogen determination, 48
    for simultaneous total protein and albumin determination, 71-72
    for uric acid determination, 54
    for zinc determination in plasma and urine using atomic absorption spectroscopy, 211

X

Xanthochromia, cerebrospinal fluid in, 323

Z

Zero-order kinetics, 215
Zinc in plasma and urine, determination of, using atomic absorption spectroscopy, 211
Zollinger-Ellison syndrome
    basal gastric secretion in, 313
    gastrin in, 311
    maximum gastric acid output in, 313
Zona fasciculata of adrenal cortex, 374
Zona glomerulosa of adrenal cortex, 374
Zona reticularis of adrenal cortex, 374
Zone in chromatography, elliptic shape of, 489
Zone electrophoresis, 528-529
Zone spreading in chromatography, 488